RESTORATION OF BOREAL AND TEMPERATE FORESTS

Integrative Studies in Water Management and Land Development

Series Editor
Robert L. France

Published Titles

Handbook of Water Sensitive Planning and Design
Edited by Robert L. France

Boreal Shield Watersheds: Lake Trout Ecosystems in a Changing Environment
Edited by J.M. Gunn, R.J. Steedman, and R.A. Ryder

Forests at the Wildland–Urban Interface: Conservation and Management
Edited by Susan W. Vince, Mary L. Duryea, Edward A. Macie, and L. Annie Hermansen

The Economics of Groundwater Remediation and Protection
Paul E. Hardisty and Ece Özdemiroğlu

Restoration of Boreal and Temperate Forests
Edited by John A. Stanturf and Palle Madsen

Forthcoming Titles

Porous Pavements
Bruce K. Ferguson

RESTORATION OF BOREAL AND TEMPERATE FORESTS

Edited by

John A. Stanturf and Palle Madsen

CRC PRESS

Boca Raton London New York Washington, D.C.

Library of Congress Cataloging-in-Publication Data

Restoration of boreal and temperate forests / edited by John A. Stanturf, Palle Madsen.
p. cm. — (Integrative studies in water management & land development ; 3)
Includes bibliographical references and index.
ISBN 1-56670-635-1
1. Reforestation. 2. Afforestation. 3. Forest ecology. 4. Restoration ecology. I. Stanturf, John A. II. Madsen, Palle. III. Integrative studies in water management and land development ; 3.

SD409.R485 2004
634.9'56—dc22 2004055355

Direct all inquiries to CRC Press, 2000 N.W. Corporate Blvd., Boca Raton, Florida 33431.

Visit the CRC Press Web site at www.crcpress.com

International Standard Book Number 1-56670-635-1
Library of Congress Card Number 2004055355
Printed in the United States of America 1 2 3 4 5 6 7 8 9 0
Printed on acid-free paper

Series statement: Integrative studies in water management and land development

Ecological issues and environmental problems have become exceedingly complex. Today, it is hubris to suppose that any single discipline can provide all the solutions for protecting and restoring ecological integrity. We have entered an age where professional humility is the only operational means for approaching environmental understanding and prediction. As a result, socially acceptable and sustainable solutions must be both imaginative and integrative in scope; in other words, garnered through combining insights gleaned from various specialized disciplines, expressed and examined together.

The purpose of the CRC Press series Integrative Studies in Water Management and Land Development is to produce a set of books that transcends the disciplines of science and engineering alone. Instead, these efforts will be truly integrative in their incorporation of additional elements from landscape architecture, land-use planning, economics, education, environmental management, history, and art. The emphasis of the series is on the breadth of study approach coupled with depth of intellectual vigor required for the investigations undertaken.

Robert L. France
Series Editor
Integrative Studies in Water Management and Land Development
Adjunct Associate Professor of Landscape Ecology
Science Director of the Center for Technology and Environment
Harvard University
Principal, W.D.N.R.G. Limnetics
Founder, Green Frigate Books

Foreword by series editor: Addressing the once and future forest

Quite simply, the history of forestry is the history of us (Harrison, R.P. 1992. *Forests: The Shadow of Civilization*). Culturally, scientifically, spiritually — we have always been and most likely always will be, shaped by those groupings of tall plants with which we share the globe. Indeed, in the world's oldest known literary creation, the *Epic of Gilgamesh*, the most significant event takes place when the hero and his companion go on an expedition in search of the trees so absent from the Summerian river plain (France, R.L. 2005. *Back to the Garden: Searching for Eden in the Mesopotamian Marshes*). It is these trees (and their logging and transport, as so brilliantly shown on the Assyrian wall carvings in the Louvre Museum) with which the world's first cities were built.

I have spent much of my professional life as an ecologist immersed within, and endeavoring to understand, the boreal forest (and accompanying lakes) of Canada (France, R.L. 2005. *Aquatic Responses to Watershed Clearcutting: Implications for Forestry and Fisheries Management*). For some of that time, before being made aware of the widespread implications of regional fire suppression, I was under a misconception about the inherent naturalness of that ecosystem. It came as a bit of a shock when I entered my first "naturally" reforested landscape in central Massachusetts. I, of course, had seen the afforested plantations of northern Europe, where the deliberate workings of humans could easily be discerned by even the most nondiscriminating of urban-attuned eyes. But I had never seen such a diversity of trees upon a reforested landscape as that within the famous Harvard Forest. Later, upon learning that in Thoreau's time, three-quarters of the surface of Massachusetts had been cleared fields with only one-quarter remaining forested, whereas today the reciprocal is true (Foster, D.R. 1999. *Thoreau's Country: Journey through a Transformed Landscape*), I recognized for the first time the power of forest regeneration, rebirth and renewal; in short, the restorative potential of nature at its strongest.

The breadth and scope of study in the present volume, edited by John Stanturf and Palle Madsen, the fourth book in the series by CRC Press — Integrative Studies in Water Management and Land Development — is truly staggering in magnitude. Over 80 authors from nearly two-dozen countries present what I believe to be the most thoroughly detailed and all-encompassing book on forest ecology and management ever produced. Underlying the entire book is a shared belief that forests today are truly dynamic ecosystems embedded within cultural landscapes. Thus, by acknowledging both the nonstatic nature of forests as well as the historical legacy of their human manipulation, the present book clearly situates the efforts of humans in ecological restoration. The chapters herein, therefore, are testament to both the reparative efforts of humans and the inherent resilience of forests.

Ecological restoration in the broadest sense, whether referred to as remediation, rehabilitation or any of the other many "r" words currently in vogue (France, R.L. 2005. *Healing Natures, Repairing Relationships: Restoring Ecological Spaces and Consciousness*), is about both return and advance (i.e., "back to the future," as the film title has it). Engaging in the sort of restoration as practiced by the authors in this book is above all a purposeful act of optimism. Rather than bemoaning the disappearance of the Earth's forests and accepting the status quo, these workers are actively involved in reversing that process, returning sections of the denuded planet to its former sylvan health and glory. The demonstrable successes of reforestation as shown in these pages or in the famous dioramas at the Fisher Museum of the Harvard Forest (Foster, D.R. and J.F. O'Keefe. 2000. *New England Forests Through Time: Insights from the Harvard Forest Dioramas*) can almost appear to be unreal. And in that light, and given that many of the chapters specifically deal with boreal forests, it is not out of place to end this foreword by quoting the lyrics of Canada's unofficial national anthem (*The Canadian Railroad Trilogy*) by folk singer Gordon Lightfoot, who referred to the landscape north of Lake Superior at the time of railroad construction as being "when the green, dark forests were too silent to be real." As the present book so clearly shows, that winsome and nostalgic "when" has become a realized "now" in many places around the world as we collectively proceed toward "the once and future forest" (aka Sauer, L.J. 1998. *The Once and Future Forest: A Guide to Forest Restoration Strategies*).

Robert L. France
Harvard University

References

Foster, D.R., *Thoreau's Country: Journey through a Transformed Landscape*, Harvard University Press, Cambridge, MA, 1999.

Foster, D.R. and O'Keefe, J.F., *New England Forests Through Time: Insights from the Harvard Forest Dioramas*, Harvard University Press, Cambridge, MA, 2000.

France, R.L., *Aquatic Responses to Watershed Clearcutting: Implications for Forestry and Fisheries Management*, CRC Press, Boca Raton, FL, in preparation.

France, R.L., *Back to the Garden: Searching for Eden in the Mesopotamian Marshes*, in preparation.

France, R.L., *Healing Natures, Repairing Relationships: Restoring Ecological Spaces and Consciousness*, MIT Press, Cambridge, MA, in press.

Harrison, R.P., *Forests: The Shadow of Civilization*, University of Chicago Press, Chicago, IL, 1992.

Sauer, L.J., *The Once and Future Forest: A Guide to Forest Restoration Strategies*, Island Press, Washington, DC, 1998.

Preface

Throughout the temperate and boreal zones, human intervention has influenced landscapes and forests for thousands of years. The degree of human disturbance has only been constrained by the technology and resources available to different cultures and by time since initial habitation. For millennia, humans have influenced forests by regulating populations of browsers; clearing for agriculture; keeping domestic livestock; cutting trees for fuel, building material, and fiber; introducing new species; using fire; and suppressing fire. Today's forests are the result of all these disturbances, along with climatic change and species migration into postglacial landscapes. The ability of humans to affect forest ecosystems increased dramatically after the Industrial Revolution. Engineering works, including mining, dams, and roads, are both more widespread and more intensive. Management has been extended to native forests over larger areas; at the same time, the switch from biomass to fossil fuels changed traditional forest management. Industrial, residential, and automotive emissions of combustion products affect forests directly and through climate change.

Forest restoration is under way in order to counteract the negative effects of human activities. Restoration is important for many interest groups in all countries, and in most countries occurs on both publicly and privately owned land. Nevertheless, the motivations for attempting forest restoration are diverse, reflecting the complexity of ecosystems, their current state and past land use, and the human context of culture, economics, and governance. Restoration goals vary, but in most countries, restoration is undertaken within the policy framework of increasing sustainability by enlarging the area of specific ecosystems, enhancing biodiversity, or repairing ecosystem functions. The most common specific objectives for restored forests include timber, wildlife habitat for game species, or aesthetics. Increasingly, other objectives are considered, including carbon sequestration, biological diversity, nongame mammals and birds, endangered animals and plants, protection of water quality and aquatic resources, and recreation.

We set out to produce a text for graduate courses in forestry and ecology. This book should also serve as a first-source reference for researchers, managers, and policymakers seeking information on restoration activities worldwide or within specific regions. Probably because of our training and experience, we take a decidedly proactive approach in this book. Our view is that to be successful, restoration requires sequential intervention and ongoing management. We believe that appropriate silvicultural operations can be designed for any forest restoration objective. Nevertheless, forest restoration requires creativity and flexibility in applying silvicultural tools. There are obvious questions that must be answered in a specific ecological, economic, and social context (What to do? How to do it? Who should pay?). The authors of the following chapters provide an understanding of how these questions are currently being answered for temperate and boreal forests.

The idea for this book, and the conference on which it was based, was hatched by a group of researchers on a field trip to view restoration research in Iceland. No doubt inspired by the majestic landscapes of that oceanic island, the authors organized the International Conference on Restoration of Boreal and Temperate Forests, held in May 2002 in Vejle, Denmark. The International Union of Forestry Research Organizations (IUFRO), the U.S. Forest Service Southern Research Station, the then Danish Forest and Landscape Research Institute, and the Swedish Agricultural University sponsored the conference. These same organizations supported the development of the book and continued to employ the editors throughout the process. The authors of the individual chapters are to be congratulated for their accomplishment and thanked for their hard work. They endured independent review of their work by at least two of their peers. After several revisions by the authors, each chapter received further revision by the editors.

The book is divided into seven parts. Part 1 provides an introduction by Stanturf to restoration terms and sets the framework for the book, particularly the case studies. Generally, *restoration* connotes some transition from a degraded state to a more "natural" condition. In the narrowest sense, restoration requires a return to an ideal natural or historic ecosystem with the same species diversity, composition, and structure as occurred before human intervention. Because this ideal state is probably impossible to attain, Stanturf adopts an alternative approach, using the term *forest restoration* broadly to describe situations where forested land use and land cover are restored (reconstruction or reclamation), as well as instances when an existing forest is rehabilitated (no change in land cover) such that structure or species composition is modified. He calls re-creation the attempt to return to a pristine natural or historic condition. All the restorative activities described (re-creation, reconstruction, reclamation, and rehabilitation) have been called forest restoration, although to the purist only re-creation would qualify as true restoration. The question of whether restoration will be guided by the desire to re-create historic or idealized natural conditions, or by specific objectives, is taken up in several chapters including those by Oliver and O'Hara, Beatty and Owen, and Van Lear and Wurtz.

Part 2 presents the ecological context for forest restoration, beginning with a chapter by Bradshaw that explores the notion of what are natural forests, and the difficulties of using the concept to drive restoration programs. Where remnants of the putative natural forest remain, or can be reconstructed from the historical record, the question of what is natural is hotly debated. The case study chapters from western North America (Beatty and Owen, Van Lear and Wurtz, Douglas and Burton, O'Hara and Waring, Jain and Graham, Kaufmann et al.) and boreal Scandinavia (Angelstam et al., Kuuluvainen et al.) are good examples of forests where extensive human alteration is historically recent. Nevertheless, even these forests have been altered, as Bradshaw points out. It would be easy to conclude that there is no scientific basis for restoring natural forests because humans and climate change have so drastically changed the whole biosphere. Indeed, some restorationists conclude that the choice of endpoint is inherently political, not scientific. Nevertheless, such choices are necessary in a restoration program and the rationale for the choice must be conveyed to the public. The task for restorationists is to interpret the scattered scientifically based knowledge of forest history, stand development, and natural processes, and combine it with practical experience to design objectives that improve sustainability. Restoration will have to proceed with very imperfect knowledge and researchers must attempt continuously to create a better foundation of documented knowledge to support restoration decisions.

Oliver and O'Hara describe the effects of restoration at the stand level, traditionally the focus of silvicultural activity. Understanding the dynamism of forest stands is critical to identifying appropriate operations for restoration. Oliver and O'Hara provide a valuable review of the changing concepts of forest ecology and how they have been applied to management and conservation. They describe the shift from viewing forest ecosystems as closed, steady-state systems with predictable development patterns to the present view of open systems that operate opportunistically, with multiple developmental pathways following disturbance. To be successful, restoration efforts must adopt this dynamic view of forests. To be effective, restorationists will have to educate the nontechnical restoration enthusiasts as well as the general public in this new paradigm.

Beatty and Owen further explore the dynamic nature of forest ecosystems by describing the role of natural disturbance. In addition to presenting some of the current concepts of disturbance, including considerations of scale and frequency, they present case studies of disturbance regimes in western subalpine and montane coniferous forests and northeastern mixed deciduous forests in the United States. They show how disturbance from fire, blowdown, and insect outbreaks occurs in both forest types, but with different effects. They offer suggestions for incorporating disturbance regimes into managed forests, paying attention to coarse-scale as well as fine-scale disturbance processes and interactions with environmental fluctuations. In many countries, forests are fragmented and rarely are large areas available where coarse-scale disturbances such as wildfire can be tolerated. Forests that are blown down by severe winds, or killed by insect and disease outbreaks, frequently are salvage logged because of their economic value. These realities pose several challenges for incorporating disturbances into restoration, illustrated here and in other chapters by attempts to restore natural or historic fire regimes (e.g., Van Lear and Wurtz, Kaufmann et al., Jain and Graham, Kuuluvainen et al., Brockway et al., and Douglas and Burton). Questions that must be faced include when to suppress wildfires, how to mimic natural fire with prescribed burning, whether there are effective surrogates to fire such as herbicides or mechanical treatments, and whether these surrogates provide the same ecological functions as fire. Managers have more opportunities for incorporating fine-scale disturbances processes such as gap creation, including altering traditional forest management as illustrated by Angelstam et al., Kuuluvainen et al., Hahn et al., and Douglas and Burton.

Vose et al. end this section on the ecological context for restoration with a discussion of watershed processes. They provide an approach to the important task of measuring successful restoration of ecological functions. Often, we assume that restoring stand structure and, if necessary, species composition will restore ecological functions. This is certainly true for afforestation; a tree plantation functions more like a forest than a cotton field. Nevertheless, even in gross terms a plantation does not offer the same degree of functioning as a multispecies forest with complex structure. Vose et al. offer a way to measure the recovery of ecological processes, specifically biogeochemical pools and processes. As they point out, restoration of biogeochemical processes is difficult to evaluate because these processes operate at multiple temporal and spatial scales. By focusing on emergent properties of ecosystem processes, such as nutrient budgets, the task is manageable. In this chapter, Vose et al. focus on nitrogen for two reasons: N pools and processes are sensitive to disturbance and restoration, and often restoring N pools and processes is a specific restoration objective. They use the resistance and resilience model for ecosystem response to disturbance and apply it to several case studies of riparian zone restoration and prescribed burning. They suggest that key indicators of ecosystem response be identified and evaluated in terms of the magnitude and duration of response. Because pools and processes that depend primarily on biological agents may vary from those dependent upon physical agents, responses will vary and evaluation of restoration success will depend upon the choice of the key indicator.

In Part 3, attention is turned to the social context for restoration in three chapters that examine restoration policy. Mercer presents an overview of policy instruments and financial methods used to encourage restoration, primarily in the United States and Western Europe. Ciccarese et al. examine the emerging potential for carbon sequestration offsets to fund restoration efforts, especially afforestation. Weber, in the final chapter in this section, explores the driving forces in Europe for afforestation, which was initiated by changing agricultural policy within the European Union. In general, the motivation for restoration is to produce or enhance ecological services, which are usually public goods or externalities. Encouraging restoration involves activity on public land and controlling or encouraging activity on private land through regulations and incentives. Because no markets exist for most of the ecological outcomes from restoration, policymakers have no easy criteria for deciding optimum levels of restoration. There is a further spatial dimension to restoration that complicates allocation decisions. For example, within a degraded watershed, restoring some areas will produce a higher level of ecological benefit than others, given the same level of restoration effort. The best combination of policies will use market solutions, government intervention, and combinations. When government intervenes, policymakers have a range of carrots (incentives) and sticks (regulation) at their disposal. Different approaches will be more feasible in some countries than others but, in general, the approach that seems the easiest, concentrating restoration on an expanded public land base, is the least feasible. Targeted voluntary and mandatory programs, combined with community-based participatory approaches, have the greatest likelihood of successfully providing long-term benefits. Aradottir and Eysteinsson, Madsen et al., Harmer et al., Lee and Suh, and Kaufmann et al. give examples in their chapters. Mercer concludes that such approaches will be complicated, costly, and will challenge our understanding of social dynamics.

One such targeted voluntary approach is payments for increased carbon sequestration. Forests are both sources and sinks for atmospheric carbon, and forestry and land-use change activities (dubbed LUCF in the climate change policy arena) can mitigate carbon emissions. Ciccarese et al. consider the role that restoration has in this arena, specifically through afforestation, rehabilitating degraded forests, and altering management of natural and plantation forests. They explain the technical barriers to considering forestry activities under present climate change treaties, including especially nonpermanence. This is the potential for carbon sequestered by restoration to be released if the forest is disturbed by natural or anthropogenic events (deforestation, harvesting, fire, pests, or storms). Other technical issues considered include how to estimate and monitor carbon sequestration at a reasonable level of precision. Ciccarese et al. point out that forest restoration for carbon sequestration is a reality; over 4 million ha have been reforested or afforested since 2001, specifically for carbon credits. The role of restoration in carbon accounting, however, is not limited to obtaining carbon offset credits or payments. All restoration activity affects carbon storage and release; increasingly, the carbon footprint of restoration projects will need to be included in national carbon accounts.

Afforestation is an important component of forest restoration, particularly in Europe. Weber provides an overview of European experience that is examined in more detail in subsequent chapters (Aradottir and Eysteinsson, Madsen et al., Balandier et al., Jõgiste et al., McCreary and Cañellas, and Renou and Farrell). Natural invasion on former agricultural land may be included in afforestation statistics, especially in mountainous areas (Balandier et al.). Forestland area has expanded in Europe in the 20th century through afforestation and natural invasion, driven initially by the need of the war industry and the fear of a wood shortage. Afforestation and agricultural policy has been examined in tandem at the European level since 1968; afforestation was a way to address problems of agricultural overproduction and rural unemployment. Increasingly, afforestation is undertaken for ecological and amenity reasons. Perhaps the most important lesson learned from the European experience is that not everyone welcomes afforestation, even when programs are voluntary. In predominantly agricultural landscapes, planting trees may be seen as an assault on rural culture and traditional landscapes. Certainly, there has been a backlash in many countries to the former afforestation practice of planting a few conifer species. (Madsen et al., Balandier et al., and Harmer et al. describe this in European countries, but it can also be seen in Asian countries, as described in Nagaike et al. and Lee and Suh.) Another important conclusion from Europe, which echoes elsewhere, is the continuing close linkage of afforestation to agriculture. The experience in the United States, described by Gardiner and Oliver, bears this out.

Part 4 provides three chapters that present overviews of key issues in restoration practice: obtaining the correct plant material, the biophysical environment faced by seedlings, and the necessary steps to take beyond the initial establishment phase. Kjær et al. point out that appropriate genetic material for restoration must be decided within a national context. In some countries, restoration may include nonnative species, but in all countries appropriate provenances will be critical for short-term success (seedling survival and stocking density) and for long-term productivity, resistance to pests, and resilience following disturbance. Forest restoration usually involves significant if not total change of species composition. The restoration process provides a unique opportunity to obtain genetically high-quality material to support multifunctional goals. This is most clearly seen when trees are planted, but genetics, in terms of plant adaptations to site conditions and disturbance regimes, is an underlying factor in all forest restoration. Other chapters clearly address these issues; for example, Bradshaw talks about species migration into postglacial landscapes and the genetic structure of natural stands. Mikułowski et al., describe how air pollution stress may reduce the genetic pool of tree species and the need to secure a genetic material for the future in a gene bank. Hahn et al., Mikułowski et al., and Hansen and Spiecker briefly touch on the role that poor choices of Norway spruce provenances played in the development of instability problems. Increasingly, we must consider the impact of future climate (whether natural fluctuation or human-induced change) on species adaptability in the restored forest.

Most restoration efforts include a regeneration phase, and choices made at this stage have far-reaching effects on restoration success. The microsite environment of the newly established propagules may result from intervention such as site preparation to enhance the rooting zone of planted seedlings, or may receive no manipulation such as passive restoration from natural invasion. Wagner and Lundqvist present the typical timeline of the regeneration process, pointing out where important strategic and practical decisions must be made. They connect the trajectory from previous silvicultural pathways, stand, and site conditions to the future species composition and stand structure. Wagner and Lundqvist emphasize experience in central and northwestern Europe; Hahn et al., Hanson and Spiecker, and Baumhauer et al. explore the issue of Norway spruce conversion in detail. Nevertheless, the principles of how life cycle stage, site, and silvicultural operations interact are universally applicable. Understanding what new plants require and what environments are available will help restorationists specify needed intervention.

Because restoration requires ongoing management, the silviculturist plays an important role in bringing to bear time-tested techniques to shape the development trajectory of a stand toward the desired condition, and to maintain the restored ecosystem. Van Lear and Wurtz discuss the challenges of fitting silvicultural operations to restoration objectives from the standpoint of what needs to be done beyond the regeneration phase. A weakness of many restoration efforts is that although success ultimately depends on a series of interventions, only the short-term operations are addressed in detail. Van Lear and Wurtz emphasize the role of fire and provide several mini-case studies from temperate and boreal forests, including types not treated in other chapters such as upland oak and Table Mountain pine in the southern United States and mixed-wood management in Canada's boreal forest.

Parts 5, 6, and 7 present regional and national case studies on afforestation, rehabilitation, and reclamation. Some 23 chapters provide documentation of current efforts to restore forest ecosystems in countries in North and South America, Asia, and Europe. Because the context for restoration varies between countries, each chapter provides a socioeconomic and policy framework for restoration. Because the legacy of past forest management and land use practices determine the major restoration needs and influence what is feasible, each case study explores these issues in some detail. Restoration is not a well-developed concept in all countries, but as far as possible, authors provide information on the extent of restoration efforts and costs.

The five chapters on afforestation present case studies from climates as varied as Iceland (Aradottir and Eysteinsson) and Spain and California (McCreary and Cañellas). Motivation for afforestation is almost as varied as the climate: in Iceland, it is to control soil erosion and restore native birch woodlands; in Spain, the problem is poor regeneration. The chapters from Denmark (Madsen et al.), the Baltic States (Jõgiste et al.), and the lowlands of the Mississippi River Valley in the United States (Gardiner and Oliver) all reflect adjustments to agricultural land use. In Denmark, the primary driver has been changing agricultural policy; in the Baltic States, more sweeping changes in governance from a centrally planned economy under the former communist regime to a more open, market-oriented economy has led to the need for alternatives to farming for large areas of land. In Mississippi, a relatively short interval of deforestation and utilization for agriculture is being reversed on some land in the floodplain of the world's third largest river. The drivers for restoration of bottomland hardwoods are enhancement of wildlife habitat and protection of water quality. In all cases, the focus is on restoring tree cover, with some consideration for directly restoring other ecosystem functions. Afforestation is touched on in other chapters (Balandier et al., Lee and Suh, Sagheb-Talebi, Paladino et al., Brockway et al., Mikułowski et al., and Renou and Farrell).

The section on rehabilitations contains 16 chapters, the most numerous in the book, reflecting the greatest interest in forest restoration. The chapters are arranged geographically, beginning in Europe and arranged from north (Scandinavia and the United Kingdom) to south (Germany and France). In boreal Sweden (Angelstam et al.), Finland (Kuuluvainen et al.), and in British Columbia (Douglas and Burton), the emphasis is on altering forest management to incorporate consideration of landscape-level concerns (biodiversity, disturbance regimes) as well as stand structure and coarse woody debris. In temperate Denmark (Hahn et al.), United Kingdom (Harmer et al.), and Germany (Hansen and Spiecker, Baumhauer et al.), the main issue is conversion from single-species conifer plantations to broadleaved or mixed stands. The predominance of conifer monocultures in Europe reflects great variation in historical development in these countries; nevertheless, a dominant theme in these chapters is that at some time, the high productivity and value of products from conifers such as Norway spruce, and the ease of their establishment, caused them to be planted even on sites where they were not adapted. Damage from windstorms, faltering timber economics, and changes in public perception have contributed to the present desire across Europe to use other species in afforestation and to change species composition of existing stands. These case studies present several approaches to addressing this issue. The chapter from southern France (Balandier et al.) addresses this same issue of conversion from monospecific conifers to mixed stands, with some additional concerns of how to accelerate development of stands that arise from spontaneous regeneration (natural invasion) on abandoned pastureland and, in all restoration efforts, how to control competing vegetation without using herbicides.

Three chapters on stand rehabilitation from Asia present very different perspectives. In Japan (Nagaike et al.), the economics of timber production have changed in favor of importing wood, rather than paying the high labor costs for managing native conifer plantations or coppiced stands of mostly oak species. Techniques for rehabilitating stands for mostly biodiversity purposes are tested in a case study in the area of Japan where heavy snowfall contributes to the difficulty of managing abandoned coppice stands. Another chapter from East Asia describes the evolution of restoration goals in the Republic of Korea in the last half of the 20th century (Lee and Suh). Following wartime turmoil and a shift from biomass to fossil fuels, early restoration goals focused on afforestation and reforestation, usually with fast-growing nonnative species; this is similar to European experience (see Weber, Madsen et al.). Later goals shifted toward timber production and the establishment of multifunctional forests. Current goals emphasize native species and sustainability. In Iran, the challenging site conditions faced by afforestation projects favor a continued use of exotic species. Sagheb-Talebi provides a fascinating example of restoration of degraded sites to secure environmental services, specifically spreading floodwater, which reduces flooding on cropland.

The chapters by Douglas and Burton (Canada) and Paladino et al. (Brazil) geographically bracket chapters on rehabilitation in the New World. Each provides an overview of restoration within a region of their respective countries. Although most of Brazil is tropical, the Atlantic Rain Forest on Brazil's coast originally stretched from the tropical to the temperate subtropical zones. Only remnants of this once vast lowland forest survives; nevertheless, interest in restoring the Mâta Atlantica is increasing. Most efforts are of limited scope and the emphasis is on using principles of ecological succession in establishing mixed-species plantations. In the western Canadian province of British Columbia, the challenge is to incorporate ecosystem restoration awareness into day-to-day forest management, which may avoid the need for restoration in the future. Douglas and Burton identify eight ecological issues and present five case studies on ground restoration, including stand structure alteration, invasive species, underrepresented habitat, and landscape fragmentation. Other Canadian experience is included in Van Lear and Wurtz, and the similarities of the situation in British Columbia to the Pacific Northwest of the United States should be noted (see O'Hara and Waring, Jain and Graham).

The four chapters from the United States cover vast areas of the west coast (O'Hara and Waring), inland west (Jain and Graham, Kaufmann et al.), and the southeastern coast (Brockway et al.). The chapters all have a common theme: restoration of fire disturbance regimes, which are touched on in Beatty and Owen, Vose et al., and Van Lear and Wurtz. Other themes are contained, including stand structures (see Oliver and O'Hara), old forest characteristics, and exotic pests. The three chapters from the western states exemplify concerns for reestablishing historic forest types and structures that dominate thinking about forest restoration in the United States and approach the notion of re-creation. They are similar to approaches in boreal Europe (Angelstam et al., Kuuluvainen et al.). Although logging, grazing, and fire suppression have heavily impacted these forests, ample remnants of relatively undisturbed conditions remain. Alternatively, such "natural" forests can be discerned from the historical record or through stand reconstruction techniques such as dendrochronology. The impact of pests as degrading agents is a topic in O'Hara and Waring and Jain and Graham, while O'Hara and Waring caution against restoration approaches that create the potential for future pest problems. In a chapter from the southeastern states, Brockway et al. similarly focus on a historic forest ecosystem, the once extensive longleaf pine forests (Gardiner and Oliver discuss another southern forest type, bottomland hardwoods). In all these case studies from the United States, however, the present forests are composed of native species although distributions and structures are not the same as they were before the influence of European settlers.

The final reclamation section of the book, comprised of only two chapters, does not attempt a comprehensive treatment of reclamation of drastically disturbed land. On the other hand, two interesting topics are presented, showing the effects of degradation at regional and local scales: reclamation of forests severely impacted by air pollution in Central Europe (Mikułowski et al.) and reclaiming industrial peatlands for forestry in the Republic of Ireland (Renou and Farrell). Both chapters present the essential dilemma of reclamation: not only must forest cover be reestablished, but the site itself has been degraded and existing forest remnants offer little guidance on how to proceed. The chapters by Renou and Farrell and Kuuluvainen et al. both treat problems of peatland forestry as reclamation and rehabilitation, respectively, primarily to repair ecological functioning.

Some similar concerns cross continental as well as national boundaries; for example, disturbance is a common theme throughout the book. Besides the obvious link of disturbance to the need for restoration brought out in the conceptual framework presented by Stanturf, Vose et al. used the disturbance response analogy of resilience and resistance in their conceptual framework for evaluating successful restoration of ecological functions. Many chapters provide examples of how prevalent disturbances are and the chapters by Bradshaw, Oliver and O'Hara, Beatty and Owen, Van Lear and Wurtz, Angelstam et al. and Kuuluvainen et al. provide an insight into temporal and spatial aspects of disturbance. Fire as a disturbance agent in a specific ecosystem is treated in these chapters as well as others mentioned above.

Another theme that emerges from several chapters is the use of nature-based silviculture, which is not simply a synonym for restoration. What is natural is examined in some fashion in almost every chapter. In addition to the effects of climate change on geological time scales brought out by Bradshaw, examples of changes in forest composition and structure due to human intervention appear throughout the many chapters. The dynamic nature of forest ecosystems, stressed by Oliver and O'Hara and Van Lear and Wurtz, shows that even without anthropogenic disturbances, it is very difficult to say what would be a natural forest in a given place and time. Nevertheless, there is a recurring theme of returning forests to more natural conditions and using more nature-based silviculture to accomplish that goal. Native species are emphasized, although several case studies present a rationale for using what are clearly exotic species under some conditions (Sagheb-Talebi, Renou and Farrell, Lee and Suh). Even native species may be discriminated against if they are not site-adapted, which is one motivation for conversion of Norway spruce in Europe (Wagner and Lundqvist, Hahn et al., Hansen and Spiecker, Baumhauer et al.) or other pines to longleaf pine in the southern United States (Brockway et al.). With the shift in emphasis from timber to ecological services in afforestation programs, even site-adapted native species may be de-emphasized for other species, especially to develop mixed-species stands (see especially Madsen et al., Gardiner and Oliver, Balandier et al., Harmer et al., Nagaike et al., Lee and Suh, Paladino et al.).

Defining a "natural" stand structure is even more arbitrary (Oliver and O'Hara). The old steady-state paradigm of succession to a climax led many to equate complex structure with old forests (O'Hara and Waring). To some, this meant that uneven-aged management was preferred to even-aged management because it led to more "natural" forests. Given the luxury of large areas of contiguous forest under a single or a few owners, it is possible to restore to a diversity of stand structures on the landscape in roughly the same proportions as occurred historically with little human influence; that is to say, under mostly natural disturbance regimes. These conditions appear to exist in western North America (Oliver and O'Hara, Beatty and Owen, Van Lear and Wurtz, Douglas and Burton, O'Hara and Waring, Jain and Graham, and Kaufmann et al.). They may also exist in boreal Europe (Angelstam et al., Kuuluvainen et al.) and Central and Eastern Europe (Weber, Jõgiste et al., and Mikułowski et al.). In more populous regions, however, nature-based silviculture that emphasizes restoring complex stand structures should stress the positive aspects of structural diversity in terms of stand stability (Hahn et al., Harmer et al., Hansen and Spiecker, Balandier et al., Nagaike et al., Lee and Suh) and biodiversity (Angelstam et al.).

A cautionary note is in order. Restoration is not necessarily a win–win effort, especially for private landowners (Mercer). As Weber points out, changing landscape conditions does not appeal to everyone. Oliver and O'Hara recommend that restoration will succeed over the long term if activities are framed in an economic perspective, with distinct, measurable objectives. Vose et al. provide one approach to measuring outcomes based on ecological functions, within a watershed context. On the other hand, Angelstam et al. use data on biodiversity at multiple scales to develop objectives with measurable outcomes. Both approaches begin to estimate outcomes in ways that will allow policymakers to compare restoration programs and to make tradeoffs (Mercer). More needs to be done.

This book is a collaboration of many scientists and practitioners from around the world. We wish to thank those who contributed to this volume and worked diligently to ensure that we produced a quality addition to the growing literature on forest restoration. First and foremost, the authors must be thanked for persevering. Editorial assistance by Lynne Breland and graphic design help from Patricia Outcalt were greatly appreciated by the editors. Institutional support from the U.S. Forest Service and the Danish Centre for Forest, Landscape and Planning, KVL, was unflagging. The patience and support of our editors at CRC, Brian Kenet and David Fausel, amazingly survived numerous delays and moving deadlines.

We relied on many external reviewers who read, commented, and offered valuable suggestions for improving the chapters. The reviewers include Joseph J. O'Brien, Thomas Waldrop, Kenneth Outcalt, Ashley Selby, Noël Lust, Lars Östland, Don English, William Mason, George Peterken, Gary Miller, Lars Edenius, Sybille Haeussler, Randy Kolka, Andy Ezell, Richard Thompson, Ilkka Vanha-Majamaa, Kevin O'Hara, Andrew Carey, Lynn Huntsinger, Edith Allen, David McGill, John Kush, Kay Kirkman, Dennis Hardin, Steve Jack, Miles Hemstrom, Penny Morgan, Ted Leininger, Inge Stupak Møller, Ralph Harmer, James Burger, Dave Borland, James Runkle, Chris Peterson, Cynthia Fowler, Mac Callaham, Brian Roy Lockhart, Robert Mickler, Andreas Brunner, Alexei Yaroshenko, Hans Christian Bruun Hansen, Jan Svejgaard Jensen, Randy Johnson, Ulrik Braüner Nielsen, Simon Bell, Ulrike Pröbstl, Sauli Valkonen, Margaret Devall, Victor Castillo, Jakob Harrekilde Jensen, Keith Kirby, Kenji Seiwa, Niels Heding, Emile Gardiner, and Michael Pedini.

Finally, we thank our families, particularly Eileen and Anne, for their patience and forbearance as we spent many evenings and weekends working on "The Book" when we rightfully should have spent time being better husbands and fathers.

John A. Stanturf
Athens, Georgia

Palle Madsen
Vejle, Denmark

About the Editors

John A. Stanturf, Ph.D., is a Project Leader and Supervisory Ecologist in the Southern Research Station, U.S. Forest Service, Athens, Georgia. Dr. Stanturf received his B.S. in plant and soil science from Montana State University and his M.S. and Ph.D. in forest soils from Cornell University. He was a Lady Davis Fellow at the Technion in Haifa, Israel and on the faculty at Cornell, the University of Pittsburgh, and the Pennsylvania State University. Currently, he is an adjunct professor at Auburn University and Mississippi State University. Before joining the Forest Service, he worked in industrial forestry research. His initial Forest Service assignment was as Project Leader, Center for Bottomland Hardwoods Research, Stoneville, Mississippi. He began his current position as Project Leader of the Disturbance Ecology unit in 2000.

His research interests focus on restoration, soil carbon, and short-rotation woody crops. Dr. Stanturf actively participates in international activities, serving as Working Party Leader for Restoration of Boreal and Temperate Forests (1.17.02) and Working Group Deputy, Restoration of Degraded Sites (1.17.00) for the International Union of Forestry Research Organizations. He is a member of the Ecosystem Management Commission of the World Conservation Union (IUCN), the Working Group on Silviculture, and the North American Forestry Commission. Dr. Stanturf is also Co-National Team Leader (U.S.) for the International Energy Agency Task 30 Short Rotation Crops for Biofuels and currently serves on the editorial board for the journal *Forest Ecology and Management*. He has authored or coauthored more than 100 technical papers in journals, proceedings, and other outlets, and has produced educational videos and slide sets.

Palle Madsen, Ph.D., is a Senior Researcher with the Danish Centre for Forest, Landscape and Planning, KVL, stationed in Vejle, Denmark. Dr. Madsen received his B.S. and Ph.D. in forestry from the Royal Danish Agricultural and Veterinary University. His silvicultural research has focused on regeneration challenges related to the forest restoration program of Danish forestry. The ultimate research goal is new and well-documented regeneration methods to support inexpensive, nature-friendly, and well-stocked regeneration for afforestation on farmland and conversion to close-to-nature silviculture on forestland. Dr. Madsen is deputy of the IUFRO Working Party 1.17.02 Temperate and Boreal Forest Restoration and he is coordinator of the IUFRO Research Group 1.10.00 Improvement and Silviculture of Beech. He has authored or coauthored more than 30 technical papers in journals, proceedings, and other outlets.

Contributors

Kaisu Aapala
Finnish Environment Institute
Helsinki, Finland

Per Angelstam
Swedish University of Agricultural Sciences
Riddarhyttan, Sweden
and
Örebro University
Örebro, Sweden

Ása L. Aradottir
Soil Conservation Service
Hella, Iceland

Philippe Balandier
CEMAGREF
Clermont-Ferrand Regional Centre
Aubière, France

Helmuth Baumhauer
Forstamt Weissenhorn
Bavaria, Germany

Susan Warren Beatty
Department of Geography
University of Colorado
Boulder, Colorado

Richard H.W. Bradshaw
Geological Survey of Denmark and Greenland
Copenhagen, Denmark

Dale G. Brockway
Southern Research Station
US Department of Agriculture Forest Service
Auburn, Alabama

Sandra Brown
Winrock International Ecosystem Services Unit
Arlington, Virginia

Philip J. Burton
Symbios Research and Restoration
Smithers, British Columbia, Canada

Isabel Cañellas
Centre for Forest Research, INIA
Madrid, Spain

Lorenzo Ciccarese
Italian Agency for Environmental Protection and Technical Services
Rome, Italy

Jorge Paladino Corrêa de Lima
Universidade Federal Rural do Rio de Janeiro
Rio de Janeiro, Brazil
and
US Department of Agriculture Forest Service
Athens, Georgia

Tanis L. Douglas
Fernhill Consulting
Vancouver, British Columbia, Canada

Tommy Ek
County Administration Board of Östergötland
Linköping, Sweden

Jens Emborg
Danish Centre for Forest, Landscape and Planning, KVL
Frederiksberg, Denmark

Thröstur Eysteinsson
Iceland Forest Service
Egilsstaðir, Iceland

E.P. Farrell
Department of Environmental Resource Management
University College Dublin
Dublin, Ireland

Søren Fodgaard
Danish Forest Association
Frederiksberg, Denmark

Peter Z. Fulé
School of Forestry and Ecological Restoration Institute
Northern Arizona University
Flagstaff, Arizona

Emile S. Gardiner
Southern Research Station
US Department of Agriculture Forest Service
Stoneville, Mississippi

Chris Geron
US Environmental Protection Agency
National Risk Management Research Laboratory
Research Triangle Park, North Carolina

Russell T. Graham
Rocky Mountain Research Station
US Department of Agriculture Forest Service
Moscow, Idaho

Lars Graudal
Danish Centre for Forest, Landscape and Planning, KVL
Copenhagen, Denmark

Maísa dos Santos Guapyassú
Fundação O Boticário de Proteção à Natureza, Curitiba
Paraná, Brazil

Jean-Luc Guitton
Auvergne Regional Centre of Private Forest
Lempdes, France

Katrine Hahn
Danish Centre for Forest, Landscape and Planning, KVL
Frederiksberg, Denmark

Christian Pilegaard Hansen
Danish Centre for Forest, Landscape and Planning, KVL
Humlebæk, Denmark

Jörg Hansen
Institute for Forest Growth
Albert-Ludwigs University
Freiburg, Germany

Ralph Harmer
Forest Research
Alice Holt Lodge
Surrey, England

Jonathan Humphrey
Forest Research Northern Research Station
Roslin, Midlothian, Scotland

Theresa B. Jain
Rocky Mountain Research Station
US Department of Agriculture Forest Service
Moscow, Idaho

Finn A. Jensen
Hedeselskabet
Viborg, Denmark

Kalev Jõgiste
Forest Research Institute
Estonian Agricultural University
Tartu, Estonia

Everett E. Johnson
School of Forestry and Wildlife Sciences
Solon Dixon Forestry Education Center
Auburn University
Andalusia, Alabama

Tomohiko Kamitani
Faculty of Agriculture
Niigata University
Niigata, Japan

Merrill R. Kaufmann
Rocky Mountain Research Station
US Department of Agriculture
Forest Service
Fort Collins, Colorado

Erik D. Kjær
Danish Centre for Forest,
Landscape and Planning, KVL
Copenhagen, Denmark

Timo Kuuluvainen
University of Helsinki
Helsinki, Finland

Lars Laestadius
World Resources Institute
Washington, D.C

J. Bo Larsen
Danish Centre for Forest,
Landscape and Planning, KVL
Frederiksberg, Denmark

Don Koo Lee
Department of Forest Science
Seoul National University
Seoul, Korea

Lars Lundqvist
Swedish University of Agricultural
Sciences
Umeå, Sweden

Palle Madsen
Danish Centre for Forest,
Landscape and Planning, KVL
Vejle, Denmark

Doug McCreary
University of California-Berkeley
Browns Valley, California

D. Evan Mercer
Southern Research Station
US Department of Agriculture
Forest Service
Research Triangle Park, North Carolina

Hideo Miguchi
Faculty of Agriculture
Niigata University
Niigata, Japan

Mikołaj Mikułowski
Forest Research Institute
Warszawa, Poland

Inge Stupak Møller
Danish Centre for Forest,
Landscape and Planning, KVL
Hørsholm, Denmark

Takuo Nagaike
Yamanashi Forest Research Institute
Masuho, Yamanashi, Japan

Tohru Nakashizuka
Research Institute for Humanity and
Nature
Kamigyo-ku, Kyoto, Japan

Kevin L. O'Hara
Environmental Science
Policy and Management
University of California
Berkeley, California

Chadwick Dearing Oliver
School of Forestry and
Environmental Studies
Yale University
New Haven, Connecticut

James M. Oliver
Ducks Unlimited Inc.
Ridgeland, Mississippi

Kenneth W. Outcalt
Southern Research Station
US Department of Agriculture
Forest Service
Athens, Georgia

Bronwyn Sigrid Owen
Department of Geography
University of Colorado
Boulder, Colorado

Gunnars Hauff Platais
World Bank
Washington, D.C.

Bernard Prevosto
CEMAGREF
Clermont-Ferrand Regional Centre
Aubière, France

Karsten Raulund-Rasmussen
Danish Centre for Forest, Landscape and Planning, KVL
Hørsholm, Denmark

Florence Renou
Department of Environmental Resource Management
University College Dublin
Dublin, Ireland

Jean-Michel Roberge
Swedish University of Agricultural Sciences
Riddarhyttan, Sweden

William H. Romme
Department of Forest Sciences
Colorado State University
Fort Collins, Colorado

Hans Roulund
Danish Centre for Forest, Landscape and Planning, KVL
Copenhagen, Denmark

Kevin C. Ryan
Rocky Mountain Research Station
US Department of Agriculture Forest Service
Missoula, Montana

Khosro Sagheb-Talebi
Research Institute of Forests and Rangelands
Tehran, Iran

Bernhard Schlamadinger
Joanneum Research
Graz, Austria

Heinrich Spiecker
Institute for Forest Growth
Albert-Ludwigs University
Freiburg, Germany

John A. Stanturf
Southern Research Station
US Department of Agriculture Forest Service
Athens, Georgia

Seung Jin Suh
Korea Forest Research Institute
Seoul, Korea

Richard Thompson
Forest Research Northern Research Station
Roslin, Midlothian, Scotland

Donald J. Tomczak
Southern Region State and Private Forestry
US Department of Agriculture Forest Service
Atlanta, Georgia

Harri Tukia
Finnish Environment Institute
Helsinki, Finland

Hardi Tullus
Faculty of Forestry
Estonian Agricultural University
Tartu, Estonia

Veiko Uri
Faculty of Forestry
Estonian Agricultural University
Tartu, Estonia

David H. Van Lear
Department of Forestry and Natural Resources
Clemson University
Clemson, South Carolina

Aivo Vares
Faculty of Forestry
Estonian Agricultural University
Tartu, Estonia

James M. Vose
Southern Research Station
US Department of Agriculture
Forest Service
Otto, North Carolina

Sven Wagner
Dresden University of Technology
Tharandt, Germany

John Walker
US Environmental Protection Agency
National Risk Management Research
Laboratory
Research Triangle Park, North Carolina

Kristen M. Waring
Environmental Science
Policy and Management
University of California
Berkeley, California

Norbert Weber
Institute for Forest and Environmental
Policy
Albert-Ludwigs University
Freiburg, Germany

Tricia L. Wurtz
Pacific Northwest Research Station
US Department of Agriculture
Forest Service
Fairbanks, Alaska

Toshiya Yoshida
Forest Research Station
Field Science Center for Northern
Biosphere
Hokkaido University
Moshiri, Horokanai, Hokkaido, Japan

Tadeusz Zachara
Forest Research Institute
Warszawa, Poland

Contents

Part 1

Introduction

chapter one

What is forest restoration?

John A. Stanturf
US Department of Agriculture Forest Service

Contents

1.1 Introduction

The need to repair habitat and restore forest structure and function is recognized throughout the temperate and boreal zones as a component of sustainable forest management (Krishnaswamy and Hanson 1999; Dobson et al. 1997). Forest restoration is a complex task, complicated by diverse ecological and social conditions, that challenges our understanding of forest ecosystems. The term restoration is used indiscriminately and it is difficult to define in a way that encompasses all situations found in the literature and in practice. Generally, restoration is seen as symmetric with degradation: an undisturbed forest in a natural or historical condition can be degraded, and a degraded forest can be restored to that natural or historical condition. As will become apparent, reality is more complicated and the fully restored state is probably unattainable (Cairns 1986; Stanturf and Madsen 2002). Terminology, however, is not merely an academic issue; definitions related to forestry and restoration are used under several international conventions such as climate change and biodiversity where distinctions and nuance have important policy implications (FAO 2002). The objective of this chapter is to provide a conceptual framework for the terms used throughout this book, in order to facilitate understanding of the diverse cultural and ecological contexts for restoration of temperate and boreal forests. This chapter has three parts: an historical context for restoration, which differs geographically; a conceptual framework for understanding the relationship between degradation and restoration; and an attempt to define restoration terms within that framework.

1.2 Historical context

Throughout history, forests have been a residual landuse; external pressures such as expanding human populations have caused forests to be cleared, usually for agriculture

(Noble and Dirzo 1997). The conversion of forest habitat to other uses has occurred at different rates and different times in history (Goudie 1986; Dobson et al. 1997). For example, most of Europe and Asia were settled millennia ago, while human occupation of the Americas is much shorter. Nevertheless, the transformation of landuse is not unidirectional; wars, plagues, population movement and fluctuations, and climate changes cause agricultural abandonment and reversion to forests. Significant changes have occurred within the last 200 years as developed nations shifted from a biomass energy economy to fossil fuels (Clawson 1979; Ericcson et al. 2000; Johan et al. 2004). Further changes are likely in the industrialized nations of the temperate zone, as changing policies for agriculture and nature conservation provide incentives for land-use shifts from agriculture to forest.

Forest restoration in the broad sense is not a new endeavor. Agricultural abandonment and natural invasion from remnant forests is a passive form of restoration that continues to occur (McIver and Starr 2001), notably in some former communist countries within the Commonwealth of Independent States (FAO 2001). Active restoration also has a long history; if the indirect effects of efforts to restore productivity to degraded land can be considered unintentional restoration. The development of secondary spruce forests in central Europe is an example of the complex pathway of degradation and restoration (Johan et al. 2004). Similarly, the loblolly pine forests of the southeastern United States were established to protect water and soil resources (USDA Forest Service 1988; Stanturf et al. 2003). In many countries, coastal dunes and heathland were planted to reclaim wasteland (e.g., Denmark; see Madsen et al. this volume). Active but unintentional restoration was motivated by the threat of timber scarcity and movements to improve nature, often with sociopolitical overtones such as providing employment and patriotic duty (Heske 1938; Orni 1969).

Forest restoration in industrialized countries at the beginning of the 21st century emerges from these earlier, more utilitarian concerns but with greater emphasis on restoring more natural forests (Farrell et al. 2000). Restoration ecology is an emerging science that developed out of restoration projects at specific sites (Hobbs and Norton 1996; Hobbs 2004). Ecological restoration "is an intentional activity that initiates or accelerates recovery of an ecosystem with respect to its health, integrity and sustainability" (SER 2002). Explicit in this definition is that ecological restoration is relative to reference conditions (Wagner et al. 2000; Perrow and Davy 2002a, b; SER 2002), which are related to a notion of natural conditions for a site (Hobbs and Norton 1996; Egan and Howell 2001). Thus, "true" restoration has the most ambitious goal of reconstructing a prior ecosystem (van Diggelen et al. 2001; SER 2002); less ambitious efforts only seek to partially restore naturalness (rehabilitation) or generally increase biodiversity (reclamation). To the purist, restoration is different from rehabilitation or reclamation because its objective is to return an ecosystem to some preexisting, natural state that often presumes an absence of human disturbance (Hobbs and Norton 1996).

Practitioners within the restoration ecology community (Hobbs 2004; Davis and Slobodkin 2004) and other resource professionals (Wagner et al. 2000; Stanturf et al. 2001) have challenged the notion of naturalness as an objective. The crux of the debate is whether naturalness represents a scientifically defensible concept (Anderson 1991) or is simply a statement of a preference for one kind of ecosystem over another (Hobbs 2004). Some restoration ecologists are moving away from the purist position, especially the more ideological views that set the goal of restoration to be an idealized pristine state, which implies a static view of ecosystems (Hobbs 2004; Davis and Slobodkin 2004). Nevertheless, a lively debate has ensued over whether the endpoint of restoration can be set in a way that is free of values (Winterhalder et al. 2004).

Crucial to the debate are the starting and ending points, and a practitioner's perception of the extent of human influence in a forested ecosystem. Forests today are human-dominated systems (Noble and Dirzo 1997; Vitousek et al. 1997; Ericsson et al. 2000), although there is a tendency in the Americas to underestimate the extent to which indigenous peoples influenced the forests described by the first European naturalists (Stanturf et al. 2002). If the starting point is a degraded forest only slightly removed in time from a natural or seminatural forest, suitable reference stands may be available for setting endpoints. Suitable reference conditions are unavailable and are likely to be unknowable if a forest minimally influenced by humans is hundreds or thousands of years distant from the starting point (Wagner et al. 2000; Hobbs 2004). Recognizing the difficulty of setting restoration goals on the basis of re-creating past conditions (Parker and Pickett 1997; Hobbs and Norton 1996; Sprugel 1991; Bradshaw, this volume) leads to the conclusion that the endpoint is a sociopolitical decision (Hobbs 2004) that can be informed by science (Keddy and Drummond 1996), but cannot be determined by science alone. What constitutes successful restoration will be defined within a cultural and ecological context, including financial costs and unexpected consequences (Anderson and Dugger 1998; Holl and Howarth 2000; Palik et al. 2000; Anand and Desrochers 2004). The appropriate intervention will be determined largely by the degree of degradation (Hobbs 2004) and the likelihood of success.

1.3 Degradation and restoration processes

Forest condition is dynamic, subject to natural developmental processes (Oliver and O'Hara, this volume) as well as natural and anthropogenic disturbances (Covington et al. 1997; Turner et al. 1998; Angelstam 1998; Stanturf et al. 2001; Beatty and Owen, this volume). Degradation results from changes to forest structure or function that lowers its productive capacity (FAO 2002), including limited biodiversity. Degradation is not synonymous with disturbance; disturbance becomes degradation, however, when it crosses a threshold beyond the natural resilience of a forest type. The simplest conceptualization of the relationship between degradation and recovery processes is to place a forest on a continuum from natural to degraded (Bradshaw 1997; Harrington 1999). Levels of state factors such as biomass or biodiversity in a forest subjected to degradation follow a linear trajectory. At any point along the trajectory, recovery toward a natural forest can be initiated once the stress or disturbance abates.

Intervention can facilitate recovery from disturbance or degradation. For convenience, intervention can be divided into three levels of increasing effort: self-renewal, rehabilitation, or reconstruction/reclamation (Stanturf et al. 2001). In the self-renewal phase, resistance and resilience mechanisms maintain or return the forest more or less to its original state, without human intervention, in a relatively short time. Sustainably managed forests rely on self-renewal processes, for example, naturally regenerated forests managed for timber. Intervention at this stage will be to ensure that composition and structure meet management objectives. Plantations of native species can be within the scope of self-renewal, where intervention (reforestation) is undertaken to control species and stocking.

At intermediate levels of disturbance, beyond the self-renewal phase, degradation occurs. If a forest is degraded but remains in forest landuse, meaning it is not deforested, it can be rehabilitated to a forest condition that is within the range of self-renewal mechanisms. Recovery to a more natural forest will take longer, but the time required can be shortened by human intervention. Rehabilitation by reforestation of forests consumed by wildfire is an example. In the most degraded state, forest cover is removed and the land is converted to another use: this is deforestation (FAO 2002). A forest degraded by acute air pollution may be deforested to a nonforest condition and the land becomes wasteland. In the most degraded conditions, after the pollution or other landuse ceases, the forest may recover to a natural forest condition in a century or longer. The recovery period may be shorter, possibly decades, with human intervention (reconstruction or reclamation). In this chapter, restoration encompasses all interventions into degraded forests, those stands disturbed beyond the range of self-renewal processes. Rehabilitation refers to restoration of degraded forests; reconstruction and reclamation encompass restoration of forests from nonforest landuses.

1.3.1 *Degradation processes*

The dynamic relationship between processes degrading and restoring forests is more easily understood if considered in light of two dimensions, changes in land cover, landuse, or both (Stanturf and Madsen 2002). Taking as the starting point the undisturbed, idealized natural mature forest (Westhoff 1983; Goudie 1986), then, conversions to other landuse such as agriculture (cultural landscape) or pasture (seminatural landscape) are through deforestation (Figure 1.1). Relatively frequent but moderate disturbance (plowing, herbicides, grazing) maintains the nonforest cover. Similarly, a change in both land cover and landuse occurs when forests are removed and the land is converted to urban areas, flooded by dams, or removed along with topsoil and overburden in mining and extractive activities. Such drastic degradation involves deforestation, usually accompanied by ongoing disturbance. The nonforest cover is maintained more or less permanently by structures, more so than by cultural activities (Figure 1.1). Agricultural land can also be converted to urban uses.

Harvesting a mature forest in a sustainable manner is a change of land cover but not landuse (FAO 2002). A new, young forest will result from natural regeneration or by reforestation (within the envelope of forest cover in Figure 1.1). Unsustainable harvesting without securing adequate regeneration, however, may degrade stand structure or diversity.

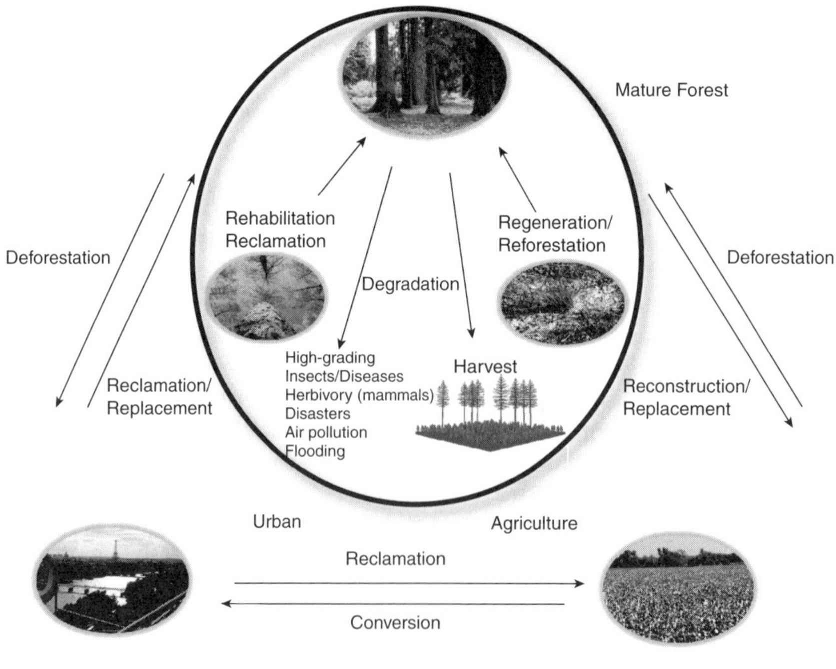

Figure 1.1 Forest restoration begins with forests that have been degraded (rehabilitation) or after deforestation and conversion to other land uses (reconstruction or reclamation). Self-renewal processes operate within forests that are disturbed but not degraded (regeneration/reforestation). (Adapted from Stanturf and Madsen, Restoration concepts for temperate and boreal forests of North America and Western Europe, *Plant Biosyst.*, 136, 143, 2002).

Outbreaks of insects or diseases (especially exotic species), fire suppression and disruption of natural fire regimes, invasion by aggressive exotic plants, or disasters such as hurricanes or wildfires can degrade forest stands and change attributes of land cover, but these stressors do not change landuse (Figure 1.1). Chronic low-level pollutant loading may degrade a forest without changing landuse, although heavy loading may deforest an area and change use into wasteland.

1.3.2 Restoration processes

Forests are resilient: given sufficient time and the cessation of disturbances, agricultural and urbanized land will revert to forest. Abandonment and reversion to forests, although secondary, seminatural, or degraded forest types, will be on a time scale of a few decades to centuries as existing forests expand into nonforest areas, or natural invasion occurs. Human intervention, however, can accelerate the reversion process (Ferris-Kaan 1995). Reconstructing forests through afforestation of agricultural land may consist of simply planting trees, although more intensive techniques are available (Stanturf et al. 2001). Reclamation of urbanized land usually requires extensive modification, including stabilization of spoil banks or removal of water control structures, followed by tree planting. Because severe site degradation may limit the possibility of restoring to natural forest condition, reclamation is sometimes called replacement (Bradshaw 1997).

1.4 Restoration terms

The dualistic notion of degradation and restoration as opposing trajectories of forest development leads to an understanding of restoration in a broader context than ecological restoration (SER 2002). In this view, the restored forest that results from reconstruction, reclamation, or rehabilitation may never recreate the original state for all functions (Cairns 1986; Bradshaw 1997; Harrington 1999). Any endpoint within the natural range of managed forests where self-renewal processes operate is acceptable as restoration. Thus, restoration to an early seral stage would be acceptable for a forest that is likely to attain a more complex structure through typical stand dynamics. How quickly the forest moves to the self-renewal phase is a function of forest type, site resources, and the amount invested to overcome the degraded conditions. This model offers a broader context for restoration on private land; landowners with management objectives other than preservation are able to contribute to ecosystem restoration (Farrell et al. 2000; Stanturf et al. 2001).

Commonly used restoration terms can be understood within a conceptual framework (Figure 1.2) that takes into account the relationships between changes in forest cover and landuse (Figure 1.1). This is not an attempt to standardize these terms, but to harmonize (FAO 2002; Hasenauer 2004) them for consistent use in the chapters that follow. The Degradation trajectory begins with the idealized forest at Ω as the starting point (Figure 1.2). This beginning point is culturally and situationally determined. In some contexts, it may represent an actual historical reality, or it may be a conceptual model of the potential natural vegetation for an area. The degradation trajectory moves toward a degraded endpoint, A in Figure 1.2. The possible endpoints are shown in Figure 1.1; the most degraded states will include deforestation and conversion to nonforest landuse. The intermediate points B_1 to B_3 represent forests degraded by air pollution, exploitive harvesting, natural disasters, etc. These degraded forests, as well as nonforest conditions (A), represent starting points for restoration trajectories. For ease of representation, the A to Ω trajectory is presented as linear; in reality it is probably more complex (Anand and Desrochers 2004).

The path extending from A to Ω, labeled Recreation, represents the strictly defined ecological or historical restoration (SER 2002). Re-creating the ideal natural or historical

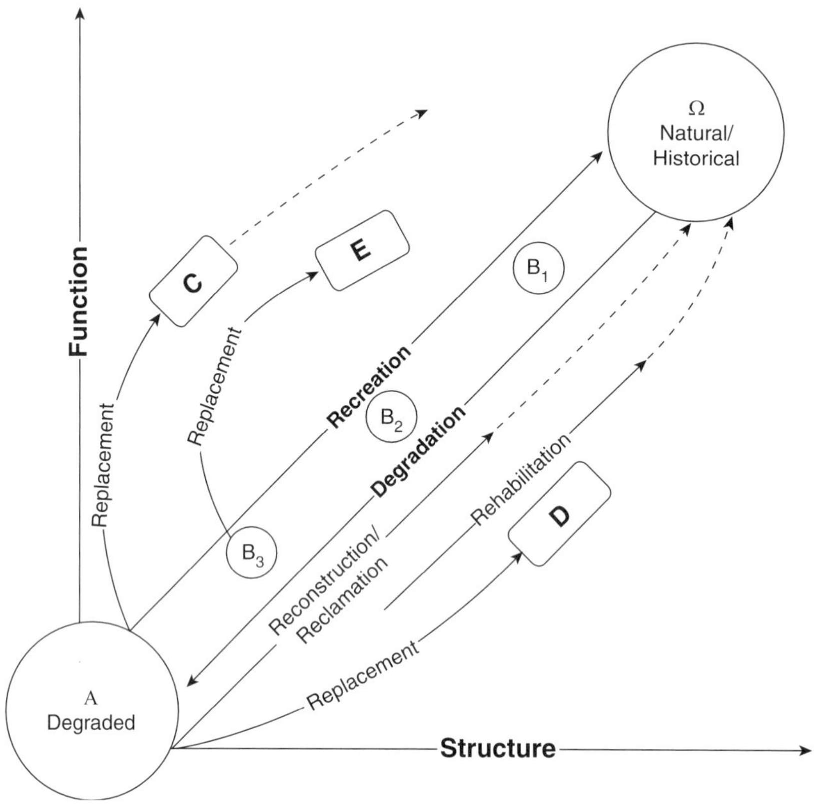

Figure 1.2 A conceptual framework for forest restoration has a starting point of a degraded forest (A) and an idealized endpoint of a forest restored to some natural or historical endpoint (Ω). The symmetric degradation/re-creation trajectories have intermediate points that represent starting/ending points (B_1 to B_3) for reconstruction or reclamation of severely degraded forests (deforested and converted to other landuse) or less severely degraded forests (rehabilitation). Replacement trajectories denote restored forests that lack the structure or species composition of native forests

forest ecosystem is unlikely to be successful over large areas (van Diggelen et al. 2001) and will certainly be expensive. Reconstruction refers to restoration of forest conditions to agricultural land (Figure 1.1), through afforestation or natural invasion. The endpoint for reconstruction (B_1 to B_3) may be a less diverse natural forest (B_2) or a mixed species plantation of native species (B_3). Alternatively, a site may be so degraded that native species are replaced by exotics; this pathway (to C) would be termed replacement.

Reclamation begins with urban or built land-use and may require land stabilization as well as afforestation. In North America and the U.K., reclamation is commonly used in the context of mined land. In the older literature, reclamation had an opposite meaning of putting derelict land to good purpose, such as draining swamplands or irrigating arid lands. For both reconstruction and reclamation, continuing intervention over time may move the forest condition closer to the natural endpoint (shown as a dashed line in Figure 1.2).

Rehabilitation of degraded forests has one of the intermediate conditions (B_1 to B_3) as a starting point; forest cover has been removed or degraded but no change to nonforest landuse has occurred. Rehabilitation encompasses many techniques to restore stand structure, species composition, natural disturbance regimes, or to remove exotic plants. Specific forms of rehabilitation are termed conversion (Nyland 2003; Spiecker et al. 2004) or transformation (Kenk and Guehne 2001).

The conditions at C, D, and E represent Replacement; these are forests that deviate from the natural range of variability but restore forestry landuse. Plantations of exotic species, for example, have a simple structure but high functioning as compared to non-forest landuse (C). Over time, with or without further intervention, even replacement stands could move toward the natural endpoint by gaining structure or additional species. For example, the conversion or transformation of Norway spruce to broadleaved forests in central Europe follows this restoration trajectory (Spiecker et al. 2004; Hanson and Spiecker, this volume; Baumhauer et al. this volume; Hahn et al. this volume). Spruce was planted on broadleaved sites for a variety of reasons (Johan et al. 2004), following the A to C trajectory (Figure 1. 2). Conversion or transformation back to mixed broadleaved forests (C to B_1 or to Ω) completes the restoration of a natural forest within the range of self-renewal processes. Starting point D represents the rehabilitation of forests with disrupted natural disturbance regimes, such as fire-suppressed conifer forests in North America (stands with high structure but low functioning). Rehabilitation requires altering structure and composition before reintroducing fire (Brockway et al. this volume; Kaufmann et al. this volume).

1.5 Conclusion

Forests are human-dominated ecosystems (Noble and Dirzo 1997). Reconstructing forests where they are now absent, as well as altering existing forests to more natural conditions, are important aspects of sustainable forest management. Important tasks for forest restorationists are to understand how ecosystems were degraded, how to reverse degradation processes, and how to efficiently initiate recovery processes (Hobbs and Norton 1996). Silviculturists and forest ecologists have important tools — diagnostic and predictive skills and effective intervention techniques — that are critical to successful restoration of complex ecosystems. These tools must be used appropriately, however, within diverse sociopolitical, ecological, and historic contexts.

A broad conception of restoration allows more diverse goals (endpoints) than the narrow construct of re-creating particular, preexisting ecosystem states (reference conditions). Greater flexibility in setting restoration objectives is not a retreat from basing restoration on ecological science (Wagner et al. 2000). Rather, it is a recognition that incomplete knowledge of past ecosystem states (Hobbs 2004), changes in the global environment (Vitousek et al. 1997), costs, and the scale of degradation argue for a pragmatic approach.

References

Anand, M. and Desrochers, R.E., Quantification of restoration success using complex systems concepts and models, *Restor. Ecol.*, 12, 117, 2004.

Anderson, J.E., A conceptual framework for evaluating and quantifying naturalness, *Conserv. Biol.*, 5, 347, 1991.

Anderson, D.H. and Dugger, B.D., A conceptual basis for evaluating restoration success, in *Transactions of 63rd North American Wildlife and Natural Resources Conference: 376-383*, Orlando, FL, Wildlife Management Institute, Washington, D.C., 111, 1998.

Angelstam, P., Maintaining and restoring biodiversity by developing natural disturbance regimes in European boreal forests, *J. Veg. Sci.*, 9, 593, 1998.

Bradshaw, A.D., What do we mean by restoration? in *Restoration Ecology and Sustainable Development*, Urbanska, K.M., Webb, N.R. and Edwards, P.J., Eds., Cambridge University Press, Cambridge, U.K., 8, 1997.

Cairns, J. Jr., Restoration, reclamation, and regeneration of degraded or destroyed ecosystems, in *Conservation Biology*, Soule, M.E., Ed., Sinauer Publishers, Ann Arbor, MI, 465, 1986.

Clawson, M., Forests in the long sweep of American history, *Science*, 204, 1168, 1979.

Covington, W.W., Fulé, P.Z., Moore, M. M., Hart, S.C., Kolb, T.E., Mast, J.N., Sackett, S.S. and Wagner, M.R., Restoring ecosystem health in ponderosa pine forests of the southwest, *J. For.*, 95, 23, 1997.

Davis, M. A. and Slobodkin, L.B., The science and values of restoration ecology, *Restor. Ecol.*, 12, 1, 2004.

Dobson, A.P., Bradshaw, A.D., and Baker, A.J. M., Hopes for the future: Restoration ecology and conservation biology, *Science*, 277, 515, 1997.

Egan, D. and Howell, E.A., Eds., *The Historical Ecology Handbook: A Restorationist's Guide to Reference Ecosystems*, Island Press, Washington, D.C, 2001.

Ericsson, S., Östlund, L., and Axelsson, A.-L., A forest of grazing and logging: deforestation and reforestation history of a boreal landscape in central Sweden, *New Forest.*, 19, 227, 2000.

Farrell, E.P., Führer, E., Ryan, D., Andersson, F., Hüttl, R., and Piussi, P., European forest ecosystems: building the future on the legacy of the past, *For. Ecol. Manage.*, 132, 5, 2000.

Ferris-Kaan, R., Ed., *The Ecology of Woodland Creation*, John Wiley & Sons, Chichester, U.K., 1995.

Food and Agriculture Organization (FAO), *Global Forest Resources Assessment 2000*, FAO Forestry Paper 140, Rome, 2001.

Food and Agriculture Organization (FAO), *Proceedings Expert Meeting on Harmonizing Forest-related Definitions for Use by Various Stakeholders*, Rome, 22–25 January 2002, FAO, Rome, 2002.

Goudie, A., *The Human Impact on the Natural Environment*, MIT Press, Cambridge, MA, 1986.

Harrington, C.A., Forests planted for ecosystem restoration or conservation, *New Forest.*, 17, 175, 1999.

Hasenauer, H., Glossary of terms and definitions relevant for conversion, in *Norway Spruce Conversion—Options and Consequences*, Spiecker, H., Hansen, J., Klimo, E., Skovsgaard, J.P., Sterba, H. and von Teuffel, K., Eds., Brill, Boston, 5, 2004.

Heske, F., *German forestry*, Yale University Press, New Haven, CT, 1938.

Hobbs, R.J., Forum: Restoration ecology: the challenge of social values and expectations, *Front. Ecol.*, 2, 43, 2004.

Hobbs, R.J. and Norton, D.A., Towards a conceptual framework for restoration ecology, *Restor. Ecol.*, 4, 93,1996.

Holl, K.D. and Howarth, R.B., Paying for restoration, *Restor. Ecol.*, 8, 260, 2000.

Johan, E., Agnoletti, M., Axelsson, A.-L., Bürgi, M., Östlund, L., Rochel, X., Schmidt, U.E., Schuler, A., Skovsgaard, J. -P., and Winiwarter, V., History of secondary spruce forests in Europe, in *Norway spruce conversion—options and consequences*, Spiecker, H., Hansen, J., Klimo, E., Skovsgaard, J.P., Sterba, H. and von Teuffel, K., Eds., Brill, Boston, 25, 2004.

Keddy, P.A. and C.G. Drummond, Ecological properties for the evaluation, management, and restoration of temperate deciduous forest ecosystems, *Ecol. Appl.*, 6, 748, 1996.

Kenk, G. and Guehne, S., Management of transformation in central Europe, *For. Ecol. Manage.*, 151, 107, 2001.

Krishnaswamy, A. and Hanson, A., Eds., *Our Forests, Our Future: Summary Report, World Commission On Forests And Sustainable Development*, Cambridge University Press, Cambridge, U.K., 1999.

McIver, J. and Starr, L., Restoration of degraded lands in the interior Columbia River basin: passive vs. active approaches, *For. Ecol. Manage.*, 153, 15, 2001.

Noble, I.R. and Dirzo, R., Forests as human-dominated ecosystems, *Science*, 277, 522, 1997.

Nyland, R.D., Even- to uneven-aged: the challenges of conversion, *For. Ecol. Manage.*, 172, 291, 2003.

Orni, E., *Afforestation in Israel*, Keren Kayemeth Leisrael, Jerusalem, 1969.

Palik, B.J., Goebel, P.C., Kirkman, L.K., and West, L., Using landscape hierarchies to guide restoration of disturbed ecosystems, *Ecol. Appl.*, 10, 189, 2000.

Parker, V.T. and Pickett, S.T.A., Restoration as an ecosystem process: implications of the modern ecological paradigm, in *Restoration Ecology and Sustainable Development*, Urbanska, K. M., Webb, N.R., and Edwards, P.J., Eds., Cambridge University Press, Cambridge, U.K. 17, 1997.

Perrow, M.R. and Davy, A.J., Eds., *Handbook of ecological Restoration 1: Principles of Restoration*, Cambridge University Press, Cambridge, U.K., 2002a.

Perrow, M.R. and Davy, A.J., Eds., *Handbook of Ecological Restoration 2: Restoration in Practice*, Cambridge University Press, Cambridge, U.K., 2002b.

Society for Ecological Restoration Science and Policy Working Group (SER), *The SER Primer on ecological restoration*, 2002. http://www.ser.org/Primer.pdf Viewed February 24, 2004.

Spiecker, H., Hansen, J., Klimo, E., Skovsgaard, J.P., Sterba, H., and von Teuffel, K., Eds., *Norway Spruce Conversion — Options and Consequences*, Brill, Boston, 2004.

Sprugel, D.G., Disturbance, equilibrium, and environmental variability: what is natural vegetation in a changing environment? *Biol. Conserv.*, 58, 1, 1991.

Stanturf, J.A., Schoenholtz, S.H., Schweitzer, C.J., and Shepard, J.P., Achieving restoration success: myths in bottomland hardwood forests, *Restor. Ecol.*, 9, 189, 2001.

Stanturf, J.A. and Madsen, P., Restoration concepts for temperate and boreal forests of North America and Western Europe, *Plant Biosyst.*, 136, 143, 2002.

Stanturf, J.A., Wade, D.W., Waldrop, T.A., Kennard, D.K., and Achtemeier, G.L., Fire in Southern Forest Landscapes, in *Southern Forest Resource Assessment*, Wear, D.M. and Greis, J.G., Eds., U.S. Department of Agriculture, Forest Service, Southern Research Station, Asheville, NC, General Technical Report. SRS-53, 607, 2002.

Stanturf, J.A., Kellison, R., Broerman, F.S., and Jones, S.B., Pine productivity: where are we and how did we get here? *J. For.*, 101, 26, 2003.

Turner, M.G., Baker, W.L., Peterson, C.J., and Peet, R.K., Factors influencing succession: lessons from large, infrequent natural disturbances, *Ecosystems*, 1, 511, 1998.

U.S. Department of Agriculture, Forest Service, *The Yazoo-Little Tallahatchie flood prevention project*, U.S. Department of Agriculture, Forest Service, Southern Region, Atlanta, GA, Forestry Report R8-FR 8, 1988.

Van Diggelen, R., Grootjans, Ab P., and Harris, J.A., Ecological restoration: State of the art or state of the science? *Restor. Ecol.*, 9, 115, 2001.

Vitousek, P.M., Mooney, H.A., Lubchenco, J., and Melillo, J.M., Human domination of Earth's ecosystems, *Science*, 277, 494, 1997.

Wagner, M.R., Block, W.M., Geils, B.W., and Wenger, K.F., Restoration ecology: A new forest management paradigm, or another merit badge for foresters? *J. For.*, 98, 23, 2000.

Westhoff, V., Man's attitude toward vegetation, in *Man's Impact On Vegetation*, Holzner, W., Werger, M.J.A. and Ikusima, I., Eds., Junk, The Hague, 7, 1983.

Winterhalder, K., Clewell, A.F., and Aronson, J., Values and science in ecological restoration — A response to Davis and Slobodkin, *Restor. Ecol.*, 12, 4, 2004.

Part 2

The Ecological Context for Forest Restoration

chapter two

What is a natural forest?

Richard H.W. Bradshaw
Geological Survey of Denmark and Greenland

Contents

2.1 Introduction

A true "natural" forest is defined here as an idealized virgin forest condition that is uninfluenced by large-scale, systematic human activity. Peterken (1996) described a continuum from natural to managed forest, and while some forest areas are almost totally managed, it is doubtful whether any truly natural, virgin forests survive in the world. Even when a "near-natural" forest has been identified, considerable research is required to verify the virgin condition, including long-term observations to establish that current dynamic processes are not the consequence of earlier human intervention (Wolf et al. 2004). Ideally, these observations are complemented by paleoecological investigations that rule out the possibility of earlier breaks in forest continuity, which may have influenced current species content (Segerström et al. 1994).

Despite the practical difficulty of positively identifying natural forest, the concept and the identification of particular near-natural sites help define the naturalness endpoint and provide a reference for use in management of natural areas (Figure 2.1) and restoration of degraded forests. All silvicultural systems are modifications of natural systems to some extent. Studies of the dynamic processes within natural forest can also serve as guidelines and a source of inspiration for the development of "close-to-nature" silvicultural systems (Bradshaw et al. 1994). As society and the forest managers seek systems that safeguard biodiversity and minimize long-term environmental impact, it seems helpful to re-examine the concept of natural and its realization in reference sites.

1-56670-635-1/05/$0.00/$1.50/

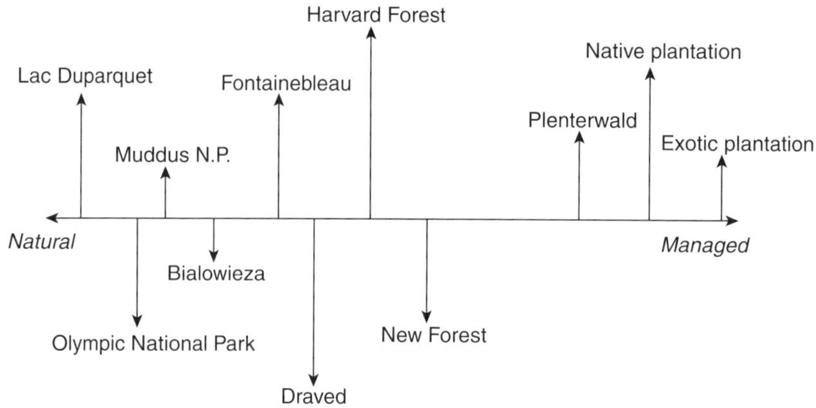

Figure 2.1 A hypothetical gradient of forests from truly natural to totally managed. A selection of forest areas and management systems are positioned on the gradient. Forests listed are: Lac Duparquet, Quebec, Canada; Olympic National Park, U.S.A.; Muddus National Park, Sweden; Bialowieza National Park, Poland, Fontainebleau Reserve, France; Draved Forest, Denmark; Harvard Forest, U.S.A.; New Forest, England.

In this chapter, I review the probable broad-scale distribution of natural forest within the boreal and temperate zones of the Northern Hemisphere and discuss some of the defining characteristics of natural forest and the fresh insight into the history of natural forest stands provided by the recent application of molecular biological techniques to forest genetics, species composition, and the genetic structure of tree populations (Petit et al. 2002). I consider, in some detail, disturbance regimes and other factors that drive dynamic processes, generate stand structure, and control continuity, and briefly describe some other important natural forest characteristics including hydrology, biogeochemistry, and deadwood.

2.2 *The distribution of natural forest within the boreal and temperate zones*

The degree of naturalness varies greatly by region among the boreal and temperate forests of the Northern Hemisphere. Old World forests have a long history of human interference. Evidence for hunting can be traced back several hundreds of thousands of years in central Europe (Thieme 1997), but major disruption of natural forests in Europe dates from the origins of organized agriculture. The Neolithic culture spread to the north and west in Europe following the first cultivation of crops and domestication of livestock in the near East, beginning around 11,000 years ago (Roberts 1998). In southeastern Europe, this process began about 10,000 years ago, but in the boreal regions of Scandinavia, only as recently as 1,000 years ago. Agriculture was introduced to Denmark about 4000 B.C. and the major structure of the cultural landscape was formed by 1000 B.C. (Odgaard and Rasmussen 2000). The advent of farming initiated forest fragmentation, altered fire regimes, affected ungulate populations and caused increasingly rapid species turnover.

The changeover from hunter–gatherer societies to settled agriculture is a convenient point to define the onset of large-scale, systematic human activity in forests that become part of organized agricultural systems, such as in Europe. Many forests were cleared for arable land or used as hay-meadows and subsequently abandoned. Larger areas were used for extensive grazing of domestic animals and collection of firewood and construction material. Thus, the least modified forest areas, and therefore best candidates for natural forest, were located in remote areas that were unsuitable for agriculture, where population densities have always been low. These included large areas of northern and eastern Europe and Asia, and the mountains of central Europe. During the 1800s, industrial forestry reached even these areas, and natural forests declined rapidly in area and natural quality. Natural forest survived only in regions and stands of little economic value or where timber extraction costs were high (Figure 2.2). Two consequences of this exploitation history are that the surviving natural forests are often inappropriate references for forests on fertile, lowland areas with a high production potential and more boreal natural forests survive than temperate natural forests.

The most natural forest areas in the Old World lie, therefore, in remote, inhospitable areas bypassed by settled agriculture or commercial forestry. In Europe, even the few large, relatively intact areas that survived have experienced disrupted ungulate–predator systems that are an integral part of the forest ecosystem (Vera 2000). The *Betula*-dominated forests along the Scandes Mountains in Scandinavia comprise a large, near-natural forest region that has escaped the attention of farmers and foresters alike, although reindeer (*Rangifer tarandus*) husbandry, particularly in recent years, has exerted an influence that is unnatural. There is a large, relatively intact mixed coniferous–deciduous forest ecosystem

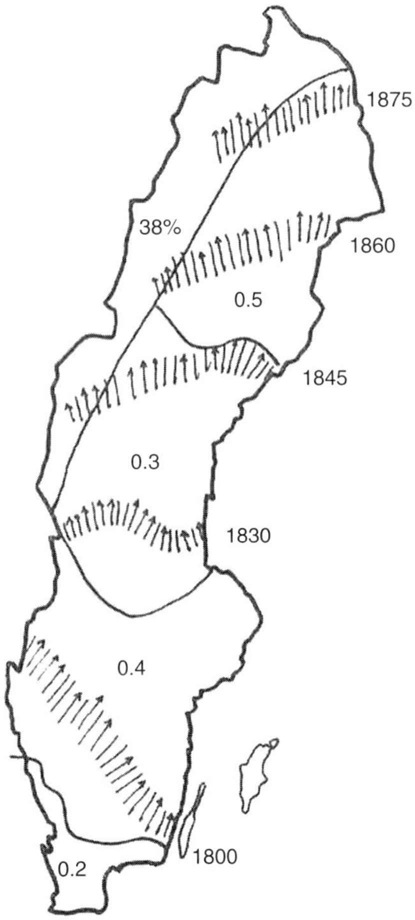

Figure 2.2 Timing of the movement of the "timber frontier" through Sweden during the 1800s, when the best-quality trees were selectively cut from the natural forests. The figures within each region are the percentage of protected forest at present (after Angelstam 1997).

in the southeast Carpathian Mountains of Romania where wolves (*Canis lupus*) are still active, but this is now being exploited for timber. The largest, unexploited forest resources in the Old World, both mixed-deciduous and boreal, lie within the northern and eastern regions of the former Soviet Union, but their future as natural forests is uncertain.

Notable smaller natural forests survive in border areas: regions where political unrest discouraged permanent settlement and agriculture, and in large private hunting reserves and estates. Of the latter, the forests that survived the best belonged to stable organizations such as monarchies or churches. Every country has examples of these areas, such as The New Forest, England; Fontainbleau, France; Neuburger Urwald, Germany; Bialowieza, Poland; and Siggaboda, Sweden (Koop 1989; Peterken 1996; Parviainen 2000).

In the New World, "presettlement" forests (pre-European settlers but not before aboriginal settlement) are often taken as a working reference for natural forest. There is widespread debate about the influence and timing of aboriginal peoples and their impact upon North American forests, particularly with regard to alteration of the fire and grazing regimes (see sections 4.1. and 4.2.), but historical documentation, among other sources, gives a useful impression of the extent and location of these forests. Large fragments of presettlement forest have survived in relatively unmodified form in many parts of Canada, Alaska, and the Pacific Northwest, and thus the widely accepted standard reference for natural forest is better described for boreal and montane North America than for Europe. As in Europe, the temperate forests have a longer history of human modification than do coniferous and boreal forests. Most of the Appalachian range in the Eastern U.S. has become widely reestablished with hardwood forest after agricultural exploitation from the early European settlers and only fragments of old-growth forest remain in remote upland areas (Foster and O'Keefe 2000).

2.3 Species composition and genetic structure of natural forest

A natural forest concept provides the regional list of native forest species and thus also defines exotic and alien species. The list of native species varies over long time-scales because natural extinctions occur and invading species become naturalized (Bradshaw 1995). There has been a loss of tree species diversity during the last 2 million years within northwest Europe (Figure 2.3) that is less apparent in North America and temperate China. The best explanation for this loss is the small and fragmented nature of the glacial refugia in Europe and the speed and severity of the climatic fluctuations (Huntley 1993; Bradshaw 1995; Hewitt 1996). The native tree species list for a country can assume an exaggerated importance in conservation debates where exotic species are regarded as potential threats to natural forest communities. This discussion may seem irrelevant to the shorter time-

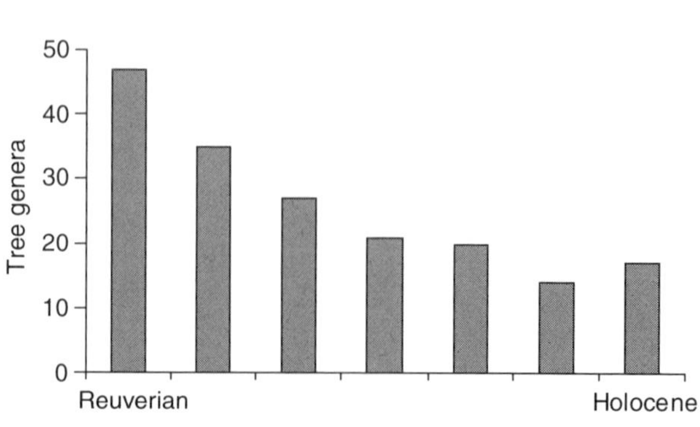

Figure 2.3 The number of tree genera recorded as fossils during some Quaternary interglacials. The time-scale is nonlinear but begins in the Reuverian c. 2.4 million years ago.

scales of forest management, but the native status of *Acer pseudoplatanus* or *Larix* are important conservation issues in some European countries, where forest conservation and management of indigenous species are prioritized. Exotics are less of an issue in the forestry of more species-rich forest regions. The take-home message is that the species content of natural forest is dynamic, particularly during periods of rapid climatic change such as the present. Most data are available for the dominant tree species, but as these structure the forest environment and act as host to many associated insect, moss, and fungal species, it is likely that a limited turnover of all forest species is a characteristic of natural forest.

Advances in our understanding of the genetic structure of forest tree populations at the continental scale in Europe emphasizes, through molecular genetics, the importance of glacial refugia as centers of genetic diversity and reservoirs for the long-term survival of forest species (Petit et al. 2002). The dynamic aspects of natural forest suggested by paleoecological analyses are also reflected in genetic structures of populations at a continental scale. For example, the pattern of chloroplast DNA variation across Europe displays the same pattern of postglacial species migration as recorded through pollen analysis (Petit et al. 2002; Brewer et al. 2002). This relationship is particularly apparent within European beech (*Fagus sylvatica*) and European white oaks (*Quercus robur* and *Quercus petraea*). *Fagus* pollen recorded from several hundred sediment profiles throughout Europe has been used to date its rise to local dominance at four time points over the last 7,000 years (Figure 2.4) (Bradshaw and Mountford 2002). The maps record the migration of beech in a northwesterly direction from glacial refugia, with the most rapid population expansion occurring 3,000 to 5,000 years ago.

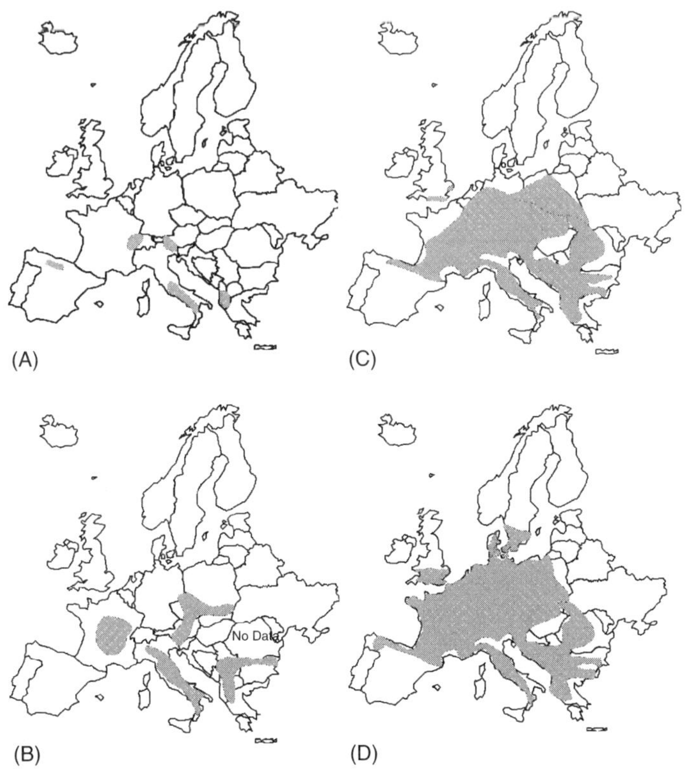

Figure 2.4 Maps based on fossil pollen data showing regions of Europe where *F. sylvatica* was a significant forest component (A) >7,000 years ago, (B) 5,000 years ago, (C) 3,000 years ago, and (D) 1,000 years ago.

A map of *Fagus* haplotypes retains the memory of earlier *Fagus* distributions; maximum diversity is recorded in presumed glacial refugia where populations may well have grown for hundreds of thousands of years (Vendramin, pers. comm.). Many contemporary *Fagus* stands in mountain areas in central Europe are among the most natural examples of surviving old growth, unmanaged forest. A *Fagus* forest in Denmark can be just as natural as one in central Italy, but their genetic structures may be quite different. As all northern populations are recently (in geological and evolutionary terms) derived from southern areas, the notion of the importance of local provenance in forest management and conservation needs careful evaluation (Haskell 2001; Wilkinson 2001).

2.4 Disturbance regimes

Natural forest dynamics are driven by infrequent but significant disturbance events such as wind, fire, drought, flooding, land-slippage, and disease (Pickett and White 1985; Peterken 1996; Gardiner and Quine 2000). These driving forces interact with the continuous disturbance of browsing and grazing animals and the localized processes of single tree replacement by seedling establishment, growth, competitive interactions, and senescence (Falinski 1986; Woods 2000). The integration of these factors in the disturbance regime, which creates the conditions necessary for regeneration, generates deadwood, influences the size and age structure of a stand, and controls the continuity of local niches and habitats. The disturbance regime is at the center of many of the dynamic processes and characteristics associated with natural forest.

2.4.1 Fire

The pivotal role of fire in many natural boreal ecosystems is widely acknowledged and has been incorporated into forest management (Johnson 1992; Goldammer and Furyaev 1996; Bradshaw et al. 1997; Angelstam 1998). Fires have always been important in boreal regions, although fire regimes have varied with prevailing climatic conditions. Individual forest fires tend to be very large in Canada and Siberia, but have been smaller in Scandinavia, at least during the last few centuries, as reconstructed by dendrochronological methods (Niklasson and Granström 2000). It is still an open question whether this difference in size of fires is due to differences in landscape form and the distribution of natural firebreaks, or whether cultural factors play a role. Road-building and fire suppression activities have impacted Fennoscandian fire regimes during the last 100 years, but the effect of these cultural factors may have been overestimated (Carcaillet et al. 2002).

Fire is one of the main disturbance factors in natural forest, but because of its long association with human societies and the ease with which a fire regime can be influenced by cultural activity, it is difficult to describe a "natural" fire regime for all but the most remote boreal forest regions of the world. The historical legacy and role of fire in temperate regions and at the boreal-temperate ecotone is less well known (Bradshaw et al. 1997), but raises issues of importance for forest restoration in these regions. Sediment records from small forest hollows throughout southern Scandinavia contain abundant charcoal from the past, but charcoal is less abundant in recent centuries and decades (Figure 2.5). For example, the record from Nissatorp in southeast Sweden records fires of probable natural origin between 4,000 and 500 B.C. (Figure 2.6). The charcoal record changes form after 500 B.C. and becomes more uniform. This can be interpreted as the onset of human management of the natural fire regime, although there was also a widespread change in climate around 500 B.C. that would also have impacted the fire regime (Lindbladh and Bradshaw 1998). *Quercus* populations, which regenerate best under somewhat open conditions, were important forest components throughout the last 6,000 years but declined in

Figure 2.5 Sediment core from a small forest hollow in southern Sweden. Note the charcoal bands indicating local fire.

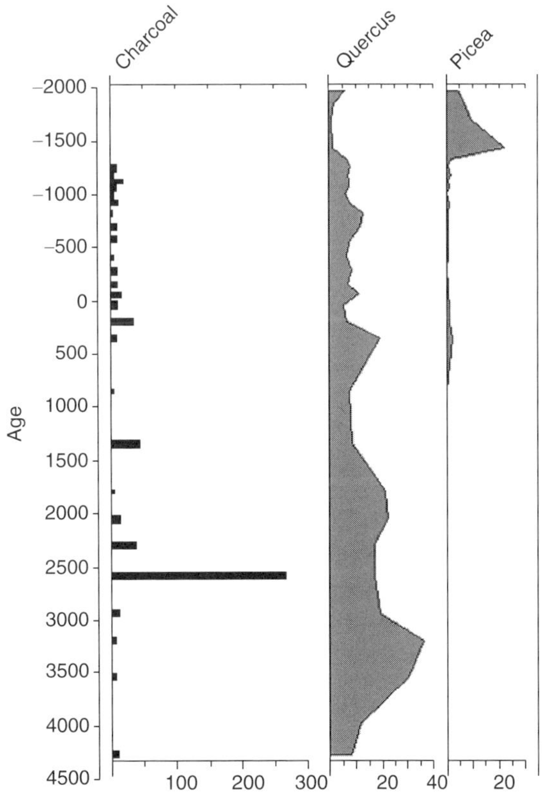

Figure 2.6 Pollen records for *Quercus* and *Picea* and charcoal fragments from a forest hollow at Nissatorp, Sweden for the last 6,000 years.

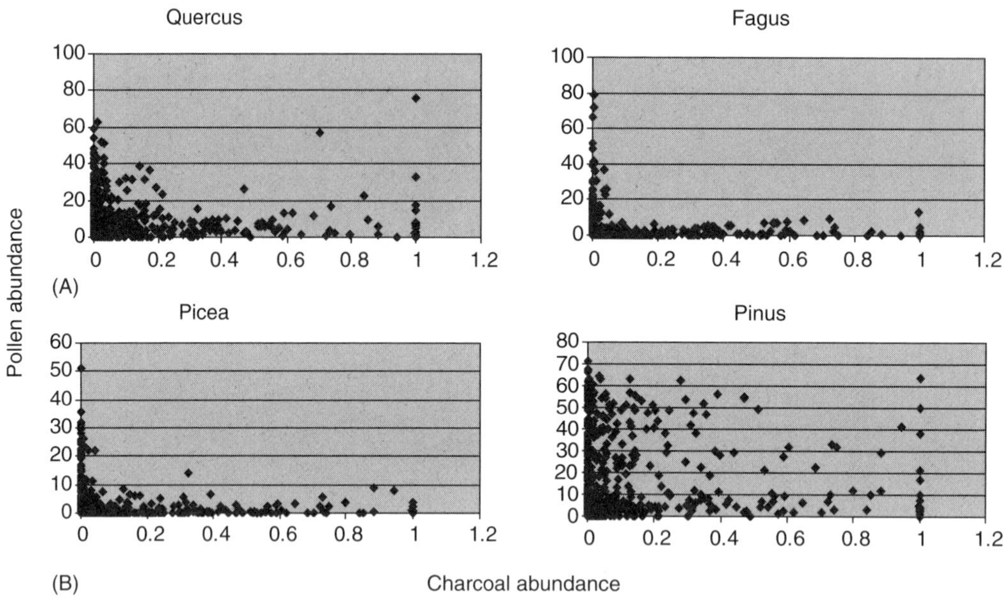

Figure 2.7 Pollen percentage values (y-axes) plotted against charcoal concentrations (x-axes) for samples from small forest hollows. The charcoal data are scaled to lie between 0 and 1. (A) *Fagus* and *Quercus* and (B) *Picea* and *Pinus*.

abundance during recent centuries. The decline of *Quercus* is associated both with an increase in the local importance of *Picea* and the cessation of charcoal in sediments. The stand-scale paleoecological record shows that *Quercus* pollen is more closely associated with charcoal particles than *Fagus* (Figure 2.7A). In a comparable manner, *Pinus* pollen is far more associated with charcoal than is *Picea* pollen (Figure 2.7B). Both *Fagus* and *Picea* have increased in abundance in unmanaged forests in southern Sweden during the last few centuries and changes in fire regime contributed to this successional development.

A comparable discussion has taken place in the eastern U.S.A. concerning the natural role of *Quercus* in temperate forests and its relationship with fire (Clark 1997; Shumway et al. 2001). The paucity of stand-scale paleoecological sites from throughout the region, however, rules out a detailed analysis. Nevertheless, the fire-sensitive *F. grandifolia* became less abundant close to the time of European settlement in the northeast U.S.A., at a time when fire frequencies were increasing; hence, it seems likely that changing fire regimes on both sides of the Atlantic has similar consequences (Gaudreau et al. 1989; Clark and Patterson 1997). Information about former fire regimes is valuable as an indication of the disturbance intensity and frequency to which the present flora and fauna are adapted. This information is valuable in conservation and the assessment of likely future fire hazard (Brown and Smith 2000).

2.4.2 *Browsing animals*

In North America, as in Europe and Asia, the megafaunal extinctions at the end of the last glaciation must have impacted forests, but little is known of the consequences (Bradshaw and Mitchell 1999; Bradshaw et al. 2003). The potential role of browsing and grazing mammals in relation to stand dynamics and ground flora in natural forest has been emphasized, but data are difficult to obtain (Vera 2000). The paleoecological record documents large changes in the fauna of boreal and temperate forests and a long history of anthropogenic modification of fauna through hunting, herding, and later domestication of large mammals. Thus, it is difficult to establish the species composition or population levels of ungulates characteristic of natural forests.

Both natural and anthropogenic factors have affected the Holocene fauna of northwest Europe. By the early Holocene (9,000 ybp), the forests of northwest Europe had been recolonized not only by species remnant in the region, such as red (*Cervus elaphus*) and roe deer (*Capreolus capreolus*), but also by elk (*Alces alces*), aurochs (*Bos primigenius*), bear (*Ursus arctos*), wolf (*Canis lupus*), lynx (*Lynx lynx*), beaver (*Castor fiber*), and others (Aaris-Sørensen 1998; Yalden 1999). However, this fauna was impoverished compared to previous interglacials, specifically in large herbivores and carnivores. Some species, such as the horse (*Equus ferus*), left the region long ago, due to natural changes in Holocene climate and vegetation; others, such as the aurochs, are now totally extinct. Most of the remaining species, however, disappeared after the Neolithic agricultural revolution or even in very recent centuries, as a result of habitat clearance and hunting. In theory, these species still belong to natural forests of the region. Models for their persistence exist in Poland and Russia, but crucial to the survival is that they are embedded in very large areas of habitat. The mammals form a semi-natural metapopulation, which can survive as a whole even if local areas become unsuitable or unavailable for various reasons (Bradshaw et al. 2002).

The Holocene records for large mammals in Sweden permit the development of models describing changing grazing regimes during the Holocene (Liljegren and Lagerås 1993) (Figure 2.8). There have been continual fluctuations in the balance between browsers and grazers, chiefly due to the early Holocene local extinction of bison (*Bison bonasus*) and aurochs (Ekström 1993), and the subsequent introduction of domestic cattle nearly 3,000 years later. The removal of domestic cattle from southern Swedish forests during the last 100 years was a significant event in long-term grazing–vegetation interactions and has

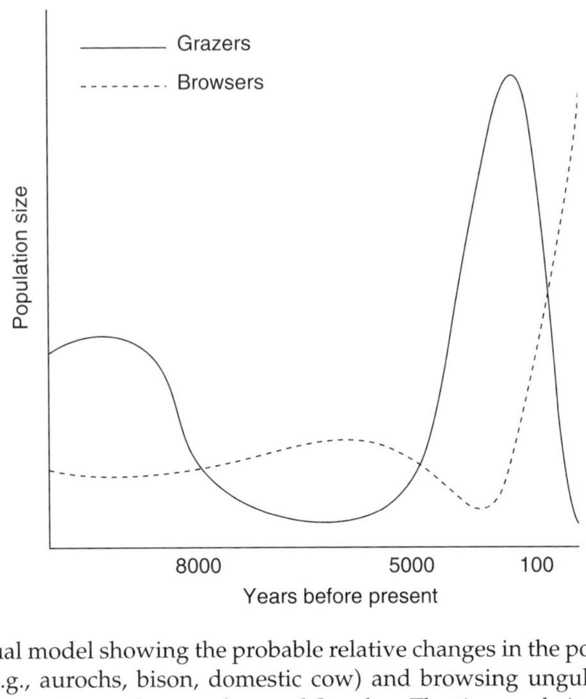

Figure 2.8 Conceptual model showing the probable relative changes in the population size of large, grazing ungulates (e.g., aurochs, bison, domestic cow) and browsing ungulates (e.g., moose, roe deer) during the Holocene in southern and central Sweden. The time-scale is nonlinear. Data taken from many sources.

had a major influence on forest composition and structure (Andersson et al. 1993). The changing balance between grazers and browsers in southern Swedish forests suggests that the present large populations of roe deer and moose are a recent development.

The much higher ungulate diversity and presumably total biomass of past interglacials strongly suggests a diverse vegetational environment that varied between interglacials, with elephants, rhinos, and many species of deer that must have been niche-separated. Climatic and edaphic factors are the major determinants of the primary habitat structure (Bradshaw et al. 2003). It is unlikely that ungulate communities caused the apparent differences in habitat diversity between past interglacials themselves (Vera 2000; Svenning 2002). Grazers, by definition, are adapted to more open conditions and can hinder woody colonization, but are unsuited for major forest clearance (Lister, pers. comm.). The large-scale, anthropogenic alteration of landscape structure in northwest Europe makes the present-day situation significantly different from previous interglacials or even from the early- to mid-Holocene. As with fire, we may gain a new insight into processes of ungulate–vegetation interactions from paleoecological data; but the current situation is unique when viewed from the long-term perspective. Studies of the past reconstruct the trajectory of events that created the present condition and provide an insight into processes, but they also show that no single set of equilibrium "base-line" conditions can be recognized in the recent geological past; hence, there is no secure reference from the past to use as a model for restoration or future management.

2.4.3 Storm

The significance of wind damage in unmanaged forests has been well reviewed for North America (Everham and Brokaw 1996; Peterson 2000; Webb 1999). In Europe, there are isolated studies from near-natural forests (Falinski 1978; Pontailler et al. 1997; Wolf et al. 2004) but most information is derived from coniferous forest or plantations (Andersen 1954; Grayson 1989; Ulanova 2000; Gardiner and Quine 2000). Tree size and species are important factors influencing the extent of storm damage. Natural forest mortality rates range between 0.5 and 3% per year in the boreal and temperate zone (Runkle 1985; Peterken 1996), and natural forests are better adapted than plantations to survive severe storm damage.

Severe storms in western Europe over the last 20 years have caused extensive damage, especially to conifer plantations, but not exclusively. In December 1999, the most powerful hurricane ever recorded in Denmark (Danmarks Meteorologiske Institut 1999) caused the greatest destruction of forest volume during the 1900s (Fodgaard 2000). Draved Forest is a seminatural deciduous forest in southern Jutland with long-term observations of tree growth and mortality. Through time storms were the major cause of mortality of large trees. Smaller trees were susceptible to competition as standing dead trees grew slower than survivors. Mortality varied by species (Figure 2.9); *Betula, Fagus,* and *Tilia* were mainly windthrown, whereas *Alnus* and *Fraxinus* were as likely to be standing dead trees (Wolf et al. 2004). In Draved Forest, 4% of all trees larger than 10 cm dbh were damaged by the storm, while damage was almost total in neighboring commercial coniferous plantations.

2.4.4 Human influences

Pests and disease, flooding and land-slippage are other natural disturbances that influence the structure and composition of natural forests; but humans are the cause of the greatest disturbances affecting forests in the boreal and temperate zones. Forests have been cleared for agriculture for over 5,000 years in Europe; even if agriculture is abandoned, the break

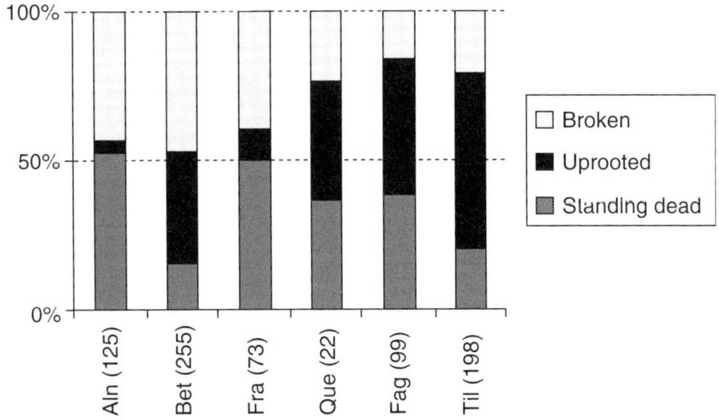

Figure 2.9 Cause of mortality for different species in Draved Forest, Denmark during the last 50 years. Sample size is shown in parentheses. Til: *Tilia cordata,* Fag: *Fagus sylvatica,* Fra: *Fraxinus excelsior,* Bet: *Betula pubescens,* Que: *Quercus robur,* Aln: *Alnus incana.*

in forest continuity can affect species composition of subsequent forests for a long time. Mattarp is a *Fagus* stand in southern Sweden that was cleared by the Vikings for crops, abandoned, and then heavily grazed during the 1700–1800s (Björkman 1996) (Figure 2.10). The present forest is poorer in natural forest species than nearby Siggaboda, one of the most natural remaining forest stands in southern Sweden (Björkman and Bradshaw 1996).

The adequate description of natural disturbance regimes and full recognition of how they have been modified, both directly and indirectly by anthropogenic activities, is one of the major challenges in the identification and study of natural forest ecosystems. Commercial forestry has altered the size structure of semi-natural forests throughout the boreal and temperate zones, removing old, large trees and shortening rotation lengths (Linder and Östlund 1992) (Figure 2.11). Densities and basal areas of large trees surviving in old North American and European temperate forests were found to be comparable (Nilsson et al. 2002).

2.5 Other important characteristics of natural forest systems

I have highlighted tree species composition and disturbance regime as defining features of natural and near-natural forests. This is, of course, a broad generalization, and many other factors and nuances will be identified in other chapters (Peterken 1996). Three other important characteristics that help characterize natural forests deserve brief mention. The hydrological regime of natural forests in Europe has not been studied in detail, but regional and local drainage of forest and agricultural areas has been relentlessly pursued in the temperate and boreal zones for many decades (Møller 2000). Regional pollen diagrams indicate a general reduction in abundance of *Alnus* and *Salix* on the landscape and field experience supports the conclusion that systematic alteration of microhabitats has decreased biodiversity in ways that we are just beginning to recognize. Landscape drainage has occurred in North America on a smaller scale relative to the natural forest resources remaining (Figure 2.12) but of great regional importance (Gardiner and Oliver, this volume). One underappreciated aspect of hydrology is the impact of reduced populations of beaver (*Castor canadensis*) since European settlement (Rudemann and Schoonmaker 1938; Naiman et al. 1986).

Forest soils and the biogeochemical cycling of their chemical constituents have received much attention, particularly with regard to the effects of atmospheric deposition

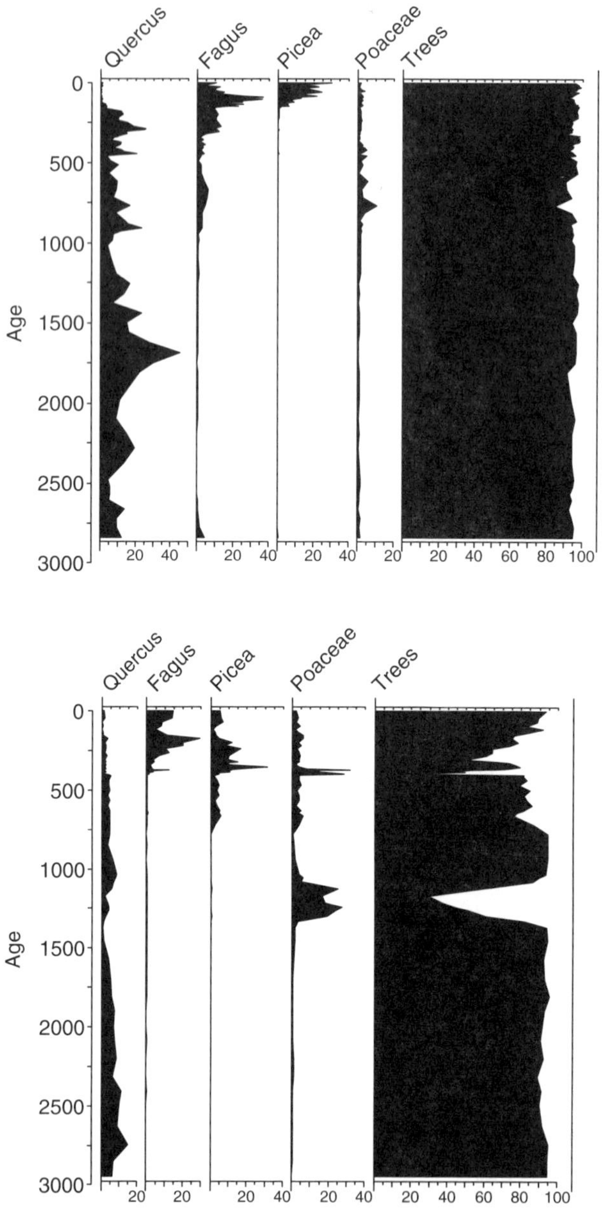

Figure 2.10 Summary pollen diagrams covering the last 3,000 years from two forest hollows in southern Sweden. Note the break in forest continuity at Mattarp 1200 years ago during the Viking period.

of first sulfates and subsequently nitrates. Nevertheless, soil change is slower than vegetation change and soil profiles may retain relic morphology, and processes such as decomposition may not be in equilibrium with current climate and litter inputs (Willis et al. 1997). Careful interpretation of soil features may identify past disturbances as breaks in forest continuity that are not apparent from present vegetation or hydrology. Windthrow mounds are a feature of many natural forests in regions that are affected by windthrow and these, together with rooting animals, such as wild boar, are important in soil turnover in natural forest. Lack of pit and mound topography may be an indicator of past agricultural use.

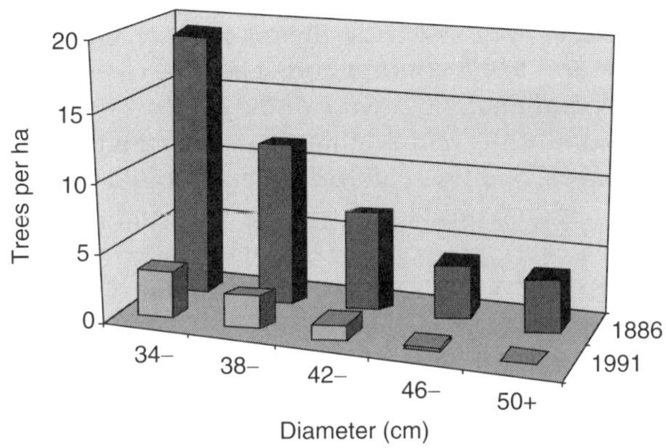

Figure 2.11 Number of trees per hectare and their diameter distribution from Orsa 'besparingsskog,' Sweden in 1886 and 1991.

Figure 2.12 *Acer macrophyllum* dripping with bryophytes in undrained, humid temperate rain forest within the Olympic National Park, U.S.A.

Standing deadwood and woody debris are important for forest biodiversity (Krankina et al. 2002; Bobiec 2002; Mountford 2002) and associated aquatic systems. The natural level and dynamics of deadwood vary greatly in space and time and deadwood is generated as a consequence of the disturbance regimes discussed above.

2.6 Conclusions

This chapter opened with the conclusion that truly natural, virgin forests probably do not exist. The many examples from western Europe of attempts to identify natural forest condition in remote forests support this conclusion. I have used the shift from hunting and gathering to sedentary agriculture to mark the change from natural to other forests. Even in areas where sedentary agriculture has not dominated, we cannot discount the effect of fire as used by nonagricultural cultures. Even in those ecosystems in North America, the influence of modern fire suppression has altered natural forests. Therefore, why is it important to create a natural forest concept and speculate over the virgin state? Firstly, it satisfies human curiosity and makes us aware of the ecological consequences of socio-economic development. Secondly, the natural forest concept provides tangible goals for management of wilderness areas, and restoration of overexploited forest. Many people would like to know what their region would look like in a truly natural state and protect a small sample for posterity. However, probably the strongest motivation is that all silvicultural systems are modifications of the natural state adapted to local needs. In a changing society and a changing environment, it is appropriate to develop these systems by adapting natural species and processes that have survived the climatic extremes of geological time. Nevertheless, we should recognize that what is natural vegetation is a range of vegetation types and disturbance regimes, and that paleoecological research can inform our understanding of the appropriate range (Sprugel 1991).

Acknowledgments

I thank my recent research students Leif Björkman, Gina Hannon, Björn Holmqvist, Matts Lindbladh, and Annett Wolf for sharing data with me. Special thanks are due to Peter Friis Møller for valuable discussion.

References

Aaris-Sørensen, K., *Danmarks forhistoriske dyreverden*, Gyldendal, Copenhagen, 232 pp., 1998 (in Danish).

Andersen, K.F., Gales and gale damage to forests, with special reference to the effects of the storm of 31 January 1953, in the northeast of Scotland, *Forestry*, 27, 97, 1954.

Andersson, L., Appelqvist, T., Bengtsson, O., Nitare, J., and Wadstein, M., *Betespräglad äldre bondesog*, Skogsstyrelsen, Jönköping, 110 pp., 1993.

Angelstam, P., Landscape analysis as a tool for the scientific management of biodiversity, *Ecol. Bull.*, 46, 140, 1997.

Angelstam, P., Maintaining and restoring biodiversity in European boreal forests by developing natural disturbance regimes, *J. Veg. Sci.*, 9, 593, 1998.

Björkman, L., Long-term population dynamics of *Fagus sylvatica* at the northern limits of its distribution in southern Sweden: a paleoecological study, *Holocene*, 6, 225, 1996.

Björkman, L. and Bradshaw, R.H.W., The immigration of *Fagus sylvatica* L. and *Picea abies* (L.) Karst. into a natural forest stand in southern Sweden during the last two thousand years, *J. Biogeogr.*, 23, 235, 1996.

Bobiec, A., Living stands and deadwood in the Bialowieza forest: suggestions for restoration management, *For. Ecol. Manage.*, 165, 125, 2002.

Bradshaw, R.H.W., The origins and dynamics of native forest ecosystems: background to the use of exotic species in forestry, *Búvísindi*, 9, 7, 1995.

Bradshaw, R.H.W., Gemmel, P. and Björkman, L., Development of nature-based silvicultural models in southern Sweden: the scientific background, *For. and Lands. Res.*, 1, 95, 1994.

Bradshaw, R.H.W., Tolonen, K., and Tolonen, M., Holocene records of fire from the boreal and temperate zones of Europe, in *Sediment Records of Biomass Burning and Global Change*, Clark, J.S., Cachier, H., Goldammer, J.G., and Stocks, B.J., Eds., Springer-Verlag, Berlin, 347, 1997.

Bradshaw, R.H.W. and Mitchell, F.J.G., The paleoecological approach to reconstructing former grazing–vegetation interactions, *For. Ecol. Manage.* 120, 3, 1999.

Bradshaw, R.H.W. and Mountford, E.P., *Report to Accompany Maps of Past European Fagus Forests*, NATMAN report to European Union, 2002.

Bradshaw, R.H.W., Hannon, G.E., and Lister, A.M., A long-term perspective on ungulate-vegetation interactions, *For. Ecol. Manage.*, 181, 267, 2003.

Brewer, S., Cheddadi, R., Beaulieu, J.L., Reille, M., and data contributors, The spread of deciduous *Quercus* throughout Europe since the last glacial period, *For. Ecol. Manage.*, 156, 27, 2002.

Brown, J.K. and Smith, J.K., Eds., *Wildland fire in ecosystems: effects of fire on flora*, U.S. Department of Agriculture, Forest Service, Rocky Mountain Research Station, Ogden, UT, General Technical Report RMRS-42-Vol. 2, 257 pp., 2000.

Carcaillet, C., Almquist, H., Asnong, H., Bradshaw, R.H.W., Carrión, J.S., Gaillard, M.-J., Gajewski, K., Haas, J.N., Haberle, S.G., Hadorn, P., Müller, S.D., Richard, P.J. H., Richoz, I., Rösch, M., Sánchez Goñi, M.F., von Stedingk, H., Stevenson, A.C., Talon, B., Tardy, C., Tinner, W., Tryterud, E., Wick, L., and Willis K.J., Holocene biomass burning and global dynamics of the carbon-cycle, *Chemosphere*, 49, 845, 2002.

Clark, J.S., Facing short-term extrapolation with long-term evidence: Holocene fire in the northeastern U.S. forests, *J. Ecol.*, 85, 377, 1997.

Clark, J.S. and Patterson, W.A., Background and local charcoal in sediments: scales of fire evidence in the paleorecord, in *Sediment Records of Biomass Burning and Global Change*, Clark, J.S., Cachier, H., Goldammer, J.G., and Stocks, B.J., Eds., Springer-Verlag, Berlin, 23, 1997.

Danmarks Meteorologiske Institut, *Rapport Orkanen over Danmark den 3–4 December 1999*, Danmarks Meteorologiske Institut, 1999 (in Danish).

Ekström, J., *The late Quaternary history of the Urus (Bos primigenius Bojanus 1827) in Sweden*, Ph.D. thesis, Lund University, Department of Quaternary Geology, 129 pp., 1993.

Everham III, E.M. and Brokaw, N.V.L., Forest damage and recovery from catastrophic wind, *Bot. Rev.*, 62, 113, 1996.

Falinski, J.B., Uprooted trees, their distribution and influence in the primeval forest biotype, *Vegetatio*, 38, 175, 1978.

Falinski, J.B.,*Vegetation Dynamics in Temperate Lowland Primeval Forests*, Kluwer, Dordrecht, 537 pp., 1986.

Fodgaard, S., Stormfald gennem tiderne, *Skoven*, 3, 144, 2000.

Foster, D.R. and O'Keefe, J.F., *New England Forests Through Time*, Harvard University Press, Cambridge, 67 pp., 2000.

Gardiner, B.A. and Quine, C.P., Management of forests to reduce the risk of abiotic damage: a review with particular reference to the effects of strong winds, *For. Ecol. Manage.*, 135, 261, 2000.

Gaudreau, D.C., Jackson, S.T. and Webb, T. III, Spatial scale and sampling strategy in paleoecological studies of vegetation patterns in mountainous terrain, *Acta. Bot. Neerl.*, 38, 369, 1989.

Goldammer, J.G. and Furyaev, V.V., *Fire in Ecosystems of Boreal Eurasia*, Kluwer, Dordrecht, 528 pp., 1996.

Grayson, A.J., *The 1987 Storm: Impacts and Responses, Forestry Commission Bulletin*, 87, 1989.

Haskell, J., The latitudinal gradient of diversity through the Holocene as recorded by fossil pollen in Europe, *Evol. Ecol. Res.*, 3, 345, 2001.

Hewitt, G.M., Some genetic consequences of ice ages, and their role in divergence and speciation, *Biol. J. Linn. Soc.*, 58, 247, 1996.

Hofmann, R.R. and Stewart, D.R.M., Grazer or browser: a classification based on the stomach-structure and feeding habits of east African ruminants, *Mammalia*, 36, 226, 1972.

Huntley, B., Species-richness in north-temperate zone forests, *J. Biogeogr.* 20, 163, 1993.

Johnson, E.A., *Fire and Vegetation Dynamics. Studies from the North American boreal forest*, Cambridge University Press, New York, 1992.

Koop, H., *Forest Dynamics*, Springer-Verlag, Berlin, 229 pp., 1989.

Krankina, O.N., Harmon, M.E., Kukuev, Y.A., Treyfeld, R.F., Kashpor, N.N., Kresnov, V.G., Skudin, V.M., Protasov, N.A., Yatskov, M., Spycher, G., and Povarov, E.D., Coarse woody debris in forest regions of Russia, *Can. J. For. Res.*, 32, 768, 2002.

Liljegren, R. and Lagerås, P., *Från mammutstäpp till kohage. Djurens historia i Sverige*, Wallin and Dalholm, Lund, 48 pp., 1993.

Lindbladh, M. and Bradshaw, R.H.W., The origin of present forest composition and pattern in southern Sweden: a study of the estate where Linnaeus was born, *J. Biogeogr.*, 25, 463, 1998.

Linder, P. and Östlund, L., *Förändringar i Sveriges Boreala Skogar 1870–1991, Report 1*, Swedish University of Agricultural Sciences, Department of Forest Ecology, 32 pp., 1992. (in Swedish).

Mountford, E.P., Fallen deadwood levels in the near-natural beech forest at La Tillaie reserve, Fontainebleau, France, *Forestry*, 75, 203, 2002.
Møller, P.F., *Vandet i skoven - hvordan får vi vandet tilbage til skoven?* GEUS, Copenhagen, 60 pp., 2000. (in Danish).
Naiman, R.J., Melillo, J.M., and Hobbie, J.E., Ecosystem alteration of boreal forest streams by beaver (*Castor Canadensis*), *Ecology*, 67, 1254, 1986.
Niklasson, M. and Granström, A., Numbers and sizes of fires: long-term spatially explicit fire history in a Swedish boreal landscape, *Ecology*, 81, 1484, 2000.
Nilsson, S.G., Niklasson, M., Hedin, J., Aronsson, G., Gutowski, J.M., Linder, P., Ljungberg, H., Mikusinski, G., and Ranius, T., Densities of large living and dead trees in old-growth temperate and boreal forests, *For. Ecol. Manage.*, 161, 189, 2002.
Odgaard, B.V. and Rasmussen, P., Origin and development of macro-scale vegetation patterns in the cultural landscape of Denmark, *J. Ecol.*, 88, 733, 2000.
Parviainen, J., *Forest Reserves Research Network*, European Commission, Brussels, 377 pp., 2000.
Peterken, G., *Natural Woodland*, Cambridge University Press, Cambridge, 522 pp., 1996.
Peterson, C.J., Catastrophic wind damage to North American forests and the potential impact of climate change, *Sci. Total Environ.*, 262, 287, 2000.
Petit R.J., Csaikl, U.M., Bordacs, S., Burg, K., Coart, E., Cottrell, J., van Dam, B., Deans, J.D., Dumolin-Lapegue, S., Fineschi, S., Finkeldey, R., Gillies, A., Glaz, I., Goicoechea, P.G., Jensen, J.S., Konig, A.O., Lowe, A.J., Madsen, S.F., Matyas, G., Munro, R.C., Olalde, M., Pemonge, M.H., Popescu, F., Slade, D., Tabbener, H., Taurchini, D., de Vries, S.G.M., Ziegenhagen, B., and Kremer, A., Chloroplast DNA variation in European white oaks—Phylogeography and patterns of diversity based on data from over 2600 populations, *For. Ecol. Manage.*, 156, 5, 2002.
Pickett, S.T.A. and White, P.S., *The Ecology of Natural Disturbance and Patch Dynamics*, Academic Press, Orlando, 1985.
Pontailler, J.-Y., Faille, A., and Lemee, G., Storms drive successional dynamics in natural forests: a case study in Fontainebleau forest (France), *For. Ecol. Manage.*, 98, 1, 1997.
Roberts, N., *The Holocene*, 2nd ed., Blackwell, Oxford, 1998, chap. 5.
Rudemann, R. and Schoonmaker, W.J., Beaver dams as geological agents, *Science*, 88, 523, 1938.
Runkle, J.R., Disturbance regimes in temperate forests, in *The Ecology of Natural Disturbance and Patch Dynamics*, Pickett, S.T.A. and White, P.S., Eds., Academic Press, Orlando, 17, 1985.
Segerström, U., Bradshaw, R.H.W., Hörnberg, G., and Bohlin, E., Disturbance history of a wet forest refuge in northern Sweden, *Biol. Conserv.*, 68, 189, 1994.
Shumway, D.L., Abrams, M.D., and Ruffner, C.M., A 400-year history of fire and oak recruitment in an old-growth oak forest in western Maryland, U.S.A., *Can. J. For. Res.*, 31, 1437, 2001.
Sprugel, D.G., Disturbance, equilibrium, and environmental variability: what is 'natural' vegetation in a changing environment? *Biol. Conserv.*, 58, 1, 1991.
Svenning, J.-C.,A review of natural vegetation openness in northwestern Europe, *Biol. Conserv.*, 104, 133, 2002.
Thieme, H., Lower Palaeolithic hunting spears from Germany, *Nature*, 385, 807, 1997.
Ulanova, N.G., The effects of windthrow on forest at different spatial scales: a review, *For. Ecol. Manage.*, 135, 155, 2000.
Vendramin, G.G., personal communication.
Vera, F.W.M., *Grazing Ecology and Forest History*, CABI Publishing, Wallingford, 506 pp., 2000.
Webb, S.L., Disturbance by wind in temperate-zone forests, in *Ecosystems of Disturbed Ground*, Walker, L., Ed., Elsevier, Amsterdam, 187, 1999.
Wilkinson, D.M., Is local provenance important in habitat creation? *J. Appl. Ecol.*, 38, 1371, 2001.
Willis, K.J., Braun, M., Sümegi, P., and Tóth, A., Does soil cause vegetation change or vice versa? A temporal perspective from Hungary, *Ecology*, 78, 740, 1997.
Wolf, A., Møller, P.F., Bradshaw, R.H.W., and Bigler, J., Storm damage and long-term mortality in a semi-natural, temperate deciduous forest, *For. Ecol. Manage.*, 188, 197, 2004.
Woods, K.D.,Dynamics in late-successional hemlock-hardwood forests over three decades, *Ecology*, 81, 110, 2000.
Yalden, D., *The History of British Mammals*, T. and A.D. Poyser, London, 305 pp., 1999.

chapter three

Effects of restoration at the stand level

Chadwick Dearing Oliver
Yale University
Kevin L. O'Hara
University of California-Berkeley

Contents

3.1 Introduction

The conservation of forests for timber, wildlife, and water became a large concern in many countries over 100 years ago, and a culture of conservation started being accepted, institutionalized, and organized. Recently, concern for forests and other ecological systems has become part of the global concern for sustainability and is building on the earlier concern for conservation (Brundtland 1987). Both conservation and sustainability contain the idea that each place and people should obtain no more or less than its fair share of values from its environment and should leave to future generations an environment that is in at least as good a condition, if not better, than at present (Oliver et al. 2002; Oliver 2003).

This chapter describes the role of forest restoration in achieving the objectives of sustainability at the stand level. The stand level is fundamental to any forest activity, since a

1-56670-635-1/05/$0.001+$1.50

stand is by definition an area of relatively uniform conditions-soil, species composition, age distribution, stand structure, and history. A stand usually varies from 2 to 200 ha in size and by definition is treated in the same way at the same times. This chapter first describes the current understanding of the dynamic nature of forest stands and the opportunities and limitations of approaching restoration at the stand level. Afterwards, it describes specific activities and pathways that can be applied to stands for restoring forest ecosystems. The chapter concludes with a brief discussion of how this restoration can be incorporated into successful management and economic frameworks.

3.2 People and forests

Forest ecosystems potentially occupy between approximately 30 and 50% of the earth's land surface area, 4.1 billion ha to 6.7 billion ha (Miller 1996; Tallis 1991) or slightly more or less, depending on how "forests" are distinguished from "marginal tree–shrub land." The total forest area was much smaller during the past glacial maximum, 18,000 years ago and much larger approximately 5,000 to 7,000 years ago (Tallis 1991).

Forests presently play an even larger role in the earth's environment than their area would suggest because most people live, obtain water, and produce foods in potentially forested ecosystems, ecosystems that would be forested were it not for human creations such as farms and cities (Tallis 1991). Nearly all forests have been impacted by people in some way. Approximately two-thirds to three-quarters of the potentially forested area is intact forests, areas presently existing in a forested state. The remaining one-quarter or one-third of the potentially forested area is in farms, residences, urban areas, or similar creations.

Restoring forests to sustainable conditions involves several considerations:

- Most people live in potentially forested areas and are highly unlikely to move to areas that are not potentially forested, even if that were desirable. Instead, restoration will need to find a way to accommodate people as part of the greater forest ecosystems, ecosystems that are forested or potentially forested.
- The world's forests area can be divided into hundreds of ecosystems (Bailey 1983), with some species unique to each ecosystem. Consequently, each forest ecosystem will need to be sustained or restored if the goal of biodiversity is to be realized; sacrificing one ecosystem to save another will not achieve the goal of sustainability and is probably not necessary.
- Not all forest ecosystems contain species that are endangered; and species become endangered for many reasons. Some are endangered because their habitats cover a small area; others, because they are severely impacted by people; and still others are endangered for natural reasons.
- Different forest ecosystems are impacted by different natural and human influences and thus will require different restoration activities. Some forest ecosystem areas contain nearly all of their intact forests, while others have been largely converted to farms, houses, and cities with much less intact forest remaining. The area of intact forests and human creations in each ecosystem area is constantly changing.

To achieve sustainability, all forest ecosystems will need to be addressed; and most "intact forests" will be in smaller areas than their potential areas, with people living within the natural boundaries of forest ecosystems, and thus preventing the intact forests from expanding to their potential. Fortunately, certain features common to forest ecosystems make it possible to restore forests to sustainable conditions under these constraints. Otherwise, people would need to make the very difficult choices of deciding what forest ecosystems to restore at the expense of others and/or how to reduce the current human population.

3.3 Changing perspectives of forest stands

3.3.1 Early ecological paradigm

Despite the hundreds of forest ecosystems (which can be subdivided into thousands of subecosystems), the wide evolutionary origins of forest plants and animals, the large variety of influences on different ecosystems, and the millions of plants and animals that live in forests, certain uniformities of processes are occurring within forest ecosystems throughout the world that make their complexity more understandable. This understanding is the result of over 100 years of scientific inquiry and a major paradigm shift, as will be described below.

Intensive scientific examination of forest processes during the late 19th and early 20th centuries led to ecological theories that governed further inquiry as well as forest management for many decades and have influenced the way forests are regarded today (Raup 1964; Boyce and Oliver 1999). Ecological communities were regarded as ecological systems (ecosystems) and were defined, bounded, and studied for both analytical behavior and their emergent properties (Odum 1971; Tansley 1935; Whittaker 1975). Forest ecosystems were assumed to be closed systems where each plant and animal had a necessary "function," which, if eliminated, would lead to the collapse of the ecosystem. At its extreme, the forest was considered similar to a "super organism," with each species coevolved with the other (for further discussion, see Boyce and Oliver 1999); elimination of one part would be similar to removing a part from a machine and the result would be complete collapse of the forest ecosystem. The "natural" forest was considered that undisturbed by people, or at least "civilized" people, and was assumed to be in a "natural," "climax," and "steady-state" condition in which the large, old trees were gradually but continuously dying and being replaced by younger trees growing from beneath (Figure 3.1).

According to this steady-state theory, if a disturbance did occur to these forests, a predictable "succession" of species would invade the disturbed site, with progressively more shade-tolerant species gradually replacing less tolerant ones (Braun 1950; Odum 1971; Oliver and Larson 1996; Whittaker 1975).

Early conservation efforts made use of this steady-state theory to attempt to protect native species. Representative tracts of these "climax" forests were set aside as preserves while forests elsewhere were exploited (Botkin 1990; Henry and Swan 1974). These forests were assumed to provide habitats for all native species, since this climax forest was allegedly the "natural" forest and therefore the only kind of forest that native species needed to survive. Since the "natural," or "climax," forest was assumed to be innately stable, these forests were expected to remain in the steady-state structure forever.

Another conservation effort was to manage forests following this theory. It was assumed that each stand was all-aged and could be sustained in its natural state through selection harvesting. This selection harvesting was endorsed by ecologists and silviculturists because they felt it was natural, scientifically based, and sustainable, and by loggers because they only had to harvest the large, valuable trees and not be bothered with cutting the others (Boyce and Oliver 1999; O'Hara 2002).

3.3.2 Changing paradigm

In the late 1940s and early 1950s, forest managers throughout the United States and Europe began to realize that the selectively harvested forests were not sustaining the species and wood quality that had been in the original forest, whereas nearby even-aged stands that had grown on old fields or following catastrophic wildfires were rapidly growing trees of high wood quality and appropriate species (Boyce and Oliver 1999). To ensure they could provide timber sustainably, foresters quickly changed to even-aged

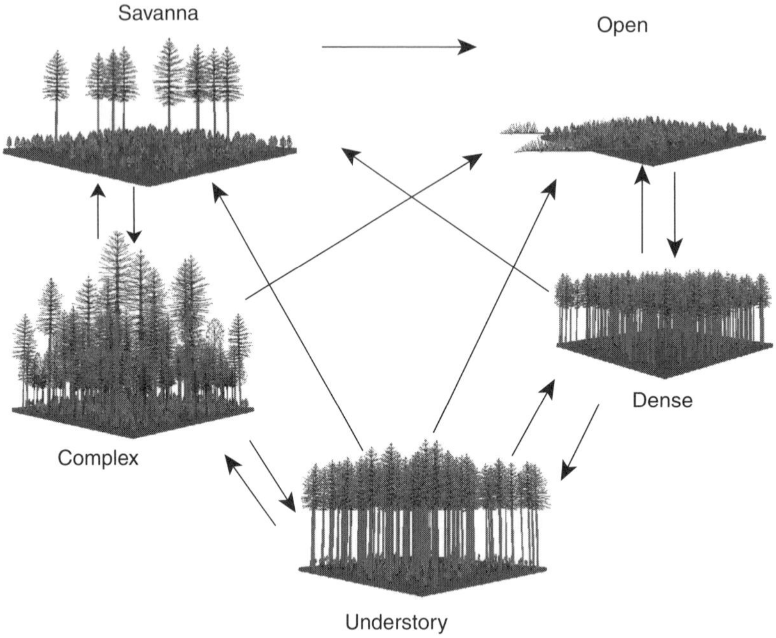

Figure 3.1 Stand structures common to many forest ecosystems in the world. The early dominant paradigm assumed "natural" forests existed only in the "complex" (a.k.a. "climax" or "old growth") structure. It is now accepted that forests change structures through growth (large arrows) and disturbances (small arrows). Different species depend on each structure within a forest ecosystem. (after Oliver and Larson, *Forest Stand Dynamics*, Updated Edition, John Wiley & Sons, New York, 1996.)

management, describing the selectively harvested forests as a mistake because the forest was not sustaining itself in tree species, growth rates, or vigor (O'Hara 2002).

As long as the steady-state forest was considered the natural forest, and forest managers were only successful in providing timber from forests that were considered artificial, managing for multiple benefits could not be considered synergistic; that is, managing for timber would presumably only be at the expense of a providing habitat and other values naturally provided by the forest, a "zero-sum" outcome. This assumed zero-sum outcome led to polarization between those wanting to sustain commodities and those wanting to sustain habitats and other natural forest values.

Accumulating evidence gradually led scientists and professionals to abandon the steady-state view of forests. The 1938 hurricane in New England blew over all trees in preserves that had been expected to remain in the "climax" condition perpetually, destroying the myth of the stable climax (Henry and Swan 1974). Other presumably climax forests that were set aside were not remaining stable in structure or species composition, but were gradually changing (Botkin 1990). During this time, studies began demonstrating that the forest was far more resilient than had been assumed; it was not a closed system, but rather was impacted by external forces such as natural disturbances, species immigrations and extinctions, and climate changes (Oliver and Larson 1996). Additionally, these studies indicated the successional sequence of younger trees and species replacing older ones within a stand was not the common pattern of forest development (Oliver 1978, 1980). Scientists began to realize that some native animal and plant species could not exist in "climax" forests, and some species became endangered, and possibly extinct, because there were not enough other forest conditions (e.g., Kirtland's warbler, *Dendroica kirtlandii* [Walkinshaw 1983], lotus blue butterfly, *Lycaeides argyrognomon lotis* [Fry and Money 1994], and others [Oliver et al. 1997; Young 1992]). By 1990, the Ecological Society of America had shifted away from the steady-state paradigm to a more dynamic paradigm of forest development (Stevens 1990).

3.3.3 *Stand development stages*

The "open system" nature of the forest ecosystem means that its behavior is an aggregation of processes that work opportunistically together to provide various, emergent properties (Johnson 2002). Some of these emergent properties can be classified as physical stand structures (Oliver and Larson 1996), and can be associated with other emergent properties (Peterken 1996). Similar stand structure patterns are repeated in many forest ecosystems throughout the world and have emergent properties similar to "strange attractors" in chaos theory (Zebrowski 1997). A common method of classifying the stand structures will be described below and will be used in this chapter (Figure 3.1). The classification is described in more detail in Oliver and Larson (1996). Different classifications have been used to describe the changing forests (e.g., Carey et al. 1999; O'Hara et al. 1996) that are helpful for specific objectives; these classifications are similar to the one described below.

A natural forest could be impacted by a stand-replacing disturbance creating an area devoid of most trees, the "stand initiation" development stage, or the "open" or "savanna" structures (Figure 3.1); during the stage of development when new trees are reoccupying the area, many tree, shrub, and herb species would invade the disturbed area (for a discussion of the relation of stand structures and development stages, see Oliver and Larson 1996; Camp and Oliver 2004). The species first expand until they occupy all above- and below-ground "growing space," at which time there are many species of plants and many animal species feeding on the diverse and short (easily accessible) plant species. When the growing space is filled, there is intense competition from the existing plants, resulting in the elimination of many species, curtailment of new species invasions, and crowding of the remaining forest. Roots generally occupy all below-ground growing space. Except on very droughty sites where soil growing space is occupied long before canopy closure, there is generally a separation of the photosynthesizing canopy layer from the ground, and very little sunlight reaches the forest floor. This stage of tight control of growing space by the canopy and/or roots is known as the "stem exclusion" stage, and creates the "dense" structure. It is characterized by a relative paucity of plant and animal species. During this stage, some tree species can out-compete others and occupy the upper canopy layer, relegating others to understory canopies (or strata) where they grow little (Figure 3.2). These understory strata species often appear younger than their overstory contemporaries because they grow so little once overtopped and shaded.

Eventually, small disturbances or natural processes of differentiation, mortality, and wind abrasion can allow more sunlight onto the forest floor and can make below-ground growing space available. Then, new herbaceous and woody plants, including trees, begin growing. These plants increase the stand's plant species diversity as well as the habitat value of the forest for animals. This time of regrowing of the understory is referred to as the "understory reinitiation" stage, and the structure created is commonly termed the "understory" structure.

When a stand is in the dense, understory, or "complex" structure, small or large disturbances can destroy some or all trees in the forest, allowing lower strata, forest floor, or newly regenerating plants to grow upward, where they either achieve the overstory or become suppressed in the shade of other, taller trees. This structure can contain very large, magnificent trees of old ages, as well as smaller trees, a diversity of plant species, and

Figure 3.2 Mixed-species, single-cohort stand with oaks in the upper stratum and maples, birches, and other species in the lower strata. Despite the large differences in sizes, virtually all trees in this stand are between 65- and 75-years-old, and grew following a clearcut 75 years ago. (Yale-Meyers Forest, CT, U.S.A.)

habitats for relatively specialized animal species. This structure is referred to as a complex structure and had previously been assumed to be the "climax" forest; however, the forest usually achieves this stage and structure as an open system, impacted by external forces such as disturbances. This condition has been referred to as the "old-growth" stage of development, but it is still in a very dynamic condition.

As each stand is disturbed, different species can invade, become dominant, and maintain the dominance for the life of the tree species, often hundreds of years. Consequently, there is no species that is naturally predetermined to dominate a site; rather, any of several species can dominate it (Henry and Swan 1974; Oliver and Stephens 1977; Oliver 1980; Oliver and Larson 1996).

3.4 Restoration and silviculture

For restoration to be effective at the stand level, it must incorporate the accepted dynamic understanding of forests. A stand cannot be restored to a steady-state and be expected to remain in that condition, even if such a steady-state were possible or desirable. Instead, the stand must be restored to a trajectory of changing structures that, in concert with other stands in the landscape, will maintain the appropriate mix of structures over time.

Forest restoration can build on the knowledge gained about silviculture and stand development for the past 100 years. At the stand level, restoration and "silviculture" will be most effective if they are considered interchangeable, with the following caveats:

- Although most definitions of silviculture consider its objectives to be broad (Daniel et al. 1979; Ford-Robertson 1971; Helms 1998; Nyland 1996; Smith et al. 1997), some definitions assume that silviculture's objective is simply to provide timber (Toumey 1947). When considered synonymous with restoration, the broader definition should be applied.
- To withstand the test of time, restoration will eventually need to adopt many of the management features learned by silviculturists, such as the need to frame the actions in an economic perspective and the need to establish distinct, measurable objectives.

By considering these terms to be interchangeable, the past century of accumulated knowledge in silviculture can immediately be applied to the practices of restoration.

A reasonable working objective of restoration can be sustainable forestry, which itself can be defined as ensuring that each forest ecosystem in the world provides its fair share of values over time and space. A reasonable working set of values has been provided by various processes and initiatives (Burley 2001) as criteria. As a beginning, the Montreal Process criteria can be used and include: biodiversity, productive capacity, forest health, soil and water, carbon sequestration, socioeconomic considerations, and the infrastructure to provide the other values (Fujimori 2001; Oliver et al. 2002; Oliver 2003).

3.5 Landscape patterns

On a landscape of many stands, the species composition depends on the pattern of disturbances and regrowth at the stand-level (Camp et al. 1997; Harris 1984; Oliver 1980; Oliver et al. 1998). If trees in many stands on the landscape are knocked over in a single disturbance, the resulting landscape can be dominated by the open structure and by species such as butterflies and deer that depend on this structure. As these stands regrow to the dense structure, the open species can be eliminated, survive in other landscapes, or become extirpated or extinct. If species depending on the understory or complex structures had survived the disturbance, their species can expand as other forests regrow to this structure.

Some animal species depend on edges between structures, while others depend on interiors of open forests, complex forests, or other structures, away from edges (Hunter 1990). As a landscape changes with time, a boundary between two stands can appear less as a sharp edge between open (or savanna) and closed (dense, understory, or complex) structures (Figure 3.3) and more as a variable canopy within an interior closed forest.

Each forested landscape could naturally maintain a variety of structures and species over time, or it could fluctuate widely from a predominance of one structure, and group of species, to another if all stands are impacted by the same disturbance. Historically and prehistorically, the large diversity of species was maintained because there were always some stands of each structure for the dependent species to utilize over the large, natural

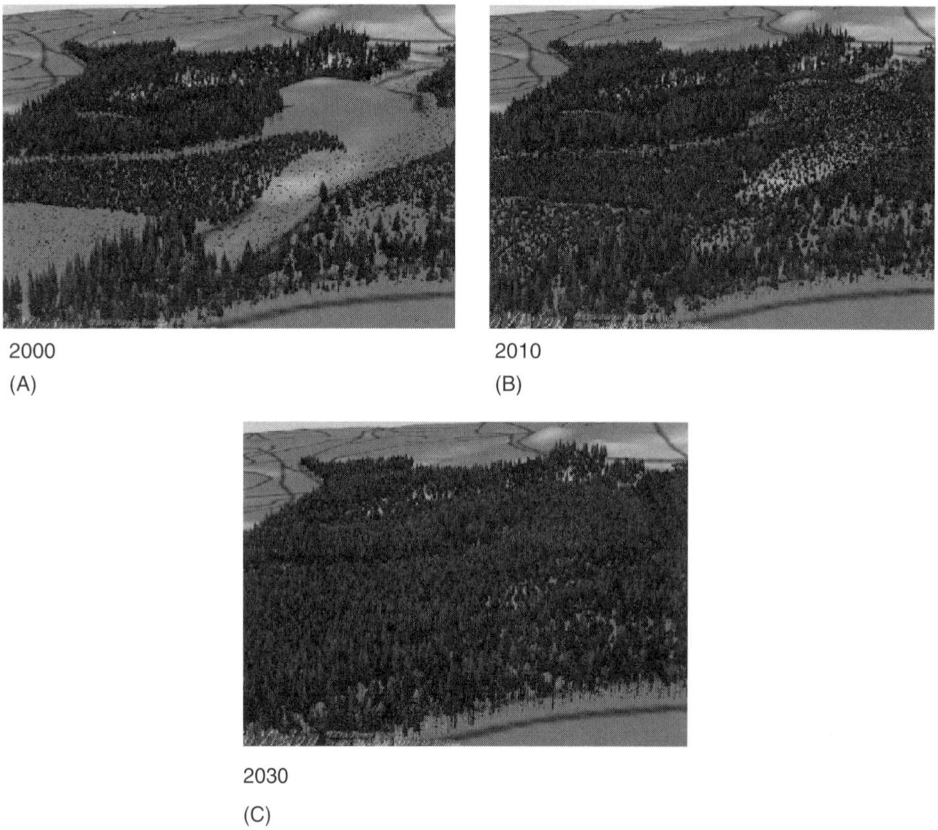

Figure 3.3 A forest with many sharp edges between stands at one time (A) can lose these sharp edges and become more hospitable to "interior closed canopy" species as the forest regrows. (B = 10 years later, and C = 30 years after A). Variability in structure within an "interior closed" forest can be achieved by edges between component stands. Simulation and visualization is of University of Washington Pack Forest inventory and management. Simulation used LMS (McCarter et al. 1998). Visualization used embedded "Envision" (Wilson and McGaughey 2000).

intact forest of each ecosystem. With people occupying part of the potentially forested area of many forested ecosystems, the area of remaining intact forests can be too small to allow the wide fluctuations in stand structures that occurred naturally and still maintain all structures and species. Consequently, proactive restoration will probably be necessary to ensure that all habitats are maintained in sufficient abundance over many of these constricted intact forests.

An important way for active management to provide the sustainability values is by restoring and maintaining a diversity of stand structures across the landscape (Hunter 1990; Oliver 1992). Using active management to mimic, avoid, and recover from natural disturbances, the variety of structures can be maintained, thus providing habitats for biodiversity, employment, and products during the harvesting to create the different structures, and maintaining the forest health and soil and water quality.

Extremely large disturbances such as meteorites hitting the earth can occur at very long time intervals (Zebrowski 1997). Evidence suggests that an event occurred 65 million years ago and resulted in extinction of the major dinosaurs and many other plant and animal species (Flannery 2001). Climatic fluctuations of the magnitude of the past 20,000 years will continue and will further contribute to the elimination of species and perhaps the emergence of others. Presently, people have no way of avoiding such disturbances, and thus the limits of human efforts to ensure sustainability need to be recognized.

3.6 Stand operations and pathways

To achieve the multiple values described above, restoration efforts will first need to determine the possible ways that a stand can change over time, both with and without human intervention. Then, the appropriate actions, or inaction, can be taken to ensure that the desired changes occur. When ecologists and silviculturists believed that a stand would develop through a predictable successional sequence of species replacing others following a disturbance, a quite straightforward set of "silvicultural systems" were developed for regenerating and therefore sustaining each species on a site. For example, to regenerate and thus sustain "pioneer" species, those assumed to regenerate immediately after a stand-replacing disturbance, clearcut harvesting was suggested. To regenerate and sustain mid-seral species, shelterwood harvesting was prescribed to leave some standing trees for protection for the first few years, was prescribed. To regenerate and sustain "climax" species, selection harvesting methods were prescribed. The difficulty in obtaining expected results when applying these harvesting methods helped lead to the paradigm shift described earlier.

Recognition of the more dynamic nature of forests has led scientists to realize that each species has regrown in a variety of partial and stand-replacing disturbances, and thus there are a variety of ways in which each species can be sustained in a stand. The variety of ways in which a stand can be successfully treated has led silviculturists to begin describing stand management through "operations" and "pathways" instead of through silvicultural systems.

3.6.1 Silvicultural operations

An operation is a specific activity done to change a stand during a relatively short time period. Operations include such activities as planting seeds or seedlings, weeding unwanted competing plants, removing some trees through thinning or various partial or complete harvesting methods, fertilizing, pruning, and others (Fujimori 2001). Successful operations generally include:

- A biological component; understanding the physiological sensitivities and needs of the plants being manipulated
- Mechanical and labor components; organizing the appropriate machines, equipment, and people to accomplish the operation
- A logistical component; ensuring that needed activities are done in the proper sequence and time for the components to converge and accomplish the targeted operation.

For example, a successful planting operation would require that the seedlings of the desired species and genetic background be grown and transported properly to the target site at the same time the planting labor was available, the weather was appropriate, and the site was properly prepared to receive the planted seedlings.

Silvicultural operations mimic natural processes to various extents, at times mimicking disturbances and other times mimicking regeneration patterns. Some operations such as burning beneath a stand mimic natural ground fires quite closely, whereas chemical applications of herbicides are quite unlike natural weeding processes. Silvicultural operations can help mimic, avoid, and recover from natural disturbances. Natural disturbances varied in frequency and magnitude in most of the world's forest types, and silvicultural operations generally lie within the "natural range of variability" of these processes. Historically, however, silvicultural operations in each forest ecosystem applied a narrower range than the natural variability of disturbances and regeneration patterns and created a narrow range of stand structures and values associated with these structures. For example, forests in some ecosystems have primarily been managed by clearcutting, and the diversity of structures and species maintained by more variable harvest operations was lost. Other forest ecosystems have primarily been managed by selective harvesting or "high grading" and there is a shift to more trees of shade-tolerant species and a decline in the open structure. Most of the world's ecosystems had been impacted by both stand-replacing disturbances and partial disturbances. Consequently, a diversity of operations will be needed not only to stay within, but to mimic the large range of natural disturbances.

Silviculturists plan operations around "windows of opportunity," times of the year when it is biologically and physically possible to accomplish an operation successfully. For example, seedlings are most easily planted when they are dormant; thus, the planting window in many temperate climates is in the spring after the soil has thawed and before the seedlings have broken dormancy. A harvesting window often depends on the soil conditions; the window for harvesting stands on easily compacted soils is commonly during dry seasons or when the soil is frozen. An additional window to be avoided when thinning stands is during the spring when the tree cambium is active; any scraping of a nonharvested tree during this period can readily scar the stem and allow stem-decay fungi to enter.

A special skill of silviculturists and others engaged in stand-level restoration is to combine the understanding of biology, windows of opportunity, and operational techniques to perform the most efficient actions to achieve the desired results. The result is a variety of operations and strategies for stand management, from minimizing regeneration costs by managing forests on a long-rotation basis (Larsen 1995), to heavily investing in regeneration but managing on a relatively short rotation (Phillips 2000).

Silviculturists have generally become quite skilled in performing operations, although some activities, such as successfully regenerating some species, still elude them. New techniques, such as creating living and dead trees with suitable shapes and cavities for use by targeted wildlife species, are being developed as new values are desired for the forest. Later chapters in this book will describe some of the successful operations, as well as new operational techniques.

3.6.2 *Silvicultural pathways*

Without disturbances, each stand will grow in a pattern or trajectory that is determined by the stand's structure (number, spatial arrangement, physical size, species, and genetic makeup of its component plants), its soil, and climate media (Oliver and Larson 1996). With time, the stand changes in structure and provides different values, as described earlier. Each large or small disturbance alters the configuration of plants, soil, or microclimate and thus causes the stand to grow along a new trajectory. Silvicultural operations planned and implemented at specific times can cause a stand to grow along a different trajectory, barring disturbances. The change in a stand over time caused by a combination of growth and specific operations at specific times is referred to as a "pathway" (Figure 3.4).

A stand can potentially follow many pathways (Figure 3.5). By understanding the potential trajectories of a stand and the ways in which various silvicultural operations can change these trajectories, a stand's pathway can be designed to provide various structures and values at different future times. Many of the values provided by a landscape will depend on the pathways followed by the component stands, with the spatial arrangement

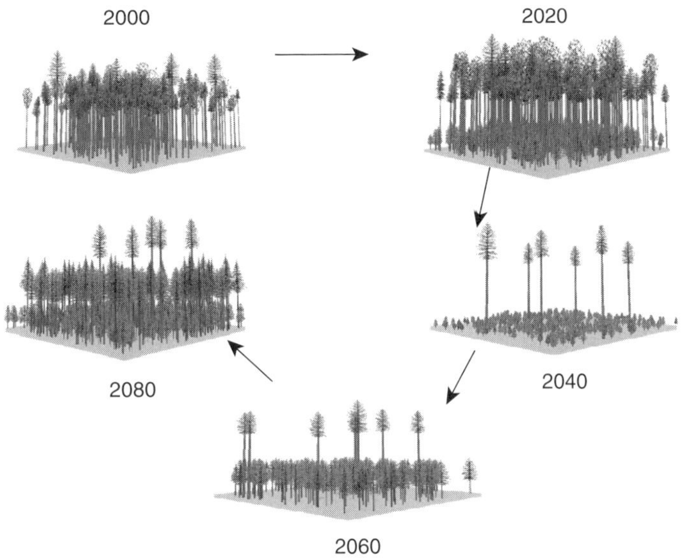

Figure 3.4 Schematic silvicultural pathway for a Douglas-fir stand. Operations were done as follows: 2005: Thinning and regeneration; 2035: Retention harvest and regeneration; 2065: Thinning of lower stratum and regeneration. (Schematic done using LMS [McCarter et al. 1998] and displayed with SVS embedded in LMS [McGaughey 1997].)

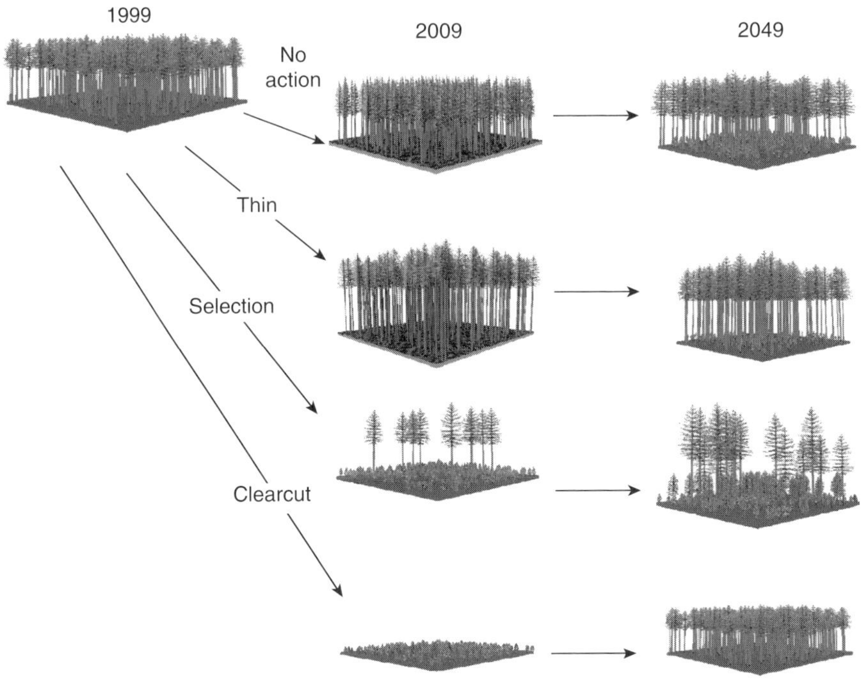

Figure 3.5 A stand can potentially follow many different pathways, depending on its present condition and the species and site conditions, as well as the type and timing of silvicultural operations. The manager decides which pathway each stand follows depending on the stand's potential, the objectives of management, and the conditions of other stands in the landscape.

of the stands' pathways influencing some values. If all stands follow similar pathways, they will pass through similar structures at the same time and provide each of the values in a "pulse" (Oliver and Larson 1996). If stands across a landscape follow different pathways, the resulting variety of structures and values can be distributed more broadly over time, creating a dynamically stable landscape. Careful coordination of stand pathways in a landscape can allow the values to be provided quite uniformly over time. The desirability of a pulse or dynamically stable landscape patterns depends largely on the regional needs and the landowners' needs and incentives (Lippke and Oliver 1993; Oliver et al. 1999).

Each stand cannot be changed to any other structure instantaneously. Instead, the potential pathways that a stand can follow and the time when it can provide future values, if at all, depend on its present structure and the soil and climate media in which it is growing. Stands also have windows of opportunity during which a pathway can be effectively changed by an appropriate operation. For example, Wilson and Oliver (2000) found that Douglas-fir (*Pseudotsuga menziesii* [Mirb.]) stands can be effectively thinned without high risk of windthrow if the trees are already stable or less than 10 m tall at the time of thinning. Above this tree height, unstable trees will not regain stability for many years and the probable result of thinning will be windthrow of the remaining trees; and the window for successful thinning will have been missed.

Silvicultural pathways can be classified in many different ways, such as according to species mix, operations done, or age distribution. The following discussion will describe various silvicultural pathways according to several, overlapping classifications: "no action," "single-species and single-cohort," "mixed-species and single-cohort," and "multiple-cohort" pathways. A "cohort" is defined as all trees beginning following a single disturbance (Oliver and Larson 1996).

3.6.2.1 No action pathway

The "no action" pathway, in which no active silvicultural operations are performed, should always be considered when prescribing pathways, to determine if any actions are really necessary to achieve the stand's objectives. The effects of this pathway depend on the condition of the stand and the objectives. A young stand with many species crowded in the same canopy stratum (i.e., "brushy stage") (sensu Gingrich 1971) may require no actions if the desired tree species will stratify naturally to the upper canopy (Oliver 1980). Alternatively, if another species will out-compete and suppress the desired species, it may be appropriate to remove the dominating species ("cleaning" operation) to restore the stand to a desired pathway. Studies in various parts of the U.S. have documented predictable patterns of stratification among oaks (Figure 3.2) (Bowling and Kellison 1983; Clatterbuck et al. 1985; Clatterbuck et al. 1987; Clatterbuck and Hodges 1988; Kittredge 1988; O'Hara 1986; Oliver 1978), Douglas-firs and associated species (Cobb et al. 1993; Stubblefield and Oliver 1978; Wierman and Oliver 1979), and Sitka spruces and western hemlocks (Deal et al. 1991); however, local professional experience is often needed to determine or ensure applicability of these patterns (O'Hara and Oliver 1999).

A young single-species stand may also differentiate readily with no intervention, allowing some trees to dominate and grow large in diameter and become wind and insect resistant while its neighbors become suppressed and commonly die (Figure 3.6A). In this case, thinning the stand would eliminate the mortality, reduce the fire risk from dead trees, and provide some timber, but would have little effect on the stand's stability. Another stand of the same age, species, and site may not readily differentiate but instead may have all trees grow nearly equally, resulting in trees with large height-diameter ratios that are susceptible to wind, snow, and insect infestations (Figure 3.6B). In this case, thinning the stand early could allow the remaining trees to become stable, while a late thinning would

(A) (B)

Figure 3.6 (A) A well-differentiating stand of Douglas-fir, Pacific Northwestern U.S.A. (B) A poorly-differentiating stand of loblolly pine (*Pinus taeda* L.), southeastern U.S.A.

simply accelerate the rate of windthrow in the stand (Wilson and Oliver 2000). Subtleties of the stand's structure such as the relative uniformity of spacing, ages, and sizes of trees can indicate the ability of a stand to differentiate naturally (O'Hara and Oliver 1999). Various tools such as density management diagrams can help indicate when a stand will become susceptible to either differentiation and mortality or stagnation (Drew and Flewelling 1979; Gingrich 1967; McCarter and Long 1986; Reineke 1933).

3.6.2.2 *Single-species and single-cohort pathways*

Single-species and single-cohort stands can develop from natural or artificial stand-replacing disturbances. Many trees and other plant species living in the region can potentially invade the area, and the type of disturbance and other conditions at the time of the disturbance help determine which species actually comprise the new stand.

The age range in single-species and single-cohort stands may be only one or two years where planting or plentiful seed sources and suitable site conditions allowed trees to reoccupy the growing space rapidly; or, it can be several decades where the stand established from wind-blown seeds on a poor site that did not occupy the growing space for many years (Oliver and Larson 1996).

Such stands with wide age ranges or irregular spacings commonly differentiate well, allowing a relatively stable stand of trees to develop with a large range of sizes (Figure 3.6A). The smaller, suppressed trees have higher height–diameter ratios than the dominant trees (Wonn and O'Hara 2001). If these stands begin at a narrow spacing, considerable mortality of small trees will occur as the stand is differentiating. The remaining trees commonly have small lower limbs or knots in their lower boles because of the initial crowding. At wide spacings, large-diameter, stable trees with large limbs will result.

Stands with narrower age distributions and more regular spacing can occur either naturally or with planting. These stands grow more readily to trees of uniform sizes with less tendency to differentiate (Figure 3.6B). Instead, they are somewhat uniformly slow in diameter growth at a predictable size for a given spacing and species (Oliver and Larson 1996); without a timely thinning, these trees will become susceptible to wind and snow damage, or insect infestations. On the other hand, stands where trees are spaced less uniformly have less need of thinnings and other operations done at precise times, but do not grow the uniform products that can be harvested very efficiently compared to more uniformly grown stands (Wilson and Oliver 2000).

If the diameter at which the trees slow is less than a merchantable size in a uniformly spaced stand, the stand will need to be thinned at a cost (precommercial thinning) to achieve large tree sizes and avoid stand health problems. Delaying this thinning will generally make the thinning less effective because the suppressed trees, the target trees for removal, will not become larger and more valuable, and the dominant trees are generally becoming less stable because they are not differentiating well. A general issue when establishing single-species plantations is tree spacing. Trying to ensure the trees are merchantable when the first thinning occurs means forecasting the minimal merchantable sizes for one or several decades in the future. Some people prefer to plant narrowly and plan on a costly precommercial thinning to ensure small branches and a choice of trees to leave for future stand composition. Others prefer to plant widely to avoid the higher cost of planting more trees and the higher cost and added commitment of a precommercial thinning. The negative aspects of wider spacing (larger limbs and less choice of trees to leave) need to be balanced against the extra costs, greater commitments to operations, and smaller sizes of trees grown at the narrower spacing. Sometimes, the problem of large limbs in plantations of wide spacings can be overcome by pruning (Hanley et al. 1995).

Thinning single-cohort, single-species stands is usually most effective if the more dominant trees are left, although some trees in the dominant crown class can be removed if necessary. The dominant trees are the most stable (against wind and snow breakage) and rapidly growing. Thinning can accomplish many objectives simultaneously (Figure 3.7):

- Depending on the intensity of the thinning, it can change the stand from the dense to the understory or savanna structures much sooner than would occur without thinning
- It sometimes provides income as well as timber for utilization
- It prevents potentially susceptible stands from succumbing to wind or snow breakage or insect attacks
- It allows the remaining trees to grow vigorously, more stably, and to large sizes.

If desirable, a stable, single-cohort stand can be converted to a multiple-cohort stand and eventually to a complex stand. It can also be clearcut to create a more open structure, or a few large trees can be retained to create a savanna structure.

The volume achieved in single-species and cohort stands varies dramatically with species (Figure 3.8). Within a species, the volume may vary at different times with stand densities and thinning pathways. The general pattern of stand growth in practically all cases is to increase tree volume and size asymptotically during the first few years or decades. In regions with subsistence rural conditions, restoring forests is difficult because local people require fuel wood constantly and often cut young trees of little volume; when such forests are protected other forests are put under even greater pressure. An alternative may be to provide these people with alternative forms of energy until the forests have begun producing high wood volumes.

3.6.2.3 *Mixed-species and single-cohort pathways*

Mixed-species, single-cohort stands can develop large ranges in tree sizes and stratify into complex vertical structures (Figure 3.2) (Oliver and Larson 1996). As a result, these stands have often been assumed to be multiple-cohort stands, with the species of smaller sizes in the understory being younger, later stages of the classical "succession" process. Many of these mixed-species stands have been shown to be single-cohort stands, in which all stems have initiated shortly after a stand-replacing disturbance and, after a period of intense competition, stratified into canopy layers by species. As the mixed-species, single-cohort

Figure 3.7 (A) Twenty-year-old Douglas-fir stand in Pacific Northwestern United States just before pruning and thinning. (B, below) Same stand 3 years later.

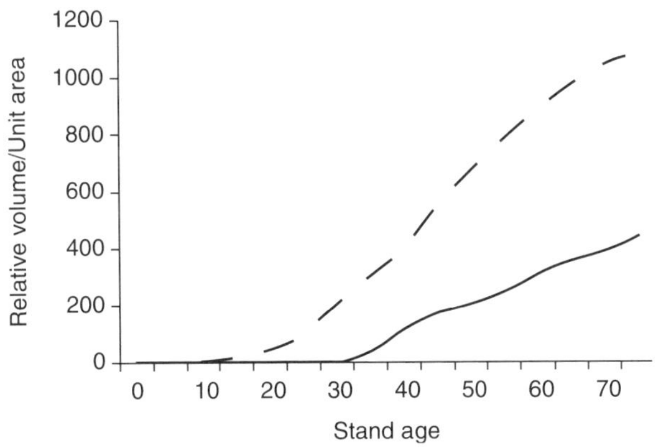

Figure 3.8 Different species can grow dramatically different volumes of wood, as shown in this simulation comparing eastern white pine (*Pinus strobus L.*, upper dotted line) and mixed hardwood stands (solid line) in western North Carolina. (Simulations done using LMS, McCarter et al. 1998.)

stand development pattern is becoming understood, it is proving to be quite useful for a variety of commodity and non-commodity objectives.

If the desired species is known to stratify naturally into the dominant canopy stratum, it is necessary only to ensure that appropriate numbers of trees of this species are in the initiating stand. If another, less desired species will out-compete the target one, either an operation will be needed to prevent the less desired species from initiating or an intermediate cleaning will be needed to eliminate it. For example, in the Pacific Northwestern U.S., red alders (*Alnus rubra* Bong.) will commonly dominate Douglas-firs if they grow close together (Newton et al. 1968; Stubblefield and Oliver 1978). To allow the Douglas-firs to dominate, either the red alders will need to be removed by early cleaning operations or they will need to be prevented from initially regenerating by minimizing mineral soil seedbeds for their light seeds.

Growing mixed-species and single-cohort stands avoids many of the concerns and commitments of single-species stands while providing additional benefits as well. The natural stratification pattern allows the target "crop" trees to begin in a crowded condition that encourages clear stems; however, instead of needing to remove the competing trees with a costly thinning to allow the crop trees to grow large and remain stable, the competing trees of other species naturally become relegated to lower canopy strata and allow the crop tree to spread their crowns above the others, grow large, and remain stable. For species that develop epicormic sprouts after being released, maintaining a crowded understory will help produce clear stems (McKinnon et al. 1935). Mixed-species stands may also help prevent some of the insect and pathogen concerns associated with single-species stands (Chandler 1990).

Afforestation or reforestation with appropriate species mixtures can rapidly create a forest with a layered canopy and a variety of tree sizes that are benefits to wildlife species, at the same time giving the advantages of efficient timber production described above. For example, planting mixtures of Douglas-firs and western redcedars (*Thuja plicata* Donn) in western Washington, U.S.A., allow the Douglas-firs to outgrow the redcedars and produce high-quality Douglas-firs and western redcedars with little thinning, at the same time creating a diverse canopy (Figure 3.9). Similarly, when a few cherrybark oaks (*Quercus pagoda* Raf.) grow with sweetgums (*Liquidambar styraciflua* L.) on old fields in the southeastern United States, a similar stratification occurs and yields high-quality oaks and a diverse canopy (Clatterbuck et al. 1985; Clatterbuck and Hodges 1988) (Figure 3.10). These mixed-species, single-cohort forests commonly have more complex forest characteristics than do single-species, single-cohort forests. In areas where there is a shortage of complex forest habitat, reforestation with such mixed-species techniques can help restore many structural features of complex stands. Conversion of single-cohort stands to multiple-cohort stands can also restore many complex structural features.

If lower stratum species are spaced correctly at an early age, they can also become merchantable. Volume growth of mixed-species stands can vary from intermediate between the volume growths of each species grown in a pure stand to greater than the volume growth of either species in a pure stand, depending on the species in the mixture (Figure 3.11) (Kelty 1986, 1989). Harvesting of mixed-species stands can occasionally be done by removing the dominant canopy, allowing lower strata trees to grow. In many cases, however, the overstory trees are so large and/or the understory trees are of such poor vigor and timber quality that removing just the overstory results in a stand of deformed trees of low vigor and a shift in species composition of the stand.

In regions with subsistence rural conditions, restoring forests to mixed-species stands may be advantageous if one species is fast growing and can be harvested while young to allow the other species to grow longer to more valuable crops.

(A) (C)

(B)

Figure 3.9 When Douglas-firs and western redcedars are planted together during afforestation or reforestation in Washington state (U.S.A.), a predictable stratification occurs that gives benefits for both wildlife and timber production. (A) Douglas-fir (foreground) and redcedar (beside forester) 3 years after planting. (B) Redcedar (left row) and Douglas-fir right row, 15 years after planting. Pruning was done for visibility. (C) Douglas-fir (center, behind forester) and redcedar (surrounding) 70 years after natural regeneration in a clearcut.

The structure of even quite young mixed-species stands has many elements of a complex forest: the large mixture of species, vertical depth and layers of foliage, large tree size range, and frequently large diameters of the dominant trees.

Many mixed-species stands were harvested through selective cutting when the all-aged succession, steady-state, and climax theories were assumed to be true. When silviculturists recognized their lack of success in managing forests through selective harvesting, they tried to grow stands through clearcutting. Loggers, however, frequently resisted clearcutting because of the extra cost of removing the understory trees of low timber value. Consequently, these loggers defined "clearcutting" as removing the most valuable trees and leaving the remaining ones because they were "unmerchantable" or considered "advance regeneration" (Figure 3.12A) (Boyce and Oliver 1999). The practice was done in mixed deciduous and coniferous forests in the eastern and western United States and in mixed deciduous forests in the Russian Far East. The remaining trees were not vigorous, often crooked and scarred, and of shade-tolerant species. They grew relatively slowly and commonly killed any shade-intolerant species trying to regenerate. The result has been a shift in species composition in these forests to more shade-tolerant species, a reduction in

Figure 3.10 Cherrybark oak (being touched by forester) with sweetgums in an old field abandoned several decades ago, Mississippi. (Clatterbuck et al., 1987; Clatterbuck and Hodges, 1988.)

the vigor and timber value of the stand, a reduction in wildlife value of the stands in the eastern U.S., and an increase in fire, insect, and disease susceptibility in these stands in the western U.S. coniferous forests. Much effort, a long time, much expertise, and much money will be needed to restore these stands to their preharvest species compositions and structures.

To distinguish between these loggers' practices and clearcutting that could sustain the species composition, timber, and structure of the forest in the long term, silviculturists termed the removal of only the best trees in a stand as "high grading," removal of only the merchantable trees while leaving the unmerchantable, usually unvigorous ones as "commercial clearcutting" or "conditional clearcutting" (in the Russian Far East), and cutting of the unmerchantable stems to replicate a stand-replacing disturbance as "silvicultural clearcutting" or "cleancutting" (Figure 3.12B) (Helms 1998). Selective harvesting, leading to multiple-cohort stands, can be done under certain circumstances in a sustainable manner as will be discussed below; however, it is subtly different in practice and dramatically different in outcomes from the common practice of "high grading," "commercial clearcutting," or "conditional clearcutting."

3.6.2.4 *Multiple-cohort pathways*

When a disturbance does not kill all trees but kills enough trees so that soil and sunlight growth factors (growing space) are made available, the remaining plants may expand and

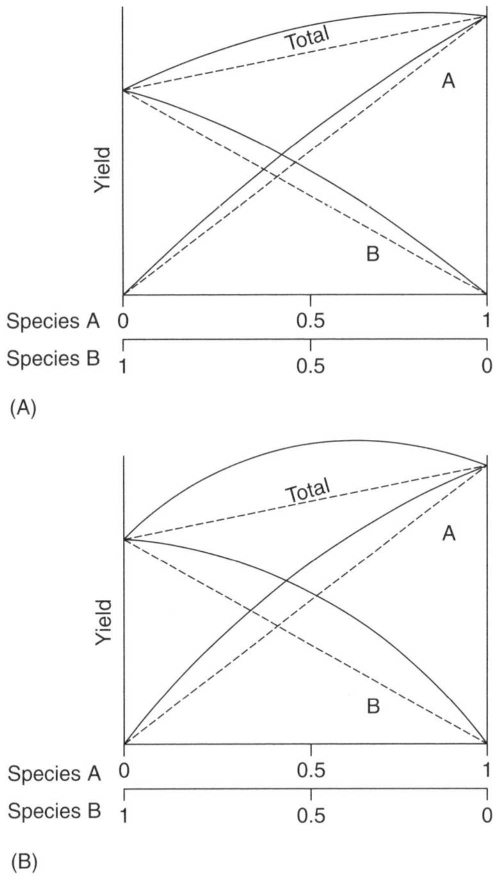

Figure 3.11 Potential relations of volume growth to proportions of species in mixed species stands. (from Kelty, *The Development and Productivity of Hemlock-Hardwood Forests in Southern New England*, Ph.D. dissertation, Yale University, New Haven, CT, 1984. With permission.) (A) Mixed volume is less than the volume of one species in a pure stand. (B) The mixed volume is greater than the volume of either species in a pure stand.

new plants may invade (Oliver and Larson 1996). If the disturbance is sufficiently large, the remaining plants are sufficiently unvigorous, and/or the new plants are sufficiently vigorous, a new cohort of plants can become established. Commonly, trees are part of this new cohort; and stands with two or more cohorts of trees are referred to as "multiple-cohort" or "multiaged" stands.

Where there is a shortage of stands of complex structure, conversion of single-cohort stands to multiple-cohort stands can help provide many complex structural features. For example, the multiple-cohort Douglas-fir stand in Figure 3.6A was similar to the single-cohort stand in Figure 3.15 before a partial cutting 17 years ago (before the photograph was taken) that allowed a new cohort to develop. As discussed earlier, reforesting or afforesting with mixed species can also provide many complex forest structural features.

A multiple-cohort stand commonly appears to have a complex structure that can be highly variable (Figures 3.13–15). A few older trees can exist in the older cohorts, with the majority of the stand being in a younger cohort. Alternatively, the opposite can occur with only a few trees in a younger cohort. Shade from much taller overstory trees (a.k.a. high

(A)

(B)

Figure 3.12 (A) "Commercial clearcutting" in which the unvigorous, crooked, unmerchantable trees of shade-tolerant species are left. They shade and exclude regeneration of shade-intolerant species. (B) "Silvicultural clearcut" in which unmerchantable trees are cut, so that new, vigorous stems of a variety of species can regenerate.

Figure 3.13 A multiple cohort stand in central Turkey, where trees invaded a former "savanna" structure when grazing was stopped (Soguksu National Park, Turkey).

Figure 3.14 Multiple cohort stand in Mississippi River floodplains. Dominance of shade-tolerant species (Sugarberry; *Celtis laevigata* Willd.) and crooked stems result from the death of shade-intolerant species and frequent suppression and release of shade-tolerant trees. Suppression and release cause sugarberry trees to leave crooks in their stems when they are released by a partial cutting. (By contrast, the dominant trees in Figure 3.2 grew without overstory shade and thus retained their straight stems.)

Figure 3.15 Douglas-fir stand in the Pacific Northwest with two cohorts. The overstory is maintained at a low density (here about 35 trees ha^{-1}) to ensure that understory grows vigorously.

shade) has different effects on tree growth than does shade from trees of the same size or only slightly taller.

The relation of overstory to understory tree growth in multiple-cohort stands is of theoretical and practical importance. For example, a larger allocation of growing space to the overstory will reduce the growth of the understory and vice versa. O'Hara (1998) presented this as a simple trade-off between the two canopy strata in two-cohort stands. Assuming that tree density by various measures is directly related to the amount of growing space occupied, various possible relationships of overstory and understory density can occur, as shown in Figure 3.16. The possible scenarios can be roughly classified into three overstory behaviors and three understory behaviors, making a total of nine stand-growth behaviors. Relationships G, H, and I are most likely to occur because of the greater growth rates of overstory trees (Kollenberg and O'Hara 1999; O'Hara 1996; O'Hara et al. 1999; Seymour and Kenefic 2002). MASAM, the Multiaged Stocking Assessment Model (O'Hara 1996; O'Hara et al. 2003), was developed to permit land managers to test the results of different growing space allocations in multiple-cohort stands.

Different species typically have different growing space response relationships that are related to their relative shade tolerance in comparison to the overstory species. The reduced growth of the understory (G, H, and I, Figure 3.16) is caused, in part, by a reduced photosynthesis at low intensities. A shade-tolerant species would probably be more productive than a less tolerant species under the same overstory light interception. The total stand productivity is probably similar for single-cohort and multiple-cohort stands, given equal total site occupancy and species composition. In ponderosa pine in the western United States, O'Hara (1996) reported that multiple-cohort stands were at least as productive as single-cohort stands on similar sites.

The correct relationships in Figure 3.16 have practical significance in determining how much forest to maintain in the complex, or multiple-cohort structure in any given forest ecosystem type. If pattern A were occurring, stands in the complex structure would be growing the most volume per ha; therefore, it would take less forest area of stands in this structure to grow and harvest a targeted volume of timber sustainably. It may be appropriate to manage the landscapes within this ecosystem with proportionately more stands following multiple-cohort pathways to create this complex structure and greater volume. Conversely, if pattern I were occurring within a forest ecosystem type, it may be appropriate to manage more stands in single-cohort trajectories with more dense and understory structures to provide the forest ecosystem type's fair share of timber commodities, leaving fewer multiple-cohort stands, but enough to provide habitat and other values.

Older cohorts in multiple-cohort stands can also affect the species composition of the younger cohorts. A very light overstory will allow all species to grow; however, even a slightly dense overstory will generally favor more shade-tolerant species (Oliver and Larson 1996). As the older cohorts regrow following a disturbance, they increasingly shade the younger cohorts, causing all trees in the younger cohorts to slow in growth. More shade-tolerant species, however, will not slow as rapidly nor die as readily as more shade-intolerant species; consequently, there is generally a shift to the more shade-tolerant species unless very few trees are left in the older cohorts (Wampler 1993). Stands having characteristics described as old growth generally contain many shade-tolerant trees because these stands have endured one or more periodic disturbances and their younger cohorts are prone to contain more shade-tolerant trees (Park 2001). When restoring a landscape to a dynamic distribution of species, maintaining a large amount of complex forests will tend to promote trees of shade-tolerant species, while more of the other structures will promote trees of less-tolerant species, even though with care and much effort, multiple-cohort stands can maintain a dominance of shade-intolerant species and single-cohort stands can maintain a dominance of shade-tolerant species. The most successful mixed species,

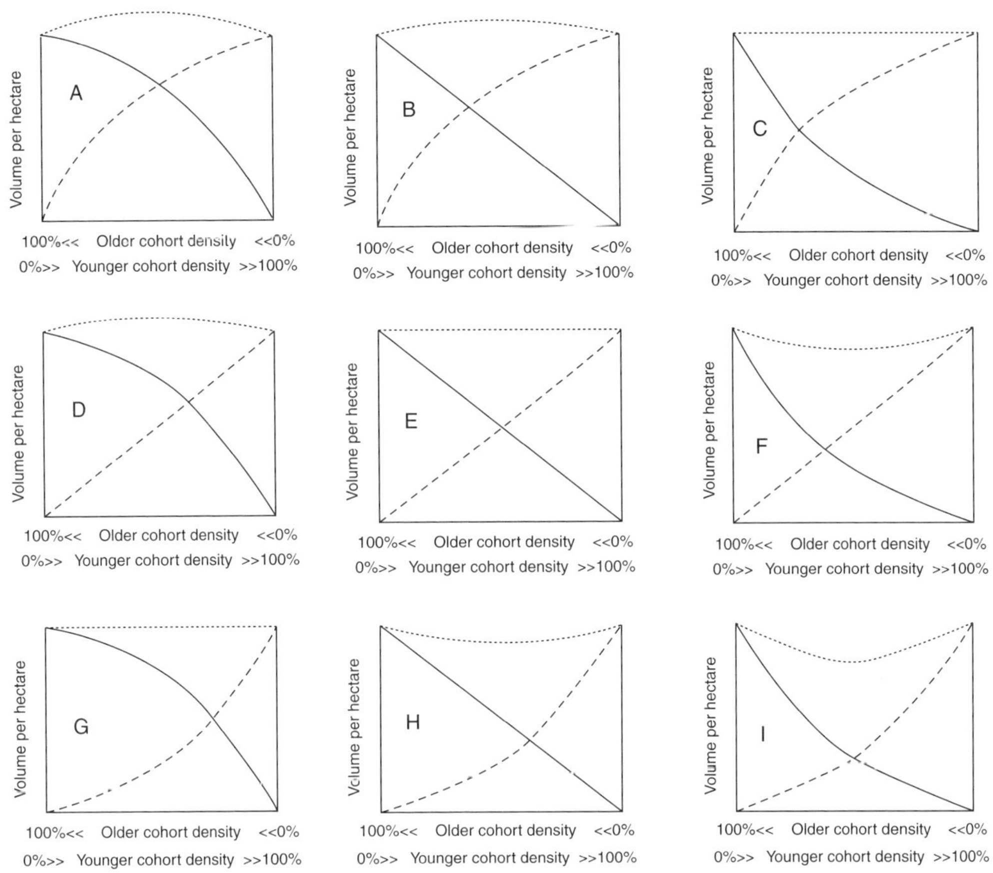

Figure 3.16 Possible theoretical relationships of older and younger cohorts in multiple-cohort stands. The relations of each cohort could be positive, neutral, or negative with respect to growth of the other cohorts, resulting in different growths for each cohort and the stand as a whole. (Solid line = older cohort; dashed line = younger cohort; dotted line = total stand)

multi-cohort stands will be those that include species with similar shade tolerance. The success with the Plenterwald in central Europe is aided by the tolerance of the three main species: Norway spruce (*Picea abies* [L.] Karst), European silver fir (*Abies alba* A.), and European beech (*Fagus sylvatica* L.).

At increasing levels of overstory shade, nearly all tree species will slow in height growth and assume a flat topped physiognomy characteristic of their species (Busgen and Munch 1929; Oliver and Larson 1996). Many conifer species will maintain a central stem but slow in height growth even more than they slow in lateral branch growth. The central stem of other conifer species and most deciduous species will also grow more slowly when shaded, but will also bend to a horizontal position and create a bush-like upper stem. Some species, such as beech (*Fagus* sp.), readily lose their upward terminal upon shading while others such as sugar maple can maintain the terminal under relatively intense shade. When released from overhead shade, all trees will resume height growth unless the initial shock of full sunlight kills them. Trees that had maintained their central vertical stem when in the understory will continue growing with a straight stem, but a dense group of branches will mark where it had been suppressed. Trees that had lost their central stem will usually have a crook, fork, or branch where the stem had turned vertical again or one or two lateral branches had assumed the dominant position (Figure 3.14).

Multiple-cohort stands often provide quite intricate structures for bird habitats similar to those provided by single-cohort mixed-species stands described earlier.

Effective management of multiple-cohort stands for timber requires several considerations:

- Multiple-cohort pathways are much easier to maintain if the species selected for management maintains a strong central stem when beneath high shade; the trees should also not readily develop wood rot if injured during partial logging operations.
- The overstory should be kept at a low enough density through thinnings to keep the younger cohorts alive and growing vigorously.
- The less vigorous trees in each cohort should be removed with each harvesting operation to prevent the result from being a high-graded stand (similar to Figure 3.12A), described earlier.
- The younger cohorts may need to be thinned to prevent overcrowding and shortening of crowns and loss of vigor.

Without extreme care, those stands comprised of species that readily become flat-topped do not provide very high-quality timber in multiple-cohort stands because of their many stem crooks. Without this care, there will need to be a balance within a forest ecosystem between stands managed in a multiple-cohort manner and those managed otherwise to provide straight trees for timber.

In regions with subsistence rural conditions, selective harvesting may be useful as a means of maintaining continuous forest cover on an area for some wildlife habitats, but at the same time obtaining some fuel wood. On the other hand, it is often quite difficult to prevent selectively harvested forests from being high graded under such circumstances; and the income and employment values from timber would not be sustained if the forest were high graded. In addition, without great care, selective harvesting tends to shift the ecosystem's species composition to shade-tolerant species.

3.7 *Coordinating restoration of many stands*

Many different stand structures can exist within a landscape, each of which has value for sustaining forests. Therefore, instead of a single pathway or structure being the target of restoration, a combination of pathways and structures, the edges between the structures, and the timber harvested during the transition from one structure to another will need to be the target of restoration (Boyce 1995; Oliver 1992). The challenge for restoration is not simply the placement of different stand structures across a landscape. These structures change with time through growth and disturbances; therefore, the changes will need to be coordinated so that some stands are developing into each structure as others are moving out of it (Figure 3.3).

Computer analytical tools have been developed that can help store and analyze the large amount of inventory, map, growth projection, and other data needed for coordination of stands across the landscape (e.g., McCarter et al. 1998; McGaughey 1997; Wilson and McGaughey 2000; http://lms.cfr.washington.edu). These management tools generate trust in forest restoration by enabling the manager to follow modern management procedures of first designing an intended outcome, and then achieving it.

3.8 *Paying for restoration*

An emerging issue in stand restoration will be paying for the operations, or paying for not doing operations that would maximize a single objective, such as high grading a stand to obtain immediate money from timber.

When a stand was managed for a single objective, such as timber, a simple microeconomic analysis could be used to determine the optimum operations and their timing. Arguments over whether to try to preserve a forest or manage it were based on the assumption that providing timber values excluded other forest values (the presumption of a zero-sum outcome described earlier). This presumption led to polarization of forest management. For example, some organizations have argued that thinning national forests in the Inland West, U.S.A., at a net cost would subsidize the timber industry, with little consideration that it would also contribute to the other values of protecting the "forest health" and soil and water quality, and providing for carbon sequestration, socioeconomic benefits, and biodiversity (Oliver and Lippke 1995).

Over time, multiple objectives have become both recognized in forest management and become economically profitable for management. For example, in parts of the southeastern United States, stands are both managed for timber and leased for hunting rights, with the hunting leases often being more profitable than the timber. Consequently, forest management is modified from maximizing timber profitability to accommodate an increased value of the forest for wildlife. Other, less obvious values are beginning to be marketed as well, such as the development rights, the right to convert a forest to another use, such as housing. Ecosystem services such as watershed protection and carbon sequestration are also beginning to be appreciated, and landowners are beginning to be rewarded for providing these values (Daily and Ellison 2002). In this way, forests are being managed as multiple-product assets (Gottfried 1992; Gottfried et al. 1996; Houghton and Mendelsohn 1996; Peters et al. 1989), with economic returns coming through providing a variety of objectives.

Managing forests as a multiple-product asset is similar to addressing economic externalities in forest management. Those changes to the environment that are considered neither a cost nor a benefit are considered externalities and are not accounted for in the relative profitability of an operation. Externalities originally included such things as air and water pollution from factories, reduction in fish habitat, reduction of complex forest habitat in logging, increased risk of fires in logging slash, and a modified scenery. Environmental and forest practice laws have forced the creators of these negative externalities to mitigate or avoid them at a cost. These costs are passed on to the consumer as part of the product cost; thus, they become internalized within the system. By internalizing the costs of these values, more values are provided by the forest, or at least the values are maintained and they no longer become externalities to the system.

The next step may be to address positive externalities in ways similar to the negative externalities. For example, forest management can create rural jobs, thus potentially reducing welfare costs for society as a whole. Different countries are beginning to consider internalizing such positive externalities in their forest management, although such practices are still being debated.

Different countries are addressing the various multiple product assets and internalizing negative and positive externalities in different ways. As these different approaches become tested and refined and the full range of sustainability criteria become internalized into the economic system, it will be increasingly easier to justify economically the restoration and sustainable management of forests.

References

Bailey, R.G., Delineation of ecosystem regions, *Environ. Manage.*, 7, 365, 1983.

Botkin, D.B., *Discordant Harmonies: A New Ecology for the Twenty-First Century*, Oxford University Press, New York, 1990.

Bowling, D.R. and Kellison, R.C., Bottomland hardwood stand development after clearcutting, *South. J. For. Res.*, 7, 110, 1983.

Boyce, S.G. and Oliver, C.D., The history of research in forest ecology and silviculture, in *Forest and Wildlife Science in America: A History*, Steen, HK., Ed., Forest History Society, North Carolina, 414, 1999.

Boyce, S.G., *Landscape Forestry*, John Wiley & Sons, New York, 1995.

Braun, E.L., *Deciduous Forests of Eastern North America*, Macmillan, New York, 1950.

Brundtland, G.H., *Our Common Future*, World Commission on Environment and Development, United Nations, Oxford University Press, New York, 1987.

Burley, J., International initiatives for the sustainable management of forests, in *Forests and Landscapes: Linking Ecology, Sustainability, and Aesthetics*, Sheppard, S.R.J. and Harshaw, H.W., Eds., CABI Publishing in association with the International Union of Forest Research Organizations, New York, 95, 2001.

Busgen, M. and Munch, E., *The Structure and Life of Forest Trees* (translated by T. Thompson), 3rd Ed., St. Giles' Works, Norwich, Great Britain, 1929.

Camp, A.E., Oliver, C.D., Hessburg, P., and Everett, R., Predicting late-successional fire refugia predating European settlement in the Wenatchee Mountains, *For. Ecol. Manage.*, 95, 63, 1997.

Camp, A.E. and Oliver, C.D., Forest Dynamics, in *Encyclopedia of Forest Sciences*, Burley, J., Evans, J., and Youngquist, J.A., Eds., Elsevier Science Ltd., Oxford, 1053, 2004.

Carey, A.B., Lippke, B.R., and Sessions, J., Intentional ecosystem management: managing forests for biodiversity, *J. Sustain. For.*, 9, 83, 1999.

Chandler, P.M., *Ecological Knowledge in a Traditional Agroforest Management System among Peasants in China*, Ph.D. Dissertation, University of Washington, Seattle, 161, 1990.

Clatterbuck, W.K. and Hodges, J.D., Development of cherrybark oak and sweetgum in mixed, even-aged bottomland stands in central Mississippi, U.S.A., *Can. J. For. Res.*, 18, 12, 1988.

Clatterbuck, W.K., Oliver, C.D., and Burkhardt, E.C., The silvicultural potential of mixed stands of cherrybark oak and American sycamore: spacing is the key, *South. J. Appl. For.*, 11, 158, 1987.

Clatterbuck, W.K., Hodges, J.D., and Burkhardt, E.C., Cherrybark oak development in natural mixed oak-sweetgum stands—preliminary results, in *Proceedings of the Third Biennial Silvicultural Research Conference*, Shoulders, E., Ed., U.S. Department of Agriculture, Forest Service, Southern Forest Experiment Station, General Technical Report SO-54, 438, 1985.

Cobb, D.F., O'Hara, K.L., and Oliver, C.D., Effects of variation in stand structure on development of mixed-species stands in eastern Washington, *Can. J. For. Res.*, 23, 545, 1993.

Daily, G.C. and Ellison, K., *The New Economy of Nature*, Island Press, Shearwater Books, New York, 2002.

Daniel, T.W., Helms, J.A., and Baker, F.S., *Principles of Silviculture*, 2nd Ed., McGraw-Hill, 1979.

Deal, R.L., Oliver, C.D., and Bormann, B.T., Reconstruction of mixed hemlock-spruce stands in coastal southeast Alaska, *Can. J. For. Res*, 21, 643, 1991.

Drew, T.J. and Flewelling, J.W., Stand density management: an alternative approach and its application to Douglas-fir plantations, *Forest Sci.*, 25, 518, 1979.

Flannery, T., *The Eternal Frontier: An Ecological History of North America and Its People*, Atlantic Monthly Press, New York, 2001.

Ford-Robertson, F.C., *Terminology of Forest Science, Technology, Practice, and Products*, Society of American Foresters, Washington, DC, 1971.

Fry, M.E. and Money, N.R., Biodiversity conservation in the management of utility rights of way, in *Proceedings of the 15th Annual Forest Vegetation Management Conference*, Redding, California, 25–27 January, 94, 1994.

Fujimori, T., *Ecological and Silvicultural Strategies for Sustainable Forest Management*, Elsevier, Amsterdam, 2001.

Gingrich, S.F., Measuring and evaluating stocking and stand density in upland hardwood forests in the Central States, *Forest Sci.*, 13, 38, 1967.

Gingrich, S.F., *Management of Upland Hardwoods*, U.S. Department of Agriculture, Forest Service, Research Paper NE-195, 26 pp., 1971.

Gottfried, R., The value of a watershed as a series of linked multi-product assets, *Ecol. Econ.*, 5, 145, 1992.

Gottfried, R., Wear, D., and Lee, R., Institutional solutions to market failure on the landscape scale, *Ecol. Econ.*, 18, 133, 1996.

Hanley, D.P., Oliver, C.D., Maguire, D.A., Briggs, D.B., and Fight, R.D., *Pruning Conifers in Northwestern North America: Opportunities, Techniques and Impacts*, University of Washington, Institute of Forest Resources, Contribution No.77, 1995.

Harris, L.D., *The Fragmented Forest: Island Biogeography Theory and the Preservation of Biotic Diversity*, University Chicago Press, Illinois, 1984.

Helms, J.A., Ed., *The Dictionary of Forestry*, Society of American Foresters, Bethesda, MD, 1998.

Henry, J.D. and Swan, J.M.A., Reconstructing forest history from live and dead plant material—an approach to the study of forest succession in southwest New Hampshire, *Ecology*, 55, 772, 1974.

Houghton, K. and Mendelsohn, R., An economic analysis of multiple-use forestry in Nepal, *Ambio*, 25, 156, 1996.

Hunter, M.L. Jr., *Wildlife, Forests and Forestry*, Regents/Prentice Hall, Englewood Cliffs, NJ, 1990.

Johnson, S., *Emergence: The Connected Lives of Ants, Brains, Cities, and Software*, Scribner, New York, 2002.

Kelty, M.J., *The Development and Productivity of Hemlock-Hardwood Forests in Southern New England*, Ph.D. Dissertation, Yale University, New Haven, CT, 1984.

Kelty, M.J., Development patterns in two hemlock-hardwood stands in southern New England, *Can. J. For. Res.*, 16, 885, 1986.

Kelty, M.J., Productivity of New England hemlock-hardwood stands as affected by species composition and canopy structure, *For. Ecol. Manage.*, 28, 237, 1989.

Kittredge, D.B., The influence of species composition on the growth of individual red oaks in mixed stands in southern New England, *Can. J. For. Res.*, 18, 1150, 1988.

Kollenberg, C.L. and O'Hara, K.L., Leaf area and tree increment dynamics of even-aged and multi-aged lodgepole pine stands in Montana, *Can. J. For. Res.*, 29, 687, 1999.

Larsen, J.B., Ecological stability of forests and sustainable silviculture, *For. Ecol. Manage.*, 75, 85, 1995.

Lippke, B. and Oliver, C.D., Managing for multiple values, *J. For.*, 91, 14, 1993.

McCarter, J.B. and Long, J.N., A lodgepole pine density management diagram, *West. J. Appl. For.*, 1, 6, 1986.

McCarter, J.M., Wilson, J.S., Baker, P.J., Moffett, J.L., and Oliver, C.D., Landscape management through integration of existing tools and emerging technologies, *J. For.*, 96, 17, 1998.

McGaughuey, R.J., Visualizing forest stand dynamics using the Stand Visualization System, in *Proceedings ACSM/ASPRS 4*, 248, 1997.

McKinnon, F.S., Hyde, G.S., and Clinie, A.C, *Cut-over Old Field White Pine Lands in Central New England*, The Harvard Forest, Harvard University, Harvard Forest Bulletin No. 18, 1935.

Miller, E.W., Forest regions of the world, in *Forests: A Global Perspective*, Majumdar, S.K., Miller, E.W., and Brenner, F.J., Eds., The Pennsylvania Academy of Science, Pennsylvania, 1996, chap. 1.

Newton, M.B., El Hassan, B.A., and Zavitkovski, J., Role of red alder in western Oregon forest succession, in *Biology of an Alder*, Trappe, J.M., Franklin, R.F., Tarrant, R.F., and Hansen, G.M., Eds., U.S. Department of Agriculture, Forest Service, Pacific Northwest Forest and Range Experiment Station, Portland, OR, 73, 1968.

Nyland, R.D., *Silviculture: Concepts and Applications*, McGraw-Hill, New York, 1996.

O'Hara, K.L., Development patterns of residual oaks and oak and yellow-poplar regeneration after release in upland hardwood stands, *South. J. Appl. For.*, 10, 244, 1986.

O'Hara, K.L., Dynamics and stocking-level relationships of multi-aged ponderosa pine stands, *Forest Sci.*, Monograph 33, 1996.

O'Hara, K.L., Silviculture for structural diversity: a new look at multi-aged systems, *J. For.*, 96, 4, 1998.

O'Hara, K.L., The historical development of uneven-aged silviculture in North America, *Forestry*, 75, 339, 2002.

O'Hara, K.L., Lähde, E., Laiho, O., Norokorpin Y. and Saksa, T., Leaf area and tree increment dynamics on a fertile mixed-conifer site in southern Finland, *Ann. For. Sci.*, 56, 237, 1999.

O'Hara, K.L., Latham, P.L., Hessburg, P. and Smith, B.G., A structural classification for inland northwest forest vegetation, *West. J. Appl. For.*, 11, 97, 1996.

O'Hara, K.L. and Oliver, C.D., A decision system for assessing stand differentiation potential and prioritizing precommercial thinning treatments, *West. J. Appl. For.*, 14, 7, 1999.

O'Hara, K.L., Valappil, N.I., and Nagel, L.M., Stocking control procedures for multi-aged ponderosa pine stands in the Inland Northwest, *West. J. Appl. For.*, 18, 5, 2003.

Odum, E.P., *Fundamentals of Ecology*, 3rd ed., Saunders, Philadelphia, 1971.
Oliver, C.D., *Development of Northern Red Oak in Mixed-species Stands in Central New England*, Yale University School of Forestry and Environmental Studies Bulletin No. 91, 1978.
Oliver, C.D., Even-aged development of mixed-species stands, *J. For.*, 78, 201, 1980.
Oliver, C.D., Forest development in North America following major disturbances, *For. Ecol. Manage.*, 3, 153, 1980.
Oliver, C.D., A landscape approach: achieving and maintaining biodiversity and economic productivity, *J. For.*, 90, 20, 1992.
Oliver, C.D., Sustainable forestry. What is it? How do we achieve it? *J. For.*, 101, 8, 2003.
Oliver, C.D. and Larson, B.C., *Forest Stand Dynamics*, Updated Edition, John Wiley & Sons, New York, 1996.
Oliver, C.D. and Lippke, B.R., Wood supply and other values and ecosystem management, in western interior forests, in *Ecosystem Management in Western Interior Forests*, Symposium held May 13–15, 1994, Spokane, Washington, Department of Natural Resource Sciences, Washington State University, Pullman, WA, 195, 1995.
Oliver, C.D. and Stephens, E.P., Reconstruction of a mixed species forest in central New England, *Ecology*, 58, 562, 1977.
Oliver, C.D., Camp, A. and Osawa, A., Forest dynamics and resulting animal and plant population changes at the stand and landscape levels, *J. Sustain. For.*, 6(3/4), 281, 1998.
Oliver, C.D., Boydak, M., Segura, G., and Bare, B.B., Forest organization, management, and policy, in *Maintaining Biodiversity in Forest Ecosystems*, ML. Hunter Jr., Ed., Oxford University, Ch.17, 556, 1999.
Oliver, C., Adams, D., Bonnicksen, T., Bowyer, J., Cubbage, F., Sampson, N., Schlarbaum, S., Whaley, R., Wiant, H., and Sebelius, J., *Report on Forest Health of the United States by the Forest Health Science Panel*, Panel chartered by Charles Taylor, Member, United States Congress, 11th District, North Carolina. (Available through U.S. House of Representatives Resources Committee at: http://www.house.gov/resources/105cong/fullcomm/apr09.97/taylor.rpt/taylor.htm) Also available as reprint through University of Washington College of Forest Resources CINTRAFOR RE43, 1997.
Oliver, C.D., Burley, J., and Maathai, W., Sustainable Forestry: What is it? How do we achieve it? in *Synopsis of May 3, 2002 Seminar*, Global Institute of Sustainable Forestry, School of Forestry and Environmental Studies, Yale University, New Haven, CT, 2002.
Park, P.S., *Forest Stand Structure Characteristics for the Cispus Adaptive Management Area, Cascade Range, U.S.A.: Implications for Old Growth, Fire Hazard, Silviculture, and Landscape Management*, Ph.D. dissertation, University of Washington, Seattle, WA, 2001.
Peterken, G.F., *Natural Woodland: Ecology and Conservation in Northern Temperate Regions*, Cambridge University Press, New York, 1996.
Peters, C., Gentry, A., and Mendelsohn, R., Valuation of an Amazonian rain forest, *Nature* 339, 655, 1989.
Phillips, Cassie, Identifying problems and challenges associated with private forest lands-overview in *Summit 2000: Washington Private Forests Forum*, University of Washington, College of Forest Resources, Institute of Forest Resources, Summer 2000.
Raup, H.M., Some problems with ecological theory and their relation to conservation, *J. Ecol.*, 52, 19, 1964.
Reineke, L.H., Perfecting a stand-density index for even-aged forests, *J. Ag. Res.*, 46, 627, 1933.
Seymour, R.S. and Kenefic, L.S., Influence of age on growth efficiency of *Tsuga canadensis* and *Picea rubens* trees in mixed-species, multi-aged northern conifer stands, *Can. J. For. Res.*, 32, 2032, 2002.
Smith, D.M., Larson, B.C., Kelty, M.J., and Ashton, P.M.S., *The Practice of Silviculture*, John Wiley & Sons, New York, 1997.
Stevens, W.K., New eye on nature: the real constant is eternal turmoil, *New York Times*, Science article, Tuesday, July 31, B5-B6, 1990.
Stubblefield, G.W. and Oliver, C.D., Silvicultural implications of the reconstruction of mixed alder/conifer stands, in *Utilization and Management of Red Alder*, Atkinson, W.A., Briggs, D., and DeBell, D.S., Eds., U.S. Department of Agriculture, Forest Service, General Technical Report PNW-70, 307, 1978.

Tallis, J.H., *Plant Community History: Long-term Changes in Plant Distribution and Diversity*, Chapman & Hall, New York, 1991.

Tansley, A.G., The use and abuse of vegetational concepts and terms, *Ecology*, 16, 284, 1935.

Toumey, J.W., *Foundations of Silviculture, Upon an Ecological Basis*, 2nd ed., New York, John Wiley & Sons, 1947.

Walkinshaw, L.H., *Kirtland's Warbler*, Cranbrook Institute of Science, Bloomfield Hills, MI, 1983.

Wampler, Marianne, *Growth of Douglas-fir under Partial Overstory Retention*, Master's thesis, University of Washington College of Forest Resources, 1993.

Whittaker, R.H., *Communities and Ecosystems*, 2nd ed., Macmillan Publishing Company, New York, 1975.

Wierman, C.A. and Oliver, C.D., Crown stratification by species in even-aged mixed stands of Douglas-fir/western hemlock, *Can. J. For. Res.*, 9, 1, 1979.

Wilson, J.S. and McGaughey, R., Presenting landscape-scale forest information: What is sufficient and what is appropriate? *J. For.*, 98, 21, 2000.

Wilson, J.S. and Oliver, C.D., Stability and density management in Douglas-fir plantations, *Can. J. For. Res.*, 30, 910, 2000.

Wonn, H.T. and O'Hara, K.L., Height:diameter ratios and stability relationships for four northern Rocky Mountain tree species, *West. J. Appl. For.*, 16, 87, 2001.

Young, M.R., Conserving insect communities in mixed woodlands, in *The Ecology of Mixed-species Stands of Trees*, M.G.R. Cannell, D.C. Malcolm, and P.A. Robertson, Eds., Blackwood Scientific Publications, London, 277, 1992.

Zebrowski, Jr. E., Death and Life, in *Perils of a Restless Planet: Scientific Perspectives on Natural Disasters*, Cambridge University Press, New York, Ch. 4, 93, 1997.

chapter four

Incorporating disturbance into forest restoration

Susan Warren Beatty and Bronwyn Sigrid Owen
University of Colorado-Boulder

Contents

4.1 Introduction

All forested ecosystems are influenced by natural disturbances. Disturbance in plant communities affects the maintenance of species richness and the process of community development (Bendix 1997; Collins and Pickett 1987; Connell 1978; Cooper-Ellis et al. 1999; Drury and Nisbet 1973; Foster 1988; Foster and Boose 1992; Grime 1979; Grubb 1977; Huston 1979; Pickett and White 1985; Platt and Weis 1977; Smith 1972; Watt 1947; Whittaker 1969; Whittaker and Levin 1977). Controls exerted by fine-scale, local disturbances interact with coarse-scale, regional disturbances in complex ways (Beatty 1991, 2003; Frelich and Reich 1995; Glitzenstein et al. 1986; Lertzman 1992; Lertzman and Fall 1998; Pastor and Broschart 1990). Disturbances are caused by both physical and biological events. Physical disturbances include generally stochastic processes that disrupt some

1-56670-635-1/05/$0.00+$1.50

aspect of the physical environment; such disturbances include wind downbursts, flooding, fire, avalanche, or landslide. Biological disturbances are often endogenous processes that stem from influences of native or exotic species in the ecosystem, and may include death from insect outbreaks, plant pathogens, or impacts caused by nonnative species (e.g., excessive herbivory, competition, or predation). In this chapter, we take a broad view of what qualifies as a disturbance, expanding on the definition of Pickett and White (1985) to include biological disturbances that often act as ongoing stress, but depending on the severity of the interaction, can also be discrete disturbance events. Both physical and biological disturbances operate at the coarse scale, such as hurricane blowdowns or gypsy moth defoliation, and at the fine scale, such as individual treefall or excessive deer browse. Coarse-scale phenomena are generally responsible for initiating successional regeneration responses that can take decades to centuries (stand-destroying fire, for example), while the fine-scale phenomena are responsible for maintenance stand regeneration responses that usually take years to decades (seedling recruitment to the canopy in a light gap, for example) (Allen and Walsh 1996; Bendix 1997; Clark et al. 1999; Kupfer and Malanson 1993; Parker and Bendix 1996; Pitman et al. 1999). The frequency of disturbance also varies with spatial scale, as coarser-scale events occur less frequently but are more intense (Drury and Nisbet 1973; Watt 1947). Disturbance, therefore, plays a different role at fine vs. coarse scale. If the incorporation of natural disturbance regimes into restoration and forest management is a long-term goal, then an understanding of forest dynamics at different spatial scales and for physical versus biological disturbances will be necessary. The many unanswered questions about the role of disturbance in forest dynamics present a challenge to those who wish to incorporate disturbance into a forest restoration plan as we will illustrate in two case studies, from needle-leaf forests in the western U.S., and northern hardwood forests of the eastern U.S.

Most disturbance events are inherently stochastic, making it difficult to predict return interval and intensity. In the western needle-leaf forests, accurate fire-history models are sought that have an ecologically meaningful fire-frequency term (Baker and Ehle 2001; Turner and Dale 1998). Managers are trying to restore these fire-maintained forest ecosystems by thinning to reduce fuel loads and by using prescription ground fires to avoid stand-replacement fires, although such stand-destroying fires may be vital to long-term ecological processes in ecosystems (Romme et al. 1995; Turner et al. 1994). In the northeastern mixed deciduous forest, a hurricane disturbance differs from the effects of forest clearing and land-use change, but restoration attempts often assume that succession following anthropogenic disturbance proceeds in a manner similar to succession after natural disturbance. Forest restoration effort needs to recognize that fine-scale heterogeneity is very important in determining species distributions within a forest, maintaining population sizes and understory richness (Beatty 1984, 1991, 2003). In addition, the finer-scale disturbances in each forest ecosystem are often affected by the coarser-scale dynamics (Webb 1988, 1989). For example, thinning to reduce fire intensity or occurrence can also modify some of the finer-scale effects, such as the role of individual treefall in forests.

Individual treefall contributes to the formation of microscale soil relief in deciduous forests, and these long-lasting features of the forest floor are important to forest species regeneration and maintenance (Beatty 1984; Beatty and Stone 1986; Lyford and MacLean 1966; Peterson and Pickett 1990). A forest regrowing on a soil that has been plowed will lack this heterogeneity and, therefore, perhaps the microsites for some forest species, particularly some tree seedlings (e.g., *Acer saccharum*) that germinate and establish in greatest abundance on old treefall mounds (Beatty 1981). Restoring the legacy effects of disturbance are as important as restoring the disturbance regime itself (Beatty 2003). Understanding the interactions between different disturbances and at different scales is vital to successful restoration of forested ecosystems. It may be easier to restore the fine-scale disturbance dynamics than the coarse-scale dynamics, since the latter is much more variable and unpredictable. Yet, much attention is focused on attempts to mimic natural disturbance on a coarse scale, such as manipulating fire extent or intensity, or the size of a harvest unit. This chapter will present arguments for paying attention to restoration of disturbance effects at both scales.

4.2 Case studies in forest disturbance regimes

To illustrate some of the challenges of incorporating disturbance into forest restoration, two case studies will compare the scale and type of disturbance. Subalpine and montane coniferous forests of the Front Range of Colorado are subject to fine-scale individual treefall disturbance (Veblen 1986). Stand-destroying coarser-scale disturbances of bark-beetle outbreaks also periodically affect these forests (Veblen et al. 1991), as well as blowdown (Veblen et al. 2001), and fire (Veblen et al. 2000). Mixed deciduous forests of the northeastern U.S. also have finer-scale treefall disturbance (Beatty and Stone 1986; Runkle 1989), but are infrequently affected by fire (Wade et al. 2000). Fires do occur, but they do not initiate a seed dispersal and germination pulse as is typical of western needle-leaf forests, but rather instigate regeneration via stump sprouting and root suckering. In the southern mixed deciduous forest, fire maintained dominance of oaks and pines, which are being replaced by maples, beech, and hemlock today (Abrams et al. 1995). This has not been reported, however, for the northeastern mixed deciduous forests, where common coarser-scale disturbances are wind storms (hurricanes or tornadoes). Recently, massive defoliations by the exotic gypsy moth have occurred periodically.

These forest systems are affected by multiple disturbance types expressed over both fine and coarse scales, including fire, blowdown, and insect outbreak. All of these disturbances can operate at a coarse scale, sometimes resulting in stand-replacement responses in the vegetation. Some of these disturbances operate at the fine scale, but the frequency and extent of coarse-scale disturbance events and the fine-scale consequences differ between forest types, and the integration of scales and types of disturbances is important to understanding forest dynamics in both ecosystems.

These case studies are based on research conducted in specific locations within Rocky Mountain needle-leaved and eastern deciduous forests in the U.S. The subalpine spruce-fir forest of the Front Range of Colorado has been studied extensively and is used to illustrate the dynamics of these three disturbances in needle-leaved forests of the Rockies. Similarly, mixed deciduous forests in upstate New York (Huyck Preserve) illustrate known disturbance dynamics in northeastern deciduous forest (also known as northern hardwoods). Clearly, the needle-leaved and deciduous forest types vary regionally, and the case studies do not represent the entire distribution of these broad forest types. The case studies, however, are representative of the specific forest types and exemplify how fire, wind, and insects disturb these ecosystems.

4.2.1 The Colorado front range subalpine spruce-fir forests

Subalpine forests in the Front Range of the Colorado Rocky Mountains occur between 2,500 and 4,000 m elevation, and mature stands are dominated by Englemann spruce (*Picea engelmannii*) and subalpine fir (*Abies lasiocarpa*), with lodgepole pine (*Pinus contorta*) and quaking aspen (*Populus tremuloides*) in some stands. This region of the southern Rocky Mountains has a continental climate with a mean annual temperature of 3.8 °C, and a range in the mean monthly temperature from −9.6 °C in January to 16.5 °C in July. The mean annual precipitation is 60.5 cm and ranges from a mean monthly precipitation of 6.3 cm in January to 3.9 cm in July. The soils are unclassified coarse-textured materials derived from Precambrian crystalline rocks and glacial deposits in upland areas and from poorly drained alluvial deposits in valley bottoms (Hoffman and Alexander 1980; Snyder et al. 1987). Much of the study area is occupied by the Mount Zirkel Wilderness Area in Routt National Forest, and had little logging impact (less than 2% of the study area) (Kulakowski and Veblen 2002). A massive blowdown occurred on 25 October 1997, caused by 200 to 250 km h^{-1} easterly winds (Wesley et al. 1998).

4.2.1.1 *Fire in subalpine forests of Colorado*

Fire is an important disturbance in needle-leaved evergreen forests, and can vary from a low-intensity ground fire to a catastrophic stand-replacing fire (Turner and Dale 1998). The former allows a reorganization of tree regeneration, without necessarily changing the canopy dominance. Stand-replacement fires usually involve regrowth by species different from those in the stand before the fire. The period before European settlement in the 19th century was characterized by a low frequency of fire. After European settlement, fire frequency greatly increased. Once technology was developed that could efficiently suppress fires, fire frequency declined greatly (Keane et al. 2002). Currently, in many needle-leaf evergreen forests on public land in the west, the policy is to let wildfires burn unless structures or people are at risk. Because of the mesic environment of the subalpine zone, fires occur less frequently than in other western forests such that differences in long return intervals are difficult to attribute to any specific anthropogenic cause (Sherriff et al. 2001). Not enough time has elapsed during the Holocene for much of the subalpine zone to exceed its natural fire-return interval. In the subalpine forests in Colorado, old-growth stands have more Engelmann spruce in the canopy and more subalpine fir in the subcanopy. Without disturbance, these two species persist as codominants. After a high-intensity stand-destroying fire, initial regeneration is by lodgepole pine, with some Engelman spruce later (Figure 4.1). Subalpine fir regenerates later, under a closed canopy environment.

4.2.1.2 *Insect outbreaks in subalpine forests of Colorado*

Spruce beetle (*Dendroctonus rufipennis*) outbreaks are regionally variable in both frequency and spatial extent. Spruce beetle outbreaks occurred in the northwest region of Colorado in 1716 to 1750, 1827 to 1845, 1860 to 1870, and 1940 to 1960 (Eisenhart and Veblen 2000), indicating their prevalence in the disturbance regime of these subalpine forests. Mild winters may trigger outbreaks, with blowdown areas possibly stimulating increased beetle populations, although this has not been a factor in the study areas (Kulakowski 2002). Old forests were much more likely to sustain a beetle outbreak than were younger stands. Spruce beetle outbreaks in an old forest stand kill the larger and older spruce trees, leaving subalpine fir as the dominant species (Figure 4.1). These fir-dominated forests are temporary in the landscape, as Engelmann spruce eventually regains dominance, but the time for recovery varies. It is not clear whether a spruce beetle outbreak makes a stand more or less susceptible to fire (Figure 4.1), as stands with recent beetle outbreaks may be more resistant to lower-intensity ground fires because surface soils are moister, presumably from less transpiration loss (Kulakowski 2002). Also, blowdowns may provide a food source that triggers a beetle outbreak (Figure 4.1), although this may not always occur.

4.2.1.3 *Blowdown in subalpine forests of Colorado*

Blowdown is less common and less extensive a disturbance in subalpine fir forests than beetle outbreaks. Old-forest stands are more susceptible to wind damage than younger stands. Blowdown susceptibility is also affected by the age of stands (younger less susceptible), which is determined in part by the fire history in the landscape; and the pattern of blowdown damage can reflect the mosaic of past disturbance history (Kulakowski

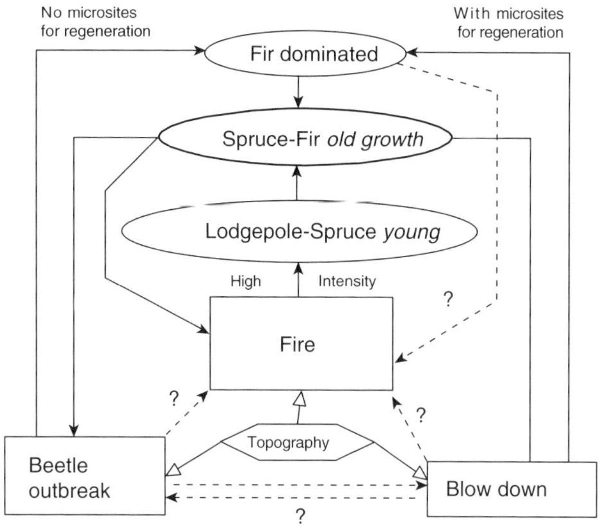

Figure 4.1 Conceptual model of the effects and interactions of three disturbance types (high-intensity fire, insect outbreak, and blowdown) on the regeneration and composition of needle-leaf subalpine forests of the Front Range of CO, USA. Rectangles indicate disturbances; ellipses indicate forest types (bold is old-growth; fine lines are successional stands). Topography is included as a modifier factor that can alter the spatial or temporal extent of a given disturbance event. Question marks indicate connections that are hypothesized.

and Veblen 2002). In blowdown areas, larger canopy spruce are removed, leaving the subcanopy and understory fir (Kulakowski and Veblen 2003). These fir-dominated stands are similar to those created by beetle outbreak with a notable exception: the majority of tree mortality comes from uprooting, which creates regeneration microsites for fir and particularly spruce (Figure 4.1). This may alter the recovery trajectory to a greater amount of spruce, as compared to the fir-dominated stand that follows beetle disturbance (Turner et al. 1994; Kulakowski and Veblen 2003).

4.2.1.4 Topography effects on disturbances in subalpine forests of Colorado

Topography influences the spatial pattern and extent of disturbances in the Colorado Front Range subalpine forests (Figure 4.1), as well as other needle-leaved forests in the western U.S. (Turner et al. 1994). Any directional disturbance (wind direction, upslope/downslope movement) can be modified by topography, which can funnel the disturbance, or act as a barrier to its propagation (Turner et al. 1994). Blowdowns are most severe at higher elevations, on easterly slopes, and close to ridgetops (Kulakowski and Veblen 2002). Fire behavior can be influenced by topography, particularly in the lower-intensity or moderate burns, although the most severe-intensity burns do not appear to be affected (Turner et al. 1994). The interaction among disturbances can influence the return interval for a given disturbance (Kulakowski and Veblen 2002).

4.2.1.5 Forest management and disturbance in subalpine forests

How forests should be managed after disturbance has generated much controversy. Arguments in favor of removing downed or dead timber after a beetle outbreak or blowdown stress the risk to the surrounding forests of a stand-replacement fire. Blowdown areas may be sources of beetle outbreaks after the blowdown (McCullough et al. 1998); the probability of an outbreak following a blowdown is about 50%, declining to 0% within

5 years (Bebi et al. 2003). Management objectives play an important role in deciding how to manage disturbed forests.

The decision of whether to salvage log after a blowdown could be based on where the blowdown is in the landscape, and the likelihood that it would carry a fire to the surrounding forest. A ridge-top blowdown could be more of a firebreak than ignition source (since fires tend to move upslope), whereas a valley or mid-slope blowdown could carry fire upslope. Allowing natural regeneration recovery after blowdown seems appropriate where a low-risk situation exists. Another approach would be to allow for natural recovery in the blowdown in any slope position, but to thin the surrounding forest to reduce fuel load and reduce the potential for carrying a fire if ignited. If recreational use is impaired by blowdown or outbreak damage (e.g., hazard of dead trees falling), a salvage operation may be indicated for that specific portion of the disturbed area, but leaving the remainder to recover naturally. Recreational uses would also have to be restricted.

4.2.2 The mixed deciduous forest of the northeastern U.S.

Broadleaf deciduous forests cover much of the eastern U.S., east of the Mississippi River and roughly between 30° and 45° latitude (Braun 1950; Delcourt and Delcourt 1996; Vankat 1979). Species composition varies greatly with latitude and altitude. Southern forests are quite diverse and are dominated by genera of *Acer, Quercus, Carya*, and a variety of other genera including *Fagus, Liriodendron, Fraxinus, Nyssa*, and *Tilia*. Northern hardwood forests have fewer canopy species, with dominants being *Acer saccharum* (sugar maple) and *Fagus grandifolia* (American beech), with other occasional occurrence of *Tsuga canadensis* (Eastern hemlock), *Fraxinus americana* (white ash), *Quercus rubra* (red oak), *Betula* spp., or *Tilia americana* (basswood). This case study will focus on the mixed deciduous forest of the Huyck Preserve located in north central New York State. It is representative of the northern hardwood forests in terms of species composition, stand ages, and disturbance history. Blowdown, insect outbreak, pathogens, and rare fires disturb northern hardwood forests. The current forests at the Huyck Preserve are dominated by *A. saccharum* and *Fagus grandifolia*, are 150 to 250 years old (McLean 2003), and have very little *Tsuga* regeneration. These mature forests have a diverse understory, closed canopy, gap phase regeneration, and no known fire history.

In northeastern mixed deciduous forests, land-use history is important in determining the trajectory of community development through succession. The clearing of forests since the 1700s to exploit wood resources and create grazing land has been widespread. Thus, nearly all old forests have been at least logged since European settlement, and many forests were grazed even if they remained unplowed. Species composition and richness will vary depending on the age of a stand, prior land-use treatment (such as having been plowed or not), and types of surrounding communities supplying propagules for colonization (McLean 2003). We use the term mature forest here instead of the term old-growth, as very little true old-growth remains in the region. In the northeastern U.S., agriculture was abandoned at different times, giving rise to stands of different ages and species composition, intermixed with stands that were farm woodlots and harvested, but not plowed. Recognition of the mosaic nature of the landscape will be important in planning for successful forest restoration in northern hardwood forests.

4.2.2.1 Fire in northern hardwood forests

Fire is an uncommon disturbance in northern hardwood forests (Figure 4.2) (Wade et al. 2000). In the western parts of the distribution of northern hardwoods (Michigan, Wisconsin, Minnesota), needle-leaf evergreen (pine or hemlock) stands are mixed with deciduous stands. In these regions, fire has been historically prevalent prior to European

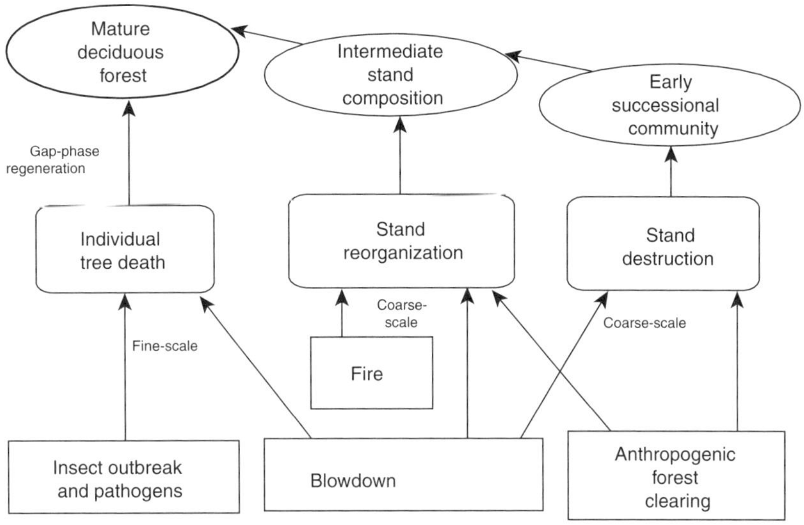

Figure 4.2 Conceptual model of the effects and interactions of four disturbance types (low-frequency fire, insect outbreak, blowdown, and anthropogenic forest clearing) on the regeneration and composition of broadleaf deciduous forests of the eastern United States. Rectangles indicate disturbances; smoothed rectangles show vegetation responses; ellipses represent forest stages (bold indicates most developed stand called "mature"). Distinction is made between disturbance events that operate at a coarse (large area) vs. a fine (small area) scale.

settlement (Heinselman 1973; Spurr 1954; Swain 1973), but it varies greatly in its frequency of occurrence (Frelich and Lorimer 1991). In the eastern portion of the forest distribution, fire is tied to anthropogenic land use (Foster 1988) and has been responsible for change in vegetation in some areas (Henry and Swan 1974). Although fires can occur naturally under extreme climatic conditions such as after 2 to 3 years of drought, or following blowdown in accumulated fuels, species in these forests are generally not adapted to recovering from fire through fire-triggered seed germination as do many needle-leaved evergreen forests in North America (Fowells 1965). Deciduous forests elsewhere, such as the Ridge and Valley Province in the southeastern U.S., do have properties such as thick bark and sprouting to withstand or recover from fire (Wade et al. 2000).

Northern hardwood forests rarely experienced fire in pre-settlement times (Foster and Zebryk 1993; Patterson and Backman 1988). Fire was so infrequent in presettlement northern hardwoods that a strong case could be made to place this forest type in the nonfire regime category (Wade et al. 2000), although fire has been noted in some areas (Lorimer 1977; Runkle 1990). Bormann and Likens (1979) consider the northern hardwoods in New Hampshire as some of the "least burnable" forests in the region. Fire response in northern hardwoods often involves sprouting or root suckering, which does not change the canopy species composition. The structure of the stand is altered, however, especially by multiple stems per individual; but fire is not considered to have any long-term ecological significance in terms of forest regeneration (Fahey and Reiners 1981). Thus, fire, similar to intense anthropogenic disturbance, introduces a pathway of forest recovery that differs from natural blowdown disturbance.

4.2.2.2 Insect outbreaks in northern hardwood forests

Northern hardwood forests recently have been subject to outbreaks of an introduced insect, the gypsy moth (*Lymantria dispar*). The species originated in Europe and in 1868 to 1869, the gypsy moth was accidentally introduced near Boston by E. Leopold Trouvelot.

The first outbreaks began in Trouvelot's neighborhood 10 years later and by 1890, the state and federal governments began their attempts to eradicate the gypsy moth. These attempts failed and the range of gypsy moth continues to spread. Although some suggest that the gypsy moth inevitably will expand its range in the future (Liebhold 1998), a fungus species (*Entomophaga maimaiga*) recently has caused considerable mortality of populations in North America (Hajek 1999; Schneeberger 1996).

The gypsy moth feeds on the foliage of hundreds of plant species in North America; but its most common hosts are oaks and aspen. Gypsy moth hosts are located throughout most of the conterminous U.S. but host trees are concentrated outside the range of northern hardwoods. Thus, insect outbreaks in northern hardwood forests are patchy in space and time (Oliver and Stephens 1977; Runkle 1998). Some areas can be defoliated multiple times within a few years, while other areas have never been impacted. Unlike the spruce beetle effects on subalpine forests, gypsy moth outbreaks generally do not kill the host trees, and in some cases benefit the understory by providing a higher light environment (Collins 1961). Massive defoliations do occur, but the trees usually leaf out again. Repeated cycles of defoliation, however, will tap resources of individual trees and lead to death. Gypsy moth outbreaks often hit most of the canopy species, therefore not leading to a shift in dominance as seen in the subalpine forests of Colorado. This, coupled with the lack of massive mortality, makes the effect of a gypsy moth disturbance in deciduous forest less dramatic and long-lasting than that of a spruce beetle outbreak in the west (Figure 4.2).

4.2.2.3 Blowdown in mixed deciduous forests of the northeastern U.S.

Blowdown can have an important effect at both coarse and fine scales. It represents the most common coarse-scale and fine-scale disturbance that occurs naturally (Foster 1988; Foster and Boose 1992; Runkle 1981; 1982; 1985; Runkle and Yetter 1987). Although patchy in occurrence, blowdown can occur in most areas of deciduous forest, usually impacting older stands more severely than young stands. Depending on the intensity of the blowdown event, stand reorganization is a common response (Figure 4.2). This response may involve the growth of remaining saplings, which gain canopy status relatively quickly, or in more intense cases requires slower regeneration from buried seed or seed dispersal. Intermediate species may regenerate quickly after blowdown, but are short-lived, such as pin cherry (*Prunus pennsylvanica*). Windstorms can also create individual treefall, resulting in gap-phase regeneration on a fine scale (Figure 4.2). Recovery to closed canopy is more rapid (about a decade) than for a coarse-scale event (centuries). In other eastern deciduous forests comprised of more shade-intolerant species, wind damage may not lead to gap-related regeneration response but rather an acceleration toward shade-tolerant species dominance (Abrams and Scott 1989; Webb 1988, 1989; Webb and Scanga 2001).

In glaciated regions of forested North America, treefall disturbance creates a remnant microrelief on the forest floor that can affect spatial heterogeneity for hundreds of years after the direct disturbance effect has disappeared (Lyford and MacLean 1966). Early regeneration patterns in large blowdowns are affected by the heterogeneity created by uprooting, such that particular forest tree species establish best in certain microsites (Peterson and Pickett 1990). In older, closed-canopy forests, this persistent mound–pit microtopography partly determines understory herbaceous species distributions at the microscale, as well as providing regeneration microsites for some canopy tree seedlings (Beatty 1984, 1991, 2003; Beatty and Sholes 1988; Beatty and Stone 1986; Canham 1984; Webb 1988). Since microtopography remains for centuries, its legacy lasts far beyond the original gap disturbance (Schaetzl and Follmer 1990).

4.2.2.4 *Anthropogenic disturbance in broadleaf deciduous forests of the eastern U.S.*

Coarse-scale anthropogenic disturbance, accompanied by severe land-use alteration, can lead to complete stand destruction. Once abandoned from such landuse (e.g., agriculture), successional development may require hundreds of years. Less intense anthropogenic disturbance, such as clearcut harvesting for wood production or selective removal of some large trees of particular species most likely will initiate a stand reorganization response. Regeneration and growth of residual populations shape the recovery of the stand to a mature state more quickly (Stanturf, this volume). The similarity of stand responses to forest clearing vs. natural blowdown is debated.

After conversion to another landuse, the nature of ongoing management will affect the trajectory of stand reorganization and development once the land-use application and restoration is initiated. For example, plowing obliterates microtopography, leaving a more uniform soil surface with less heterogeneity in the upper horizons (Figure 4.3). Several generations of treefall activity may be required for the plowed site to begin to redevelop the heterogeneity of the original forest. Grazing, haying, or forest cutting alone (no subsequent land-use change) preserves some of the previous microtopography, so that a young successional forest will have a degree of spatial heterogeneity from the beginning (Figure 4.3). In both cases, the forest regrows and trees reach sufficient age and size to uproot and create new mound–pit microsites. Microrelief accumulates with stand age, adding to and reworking existing microtopography. Although the diversity of the understory varies with stand age, stands developing on plowed substrates have a lower diversity than those with microrelief (Beatty 2003; McLean 2003). One goal of forest restoration might be to preserve

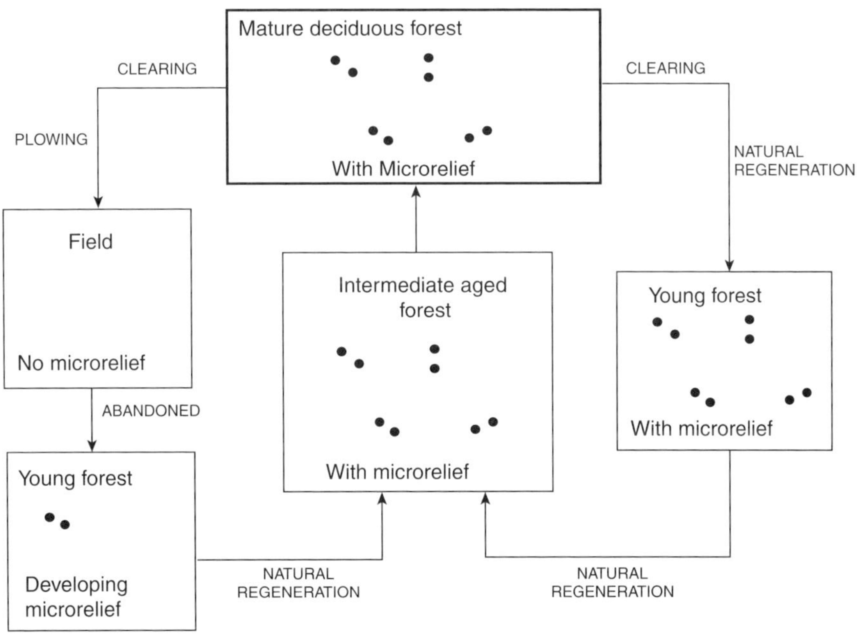

Figure 4.3 Conceptual model of the effects of anthropogenic forest clearing on forest regeneration. Forest floor microrelief is important for tree seedling and understory species colonization and establishment. Rectangles are vegetation types (field, forest successional stages). Fine lines indicate young stands; bold lines indicate more developed stands. Filled circles in rectangles show mound–pit microrelief from treefalls. Arrows indicate successional pathways.

habitat for rare or endangered species, while another might be to assure adequate forest regeneration. Maintenance of fine-scale heterogeneity in microsites could be important for either goal, and in the case of restoration of severely disturbed sites, artificial creation of microrelief might be warranted.

4.2.2.5 *Forest management and disturbance in northern hardwood forests*

Interactions between disturbances to deciduous forest can alter regeneration responses and species composition of the forest; for example, increased mortality from repeated insect outbreaks accompanied by drought increases the chance of fire. Since northern hardwood forests do not have seed regeneration modes triggered by fire, some species will be favored over other species following fire that would regenerate after blowdown. Insect outbreaks, coupled instead with extremely wet conditions, increase the likelihood of blowdown from saturated soils and lower root stability from repeated defoliation. An increase in these events would not necessarily alter species composition, but would certainly shift age structure over a landscape. Spatial patterning of disturbance patches within the landscape can change the susceptibility of undisturbed stands to perturbation as well. A cleared or burned area adjacent to mature stands could increase wind speeds sufficiently to trigger a massive blowdown in the adjacent forest. On the finer scales, successful regeneration of a stand that has been anthropogenically cleared may depend on maintaining microsites for seedling establishment. Restoration of afforested land, artificially creating soil disturbances to mimic mound and pit microtopography, would promote forest understory colonization and establishment (Beatty 2003), and may enhance overstory recovery.

4.3 Conclusions

In both subalpine spruce-fir and northern hardwood forests of North America, numerous disturbances affect species composition, age, and spatial distribution of forest stand structures. Fire, blowdown, and insect outbreaks occur in these forest types; however, the ultimate effect of these disturbances often is very different between the two types. Colorado subalpine forests are adapted to fire, whether a low-intensity surface fire or a stand-destroying crown fire. Fire in northern hardwood forests of the northeastern U.S. does not stimulate seed or seedling regeneration, but instead promotes vigorous sprouting. Fire in subalpine spruce-fir forests has a long-term natural cycle, whereas most fire in northern hardwoods is more recent and is related to human occupation of the landscape.

Blowdown disturbances are common in both forest types, and have similar effects on forest regeneration. Blowdown recovery is generally faster in deciduous than evergreen forests; however, the extent of blowdown is often greater and more frequent in deciduous forests. Depending on the scale of the blowdown, forest recovery varies from gap phase regeneration (fine scale) to stand reorganization or succession (coarse scale). In both forest types, microsites created by uprooting of trees are important for species regeneration. In deciduous forests, anthropogenic disturbance has been very widespread, such that very little old forest exists in the eastern U.S. Forest clearing, followed by long-term manipulation of the soil surface, destroys forest floor microsites. Passive restoration of forest conditions follows an extended pathway of succession, often requiring decades in a woody, old-field stage before forest regrowth initiates. The developing forest likely follows a different regeneration pathway due to lack of microtopography than a forest that has been clearfelled and immediately released to regrowth. In the latter forest, microsites are still available for regeneration.

Insect outbreaks are also common in both forest types, but the spruce beetle that affects subalpine spruce-fir forests in Colorado is native to North America, while the gypsy moth that defoliates deciduous forests in the eastern U.S. was introduced from

Europe in the 1900s. The spruce beetle generally kills a significant portion of spruce in a stand, leaving the secondary species (fir) to achieve dominance. In the case of gypsy moth defoliation, trees are usually not killed, but defoliation temporarily enriches the light environment of the understory. An insect outbreak may be connected to the probability that a stand suffers from another type of disturbance. For example, spruce beetle damage may lead to increased fire susceptibility, or a blowdown area may trigger a beetle outbreak. In deciduous forests, defoliation may increase susceptibility to blowdown or the rare fire if climatic conditions are right (dry).

Recently, disturbances from invasive exotic species have gained the attention of ecologists, managers, and conservationists. Exotic species are increasingly a problem in the arid west; however, there are successful exotic species in deciduous forests as well, such as a honeysuckle shrub (*Lonicera maackii*) in forests of Ohio and Japanese honeysuckle vine (*Lonicera japonica*) in the southeastern forests of the U.S. Treefall gaps in northeastern forests can be colonized by some exotic species (Goldblum and Beatty 1999), perhaps gaining a foothold in forested communities. A recent study (Beatty, unpublished data) of forested wilderness areas across the U.S. found that exotic herbaceous species were capable of invading forest interiors away from trails via treefall gap openings that provided sufficient light for the exotic species to survive, reproduce, and disperse. This interaction between treefall and exotic invasions could potentially affect native diversity and the ecosystem's ability to respond to natural disturbance. In any restoration project, attention to the possibility of creating conditions conducive to exotic invasion is crucial.

Successful incorporation of disturbance into forest restoration will need to be taken into account: the interaction of different disturbance types; the scale at which a disturbance operates; and the different forms of disturbance, such as physical and biological perturbations. Goals may be achieved through active and/or passive management, and should attempt to satisfy a mix of consumptive and nonconsumptive outputs.

Incorporating disturbance regimes and legacies into forest restoration presents a challenge at several levels. First, different forest types will have species that vary in life-history characteristics, such as shade tolerance, likelihood for seed vs. clonal reproduction, sensitivity to browse, and environmental fluctuations. These differences will affect the susceptibility to disturbance and rate of recovery. Regeneration requirements for needle-leaved species differ from requirements of deciduous species (fire-triggered seed dispersal and germination vs. episodic masting and caching by animals). Competition affects seedlings of some species more than others (gap-phase regeneration vs. seedling-bank shade-tolerant regeneration). In restoration that requires active management to provide optimum conditions for regeneration, knowing the species life history and interactions with other species will be critical. These life-history characteristics have often been shaped by the disturbance regimes affecting these forests. Therefore, knowing how disturbance affects the species and forests at both fine- and coarse-scales will contribute to understanding the species requirements.

Second, younger stands of almost any forest type are less susceptible to disturbances that cause stand destruction. Young subalpine spruce-fir forests are less susceptible to fire (insufficient fuel load), beetle outbreak (oldest trees are killed), and blowdown (lower profile, greater flexibility). Mixed deciduous forests rarely experience fire. Young stands also often do not have species in them that are favored food for the gypsy moth, and are less susceptible to uprooting or snapping in a blowdown. In fact, the creation of treefall gaps and microtopography does not begin until the forest reaches a substantial stage of development where trees are large enough to uproot, and the canopy is closed enough for a gap to make a difference in the light environment. Therefore, in early stages of forest restoration (assuming a successional series is the goal), there is much less need to incorporate the coarse-scale disturbance regimes, and attention should be paid to managing the fine-scale disturbances and their legacies. In some cases, planting is a technique used to hasten forest recovery, particularly with respect to economically valuable species. Accumulating evidence of the importance of microsite to seedlings suggests that restoration action might include creating microsites in conjunction with the planting. In a northern hardwood forest, a restoration goal to hasten development toward a "mature" species composition that included sugar maple could include in restoration plans creating artificial mound–pit formations to provide microsites conducive to regeneration. Where seedlings germinate and establish best on fallen logs, which is often the case for birch or hemlock, restoring microsites to younger stands would mimic the regeneration sites commonly found in more developed stands. In particular, if the forest restoration project is applied to cultivated or otherwise intensely used land, restoration of these microsites could be important for regaining herbaceous plant diversity.

In rehabilitation projects, a first step should be to evaluate whether fine-scale processes are occurring (treefall, microtopography, logs, episodic regeneration, and sapling attainment of canopy status), then no restoration is needed at this scale. However, the stand may be at a stage of development where coarse-scale disturbances become likely. A decision needs to be made about the response if a natural event occurs (fire, windstorm, outbreak). In the case of fire, a let-burn approach is often taken where nonconsumptive outputs are most important. Similarly, recovery after an outbreak or blowdown would be through passive management. In the case of consumptive needs being important, salvage logging or thinning may be appropriate in postdisturbance stands and surrounding undisturbed forest.

Restoration of disturbance regimes may require artificial means, such as controlled burns to mimic wildfire or thinning to mimic treefall gaps. In the case of thinning, leaving some downed wood may also provide additional fine-scale disturbance.

In any restoration project, there will be natural environmental fluctuations that cannot be controlled. A sequence of drought years coupled with an insect outbreak would signal conditions favorable for a fire, particularly in needle-leaved forest. A restoration manager would have to decide whether to let the forest experience any natural event, or take steps to prevent such an occurrence. By knowing the interactions between such coarse-scale disturbance events, a manager is in a better position to make an informed decision. If management has provided for maintenance of fine-scale disturbance events and legacies, the forest may better withstand environmental fluctuations. The maintenance of particular species in the understory of northern hardwood forests is closely tied to the availability of several kinds of microsites that provide refugia during stressful years. Although active management at the fine scale is often overlooked, it may be a very important part of restoring a forest to a state where it can withstand stresses and disturbances, or recover rapidly from them.

References

Abrams, M.D. and Scott, M.L., Disturbance-mediated accelerated succession in two Michigan forest types, *Forest Sci.*, 35, 42, 1989.

Abrams, M.D., Orwig, D.A., and DeMeo, T.E., Dendroecological analysis of successional dynamics for a presettlement origin white pine-mixed oak forest in the southern Appalachians, USA, *J. Ecol.*, 83, 123, 1995.

Allen, T.R. and Walsh, S.J., Spatial and compositional pattern of alpine treeline, Glacier National Park, Montana, *Photogramm. Eng. Rem. S.*, 62, 1261, 1996.

Baker, W.L. and Ehle, D., Uncertainty in surface-fire history: the case of ponderosa pine forests in the western United States, *Can. J. For. Res.*, 31, 1205, 2001.

Beatty, S.W., *The Role of Treefalls and Forest Microtopography in Pattern Formation in Understory Communities*, Dissertation, Cornell University, Ithaca, NY, 143 pp., 1981.

Beatty, S.W., Influence of microtopography and canopy species on spatial patterns of forest understory plants, *Ecology*, 65, 1406, 1984.

Beatty, S.W., Colonization dynamics in a mosaic landscape: the buried seed pool, *J. Biogeogr.*, 18, 553, 1991.

Beatty, S.W., Habitat heterogeneity and maintenance of species in understory communities, in *The Herbaceous Layer in Forests of Eastern North America*, Gilliam, F. S. and Roberts, M. R., Eds., Oxford University Press, New York, 177, 2003.

Beatty, S.W. and Sholes, O.D.V., Leaf litter effect on plant species composition of deciduous forest treefall pits, *Can. J. For. Res.*, 18, 553, 1988.

Beatty, S.W. and Stone, E.L., The variety of soil microsites created by treefalls, *Can. J. For. Res.*, 16, 539, 1986.

Bebi, P., Kulakowski, D., and Veblen, T.T., Interactions between fire and spruce beetles in a subalpine Rocky Mountain forest landscape, *Ecology*, 84, 362, 2003.

Bendix, J., Flood disturbance and the distribution of riparian species diversity, *Geogr. Rev.*, 87, 468, 1997.

Bormann, F.H. and Likens, G.E., Catastrophic disturbance and the steady state in northern hardwood forests, *Am. Sci.*, 67, 660, 1979.

Braun, E.L., *Deciduous Forests of Eastern North America*, Blakiston, Philadelphia, PA, U.S.A., 1950.

Canham, C.D., *Canopy Recruitment in Shade Tolerant Trees: The Response of Acer saccharum and Fagus grandifolia to Canopy Openings*, Dissertation, Cornell University, Ithaca, NY, 1984.

Clark, D.B., Palmer, M.W., and Clark, D.A., Edaphic factors and the landscape-scale distributions of tropical rain forest trees, *Ecology*, 80, 2662, 1999.

Collins, B.S. and Pickett, S.T.A., Influence of canopy opening on the environment and herb layer in a northern hardwoods forest, *Vegetatio*, 70, 3, 1987.

Collins, S., Benefits to understory from canopy defoliation by Gypsy Moth larvae, *Ecology*, 42, 836, 1961.

Connell, J.H., Diversity in tropical rainforests and coral reefs, *Science*, 199, 1302, 1978.

Cooper-Ellis, S., Foster, D.R., Carlton, G., and. Lezberg, A., Forest response to catastrophic wind: results from an experimental hurricane, *Ecology*, 80, 2683, 1999.

Delcourt, H.R. and Delcourt, P.A., Eastern deciduous forests, in *North American Terrestrial Vegetation*, Barbour, M.G. and Billings, D.W., Eds., Cambridge University Press, Cambridge, U.K., 357, 1996.

Drury, W.H. and Nisbet, I.C.T., Succession. *J. Arnold Arbor. Harv. Univ.*, 54, 331, 1973.

Eisenhart, K. and Veblen, T.T., Dendrochronological identification of spruce bark beetle outbreaks in northwestern Colorado, *Can. J. For. Res.*, 30, 1788, 2000.

Fahey, T.J. and Reiners, W.A., Fire in the forests of Maine and New Hampshire, *B. Torrey Bot. Club*, 8, 362, 1981.

Foster, D.R., Disturbance history, community organization and vegetation dynamics of the old-growth Pisgah forest, southwestern New Hampshire, U.S.A., *J. Ecol.*, 76, 105, 1988.

Foster, D.R. and Boose, E.R., Patterns of forest damage resulting from catastrophic wind in central New England, U.S.A., *J. Ecol.*, 80, 79, 1992.

Foster, D.R. and Zebryk, T.M., Long-term vegetation dynamics and disturbance history of a Tsuga-dominated forest in New England, *Ecology*, 74, 982, 1993.

Fowells, H.A., Silvics of forest trees of the United States, *Ag. Handb.* 271, U.S. Department of Agriculture, Washington, DC, 762 pp., 1965.

Frelich, L.E. and Lorimer, C.G., Natural disturbance regimes in hemlock-hardwood forests of the Upper Great Lakes region, *Ecol. Monogr.*, 61, 145, 1991.

Frelich, L.E. and Reich, P.B., Spatial patterns and succession in a Minnesota southern-boreal forest, *Ecol. Monogr.*, 65, 325, 1995.

Glitzenstein, J.S., Harcombe, P.A., and Streng, D.R., Disturbance, succession, and maintenance of species diversity in an east Texas forest, *Ecol. Monogr.*, 6, 243, 1986.

Goldblum, D. and Beatty, S.W., Influence of an old field/forest edge on a northeastern United States deciduous forest understory community, *B. Torrey Bot. Club*, 126, 335, 1999.

Grime, P., *Plant Strategies and Vegetation Processes*, John Wiley and Sons, New York., 418 pp., 1979.

Grubb, P.J., The maintenance of species-richness in plant communities: the importance of the regeneration niche, *Biol. Rev.*, 52, 107, 1977.

Hajek, A.E., Pathology and epizootiology of *Entomophaga maimaiga* infections in forest Lepidoptera, *Microbiol. Mol. Biol. Rev.*, 63, 814, 1999.

Heinselman, M.L., Fire in the virgin forests of the Boundary Waters Canoe Area, Minnesota, *Quaternary Res.*, 3, 329, 1973.

Henry, J.D. and Swan, J.M.A., Reconstructing forest history from live and dead plant material — an approach to the study of forest succession in southwestern New Hampshire, *Ecology*, 55, 773, 1974.

Hoffman, G.R. and Alexander, R.R., *Forest Vegetation of the Routt National Forest in Northwestern Colorado: A Habitat Classification*, U.S. Department of Agriculture, Forest Service, Rocky Mountain Forest and Range Experiment Station, Fort Collins, CO, Research Paper RM-221, 1980.

Huston, M., A general hypothesis of species diversity, *Am. Nat.*, 113, 81, 1979.

Keane, R.E., Ryan, K.C., Veblen, T.T., Allen, C., Logan, J., and. Hawkes, B., The cascading effects of fire exclusion in Rocky Mountain ecosystems, in *Rocky Mountain Futures: An Ecological Perspective*, Baron, J., Fagre, D., and Hauer, R., Eds., Island Press, 2002.

Kulakowski, D., *Interactions Among Natural Disturbances in Subalpine Forests in Northwestern Colorado*, Dissertation, University of Colorado, Boulder, CO, USA, 2002.

Kulakowski, D. and Veblen, T.T., Influences of fire history and topography on the pattern of a severe wind blowdown in a Colorado subalpine forest, *J. Ecol.*, 90, 806, 2002.

Kulakowski, D. and Veblen, T.T., Subalpine forest development following a blowdown in the Mount Zirkel Wilderness, Colorado, USA, *J. Veg. Sci.*, 14, 653, 2003.

Kupfer, J.A. and Malanson, G.P., Structure and composition of a riparian forest edge, *Phys. Geogr.*, 14, 154, 1993.

Lertzman, K.P., Patterns of gap-phase replacement in a subalpine, old-growth forest, *Ecology*, 73, 657, 1992.

Lertzman, K. and Fall, J., From forest stands to landscapes: spatial scales and the roles of disturbances, in *Ecological Scale: Theory and Applications*, Peterson, D.L. and Parker, V.T., Eds., Columbia University Press, New York, 339 pp., 1998.

Liebhold, S., *Gypsy Moth in North America*, U.S. Department of Agriculture, Forest Service, Northeastern Research Station, http://www.fs.fed.us/ne/morgantown/4557/gmoth/, 1998.

Lorimer, C.G., The presettlement forest and natural disturbance cycle of northeastern Maine, *Ecology*, 58, 139, 1977.

Lyford, W.H. and MacLean, D.W., Mound and pit microrelief in relation to soil disturbance and tree distribution in New Brunswick, Canada, *Harv. For. Pap.*, 15, 1, 1966.

McCullough, D.G., Werner, R.A., and Neumann, D., Fire and insects in northern and boreal forest ecosystems of North America, *Annu. Rev. Entomol.*, 43, 107, 1998.

McLean, M.B., *Spatial and Temporal Pattern in the Understory: Disturbance Legacy and Environmental Variation*, Dissertation, University of Colorado, Boulder, CO, USA, 2003.

Odum, E.P., Vegetation of the Edmund Niles Huyck Preserve, New York, *Am. Midl. Nat.*, 29, 72, 1943.

Oliver, C.D. and Stephens, E.P., Reconstruction of a mixed-species forest in central New England, *Ecology*, 58, 562, 1977.

Parker, K.C. and Bendix, J., Landscape-scale geomorphic influences on vegetation patterns in four environments, *Phys. Geogr.*, 17, 113, 1996.

Pastor, J. and. Broschart, M., The spatial pattern of a northern conifer-hardwood landscape, *Landscape Ecol.*, 4, 55, 1990.

Patterson, W.A. and Backman, A.E., Fire and disease history of forests, in *Handbook of Vegetation Science — Vegetation History, Vol. 7*, Huntley, B. and Webb, T. III, Eds., Kluwer Academic Publishers, Netherlands, 603, 1988.

Peterson, C.J. and Pickett, S.T.A., Microsite and elevational influences on early forest regeneration after catastrophic windthrow, *J. Veg. Sci.*, 1, 657, 1990.

Pickett, S.T.A. and White, P.S., *The Ecology of Natural Disturbance and Patch Dynamics*, Academic Press, New York, 1985.

Pitman, N.C.A., Terborgh, J., Silman, M.R., Nuñez and V.P., Tree species distributions in an upper Amazonian forest, *Ecology*, 80, 2651, 1999.

Platt, W.J. and Weis, I.M., Resource partitioning and competition within a guild of fugitive prairie plants, *Am. Nat.*, 111, 479, 1977.
Romme, W.H., Turner, M.G., Wallace, L.L., and Walker, J., Aspen, elk and fire in northern Yellowstone National Park, *Ecology*, 76, 2097, 1995.
Runkle, J.R., Gap regeneration in some old-growth forests of the eastern United States, *Ecology*, 62, 1041, 1981.
Runkle, J.R., Patterns of disturbance in some old-growth mesic forests of eastern North America, *Ecology*, 63, 1533, 1982.
Runkle, J.R., Disturbance regimes in temperate forests, in *The Ecology of Natural Disturbance and Patch Dynamics*, Pickett, S.T.A. and White, P.S., Eds., Academic Press Inc., San Diego, CA, 17, 1985.
Runkle, J.R., Synchrony of regeneration, gaps, and latitudinal differences in tree species diversity, *Ecology*, 70, 546, 1989.
Runkle, J.R., Gap dynamics in an Ohio Acer-Fagus forest and speculations on the geography of disturbance, *Can. J. For. Res.*, 20, 632, 1990.
Runkle, J.R., Changes in southern Appalachian canopy tree gaps sampled thrice, *Ecology*, 79, 1768, 1998.
Runkle, J.R. and Yetter, T.C., Treefalls revisited: gap dynamics in the southern Appalachians. *Ecology*, 68, 417, 1987.
Schaetzl, R.J. and Follmer, L.R., Longevity of treethrow microtopography: implications for mass wasting, *Geomorphology*, 3, 113, 1990.
Schneeberger, N.F., Gypsy moth populations plummet in 1996 while "The Fungus" skyrockets, *Gypsy Moth News*, 42, 1, 1996.
Sherriff, R.L., Veblen, T.T., and Sibold, J.S., Fire history in high elevation subalpine forests in the Colorado Front Range, *Ecoscience*, 8, 369, 2001.
Smith, F.E., Spatial heterogeneity, stability, and diversity in ecosystems, *Trans. Conn. Acad. Arts Sci.*, 44, 309, 1972.
Snyder, G.L., Patten, L.L. and Daniels, J.J., *Mineral Resources of the Mount Zirkel Wilderness and Northern Park Range Vicinity, Jackson and Routt Counties, Colorado*, U.S. Geological Survey, Washington, DC, 1554, 1987.
Spurr, S.H., The forests of Itasca in the nineteenth century as related to fire, *Ecology*, 35, 21, 1954.
Stearns, F.W., Ninety years change in a northern hardwood forest in Wisconsin, *Ecology*, 30, 350, 1949.
Swain, A.M., A history of fire and vegetation in northeastern Minnesota as recorded in lake sediments, *Quaternary Res.*, 3, 383, 1973.
Turner, M.G. and Dale, V.H., Comparing large, infrequent disturbances: what have we learned? *Ecosystems*, 1, 493, 1998.
Turner, M.G., Dale, V.H., and Everham, E.E. III, Fires, hurricanes and volcanoes: comparing large-scale disturbances, *BioScience*, 47, 758, 1997.
Turner, M.G., Hargrove, W.H., Gardner, R.H., and Romme, W.H., Effects of fire on landscape heterogeneity in Yellowstone National Park, Wyoming, *J. Veg. Sci.*, 5, 731, 1994.
Turner, M.G., Romme, W.H., and Gardner, R.H., Landscape disturbance models and the long-term dynamics of natural areas, *Nat. Areas J.*, 14, 3, 1994.
Vankat, J.L., *The Natural Vegetation of North America*, John Wiley and Sons, Inc., New York, NY, USA, 1979.
Veblen, T.T., Treefalls and the coexistence of conifers in subalpine forests of the Central Rockies, *Ecology*, 67, 644, 1986.
Veblen, T.T., Hadley, K.S., Reid, M.S., and Rebertus, A.J., Stand response to spruce beetle outbreak in Colorado subalpine forests. *Ecology*, 72, 213, 1991.
Veblen, T.T., Kitzberger, T., and Donnegan, J., Climatic and human influences on fire regimes in Ponderosa pine forests in the Colorado Front Range, *Ecol. Appl.*, 10, 1178, 2000.
Veblen,T.T., Kulakowski, D., Eisenhart, K.S., and Baker, W.L., Subalpine forest damage from a severe windstorm in northern Colorado, *Can. J. For. Res.*, 31, 2089, 2001.
Wade, D.D., Brock, B.L., Brose, P.H., Grace, J.B., Hoch, G.A., and Patterson, W.A. III, Fire in eastern ecosystems, in *Wildland Fire in Ecosystems: Effects of Fire on Flora*, Brown, J.K. and Smith, J.K., Eds., General Technical Report RMRS-GTR-42, Vol. 2, Ogden, UT, 53, 2000.

Watt, A.S., Pattern and process in the plant community, *J. Ecol.*, 35, 1, 1947.
Webb, S.L., Windstorm damage and microsite colonization in two Minnesota forests, *Can. J. For. Res.*, 18, 1186, 1988.
Webb, S.L., Contrasting windstorm consequences in two forests, Itasca State Park, Minnesota, *Ecology*, 70, 1167, 1989.
Webb, S.L. and Scanga, S.E., Windstorm disturbance without patch dynamics: twelve years of change in a Minnesota forest, *Ecology*, 82, 893, 2001.
Wesley, D.A., Poulos, G., Meyers, M., Snook, J., and Judson, A., Observations and forcing mechanisms during the October 1997 Front Range blizzard and forest destruction, in *American Meteorological Society, Eighth Conference on Mountain Meteorology*, Flagstaff, AZ, USA, 25, 1998.
Whittaker, R.H., Evolution of diversity in plant communities, *Brookhaven Sym. Biol.*, 22, 178, 1969.
Whittaker, R.H. and Levin, S.A., The role of mosaic phenomena in natural communities, *Theor. Popul. Biol.*, 12, 117, 1977.

chapter five

Restoration effects on N cycling pools and processes

James M. Vose
US Department of Agriculture Forest Service
Chris Geron and John Walker
US Environmental Protection Agency National Risk Management Research Laboratory
Karsten Raulund-Rasmussen
Danish Centre for Forest, Landscape and Planning, KVL

Contents

5.1 *Introduction*

Over the past several years, there has been an acceleration of restoration efforts to mitigate the consequences (i.e., ground and surface water chemical pollutants, erosion, etc.) of degraded ecosystems and enhance structural and functional components of watershed ecosystems that regulate biogeochemical cycling and associated aquatic components. Biogeochemical processes are complex because they operate at a variety of spatial and temporal scales (e.g., near instantaneous soil chemical reactions vs. bedrock weathering). Large-scale assessments of the integrity of these fine-scale processes would be a daunting task; however, our knowledge of ecosystem processes at larger scales (e.g., stands, watersheds, basins) suggests emergent (and measurable) properties that integrate these finer-scale processes. For example, nutrient budgets are key indicators of watershed health and have proven useful for evaluating response and recovery to a variety of disturbances and management activities (Swank and Vose 1997; Swank et al. 2001). Since water moves nutrients through the system, hydrologic, aquatic, and biogeochemical processes are tightly linked. Nutrient budgets are comprised of inputs (e.g., atmospheric deposition, fixation), internal pools and cycling (e.g., soil, litter, and vegetation pools and cycling rates), and outputs (e.g., stream chemistry, seepage or groundwater chemistry, denitrification) (Figure 5.1). The nutrient cycling components shown in Figure 5.1 serve as a focal point for identifying ecosystem nutrient pools or processes requiring restoration to enhance biogeochemical processes and aquatic systems. However, pool sizes and cycling rates can vary considerably among ecosystems (Table 5.1). This variation makes general assessments of restoration effects on biogeochemical cycling difficult because similar types of ecosystem degradation (e.g., insect-related mortality) may not have similar impacts on

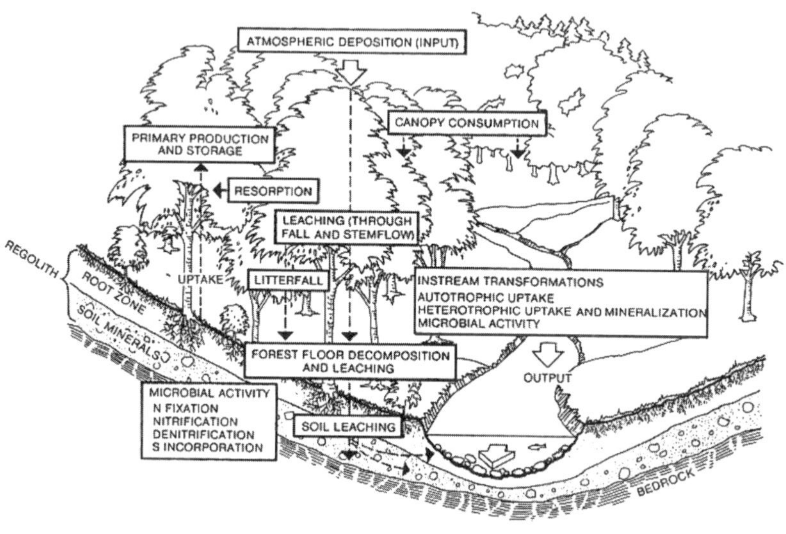

Figure 5.1 Components of ecosystem biogeochemical cycling in a watershed context.

Table 5.1 Comparison of Pool Sizes and Cycling Rates Among Three Contrasting Forest Ecosystem Types in North America (adapted from Swank 1986)

	Oak/Hickory (North Carolina)			Northern Hardwood (New Hampshire)			Western Conifer (Oregon)		
	N	Ca	K	N	Ca	K	N	Ca	K
Pool Size (kg ha^{-1})									
Vegetation	995	830	400	532	484	218	560	750	360
Forest Floor	140	130	20	1,256	372	66	740	570	90
Mineral Soil	6,800	2,500	124,000	4,890	9,600	—	4,500	—	—
Cycling Rate (kg ha^{-1} year^{-1})									
Litterfall	33	44	18	54	41	18	21	41	9
Woody Increment	13	23	13	9	8	6	−2	−4	−1

nutrient pools and processes among ecosystems. Similarly, this variation also implies that restoration treatments applied in one ecosystem may not have similar results in another. Despite these complexities, it is important to develop and apply restoration practices that restore biogeochemical cycling processes in degraded ecosystems, and to develop methods, metrics, and approaches for assessing the recovery of biogeochemical cycling pools and processes among contrasting ecosystem types. This chapter focuses primarily on developing metrics and approaches for assessing the recovery of biogeochemical cycling pools and processes.

The degree of biological vs. physiochemical regulation of pool sizes and flux rates is element dependent. For example, nitrogen (N) cycling is especially dependent on microbially driven biological processes such as mineralization, nitrification, and denitrification. Most of the stored N pool is contained in soil organic matter and vegetation, and inputs are almost exclusively driven by atmospheric contributions. In contrast, calcium (Ca) cycling is especially dependent on soil and bedrock weathering for inputs and often most ecosystem Ca is stored in the mineral soil. These differences are especially relevant when evaluating restoration alternatives and evaluating the rate and magnitude of potential responses.

The complexity and interdependency of nutrient cycles require focusing on "key indicators" that are responsive to restoration activities. In some cases, the choice of the "key indicator" is obvious, especially if the overall goal of restoration is to alter the pool size or cycling rate of a particular element or combination of elements. For example, examining restoration success to offset or mitigate the effects of elevated sulfur (S) deposition (e.g., liming) would focus primarily on the S cycle, with additional measurements of base cations and acidity as codependent variables. In contrast, restoration goals may not be directly focused on biogeochemical cycling *per se*. In this case, examining nutrient cycling processes may provide an integrated estimate of ecosystem-level responses or indicate undesirable (or desirable) consequences to other "nontargeted" ecosystem components. For example, prescribed burning has been advanced as a restoration tool to improve the structure and diversity of forest ecosystems. Because fire alters nutrient pools and cycling rates, many of the functional aspects of biogeochemical cycling are either directly or indirectly affected by restoration burning. The degree of departure (in terms of both magnitude and duration) of biogeochemical pools or processes from baseline or desired conditions may serve as a tool for evaluating restoration success or comparing restoration alternatives. Because biogeochemical cycling processes are complex and involve multiple elements, several factors need to be considered when choosing an appropriate metric(s) for evaluating biogeochemical responses to restoration. First, the indicator (or response variable) needs to be responsive to the restoration activity. For example, if restoration involves manipulation of forest floor and/or surface soils, then pools (e.g., forest floor biomass and nutrient content, available nutrients, soil solution chemistry) or processes (e.g., litter decomposition, mineralization, microbial activity) tightly linked with those ecosystem components should be monitored. Second, due to the inherent variability in biogeochemical cycling processes at the watershed scale, response variables need to be selected such that their response magnitude exceeds the natural variability of the system. For example, seasonal and interannual variation in stream chemistry can result in nearly twofold differences in NO_3^-, Ca, and K concentrations (Swank et al. 2001). Restoration practices that cause only minor changes in pools or processes may not result in measurable responses (+ or −) in terrestrial or aquatic biogeochemical cycling. For example, in ecosystems where baseline stream chemistry is inherently high and variable, integrated responses reflected by changes in stream nutrient concentration may be too subtle to detect (Likens et al. 1970).

Data from long-term watershed studies provide some insight into elements that meet the criteria outlined above (Swank and Vose 1997). Analyses of nutrient budgets suggest that terrestrial ecosystems are inherently conservative (i.e., outputs< inputs) for nutrients that are in limiting supply and whose availability is largely controlled by biological processes (Swank 1986). For example, when examining N budgets across a range of ecosystem types in North America, Henderson et al. (1978) found comparable rates of nitrogen retention, despite large differences in N inputs. The most significant pathway for N loss from terrestrial ecosystems is solution loss in groundwater or streamwater. Undisturbed ecosystems that receive low amounts of N inputs generally lose very little inorganic N via stream or groundwater export. This high degree of N conservation is indicative of tightly linked N pools and cycling processes such as storage, uptake, immobilization/mobilization, N fixation, and microbial processes. When some (or many) of these pools and processes are altered due to disturbance or restoration, measurable changes in N pools, processes, and outputs are likely. For example, in the southern Appalachian mountains of Western North Carolina, stream NO_3 is a strong indicator of ecosystem response to disturbances such as insect defoliation (Swank 1981), clearcutting (Swank et al. 2001), or hardwood forest conversion to grass cover or pines (Swank and Vose 1997).

In this chapter, we present a conceptual framework for evaluating biogeochemical responses to restoration and synthesize data from several restoration studies to illustrate this conceptual framework. Our chapter focuses specifically on N cycling pools and processes for two reasons. First, as explained above, N pools and processes are sensitive to disturbance (and restoration) and may serve as "key indicators" of ecosystem responses to restoration. Second, restoring N cycling pools and processes is a primary goal of many restoration efforts because N commonly limits terrestrial ecosystem productivity, and N (e.g., NO_3^-) losses to groundwater and surface water can have important consequences for human health and for the ecological integrity of aquatic ecosystems. The studies contrast the potential uses of resistance/resilience analogy. In the first set of studies, the primary goal of restoration was to restore N cycling processes in degraded riparian zone ecosystems. In the others, the primary goal was to alter vegetation structure and composition using a variety of prescribed burning techniques. Here, we use the resistance/resilience analogy to evaluate restoration impacts on other "nontarget" ecosystem components.

5.2 Conceptual framework for evaluating ecosystem responses to restoration

The resistance and resilience model of ecosystem response to disturbance (Holling 1973) is a useful conceptual framework for evaluating the restoration success of enhancing biogeochemical cycles and aquatic systems (Waide 1988; Carpenter et al. 2001). In the original development of resistance/resilience theory, resilience was defined as the amount of change a system can undergo before moving into a different state space and a different set of controls (Holling 1973). Resistance has been defined as the amount of external pressure needed to bring about a given amount of disturbance in the system (Carpenter et al. 2001). In further revisions of resistance/resilience theory, organic matter pool sizes and turnover rates were key indices of disturbance responses (Webster et al. 1975; Waide 1988), and it has been further modified to evaluate ecosystem responses to restoration burning (Vose 2000). The resistance–resilience model can be used as a conceptual construct for evaluating spatial and temporal aspects of current ecosystem condition, desired ecosystem condition, and ecosystem response to restoration (Figure 5.2). Here, we define the y-axis as the response magnitude (i.e., percent change from initial condition) and the x-axis as the response duration. The line labeled (a) represents an ecosystem where the initial condition

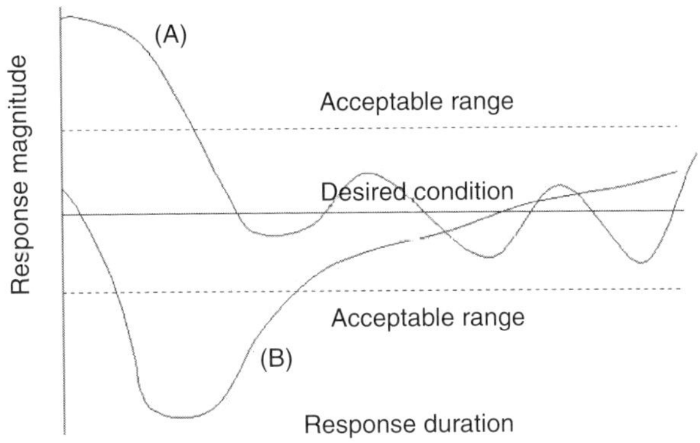

Figure 5.2 Conceptual model of approach to evaluating ecosystem responses to restoration. The approach focuses on determining the magnitude (y-axis) and duration of response (x-axis) with respect to desired conditions and acceptable ranges. Line (a) represents the hypothetical response of a parameter outside the range of acceptable conditions and the response to restoration. Line (b) represents the hypothetical response of a nontargeted parameter that may respond negatively to the restoration of other ecosystem components.

of the pool or process is outside the boundary of acceptable conditions and restoration treatments are used to change the pool/process into an acceptable range. The line labeled (b) represents potential responses of nontarget ecosystem pools/processes. Here, the initial condition may be within acceptable limits; however, restoring other ecosystem components may directly (or indirectly) alter nontargeted ecosystem components. In application, the historical and contemporary disturbance legacy of an ecosystem (or watershed) determines its degree of departure from desired conditions, influences the rate of attainment of desired ecosystem conditions or response duration, and influences the magnitude of response to the restoration treatment. We emphasize that the desired conditions (i.e., processes or pools) are defined as existing within a boundary of acceptable conditions reflective of the range of natural variability or some other (e.g., land manager defined) acceptable range of variability. This point is especially important in evaluating biogeochemical responses to restoration because nutrient cycling processes are not constant, but are more likely to persist in a dynamic equilibrium. In contrast, severely degraded ecosystems are characterized as having initial pools and processes that are farthest from desired conditions, consistently remain outside the boundary of acceptable conditions, require substantial restoration efforts to move toward desired conditions, and take the longest time to attain desired conditions.

5.3 Restoration of N cycling processes in degraded riparian zones

Riparian zones are the key linkage between terrestrial and aquatic ecosystems. While several aspects of riparian zone structure and function remain unknown, the importance of riparian zones for reducing nutrient inputs to aquatic systems is well documented (Lowrance et al. 1984; Gilliam et al. 1997). For example, several studies have demonstrated substantial (e.g., 60 to 100% of inputs) reductions in N concentration in subsurface waters by both grass and forest riparian zones (Daniels and Gilliam 1996; Cooper et al. 1987; Osborne and Kovacic 1993). Riparian zone nutrient cycling processes that reduce nutrient fluxes to aquatic systems include physical entrapment of organic matter and sediment, uptake by riparian vegetation and microbes, and microbial processes such as denitrification (Figure 5.3). Factors that influence nutrient uptake efficiency include the type and

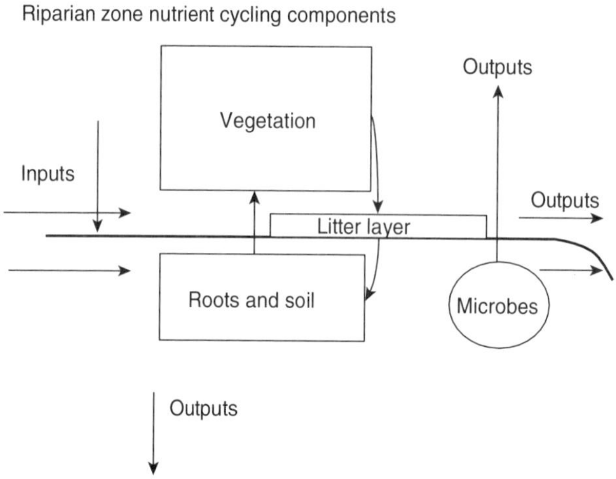

Figure 5.3 Key nutrient cycling pools and processes in riparian zones; inputs come from uplands and outputs are to the stream.

density of riparian vegetation, soil surface conditions (i.e., the presence or absence of a litter layer; the presence or absence of compacted soil), and soil physical characteristics (e.g., infiltration rates) (Osborne and Kovacic 1993).

In many ecosystems, the structure and function of riparian zone ecosystems have been severely degraded. The most dramatic changes occur where the natural landscape has been converted to alternative uses, such as urban and suburban development, livestock grazing, hayfields, or row crop agriculture. More subtle changes include altered disturbance regimes (e.g., changes in flooding regimes, timber harvesting), changes in species composition (e.g., introduction of exotic species, loss of native species), or lower-intensity disturbances such as concentrated recreation use or prescribed fire. Natural resources managers, farmers, and local citizen-based groups are becomingly increasingly aware of the importance of riparian zones for protecting water quality. In the U.S., many regions have set goals (i.e., linear stream miles of riparian zones to be restored) for the restoration of degraded riparian zones and long-term protection. In this example, we quantify the short-term recovery of N cycling processes in restored riparian zones in agricultural (pasture and hayfields) watersheds and use the resistance/resilience analogy to evaluate responses.

5.3.1 *Study sites*

Two study sites were selected for the study of N cycling recovery following restoration. The Killian Farm (KF) site was located on Cartoogechaye Creek, a tributary of the Little Tennessee River in Macon County, North Carolina. Livestock grazing for several years heavily impacted the site and cattle were allowed creek access along the entire restoration treatment area (approximately 400 m). Unstable streambanks, heavily compacted areas, trails leading into and within the riparian area, and sparse vegetative cover characterized the site. A 100 m section upstream from (but adjacent to) the degraded area served as the control. The control area was fenced and had a lush understory and fully developed overstory along the riparian zone. The topography of area adjacent to the creek was gently sloping (~5%) toward the riparian area for approximately 30 m, and steeply sloping (20%) thereafter.

The Rabun Farm (RF) site was located on Sutton Branch, a small tributary of Betty's Creek in Rabun County, Georgia. The site was heavily impacted by cattle, resulting in significant compaction and bare areas in the riparian zone over about 75% of the stream (approximately 300 m) reach. The other 25% (approximately 100 m) had a thick vegetative cover of alder and rose and had no evidence of current or historical cattle use. Since this area was unused by cattle, it served as a control. The streambank was generally stable, with only a few areas showing signs of excessive erosion. The topography of area adjacent to the branch was gently sloping (~5%) toward the riparian area for approximately 20 m, and steeply sloping (25%) thereafter.

5.3.2 *Restoration treatments*

Restoration treatments were initiated in late fall 1997 at the KF site, and early spring 1998 at the RF site. Cattle were excluded from the area during restoration treatments and fencing was installed to permanently exclude cattle from the riparian zone. Since cattle were active at both sites, the distance from the stream edge to the fence defined the buffer width. At the KF site, the distance from the stream edge to the fence line ranged from 7 to 28 m, and averaged 11 m. At the RF site, the distance from the stream edge to the fence line ranged from 2 to 9 m, and averaged 6 m. Tree seedlings of a variety of riparian zones species and species with potential wildlife benefits (e.g., *Cornus florida*, *Salix* spp., and *Platanus occidentalis*) were planted (~10 m spacing) at all sites. Riparian areas outside the fenced area continued to be used by livestock and served as unrestored sample areas for some parameters.

To illustrate the effects of riparian zone restoration on N cycling pools and processes, comparisons were made using several approaches. Comparisons were made between restored areas and nearby controls, between restored areas and undisturbed forested riparian zones, and between restored and unrestored areas.

5.3.3 *Overland flow sampling*

Total suspended solids (TSS) entering the streams in overland flow were sampled on a storm-event basis. TSS included the combined contributions of sediment and organic matter. We did not measure the N content of the overland flow samples; however, both sediment and organic matter contain N and we can infer that changes in N amount in overland flow are proportional to changes in TSS. Collectors were installed in locations where overland flow was obvious within 1 week after restoration treatments were completed. Overland flow collectors were installed at five locations in the degraded riparian zones at the KF and two locations in the controls. Four collectors were installed in degraded areas at the RF site and two in the controls. Collectors were checked after all storms, but typically only collected overland flow when storms exceeded 1.25 cm during the first month following restoration treatments, and 5.0 cm thereafter. When overland flow was present in the carboys, they were removed from the overland flow collector and vigorously agitated to resuspend solids, and then quickly subsampled with a 1,000 ml nalgene bottle.

5.3.4 *Soil solution sampling*

Porous cup lysimeters (30 cm depth) were used to sample soil solution NO_3^- concentration. Lysimeters were installed at 3-m spacing from the stream edge to just inside the unrestored area along transects (nine transects at RF, eight transects at KF) equally spaced throughout the riparian zone. After an equilibration period of 3 to 4 weeks, lysimeters were sampled weekly for approximately 2 years at each site. Samples were composited and analyzed monthly.

5.3.5 *Nitrogen trace gas sampling*

Trace gas fluxes were determined using a flow-through chamber mass balance technique. Details of the sampling and analytical procedures are presented in Walker et al. (2002); hence, only a summary is presented here. Nitric oxide (NO) and NH_3 fluxes were measured approximately every 3 months between July 1998 and August 2000, with sampling intervals roughly corresponding to seasonal midpoints. Nitrous oxide (N_2O) measurements began in June 1999. Fluxes were measured in two locations within both the restored and unrestored riparian zone during each measurement period. Chambers were placed in the same general area from season to season. Reported fluxes are for daytime values only. Nighttime values were generally depressed, largely due to the exponential temperature dependence of emissions.

5.3.6 *Soils and vegetation characterization*

Bulk density (g cm^{-3}), infiltration rates (cm h^{-1}), and plant biomass (g m^{-2}) were measured at both the KF and RF sites. Bulk density was measured at 20 random locations within the restored riparian zone (n = 20 at KF and n = 20 at RF) and ten random locations in the controls. Bulk density samples were taken during the winter (soils were not frozen) in 1997, 1998, and 1999 at KF and in 1998 and 1999 at RF. Infiltration rates were measured with an infiltrometer at random locations within the restored riparian zone (n = 16 at KF and n = 20 at RF) and five random locations in the controls. The peak standing plant biomass (midsummer) was measured on 20 randomly located 0.84 m^{-2} clip plots within the restored riparian zone and 20 locations in the degraded riparian zone. Both woody and herbaceous vegetation were sampled. Clipped vegetation was dried at 70°C to a constant weight. Samples were taken in 1998 and 1999 at KF and in 1999 and 2000 at RF.

5.3.7 *Nitrogen cycling responses to riparian zone restoration*

Several components of the riparian N cycle were improved by restoration. The "initial condition" of soil solution NO_3^- in the degraded riparian zone was approximately tenfold higher than the baseline value of 0.01 mg l^{-1} (i.e., typical for soil solution NO_3^- concentrations in undisturbed riparian zones in the southern Appalachians) (Yeakley et al. 2003). Restoration resulted in a considerable reduction in soil solution NO_3^-, with values in near-stream areas comparable to baseline levels (Figure 5.4). We attribute this decrease in soil solution NO_3^- to a combination of reduced inputs and increased storage in the recovering vegetation, litter layer, and recovering microbial populations and transformation rates. For example, vegetation biomass increased two- to fourfold on the restored riparian area compared to the degraded riparian zone and was comparable to baseline levels from control areas (Table 5.2). Increased vegetation biomass, in combination with changes in bulk density and infiltration, also appears to have played a role in decreasing the movement of

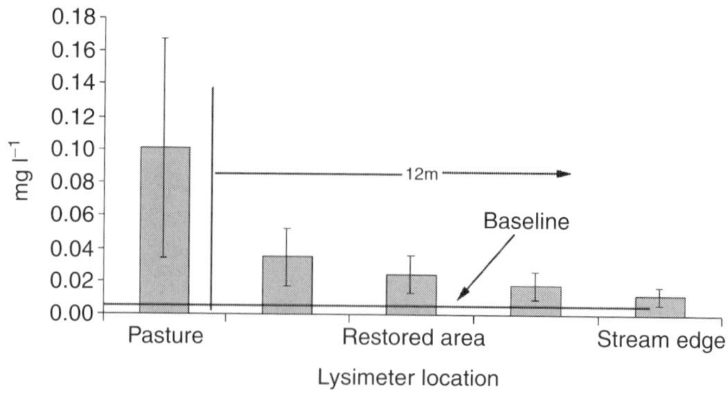

Figure 5.4 Soil solution NO_3^- concentrations (30 cm soil depth) along a transect from stream edge to unrestored pasture two years after restoration.

Table 5.2 Mean Vegetation Biomass (g m^{-2}) 2 Years after Restoration Compared to a Nearby Degraded Riparian Zone

Site	Restored (g m^{-2})	Degraded (g m^{-2})
RF	394 (133)	154 (95)
KF	399 (135)	91 (51)

Data in parentheses are standard errors ($n = 20$).

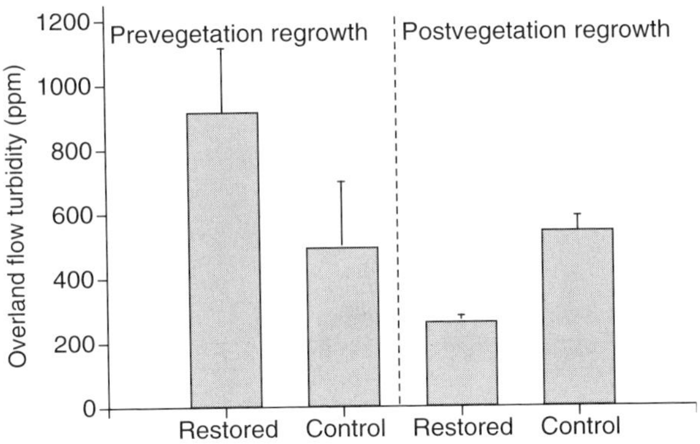

Figure 5.5 Effectiveness of vegetation regrowth in restored riparian zones for reducing overland flow turbidity.

nutrients (and sediment) through the riparian zone in overland flow (Figure 5.5). We observed a general decline in the amount of TSS in overland flow over time (Figure 5.6); however, since baseline values for TSS in overland flow in undisturbed riparian zones in the southern Appalachians are extremely low due to high infiltration rates (e.g., 25 cm h^{-1}) (Swift 1986), restoration treatments have not resulted in a full recovery of soil surface/vegetation conditions that promote rapid infiltration and TSS trapping. For example, the infiltration rate on our study sites (mean of the KF and RF sites) was only 8.5 cm h^{-1} 1 year after restoration, well below the values observed in undisturbed ecosystems. It should be noted that the size and severity of disturbed upland area, as well as the width of the riparian zone, have a significant influence on the capacity of the riparian zone to reduce TSS. Hence, small "fully restored" riparian zones with large, highly disturbed upland areas may still not have the functional capacity to trap and store all of the nutrients or sediment in overland flow. In our study, restoration did result in values comparable to the control (Figure 5.5), indicating that recovery was achieved within the constraints of the upland-riparian complex. In general, microbial processes regulating N trace gas emissions also increased after restoration (Walker et al. 2002). In this case, the desired condition is high levels of trace gas N flux to the atmosphere to prevent transport of subsurface N into surface water or groundwater. This contrasts with previous desired conditions, where we have used baseline values (which are typically quite low) (Davidson and Swank 1986) from undisturbed riparian zones. Here, NO fluxes were five- to sixfold greater in the restored riparian zone the first year following restoration (Figure 5.7), and available soil N was more efficiently converted to N trace gases (i.e., flux per unit soil [N]) in the restored riparian zone relative to the degraded areas (Walker et al. 2002). This pattern is likely due to the recovery of microbial communities following changes in soil physical characteristics (i.e., decreased bulk density [−5% from pre-restored conditions] and increased infiltration [+73% from pre-restored conditions]) that promote aeration.

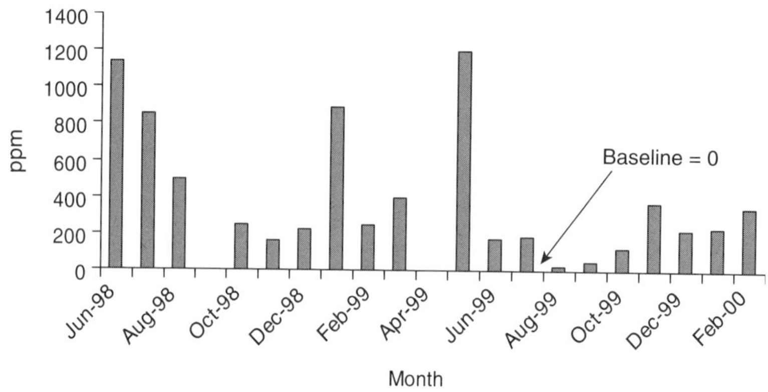

Figure 5.6 Temporal patterns in overland flow turbidity (TSS in ppm) immediately after restoration (June 1998) and 1.5 years later (February 2000).

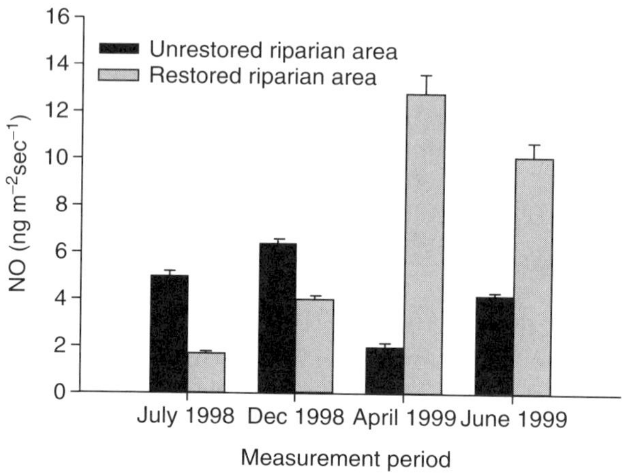

Figure 5.7 Temporal patterns in NO flux immediately after riparian zone restoration (July 1988) and 1 year later.

Evaluation of restoration of N cycling processes using the resistance/resilience analogy suggests a rapid and efficient recovery of many key N cycling pools, processes, and resultant fluxes (Figure 5.8). For example, soil solution NO_3^-, which was elevated tenfold as a result of riparian zone degradation and increased inputs from animal wastes, decreased to near-baseline conditions by the second year of recovery. Nitrogen trace gas fluxes and microbial N processing efficiency increased substantially in the first year of recovery, suggesting a rapid recovery of microbial populations and activity to desired conditions (i.e., significant and efficient conversion of subsurface N to gaseous forms). Both soil solution NO_3^- and N trace gas fluxes exhibited considerable resilience and low resistance (i.e., the return to desired conditions occurred rapidly and with minimal restoration intensity). In contrast, TSS (and the N it contains) in overland flow did not achieve desired conditions during the period of study. In part, this may be related to an incomplete recovery of infiltration rates and vegetation regrowth. This pattern of responses indicates that physical changes in soil surface characteristics may be more resistant and less resilient than N cycling-related biological responses. However, because the processing and storage capacity for TSS in overland flow is related to riparian zone width and the amount of material transported from upland areas in TSS (which is related to size, amount of

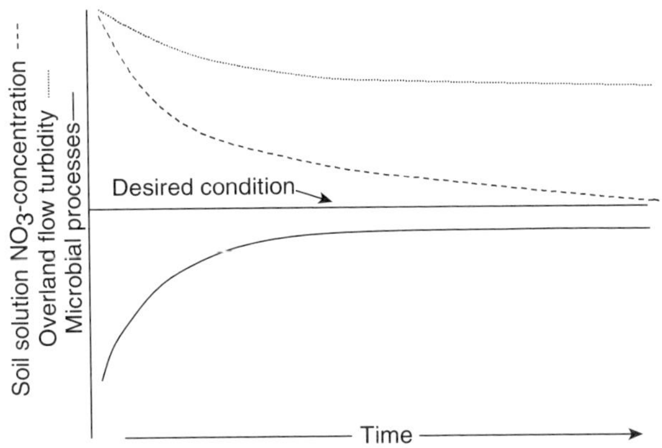

Figure 5.8 General patterns of changes in nutrient cycling pools and processes 1 to 1.5 years following riparian zone restoration.

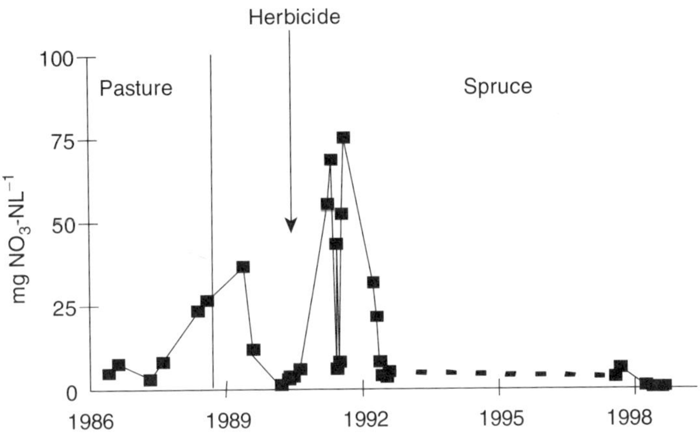

Figure 5.9 Effects of afforestation and herbicide treatment on groundwater N.

disturbance, and nutrient inputs from external sources such as animal waste and fertilizer), recovery in these restored riparian zones may never be sufficient to achieve conditions comparable to undisturbed forest ecosystems.

Results from a 2-year study provide only a snapshot of potential longer-term responses and very little information concerning year-to-year variability. For example, while results from the riparian zone restoration study suggest rapid recovery, we can make few conclusions about how ecosystem pools and processes will respond to factors such as extremes in abiotic variables, disturbance, or successional changes as the soils and vegetation continue to develop. As an example, results from longer-term afforestation studies in Denmark suggest that post-restoration vegetation dynamics are very important in regulating NO_3^- - N concentration in seepage water (Figure 5.9) (Callesen et al. 1999). Prior to afforestation, groundwater NO_3^- - N ranged from 10 to 25 mg l^{-1}, and then decreased substantially in the first few years after afforestation with Norway spruce (*Picea abies*). Herbicide treatment to release the spruce resulted in a significant spike (> 75 mg l^{-1}) in seepage water NO_3^- -N, followed by a reduction to levels around 5 mg l^{-1} in the fourth year and <1 mg l^{-1} by the tenth year.

5.4 Effects of stand restoration burning on stream nitrogen

The onset of fire suppression early in the 20th century significantly reduced the role of fire in shaping the structure and function of ecosystems in the southern Appalachians. A century of fire exclusion has resulted in a buildup of woody and fine fuels, increasing the risk of catastrophic wildfires, as well as altering ecosystems and wildlife habitat historically dependent on periodic wildfire for maintenance (Van Lear and Waldrop 1989). Prescribed fire has been used as a tool in many forest ecosystems to enhance overall stand health and productivity and reduce fuel loads (Sanders and Van Lear 1987; Van Lear and Waldrop 1989). While often satisfying the primary goal of reducing competition to commercially desirable tree species, prescribed burning also improves habitat for both avian and terrestrial wildlife (Cooper 1971). The continued (and, perhaps, expanded) use of fire in the southern Appalachians has generated considerable interest about the effects of prescribed fire on ecosystem integrity, particularly how it influences losses of key plant nutrients such as nitrogen. For example, total ecosystem nitrogen may decrease in forested systems following fire (Neary and Currier 1982; Rapp 1990) due to volatilization of nitrogen stored in coarse and fine fuels, and/or increased leaching of released NO_3^- from the system (Knoepp and Swank 1993). While the objective of the restoration burning did not focus on biogeochemical cycling, we propose that the magnitude (e.g., resistance) and duration (e.g., resilience) of NO_3^- response in stream water may be a useful metric for assessing ecosystem response to restoration burning. Here, we summarize and compare stream NO_3-N responses from four southern Appalachian watersheds receiving different burn prescriptions and use the resistance/resilience analogy to evaluate ecosystem responses.

5.4.1 Study sites

All four burned areas are on the Nantahala National Forest in western North Carolina, U.S. Three of the four sites are characterized as the xeric oak-pine forest type, in a generally degraded condition due to the combined effects of land management practices (e.g., high-grade logging), fire exclusion, and drought. Stands typically consisted of mixtures of pitch pine (*Pinus rigida*), Virginia pine (*Pinus virginiana*), shortleaf pine (*Pinus echinata*), scarlet oak (*Quercus coccinea*), chestnut oak (*Q. prinus*), and red maple (*Acer rubrum*), and have dense understories of the ericaceous shrub, mountain laurel (*Kalmia latifolia*). Much of the yellow pine overstory on these sites was dead because of recent drought-related southern pine beetle (*Dendroctonus frontalis*) infestations (Smith 1991).

The fourth site is within the Joyce Kilmer-Slickrock Wilderness Area and Memorial Forest, located approximately 60 km west of the other three sites. The area has an old-growth designation, and is one of the few remaining large uncut forested tracts in the southern Appalachians. This high-elevation site contains large areas of both mesic and xeric forest types.

5.4.2 Burning methods and sample collection and analysis

Due to the degraded condition of many xeric oak-pine stands in the southern Appalachians, the fell-and-burn site preparation prescription has commonly been used to increase commercial viability and enhance other attributes of ecosystem health (Vose et al. 1999). In 1990, a 5.25-ha watershed, designated Jacobs Branch (JB), was selected to examine ecosystem-level effects of the fell-and-burn treatment. Initial and long-term results have been published previously (Clinton et al. 1993; Elliott and Vose 1993; Knoepp and Swank 1993; Swift et al. 1993; Vose and Swank 1993; Elliott and Vose 1995; Clinton et al. 1996; Clinton and Vose 2000), and a complete burning characterization has been described by Swift et al. (1993). All woody vegetation on JB was felled in the summer of 1990. The downed material was allowed to cure for approximately 10 weeks before a headfire was set at the base of the slope. The entire slope burned in less than 30 minutes. The fire was characterized as a high-intensity, low-severity fire that consumed most of the litter layer and a small portion of the humus layer (Vose and Swank 1993). Stream samples were collected weekly from a small stream at the base of the burned watershed beginning 6 months prior to burning and continued for more than 12 months after burning. Weekly samples were composited to obtain a monthly mean.

The Wine Spring Creek (WSC) burn was implemented in April 1995 to simulate a stand-replacement wildfire covering approximately 82 ha. The fire prescription was designed to examine the effects of wildfire on responses of vegetation, soil processes, and stream chemistry and turbidity (Major 1996; Elliott et al. 1999; Vose et al. 1999). The site was ignited by a helicopter and helitorch using strip headfires along the contour beginning at the lower and mid-slope positions. Crown fires were frequent at mid- and upper-slope positions due to the combination of heavy fuels, steep slopes, thick mountain laurel (*Kalmia latifolia*) understories, and firing techniques. Streams and roads provided natural firebreaks, and a backing fire was set along the top of the ridge. Automated time-increment proportional stream samplers (American Sigma, Inc.) were installed in 1993 at the base of the burned slope and upstream of the burned area, as well as in several streams in other unburned areas to be used as references, for the purpose of characterizing dissolved inorganic chemistry. In addition, grab samples (e.g., 250 ml collections at designated locations) were collected weekly at each stream sampler location.

The Joyce Kilmer-Slickrock Wilderness Area (JK) wildfire occurred in November 1999 and burned approximately 970 ha or about two thirds of the Little Santeetlah Creek Watershed within the JK site. The fire was an understory burn that traveled well into riparian zones of many small streams draining the watershed. During the time of the burn, stream samples were being collected for an unrelated study of stream water chemistry. On the JK site, weekly grab samples were taken beginning in August 1999 and continued after the burn for 6 months.

The Hickory Branch (HB) site was prescribed-burned (stand restoration burning) in March 1999. HB site vegetation was predominantly dry-site mixed-pine hardwood in the overstory and mountain laurel in the understory. Using a helitorch, the site was strip-burned along the contour beginning at the ridge and continuing to the base of the watershed. Crown fires frequently occurred on steep, south-facing slopes containing standing dead yellow pine snags and thick mountain laurel understory. Approximately 75% of the watershed burned, with little or no burning occurring in the more mesic areas around riparian zones. Stream water grab samples were taken weekly, beginning approximately 12 months before the burn at the watershed outlet on Pine Branch (hereinafter referred to as Lower Pine Branch) and at two locations upstream immediately above the confluence of Pine Branch (Upper Pine Branch) and Matheson Branch. Approximately 2 weeks before the burn, an automated flow-increment proportional sampler (American Sigma, Loveland, CO, U.S.A.) was installed at the watershed outlet (Lower Pine Branch), which collected samples for 2 months after the burn. Grab samples continued to be collected at all three collection sites.

All stream water samples were analyzed at the Coweeta Hydrologic Lab using established protocols (Deal et al. 1996) for the determination of NO_3-N concentrations using a Dionex ion chromatograph (Dionex Inc.). To account for seasonal differences and provide long-term baseline values, long-term monthly mean stream NO_3 concentration for a south-facing mixed-hardwood control watershed at Coweeta (Watershed 2) was used as an undisturbed reference for stream NO_3 concentration.

5.4.3 *Effects of fire on stream NO_3^-*

Stream NO_3-N responses to burning varied among the four studies. Two of the studies exhibited increased stream NO_3^-. After the JB burn, NO_3-N in stream water increased from <0.01 to 0.055 mg N l^{-1} after 3 months and up to a maximum of 0.075 mg l^{-1} after 7 months, and remained elevated until June of the following year (Figure 5.10). Similarly, stream water NO_3^- concentration following the JK wildfire showed a marked increase over preburn values. Compared to the values for the previous 2 months, stream NO_3 increased from a low of 0.02 to 0.12 mg N l^{-1} within 6 weeks of the burn, and then quickly returned to near preburn values (Figure 5.9). In contrast, on both the HB and WSC burns, there were no measurable fire effects on stream NO_3-N (Figure 5.9). For example, using pre- and postburn regressions of stream NO_3, Vose et al. (1999) found no difference in NO_3 concentrations between the streams directly below the WSC burned area and an unburned reference stream. Similarly, there were strong similarities in monthly mean stream water NO_3 concentrations between the HB burn site and an undisturbed reference watershed both in terms of the magnitude and the seasonality before and after the burn (Figure 5.10).

The initial fuel condition and the type and season of burning appeared to be the primary factors determining the observed differences in the amount and duration of stream water NO_3^- responses (Table 5.2). Large differences were observed between fall burns and spring burns. The two sites that burned in the fall (JB and JK) showed large increases in NO_3^- concentration compared to the spring burns (WSC and HB) (Table 5.2). For example, the two sites that showed an NO_3^- response were burned in early and late fall when vegetation uptake and microbial immobilization were low. On the JB site, Knoepp and Swank (1993) observed significant increases in soil water NO_3^- concentration at 30 and 60 cm depths soon after the burn in September and a sharp increase again in June the next year. The initial postburn increase (Figure 5.10) was likely due to leaching of excess NO_3^- resulting from reduced uptake and increased mineralization (Knoepp and Swank 1993), and the later increase was probably related to increased decomposition of exposed surface organic layers and other fine organic material, and delayed increases in rates of nitrification often observed following fire (Knoepp and Swank 1993; White 1986). Because virtually no uptake of soil nutrients was taking place during the dormant season, and because of vegetation removal to within a few meters of the stream, excess NO_3-N was leached to the stream. To some extent, the same explanation can be used for observed increases in stream NO_3^- concentration following the JK burn. In late fall, no vegetation uptake was occurring to immobilize excess NO_3^-. Additionally, some of the riparian areas were burned, thus providing a potential direct input of NO_3^- to the stream.

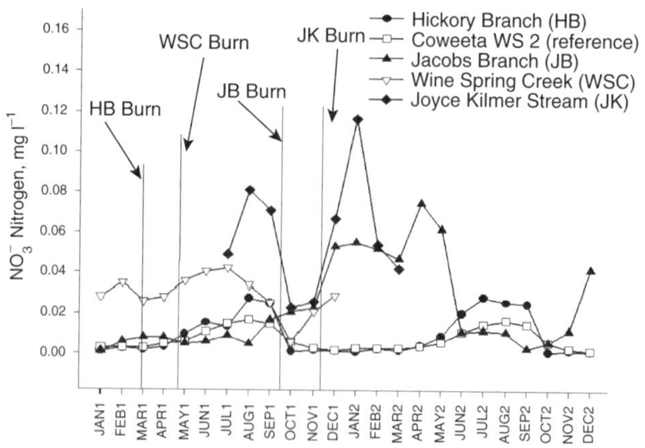

Figure 5.10 Response of streamwater nitrate (NO_3-N) to a variety of burning prescriptions in the southern Appalachians, U.S. Coweeta WS 2 represents unburned reference conditions.

The streams that showed no measurable response (HB and WSC) drained watersheds that burned in early spring at the onset of leaf growth and nutrient uptake. Hence, vegetation that was not killed by the fire may have immobilized NO_3^- via uptake. In addition, Vose et al. (unpublished data) have demonstrated the effectiveness of riparian zones in reducing NO_3^- delivery to streams through microbial uptake. On the WSC site, the riparian zone served to buffer fire effects because it did not burn (Vose et al. 1999). Similarly, the riparian zone on the HB site, although burned in mid-March, may have immobilized most of the NO_3^- mobilized during the burn, since only a weak stream water NO_3^- response in the upper reaches of the watershed was observed on the HB site 2 weeks after burning.

Some of the variation in stream NO_3^- losses among burns may be related to differences in the initial condition (i.e., amount and quality) of forest floor fuels. For example, the fell and burn site (JB) had extremely high woody fuel loadings because all standing vegetation was felled prior to burning. Higher severity fires resulting from the heavy fuel loads have the potential to reduce the vigor of regrowing vegetation and reduce nutrient uptake (Elliott et al. 2002). For example, stream NO_3^- response on the JB site suggested low resistance (+750%) and resilience (30 week duration). This is likely due to high intensity and severity fires, which reduce the ability of vegetation regrowth to mediate the effects of burning. In contrast, the stand-replacement burns showed high resistance and resilience to disturbance due in part to lower fuels (hence, lower intensity and severity fires), an intact riparian buffer, and season of the burn. The JK wildfire was intermediate in resilience but less resistant to disturbance, possibly due to large accumulations of recalcitrant nitrogen in old-growth forest floor. Another possible explanation for the pulse of NO_3^- following the JK wildfire is low resistance to disturbance in old-growth forest ecosystems. There is some evidence to suggest that old-growth forests are less capable of sequestering nutrients mobilized by disturbance than actively growing, aggrading forest ecosystems. In addition, stand age, in the absence of fire, determines the accumulation of recalcitrant forms of N that are released during burning and become subject to volatilization and leaching from the system.

Applying the restoration response graph to the data (Figure 5.11) suggests that the "fell and burn" restoration treatment has the greatest potential for undesirable effects on stream NO_3^- because the magnitude (i.e., +750%) and duration (i.e., 30 weeks) of response exceed that observed in the old-growth wildfire. However, it should be noted that even the "worst-case scenario" of peak stream water NO_3^- concentration (i.e., 0.12 mg l^{-1} for the JK wildfire) is still quite low and within acceptable limits for surface water

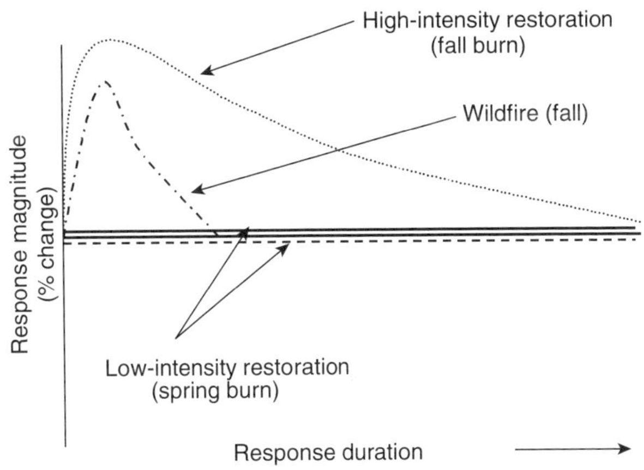

Figure 5.11 General patterns of changes in stream NO_3^- in response to different burning prescriptions.

quality. Using the resistance/resilience analogy, we conclude that while low baseline stream NO_3^- levels allow for a high-resolution assessment of response patterns across a range of fire types, using any of these prescribed fire restoration techniques to alter vegetation structure does not result in unacceptable effects on stream NO_3^-.

5.5 Summary

We have presented two different aspects of evaluating the effects of restoration on biogeochemical cycling and aquatic ecosystems. The first examined the direct responses of nutrient cycling pools and processes to restoration efforts aimed at improving nitrogen retention and processing in degraded riparian zones. The second, stream NO_3^- responses to restoration burning, examined responses to restoration efforts that do not target biogeochemical cycling *per se*, but directly or indirectly influence nutrient cycling pools and processes. We presented an approach to evaluate short- and long-term responses of restoration on biogeochemical cycling using the resistance and resilience analogy. Our adaptation of the resistance/resilience analogy to evaluate restoration responses focused on three key attributes. First, the complexity of ecosystem biogeochemical cycling requires the determination of key indicators of response. In our example, we focused on N cycling pools and processes because N is often tightly conserved, responds to changes in physical and biological conditions, limits productivity in many ecosystems, and is toxic in some forms and concentrations. Second, evaluation of responses (direct or nontarget) can be approached by assessing response magnitude and duration. Restoration effectiveness can be assessed in terms of how quickly pools and processes return to desired conditions and how long they stay with desired boundary conditions. Finally, variable responses should be expected since the impacts of degradation may impact pools and processes differently. For example, pools and processes that depend primarily on changes in biological parameters may respond quite differently than those dependent on changes in physical parameters.

References

Callesen, L., Rauland-Rasmussen, K., Gundersen, H., and Stryhn, H., Nitrate concentration in soil solution below Danish forests, *For. Ecol. Manage.*, 114, 71, 1999.

Carpenter, S., Walker, B., Anderies, J.M., and Abel, N., From metaphor to measurement: resilience of what to what? *Ecosystems*, 4, 765, 2001.

Clinton, B.D. and Vose, J.M., Plant succession and community restoration following felling and burning in the southern Appalachian mountains, in *Fire and Forest Ecology: Innovation Silviculture & Vegetation Management*, Tall Timbers Fire Ecology Conference Proceedings, No. 21, Moser, K.W. and Moser, C.F., Eds., Tall Timbers Research Station, Tallahassee, FL, 22, 2000.

Clinton, B.D., Vose, J.M., and Swank, W.T., Site preparation burning to improve southern Appalachian pine-hardwood stands: vegetation composition and diversity of 13-year-old stands, *Can. J. For. Res.*, 23, 2271, 1993.

Clinton, B.D., Vose, J.M., and Swank, W.T., Shifts in aboveground and forest floor carbon and nitrogen pools after felling and burning in the southern Appalachians, *Forest Sci.*, 42, 431, 1996.

Cooper, J.R., Gilliam, J.W., Daniels, R.B., and Robarge, W.P., Riparian areas as filters for agricultural sediment, *Soil Sci. Soc. Am. J.*, 51, 416, 1987.

Cooper, R.W., The pros and cons of burning in the south, *For. Farmer*, 31, 10, 1971.

Daniels, R.B. and Gilliam, J.W., Sediment and chemical load reduction by grass and riparian filters, *Soil Sci. Soc. Am. J.*, 60, 246, 1986.

Davidson, E.A. and Swank, W.T., Environmental parameters regulating gaseous nitrogen losses from two forested ecosystems via nitrification and denitrification, *Appl. Environ. Microb.*, 52, 1287, 1986.

Deal, J.M., Brown, C., and Holland, C., *Procedures for Chemical Analysis at the Coweeta Hydrologic Laboratory*, Coweeta Files, 1996.

Elliott, K.J. and Vose J.M., Site preparation burning to improve southern Appalachian pine-hardwood stands: photosynthesis, water relations, and growth of planted *Pinus strobus* during establishment, *Can. J. For. Res.*, 23, 2278, 1993.

Elliott, K.J. and Vose, J.M., Evaluation of the competitive environment for white pine (*Pinus strobus* L.) seedlings planted on prescribed burn sites in the southern Appalachians, *Forest Sci.*, 41, 513, 1995.

Elliott, K.J., Hendrick, R.L., Major, A.E., Vose, J.M., and Swank, W.T., Vegetation dynamics after a prescribed fire in the southern Appalachians, *For. Ecol. Manage.*, 114, 199, 1999.

Elliott, K.J., Vose, J.M., and Clinton, B.D., Growth of eastern white pine (*Pinus strobus* L.) related to forest floor consumption by prescribed fire in the southern Appalachians, *South. J. Appl. For.*, 29, 18, 2002.

Gilliam, J.W., Osmond, D. L., and Evans, R. O., *Selected Agricultural Best Management Practices to Control Nitrogen in the Neuse River Basin*, North Carolina Agricultural Research Technical Bulletin 311, North Carolina State University, Raleigh, NC, 1997.

Henderson, G.S., Swank, W.T., Waide, J.B., and Grier, C.C., Nutrient budgets of Appalachian and Cascade region watersheds: a comparison, *Forest Sci.*, 24, 385, 1978.

Holling, C.S., Resilience and stability of ecological systems, *Annu. Rev. Ecol. Syst.*, 4, 1, 1973.

Knoepp, J.D. and Swank, W.T., Site preparation burning to improve southern Appalachian pine-hardwood stands: nitrogen responses in soil, soil water, and streams, *Can. J. For. Res.*, 23, 2263, 1993.

Lowrance, R.R., Todd, J., Fail, J., Hendrickson, O., Leonard, R., and Asmussen, L., Riparian forests as nutrient filters in agricultural watersheds, *Bioscience*, 34, 374, 1984.

Likens, G.E., Bormann, F.H., Johnson, N.M., Fisher, D.W., and Pierce, R.S., Effects of forest cutting and herbicide treatment on nutrient budgets in the Hubbard Brook watershed ecosystem, *Ecol. Monogr.*, 40, 23, 1970.

Major, A.E., *The Effects of Stand-Replacement Fires on Pine and Oak Communities in the Southern Appalachians*, M.S. thesis, University of Georgia, Athens, 1996.

Neary, D.G. and Currier, J.B., Impact of wildfire and watershed restoration on water quality in South Carolina's Blue Ridge Mountains, *South. J. Appl. For.*, 6, 81, 1982.

Osborne, L.L. and Kovacic, D.A., Riparian vegetative buffer strips in water quality restoration and stream management, *Freshwater Biol.*, 29, 243, 1993.

Rapp, M., Nitrogen status and mineralization on natural and disturbed Mediterranean forests and coppices, *Plant Soil*, 128, 21, 1990.

Sanders, B.M. and Van Lear, D.H., Pre-and post-burn photo series for pine-hardwood logging slash in the southern Appalachians, in *Proceedings of the Ninth Conference on Fire and Forest Meteorology*, American Meteorological Society, Boston, MA, 41, 1987.

Smith, R.N., *Species Composition, Stand Structure, and Woody Detrital Dynamics Associated with Pine Mortality in the Southern Appalachians*, M.S. thesis, University of Georgia, Athens, 1991.

Swank, W.T., Review of biogeochemical cycling of mineral-forming elements: studies in environmental science, *Q. Rev. Biol.*, 56, 91, 1981.

Swank, W.T., Biological control of solute losses from forest ecosystems, in *Solute Processes*, Trudgill, S. T., Ed., John Wiley & Sons, New York, 85, 1986.

Swank, W.T. and Vose, J.M., Long-term nitrogen dynamics of Coweeta forested watersheds in the southeastern United States of America, *Global Biogeochem. Cy.*, 11, 657, 1997.

Swank, W.T., Vose, J.M., and Elliott, K.J., Long-term hydrologic and water quality responses following commercial clearcutting of mixed hardwoods on a southern Appalachian catchment, *For. Ecol. Manage.*, 143, 163, 2001.

Swift, L.W. Jr., Filter strip widths for forest roads in the southern Appalachians, *South. J. Appl. For.*, 10, 27, 1986.

Swift, L.W. Jr., Elliott, K.J., Ottmar, R.D., and Vihnanek, R.E., Site preparation burning to improve southern Appalachian pine-hardwood stands: fire characteristics and soil erosion, moisture, and temperature, *Can. J. For. Res.*, 23, 2242, 1993.

Van Lear, D.H. and Waldrop, T.A., *History, Uses, and Effects of Fire in the Appalachians*, U.S. Department of Agriculture, Forest Service, Southeastern Research Station, Asheville, NC, General Technical Report SE-54, 1989.

Vose, J.M., Perspectives on using prescribed fire to achieve desired ecosystem conditions, in *Fire and Forest Ecology: Innovation Silviculture & Vegetation Management, Tall Timbers Fire Ecology Conference Proceedings*, No. 21, Moser, K.W. and Moser, C.F., Eds., Tall Timbers Research Station, Tallahassee, FL, 12, 2000.

Vose, J.M. and Swank, W.T., Site preparation burning to improve southern Appalachian pine-hardwood stands: above-ground biomass, forest floor mass, and nitrogen and carbon pools, *Can. J. For. Res.*, 23, 2255, 1993.

Vose, J.M., Swank, W.T., Clinton, B.D., Knoepp, J.D., and Swift, L.W. Jr., Using stand replacement fires to restore southern Appalachian pine-hardwood ecosystems: effects on mass, carbon, and nutrient pools, *For. Ecol. Manage.*, 114, 215, 1999.

Waide, J.B., Forest ecosystem stability: revision of the resistance–resilience model in relation to observable macroscopic properties of ecosystems, in *Forest Hydrology and Ecology at Coweeta. Ecological Studies*, Vol. 66, Swank, W.T. and Crossley, D.A. Jr., Eds., Springer-Verlag, New York, 383, 1988.

Walker, J.T., Geron, C.D., Vose, J.M., and Swank, W.T., Nitrogen trace gas emissions from a riparian ecosystem in southern Appalachia, *Chemosphere*, 49, 1389, 2002.

Webster, J.R., Waide, J.B., and Patten, B.C., Nutrient recycling and the stability of ecosystems, in *Mineral Cycling in Southeastern Ecosystems*, Howell, F.G., Gentry, J.B. and Smith, M.H., Eds., Energy Research and Development Administration, Technical Information Center, Washington, D.C., 1, 1975 [Available from National Technical Information Service, 5285 Port Royal Roads, Springfield, VA 22161].

White, C.S., Effects of prescribed fire on rates of decomposition and nitrogen mineralization in a ponderosa pine ecosystem, *Biol. Fert. Soils*, 2, 87, 1986.

Yeakley, J.A., Coleman, D.C., Haines, B.L., Kloeppel, B.D., Meyer, J.L., Swank, W.T., Argo, B.W., Deal, J.M., and Taylor, S.F., Hillslope nutrient dynamics following upland riparian vegetation disturbance, *Ecosystems*, 6, 154, 2003.

Part 3

The Sociopolitical Context for Restoration

chapter six

Policies for encouraging forest restoration

D. Evan Mercer
US Department of Agriculture Forest Service

Contents

6.1 Introduction

Throughout the 20th century, many countries created national parks, forests, nature reserves, and sanctuaries to provide benefits that are underproduced on private lands. Private lands are now especially valuable for providing ecological services that public lands cannot provide, due to the increasing demands for all uses and the political and economic conflicts associated with allocating public lands between competing uses (e.g., recreation, watershed protection, biodiversity conservation, wildlife habitat, commodity production) (Kline et al. 2000). In many countries, the supply of public lands may not be adequate to ensure desirable flows of beneficial ecosystem services. Furthermore, because many ecological processes cross ownership borders, enhancing the flow of benefits requires management at a broader, landscape scale and with the participation of both public and private landowners (Kline et al. 2000; Gottfried et al. 1996; Boyd and Wainger 2002a; Johnston et al. 2002).

Many ecological services are public goods (goods or services for which it is hard or impossible to exclude from benefiting those who do not contribute to paying the costs of producing the good) or externalities (costs or benefits that are inflicted on or received by others, that are not reflected in market transactions). Ecological services typically are underproduced on private lands because the landowners bear the costs of restoring, preserving, and managing their lands for ecological services as well as the opportunity costs of foregoing alternative income-producing activities, while all members of the community enjoy the benefits for free. As a result, governments worldwide have been experimenting with policies and programs to encourage forest restoration, including a combination of activities on public lands and regulations and incentives to private landowners.

Deciding on the best combination of policies for encouraging restoration requires determining which ecosystem services are amenable to market solutions, which require government intervention, and which require a combination of government and market approaches. Ecosystem services that can be efficiently supplied by markets must provide goods or services with some commercial value or for which a commercial value can be attached. Unless the landowner can appropriate some of that value, however, market solutions are infeasible and government intervention is required (Heal 1999).

Ecosystem services are frequently a combination of public (if provided for one, they are provided for all) and private goods. The extent to which they are public goods determines the necessity for government intervention. Watershed restoration projects provide an example of mixed public and private goods. Water quality is a public good because if it is produced for one user of the watershed, it is produced for all, no matter who bears the cost. Because some individuals and communities can be excluded from consuming water, it can also be considered a private good (Heal 1999). Indeed, during the recent droughts that have inflicted the western and southeastern U.S., it was not uncommon to see news accounts of communities selling water and water rights to other communities with shortages of water. Likewise, the protective role of forest ecosystems in a watershed produces public goods (e.g., biodiversity, carbon sequestration, wildlife habitat), whose production may or may not conflict with private goods produced from the ecosystem such as timber.

The situation becomes even more complicated when considering ecological processes that produce ecosystem services at a landscape scale. Individual owners acting alone are unlikely to produce the socially optimal amount of ecosystem services. Just as prices in a perfectly competitive market are determined by the interactions of all buyers and sellers in a market, the mix of commodities and services produced from an ecosystem or watershed depends on the spatial pattern of land-use decisions made by all the landowners in that ecosystem. These "economies of configuration" (Gottfried et al. 1996) suggest that policies and programs designed and implemented at a landscape level rather than with individual ownerships are more likely to produce optimal quantities of ecosystem services, unless the effects of landowners' decisions can be separated spatially.

Gottfried et al. (1996) analyzed the ability of markets to create optimal landscapes and demonstrated that market forces in decentralized, unregulated economies are inadequate for optimizing ecological services at the landscape level. An optimal market solution would require that all public and private landowners compensate each other for the production of all externalities, both positive and negative. Unfortunately, this optimal situation requires intimate, quantifiable information, knowledge, and monitoring of the interrelationships within and between ecosystems and between ecosystems and economic systems, a level of knowledge that is not yet possible. Even if possible, implementation would likely be prohibitively expensive.

Incentives to individuals (e.g., taxes or subsidies) for restoration will usually suffer from scale problems (Gottfried et al. 1996) because the location of each ownership within a landscape determines, to a large extent, the production of a landscape's mix of goods and services, and landowners jointly (but in different ways) affect the landscape's ecological functions. Therefore, in landscapes with multiple ownerships that have spatially varied impacts on the ecosystem, markets will most likely fail to provide adequate ecosystem services, even in the presence of traditional methods of internalizing externalities (e.g., taxes and subsidies). This suggests that a combination of government and market interventions will usually be required to ensure successful and efficient ecosystem restoration efforts across a landscape (Gottfried et al. 1996).

6.2 Costs and benefits of restoration policy options

Governments have used a variety of mechanisms to encourage ecosystem restoration. These include tax incentives, subsidies, and cost-share programs; purchase of conservation easements; fee-simple purchases; regulations restricting landuse; tradable development rights; and cooperative/collective efforts (Alberini and Segerson 2002; Boyd and Simpson 1999; Cubbage et al. 1993; Granskog et al. 2002; OECD 1999). Table 6.1 provides a qualitative overview of the relative costs and benefits of alternative restoration policies, which are elaborated in this section.

The fundamental decision for policymakers is whether the government has both the ability and will to impose unwanted costs on landowners through either mandatory or voluntary approaches to encourage restoration. With mandatory approaches, governments impose net costs on landowners, who are therefore worse off financially than in the absence of the policy. Landowners will participate in voluntary programs if the total amount of benefits they receive from the land and government (both financial and nonfinancial) are at least as high as they receive without participating. For voluntary approaches to succeed, the landowner must perceive some gain or at least no net loss from participating (Alberini and Segerson 2002).

No matter which mechanism is used, the social cost of restoring and preserving ecosystems is the value of the lost income from whatever economic activities are foregone as a result of restoring and managing the land for ecological services. In other words, the social cost of restoration is the difference between the value of the land in its highest and best private use and its value following restoration. This cost is constant for individual properties no matter which type of policy or program is used to encourage restoration (Boyd and Simpson 1999). However, these opportunity costs will vary with the options provided to landowners to choose which parcels of their lands will be used for restoration and which for economic activities.

Available policy and program options, however, differ in the size of the transaction costs associated with the institutional, organizational, and informational requirements for implementing the policy. They also differ in who pays for the restoration of ecological services that benefit the whole society. If restoration is mandated by regulations that restrict and/or require certain activities on private lands, the private landowners pay. The costs of outright (fee-simple) purchase and purchase of conservation easements on private lands by the government are paid by taxpayers. Developers and future landowners pay under systems of tradable development rights. Under voluntary tax incentive and cost-sharing programs, both the landowner and taxpayers bear the costs. As Boyd and Simpson (1999) state,

> Reasonable persons may differ regarding which groups are more morally deserving of bearing or escaping the burden of payment. But someone must pay.

6.2.1 *Mandatory approaches*

Mandatory (or regulatory) approaches use laws and policies to either dictate specific land management actions or otherwise limit or control how landowners (private and public) manage their lands. Granskog et al. (2002) list 13 federal statutes and 6 types of state laws regulating forest management in the southern U.S., while a 1991 survey identified 359 local ordinances regulating forestry activities in the eastern U.S. (Cubbage et al. 1993). Forest and land-use regulations vary considerably across Europe, but most European countries have stringent laws regulating timber harvesting, environmental protection, forest conversion, and restoration. In contrast to the U.S., however, most private lands in Europe are required to allow public access (Cubbage et al 1993). Regulation is sometimes viewed as an efficient way of producing or protecting ecological services, because, in contrast to other approaches, fewer intervening institutions (e.g., tax assessment and collection agencies or markets) are usually required. Monitoring and enforcement costs,

Table 6.1 Summary of Qualitative Levels of Costs and Benefits of Alternative Restoration Tools

Policies	Costs					Who Pays?	Benefits
	Transactions	Monitoring	Enforcement	Opportunity	Political		
Mandatory							
Land-use regulations	Medium	High	High	High	High	Landowners	High
Compensatory mitigation	High	High	High	High	Medium	Landowners and developers	Low–medium
Mitigation banking	High	Medium	Medium	Low	Low	Landowners and developers	Medium
In-lieu mitigation fees	Medium	High	High	Low	Medium	Landowners and developers	Low
Fee-simple purchase	High	Low	Low	High	High	Taxpayers	High
Easements	High	Medium	Medium	High	High	Taxpayers	High
Voluntary							
Subsidies/cost-share	Medium	Medium	Medium	Medium	Low	Taxpayers and landowners	Low–medium
Tax credits	Medium	Medium	Medium	Medium	Low	Taxpayers and landowners	Low–medium
Fee-simple purchase	High	Low	Low	Low	Medium	Taxpayers	Low–medium
Easements	High	High	High	Low	Medium	Taxpayers	Low–medium
Cooperative-collective actions	High	High	High	Low	Low	Landowners	Medium–high

however, are potentially very large, and regulations can often be politically difficult to pass and implement as landowners may resist relinquishing rights to how they manage their land.

Among the most common regulatory approaches for mandating restoration and preservation of ecosystems are the so-called compensatory mitigation regulations, also referred to as "no-net-loss" policies. Mitigation is the creation, enhancement, or restoration of habitats or ecosystems in response to an action by the landowner that impacts negatively on the services provided by the ecosystem. Although the U.S. initiated compensatory mitigation in 1972 (Clean Water Act: 33U.S.C. 1344 [1972]), the European Union and its member states have also recently implemented a variety of no-net-loss policies (Ledoux et al. 2000). Compensatory mitigation regulations have been primarily directed at wetlands, but they are also applicable to a wide variety of ecosystems and habitats (Fernandez 1999; Ledoux et al. 2000).

The three general mechanisms for implementing compensatory mitigation actions are landowner or permittee-responsible, mitigation banks, and in-lieu fees (NRC 2001). Early compensatory mitigation actions in the U.S. encouraged on-site mitigation under the direct responsibility of the landowner or permittee and allowed developers to simultaneously develop and attempt restoration on the same site. Critics argued that on-site mitigation produced scattered small islands of restored ecosystems that were too small to adequately compensate for the loss of landscape-level ecological services, and that temporal losses of wetland functions were too large when developers were allowed to mitigate contemporaneously with development. The early mitigation efforts were also criticized for failing to protect high value wetlands, to enforce minimum mitigation requirements for replacement habitats, and to enforce and monitor compliance (Ledoux et al. 2000; NRC 1992, 2001; Race 1985; Race and Fonseca 1996; Reppert 1992; Roberts 1993; Zedler 1996).

Mitigation banking and in-lieu fee programs were developed in response to problems with individual on-site, postproject, landowner-responsible mitigation. Private firms began producing wetland credits for sale in the early 1990s, and by 1995 U.S. agencies issued guidelines for approving the sale of wetland mitigation credits (Fernandez and Karp 1999). Mitigation banking provides for advanced compensation of unavoidable habitat loss by creating, restoring, or enhancing large areas in the same watershed prior to future development and as a precondition for future development. Mitigation banks are usually relatively large blocks of restored, created, or enhanced ecosystems that compensate for ecological impacts from particular development projects, but also act as a repository with credits available for compensating future development projects (Reppert 1992).

When developers or landowners want to develop an environmentally sensitive area, they are required to have credits in hand from investing in a completed restoration site. Credits are denominated as Habitat Units (HU), which are calculated as the product of the number of species or ecological functions per ha times the number of ha being developed. The number of credits required depends on the success of the restoration project (Fernandez and Karp 1999). In essence, mitigation banking sets up a market in which restored ecosystem values and functions are quantified as credits, deposited into an account, and later purchased by developers or landowners when regulations require compensation for authorized losses of ecosystem function (Ledoux et al. 2000).

Mitigation banks offer several advantages over individual on-site, postproject mitigation. First, failure rates should be lower because temporal losses of ecosystem function and uncertainty over the success of mitigation efforts are reduced when restoration occurs prior to implementation of the development project. Second, it is usually ecologically advantageous to have larger, consolidated mitigation sites instead of the smaller, isolated islands of restored ecosystems that individual on-site mitigation tends to produce. Third, economies of scale (financial, regulatory, ecological, scientific) associated with mitigation banking should lead to more cost-effective and superior mitigation projects (NRC 2001).

The third approach to compensatory mitigation involves the use of in-lieu fees as a payment to natural resource managers for implementing specific or general restoration projects. U.S. agencies require the sponsors of in-lieu fee accounts to enter into agreements similar to banking instruments to define appropriate conditions for in-lieu fee mitigation.

Historically, U.S. agencies have generally preferred on-site mitigation to off-site mitigation. For example, 75% of compensatory mitigation projects in 1998 were implemented on-site, while 9% occurred through mitigation banks and the remainder through other mechanisms such as in-lieu fees and in-kind exchanges (NRC 2001). However, when off-site mitigation is approved, U.S. agencies usually prefer mitigation banks to in-lieu fees as long as there are credits available from an approved mitigation bank in the service area. This is based on criticisms that in-lieu fee programs were allowing compensation outside the impacted watershed, funds were being used for nonmitigation activities, programs resulted in out-of-kind compensation, and preservation was too often allowed as a compensatory action. Nevertheless, in-lieu fees are preferred when in-kind mitigation is not available from a bank or the bank only provides preservation credits and the in-lieu fee arrangement offers in-kind restoration (NRC 2001).

The National Research Council (NRC 2001) recently completed a thorough review of restoration projects in the U.S. required under the Clean Water Act (CWA) Section 404. Success was spotty. The NRC found some sites meeting established criteria and showing promise of becoming fully functional ecosystems and watersheds. Others were never initiated, or if initiated, were poorly designed, carelessly implemented, or both. Compensation sites were often placed in landscapes that did not provide enough hydrological impacts and/or associated communities to adequately compensate for the loss of ecological services. At most sites, the absence of long-term monitoring and legal and financial assurances for long-term protection suggest that the mitigation efforts may not be sustainable over time (NRC 2001).

Based on this analysis, the NRC recommended that the U.S. mitigation programs develop and implement a watershed-based approach. The recommended approach would use watershed assessments to determine existing and reference conditions, incorporate results from these assessments into resource management planning, and foster collaboration among landowners in a watershed. Some state and federal agencies have experimented with a process called Advanced Identification (ADID) to plan mitigation sites. The ADID process sets watershed restoration priorities and designs mitigation strategies by assessing the functions and values of sites in a watershed and identifying the most degraded, least valuable sites for future development and the more valuable sites for restoration. Since the public (rather than the developer) pays for the substantial costs that are incurred, ADID can be considered a subsidy to landowners to produce public benefits while satisfying the constraint of no-net-loss criteria (Fernandez and Karp 1999).

Because on-site, permittee-responsible mitigation will likely continue to dominate mitigation efforts in the near future (due in large part to the scarcity of mitigation banks and in-lieu fee programs), the NRC recommended that agencies establish and enforce clear compliance requirements to assure that projects are initiated at least concurrently with the development action, science-based design criteria and adaptive management are used to implement and construct mitigation projects, performance standards are specified and attained before permits are approved, and the permittee provides a government or nonprofit stewardship organization with an easement on or title to the restored site and cash payments large enough to ensure long-term monitoring, management, and maintenance of the site (NRC 2001).

In Austria, Weiss (2000) compared the use of regulations, subsidies, and an educational program to promote mountain forest restoration. Although the regulations (Forest Act of 1987) require the landowner to bear the cost of restoring and ensuring the existence of mountain forests under a polluter-pays principle, it also requires downstream and downhill beneficiaries to pay for additional costs to implement measures or management actions for specific protective functions such as control of avalanches or soil erosion. However, to avoid the transactions costs of resolving conflicts over who owes whom what, the regulations were rarely implemented. Instead, the forest authorities switched primarily to forest restoration subsidies, which are politically more palatable, less conflictive, and therefore less costly to implement.

6.2.2 *Voluntary approaches*

Voluntary approaches for environmental policy and programs have become increasingly popular throughout the world (OECD 1999). In many countries, landowner resentment toward increasing regulation of private property has driven governments to initiate policies and programs that utilize positive incentives to encourage cooperation with government agencies, or that appeal to landowners' sense of shared responsibility toward environmental stewardship, in order to encourage voluntary, collective restoration projects (Gottfried et al. 1996; Hodge and McNally 2000). The increased flexibility of voluntary approaches for both landowners and government agencies may result in cost savings. Landowners may choose the most efficient and cost-effective means to restore ecosystems for their particular site rather than implementing dictated restoration strategies under mandatory approaches. In addition, enforcement and monitoring costs should be lower when landowners volunteer to participate. Critics suggest, however, that relying on voluntary approaches may result in the restoration of the lowest quality lands that would not necessarily produce the highest ecological and social benefits.

Landowners or firms may voluntarily comply with environmental regulations and programs for a number of reasons. The combined financial incentives and value of improved environmental services to the landowner may be sufficiently high to offset the costs of participating. Individuals or firms may be motivated to voluntarily comply in order to project a positive image as responsible environmental stewards with their neighbors, customers, or communities. Fear of stricter regulations in the future may also motivate voluntary compliance or participation (Arora and Cason 1996; Kline et al. 2000). However, voluntary approaches based on exemptions from existing regulations or taxes probably provide greater and more credible incentives than threats to enact new regulations or taxes should voluntary approaches fail. Several authors have suggested that voluntary approaches backed by a strong regulatory framework are more likely to succeed (Alberini and Segerson 2002; Khanna 2001).

Voluntary participation in ecosystem restoration, preservation, and conservation occurs through unilateral and bilateral government programs, individual or collective initiatives, or both. The most common are unilateral government programs in which the government offers financial incentives, such as tax credits or subsidies, in exchange for certain land management actions by landowners. The second type consists of bilateral agreements between a government agency (or nonprofit organization) and landowners (either individually or as groups or communities). Conservation easements and nonmandatory, fee-simple land purchases are examples of this second type. The third type, unilateral action by one or more landowners with or without government involvement, includes collective/community action and individual efforts at environmental stewardship.

Tax incentives, financial subsidies, or cost-sharing schemes reduce private costs of restoration and increase the likelihood that the benefits to the landowner will exceed the private costs. By using positive incentives, the government attempts to guarantee the landowner a net benefit level from participating that is at least as high as without participating. This "carrot" approach has been used extensively throughout the U.S. and Europe to induce farmers and other landowners to adopt conservation practices, retire environmentally fragile lands, and restore degraded ecosystems (Cooper and Keim 1996; Cubbage et al. 1993; Granskog et al. 2002).

In the U.S., a number of federal income tax incentives and cost-share programs have been directed at improving forest management and encouraging reforestation and ecosystem restoration. These include the Forestry Incentive Program (FIP), Conservation Reserve Program (CRP), Wetlands Reserve Program (WRP), Stewardship Incentives Program (SIP), Environmental Quality Incentives Program (EQIP), and the Wildlife Habitat Incentives Program (Cubbage et al. 1993; Granskog et al. 2002). An equally wide variety of similar programs exist under the European Union and its member states (Cubbage et al. 1993; Hodge and McNally 2000; Ledoux et al. 2000; McCarthy et al. 2003; OECD 1999; Terstad 1999; Weiss 2000; Whitby and Saunders 1996).

As always, a central problem for policymakers is determining the level of subsidy or incentive that encourages the optimal level of participation. Another critical problem with these programs concerns the ability to target the sites and elicit participation from landowners that are most crucial for maximizing the benefits associated with ecosystem restoration. Obviously, landowners vary significantly in terms of their likelihood of participation and the benefits available from restoring ecosystems on their individual sites. A large literature has developed that analyzes the characteristics of landowners that affect their likelihood of participation. The most important factors include income, personal values, tract size, residence, long-term plans, knowledge of management options and benefits, tax policies, available capital, and resource commodity values (Birch 1996; Bliss et al. 1997; de Steiguer 1984; English et al. 1997; Nagubadi et al. 1996; Wicker 2002).

If targeting is difficult or impossible because government agents are unable to observe the specific characteristics of individual landowners, a second-best approach uses a policy "menu" that offers different financial incentives in return for providing different levels of restoration. This would essentially involve paying additional informational subsidies that exceed the minimum amount to induce participation if landowner characteristics were observable by the government agent (Alberini and Segerson 2002; Wu and Babcock 1999).

Conservation easements are the most common type of bilateral, voluntary mechanism for encouraging ecosystem restoration. Conservation easements are legally binding agreements between landowners and governments or nonprofit organizations that are effectively a form of shared ownership. The landowner relinquishes the right to certain landuses (or agrees to manage the land in certain ways) for a given period of time (often in perpetuity) in exchange for tax benefits or direct monetary compensation.

Conservation easements are usually less costly to acquire than fee-simple purchase of entire properties so that with a fixed amount of public or nonprofit funds, more land can typically be restored and preserved with conservation easements. Compared to tax incentives, subsidies, or tradable development rights, easements carry less administrative burden and cost and typically require fewer changes in property or environmental laws or regulations. Because they are voluntary, easements tend to be more palatable than direct land-use regulation to private landowners. However, the long-term monitoring and enforcement of easement contracts may entail large administrative costs; nevertheless, the use of conservation easements has grown rapidly. Because conservation easements often result in a fragmented pattern of restored ecosystems, they should be combined with other approaches that identify broader areas for restoration and protection (Granskog et al. 2002).

A growing group of scientists, land managers, and policymakers suggest that effective ecosystem restoration must be planned at a landscape or watershed level (Gottfried et al.

1996; Hodge and McNally 2000; Kline et al. 2000; NRC 2001). Successful restoration efforts will depend not only on how landowners individually react to government incentives but also whether neighboring landowners and communities participate and can work together to maximize the benefits from restoration. For example, Hodge and McNally (2001) state that given the technical constraints and costs of water quality management, "wetland restoration will not be possible unless all producers cooperate (within a specific location)."

Watershed councils are used widely throughout the Midwestern U.S. to develop and implement watershed restoration plans and to coordinate management of riparian areas (Gottfried et al. 1996). In Ecuador, fishers, shrimp aquaculturists, and a variety of landowner and conservation groups cooperate to develop agreements to improve riparian areas and water quality, such as restricting harvesting of mangroves. In Costa Rica, landowners cooperate to develop ecotourism reserves in coastal mountain regions (Gottfried et al. 1996). In the Philippines, socioeconomic factors such as a knowledge of trees and tree planting, land-use patterns and ownership, and social and community organizations were more important than ecological factors in determining the relative success of forest restoration efforts (Walters 1997).

Individuals or communities may be motivated to unilaterally initiate restoration activities by environmental stewardship, personal satisfaction, or utility gained from improving the environment. Recent research in experimental economics suggests that effective communication and high marginal payoffs are essential to achieving voluntary provision of public goods and cooperative actions within a landscape (Ledyard 1995). Therefore, successful cooperative restoration efforts will most likely require some external agent (government or nongovernmental) that facilitates communication between landowners, provides information, incurs transactions costs, and provides resources to lower the costs to individual landowners for participating (Hodge and McNally 2000).

One would expect that if incentives such as tax reductions and subsidies are high enough, most landowners in a watershed or landscape would participate in coordinated efforts at ecosystem restoration. Nevertheless, collective provision of ecological services requires significant and costly information on technical requirements for ecosystem restoration. Indeed, the transaction costs involved in acquiring information and negotiating agreements between landowners may be so high that collective restoration efforts will not succeed, even with government incentives (Hodge and McNally 2000).

An example of such an approach in the U.S. is the Oregon Coastal Salmon Restoration Initiative (OCSRI). A coalition of state agencies and private interest groups developed OCSRI in order to avoid the consequences of the Pacific Northwest Coho salmon being listed as threatened under the Endangered Species Act (Kline et al. 2000). By appealing broadly to Oregonians' collective responsibility to protect salmon habitat, the initiative relied on community-based, voluntary efforts by private landowners, local interest groups such as watershed councils, and soil and water conservation districts. The actions by forest landowners to reduce timber harvests in critical riparian buffer areas were crucial for successful restoration. Apparently, a high proportion of forest landowners in Oregon are motivated to own land for protecting and enhancing habitat for threatened or endangered species, as well as timber production (Kline et al. 2000), suggesting a high potential for success. However, the owners with the largest tracts of forestland and the largest percentage of forestland were motivated primarily by timber production. Therefore, a combined policy that provides economic incentives such as tax relief or cost sharing may be required to induce cooperation of a sufficient number of landowners to achieve success (Kline et al. 2000).

Internal Drainage Boards (IDB) in the U.K. evolved from informal associations (clubs) of local landowners organized to improve land drainage into formal institutions with statutory responsibilities (established in 1976). They now act like a local government organization representing the competing interests of local residents. Although the IDBs are well positioned to promote watershed-based collective action for restoring wetland ecosystems, several reforms are needed (Hodge and McNally 2000). The current system of using standard contracts to landowners severely limits flexibility and reduces the capacity to develop new cooperative actions by landowners. Even with institutional reform of the IDBs, collective action will be limited unless financial incentives are sufficiently high to encourage participation of adjacent farmers. Hodge and McNally (2000) recommended allowing groups of landholders to compete for restoration grants and contracts in order to promote larger cooperative restoration projects.

6.3 Measuring success

A variety of criteria have been used or suggested for evaluating the success of government policies and programs for forest restoration (Cubbage et al. 1993). These include physical measures (e.g., number of ha restored), measures of ecological function (habitat, water quality, biodiversity), and socioeconomic measures (efficiency, cost effectiveness, and equity). All of these measures involve different ways of determining and comparing benefits and costs (ecological, economic, and social) of program and policy alternatives. Although measuring the direct costs of restoration programs is relatively straightforward (Boyd and Simpson 1999), measuring the benefits is more difficult and is the issue addressed here.

In addition to producing and supporting species and biological and ecological functions, forest ecosystems produce and support a wide variety of socially valuable services such as water quality, erosion and flood control, wildlife habitat, recreation, biodiversity, and aesthetics. When ecosystems are degraded and subsequently restored, the social value of the services produced also changes. Ecosystem services are the beneficial outcomes of ecosystem functions. For example, reducing runoff, erosion, and flood peaks are biophysical functions of a watershed restoration project. Examples of the services they provide include reducing damage to agriculture, water quality, buildings, and roads. Even though a restored ecosystem may have high rates of ecosystem function, it may not necessarily produce ecosystem services with high social values (Boyd and Wainger 2002b). Therefore, evaluating the effectiveness of ecosystem restoration requires more than just good ecological analyses. Evaluating the success of alternative policies and programs for restoring ecosystems requires comparing the social value of the change in ecosystem services to the costs of implementation, management, monitoring, and enforcement.

The problem of how to estimate benefits can be divided into supply and demand components (Johnston et al. 2002). The supply component consists of estimating how policy and program alternatives affect the quantities of ecosystem services and the demand side establishes how people and society value the various services. Most current governmental programs rely on purely biophysical descriptions to quantify the success of supplying ecological functions from restoration projects. Further, these estimates are often very simple, such as area treated. The demand or value side is rarely assessed (Polasky 2002). In the absence of adequate assessments of the value of the services produced by restored ecosystems, regulators and program administrators tend to use differences in the program costs as the primary criteria for success. However, if restoration efforts are evaluated solely on area restored (or treated) and cost effectiveness, most restoration will occur on the least expensive, least ecologically valuable lands. The result would likely be a migration of restored habitats and ecosystems to remote, cost-effective sites, an outcome that is unlikely to be ecologically or economically optimal (Boyd and Wainger 2002b).

Current regulatory land-use programs in the U.S. under analyze the social value of both the lost and restored ecosystems under compensatory mitigation programs (Boyd and Wainger 2002a, 2002b). Usually, mitigation decisions are based solely on requiring restoration of an "acre for an acre" of a biophysically similar site when the landowner seeks to develop or otherwise negatively impact an ecosystem (usually wetlands). Unfortunately, this simplistic approach fails to consider how society values the destroyed and restored ecosystems. Indicators of social value include the location in the greater landscape, availability and importance of local substitutes and complements to the site, and future risks to the restored site's ability to continue to provide services.

Surveys and econometric analyses are used to develop evidence and estimates of the demand for and resulting social value of different ecosystem services (Costanza et al. 1998; Holmes et al. 2004; Kenyon and Nevin 2001; Kline et al. 2000; Loomis et al. 2000; Lupi et al. 2002; Schaberg et al. 1999). Unfortunately, these analyses tend to be difficult, costly, and incomplete; rarely do they value the full range of ecosystem services at a site. In response, researchers are beginning to develop alternative approaches by combining economic and ecologic indicators and attributes that identify potential differences in social benefits generated by restored ecosystems on a landscape level (Boyd and Wainger 2002a, 2002b; Johnston et al. 2002). These efforts require credible models of the production of ecosystem services that link models from the natural sciences (which predict how changes in ecosystems impact the services produced by those ecosystems) with economic models (which estimate the value of those services and predict landowner participation). Since the value of services from any particular site depends on both the conditions at the site as well as its location on the landscape, these models will need to operate at landscape level (Polasky 2002).

6.4 Conclusions

Different approaches taken by governments to encourage restoration will differ in their implementation costs (transactions costs), in who ultimately bears the cost of restoration, and in the ability to select and maintain appropriate habitats and ecosystems. Policymakers have the difficult task of making trade-offs between how much to spend, who pays, and how to ensure successful restoration. For example, fee-simple purchase by the government of lands for restoration provides the lowest risk of failure and the best targeting of ecosystem restoration. However, the costs are very high and political perils may be even higher if the government begins a process of requiring large numbers of private landowners to sell their lands to the government. In contrast, a broad-based program of tradable development rights (e.g., compensatory mitigation) may result in many restoration commitments on paper, but significant ongoing problems with compliance. Voluntary approaches tend to be more politically palatable and usually less expensive, but often result in a fragmented landscape of restored and unrestored sites with less than optimal production of ecological services.

All this suggests that there is no restoration policy panacea. All restoration policies and programs involve trade-offs. A combined approach of carrots (voluntary approaches) and sticks (mandatory or regulatory approaches) will usually be required to achieve maximum production of ecological services at the lowest cost. The exact combination of efforts needed to optimize restoration efforts will depend on local political, economic, social, and ecological conditions and require significant and difficult economic and ecologic modeling and analysis.

Perhaps the greatest unresolved policy issue, no matter which policy or combination of policies is used, concerns targeting and choosing the properties for restoration that will produce the greatest ecological benefits at the lowest societal cost. The use of a spatial, landscape, or watershed basis for planning and implementing restoration policies and programs is very complicated and difficult due to the interdependence of habitats and properties across a landscape, scientific debate over restoration priorities, and the inherent difficulties in predicting human development patterns across a landscape (Boyd and Simpson 1999).

The easiest approach to the problem of ensuring sufficient restoration across a landscape would be through public lands. However, public lands are generally too small to provide the necessary economies of configuration (Gottfried et al. 1996). Thus, creative policy and program approaches are still needed and usually require a combination of targeted voluntary and mandatory policy instruments combined with community-based, cooperative approaches between government agencies, nonprofit organizations, and landowner groups and organizations to effectively encourage the majority of landowners in a landscape or watershed to participate. Obviously, this type of approach is complicated, costly, and challenging; it requires considerable understanding of the social dynamics that promote or obstruct alternative institutional solutions to ecosystem restoration at a landscape level. Considerable social, economic, and institutional research and political support will be necessary to turn this dream into a reality.

References

Alberini, A. and Segerson, K., Assessing voluntary programs to improve environmental quality, *Environ. Resour. Econ.*, 22, 157, 2002.

Arora, S. and Cason, T., Why do firms volunteer to exceed environmental regulations? Understanding participation in EPA's 33/50 program, *Land Econ.*, 72, 413, 1996.

Birch, T.W., *Private Forest Landowners in the Southern United States, 1994*, U.S. Department of Agriculture, Forest Service, Northeastern Forest Experiment Station, Radnor, PA, Resource Bulletin NE-138, 195 pp., 1996.

Bliss, J.C., Nepal, S.K., Brooks, R.T., and Larsen, M.D., In the mainstream: environmental attitudes of Midsouth forest owners, *J. For.*, 21, 37, 1997.

Boyd, J. and Simpson, R.D., Economics and biodiversity conservation options: an argument for continued experimentation and measured expectations, *Sci. Total Environ.*, 240, 91, 1999.

Boyd, J. and Wainger, L., Landscape indicators of ecosystem service benefits, *Am. J. Agric. Econ.*, 84, 1371, 2002a.

Boyd, J. and Wainger, L., Measuring ecosystem service benefits for wetland mitigation, *Nat. Wetlands Newsl.*, 24, 1, 11, 2002b.

Cooper, J.C. and Keim, R.W., Incentive payments to encourage farmer adoption of water quality protection practices, *Am. J. Agric. Econ.*, 78, 54, 1996.

Costanza, R., d'Arge, de-Groot, R., Farber, S., Grasso, M., Hannon, B., Limburg, K., Naeem, S., O'Neil, R., Paruelo, J., Raskin, R., Sutton, P., and van den Belt, J., The value of ecosystem services and natural capital, *Ecol. Econ.*, 25, 3, 1998.

Cubbage, F.W., O'Laughlin, J., and Bullock, C.S., *Forest Resource Policy*, John Wiley, New York, 562 pp., 1993.

De Steiguer, J.E., Impact of cost-share programs on private reforestation investment, *Forest Sci.*, 30, 697, 1984.

English, B.C., Bell, C.D., Wells, G.R., and Roberts, R.K., Stewardship incentives in forestry: participation factors in Florida, *South. J. Appl. For.*, 21, 5, 1997.

Fernandez, L., An analysis of economic incentives in wetlands policies addressing biodiversity, *Sci. Total Environ.*, 240, 107, 1999.

Fernandez, L. and Karp, L., Restoring wetlands through wetlands mitigation banks, *Environ. Resour. Econ.*, 12, 323, 1999.

Gottfried, R., Wear, D., and Lee, R., Institutional solutions to market failure on the landscape level, *Ecol. Econ.*, 18, 133, 1996.

Granskog, J.E., Haines, T., Greene, J.L., Doherty, B.A., Bick, S., Haney, H.L., Moffat, S.O., Speir, J., and Spink, J.J., Policies, regulations, and laws, in *Southern Forest Resource Assessment*, Wear, D.N. and J.G. Greis, Eds., U.S. Department of Agriculture, Forest Service, Southern Research Station, Asheville, NC, General Technical Report SRS-53, 189, 2002.

Heal, G., Markets and sustainability, *Sci. Total Environ.*, 240, 75, 1999.

Hodge, I. and McNally, S., Wetland restoration, collective action, and the role of water management institutions, *Ecol. Econ.*, 35, 107, 2000.

Holmes, T., Bergstrom, J.C., Huszar, E., Kask, S.B., and Orr, III, F., Contingent valuation, net marginal benefits, and the scale of riparian ecosystem restoration, *Ecol. Econ.*, 49, 1, 2004.

Johnston, R.J., Magnusson, G., Mazzotta, M.J., and Opaluch, J.P., Combining economic and ecologic indicators to prioritize salt marsh restoration actions, *Am. J. Agric. Econ.*, 84, 1362, 2002.

Kenyon, W. and Nevin, C., The use of economic and participatory approaches to assess forest development: a case study in the Ettrick Valley, *Forest Policy Econ.*, 3, 69, 2001.

Khanna, M., Non-mandatory approaches to environmental protection, *J. Econ. Surv.*, 15, 291, 2001.

Kline, J.D., Alig, R.J., and Johnson, R.L., Forest owner incentives to protect riparian habitat, *Ecol. Econ.*, 33, 29, 2000.

Ledoux, L., Crooks, S., Jordan, A., and Turner, R.K., Implementing EU biodiversity policy: UK experiences, *Land Use Policy*, 17, 257, 2000.

Ledyard, J.O., Public goods: a survey of experimental research, in *The Handbook of Experimental Economics*, Kagel, J.H. and A.E. Roth, Eds., Princeton University Press, Princeton, NJ, 1995.

Loomis, J., Kent, P., Strange, L., Fausch, K., and Covich, A., Measuring the total economic value of restoring ecosystem services in an impaired river basin: results from a contingent valuation survey, *Ecol. Econ.*, 33, 103, 2000.

Lupi, F., Kaplowitz, M.D., and Hoehn, J.P., The economic equivalency of drained and restored wetlands in Michigan, *Am. J. Agric. Econ.*, 84, 1355, 2002.

McCarthy, S., Matthews, A., and Riordan, B., Economic determinants of private afforestation in the Republic of Ireland, *Land-Use Policy*, 20, 51, 2003.

Nagubadi, V., McNamara, K.T., Hoover, W.L., and Mills, Jr., W.L., Program participation behavior of nonindustrial forest land owners: a probit analysis, *J. Agric. Appl. Econ.*, 28, 323, 1996.

NRC, *Restoration of Aquatic Ecosystems: The Status of Science, Technology, and Public Policy*, National Research Council, National Academy Press, Washington, D.C., 1992.

NRC, *Compensating for Wetland Losses under the Clean Water Act*, Committee on Mitigating Wetland Losses, National Research Council, National Academy Press, Washington, DC, 305 pp., 2001.

OECD, *Voluntary Approaches for Environmental Policy: An Assessment*, Organisation for Economic Cooperation and Development, Paris, 143 pp., 1999.

Polasky, S., The economics of wetland ecosystem restoration and mitigation: discussion, *Am. J. Agric. Econ.*, 84, 1379, 2002.

Race, M.S., Critique of present wetland mitigation policies in the United States based on analysis of past restoration projects in San Francisco Bay, *Environ. Manage.*, 9, 71, 1985.

Race, M.S. and Fonseca, M.S., Fixing compensatory mitigation — what will it take? *Ecol. Appl.*, 6, 94, 1996.

Reppert, R., *National Wetland Mitigation Banking Study: Wetlands Mitigation Banking Concepts*, Institute of Water Resources, Water Resources Support Center, U.S. Army Corps of Engineers, IWR Report 92-WMB-1, 25 pp., 1992.

Roberts, L., Wetland trading is a loser's game, says ecologist, *Science*, 260, 1890, 1993.

Shaberg, R.H., Holmes, T.P., Lee, K.J., and Abt, R.C., Ascribing value to ecological processes: an economic view of environmental change, *For. Ecol. Manage.*, 114, 329, 1999.

Terstad, J., Swedish experiences of incentives for the protection of nature, *Sci. Total Environ.*, 240, 189, 1999.

Walters, B.B., Human ecological questions for tropical restoration: experiences from planting native upland trees and mangroves in the Philippines, *For. Ecol. Manage.*, 99, 275, 1997.

Weiss, G., Evaluation of policy instruments for protective forest management in Austria, *For. Policy Econ.*, 1, 243, 2000.

Whitby, M. and Saunders, C., Estimating conservation goods in Britain, *Land Econ.*, 72, 313, 1996.

Wu, J. and Babcock, B.A., The relative efficiency of voluntary vs. mandatory environmental regulations, *J. Environ. Econ. Manage.*, 38, 158, 1999.

Wicker, G., Motivation for private forest landowners, in *Southern Forest Resource Assessment*, Wear, D.N. and J.G. Greis, Eds., U.S. Department of Agriculture, Forest Service, Southern Research Station, Asheville, NC, General Technical Report SRS-53, 225, 2002.

Zedler, J.B., Coastal mitigation in Southern California: the need for a regional restoration strategy, *Ecol. Appl.*, 6, 84, 1996.

chapter seven

Carbon sequestration through restoration of temperate and boreal forests

Lorenzo Ciccarese
Italian Agency for the Environmental Protection and Technical Services
Sandra Brown
Winrock International Ecosystem Services Unit
Bernhard Schlamadinger
Joanneum Research

Contents

7.1 Introduction

Both human and natural disturbances affect carbon storage and flux in forested ecosystems, including harvesting, conversion to agriculture or pasture, wildfire and fire suppression, and pest and disease outbreaks. Forests are sources of atmospheric carbon dioxide (CO_2) when total respiration or oxidation of plants, soil, and dead organic matter exceed net primary productivity. Conversely, forests can become carbon sinks when agricultural land and pasture are abandoned and revert naturally to forests, or are restored to native forests or plantations through afforestation. Aggrading forest ecosystems also fix more carbon than they respire. Forests play an important role in the global carbon cycle because they store large quantities of carbon in vegetation and soil, exchange carbon with the atmosphere through photosynthesis and respiration, are sources of atmospheric carbon when they are disturbed, and become atmospheric carbon sinks (i.e., net transfer of CO_2 from the atmosphere to the land) during regrowth after disturbance (Brown et al. 1996a). Humans have the potential to alter the magnitude of forest carbon stocks and the

1-56670-635-1/05/$0.00+$1.50

direction of forest carbon fluxes through changes in landuse and management, and thus alter the role of forests in the global carbon cycle.

The global carbon cycle is one of the major biogeochemical cycles because it regulates the concentration of CO_2, the most important greenhouse gas (GHG) in the atmosphere. The concentration of atmospheric CO_2 has increased from around 280 parts per million by volume (ppmv) since the start of the industrial era, to about 370 ppmv currently. The observed global warming of about 0.75°C in the past century is attributed to the increase in CO_2 and other GHGs (Hansen et al. 2000; Hansen and Sato 2001). During the 1990s, fossil fuel combustion emitted 6.3 ± 0.4 gigatons of carbon (Gt C) per year (Prentice et al. 2001) and was the main cause of increased atmospheric CO_2. Nevertheless, degradation and loss of forests caused significant CO_2 flux to the atmosphere (Dixon et al. 1994; Fearnside 2000; Prentice et al. 2001; Houghton 2003). During the 1990s, tropical deforestation and forest degradation resulted in an estimated average annual CO_2 release to the atmosphere of 1.6 ± 0.8 Gt C, or about 25% of total global anthropogenic carbon emitted. On the other hand, terrestrial ecosystems sequester globally 0.7 ± 1.0 Gt C; the carbon residual without deforestation was 2.3 ± 1.3 Gt C (IPCC 2000), due in part to effects of CO_2 or nitrogen fertilization largely in the temperate and boreal zones.

Landuse change and forestry activities (dubbed LUCF in the shorthand of international climate change policy) can mitigate carbon emissions by conserving or expanding existing carbon storage in forest ecosystems and by substituting sustainably grown wood for energy-intensive and cement-based products. Forest restoration clearly has a role in these efforts; a variety of specific actions that expand carbon stocks have a restoration component. Examples include increasing the area or carbon density of forests by afforestation, rehabilitating degraded forests, and altering the management of natural forests and plantations to sequester more carbon longer by lengthening rotations and reducing logging impacts (Myers and Goreau 1991; Brown et al. 1996b; Kauppi and Sedjo 2001). Most of these practices have positive social, economic, and ecological benefits in addition to their possible role in mitigating climate change (Brown et al. 1996b), and payments for the carbon conserved and sequestered may provide incentives for restoration.

Implementing an aggressive program to change forestry practices on about 700 million ha over the next 50 years could avoid and remove emissions of about 60 to 87 billion tons of carbon (Brown et al. 1996a; Kauppi and Sedjo 2001) or the equivalent of 12 to 15% of the "business-as-usual" fossil fuel emissions over the same period (Table 7.1). For the temperate and boreal zone forests, only activities that sequester new carbon are included in this estimate; the potential for conserving carbon through changes in management has not been estimated for these regions. Some 215 million ha of the temperate and boreal zone are technically suitable for afforestation and reforestation activities, including agroforestry on some agricultural lands. These lands are potentially capable of sequestering about

Table 7.1 Global Estimates of the Potential Amount of Carbon that could be Sequestered and Conserved by Forestry Practices from 1995 to 2050 (Brown et al. 1996a)

Latitudinal belt	Practice	Area (million ha)	Amount of carbon (Gt)
Boreal	Afforestation/reforestation	95	2.4
Temperate	Afforestation/reforestation	113	11.8
	Agroforestry	7	0.7
Tropics	Afforestation/reforestation	67	16.4
	Agroforestry	63	6.3
	Regeneration	217	11.5–28.7
	Slow deforestation	138	10.8–20.8
Total		700	60–87

15 billion t C over the 50-year period, which provides a rough estimate of the potential contribution of forest restoration in temperate and boreal forests. Our objective in this chapter is to provide an overview of the relationship between carbon sequestration and forest restoration. The specific purposes are to summarize policy and technical issues surrounding implementation of land-use change and forestry activities for mitigating GHG emissions, to describe the tools and techniques used for measuring, monitoring, and estimating the carbon benefits, and to describe some existing forest restoration carbon projects.

7.2. United Nations climate convention and Kyoto Protocol

The United Nations Framework Convention on Climate Change (UNFCCC) was adopted in 1992 and ratified by almost 200 countries. The UNFCCC recognized climate change as a serious threat, established a framework to address the increasing concentrations of GHGs in the atmosphere and to mitigate its adverse effects, but does not commit signatories to limit GHG emissions. The UNFCCC urged parties to adopt measures to enhance and conserve ecosystems such as forests that act as sinks and reservoirs of GHGs. The Kyoto Protocol (KP), adopted in 1997, supplements the UNFCCC with an enforceable agreement with quantitative targets for reducing GHG emissions. This protocol limits emissions of six major GHGs (CO_2, CH_4, N_2O, and three industrial gases) by industrialized countries, on average by 5.2% below the 1990 level, for the first commitment period 2008 to 2012.

Forestry activities were included in the KP to mitigate carbon emissions and as a mechanism for countries to meet their commitments to reduce net emissions to the atmosphere. Several Articles in the KP refer to human-induced land-use change and forestry activities: Article 3.3 refers to emissions by sources and removals by sinks resulting from afforestation, reforestation, and deforestation activities since 1990; Article 3.4 refers to additional activities in forest management (among other types of landuse); Article 6 (Joint Implementation [JI]) refers to the trading of emissions reduction units in any sector of the economy; and Article 12 (Clean Development Mechanism [CDM]) refers to emission-offset trading between developed and developing countries. CDM limits forestry activities to afforestation and reforestation and JI and CDM projects must lead to emission reductions or carbon sequestration that are additional to what would have occurred without the JI or CDM funding. All Articles require real, measurable, and verifiable changes in carbon stocks; for JI and CDM projects, this means a comparison of actual emissions or sequestration with and without the project.

Including forestry activities has proven complex and controversial, requiring political negotiation and technical investigation (Schneider 1998; Schlamadinger and Marland 2000). For example, very basic questions, such as what constituted a forest, had to be negotiated, as this could heavily affect consistency of methodologies and comparability among nations. Other major issues were the extent to which management of existing forests could be included, and the rules for the accounting and inventorying emissions and removals from land-use activities. To help clarify the science behind these issues and aid negotiators, the Intergovernmental Panel on Climate Change (IPCC) prepared a Special Report on Land-Use, Land-Use Change and Forestry (LULUCF) that was used to reach a final agreement on land-use activities in the KP (IPCC 2000). Decisions reached in Marrakech in 2001 have implications for forest restoration. For example, countries were given some flexibility in defining the term forest, to account for national circumstances and data availability. What constitutes a forest determines the area potentially available for afforestation, reforestation and deforestation, but also forest management. A country may choose a minimum area of land that qualifies as forest between 0.05 and 1.0 ha, with a minimum tree crown cover between 10 and 30%, and a minimum tree height between 2 and 5 m at maturity. Lands that temporarily fail to meet these thresholds, but are expected to do so under continuation of current management, are included under "forest" (e.g., if they were clearcut-harvested and regenerated).

Once a parcel of land enters the accounting system because it is subject to one of these activities, all future GHG emissions and removals on this land have to be accounted for. The accounting of GHGs in KP's first commitment period encompasses the time period from 2008 to 2012. Indirect human-induced and natural carbon emissions and removals, including those due to CO_2 or nitrogen fertilization of forests, are excluded from crediting. For now, this is met by a cap on the amount of credits a country can claim from its managed forests. Researchers are investigating whether indirect and natural effects can be separated from anthropogenic impacts on forest carbon stocks.

7.3 Scientific and technical issues surrounding carbon sequestration through forest restoration

Several technical and operational concerns have been raised about landuse and land-use forestry activities and projects and how to properly account for carbon emissions and sequestration. Some of these concerns are specific to land-use projects, but others apply generally to mitigation efforts to offset or reduce GHG emissions.

7.3.1 Leakage

Leakage refers to indirect effects of a mitigation project on GHG emissions outside the project boundary, due to actions that are usually beyond the control of project participants. Leakage includes emissions caused by the displacement of activities, such as agricultural production, or increases of carbon stocks in one place caused by decreases in another. Leakage is likely to occur when a project prevents an activity that provided products or employment locally, without offering an alternative (IPCC 2000). For example, afforestation can displace agricultural use of the land without decreasing demand for agricultural products. Thus, production simply shifts elsewhere, possibly leading to deforestation or lower rates of forest establishment. Similarly, timber produced from reforestation projects increases the global supply of timber, thus leading to lower timber prices, and potentially reducing the attractiveness of forestry uses of land.

Project design or accounting as part of project monitoring can address the risk of leakage (Brown et al. 1997). Leakage can be accounted for by project-specific research, or through default values such as discounting carbon credits. Project design can address leakage by ensuring that the project includes provisions for producing the displaced goods and services. For example, an afforestation project on marginal agricultural land could include retaining part of the project land in agriculture and improving production methods in order to maintain the same overall yields. However, in many developed countries of the boreal and temperate zones, forest restoration on agricultural land set aside to reduce agricultural outputs would likely lead to minimal leakage.

7.3.2 Additionality

Additionality can be a concern for any type of carbon-offset project, not only landuse projects. Additionality is the requirement that carbon credits represent real improvements over what would have happened anyway, in the absence of carbon financing. Thus, carbon sequestration in a reforestation or restoration project must be quantified by comparing it to the carbon uptake or carbon emissions of a baseline. Additionality is explicitly required for JI and CDM projects. It is less of an issue for restoration of natural forests, where the requirement can easily be fulfilled and verified. Restoration may yield environmental, socioeconomic, and other benefits that are not captured financially by the landowner, thus making restoration unlikely without the additional income from carbon credits. The situation is more difficult for commercial plantations, which would receive income from wood production, thus making them economically viable without carbon financing and a likely baseline scenario. Leakage and additionality pose no significant differences between land-use projects and efforts to reduce fossil-fuel emissions.

7.3.3 *Nonpermanence*

Carbon sequestered in a forest restoration project can be released back to the atmosphere if the forest is disturbed by natural or anthropogenic events such as harvesting or deforestation, fires, pests and diseases, or storms. This is frequently referred to as "nonpermanence." Reductions in emissions from fossil fuels through fuel switching or energy efficiency improvements, on the other hand, do not lead to increased emissions later, should the project be discontinued. For example, replacing coal-based electricity with electricity from woody biomass for 20 years will result in a certain amount of emission savings. Even if energy production switches back to coal from a renewable fuel, a permanent savings will have been made of 20 years worth of emissions from coal combustion.

Forest-level management and planning can ensure that there is no net release of carbon, although fluxes of carbon between the forest and the atmosphere will certainly occur. In a regulated forest, carbon stocks of the entire forest remain constant or even increase, and a few forest units are harvested each year. Just as annual volume harvests are scheduled to remove less than annual volume growth, managers can plan regular harvesting and tending activity to maintain carbon stocks. Because natural and anthropogenic disturbances can differ in severity, frequency of occurrence, and predictability, they may affect this balance in unforeseen ways. A manager may wish to reduce the risk of noncompliance by planning to harvest less than growth.

Nonpermanence is an issue for CDM afforestation and reforestation projects. A contentious issue is how to account for the reemission of carbon from new forests in developing countries. In December 2003, at the ninth session to the Conference of the Parties to the UNFCCC (COP9), an important decision was agreed to for addressing nonpermanence — as well as additionality, leakage, uncertainties, and socioeconomic and environmental impacts — associated with afforestation and reforestation project activities under CDM. Carbon credits from CDM projects are considered temporary, so that upon expiration of the credits, or when the carbon is released back to the atmosphere prematurely, the investor must renew the credits from the same project, acquire credits from another project/seller, or reduce internal emissions correspondingly.

7.3.4 *Saturation*

Ultimately, the scope for increasing carbon stocks in vegetation will reach ecological or practical limits, a phenomenon termed saturation. The global potential for carbon sequestration through forest restoration (or any other type of land-use activities) is limited by the amount of available land and by the amount of carbon that can be stored on that land (Koskela et al. 2000). By contrast, renewable energy sources can displace emissions from fossil fuel in electricity production as long as a power station operates. In principle, this displacement is not limited by how long fossil fuels are available. Afforestation and reforestation projects can effectively address saturation constraints by producing biofuels and wood products that replace fossil fuels and energy-intensive nontimber building materials such as steel or concrete, thereby expanding the scope for emission reductions and allowing repeated CO_2 emissions savings.

7.4 Monitoring and estimation of carbon uptake through forest restoration

Much experience has been gained from forest-based carbon-offset projects with respect to measuring, monitoring, and accounting for carbon benefits (Brown et al. 2000a). Identifying the correct carbon stocks to quantify, accurately measure them to an often predetermined level of precision, and monitor them over the length of the project are significant challenges. The major carbon pools in forestry projects are above-ground biomass, below-ground biomass, litter, deadwood, and soil organic carbon. Techniques and methods for measuring and monitoring terrestrial carbon pools based on commonly accepted principles of forest inventory, soil sampling, and ecological surveys are well established (Pinard and Putz 1997; MacDicken 1997a; Post et al. 1999; Brown et al. 2000a, 2000b; Brown 2002a; Segura and Kanninen 2002). These methods can all be applied to project-level measuring and monitoring of the key carbon pools in forestry projects.

Landuse and forestry projects are generally easier to quantify and monitor than national inventories, due to clearly defined boundaries for project activities, relative ease of stratification of project area, and choice of carbon pools to measure (Brown et al. 2000b). Criteria affecting the selection of carbon pools to inventory and monitor include the type of project, pool size and its rate of and direction of change, availability of appropriate methods to attain required accuracy and precision, and the cost to measure (MacDicken 1997a, 1997b). The carbon credits from a project for all measured pools are essentially the difference between the carbon pools in the with-project case minus the without-project or baseline case.

The difference for some pools clearly will be positive, while for other situations the difference will be negative, depending on the starting point. For example, lengthening a forest rotation will store more carbon in trees on average (with-project) than a shorter rotation (without-project); the deadwood pool in a reduced-impact logging project will be less than in a conventional logging. To make measurements tractable, a manager can use a selective or partial accounting system that includes all pools expected to decrease and a choice of pools expected to increase as a result of the project, relative to the without-project baseline. Only pools that are measured (or estimated from a measured parameter) and monitored should be incorporated into the calculation of carbon benefits.

The desired precision in measuring and monitoring carbon must be balanced with measurement cost, and this is affected by the variability in carbon stocks on the project lands and the magnitude of expected change. More plots are needed to attain a desired precision level if variability is high, increasing the cost of measuring and monitoring. Stratifying the project land into a reasonable number of relatively homogeneous units can reduce the number of plots needed for a given level of precision. Generally, costs are a function of the number of pools to be measured and monitored, the frequency of monitoring, targeted precision level, and the complexity of monitoring methods. The frequency of monitoring is related to the rate and magnitude of change; less frequent monitoring is appropriate if only a small change is expected.

Experience to date with large projects implemented by The Nature Conservancy in Belize, Brazil, and Bolivia (Brown et al. 2000a; Brown 2002a) has shown that for a modest cost, a properly designed project with the number of plots needed to attain a given precision level can measure the carbon stocks in all components within a 95% confidence interval that is 5 to 7% of the mean. For example, a tropical forest conservation project in Bolivia measured six strata on 625 plots in a 634,000 ha project area and obtained an estimated total carbon stock of about 115 M t C, with a 95% confidence interval of 4.2% of the mean at a cost of less than one cent per t carbon (Brown et al. 2000a). Similar results have been obtained for other projects (e.g., 150 plots for approximately 14,000 ha of tropical forests in Belize and 140 plots for a 4,000 ha project in Brazil). In sum, well-designed studies (stratified with a predetermined number of plots installed based on variance of the strata) can measure carbon stocks and their change, to within <10% of the mean with 95% confidence at a modest cost of no more than US$0.25/t C. Advances in remote sensing offer another reliable and readily applicable source of information for the measurement, monitoring, and verification of forest area, as well as above-ground biomass and thus carbon (Brown 2002b).

In 2003, the IPCC finalized the "Good Practice Guidance for LandUse, Land-Use Change and Forestry," which shall be used by Annex I countries (signatories to the United Nations Framework Convention on Climate Change; list available at <http://www.ictp.trieste.it/~moltenif/UN_climate/annex1.htm>)for preparing inventories under the Convention and the KP for the first commitment period. It contains methodologies for measuring, estimating, and reporting GHG emissions and removals from forest restoration activities under Articles 3.3 and 3.4, as well as forestry projects under JI and CDM.

7.5 Restoration with a carbon component

More than 40 reforestation projects were established by 2001, totaling more than 4 million ha (Moura-Costa 2001). These projects were voluntarily established in anticipation of expected changes in national environmental legislation that would require polluters to reduce GHG emissions. Third-party independent organizations have certified the carbon sequestered by many of these voluntary initiatives. GHG emitters have funded many of these projects to gain experience, in light of future prospects of offsetting their GHG emissions (Trexler et al. 1999). Sometimes the investments of individual companies or investors are pooled into carbon investment funds. For example, the World Bank has created the Prototype Carbon Fund (PCF) and the BioCarbon Fund (BCF) (www.biocarbonfund.org). The PCF includes mostly energy-related projects with a couple of landuse and forestry projects. The new BCF provides carbon financing for projects that sequester or remove GHGs in forest and agricultural ecosystems. The BCF aims to deliver cost-effective carbon emission reductions, while promoting biodiversity, conservation, and sustainable development. The target size of the BCF is US$100 million.

Forest restoration projects for carbon sequestration are a reality and are already being implemented. Some examples of forest restoration projects for carbon sequestration include:

1. The Italian Inter-Ministerial Committee for Economic Planning adopted a National Plan for reducing GHGs in December 2002, establishing a national strategy for fulfilling its KP commitments. The policy allocated landuse and forestry options a sequestration capability of 10.2 Mt CO_2 equivalents, which corresponded to 11% of Italy's emissions reduction commitments. Forest management will play a major role (4.1 Mt CO_2), as well as afforestation and reforestation (3 Mt CO_2). Almost 110,000 ha of new stands will be established by the end of the first commitment period, funded by European Union forestry programs and by the Soil Conservation Act (300m Euro).
2. Rusafor-Sap is a Russia–U.S.A. collaborative effort, established in the Saratov region of Russia. The project consists of afforestation and reforestation of four areas of marginal agricultural land and burned forest stands, totaling 900 ha. The expected lifetime of the project is 60 years; carbon sequestration of 200 Mt C is expected. Other objectives of the project are wildlife habitat preservation through the restoration of native trees, encouragement of ecotourism, and local sustainable timber production.
3. The Scottish Forest Alliance launched a forest reforestation program in 2002. The Alliance is collaborating with British Petroleum, Forest Enterprise (a woodland management agency that manages the nationally owned forest estate), the Royal

Society for the Protection of Birds Scotland, and the Woodland Trust Scotland. This program aims to enhance forest biodiversity and regenerate and expand native woodland in Scotland, promote social and economic gains for local communities, and contribute to carbon sequestration in Scotland. BP has committed £1m per annum for 10 years to the project (www.scottishforestalliance.org.uk).

4. The State Forests of the Forestry Commission of New South Wales, Australia signed an agreement in 2002 with an overseas-based company to establish 1,000 ha of new forest in order to reduce GHG reduction. An ongoing program of between 10,000 and 40,000 ha is planned over the next 10 years. To date, some $2.3m has been received from the company under this initiative.
5. PRIMAKLIMA has successfully reforested and implemented CO_2-related ecological measures on more than 2,600 ha of land all over the world, of which 1,800 ha are located in Germany (http://www.prima-klima-weltweit.de/english/intro.php3?top=english).
6. Significant reforestation activities are taking place in the Lower Mississippi Alluvial Valley (LMAV) of the U.S.A. Over the last hundred years, millions of ha of native bottomland hardwood forest were converted to soybean and cotton production

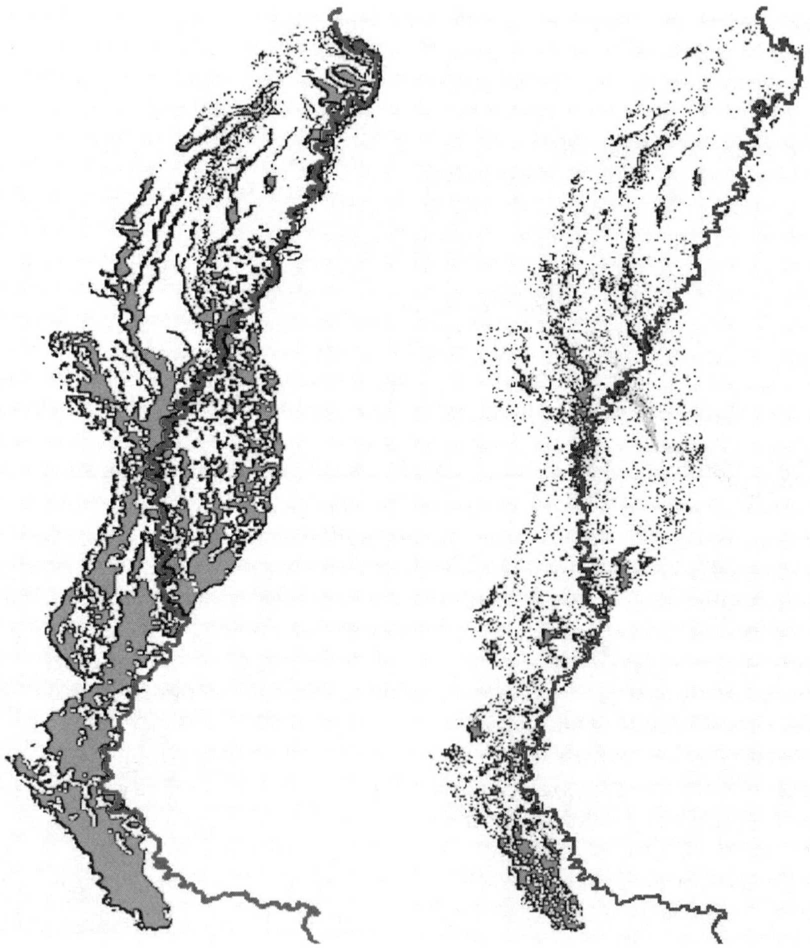

Figure 7.1 Forest cover in LMAV in 1950 (left) and 1990 (right). *Source*: USGS and USFWS (1999), Black Bears and Songbirds of the Lower Mississippi River Valley CD-ROM (www.lacoast.gov/freestuff/cd-roms/blackbears.htm).

> (Figure 7.1). Frequent flooding in some areas reduces the income from agriculture, rendering farming economically marginal. Electric utility companies have funded some restoration, although most is by federal programs (Gardiner and Oliver, this volume). Even though there is no cap on GHG emissions from industry in the U.S., these companies have voluntarily invested in GHG offset projects. Projections are that 900 t CO_2 per ha can be sequestered over a 70-year period. To date, about 24,000 ha have been planted with about 18 million hardwood seedlings for carbon sequestration projects with funding from electric utility and oil companies. The majority of the land is in National Wildlife Refuges managed by the U.S. Fish and Wildlife Service. About 150 permanent plots have been installed to determine carbon sequestration over time.

The potential expansion of the market for carbon-driven restoration projects is dependent on a host of factors, including how forestry carbon offsets will be counted; how additionality, baselines, leakage, and other issues described in this chapter are treated; how project implementation and transaction costs change in the future as more and more projects are carried out; and whether clear and simple rules will be adopted (Danish et al. 1999) that allow the private sector to take cost-effective action. Interest in forest restoration projects for carbon offsets will also depend on the extent other environmental and socioeconomic benefits can be fulfilled through forest restoration.

Climate change will affect ecosystems and impact biological diversity, and some actions proposed in the UNFCCC and its KP to mitigate climate change will influence biodiversity. Forest restoration projects for carbon offsets will prove most successful if they meet a host of other environmental and socioeconomic objectives. However, like the carbon that is being sequestered and measured and monitored, methods need to be developed to quantify these other co-benefits for their full value to be realized. Projects that are worthwhile even without carbon considerations are also more likely to provide permanent carbon offsets.

References

Brown, S., Measuring, monitoring, and verification of carbon benefits for forest-based projects, *Philos. Trans. R. Soc. Lond.*, A, 360, 1669, 2002a.

Brown, S., Measuring carbon in forests: current status and future challenges. *Environ. Pollut.*, 116, 363, 2002b.

Brown, S., Burnham, M., Delaney, M., Vaca, R., Powell, M., and Moreno, A., Issues and challenges for forest-based carbon-offset projects: a case study of the Noel Kempff Climate Action Project in Bolivia, *Mitigation Adaptation Strategies Global Change*, 5, 99, 2000a.

Brown, P., Cabarle, B., and Livernash, T., *Carbon Counts: Estimating Climate Change Mitigation in Forestry Projects*, World Resources Institute, Washington, DC., 25 pp., 1997.

Brown, S., Masera, O., and Sathaye, J., Project-based activities, in *LandUse, Land-Use Change, and Forestry. A Special Report of the Intergovernmental Panel on Climate Change (IPCC)*, R.T. Watson, I.R. Noble, B. Bolin, N.H. Ravindranath, D.J. Verardo, and D.J. Dokken, Eds., Cambridge University Press, Cambridge U.K., 283, 2000b, chap. 5.

Brown, S., Sathaye, J., Cannell, M., and Kauppi, P., Management of forests for mitigation of greenhouse gas emissions, In *Climate Change 1995. Impacts, Adaptations and Mitigation of Climate Change: Scientific–Technical Analyses. Contribution of Working Group II to the Second Assessment Report of the Intergovernmental Panel on Climate Change (IPCC)*, R.T. Watson, M.C. Zinyowera, and R.H. Moss, Eds., Cambridge University Press, Cambridge, U.K., 773, 1996a.

Brown, S., Sathaye, J., Cannell, M., and Kauppi, P.E., Mitigation of carbon emissions to the atmosphere by forest management, *Commonw. For. Rev.*, 75, 80, 1996b.

Danish, K., Brenes, E., and Rotter, J., Legal dimensions of AIJ project development activities, in *The U.N. Framework Convention on Climate Change Activities Implemented Jointly (AIJ) Pilot: Experiences and Lessons Learned*, R.K. Dixon, Ed., Kluwer Academic Publishers, Dordrecht, 1999.

Dixon, R.K., Brown, S., Houghton, R.A., Solomon, A.M., Trexler, M.C., and Wisniewski, J., Carbon pools and flux of global forest ecosystems, *Science*, 263, 185, 1994.

Fearnside, P.M., Global warming and tropical land-use change: greenhouse gas emissions from biomass burning, decomposition and soils in forest conversion, shifting cultivation and secondary vegetation, *Climatic Change*, 46, 115, 2000.

Hansen, J., Sato, M., Ruedy, R., Lacis, A., and Oinas, V., Global warming in the twenty-first century: an alternative scenario, *Proc. Natl. Acad. Sci.*, 97, 9875, 2000.

Hansen, J.E. and Sato, M., Trends of measured climate forcing agents, *Proc. Natl. Acad. Sci.*, 98, 14778, 2001.

Houghton, R.A., Revised estimates of the annual net flux of carbon to the atmosphere from changes in landuse and land management 1850–2000, *Tellus*, 55, 378, 2003.

IPCC, Landuse, land-use change, and forestry, in *A Special Report of the Intergovernmental Panel on Climate Change (IPCC)*, R.T. Watson, I.R. Noble, B. Bolin, N.H. Ravindranath, D.J. Verardo, and D.J. Dokken, Eds., Cambridge University Press, Cambridge, U.K., 2000.

Kauppi, P. and Sedjo, R., Technological and economic potential of options to enhance, maintain and manage biological carbon reservoirs and geo-engineering, in *Climate Change 2001, Mitigation, Contribution of Working Group III to the Third Assessment Report of the Intergovernmental Panel on Climate Change (IPCC)*, B. Metz, O. Davidson, R. Swart, and J. Pan, Eds., Cambridge University Press, Cambridge, U.K., 2001.

Koskela, J., Nygren, P., Berninger, F., and Luukanen, O., Implications of the Kyoto Protocol for tropical forest management and landuse: prospects and pitfalls, *Tropical Forestry Reports*, University of Helsinki, 22, 103 pp., 2000.

MacDicken, K., *A Guide to Monitoring Carbon Storage in Forestry and Agroforestry Projects*, Winrock International, Arlington, VA, 87 pp., 1997a.

MacDicken, K., Project specific monitoring and verification: state of the art and challenges, *Mitigation Adaptation Strategies Global Change*, 2, 191, 1997b.

Moura-Costa, P., The climate convention and evolution of the market for forest carbon offsets, *Unasylva*, 52, 34, 2001.

Myers, N. and Goreau, T.J., Tropical forests and the greenhouse effect: a management response, *Climatic Change*, 19, 215, 1991.

Pinard, M. and Putz, F., Monitoring carbon sequestration benefits associated with reduced-impact logging project in Malaysia, *Mitigation Adaptation Strategies Global Change*, 2, 203, 1997.

Post, W.M., Izaurralde, R.C., Mann, L.K., and Bliss, N., Monitoring and verification of soil organic carbon sequestration, in *Symposium: Carbon Sequestration in Soils Science, Monitoring and Beyond*, N.J. Rosenberg, R.C. Izaurralde, and E.L. Malone, Eds., Batelle Press, Columbus, OH, 41, 1999.

Prentice, I.C., Farquhar, G., Fashm, M., Goulden, M., Heimann, M., Jaramillo, V., Kheshgi, H., Le Quéré, C., and Scholes, R.J., The carbon cycle and atmospheric carbon dioxide, in *Climate Change 2001: The Scientific Basis. Contribution of Working Group I to the Third Assessment Report of the Intergovernmental Panel on Climate Change*, J.T. Houghton, Y. Ding, D.J. Griggs, M. Noguer, P.J. van der Linden, X. Dai, K. Maskell, and C.A. Johnson, Eds., Cambridge University Press, Cambridge, U.K., 183, 2001.

Schlamadinger, B. and Marland, G., *LandUse and Global Climate Change: Forests, Land Management, and the KP*, Pew Center on Global Climate Change, Arlington, VA, available on Internet at www.pewclimate.org/proje cts/index_policy.cfm, 54 pp., 2000.

Schneider, S.H., Kyoto Protocol: the unfinished agenda. An editorial essay, *Climatic Change*, 39, 1, 1998.

Segura, M. and Kanninen, M., Inventario para estimar carbono en ecosistemas forestales tropicales, in *Inventarios Forestales Para Bosques Latifoliados en America Central CATIE — Centro Agronómico Tropical de Investigación y Enseñanza*, L. Orozco and C. Brumér, Eds., 202, 2002.

Trexler, M., Kosloff, L., and Gibbons, R., Forestry and land-use change in the AIJ pilot phase: the evolution of issues and methods to address them, in *The UN Framework Convention on Climate Change Activities Implemented Jointly Pilot: Experiences and Lessons Learned*, R. Dixon, Ed., Kluwer Academic Publishers, Dordrecht, 121, 1999.

chapter eight

Afforestation in Europe: lessons learned, challenges remaining

Norbert Weber
Institute for Forest and Environmental Policy, Albert-Ludwigs University

Contents

8.1 Introduction

Afforestation is an essential component of forest restoration and can be described as the change from agriculture to forest (Stanturf and Madsen 2002). Worldwide, there are more than 80 definitions of afforestation (Lund 2000); in the context of this contribution, afforestation is viewed as the "establishment of forest by planting or regeneration on land recently lacking forest cover" (Mather 1993).

Afforestation in Europe is as varied as is forestry, generally, in the respective countries, and there are great spatial and temporal variations, even within the nations. Afforestation policies of European countries are a product of their respective amount of forest area, site conditions, ownership structures, human density population, and general economic situation. There is a steady transfer between four major types of land-use: agricultural, wooded, unused lands, and urban uses (Buttoud et al. 1993; FAO 2000). Consequently, it is difficult, on closer scrutiny, to quantify "true" afforestation areas. Especially in France (Buttoud et al. 1993), but also in mountainous regions of Austria, Switzerland, Italy, and

1-56670-635-1/05/$0.00+$1.50

the Scandinavian states, natural invasion on former agricultural land is an important element of land-use change (Plochmann and Thoroe 1991; Piussi and Pettenella 2000) and may be included in afforestation statistics. Nevertheless, some general tendencies distinguish the European situation in contrast to other regions in the world.

Although the total forest area has been decreasing elsewhere, in Europe a remarkable expansion is apparent. From a low level in the mid-19th century, forest area has increased 2 million ha per decade, excluding Russia (Steinlin 1994). The loss of forest for urban development and transportation infrastructure is more than offset by expansion of forests onto agricultural and other land (UN-ECE/FAO 1999). The average annual increase of European forest area is about 512,000 ha for 38 European countries, of which 340,000 ha are within the member states of the European Union (EU) (UN-ECE/FAO 2000). It is notable that this includes changes from other wooded land into closed forests.

The current area of European forests is 175 million ha, of which 113 million ha belong to member states of the EU (UN-ECE/FAO 2000). The corresponding figures for other wooded land are 39 and 22 million ha, respectively.

More than 50% of the forests are predominantly coniferous; the figure for the EU is 70%. About one third consists of predominantly broadleaved species and the rest are mixed forests. Norway spruce (*Picea abies* (L.) Karst.) and Scots pine (*Pinus sylvestris* L.) are the main coniferous tree species in Europe while various oaks (*Quercus* sp.) and beech (*Fagus sylvatica* L.) are the main broadleaved trees (UN-ECE/FAO 2000; Päivinen et al. 1999). Eucalypts, poplars, and pine, especially *P. radiata*, are frequently grown as short-rotation woody crops in southern European countries, sometimes resulting in environmental problems (Pereira 1993).

Removals from the forests of Europe total 360 million cubic meters underbark. Fellings of growing stock accounted for 52.6% of net annual increment (UN-ECE/FAO 2000). In the mid-1990s, it was shown that the increment of European forests had grown dramatically, in spite of dire predictions of air pollution effects (Spiecker et al. 1996). In economic terms, forestry in the EU is an important factor, with 82,000 enterprises, 3 million employees, and a contribution of one-eighth of the added value of all EU industries (Laitat et al. 2000). In all Europe, about 77,000 parcels are in public ownership, whereas 10.7 million parcels are in private hands. The average size of public holdings is 1,200 ha; private holdings are smaller, on average 10.6 ha (UN-ECE/FAO 2000).

8.2 History of afforestation in Europe

The distribution of forests and fields across the landscape results from the combination of natural factors (soil, climate, topography) and societal factors (political, economic, and property relationships) through the historical development of a region (Brünig and Mayer 1980). Long ago, forests covered more than 90% of some areas of Europe. Increased population and dependence on wood for many purposes led to exploitation that exceeded the capability of forests to regenerate. In the 18th century, as a consequence of long-lasting and intensive use of the resource, the forested area reached its lowest level in Europe. At that time, European rulers began to recognize the economic value of the forest that had to be protected. During the 19th century, gradual transitions between dense and less dense forested areas gave way to strict separation between forest, meadow, and field. Forest grazing was prohibited; forests and farmland became separated so that borderlines between these land-uses could be drawn on maps. Forest sciences were developed rapidly, resulting in academically trained foresters. Under their guidance, large forest areas were designated and afforested by seed or planting; degraded forest areas were replanted with spruce or pine. Often criticized today, the widespread conifer plantings were in many cases the only option on degraded soils and under high browsing pressure from hoofed game animals. Large-scale afforestation in Central Europe in the second half of the 19th century was only possible after the use of wood had been substituted by coal, brickstones, and steel. In addition, idle heathland was afforested after imported cotton partially replaced wool. Mineral fertilizers increased productivity on better soils, releasing marginal agricultural land such as steep slopes, flat soils, dunes, and overexploited wasteland to afforestation (Küster 1999).

Native forest had almost completely disappeared by the 18th or 19th century, especially in northwest Europe. Most of the present forest area resulted from afforestation (Mather 1993). In some Nordic countries, afforestation assumed the character of national movements: Denmark had its "Heath Cause," and Sweden and Norway had their "Cover the Mountains" (Helles and Linddal 1996). Afforestation efforts late in the 18th and early in the 19th century appeared to be success stories: improving employment, defeating erosion, and increasing wood production. However, social tensions arose in some countries when governments took away pastureland for afforestation (van Maaren 1993). In Spain and Portugal, conflicts with shepherds and goatherds emerged over traditional grazing rights, but also other rights were lost; for example, the possibility to gather firewood or material for producing charcoal and secondary products, such as mast, honey, and mushrooms (Groome 1993; Brouwer 1995). Even in the late 20th century, afforestation can be cast as a "people vs. forestry issue" and may cause serious conflicts. In Ireland, attitudes inimical to afforestation have hindered development at all stages, sometimes leading to violent protests (Gillmor 1993).

Despite the long history of afforestation in Europe, the scale of planting generally remained small until the 20th century (Mather 1993). Afforestation of arable land and meadows on a large scale only started in the 1960s, when member states of the former European Economic Community (EEC), especially in Germany, France, and Italy, had begun or were planning afforestation of several million ha. Direct or indirect involvement by the state was responsible for afforesting wasteland and marginal farmland (Mantel 1965).

In Eastern Europe, the area of forests increased in most countries in the postwar period. Especially in the Baltic region, the extension of forests was most remarkable: 125% in Estonia, 75% in Lithuania, and 61% in Latvia. Forest area has also increased in other Eastern European countries: Macedonia showed a growth of 117% and Hungary 56%. The highest absolute increase of forest area has occurred in Poland with 2.3 million ha, corresponding to 38%. In most of these countries, the trend of expanding forest areas has continued (PHARE 2000).

Two main developments have been decisive for the increase of forest areas in Europe in the 20th century. The more obvious development has been large-scale plantation programs and active conversion of agricultural land or other wooded land into closed forest, for example, in France, Ireland, Turkey, and Spain. This phenomenon is a consequence of the continuing changes in European agriculture and the resulting depopulation of rural areas. A more subtle development has been the new forests that originated from natural invasion onto abandoned farmland, mostly in mountainous regions on low-quality soils. The most important increase in natural forest extension over the last 10 years can be found in France and Switzerland, totaling 225,000 and 127,000 ha, respectively (EFIDAS 2002). However, changes of land-use between agriculture and forestry are sometimes hard to monitor because of the dynamics of forest cover; every year new fallow land turns spontaneously to woodland, but also forest stands evolve toward pastures with scattered trees or other land cover types because of fires or excessive grazing. Sometimes, young woodlots on former farmland are taken back to agricultural use, especially in hilly regions. In some cases, even the owners or managers of the land are unknown, making land-use statistics unreliable (Piussi and Pettenella 2000).

8.3 Afforestation policy of the EU

Afforestation has been a central concern of agricultural policy on the European level since the publication of the so-called Mansholt Memorandum in 1968. Mansholt, then Commissioner for Agriculture of the European Community, intended to streamline food production and increase the competitiveness of agricultural enterprises. He recommended reducing the agricultural area of the European Community by at least 5 million ha and afforesting about 4 million ha of these lands (Le Chatelier 1969). Between 1973 and 1988, about 9 million ha were abandoned by agriculture, but less than 3% of the released area was subject to afforestation grants (Eurofor 1994). In 1990, there was potentially 15 million to 20 million ha of afforestation area in the EC (Lust and Muys 1993). Weisgerber (1988) estimated that 1 million to 5 million ha was suitable for short-rotation forests. A recent estimate is that another 6 million to 16 million ha could be released due to further technical and economic developments.

Although a common forest policy has no legal basis, the European Commission, since 1966, has undertaken to improve the general condition for afforestation and to harmonize national legal provisions. In 1979, a community action program in the Mediterranean region was initiated (Reg. [EEC] 269/1979), and afforestation and restoration of degraded forests were included in this program (Eurofor 1994). Several further regulations, valid for the whole community area, explicitly dealt with the afforestation question; the most important ones are Reg. 797/85, 1096/88, 1609/89, 1610/89, 2328/91, 2080/92, 1257/99, and 1750/99.

Within the framework of Reg. 797/85, only 69,775 ha of new forests were created in the period between 1986 and 1991 (Baillieux 1992). A community aid scheme for forestry measures in agriculture was introduced in 1992 (Reg. 2080/92), which resulted in nearly 900,000 ha afforested up to 1999 (Table 8.1); Spain, Portugal, the U.K., and Ireland were the main beneficiaries. The provisions of this scheme included an afforestation aid up to 4,000 €ha^{-1}, a maintenance premium for five years up to 500 €ha^{-1}, and a compensatory premium for loss of income up to 600 €$ha^{-1}year^{-1}$. This scheme has been subject to several evaluations, starting with an internal report of the European Commission followed by the

Table 8.1 Agricultural Area, Forest Land, and Areas Afforested within the Framework of Reg. (EC) 2080/92

Nation State	Utilized Agricultural Area (ha)	Forest and Other Wooded Land (ha)	Areas Afforested (ha)	Agricultural Area Afforested (%)
Spain	22,984,000	25,984,000	406,211	1.76
Portugal	3,822,000	3,467,000	123,998	3.24
United Kingdom	16,169,000	2,489,000	101,281	0.62
Ireland	4,342,000	591,000	98,258	2.26
Italy	14,833,000	10,842,000	60,235	0.40
France	28,331,000	16,989,000	45,147	0.15
Germany	17,160,000	10,740,000	27,067	0.15
Greece	3,499,000	6,513,000	16,401	0.46
Finland	2,172,000	22,768,000	15,920	0.73
Denmark	2,689,000	538,000	2,317	—
Netherlands	2,011,000	339,000	2,271	—
Austria	3,415,000	3,924,000	410	—
Belgium	1,383,000	672,000	302	—
Sweden	3,109,000	30,259,000	27	—
Luxembourg	127,000	89,000	12	—
EU total	126,045,000	136,204,000	899,857	—

Source: Modified from du Breil de Pontbriand (2000) and UN-ECE/FAO (2000).

so-called Otila Report for the European Parliament, which criticized the high total funding per ha:

> With this sort of money available, it is not surprising to hear that significant numbers of traditional poplar growers in a certain member state put their plantations back into agriculture for the minimum three years necessary to qualify as agricultural land, and thus make it eligible for replanting trees, using the different afforestation measures, all perfectly legally (European Parliament 1998).

The French organization Institut pour le Développement Forestier (IDF) provided a comprehensive external evaluation of Reg. 2080/92 on behalf of the European Commission, which stated that the afforestation grants and premiums encouraged the predominant use of broadleaved species that represent 56.8% of the areas planted, with cork oak and evergreen oak predominating. The share of conifers amounts to 32.1% of the plantations. It concluded that

> The "2080" forester is a Mediterranean farmer, more Spanish than anything else, over 50 years of age, who has been growing 45.8% broadleaf species…on an area of 8 hectares very probably devoted to low productivity grazing.

According to IDF estimates, about 150,000 full-time equivalent jobs have been created by the afforestation operations, but forestry activities do not replace agricultural ones. They require less labor and a different know-how (IDF 2001). Although Reg. 2080/92 also promoted the improvement of woodlands, this provision has been little-used, expending only 4% of the aid budget allocated. The main recipients were Germany, Finland, and Austria (du Breil de Pontbriand 2000).

Overall, Reg. 2080/92 has had little impact on reducing agricultural surpluses (Table 8.1). The contribution of Reg. 2080 to biodiversity will be generally positive and relatively significant. The effect on carbon sequestration by 2012 will be negligible, and will hardly be significant until 2030, when it should represent approximately 2 to 3% of the total amount of carbon fixed by European forests, an estimated 2.7 million tons of carbon per year (Seoane 2002). Afforestation under Reg. 2080 will protect against soil erosion and desiccation, but this cannot be quantified. Revised provisions for afforestation have been anchored in the new Reg. 1257/99 and 1750/99. These regulations emphasize that afforestation should not be seen as the goal *per se*, but as a means to improve living and working conditions in rural areas. "Support shall be granted for the afforestation of agricultural land provided that such planting is adapted to local conditions and is compatible with the environment" (Reg. 1257/99, Art. 31, § 1; CEC 1998).

8.4 National afforestation policies

8.4.1 Afforestation objectives and targets

Because they are closely linked, objectives, means, and implementation of afforestation in the respective states of Europe have to be analyzed in combination. An assessment of afforestation programs also has to take into account the natural, political, and socioeconomic conditions prevailing at the time programs were formulated (Helles and Linddal 1996).

In many countries, national forest policies and afforestation measures originated from the need for wood for the war industry. Even after World War II, wood production to increase self-sufficiency was the main objective of afforestation in European countries and was driven by strategic and economic reasons. Predictions of wood shortages, however, have not materialized (Mather 1993). In many cases, agricultural considerations such as the retrenchment of farms from marginal lands or overproduction were decisive for furthering afforestation programs. When the Danish government adopted a resolution in 1989 to double the area presently covered with forest over the span of a tree generation, that is, by the year 2100, the reduction of overproduction from agriculture within the EU was a main objective (Madsen et al. 2002). Often, improving employment was an important objective for creating new forests, especially in Spain, Denmark, and Portugal. Case studies from several countries show that the employment effect had been overestimated (Gillmor 1993), mostly due to postwar mechanization of forest work and decreased need for labor when the forests grow older (Bulfin 1993; Peck and Descargues 1995; Anderson et al. 1996; Brouwer 1995). In Spain, afforestation was sometimes accused of being "bread for today and hunger for tomorrow" (Arias 1993). Even today, there are fears that afforestation may hasten the depopulation of rural areas (Helles and Linddal 1996), although forestry provides more jobs in comparison to alternative land-uses, such as sheep farming, if downstream employment is considered. But large-scale afforestation, especially as carried out by corporations or absentee owners, may use contractors rather than local residents (Mather 1993). In some parts of Europe, forestry is regarded as a threat to rural culture and of little economic importance (Elands and O'Leary 2002). Nevertheless, afforestation is still regarded by planners as an appropriate means for mitigating rural exodus, if applied in combination with other instruments.

Afforestation for ecological reasons is an issue of increasing importance. For instance, large-scale plantings in Spain are used as a means to combat soil erosion and desertification (RDP Spain 2000 to 2006). The creation of new woodlands in densely populated areas, taking place in the U.K., Belgium, and the Netherlands, is seen as a means of increasing ecological and amenity value of the landscape (Lust and Muys 1993; van Maaren 1993). The creation of forests to fix CO_2 emissions, for example, comprising the Dutch foundation *Face*, the Swiss *Factor* organization, or the German foundation *Prima Klima*, is gaining importance. Afforestation, especially for recreational purposes, is not yet common, although it could be an important theme for the establishment of future forests.

Several shifts in afforestation policies have occurred in recent years in Europe as a whole and in the respective countries. The first shift was described at some length above, and can be described as a change of underlying objectives, from increasing forest production to reducing agricultural overproduction. More recently, the discussion has focused on environmental policy with a discussion of climate change and the global character of forests. Throughout these changes in objectives, rural development has been an additional concern. The second policy shift comes from overlapping national afforestation aims with international obligations. Expansion of the forest area of Europe has become a task of global and European importance (Forest Principles 8 [a] and 8 [b]; Chapter 11 of Agenda 21 and the Pan-European Resolutions Helsinki H 1 and Lisbon L 2). As signatory parties to the Framework Convention on Climate Change and the Kyoto agreement, European governments were to enhance sinks of greenhouse gases (Adger et al. 1996; IPCC 2000). The third shift is from state afforestation to private afforestation. In many countries, public agencies are no longer the leading actors in afforestation activities, mirroring the partial retreat, generally, of government, from policy and development arenas in the forest sector (Slee and Wiersum 2001). Nevertheless, countries are still instrumental in initiating afforestation either directly (on land acquired for that purpose) or indirectly (through incentives offered to the private sector). The fourth shift derives from the third, a change in means of afforestation policy. Regulative instruments, which are characterized by the use of government power and the application of top-down policy and implementation, are decreasing in importance. Economic instruments, such as financial incentives for farmers or private forest investors, and information programs, have superseded large-scale state afforestation programs. Some countries continue to plan from the top, however. For example, in Hungary, an overall total of 335,000 ha will be afforested and targets have been designated (Keresztesi 1993) for veneer plantations (30,000 ha), conifer plantations (140,000 ha), fodder forests (65,000 ha), beekeeping forests with *Robinia pseudoacacia* (L.) (60,000 ha), and energy forests (40,000 ha).

8.4.2 *What can we learn from previous afforestation activities in Europe?*

8.4.2.1 *Success is not assured*

Afforestation policies should be judged by their long-term success rather than by the number of ha planted per year. Success can be judged by achieving a balanced mix of species, the techniques used, the multiple interests the plantings satisfy, and positive externalities generated by afforestation (Groome 1993). Afforestation success is elusive, especially in Mediterranean areas where young forest stands are at risk from fire or browsing by goats and sheep. For instance, in Spain between 0.85 million and 1.2 million ha of forest plantations failed (Groome 1993). Additionally, changes between land-uses are reversible and show considerable dynamics over time (Buttoud et al. 1993; Jensen 1993; Peck and Descargues 1995; Crabtree 1996).

8.4.2.2 *Many factors influence afforestation programs*

Afforestation occurs within a political and economic context and activities are embedded within a framework of cultural beliefs, values, and attitudes of landowners (Selby and Petäjistö 2000; Phelan 2001). Active drivers of afforestation policy and programs include a shortage of wood, whether real, perceived, or feared; key personalities with a strong interest in afforestation; or fortuitous events, such as the change in national boundaries in Hungary. Passive inducements include the recognition that forests are undersupplied, changes in agricultural policy releasing land to other uses, availability of suitable species, proven technology, and financial incentives (Mather 1993). Afforestation levels are the result of push–pull inducements; decreased agricultural supports push, and subsidies pull afforestation levels up (Selby and Petäjistö 1994). Passive inducements, especially the retrenchment of agriculture, have been the most important driving forces for afforestation, even though no direct relationship can be observed at the national level between forest expansion and agricultural contraction (Mather 2000). Grants and premiums for private afforestation can induce afforestation, but waves soon flatten if the incentives are reduced or abolished.

8.4.2.3 *Social context is increasingly important*

Not everyone welcomes afforestation. Sometimes, afforestation of marginal lands is viewed negatively as an assault on rural culture and rural landscapes (Lanly 1997). Afforestation may revert a landscape back to climax biotypes, for example, taiga forests in Finland. Depending on one's point of view, this can be regarded as ecosystem rehabilitation or destruction of the cultural landscape, especially in early-seral stages of scrub vegetation (Selby 2003, pers. comm.). In contrast to former times when there was little resistance to establishing new forests, sometimes resistance comes from environmental authorities, NGOs, farmers, or local residents. Obligatory application procedures are common and some legislation had introduced required assessments of the effects of afforestation projects. For instance, in Ireland large-scale afforestation has been subject to an Environmental Impact Assessment (EIA) if initial afforestation is 200 ha or more. An EIA is triggered if 10 ha or more of broadleaved high forest is replaced with conifers. The afforestation threshold has been reduced from 200 to 70 ha, and a new provision has been added that triggers an EIA if the cumulative afforestation by a single owner within a 3-year period exceeds 70 ha in aggregate (Nieuwenhuis 1999).

8.4.2.4 Inconsistent policies

Despite efforts to harmonize afforestation policy within the EU, in several countries inconsistencies can be observed between afforestation and other land-use policy. For example, in Ireland, funding for forestry conflicted with the Rural Environment Protection Scheme, an environmental scheme that provided support payments to farmers who complied with environmental protection requirements. This resulted in land being withheld from forestry, even though forestry would be the optimal land-use (Phelan 2001). In Spain, premiums for goats and sheep were sometimes granted for the same regions that were subject to afforestation subsidies (Herkendell 2001, pers. comm.).

8.5 Opportunities and threats for future afforestation

8.5.1 Further increase of European forest areas

There has been much progress in the creation of new forests in Europe during recent decades; however, it is difficult to explain the underlying causes and patterns of this development confidently and credibly, so that we can forecast future developments. One reason for this unsatisfactory situation is that any estimate of potential afforestation has to take into account many particularities of land-use change in each country. In addition, there are several contrary tendencies in the policy sectors of agriculture, forestry, environment, and economy, which complicate the situation. Nevertheless, further increases in forest area in Europe can be expected.

The goal of planting new forests is fixed within regional, national, and international political programs. Active afforestation efforts will take place to fulfill the political promises made in the field of climate policy, both in the respective countries and in the form of Joint Implementation partnerships (Ciccarese et al. this volume). Afforestation also enables countries to fulfill their obligations toward international and supranational agreements on nature conservation. In countries with low levels of forest cover, forests will be created to restore biodiversity, even though biodiversity preservation is sometimes used as an argument against new forests (Government of Ireland 2002). Many national forest laws and political programs contain the objective to establish new forests. For example, in Germany, a program aimed at coastal protection and improving agriculture stated that afforestation was a reasonable alternative land-use for abandoned agricultural areas and that it provided desirable environmental effects, especially the contribution to climate protection. Promotion of afforestation is treated as a matter of high significance (Deutscher Bundestag 2002).

Afforestation increasingly is regarded as integral to improving the conditions of rural areas. Most EU members have integrated afforestation into their rural development programs through 2006. In Scotland, discretionary payments are available for short-rotation coppice or for afforestation of nonagricultural land. Only a few countries have quantitative afforestation targets within their rural development plans; France, for example, intends to afforest 9,500 ha of agricultural land per year. In the Belgian region of Flanders, during the period 1999 to 2007, as many as 1,000 ha per year are to be afforested (Lust 2003, pers. comm.).

New forests will develop as a consequence of natural invasion, especially in Alpine regions (Piussi and Pettenella 2000) and Scandinavia. These processes are loosely connected with the retrenchment of agriculture and the depopulation of rural areas.

The expected retrenchment of agriculture in former East Bloc countries will provide many suitable areas for afforestation. Some long-term expectations (until 2050) indicate up to 50,000 ha for the Czech Republic, 300,000 to 400,000 ha for Estonia, 700,000 to 800,000 ha for Hungary, 300,000 ha for Latvia, as well as for Lithuania, one million ha for Poland, 225,000 ha plus 10,000 km of shelter belts for Romania, and 250,000 to 300,000 ha for Slovenia (PHARE 2000). Bulgaria, Hungary, Lithuania, Poland, and Romania have already formulated afforestation programs (Marghescu 2001).

Afforestation can also mitigate degradation in regions suffering from erosion and deflation. For instance, in Ukraine over 13 million ha of lands have been damaged by water erosion and 6 million ha by wind erosion. The area eroded has increased by 70,000 to 100,000 ha per year during the last decades. Ukrainian scientists have developed sophisticated goals for the increase of the forest area in different regions (Soloviy 2000).

8.5.2 *Preconditions for successful afforestation programs*

In the past, some foresters thought that afforestation was the automatic consequence of agricultural retrenchment. Experience has shown, however, that afforestation is one of many alternative uses for agricultural land (Weber 1993). Thus, it is important to be aware of divergent opinions and to identify the most powerful arguments for afforestation. The notion that everybody loves the forest seems to be a myth (Elands and O'Leary 2002). In some cases, local people view afforestation as creating isolation, deteriorating landscapes, and threatening traditional land-uses. Severe resistance to afforestation can be expected in regions where forestry is not a local tradition or the afforestation program is imposed on local people by large landowners, external investors, or the state (Brouwer 1995; Selby 2003, pers. comm.) In France, for example, the political discourse on afforestation in a mountainous area contained two opposing views, which could be identified as "pro-forest" vs. "pro-landscape." The first view focused on open space as a vacant area to afforest and suitable for making profit from timber wood and hunting, and the second view identified open space as a landscape to preserve (Le Floch and Deuffic 2002). In France, as elsewhere, afforestation is questioned economically and socially (Terrasson 2001). In Britain, the term "afforestation" is being dropped in favor of "woodland creation" (Pryor 2000). Some German foresters speak of "*Neuwaldbildung*" (i.e., creating new woodlands) and "*Wiederbewaldung*" (i.e., bringing woodlands back) instead of using the word "*Erstaufforstung*" (corresponding to afforestation) to raise the acceptance for the change of land-use in favor of forests.

The mistakes of the past that resulted in skepticism and antagonism toward afforestation, and forestry as a whole, can be avoided. Planning and designing new forests can be done to adequately meet the needs of the population. Modern planning approaches, such as those in Scotland and Ireland (Bulfin 2000), are applying remote sensing and GIS techniques to take into account regional and local conditions. Many options are available to adapt new plantations to the landscape. For instance, the shape and edge of forests can be designed using roads, rides, and fire breaks that are adapted to the respective peculiarities of the landscape (McCormack and O'Leary 2000).

More attention should be given to local conditions and public opinion. Case studies have identified potential scenarios for afforestation at the local level based on two variables: the proportion of landowners who intend to plant and the share of community inhabitants who support planting. The *status quo* scenario is characterized by low support for afforestation from both sides. In an underplant scenario, public support for afforestation is much higher than the likelihood of planting by farmers. A plant-conflict situation is to be expected if farmers have a high interest in afforestation that meets with low public support. In the strategic scenario, many landowners, as well as the public, are in favor of the establishment of new forests (O'Leary and Elands 2002). Knowing the social context in a candidate area for afforestation can help overcome low support and acceptance, or suggest that funds are better spent elsewhere.

8.5.3 *New approaches*

Emphasis today in afforestation programs is on outcomes rather than activity levels (Groome 1993; Pryor 2000). Rather than measuring area planted, policymakers are interested in the numbers of people visiting woodlands, expenditures by visitors to woodland that indirectly produces employment, employment directly associated with planting, forest management, timber harvesting and processing, increases in populations of key species or areas of key habitat, or increase in property values and inward investment, which can be attributed to environmental improvement (Pryor 2000).

Future afforestation offers scope for a variety of designs, including afforestation to create forests for special purposes, which could reduce pressure on existing forests. In addition to plantations for producing wood, forests could be dedicated especially for mountain bikers, horsemen, labyrinth walkers, picnic sites, playgrounds, cemeteries, adventurers, art, etc. Instead of degrading existing forests by reintroducing forest pasture, as proposed by some environmentalists, parts of new forests could be designated for that purpose in the medium term.

One innovative approach is the *NeighbourWood* forests in Britain and Ireland. Based on increasing demand for forests close to urban areas, such as villages, towns, or cities, the NeighbourWood scheme provides support for the development of new and existing forests designated for public access, recreation, and enjoyment. Mostly, this is on land owned or managed by local authorities, but there are some private initiatives. NeighbourWoods should be accessible, attractive, and form an integral part of the locality. They can range in scale from small pockets, in and around residential areas and public parks, to larger-scale woodland amenities, situated on greenbelts and along the urban fringe. It would also be possible to include existing woodlands outside towns and villages, used by local communities for amenity and recreation (Department of the Marine and Natural Resources 2001).

8.6 Conclusions

At the beginning of the 21st century, three major features characterize afforestation in Europe. First, afforestation is occurring over most of the continent. Though not closely, it is linked to agricultural retrenchment. Second, most afforestation is directed toward providing services, rather than primarily toward timber production. Third, there is an increasing degree of democratization and public participation involving landowners, as well as the general public, in the planning processes (Mather 2000). The results of afforestation research and regional case studies (Merlo et al. 1987; Bock and Rondeux 1990; Mather 1993; Volz and Weber 1993; Weber 2000) can improve the formulation and implementation of afforestation objectives and targets. However, some socioeconomic aspects of afforestation have not received sufficient attention, for instance, the multifunctional role of forests in the context of rural enterprises (Selby and Petäjistö 2000).

The forest area of Europe could increase by 4 million ha by 2050 (Nabuurs 2001). A large part of the expansion of exploitable forests is expected to take place in Spain, France, and Poland (UN-ECE/FAO 1996). However, many uncertainties make it difficult to forecast the extent of future afforestation. Earlier estimates of afforestation potential have proven to be too high and future estimates must recognize many constraints in infrastructure, social aspects, and economic aspects (Weber 1993; Ciesla 1996). Environmental issues will have an uncertain influence; afforestation is welcome in the context of climate policy or the protection against erosion, but expanding conservation constraints at the local level may limit suitable areas for new forests in most European countries.

Agricultural policy and future development remain crucial for afforestation programs. Further increases in forest area will utilize abandoned agricultural land in rural areas rather than on other land categories. In some countries with mountainous regions, including France, Austria, Switzerland, and Italy, invasion and old-field succession will continue. Degraded lands, such as cutaway peatlands in Ireland, quarries in Great Britain, brown coal opencast mining areas in Germany, and eroded lands in Spain or Ukraine, will deliver additional potential afforestation areas. Due to high land prices and extensive infrastructure in place, the realization of afforestation in urban areas will be difficult. However, new approaches similar to NeighbourWood forests could be a solution for concentrating efforts.

Nation states and their subunits will continue to be instrumental in initiating afforestation, both directly, by acquiring land (although to a decreasing extent), and indirectly, by offering incentives for the private sector. However, it will be more difficult to justify long-term subsidies, and the objectives for funding afforestation as well as the conditions for receiving subsidies will have to be revised. Landowners increasingly will be part of afforestation policy rather than simply recipients of a subsidy. Greater effort will be needed to inform landowners about afforestation as an alternative landuse and the financial subsidies available. Implementation of well-developed afforestation plans is not feasible unless planners know the attitudes and preferences of landowners and the public, and programs are designed to overcome opposition by addressing local needs. In some countries, foundations encouraging afforestation for protecting the climate and other environmental values will play an important role. Although they do not always welcome new forests, some environmental NGOs encourage the planting of special forests; for example, wedding trees planted on the occasion of marriage celebrations.

Future afforestation and restoration approaches must be seen in the context of many challenges that can only be met by holistic strategies and planning approaches. As a consequence of the shifts in actors of afforestation policy, efforts toward an integrated landuse policy should be strengthened (Selby and Petäjistö 2000). It will be necessary to steer the difficult course between land-use change, the limits of conventional afforestation planning, values and attitudes of landowners, and different opinions of the public about forests (Ní Dhubháin and Wall 1998; Tyrväinen and Tahvanainen 2000; Præstholm and Jensen 2001).

Even if afforestation activity slows, Europe in the future will be characterized as a continent with a net increase in forest area. Illustrating the characteristics of transition from declining to expanding forest resources (Palo 2000), the European development of forest areas raises hopes for the rest of the world.

References

Adger, N., Pettenella, D., and Whitby, M., Land-use in Europe and the reduction on greenhouse gas emissions, in *Draft Proccedings Workshop on Instruments for Global Warming Mitigation: The Role of Agriculture and Forestry*, Viote des Monte Bondone, Trento, 1, 1996.

Anderson, C., Thomson, K.J., and Psaltopoulos, D., Forestry contracting in northern Scotland: a survey report, *Forestry*, 69, 143, 1996.

Arias, M.L., Experiences of Eucalyptus-afforestation in Spain, in *Afforestation of Agricultural Land*, Volz, K.-R. and Weber, N., Eds., Commission of the European Communities, Luxembourg, 77, 1993.

Baillieux, P., Die EG-Verordnung 2080/92 aus der Sicht der EG-Kommission, *EG-Regionalseminar über "Die Aufforstung von Landwirtschaftsflächen"*, 16–18 November 1992, CEPFAR, Bruxelles, 9, 1992.

Bock, J. and Rondeux, J., Eds., *Marginal Agricultural Land and Efficient Afforestation, Proceedings of a Workshop in the CEC Land and Water Use Research Programme*, Gembloux, Belgium, 20–21 October 1988, Commission of the European Communities, Luxembourg, 1990.

Brouwer, R., *Planting Power: The Afforestation of the Commons and State Formation in Portugal*, eburon P&L, Delft, 1995.

Brünig, E. and Mayer, H., *Waldbauliche Terminologie. Fachwörter der Forstlichen Produktion*, IUFRO Gruppe Ökosysteme, Institut für Waldbau, Universität für Bodenkultur, Wien, 1980.

Bulfin, M, Private forestry in Ireland: progress and problems, in *Afforestation of Agricultural Land*, Volz, K.-R. and Weber, N., Eds., Commission of the European Communities, Luxembourg, 59, 1993.

Bulfin, M., Physical planning for forestry in Ireland, in *NEWFOR — New Forests for Europe: Afforestation at the Turn of the Century*, Weber, N., Ed., EFI Proceedings No. 35, European Forest Institute, Joensuu, 89, 2000.

Buttoud, G., Cavailhes, J., and Normandin, D., Afforestation of agricultural land in France: its dynamics and related policies, in *Afforestation of Agricultural Land*, Volz, K.-R. and Weber, N., Eds., Commission of the European Communities, Luxembourg, 23, 1993.

CEC — Commission of the European Communities — Directorate General VI, *Towards a Greening of the Common Agricultural Policy*, Fact-Sheet Environment, Brussels, 1998.

Ciesla, W.M., Forestry options for mitigating predicted effects of climate change, in *Workshop on Instruments for Global Warming Mitigation*, Viote des Monte Bondone, Trento, 32, 1996.

Crabtree, B., Policy instrument design for multi-objective environmental forestry: the case of carbon retention by farm woodlands, in *Draft Proceedings Workshop on Instruments for Global Warming Mitigation: The Role of Agriculture and Forestry*, Viote des Monte Bondone, Trento, 182, 1996.

du Breil de Pontbriand, L., European experiences with regulation 2080/92 and the new afforestation policy under agenda 2000, in *NEWFOR — New Forests for Europe: Afforestation at the Turn of the Century*, Weber, N., Ed., EFI Proceedings No. 35, European Forest Institute, Joensuu, 23, 2000.

Department of the Marine and Natural Resources, NeighbourWood Scheme, Forest Service, Wexford/Ireland, 2001.

Deutscher Bundestag, Rahmenplan der Gemeinschaftsaufgabe "Verbesserung der Agrarstruktur und des Küstenschutzes" für den Zeitraum 2002–2005, Drucksache 14/9009, 2002.

EFIDAS — European Forestry Information and Data Analysis System, data table Long-term European Forest Resources, generated 20 July, 2002.

Elands, B.H.M. and O'Leary, T.N., The myth of forests; a reflection of the variety of rural identities in Europe and the role of forests in it, in *The Changing Role of Forestry in Europe: Perspectives for Rural Development*, Wiersum, K.F. and Elands, B.H.M., Eds., Proceedings 2002-02, Forest and Nature Conservation Policy Group, Wageningen University, Wageningen, 25, 2002.

Eurofor, L'Europe et La Foret, Etude pour le Parlement Européenne, Tome 2, Office des publications officielles des Communautés Européennes, Luxembourg, 1994.

European Parliament, *Report on the Report from the Commission to the Council and the European Parliament on the Application of Regulation (EEC) No 2080/Instituting a Community Aid Scheme for Forestry Measures in Agriculture*, Session Documents A4-0346/98, 30 September 1998, 1998.

FAO, FRA 2000, *On Definitions of Forest and Forest Change, Forest Resources Assessment Programme*, Working Paper 33, Rome, 2000.

Gillmor, D.A., Afforestation in the Republic of Ireland, in *Afforestation. Policies, Planning and Progress*, Mather, A.S., Ed., Belhaven Press, London and Florida, 34, 1993.

Government of Ireland, *CBD First National Report Ireland*, online available at http://www.biodiv.org/doc/world/ie/ie-nr-01-en.pdf, 2002.

Groome, H.J., Afforestation policy and practice in Spain, in *Afforestation. Policies, Planning and Progress*, Mather, A.S., Ed., Belhaven Press, London and Florida, 72, 1993.

Helles, F. and Linddal, M., *Afforestation Experience in the Nordic Countries*. Executive Summary of a Scientific Report, Department of Economics and Natural Resources, Royal Veterinary and Agricultural University, Frederiksberg, 1996.

Herkendell, J., personal communication, 2001.

IDF, *Evaluation du système d'aide communautaire pour les mesures forestières en agriculture du Règlement 2080/92*, Rapport final, Auzeville, 2001.

IPCC, *Land-Use, Land-Use Change, and Forestry*, Special Report of the Intergovernmental Panel on Climate Change, Summary for Policymakers, Geneva, 2000.

Jensen, K.M., Afforestation in Denmark, in *Afforestation. Policies, Planning and Progress*, in Mather, A.S., Ed., Belhaven Press, London and Florida, 49, 1993.

Keresztesi, B., Hungary, in *Afforestation, Policies, Planning and Progress*, Mather, A.S., Ed., Belhaven Press, London, 1993.

Küster, H., *Geschichte der Landschaft in Mitteleuropa. Von der Eiszeit bis zur Gegenwart*, C.H. Beck, München, 1999.

Laitat, E., Karjalainen, T., Loustau, D., and Lindner, M., Introduction: towards an integrated scientific approach for carbon accounting in forestry, *Biotechnol. Agron. Soc. Environ.*, 4, 241, 2000.

Lanly, J.-P., Forestry and woodland resources, in *Proceedings of the XI World Forestry Congress Antalya*, 1A, 3, 1997.

Le Chatelier, X., Die forstlichen Massnahmen des Mansholt-Memorandums, *AllgemeineForstZeitschrift*, 24, 471, 1969.

Le Floch, S. and Deuffic, P., Social relationships to forests as an indication of present issues regarding rural areas; content analysis of in-depth interviews carried out in the Monts D'Arrée area, France, in *The Changing Role of Forestry in Europe: Perspectives for Rural Development*, Wiersum, K.F. and Elands, B.H.M., Eds., Proceedings 2002-02, Forest and Nature Conservation Policy Group, Wageningen University, Wageningen, 75, 2002.

Lund, H.G., Coord., Definitions of forest, deforestation, afforestation and reforestation [Online], Manassas, VA: Forest Information Services, http://home.att.net/~gklund/DEFpaper.html, 2000.

Lust, N. and Muys, B., Conflicts caused by afforestations of agricultural land in densely populated areas, in *Afforestation of Agricultural Land*, Volz, K.-R. and Weber, N., Eds., Commission of the European Communities, Luxembourg, 115, 1993.

Lust, N., personal communication, 2003.

Maaren, A. van, Afforestation experiences: opportunities and constraints, in *Afforestation of Agricultural Land*, Volz, K.-R. and Weber, N., Eds., Commission of the European Communities, Luxembourg, 141, 1993.

Madsen, P., Matthesen, P. and Thormann, A., Afforestation in Denmark, in *Proceedings of the IUFRO Conference on Restoration of Boreal and Temperate Forests*, Gardiner, E.S. and Breland, L.J., Eds., Danish Centre for Forest Landscape and Planning, Hørsholm, 40, 2002.

Mantel, K., Der Standort der Forstwirtschaft im Wettbewerb um den Raum, *Der Forst- und Holzwirt*, 20, 530, 1965.

Marghescu, T., Ed., *Nature Conservation in Private Forests of Selected CEE Countries*, Opportunities and Constraints, Programme paper, IUCN, Tilburg, 2001.

Mather, A.S., Ed., *Afforestation. Policies, Planning and Progress*, Belhaven Press, London and Florida, 1993, chaps. 1 and 14.

Mather, A.S., Afforestation: progress, trends and policies, in *NEWFOR — New Forests for Europe: Afforestation at the Turn of the Century*, Weber, N., Ed., EFI Proceedings No. 35, European Forest Institute, Joensuu, 11, 2000.

McCormack, A.C. and O'Leary, T.N., Towards a holistic and balanced forest landscape strategy in Ireland, in *NEWFOR — New Forests for Europe: Afforestation at the Turn of the Century*, Weber, N., Ed., EFI Proceedings No. 35, European Forest Institute, Joensuu, 101, 2000.

Merlo, M., Stellin, G., Harou, P., and Whitby, M., Eds., Multipurpose agriculture and forestry, in *Proceedings of the 11th Seminar of the EAAE*, 28 April–3 May 1986, Wissenschaftsverlag Vauk, Kiel, 1987.

Nabuurs, G.-J., European forests in the 21st century: impacts of nature-oriented forest management assessed with a large-scale scenario model, *Alterra Scientific Contributions* 2, Alterra Green World Research, Wageningen, 2001.

Ní Dhubháin, Á. and Wall, S., Farmers and rural attitudes to forestry, in *Public Perception and Attitudes of Forest Owners to Forests in Europe*, Terrasson, D., Ed., Cemagref Editions, Antony Cedex, 165, 1998.

Nieuwenhuis, M., The Republic of Ireland, in *Forestry in Changing Societies in Europe. Information for Teaching Module. Part 2*, Pelkonen, P., Pitkänen, A., Schmidt, P., Oesten, G., Piussi, P., and Rojas, E., Eds., SILVA Network, Joensuu, 167, 1999.

O'Leary, T. and Elands, B.H.M., Anyone for more forests? Current perspectives and future expectations on afforestation and forest functions across Europe, in *The Changing Role of Forestry in Europe: Perspectives for Rural Development*, Wiersum, K.F. and Elands, B.H.M., Eds., Proceedings 2002-02, Forest and Nature Conservation Policy Group, Wageningen University, Wageningen, 51, 2002.

Päivinen, R., Lin, C., Ottitsch, A., Schuck, A., and Moiseyev, A., The European dimension, in *Forestry in Changing Societies in Europe. Information for Teaching Module. Part 2*, Pelkonen, P., Pitkänen, A., Schmidt, P., Oesten, G., Piussi, P. and Rojas, E., Eds., SILVA Network, Joensuu, 5, 1999.

Palo, M., Global prospects on deforestation and forest transition, in *World Forests from Deforestation to Transition?* Palo, M. and Vanhanen, H., Eds., Kluwer Academic Publishers, Dordrecht, 3, 2000.

Peck, T. and Descargues, J., The policy context for the development of the forest industries sector in Europe, *Beiträge der Professur Forstpolitik und Forstökonomie* der ETH Zürich, Zürich, 14, 1995.

Pereira, J.S., Environmental impact assessment in afforestation with Eucalyptus in Portugal, in *Afforestation of Agricultural Land*, Volz, K.-R. and Weber, N., Eds., Commission of the European Communities, Luxembourg, 179, 1993.

PHARE (civilian aid programme for the Central and Eastern European Region), *Conservation and Sustainable Management of Forests in Central and Eastern European Countries*, PHARE Multi-Country Forestry Programme, 2000.

Phelan, J., *Promotion of Afforestation through Relevant EU Funds*, paper for the Conference *"Afforestation in the Baltic Sea Region"*, Vilnius, 2–3 October, http://redep.gamta.lt/phelan.html, 2001.

Piussi, P. and Pettenella, D., Spontaneous afforestation of fallows in Italy, in *NEWFOR — New Forests for Europe: Afforestation at the Turn of the Century*, Weber, N., Ed., EFI Proceedings No. 35, European Forest Institute, Joensuu, 151, 2000.

Plochmann, R. and Thoroe, C., Förderung der Erstaufforstung, *Schriftenreihe des Bundesministers für Ernährung, Landwirtschaft und Forsten*, Reihe A — angewandte Wissenschaft — Heft 397, Bonn, 1991.

Præstholm, S. and Jensen, F.S., *Attitudes and Practices in Relation to the Landscape and Forest in Haderslev and Hvorslev Municipalities*, Denmark, Danish Forest and Landscape Research Institute, Hørsholm, 2001.

Pryor, S., Afforestation in Britain: opportunities and obstacles, strategies and targets, in *NEWFOR — New Forests for Europe: Afforestation at the Turn of the Century*, Weber, N., Ed., EFI Proceedings No. 35, European Forest Institute, Joensuu, 77, 2000.

Selby, A., personal communication, 2003.

Selby, A. and Petäjistö, L., *Field Afforestation in Finland in the 1990s: Objections, Preconditions and Alternatives*, Finnish Forest Research Institute, Research Papers 502, Helsinki, 1994.

Selby, A. and Petäjistö, L., A critical appraisal of afforestation programmes in the light of Finnish and Irish experiences, in *NEWFOR — New Forests for Europe: Afforestation at the Turn of the Century*, Weber, N., Ed., EFI Proceedings No. 35, European Forest Institute, Joensuu, 52, 2000.

Seoane, I., European Community Aid Schemes, ECCP Working Group on Forest Carbon Sinks, unpublished presentation, 2002.

Slee, B. and Wiersum, K.F., New opportunities for forest-related rural development in industrialized countries, *For. Policy Econ.*, 3, 1, 2001.

Soloviy, I.P., Afforestation in Ukraine — potential and restrictions, in *NEWFOR — New Forests for Europe: Afforestation at the Turn of the Century*, Weber, N., Ed., EFI Proceedings No. 35, European Forest Institute, Joensuu, 195, 2000.

Spiecker, H., Mielikäinen, K., Köhl, M., and Skovsgaard, J.P., Eds., *Growth Trends in European forests — Studies from 12 Countries*, Springer, Heidelberg, 1996.

Stanturf, J.A. and Madsen, P., Restoration concepts for temperate and boreal forests of North America and Western Europe, *Plant Biosyst.*, 136, 143, 2002.

Steinlin, H., Globale und volkswirtschaftliche Aspekte der Forstwirtschaft, *Mainauer Gespraeche*, Mainau, Band 10, 15, 1994.

Terrasson, D., Introduction to the parallel session 2, *International Policy Research Symposium on The Changing Role of Forestry in Europe*, 13 November, Wageningen, The Netherlands, 2001.

Tyrväinen, L. and Tahvanainen, L., Impacts of afforestation on the scenic value of rural countryside, in *NEWFOR — New Forests for Europe: Afforestation at the Turn of the Century*, Weber, N., Ed., EFI Proceedings No. 35, European Forest Institute, Joensuu, 141, 2000.

UN-ECE/FAO, *Main Findings and Implications of the Study European Timber Trends and Prospects: Into the 21st Century (ETTS V)*, Timber Committee 54th Session, 1996.

UN-ECE/FAO, *State of European Forests and Forestry 1999*, ECE/TIM/SP/16, United Nations, Geneva, 1999.

UN-ECE/FAO, *Forest Resources of Europe, CIS, North America, Australia, Japan and New Zealand*, Geneva Timber and Forest Study Papers, No. 17, United Nations, New York and Geneva, 2000.

Volz, K.-R. and Weber, N., Eds., *Afforestation of Agricultural Land*. Proceedings of a workshop held in Brussels, 12–13 December 1991, Commission of the European Communities, Report EUR 14804 EN, 1993.

Weber, N., Empirical investigations on afforestation in Germany, in *Afforestation of Agricultural Land*, Volz, K.-R. and Weber, N., Eds., Commission of the European Communities, Luxembourg, 47, 1993.

Weber, N., Ed., *NEWFOR — New Forests for Europe: Afforestation at the Turn of the Century*, EFI Proceedings No. 35, European Forest Institute, Joensuu, 2000.

Weisgerber, H., Aufforstung landwirtschaftlicher Flächen aus forstlicher Sicht — Neue Anbauformen mit raschwüchsigen Baumarten in kurzen Umtriebszeiten, *Agrarspectrum* 14, 213, 1988.

Part 4

The Practice of Forest Restoration

chapter nine

Procurement of plant material of good genetic quality

Erik D. Kjær, Christian Pilegaard Hansen, Hans Roulund, and Lars Graudal
Danish Centre for Forest, Landscape and Planning, KVL

Contents

9.1 Introduction

Trees are planted for many reasons and have a wide range of uses, including production of timber, pulpwood, environmental protection, recreation, urban forests, and nature recreation. Many factors influence the success of such plantings, but in this chapter only one factor of major importance will be considered: the importance of selecting the right genetic material. Selection of appropriate genetic material is an important decision, as it will influence all phases of a plantation from establishment to harvest. If trees are allowed to regenerate naturally, or if stands are subsequently used for seed collection, the selection of planting material will also influence the success and quality of future forests.

Appropriate genetic material can increase the amount and quality of the products deriving from the plantations, economic output, resistance to pests and diseases, and stability of the trees. Using inappropriate genetic material, on the other hand, may lead to suboptimal timber outputs or poor stability and resistance to windthrow. Some examples will illustrate the importance of selecting the right genetic material, the benefits that can

1-56670-635-1/05/$0.00+$1.50

be derived, and the problems that may arise from using suboptimal genetic material. The North American lodgepole pine (*Pinus contorta* Dougl.) has been planted as an exotic species in Denmark on wind-exposed, infertile coastal sites. Initially, provenances from western coastal areas of the U.S. were assumed to fit Danish maritime conditions. Seed from these origins, however, caused many total stand failures. However, plantings of material from seed of other, more continental sources grow very well in Denmark (Larsen 1997; Wellendorf and Feilberg 1984; Larsen and Nielsen 1982). The success or failure of planting *Pinus contorta* in Denmark is therefore a question of carefully selecting planting material of optimal origin.

Tree improvement has long been important to commercial forest plantations. Breeding programs focus on increasing yields, wood quality, and pest resistance. For example, the growth rate of radiata pine (*Pinus radiata* D. Don) in New Zealand has been improved by 30% following 30 to 40 years of tree improvement (Carson et al. 1990). In Australia, increases in productivity of 14 to 24% following one generation of tree breeding have been reported (Cotterlill et al. 1989). In the southern U.S., slash pine (*Pinus elliottii* Engelm.) is planted on approximately 4 million ha. The gain from the first generation of tree improvement has been approximately 7% in volume production, and is expected to increase to 17% after the second generation of breeding (White et al. 1993). At the same time, resistance to rust has also been improved. Gains from growing improved trees on such a large area are obviously substantial. Loblolly pine (*Pinus taeda* L.) is even more widely planted. Family plus within-family selection was the most effective to achieve genetic gain for early selection on both height and volume (Xiang et al. 2003) and an additional gain of 10 to 40% can be achieved by capturing the nonadditive genetic component through mass production of full-sib crosses or vegetative propagation.

The economic benefits of breeding Norway spruce (*Picea abies* (L.) Karst.) and Scots pine (*Pinus sylvestris* L.) in Sweden have been estimated at 600 million SKr (approximately US$75 million) for a low-input breeding program to 900 million SKr (US$110 million) for an intensive program. The internal rate of return in the intensive tree improvement program is calculated as 4.1% (SkogsForsk 1995).

These examples all relate to volume production where gains can be considerable. The meaning of "appropriate" is, however, different for different objectives. For example, adaptability (being adapted to the planting location) is crucial for all plantings, but it is more important for restoring natural ecosystems than for plantations. Restoration requires sufficient genetic diversity so that natural regeneration will continue for generations. Plantations need only survive one generation and do not even need to reproduce. Some examples from Denmark of appropriate genetic material are presented in Table 9.1, which considers the objective of planting.

9.2 The seed procurement context

A prerequisite for any planting program is the availability of seed (or other reproductive material) in sufficient quantity and quality. Quality relates to physiological vigor as well as genetic constitution. The physiological condition will determine whether a seed can germinate and survive the early stages of establishment. The genetic constitution will often have impact on all phases of a planting, from establishment to harvest (Foster et al. 1995). However, a given seed source may not be superior at all sites, or for all purposes. Genetic quality must therefore be defined relative to a given planting site and planting purpose (Table 9.1). Seed sources are rarely universally "good" or "bad" but may be inappropriate for some objectives. Thus, procurement of good seed is complicated and is always an important part of any planting program. Unfortunately, seed sources and procurement systems are rarely developed in the initial phases of larger planting programs, which leads to

Table 9.1 Examples of Objectives for Different Types of Plantings on Different Sites in Denmark, Showing the Requirements for the Planting Material, and Some Possible Seed Sources

Objective of the Planting	Requirements for the Planting Material	Examples of Seed Sources Which Fulfil the Requirements
Production of *Abies procera* decoration greenery in approx 35-year rotation	Few needle damages, good decoration quality, color, high production (amount of decoration). High survival in establishment	Seed orchards: C.E. Flensborg (FP 623) Approved stands: Frisenborg (F449)
Economic production of Norway spruce (*Picea*	Wood quality, production, health, and stability.	Seed orchards: Hoffmansgave (FP 240)
abies) timber in 50-year rotation. Site with poor soil and spring frost. No requirements of natural regeneration	High survival in establishment	Approved stands: Fårhus (F 463) Cuttings: E.g., cuttings from progeny of plus trees, where the clonal mix has the desired traits.
Economic production of oak timber in East Denmark in 130-year rotation	Stem form, growth rate, and heartwood percentage. Establishment security	Approved stands of foreign origin: *Quercus robur*, Bidstrup (F148)
Oak forest on poor soil in Western Denmark. Protection and recreation forest; timber production in 150-year rotation	Health and adaptation in the long term Stem form, growth rate. Establishment security	Approved stands of Danish origin: *Q. petraea*, Løvenholm (F212)
Prunus padus for city plantings	Homogenous. Health	Controlled breeding: Para Dafo
Prunus padus for planting in the forest edges for long-term survival and natural regeneration. Protection and recreation forest, biodiversity (insects, birds, game)	Health and adaptation in the long term Establishment security. Genetic diversity	Selected seed sources: Rådmandshaven
Economic production of noble ash (*Fraxinus excelsior*) timber in 50-year rotation	Stem form, growth rate. Establishment security	Seed orchards: Birkemarken (FP202)
Ash forest (*Fraxinus excelsior*) established for production of quality wood and introduction of a continuous and stabilising broadleaved element for later natural regeneration, possibly in mix with beech (*Fagus sylvatica*)	Health and adaptation in the long term. Stem form, growth rate. Establishment security	Approved stands: Gråsten (F615)

[*Source*: Graudal, E.D. Kjær and Thomsen, A., Eds., Brug af frøkilder af træer og buske—Bidrag til en bæredygtig anvendelse af de genetiske ressourcer [Sustainable use of genetic resources of trees and shrubs in Denmark]. The Tree Improvement Station, The Danish Forest and Nature Agency, Copenhagen, Denmark, 1997.].

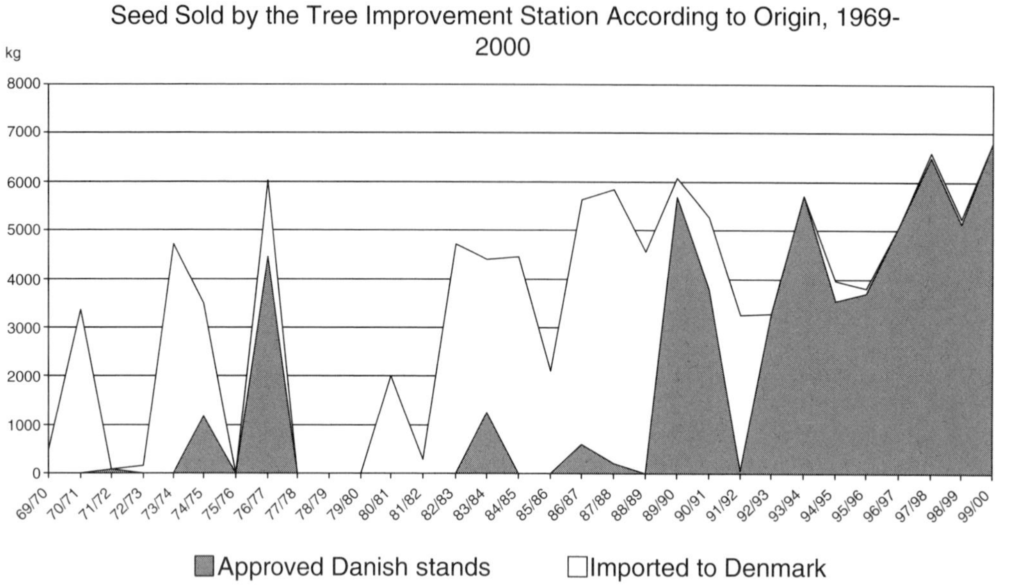

Figure 9.1 Change in origin of beech seeds (*Fagus sylvatica*) sold from the Tree Improvement Station in Denmark over a 30-year period. Improved storage technology for beech has increased the share of local seed sources. (*Source*: The Tree Improvement Station of the Danish Forest and Nature Agency, Krogerupvej 21, DK3050 Denmark (unpublished data.))

importing seed from long distances (Figure 9.1). Because it is often difficult to obtain sufficient amounts of seeds at the beginning of a planting program, it may be best to rely on robust strategies such as procuring seed of local origin, which often presumes a definition of seed deployment zones or planting zones.

9.2.1 Seed source management

Establishing and managing good seed sources is necessary in order to increase the supply of good quality seed. Improved genetic material can only be obtained by applying tree improvement (selection and breeding) activities. Even in the short run, decisions on seed sources should consider genetic implications. Heavy thinning regimes generally promote flowering, but can also lead to directional selection if selection is targeting a character under modest or high genetic control at the site. Collection of seed from only a few trees will also have genetic implications, because low population number is associated with genetic drift and inbreeding. Therefore, tree improvement and seed source management are overlapping activities.

Tree improvement will only benefit plantings if improved genetic material can be successfully propagated on a commercial scale and deployed in the actual planting programs according to its genetic features. Tree improvement programs must be designed and implemented in ways that make multiplication and large-scale use both economical and technically feasible. Tree improvement decisions must therefore include seed procurement and propagation considerations in order to form an integrated program.

9.2.2 Conservation of genetic resources

The purpose of conservation of genetic resources is to secure the ability for a species to adapt to environmental change and to maintain the basis for future selection and

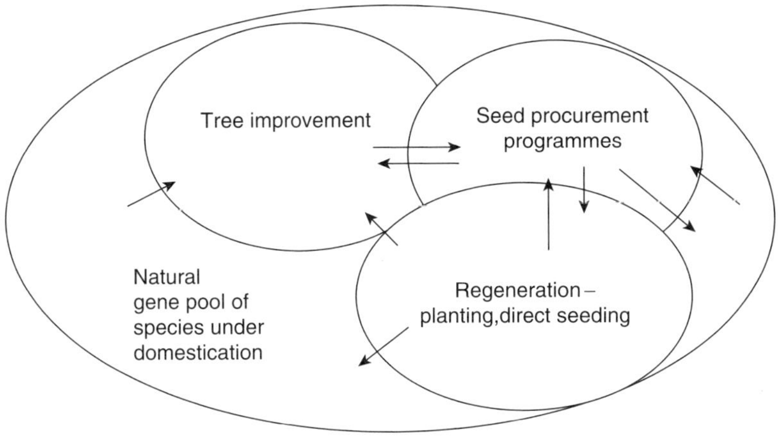

Figure 9.2 The seed procurement context.

breeding activities (Graudal et al. 1995). Seed procurement should therefore be seen in an integrated context (Figure 9.2). Tree improvement, seed procurement, and planting are activities on a scale that often influences the genetic resource. For example, large-scale plantings of a native species may be based on imported seed that is poorly adapted to local conditions (Figure 9.1). However, problems also arise when plantings of rare or uncommon species are based on highly selected, but genetically very narrow seed sources or clonal mixtures. A few alleles (genes) from selected genotypes may crowd out the native gene pool by genetic drift alone, or in combination with selective advantages. The result can be unintended genetic erosion of the native gene pool. Breeding programs can be an effective means of conserving and developing the genetic resources of domesticated species, if based on sound genetic principles in a multiple population context (Namkoong et al. 1988), for example, the Swedish program for Norway spruce (Danell 1993a).

9.3 From wild populations to improved material

Humans have always used natural forests, and for centuries afforestation with exotic or native species has taken place to produce wood, prevent erosion, or create a more sheltered environment for people, livestock, or crops. Trees are genetically diverse organisms and better growth, quality, and adaptability can be achieved by carefully selecting the best seed sources for regeneration. Tree improvement can therefore be seen as a process from using material from wild populations to using identified seed sources and selected material that enhances the economic and social benefits of the plantings.

Tree improvement covers a continuum of activities from simple selection and seed collection to advanced breeding schemes with supportive research. Activities may include multiple test sites and propagation in intensely managed seed orchards, by cuttings, or *in vitro*. In less advanced programs, tree improvement activities may merely concentrate on the establishment of a few well-adapted seed sources based on the selection of a number of phenotypically selected mother trees.

Tree improvement in a broad sense may consist of the following activities:

- Seed source identification and selection of seed sources (provenances) based on seed zone systems without testing

- Seed source (provenance) testing and use of best seed sources
- Breeding based on selection of superior trees (plus trees), establishment of seed orchards (from seed or by grafting), and progeny trials at multiple sites
- Mass propagation of improved material for commercial plantations.

Identification and selection of seed sources based on an untested seed zone system (Activity 1) is not considered tree improvement in a narrow sense, but it is the starting point for domestication. Commercial mass propagation of improved genetic material (Activity 4) is, strictly speaking, not a tree improvement activity, but another important part of the domestication process.

9.3.1 *Selection provenances without testing (seed zone systems)*

The simplest form of improvement is to select seed sources based on their appearance at their respective locations and use them only in similar environments, that is, match seed source to planting site based on ecological similarity. Seed deployment zones or planting zones are ecologically similar areas (Lillesø, et al. 2000). Seed should not be transferred between zones, but transfer within a zone is unrestricted, in order to avoid establishing plantations with untested planting material that has been adapted to completely different climatic conditions. Using a seed zone system does not guarantee that the best seed sources will be used at a given location (Namkoong 1969), but it does reduce the risk of using nonadapted genetic material, and therefore the risk of plantation failure. Seed deployment zones can be developed based on results from actual testing (Campbell 1986, 1991; Westfall 1992; Illingworth and Szalkai 1993; Randall 1996; Lindgren and Ying 2000; Haman et al. 2000; Schmidtling 2001; Randall and Berrang 2002), in which case gains from using such zones can be qualified and quantified.

9.3.2 *Breeding*

Conventional breeding consists of selection in defined breeding populations, combined with testing. Testing is typically done in progeny trials; seed is produced in separate clonal seed orchards. Results from progeny trials enable the removal of inferior clones from the seed orchard and selection of clones for the next breeding cycle. There are many combinations of testing, selection, crossing of material, and seed orchard management (Namkoong et al. 1988).

9.3.3 *Multiple population breeding strategy based on seed zones*

Because trees are planted under variable conditions, seed sources must be matched to sites. Seed sources developed in one region may not be suited to different environmental conditions. To cope in tree improvement with this problem of "genotype by environment interaction," the Multiple Population Breeding Strategy (MPBS) was developed (Namkoong et al. 1980, 1988). In MPBS, seed sources are raised, tested, and managed separately in a number of subregions. This strategy allows faster progress and more flexibility in tree improvement, not only to cope with differences in environment but also to direct the breeding toward improvement in different traits.

9.3.4 *Gains of tree breeding*

Tree breeding is normally based on genetically diverse material, in contrast to the strategy of breeding agricultural crops (NRC 1991a, 1991b). This is because trees are grown in long rotations and under less controlled environments, usually without irrigation or fertiliza-

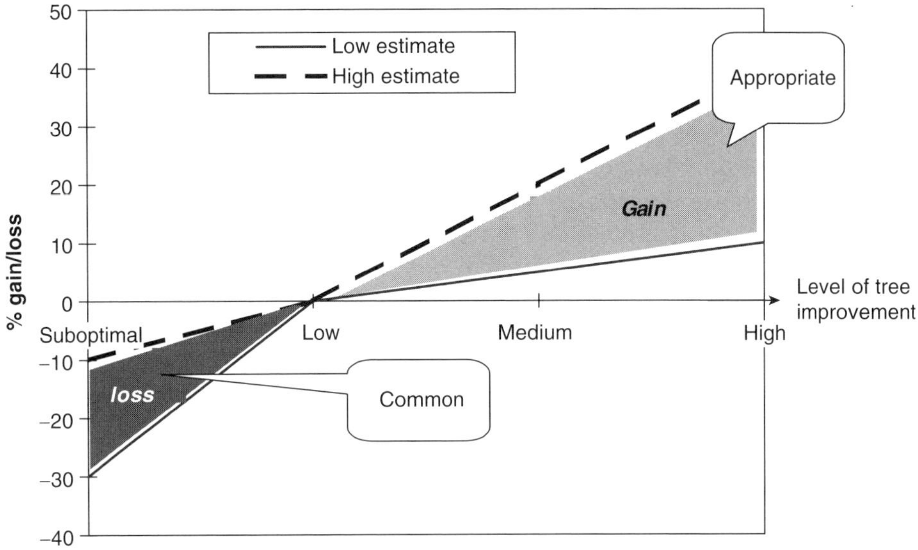

Figure 9.3 Possible gains from the first cycle of tree improvement, based on realized experience with volume production. Gains possible from tree improvement increase from left to right. The suboptimal area (negative gain) is where many tree plantings originate and the "gain" area above the line is where they ought to originate. The use of tested seed orchards (high level of improvement) requires establishment of field trials, whereas the untested seed orchards (medium level of improvement) are based on simple selection and propagation of old, superior trees (*Source:* modified after: Graudal and Kjær (2001) mainly based on data from Foster *et al.*, 1995.)

tion. This will, in general, lead to lower rates of gain compared to breeding agricultural crops. Typical gains in vigor from domestication programs for forest tree species are in the range of 10 to 40% for each breeding cycle (Foster et al. 1995) (Figure 9.3). The duration of breeding cycles varies from species to species. In the boreal and temperate zones, it will often be 15 to 30 years. As a rule of thumb, the productivity of planting material may be increased approximately 1% annually by breeding (Danell 1990). Gains in wood quality may be as large as volume gains and certainly are equally as important.

9.3.5 *Management of diversity in breeding programs*

Human activities influence planting material in ways other than simply improving desired characters. Genetic diversity will be reduced if plantings are established based on only one or a few clones. This will be especially problematic if the entire breeding program is based on a few selected trees (clones), and therefore not designed for long-term breeding (Roulund 1981). Narrowing the genetic base may cause future problems because diversity serves as a buffer against environmental fluctuations, pests, and inbreeding depression.

Tree improvement will not result in the loss of any substantial amount of genetic variance if breeding is based on a sufficient number of genetically distinct individuals (Danell 1993b; Falconer 1989). The minimal acceptable diversity in individual plantings has been discussed intensely (Roberds et al. 1990; Lindgren 1992; Libby and Ahuja 1992; White 1992; Johnson et al. 2001). It will be lower than the number of individuals required for the whole breeding population and will depend on the species, silvicultural regime, and expected length of rotation. Factors that increase the recommended amount of genetic variation needed in planting stock include a long rotation, use of untested material, and intent to use natural regeneration. Less diversity is acceptable for short rotation regimes or if the planting material has been intensively tested. Narrow genetic diversity may be suitable only for species with very short rotation and therefore rapid replacement. In all cases, it is important that genetic diversity is maintained in long-term breeding programs.

Other traits may unintentionally change in response to breeding. Fast-growing trees, for example, may have low wood density, bad stem form, or be less fertile. Trees without epicormic branches are attractive to sawmills but may suffer more if defoliated. It is therefore important to investigate for correlations between desired and undesired traits. Using only tested stock can reduce the risk of introducing undesirable traits. Intensive selection should only be performed as part of intensive breeding programs. Less intensive breeding programs should work with broad genetic populations. The importance of maintaining sufficient genetic diversity in the breeding populations should always be recognized (El-Kassaby 2000; Johnson and Lipow 2002).

New seed sources should be tested carefully, as local seed sources are not necessarily superior (Namkoong 1969). Some examples of nonlocal seed sources that were superior, include the Livingston Parrish source of loblolly pine for fusiform rust resistance and growth (Wells 1985), the Haney provenance of Sitka spruce for weevil resistance (Ying 1991; 1997), and the Guadalupe source of *Pinus radiata* for wood density (Burdon and Low 1992; Low and Smith 1997).

Nevertheless, using untested, introduced seed sources may add additional risk. Adaptation problems may only be apparent after a number of years, such as following a year with extreme climatic conditions. Testing therefore requires a sufficient time span to encompass the full range of climatic conditions. Experience with plantings of Norway spruce in Denmark illustrates this problem. The species is exotic to Denmark, but has been widely planted for a century because of its easy establishment, high productivity, and high wood quality. Planting in Denmark has historically been based on seed imported from western Germany. However, this situation changed in the early 1980s when results from testing in southern Sweden and Denmark suggested that provenances from eastern Europe (Poland and Romania) were superior to traditional sources. From the early 1980s, Romanian and Polish seed sources were used over large areas in both Sweden and Denmark. Later assessments of field trials indicated increasingly poor adaptability of the Eastern European provenances to harsh sites, especially sites influenced by wind (Larsen and Wellendorf 1997; Wellendorff 1988). Stands established with seed from eastern Europe have developed very poorly, often deteriorating long before rotation age. This illustrates the importance of careful selection of genetic material and the large, adverse effects that may arise from premature use of insufficiently tested material.

9.3.6 *Role of biotechnology*

Biotechnology has produced techniques for genetic studies and mass propagation of plants. Genetic markers at the level of DNA have been used for genome mapping and other genetic studies (Wang and Szmidt 2001) such as marker-aided selection. Herbicide resistance has been introduced into some species by genetic engineering and progress has been made on insect resistance. Micropropagation from buds and somatic embryogenesis have been developed to rapidly multiply improved material. Advances from biotechnology, so useful for agricultural crops, have so far proven less useful for tree species. Presently, there are no large-scale examples where biotechnology has been applied to commercial advantage in forest plantation programs (Haines and Martin 1997). To date, few, if any, commercial plantation programs have been based on transgenic plants.

Because many genes regulate most of the desired characteristics in trees, the use of biotechnology in breeding and propagation will probably be limited for the near future, except for a few very important commercial species. The necessary conditions needed for safe and advantageous use of biotechnology for improved tree planting are discussed by Carson et al. (1996), Davey et al. (1998), and El-Kassaby (2003).

9.3.7 Matching environment and climate change

Many years' experience domesticating exotics and improving important native species have shown that plant lots (provenances, seed orchard, clone mixtures, individual clones) change rank characters when planted on different soil types and in different climates. This interaction between genotype and the environment is recognized both in the use of homogeneous seed zones and breeding programs based on such zones (Matheson and Cotterill 1990; Lynch and Walsh 1998). The prospect of future climate change, however, may negate the usefulness of these zones. Climate change is a difficult concept to plan for, first because it is a very uncertain prospect, and second because it is difficult to agree upon a timeframe.

One strategy is to breed for general adaptability (Larsen 1990), and to ensure the presence of a high level of genetic variability in planting stocks in order to adapt to whichever direction climate changes (Larsen 1995). Another strategy is to network and exchange breeding material of a species between zones representing various climatic conditions, such as the Swedish program for Norway spruce (Danell 1993a). This will result in improved material adapted to many different climatic and edaphic conditions. Recently, focus has increased on low-input breeding models that include the ability to provide gain in many species and populations simultaneously, and to maintain high genetic diversity at low costs (Lindgren 1999, 2000).

9.4 Legislation, regulation, and control of genetic quality

Tree planting has taken place in nearly all countries in the boreal and temperate zones in the last several hundreds of years. Over the last 50 to 70 years, most countries developed legislation and regulation for plant material moving in international trade. However, it is not easy to control seed deployment through legislation, because a given seed source may be valuable in one context, but inferior in another. Three strategies can be pursued for regulating deployment of seed sources:

1. Strict legislation focused on defining which seed sources can be used where, and for what purpose. Although this kind of regulation is inherent in the seed transfer guidelines developed for some undomesticated species, it is rarely based on strict legislation. Very "fine tuned" legislation would be needed if this approach was applied to improved species.
2. Legislation focused on allowing only "superior" (approved) seed sources on the commercial market. This kind of legislation has been applied for a number of species within the European Union, where each country has a designated authority that approves seed sources at different genetic levels (identified, selected, or tested). The authorities publish official lists of seed sources, and control seed harvest, sale, imports, and exports. This kind of legislation may discriminate against the use of local seed sources adapted to harsh conditions because such sources may not qualify for wider use. Another drawback with this system is that a given seed source may be well adapted in one part of Europe, but poorly adapted in another part. In the European system, it is possible to exclude the use of seed of specific origin, but this option is rarely exercised.

3. Regulation through information and development of specific seed source recommendations. This strategy requires that the end users realize the value of using appropriate planting material, are able to identify the origin and trust the quality of this information, and have access to preferred seedlings. Official control of seed collection, processing, and marketing by a public authority (EU 1999; OECD 1974) results in a system where each seed lot has an informative label stating the origin and genetic features, combined with public access to information about seed sources and their utilization. Market access to seedlings of preferred origin will usually be driven by demand, but coordination and planning by some public agency will be required. The time lag between initial demands until the seed can be effectively procured is one complication that requires a proactive effort.

An important practical aspect is that the cost of these regulations does not exceed the expected gain from using improved and more adapted seed sources.

9.5 Conclusions

Tree planting offers the opportunity to carefully consider the choice of species and genetic material to be used. Significant benefits will arise from using appropriate genetic material, such as enhanced production and quality of products, more stable plantations, reduced damage from pests and diseases, or better environmental services. Consequently, suboptimal planting material will reduce benefits, or may, in worst cases, lead to unstable plantings or stand degradation. Breeding programs built on sound genetic principles can maintain, and may even increase, genetic diversity. Great care should be taken in domestication programs to secure sufficient genetic diversity in the planting material. This is important in order to encompass the range of conditions presented by heterogeneous planting sites and to take into account the effects of climatic fluctuations and the possibility of climate change.

References

Burdon, R.D. and Low, C.B., Genetic survey of *Pinus radiata*, 6: Wood Properties; Variation, heritabilities, and interrelationships with other traits, *N. Z. J. For. Sci.*, 2, 228, 1992.

Campbell, R.K., Mapped genetic variation of Douglas-fir to guide seed transfer in southwest Oregon, *Silvae Genet.*, 35, 85, 1986.

Campbell, R.K., Soils, seed-zone maps, and physiography: guidelines for seed transfer in SW Oregon, *For. Sci.*, 37, 973, 1991.

Carson, M.J., Burdon, R.D., Carson, S.D., Firth, A., Shelbourne, C.J.A., and Vincent, T.G., Realising genetic gains in production forests, in *Joint Working Parties Handbook*, Joint Meeting of Western Forest Genetics Association and IUFRO Working Parties S2.02-05, 06, 12 and 14, Douglas-fir, Lodgepole pine, Sitka and Abies spp., Breeding and Genetic resources, 1990.

Carson, M.J., Carson, S.D., Richardson, T.E., Walter, C., Wilcox, P.L., Burdon, R.D., and Gardner, R.C., Molecular biology application to forest trees: fact or fiction? in *Tree Improvement for Sustainable Tropical Forestry, QFRI-IUFRO Conference*, Dieters, M.J., Matheson, A.C., Nikles, D.G., Harwood, C.E., and Walker, S.M., Eds., Caloundra, Queensland, Australia, 27 Oct.– 1 Nov. 1996, Queensland Forest Research Institute, Gympie, 1996.

Cotterlill, P.P., Dean, C., Cameron J., and Brindbergs, M., Nucleus breeding: a new strategy for rapid improvement under clonal forestry, in *Breeding Tropical Trees, Proceeding from IUFRO Conference*, Gibson, G.I., Griffin, R., and Matheson, A.C., Eds., Pattaya, Thailand, 1989.

Danell, Ö., Possible gains in initial stages of a national tree improvement programme using different techniques, in *Consequences of Modern Techniques in Tree Breeding Strategies, Forest Tree Improvement*, 23, 11, 1990.

Danell, Ö., Breeding programmes in Sweden — General approach, in *Progeny Testing and Breeding Strategies, Proceedings of the Nordic Group for Tree Breeding*, Edinburgh, Scotland, 6–10 Oct. 1993, Supplement, Lee, S.J., Ed., Forestry Commission, Edinburgh, 1993a.

Danell, Ö., Tree breeding strategies: are we too concerned conservationists but inefficient breeders? in *Progeny Testing and Breeding Strategies, Proceedings of the Nordic Group of Tree Breeding*, Lee, S.J., Ed., Edinburgh, Oct. 1993, Forestry Commission, United Kingdom, 1993b.

Davey, M.R., Alderson, P.G., Lowe, K.C., and Power, J.B., Eds., *Tree Biotechnology: Towards the Millennium*, Nottingham University Press, U.K., 337, 1998.

El-Kassaby, Y.A., Effect of forest tree domestication on gene pools, in *Forest Conservation Genetics, Principles and Practice*, Young, A., Boshier, D., and Boyle, T., Eds., Collingwood, CSIRO Publishing, VIC, Australia, 197, 2000.

El-Kassaby, Y.A., *Feasibility and Proposed Outline of a Global Review of Forest Biotechnology*, Working Paper, FAO Panel of Experts on Forest Gene Resources, 13th Session 10–12 November 2003, FAO, Rome, Italy, 2003.

European Union, *Council Directive 1999/105/EC of 22 December 1999 on the Marketing of Forest Reproductive Material*, Official Journal L011, 15/01/2000, 17, 1999.

Falconer, D.S., *Introduction to Quantitative Genetics*, 3rd ed., Longman Scientific and Technical, London, 1989.

Foster, G.S., Jones, N., and Kjær, E.D., Economics of tree improvement in development projects in the tropics, in *Environmental & Economic Issues in Forestry: Selected Case Studies in Asia*, Shen, S. and Contreras-Hermosilla, A., Eds., World Bank Technical Paper No. 281, The World Bank, Washington, DC, 95, 1995.

Graudal, L., and Kjær, E. D., Can national tree seed programmes generate economic, social and/or environmental benefits that cover their costs - considerations on economics, sustainability and challenges ahead for tree seed centres in tropical countries, in Eyog-Matg, Kigomo, B., and Boffa, J.M., Eds., *Recent Research and Development in Forest Genetic Resources*, Proceedings from the regional workshop, Nairobi 6-11 December 1999, IPGRI, Nairobi, Kenya, 2000.

Graudal, L., and Kjær, E.D., Tree improvement priorities and strategies in *Recent Research and Development in Forest Genetic Resources*, Eyog-Matg, Kigomo, B., and Boffa, J-M., Eds., Proceedings from the regional workshop in Nairobi 6–11 December, 1999. International Plant Genetic Resource Institutte, Nairobi, Kenya, 2001.

Graudal, L., Kjær, E.D., and Canger, S.C., A systematic approach to the conservation of genetic resources of trees and shrubs in Denmark, *For. Ecol. Manage.*, 73, 117, 1995.

Graudal, L., Kjær, E., and Thomsen, A., *Proposal for a strategy for the use of seed sources of trees and shrubs for forest and landscape plantings in Denmark* [In Danish, *Oplæg til en strategi for brug af frøkilder af træer ogbuske til skov og landskabsformål i Danmark*], Skov- og Naturstyrelsen, Statskovenes Planteavlsstation, 100 pp., 1997.

Haines, R. and Martin, B.E., Biotechnology and sustainable production of tropical timber, *For. Gene. Resour.*, 25, 52, 1997.

Haman, A., Koshy, M.P., Namkoong, G., and Ying, C.C., Genotype × environment interactions in *Alnus rubra*: developing seed zones and seed transfer guidelines with spatial statistics and GIS, *For. Ecol. Manage.*, 136, 107, 2000.

Illingworth, K. and Szalkai, A., Summary, in *British Columbia Interior Tree Improvement Council Second Progress Report 1985–1990*, Albricht, M., Ed., B.C. Ministry of Forests, 52 pp., 1993.

Johnson, R., St. Clair, J.B., and Lipow, S.R., Genetic conservation in applied tree breeding programs, in *Proceedings: International Conference on Ex situ and In situ Conservation of Commercial Tropical Trees* (http://www.fs.fed.us/pnw/pubs/journals/Johnson_StClair_Lipow_2001.pdf) 215, 2001.

Johnson, R., and Lipow, S.R., Compatibility of breeding for increased wood production and long-term sustainability: the genetic variation of seed orchard seed and associated risks, in *Congruent Management of Multiple Resources: Proceedings from the Wood Compatibility Initiative Workshop*, Adelaide C. Johnson, Richard W. Haynes and Robert A. Monserud, Eds., http://www.fs.fed.us/pnw/pubs/gtr563/PNWGTR563B.pdf, 2002.

Larsen, J.B., Breeding for physiological adaptability in order to counteract an expected increase in environmental heterogeneity, in Consequences of Modern Techniques in Tree Breeding Strategies, *For. Tree Improvement*, 23, 5, 1990.

Larsen, J.B., Ecological stability of forests and sustainable silviculture, *For. Ecol. Manage.*, 73, 85, 1995.

Larsen, J.B., Contorta: provenance variation and choice of seed source, in *Selection of Tree Species and Provenances in Sustainable Forestry*, Larsen, J.B., Ed., *Dansk Skovbrugs Tidsskrift*, Copenhagen, (87), 240, 1997 (in Danish).

Larsen, J.B. and Nielsen, C.N., Proveniensforsøg med Contortafyr (*Pinus contorta* Dougl.) i Danmark, *Forstl. Forsøgsv. Danm.*, 38, 239, 1982.

Larsen, J.B. and Wellendorf, H., Norway Spruce: provenance variation, tree improvement and choice of seed source, in *Selection of Tree Species and Provenances in Sustainable Forestry*, Larsen, J.B., Ed., *Dansk Skovbrugs Tidsskrift*, Copenhagen, 87, 144, 1997 (in Danish).

Libby, W.J. and Ahuja, M.R., Clonal forestry, in *Clonal Forestry II. Conservation and Application*, Ahuja, M.A. and W.J. Libby, Eds., Springer-Verlag, Berlin, 1, 1992.

Lillesø, J-P.B., Dhakal, L.P., Shrestha, T.B., Tamrakar, P.R., Shrestha, R., Nayaju, R.P., Shrestha R., and Kjær, E.D., *Tree Planting Zones in Nepal*, DFSC Case Study 1, TISC Technical Paper 103, Danida Forest Seed Centre, Denmark, Tree Improvement and Silviculture Component, Kathmandu, 2001.

Lindgren, D., The population biology and clonal deployment, in *Clonal Forestry II. Conservation and Application*, Ahuja, M.A. and W.J. Libby, Eds., Springer-Verlag, Berlin, 1, 1992.

Lindgren, D., Long-term forest tree improvement while maintaining diversity: goal and methods, in *Proceedings from the 1998 Meeting of the Nordic Group for the Management of Genetic Resources of Trees*, Skrøppa, T., Ed., 13, 1999.

Lindgren, D., Low-intensity tree breeding, in *Rapid Generation Turnover in the Breeding Population and Low-Intensity Breeding*, Lundkvist, K., Ed., Department of Forest Genetics, Uppsala, SLU, Sweden, Research Notes 55, 37, 2000.

Lindgren, D. and Ying, C.C., A model integrating seed source adaptation and seed use, *New For.*, 20, 87, 2000.

Low, C.B. and Smith, T., Use of the Guadalupe provenance in *Pinus radiata* improvement in New Zealand, in *IUFRO 97 Genetics of Radiata Pine, Proceedings of NZ FRI-IUFRO Conference*, R.D. Burdon and J.M. Morre, Eds., Rotorua, New Zealand, 1–4 Dec. 1997, FRI Bulletin 203, Rotorua, New Zealand, 57, 1997.

Lynch, M. and Walsh, B., *Genetics and Analysis of Quantitative Traits*, Sinauer Associates Inc., Massachusetts, 980 pp., 1998.

Matheson, A.C. and Cotterlill, P.P., Utility of genotype × environment interactions, *For. Ecol. Manage.*, 30, 159, 1990.

Namkoong, G., Non-optimality of local races, in *Proceedings of the 10th Southern Forestry Tree Improvement Conference*, U.S. National Tree Improvement Society, Springfield, VA, 149, 1969.

Namkoong, G., Barnes, R.D., and Burley, J., *A Philosophy of Breeding Strategy for Tropical Trees*, Tropical Forestry Paper No. 16, Oxford Forestry University, 67 p., 1980.

Namkoong, G., Kang H.C., and Brouard, J.S., *Tree Breeding: Principles and Strategies*, Springer-Verlag, Berlin, 1988.

NRC, *Managing Global Genetic Resources*, The U.S. National Plant Germplasm System, National Research Council (Authors: Allard, R.W., Alvim, P. de T., Barton, J.H., Buttel, F.H., Chang, T.-T., Day, P.R., Evenson, R.E., Fitzhugh, H.A., Goodman, M.M., Hardon, J.J., Marshall, D.R., Sastrapradja, S., Smith, C., and Spence, J.A.), National Academy Press, Washington, DC, 1991a.

NRC, *Managing Global Genetic Resources, Forest Trees*, National Research Council (Authors: Allard, R.W., Alvim, P. de T., Amram, A., Barton, J.H., Bawa, K.S., Burley, J., Buttel, F.H., Chang, T.-T., Day, P.R., Evenson, R.E., Fitzhugh, H.A., Goodman, M.M., Hardon, J.J., Marshall, D.R., Namkoong, G., Paroda, R.S., Sastrapradja, S., Shen, S., Smith, C., Spence, J.A., Wilkes, H.G., and Withers, L.A.), National Academy Press, Washington, DC, 1991b.

OECD, *OECD Scheme for the Control of Forest Reproductive Material Moving in International Trade*, Organisation for Economic Cooperation and Development, Directorate for Agriculture and Food, Paris, France, 1974.

Randall, W.K., *Forest Tree Seed Zones for Western Oregon*, Oregon Department of Forestry, Salem, OR, 1996.

Randall, W.K. and Berrang, P., *Washington Tree Seed Transfer Zones*, Washington Department of Natural Resources, Olympia, WA, 2002.

Roberds, J.H., Namkoong, G., and Skrøppa, T., Genetic analysis of risk in clonal population of forest trees, *Theor. Appl. Genet.*, 79, 841, 1990.

Roulund, H., Problems of clonal forestry in spruce and their influence on breeding strategy, *For. Abst.*, 42, 10, 457, 1981.

SkogsForsk, *Strategy for Future Tree Improvement in Sweden*, SkogForsk, Uppsala, Sweden, 1995 (in Swedish).

Schmidtling, R.C., *Southern Pine Seed Sources*, U.S. Department of Agriculture, Forest Service, Southern Forest Research Station, Asheville, NC, General Technical Report SRS-44 (http://www.srs.fs.usda.gov/pubs/gtr/gtr_srs044.pdf), 2001.

Wang, X.R. and Szmidt, A.E., Molecular markers in population genetics of forest trees, *Scand. J. For. Res.*, 16, 199, 2001.

Wellendorf, H., A Danish Norway spruce breeding plan from 1972 — A retro perspective review 15 years later, in *Proceedings of the IUFRO Working Party Meeting, S2.02-11, Norway Spruce; Provenances, Breeding and Genetic Conservation*, Stener, L.-G., and Werner, M., Eds., Report no. 11, The Institute for Forest Tree Improvement, Uppsala, 279, 1988.

Wellendorf, A.H. and Feilberg, L., Needle damage on *Pinus contorta* during cold winters, Dansk Skovforening, Copenhagen, *Skoven*, 4, 116, 1984 (in Danish).

Wells, O.O., Use of Livingston Parish, Louisiana loblolly pine by forest products industries in the southeast, *South. J. Appl. For.*, 9, 180, 1985.

Westfall, R.D., Developing seed transfer zones, in *Handbook of Quantitative Forest Genetics*, Fins, L., Friedman, S.T., and Brotschol, J.V., Eds., Kluwer Academic Publishers, Dordrecht, 313, 1992.

White, T., Advanced-generation breeding populations: size and structure, in *Proceedings of IUFRO Conference S2.02-08, Breeding Tropical Trees — Resolving Tropical Forest Resource Concerns through Tree Improvement, Gene Conservation and Domestication of New Species*, 9–18 Oct. 1992, Cali, Colombia, CAMCORE, Raleigh, NC, 208, 1992.

White, T.L., Hodge, G.R., and Powel, G.L., An advanced generation tree improvement plan for slash pine in Southeastern USA, *Silvae Genet.*, 42, 359, 1993.

Xiang B., Li, B., and McKeand, S., Genetic gain and selection efficiency of loblolly pine in three geographic regions, *For. Sci.*, 49, 196, 2003.

Ying, C.C., *Genetic Resistance to White Pine Weevil in Sitka Spruce*, British Columbia Ministry of Forests, Research Note No. 106, Victoria, BC, Canada, 1991.

Ying, C.C., Effects of site, provenance and site interaction in Sitka spruce in Coastal British Columbia, *For. Genet.*, 4, 99, 1997.

chapter ten

Regeneration techniques and the seedling environment from a European perspective

Sven Wagner
Dresden University of Technology
Lars Lundqvist
Swedish University of Agricultural Sciences

Contents

10.1 Introduction

Regeneration is key to sustainable forest management because measures taken by the manager at this stage affect development of the next stand and even beyond. Regeneration choices must be appropriate for current management objectives and flexible enough that future options are retained. Because forests can be long-lived, forest managers continuously face the risk that changing conditions in the future will negate the assumptions on which they based their current decisions. The importance of actively intervening to restore forests, using standard and novel regeneration techniques, is not universally recognized (Keddy and Drummond 1996). Planting or direct seeding under afforestation may be the only appropriate method to restore degraded land, but natural invasion may also play a role. In the rehabilitation of forests, some groups advocate relying on natural regeneration

1-56670-635-1/05/$0.00+$1.50

or simply letting successional dynamics do the job of regeneration and restoration (Young 2000). Natural regeneration techniques, however, are not innately superior or always more appropriate for restoring forest ecosystems than artificial techniques. Where management objectives are best served by controlling the timing of restoration and the species composition of the restored forest, active intervention at the regeneration stage is critical. Intervention must be based on management objectives, informed by evaluation and interpretation of site conditions, and incorporate silvicultural knowledge and skills.

In traditional forest management, regeneration success is defined by a minimum stocking density of genetically adapted, vigorous plants with appropriate leader shoot growth (Ferguson et al. 1986; McWilliams et al. 1995; Leinonen et al. 1989; Ponder 1995). These aims also apply to regeneration for restoration. Although stocking level may differ depending on management objectives, in many cases financial return to the landowner is an important objective of forest restoration, together with other objectives such as biodiversity conservation. Restoration often aims to alter species composition, sometimes placing additional requirements for a minimum number of individuals of several species, and stocking levels may be different than for maximum timber production. Restoration goals may also call for a change in stand structure. When the target is an uneven-age stand structure, an appropriate stocking density is somewhat more complex, because there may not be a straightforward relationship between seedling density at a given time and the recruitment of trees for the upper tree layers (Lundqvist 1995).

Regeneration involves plant life cycles and developmental stages (Harper 1977; Clark et al. 1998) and their interactions with environment (Smith et al. 1997). Because each developmental stage has critical environmental limits, these interactions must be evaluated for each stage (Schupp 1995). Important life cycle stages for regeneration include flowering, fruiting, seed or vegetative diaspores, storage, germination, succulent stage, and establishment (Figure 10.1). The forest manager can influence these stages to further management aims. Forest regeneration methods generally are separated into natural vs. artificial methods. Natural regeneration refers to reproductive as well as vegetative material, which is dispersed from the mother trees. Reproductive diaspores are seeds, whereas vegetative

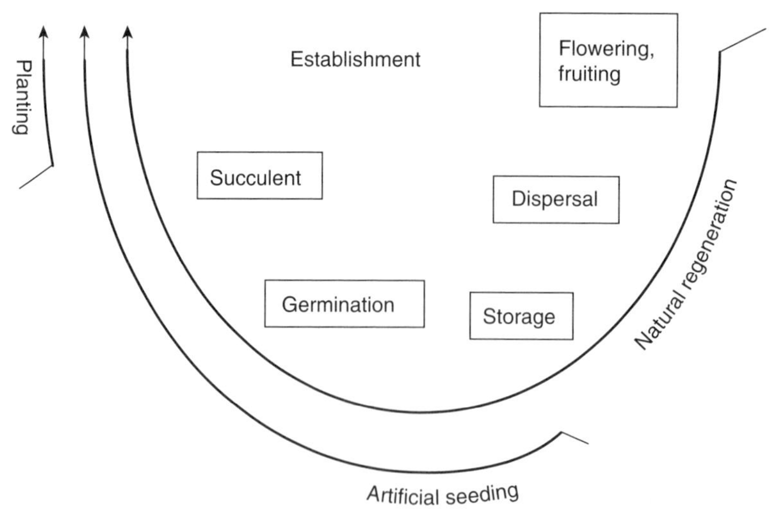

Figure 10.1 Regeneration stages through which organisms must traverse. A propagule will be at risk for different biotic and abiotic threats for varying lengths of time, depending upon the regeneration technique used.

diaspores may be pieces of branches or roots. Vegetative natural regeneration may also include root suckers and sprouts from stumps, which are extensively used in coppice systems. Artificial methods are based on planting seedlings or cuttings and direct seeding.

Our aim in this chapter is to illustrate the interactions between life cycle stages, site factors, and the techniques available to intervene in the regeneration process. We restrict the scope of the chapter to ecological aspects; for more comprehensive treatments that consider economics and landowner objectives, the reader is referred to Cubbage et al. (1991) and Dougherty and Duryea (1991). The complex task of choosing regeneration measures can be approached in two steps: analysis and decision. Environmental conditions at several scales, from stand to microhabitat, must be evaluated for different regeneration measures to properly predict options and obstacles over the entire regeneration phase (Nyland 1996). This has to be done for each stage of the life cycle, in order to identify potential environmental hazards as well as environmental prerequisites for each stage. Recommended regeneration measures are derived from these evaluation options (Perala and Alm 1990a; Jobidon 1994; Van Der Meer et al. 1999).

Analysis can be a daunting task, and oftentimes practical considerations force a manager to compromise and make a decision with incomplete information. For example, site conditions may be surveyed and analyzed at different scales and levels of intensity (Barnes et al. 1982; Kimmins 1987; Otto 1996; Gholz and Boring 1991; Smith et al. 1997). In practice, the quality and intensity of site surveys depend on available resources and may be constrained by factors such as the available staff (both numbers of individuals and their competence) as well as the willingness of the forest owner to invest in site mapping and analyses. The manager's knowledge and experience of local sites and the performance of species may offset the lack of complete information. Such experience may be critical in accounting for changes that may occur after the regeneration intervention, such as potential competing vegetation. Once the aims for the future provenance, species composition, and stand and age structure have been set, the manager must decide on the correct regeneration method or combination of methods.

Despite the complicated and theoretical approach we have described, to a great extent this is the process behind much of the practical experience and silvicultural tradition employed by managers. Forest restoration is in some ways a new application of silviculture; therefore, we must remain mindful that widely-used regeneration techniques may be underlain by different underlying objectives and assumptions. We may carefully use existing knowledge but should be prepared to seek new information or analyses. The various regeneration methods have advantages and disadvantages, which will be outlined in the following sections. The presentation in separate sections does not mean that these techniques are mutually exclusive; combined measures may well be effective, as in the example of integrated regeneration methods used in the Weissenhorn Forest District, Germany (Baumhauer, this volume). Here, beech (*Fagus sylvatica* L.), European silver fir (*Abies alba* Mill.), and to some extent sycamore maple (*Acer pseudoplatanus* L.) were directly sown under Norway spruce shelterwood to convert pure spruce stands to mixed stands, relying on natural spruce regeneration to maintain spruce in the future stand. If management aims had included more species, supplemental plantings of seedlings in small groups would have illustrated regeneration using natural, planting seedlings, and direct seeding methods.

Regeneration conditions are linked to past silvicultural practices in a stand, including past and present species composition, stand structure, water regime (artificial drainage), and game population density. These management practices interact with site characteristics (soil type, climate, natural drainage) that affect important factors for regeneration such as natural seed rain, ground flora competition, soil compaction, humus layer, browsing regime, rodent populations, and microclimate. This is yet another chain or timeline for the forest manager to observe in the integrated analyses of past silviculture, present site characteristics, and choice of regeneration method to fulfill management objectives. Additionally, it is a plentiful source of challenges for forest research to create new and relevant knowledge.

10.2 Regeneration stages and their silviculture

The following section gives an overview of the limitations to the regeneration process by stage and the measures available to counteract them. An acceptable result for each stage is taken to be a minimum density of seeds, seedlings, or sprouts required to regenerate the site; the limitations are those factors that result in lower densities than required. In this context, a biological phenomenon will be called effective if adequate densities result. This is particularly important in seed dispersal; the maximum distance seeds travel may be large, but effective distances to regenerate a site are shorter.

10.2.1 Flowering and fruiting

The intensity and frequency of flowering and fruiting depend on weather, species, and genetic variability within a species, age and vitality of individual trees, and presence or absence of pollinators or seed and flower predators. Light-demanding pioneer species such as birch usually produces an acceptable fruit crop almost every year (Perala and Alm 1990a). Heavy-seeded species, which often, but not always, are shade-tolerant climax species, have good mast years (seed production years) that usually occur at intervals longer than 5 years (Watt 1923; Burschel et al. 1964; Sork 1993).

Fruiting of species often differs from one year to the next and is not synchronized between species, although seed rain may be synchronized among co-occurring species of a genus (Shibata and Nakashizuka 1995). Alternating abundance and composition in seed rain (Lässig et al. 1995) may favor some species on disturbed sites. In a stand without advance regeneration, for example, a heavy-fruiting species will be favored over a nonfruiting species by sudden natural disturbances such as windthrow, as well as harvesting or site preparation, because it occupies the local regeneration niches first. Because seeds disperse over limited distances (Johnson 1988; Ribbens et al. 1994), the availability of seeds depends greatly on the distance to and strength of the nearest seed source (Clark et al. 1998). Selectively cutting seed trees of undesired species prior to site preparation or other regeneration measures can favor a desirable species. For dioecious species, cuttings can be concentrated on female trees.

Once a tree reaches the minimum age for seed production, its contribution depends on its position in the canopy, vigor, and genetics. Preparatory cuttings or fertilization to induce flowering and fruiting can improve the vitality of individual trees. The seed crop on an area is maximized with an optimal combination of individual tree vitality and tree density (Nyland 1996). This optimum is seldom present in stands of maximum density, due to poorly developed individual tree crowns. Thinning to promote trees of good phenotype has been suggested to improve the probability of passing desirable characters to the new stand (Nyland 1996), although losses in genetic variability in one generation in such selected stands are barely detectable (Muona and Harju 1989).

10.2.2 Dispersal

Density of seeds on regeneration sites is determined by the size of the seed crop and distance from the seed bearers. The effect of distance varies by species and the mode of seed dispersal. Dispersal by wind (anemochory) and birds, mammals, and insects (zoochory) are the most important in the temperate and boreal zone, but gravity and water also have to be taken into account (Smith et al. 1997; Bonn and Poschlod 1998). Seed density on the ground generally decreases monotonically with distance from seed source (Clark et al. 1999), resulting in very low densities at long distances (Bullock and Clarke 2000). Effective dispersal distances are seldom greater than a few times the height of the seed bearers (Nyland 1996), but differences among species are considerable (Ribbens et al. 1994). Maximum dispersal distances for different species (Table 10.1) are important for long-term dispersal and invading scenarios, particularly in fragmented landscapes (Malanson and Armstrong 1996), and natural invasion (Pacala et al. 1996). The rate of colonization slows with increasing distance from seed sources (Harper 1977). Planting or direct seeding later successional species of limited dispersal ability can supplement natural regeneration (McClanahan 1986).

Regular invasions of stands of other species can be observed for some species, for example, oaks invading pines (Gholz and Boring 1991) by bird dispersal (zoochory). Pure pine stands are suitable regeneration sites beginning in the pole stage, when light availability on the ground increases steadily (Messier and Puttonen 1995). Several bird species bring in acorns; jays are particularly important and far flying (Bossema 1979). Densities of more than 2,000 stems per ha of acceptable quality can develop if browsing by deer does not stunt or kill the oak seedlings (Mosandl and Kleinert 1998). Fencing deer out of a stand can accelerate this process. The European jay has regenerated several thousand ha of pine stands with oak (Figure 10.2) in Lower Saxony in this way (Otto 1996). Some foresters deliberately offer acorns to jays in special boxes instead of merely observing the succession. Similar observations of succession have been made with rowan (*Sorbus aucuparia* L.), well known in central Europe for its ability to invade pure Norway spruce stands (Küßner 1997), as well as Scots pine (*Pinus sylvestris* L.) stands. In spruce stands, however, gaps of minimum area are essential for rowan to succeed. Wind-dispersed birch species (anemochory) are able to regenerate in pure conifer stands, but are particularly favored by heavy disturbances (Perala and Alm 1990b).

A knowledge of dispersal distances is required to determine the effective density of seeds, which should not focus on maximum-recorded distances. In mixed stands, particular interest is in the effective distance to seed trees of all species in order to maintain the mixture. In wind dispersal, effective distance is influenced by wind speed and direction,

Table 10.1 Mean Dispersal Distances (MDD) for Some Temperate Tree Species as Reported by Different Authors

Species	MDD (m)	Site Conditions	Reference
Fagus sylvatica	8.0	Into clearcut	Karlsson (2001)
Fagus sylvatica	9.0	In forest	Wagner (1999)
Acer platanoides	42.5	Into clearcut	Karlsson (2001)
Acer saccharum	66.7	Into clearcut	Hughes and Fahey (1988) (cf. Willson 1993)
Acer saccharum	22.3	In forest	Clark et al. (1998)
Fraxinus excelsior	52.3	In forest	Wagner (1997)
Fraxinus americana	66.7	Into oldfield	Burton (cf. Willson 1993)
Fraxinus americana	17.1	In forest	Clark et al. (1998)
Betula pendula/pubescens	49.3–81.0	Into clearcut	Karlsson (2001)
B. pendula	35.9	In forest	Zeibig and Wälder (2002)
B. alleghaniensis	66.7	Into clearcut	Hughes and Fahey (1988) (cf. Willson 1993)
B. alleghaniensis	32.9	In forest	Clark et al. (1998)

Figure 10.2 Pedunculate oak (*Q. petraea* Liebl.) directly seeded under a light shelter of beech (*Fagus sylvatica* L.) in the Spessart area (Germany). Sowing is done in rows, using approximately 750 kg ha^{-1} of seed.

which can be influenced by the density and height of the remaining trees (Alexander 1986; Greene and Johnson 1996; Karlsson 2001). In animal dispersal, foresters should have a clear idea of the factors involved such as vector species, preferred habitats, and distances covered. Silvicultural techniques rely on seed dispersal by gravity or wind, especially when sources for the desired species (including beech, maple, spruce, birch, and pine) are on site nearby (LePage et al. 2000). Animal dispersal can be an important complement to other vectors if desired species are not present, especially for heavy-seeded species that are not wind dispersed. Zoochory may be an important silvicultural tool for transformation when relying on natural regeneration (Stimm and Böswald 1994; Ghazoul and Simmons 2001).

10.2.3 Storage, germination, and succulent stage

Seeds of most temperate and boreal tree species are stored on or in the soil because of unfavorable conditions for germination and growth. This resting period is short for species that germinate within the current growing season (*Acer rubrum* L., *Ulmus laevis* Pall.) and may be several years long in seed banks for species with dormancy adaptations (black cherry, *Prunus serotina* Ehrh.). Winter storage is an adaptation of species development to the annual climate cycle in medium and high latitudes (Runkle 1989). Extended periods of opportunistic dormancy (i.e., enforced and induced dormancy) (Harper 1977) can be seen as an alternative to dispersal in space, sometimes called dispersal in time (Willson 1993). Some species build up seed banks in the litter layer after dispersal (*Prunus serotina*), which is an important strategy, but falling to the ground exposes the seed to new risks, such as predation and parasitism. Losses due to mice or birds are common (Madsen 1995), and are determined partly by the duration of storage. Aerial seed banks, where seed is retained in tree crowns, are observed in some fire-adapted species such as *Pinus contorta* (Richardson 1998).

Seeds germinate only if certain conditions are met, including breaking dormancy. The period from germination through the succulent stage is very risky as germinants expand leaves, develop roots, begin to grow, and pass from the succulent stage to harden sufficiently to withstand environmental conditions (Nyland 1996). This is the most vulnerable stage in the entire lifetime of a plant, and is therefore crucial for any natural regeneration strategy. Unfortunately, pathogens often benefit from the same environmental conditions that are favorable for germinating seeds or succulent seedlings; thus, fungal attack and death by damping-off disease is a common fate (Smith et al. 1997).

Environmental conditions can change dramatically within a few cm, and a microsite where the germination process can be successfully completed is termed a safe site (Harper 1977). Safe sites generally provide a seed with sufficient moisture, warmth, oxygen, and sometimes light of an appropriate quality (Smith et al. 1997). Pathogens are absent or ineffective at safe sites. Characteristics of the soil surface determine microsite conditions, acting as a boundary layer (a transition zone between the atmosphere and soil). The characteristics of the boundary layer are determined primarily by litter quality and quantity and the development of a humus layer. Existing vegetation can alter boundary layer conditions by intercepting radiation, affecting moisture content, and determining the thickness of the litter and humus layers (Wardle 1959; George and Bazzaz 1999). Vegetation also provides shelter for herbivorous mammals from predators.

The humus layer is particularly crucial for moisture supply. The surface materials and the vertical depth of accumulation in relation to the relative size of seeds and germinants often determine what a safe site for a particular species is. Acorns often germinate best when buried 2 to 3 cm, regardless of the covering material, but birch seeds require physical contact with fine humus or the uppermost mineral soil to germinate successfully (Carlton and Bazzaz 1998; Karlsson 2001). In general, exposed mineral soil usually is the best material for storage, germination, and initial growth because of its favorable moisture supply. Thick humus layers often lead to establishment failure, particularly for species with small seeds. In contrast, many species are favored by woody debris (Dimbleby 1953; LePage et al. 2000), especially on harsh sites, for example, forests in mountainous areas or with a grass cover. Successful establishment of species such as Norway spruce, silver fir, sycamore, and rowan could be enhanced in high-elevation grass-covered areas by supplying woody debris (Stöckli 1995). Thick humus layers may need to be reduced to promote the survival of seeds and to facilitate germination. This can be done indirectly by preparatory fellings, but sometimes this risks promoting advance regeneration of undesired species if thinning intensity is too heavy. The undesired species often are more shade-tolerant than the desired ones, for example, Norway spruce on wet soils where pedunculate oak (*Quercus robur* L.) is preferred. In mixed stands, early thinning that selectively removes undesirable species can alter these features.

Seed densities can increase rapidly following a major event (masting in heavy-seeded oak and beech) or progressively by lesser dispersal events and seed storage, building up a seed bank by dormancy (*Prunus serotina* Ehrh.). Foresters can use a range of regeneration methods that rely on seed banks, but must know about the particular mechanisms needed to break the dormancy of desired species. These mechanisms can be light or heat, leading to particular cutting or prescribed burning activities, for example, in eucalypts or serotinous pines and spruces (Smith et al. 1997). Wildlife management is often essential for successful regeneration. Browsing by deer can selectively remove succulent plants of preferred species (Shimoda 1994; Madsen 1995; Harmer 2001), altering species composition toward more unpalatable conifers. Fencing is an expensive alternative to hunting.

10.2.4 Establishment

Because all autotrophic plants require the same resources (water, nutrients, and light), competition is the most important process during the establishment stage. Competition between individual plants can be partitioned into the effect on resources and the response to resources (Goldberg 1990). In recent years, our knowledge of the strategies used by plants to acquire and allocate resources has improved, particularly for photosynthetically

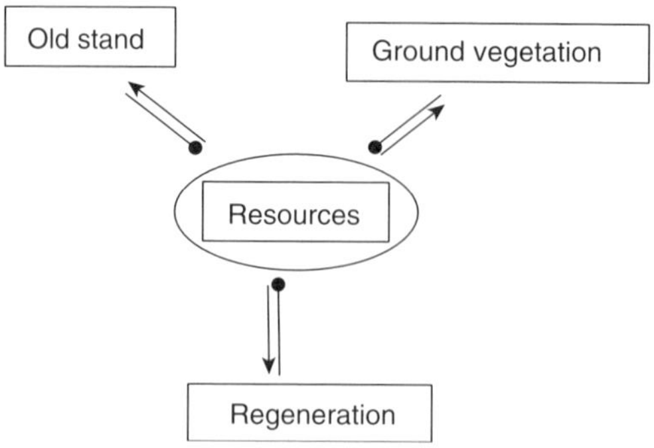

Figure 10.3 Different functional groups and their impact on the resource pool.

active radiation (PAR) (Lüpke 1987; Pacala et al. 1996; Coates and Burton 1999). In contrast, our knowledge of belowground mechanisms and their importance for seedling vitality has scarcely increased (Havranek and Benecke 1978; Flaig and Mohr 1990). There is large variation in belowground resources in the forest, at both micro- and macroscales, and this variation is difficult to control and separate from variation in PAR (Walters and Reich 1997; Finzi and Canham 2000; Huss and Stephani 1978; Reed et al. 1983). Nevertheless, the species-specific response to the complex resource pool is key to understanding the coexistence of species (Tilman 1982).

The old stand (overstory and ground vegetation) reduces available resources for seedlings (Figure 10.3), and seedling mortality is common during establishment. Some species survive prolonged periods of resource shortage by building up seedling (*Fraxinus* spp., *Acer* spp.) or sapling banks (beeches, firs) under closed canopies. Most of these species produce less seed per tree and lower densities of seeds on the ground over time than more light-demanding species (Ribbens et al. 1994; Clark et al. 1998); seedling and sapling banks are strategies by which low seed production is compensated by prolonged time. Underlying mechanisms include architectural adaptations such as the relative portion of leaf and branch biomass (Kohyama 1987), high morphological plasticity such as opportunistic plagiotropy in beeches (Brown 1951), and metabolism adaptations such as low respiration rates (Walters and Reich 2000). Individual seedlings are often capable of vigorous responses to sudden improvements in resource supply, and may thus occupy promising niches in advance of other less tolerant species.

10.3 Natural regeneration techniques

Natural regeneration is appropriate if the provenance and species already at the site match the site conditions in time and space. The major advantage of natural over artificial regeneration, particularly planting, is low regeneration costs. Successful natural regeneration provides inexpensive and densely stocked stands of potentially high future quality. Dense stocking, however, may create a need for expensive precommercial thinnings. If less successful, natural regeneration may require expensive supplementary planting to fill in failed spots. Another advantage over planting may be the development of natural root systems, undisturbed by root undercutting in nursery and transplanting (intact taproot). Poorly planted seedlings with deformed root systems may develop stability problems. Natural regeneration is the dominant regeneration method within close-to-nature forestry (Hahn et al., this volume). Management aims within this silvicultural approach include the development of stands mixed in various ways (mixed species, age and tree size), and natural regeneration is more likely to produce such stands unless complicated plantings are attempted.

The tree canopy in old stands of forest ecosystems control most of the resources required by other organisms for growth (Figure 10.3). Felling regimes are a powerful silvicultural tool for guiding the regeneration process. Opening up the canopy and rooting zone dramatically changes environmental conditions on the forest floor by promoting humus mineralization, growth of tree regeneration and competing vegetation, and attracting herbivores to the site. Individual trees and the entire stand affect most environmental factors such as temperature and wind speed (Kimmins 1987); because forest ecosystems are organized hierarchically (Ulrich 1994), the mature stand in most cases determines the resource availability for both the tree regeneration and competing ground vegetation.

Different felling (cutting) methods create favorable conditions for preferred species, particularly in regard to resources, affecting growth (Lust 1994) and mortality (Kobe et al. 1995) of the regeneration. In plant communities, space can be used as a surrogate for resource supply (Tilman 1982). The concept of growing space (Oliver and Larson 1996) provides a way to integrate the underlying mechanisms important in regeneration. In the regeneration stage, seedlings profit from available growing space that was previously occupied by one or several old trees that were killed by a natural disturbance or were removed by cutting or deadening. Seedlings can utilize this growing space but it is also available to the remaining old trees and competing ground vegetation.

Light-demanding species have difficulty germinating, establishing, and maintaining in small gaps. Traditionally, this has been regarded as a problem of insufficient light (Lüpke 1998; Van Der Meer et al. 1999). However, root competition from the surrounding overstory might be more important than a shortage in PAR (Wagner 1996; Brockway and Outcalt 1998; Coomes and Grubb 2000). Regardless, the lateral expansion of roots (Ammer and Wagner 2002) and crowns (Le Goff and Ottorini 1995; Guericke 2001) of older trees along the edges of gaps may impact the future growth of seedlings. Additional felling may be necessary to avoid the increased mortality that results after prolonged periods of suppression due to gap reduction (Canham 1989; Emborg 1996). This problem is less pronounced in shade-tolerant species.

Four high-forest felling methods are available for regeneration (Smith et al. 1997): clearcut, seed tree, shelterwood, and selection. These strategies differ in the space allocated to the new age class at the time of cutting. Except for clearcutting, the methods are further differentiated by the way and the speed with which subsequent fellings are performed until the regeneration is finally released. As a rule of thumb, selection cuttings of low intensity and over long regeneration periods favor seedlings of species tolerant of shade and root competition from old trees, while intolerant species are favored by seed-tree or clearcutting methods with short periods until release. However, the local microenvironment around the seedling is most important. A group selection system with an appropriate gap size might maintain a considerable proportion of moderately tolerant to intolerant species (Abe et al. 1995) while a single-tree selection system with a comparable residual basal area will fail to do so (Nyland 1996).

10.3.1 Shelterwood and seed-tree systems

A single cutting event opens up the canopy of a mature stand in shelterwood and seed-tree systems, often in combination with heavy fruiting of the desirable species (masting). Seedlings from succeeding years contribute to the regeneration cohort when conditions are favorable for their establishment and growth, but frequently rapid establishment of competing ground vegetation and the competitive advantage of older seedlings interfere. Consequently, these systems establish temporary two-cohort stands and do not substantially alter age structure. In both shelterwood and seed-tree systems, structures might become more complex if there is a long period between the seeding felling and the final

Figure 10.4 Natural regeneration of pedunculate oak (*Quercus petraea* Liebl.) under a light shelter of old con-specific trees south of the Harz area (Germany).

removal of the old stand. The longer the residual trees of the old stand remain, and the more clumped they are, the more diverse the vertical and horizontal structures of the regeneration in the area will become (Kuuluvainen et al. 1993). Selection fellings or utilization of spontaneously created gaps is necessary to create more lasting, complex uneven-aged structures (Figure 10.4).

Close-to-nature silviculture using natural regeneration may be based mainly on one mast year followed by light to moderate shelterwood cuts to open the canopy up for the regeneration to gain a foothold, while still protecting against excessive weed competition. For the more light-demanding Scots pine (*Pinus sylvestris* L.), shelterwood cuts following the mast year may only leave scattered seed trees. Regeneration may be based on preparatory thinning to promote the seed production in dense and dark stands. Such measures may include selection fellings to regenerate shade-tolerant silver fir in mixed stands with Norway spruce and beech, or gradual development of advance regeneration of pedunculate oak (*Q. robur* L.) by thinning in Scots pine stands.

10.3.2 Selection systems

Central Europe has a rich experience in transforming forests from regular to irregular communities, starting in 1886 with Femelwald (Gayer 1886), Plenterwald in 1901 (Biolley cited in Schütz 2001), and Dauerwald in 1920 (Möller cited in Guldin 1996). The underlying idea of these concepts is to use gaps (Runkle 1982) to initiate new regeneration cohorts in groups, leaving parts of the stand undisturbed until new gaps are created in the next establishment phase (Nyland 1996). The same principle applies when relying on gaps spontaneously created by natural disturbances. Multiple entries into the stand are needed over time to affect the entire area. For example, long-term efforts to establish the shade-tolerant European beech (*Fagus sylvatica* L.) in younger Norway spruce stands are constrained by the edge trees of the spruce stand, which slowly expand into the openings (Müller and Wagner 2003). Continued thinnings are necessary to safeguard the development of the beech.

Despite this long tradition, many management questions remain, including the horizontal distribution of age cohorts, desirable variation in age classes, and maintenance of the age structure in the future. The importance of gaps has been questioned in the boreal zone; some stress the importance of gap formation for regenerating boreal uneven-aged Norway spruce forests (Hytteborn and Packham 1985; Skarpe et al. 1989; Leemans 1991; Qinghong and Hytteborn 1991), but although the regeneration is spatially aggregated (Lundqvist 1995), the groups are not associated with canopy openings or patches with low overstory density (Lundqvist and Fridman 1996).

The most important consideration is the best time for initiating regeneration and the accompanying cutting measures. Besides economic concerns (Nyland 1996; Bachmann 1988), early establishment of regeneration under the old stand offers the opportunity to develop age and dimensional variation within the regeneration stratum as well as the entire stand. Aggregation of regeneration in small patches created by cuttings (group selection) accomplishes this structuring of the regeneration stratum as suggested for sessile oak (*Quercus petraea* Liebl.) in Scots pine stands (Mortzfeld 1896).

10.3.3 Clearcut system

Artificial regeneration after clearcutting can be considered a restoration activity, for example, replacing a pure Norway spruce stand on a poorly drained soil by pedunculate oak, a more adapted species on the site. Because of the high risk of windthrow in recently cut spruce stands and the low shade tolerance of oak, clearfelling is an appropriate restoration tool. An even more close-to-nature method would be to allow the natural regeneration of pioneers like birch and aspen on these clearcuts, in order to create a mixture of species.

10.3.4 Site preparation

Direct treatments of the site to make the physical environment suitable for germination are more common than preparatory fellings. These treatments are intended to alter site microclimate, improve access to water supply in exposed mineral soil, or eliminate competing vegetation (Morris et al. 1993). In the context of restoration, techniques chosen should minimize the surface area affected by the treatment. Strips, bands, and spots should be preferred to broadcast treatments. Prescribed burning seems appropriate in naturally fire-dependent communities such as Scots pine in the boreal zone.

Cutting treatments are often accompanied by intensive cultivation that exposes mineral soil (Kuuluvainen and Pukkala 1989), and harrowing after seed dispersal for some species (Madsen 1995). Although site preparation may interfere with natural succession (Nyland 1996), benefits include eliminating nonpreferred tree species (Gordon et al. 1995; Lautenschlager 1995) and eliminating grass and herbaceous cover for seed predators such as mice. This is usually directed against species that have established as advance regeneration and control by weeding or precommercial thinning is not viable. Effective site preparation techniques include mechanical, chemical (Nyland 1996), and prescribed burning (Van Lear and Waldrop 1991). Mechanical treatments such as brush saws are appropriate for non-sprouting species, including most conifers. For species that sprout readily, a combination of chemical and mechanical techniques is used (Röhrig and Gussone 1990). Mechanical site preparation may facilitate short-term seed storage and germination of desired species.

10.3.5 Competing vegetation

Controlling competing vegetation during the establishment stage can increase soil temperature, PAR availability, and nutrient availability, thereby improving survival and growth of seedlings under limiting site conditions (Brand 1991; Madsen 1995; Groot 1999). Low soil temperature and nitrogen availability are often limiting in the boreal zone (Brand 1991). Limitations in PAR availability due to competing vegetation are species dependent (Küßner et al. 2000), but also depend on interactions with other environmental resources (Lautenschlager 1999; Küßner et al. 2000). Specific recommendations for vegetation control depend on desirable tree species (Löf 2000), weed species (Lautenschlager 1995), and site. Specifying an appropriate level is not easy (Tappeiner and Wagner 1987; Cain 1989), and there is a difference between a competition threshold and a critical-period threshold (Wagner 2000). This differentiates between the vegetation densities at which yield loss occurs (Jobidon 1994) and the time when vegetation control must begin to prevent yield loss (Wagner et al. 1999).

Mechanical and chemical methods are available for vegetation control (Wagner 1993). Livestock grazing and cover cropping with desirable noncrop plants are alternatives to conventional herbicides (Wagner 1993). Direct vegetation management is often unnecessary under the shelter of old trees when the crop species are more shade-tolerant than species of competing vegetation (Lautenschlager 1999). This is used in central Europe with plantings of beech under shelter of Norway spruce; beech is shade tolerant, but grasses such as *Avenella flexuosa* Parl. or *Calamagrostis villosa* Gmel. are not. Nevertheless, vegetation management is essential in beech plantings on clearcut sites or exposed strip cuttings, in order to reduce the risk of mice damage.

10.4 Artificial regeneration techniques

10.4.1 Direct seeding

The prime advantages of direct seeding over planting are potentially much lower costs and natural root system development, which may be important for the stability of single trees of oak, pine, and fir (Sutton 1991). The prime disadvantage is that the plant has to spend its most vulnerable stages (seed, germinant, and succulent) of its life in the hostile field environment. This may lead to failure and thereby not only increased regeneration costs, but also waste of limited seed lots. Other problems of sowing are determining the best time and the handling of huge amounts of seeds.

Figure 10.5 Natural regeneration of pedunculate oak (*Quercus petraea* Liebl.) under a shelter of Scots pine (*Pinus sylvestris* L.) in the lowlands of northern Germany. The acorns were transported to the stand by the European jay (*Garrulus glandarius* L.).

Seeding by broadcast, strip, or spot techniques (Nyland 1996) is generally recommended at the beginning of the growing season in order to reduce risk from seed predators, fungi, or unfavorable environment conditions such as excess water or heavy frost (Figure 10.5). The probability of success increases if seeds are pretreated (stratification, incubation) to obtain prompt and vigorous germination (Bergsten 1987). Additionally, seeds may be treated with repellents. Soil scarification is one way to prepare a site for artificial seeding. Sowing can be mechanized, such as beechnuts or acorns in drills or spots (Leder and Wagner 1996), as well as sowing of acorns in hoe strips (Preuhsler and Pinto da Costa 1994). Small-seeded, shade-tolerant silver fir (*Abies alba* Mill.) can also be sown by machine. Machine sowing can cause injuries in shallow-rooted old stands, particularly in spruce, but the problem can often be ignored in pine.

10.4.2 Planting

Planting is the most common technique, while seeding is usually limited to large-seeded species such as oaks and beech, or to particular forest districts where experience in this technique has accumulated. Planting desired species is a standard regeneration technique in forestry and will, therefore, not be discussed here in detail (Dougherty and Duryea 1991). The ideal planting material is adapted to the soil, site, and method of planting and time of planting. The two types of nursery material are bareroot and container seedlings. Either type may provide good results when handling and cultural practices are appropriate. In general, more intensive management is necessary in more extreme situations caused by competing vegetation, herbivores, drought, or high water tables. Special spot preparation methods and a careful choice of high-quality material might prove necessary.

Planting wildings is another way to establish desirable species. Wildings may have fewer establishment problems than standard nursery stock under the canopy of an old stand, as they have adapted to shade by physiological conditioning under an overstory (Nörr et al. 2002). Deer may preferentially browse well-watered and fertilized nursery stock (Suchant et al. 2000). Wildings need careful selection and handling, however, to protect leader shoots and obtain a sufficient mass of fine roots.

The number of seedlings to plant under partial canopies remains an open question. Numbers vary, depending on the objectives, risks, species, and the light environment. If timber production is an objective, the number of stems desired at maturity and an estimate of mortality provide a rough estimate. Numbers for broadleaves can be much higher in order to have side competition to induce self-pruning in the pole stage. For enrichment plantings, it is unclear as to how the planted seedlings should be protected from competition by regeneration of unpreferred species or provenances.

Although species differ widely in their height growth response to environmental conditions like radiation (Coates and Burton 1999), there are only a few examples of mixture development guided by felling regime (Mosandl and El Kateb 1988; Lüpke 1998). Suppression, and finally death (Kobe and Coates 1997), of more light-demanding species in the shaded understory of old stands is sometimes desirable in restoration activities. In even-aged stands of pioneer species, the light climate of the succeeding pole stage often allows for advance regeneration of shade-tolerant species (Figure 10.6). This can be an advantage for restoration by preventing the establishment of an undesirable pioneer in the succeeding regeneration stratum. Black cherry (*Prunus serotina* Ehrh.) is a strong exotic invader in the lowland forests of central Europe. Planting beech in cherry stands helps to retard the development of advance black cherry regeneration. Similarly, planting pedunculate oak in mixtures with Scots pine can minimize stocking of the light-demanding pine if the oaks regenerate first. This task is more complicated when the undesirable species is intermediate in shade tolerance. The application of very light cuttings and a lengthy

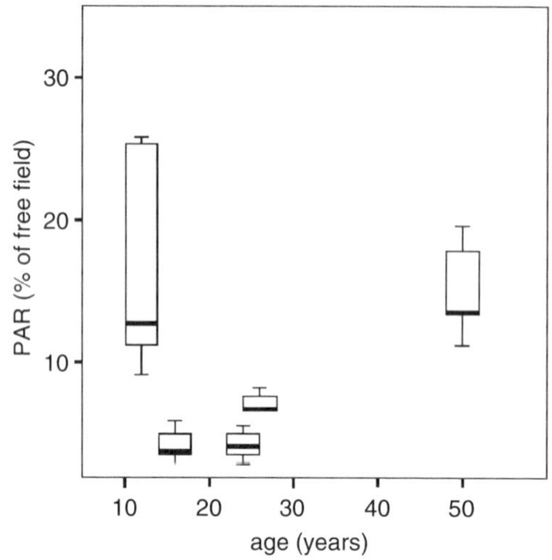

Figure 10.6 Boxplots of photosynthetically active radiation (PAR) at the forest floor in pure even-aged stands of *Prunus serotina* in Lower Saxony. Note the decline in light reaching the forest floor after crown closure at about age 15.

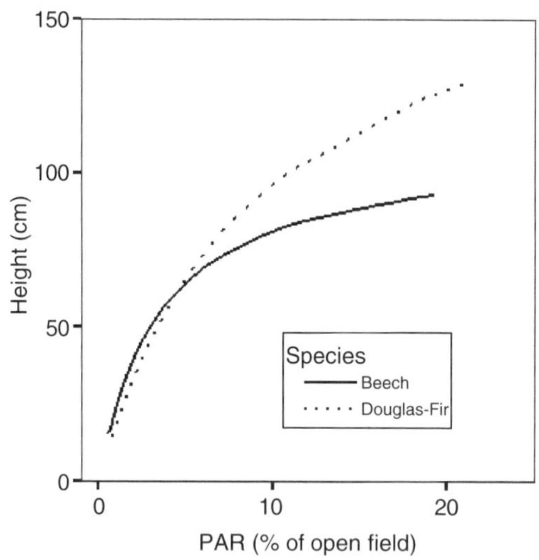

Figure 10.7 Height of 4-year-old European beech (*Fagus sylvatica* L.) and Douglas-fir (*Pseudotsuga menziesii*) under the same light regimes. Under lower light, the beech is better able to compete with the Douglas-fir.

regeneration period in shade will favor shade-tolerant species only if the old stand itself is primarily made up of such species. Mixtures of European beech and Douglas fir can be managed in this way to promote beech (Knoerzer 1999; Guericke 2000). To keep the Douglas fir disadvantaged, light levels have to be so low that cuttings can hardly be applied and the stand should simply be left alone (Figure 10.7).

References

Abe, S., Masaki, T., and Nakashizuka, T., Factors influencing sapling composition in canopy gaps of a temperate deciduous forest, *Vegetatio*, 120, 21, 1995.

Alexander, R.R., *Engelmann Spruce Seed Production and Dispersal and Seedling Establishment in the Central Rocky Mountains*, U.S. Department of Agriculture, Forest Service, Rocky Mountain Research Station, General Technical Report RM-134, 1986.

Ammer, C. and Wagner, S., Problems and options in modelling fine root biomass of single mature Norway spruce trees at given points from stand data, *Can. J. For. Res.*, 32, 581, 2002.

Bachmann, P., Wertzuwachsuntersuchungen für die Verjüngungsplanung, *Allgemeine Forst- u.Jagd-Zeitung*, 62, 1988.

Barnes, B.V., Pregitzer, K.S., Spies, T.A., and Spooner, V.H., Ecological forest site classification, *J. For.*, 80, 493, 1982.

Bergsten, U., *Incubation of Pinus sylvestris L. and Picea abies L. (Karst.) Seeds at Controlled Moisture Content as an Invigoration Step in the IDS Method*, Ph.D thesis, Depart. of Silviculture, Swedish University of Agricultural Sciences, 1987.

Bonn, S. and Poschlod, P., *Ausbreitungsbiologie der Pflanzen Mitteleuropas*, Quelle & Meyer, Wiesbaden, 1998.

Bossema, I., Jays and oaks: an eco-ethological study of a symbiosis, *Behaviour*, 70, 1, 1979.

Brand, D.G., The establishment of boreal and sub-boreal conifer plantations: an integrated analysis of environmental conditions and seedling growth, *For. Sci.*, 37, 1, 68, 1991.

Brockway, D.G. and Outcalt, K.W., Gap-phase regeneration in longleaf pine wiregrass ecosystems, *For. Ecol. Manage.*, 106, 125, 1998.

Brown, J.M.B., *Influence of Shade on the Height Growth and Habit of Beech*, Report on Forest Research, Forestry Commission, London, 62, 1951.

Bullock, J.M. and Clarke, R.T., Long distance seed dispersal by wind: measuring and modelling the tail of the curve, *Oecologia*, 124, 506, 2000.

Burschel, P., Huss, J., and Kalbhenn, R., Die natürliche Verjüngung der Buche. *Schriftenreihe der Forstlichen Fakultät der Uni Göttingen und der Niedersächsischen Forstlichen Versuchsanstalt Göttingen*, 34, J.D. Sauerländer's Verlag, Frankfurt, 1964.

Cain, M.D., A simple competition assessment system associated with intensive competition control in natural loblolly-shortleaf pine seedling stands, *South. J. Appl. For.*, 13, 8, 1989.

Canham, C.D., Different responses to gaps among shade-tolerant tree species, *Ecology*, 70, 3, 548, 1989.

Carlton, G.C. and Bazzaz, F.A., Regeneration of three sympatric birch species on experimental hurricane blowdown microsites, *Ecol. Monogr.*, 68, 1, 99, 1998.

Clark, J.S., Macklin, E., and Wood, L., Stages and spatial scales of recruitment limitation in southern Appalachian forests, *Ecol. Monogr.*, 68, 2, 213, 1998.

Clark, J.S., Silman, M., Kern, R., Macklin, E., and HilleRisLambers, J., Seed dispersal near and far: patterns across temperate and tropical forests, *Ecology*, 80, 5, 1475, 1999.

Coates, K.D. and Burton, P.J., Growth of planted tree seedlings in response to ambient light levels in northwestern interior cedar-hemlock forests of British Columbia, *Can. J. For. Res.*, 29, 1374, 1999.

Coomes, D.A. and Grubb, P.J., Impacts of root competition in forests and woodlands: a theoretical framework and review of experiments, *Ecol. Monogr.*, 70, 2, 171, 2000.

Cubbage, F.W., Gunter, J.E., and Olson, J.T., Reforestation economics, law, and taxation, in *Forest Regeneration Manual*, Duryea, M.L. and Dougherty, P.M., Eds., Kluwer Academic Publishers, Dordrecht, 9, 1991.

Dimbleby, G.W., Natural regeneration of pine and birch on the heather moors of northeast Yorkshire, *Forestry*, 26, 41, 1953.

Dougherty, P.M. and Duryea, M.L., Regeneration: an overview of past trends and basic steps needed to ensure future success, in *Forest Regeneration Manual*, Duryea, M.L. and Dougherty, P.M., Eds., Kluwer Academic Publishers, Dordrecht, 3, 1991.

Emborg, J., *The Structure, Dynamics and Light Conditions of Suserup Skov, a Semi-Natural Temperate Deciduous Forest in Denmark*, Ph.D. thesis, Depart. of Economics and Natural Resources, The Royal Veterinary and Agricultural University, Copenhagen, 1996.

Finzi, A.C. and Canham, C.D., Sapling growth in response to light and nitrogen availability in a southern New England forest, *For. Ecol. Manage.*, 153, 2000.

Ferguson, D.E., Stage, A.R., and Boyd, R.J., Predicting regeneration in the Grand Fir–Cedar–Hemlock ecosystem of the northern Rocky Mountains, *For. Sci. Monogr.*, 1, 1986.

Flaig, H. and Mohr, H., Auswirkungen eines erhöhten Ammoniumangebots auf die Keimpflanzen der gemeinen Kiefer (*Pinus sylvestris* L.), *Allgemeine Forst- u. Jagd-Zeitung*, 35, 1990.

Gayer, K., *Der gemischte Wald*, Parey, Berlin, 1886.

George, L.O. and Bazzaz, F.A., The fern understory as an ecological filter: emergence and establishment of canopy-tree seedlings, *Ecology*, 833, 1999.

Ghazoul, J. and Simmons, E., Biological interactions and disturbance: plants and animals, in *The Forests Handbook*, Evans, J., Ed., Blackwell Science Ltd., Oxford, 105, 2001.

Gholz, H.L. and Boring, L.R., Characterizing the site: environment, associated vegetation, and site potential, in *Forest Regeneration Manual*, Duryea, M.L. and Dougherty, P.M., Eds., Kluwer Academic Publishers, Dordrecht, 163, 1991.

Goldberg, D.E., Components of resource competition in plant communities, in *Perspectives on Plant Competition*, Grace and Tilman, Eds., Academic Press, New York, 27, 1990.

Gordon, A.M., Simpson, J.A., and Williams, P.A., Six-year response of red oak seedlings planted under a shelterwood in central Ontario, *Can. J. For. Res.*, 603, 1995.

Greene, D.F. and Johnson, E.A., Wind dispersal of seeds from a forest into a clearing, *Ecology*, 595, 1996.

Groot, A., Effects of shelter and competition on early growth of planted white spruce, *Can. J. For. Res.*, 29, 1002, 1999.

Guericke, M., Waldwachstumskundliche Untersuchungen in gleichaltrigen Buchen-Douglasien-Mischbeständen im Westniedersächsischen Tiefland, *Forst Holz*, 718, 2000.

Guericke, M., *Untersuchungen zur Wuchsdynamik von Mischbeständen aus Buche und Europ, Lärche als Grundlage für ein abstandsabhängiges Einzelbaumwachstumsmodell*, Cuvillier-Verlag, Göttingen, 2001.

Guldin, J.M., The role of uneven-aged silviculture in the context of ecosystem management, *West. J. Appl. For.*, 4, 1996.

Harmer, R., The effect of plant competition and simulated summer browsing by deer on tree regeneration, *J. Appl. Ecol.*, 1094, 2001.

Harper, J.L., *Population biology of plants*, Academic Press, London, 1977.

Havranek, W.M. and Benecke, U., The influence of soil moisture on water potential, transpiration and photosynthesis of conifer seedlings, *Plant Soil*, 91, 1978.

Huss, J. and Stephani, A., Lassen sich angekommene Buchennaturverjüngungen durch frühzeitige Auflichtung, durch Düngung oder Unkrautbekämpfung rascher aus der Gefahrenzone bringen? *Allg. Forst-u. Jagd-Zeitung*, 133, 1978.

Hytteborn, H. and Packham, J.R., Left to nature: forest structure and regeneration in Fiby urskog, central Sweden, *Arboricultural J.*, 9, 1, 1985.

Johnson, W.C., Estimating dispersibility of Acer, Fraxinus and Tilia in fragmented landscapes from patterns of seedling establishment, *Landscape Ecol.*, 175, 1988.

Jobidon, R., Light threshold for optimal black spruce (*Picea mariana*) seedling growth and development under brush competition, *Can. J. For. Res.*, 1629, 1994.

Karlsson, M., *Natural Regeneration of Broadleaved Tree Species in Southern Sweden*, Acta Universitatis Agriculturae Sueciae, *Silvestria* 196, Swedish University of Agricultural Sciences, Alnarp, 2001.

Keddy, P.A. and Drummond, C.G., Ecological properties for the evaluation, management, and restoration of temperate deciduous forest ecosystems, *Ecol. Appl.*, 748, 1996.

Kimmins, J.P., *Forest ecology*, Macmillan Publishing Company, New York, 1987.

Knoerzer, D., Zur Naturverjüngung der Douglasie im Schwarzwald, *Dissertationes Botanicae*, No. 306, J. Cramer in der Gebrüder Borntraeger Verlagsbuchhandlung, Berlin, Stuttgart, 1999.

Kobe, R.K. and Coates, K.D., Models of sapling mortality as a function of growth to characterize interspecific variation in shade tolerance of eight tree species of northwestern British Columbia, *Can. J. For. Res.*, 227, 1997.

Kobe, R.K., Pacala, S.W., Silander, J.A., and Canham, C.D., Juvenile tree survivorship as a component of shade tolerance, *Ecol. Appl.*, 517, 1995.

Kohyama, T., Significance of architecture and allometry in saplings, *Funct. Ecol.*, 1, 399, 1987.

Küßner, R., Sukzessionale Prozesse in Fichtenbeständen des Osterzgebirges — Möglichkeiten ihrer waldbaulichen Beeinflussung und ihre Bedeutung für einen ökologisch begründeten Waldumbau, *For. Centralbl.*, 359, 1997.

Küßner, R., Reynolds, P.E., and Bell, F.W., Growth response of *Picea mariana* seedlings to competition for radiation, *Scand. J. For. Res.*, 15, 334, 2000.
Kuuluvainen, T. and Pukkala, T., Effect of Scots pine seed trees on the density of ground vegetation and tree seedlings, *Silva Fenn.*, 159, 1989.
Kuuluvainen, T., Hokkanen, T.J., Järvinen, E., and Pukkala, T., Factors related to seedling growth in a boreal Scots pine stand: a spatial analysis of a vegetation-soil system, *Can. J. For. Res.*, 2101, 1993.
Lässig, R., Egli, S., Odermatt, O., Schönenberger, W., Stöckli, B., and Wohlgemuth, T., Beginn der Wiederbewaldung auf Windwurfflächen, *Schweiz. Z. For.*, 893, 1995.
Lautenschlager, R.A., Competition between forest brush and planted white spruce in north-central Maine, *North. J. Appl. For.*, 163, 1995.
Lautenschlager, R.A., Environmental resource interactions affect red raspberry growth and its competition with white spruce, *Can. J. For. Res.*, 29, 7, 906, 1999.
Leder, B. and Wagner, S., Bucheckern/Streu-Voraussaat als Alternative beim Umbau von Nadelholzbeständen in Mischbestände, *Forstarchiv*, 7, 1996.
Leemans, R., Canopy gaps and establishment patterns of spruce (*Picea abies* (L.) Karst.) in two old-growth coniferous forests in central Sweden, *Vegetatio*, 93, 157, 1991.
Le Goff, N. and Ottorini, J.M., Crown development and growth of beech (*Fagus sylvatica* L.) after thinning in an experimental stand, Genetics and silviculture of beech, in *Proceedings from the 5th beech Symposium of the IUFRO Project Group P1.10-00*, S.F. Madsen, Ed., *Forskningsserien*, No. 11, Danish Forest and Landscape Research Institute, Hörsholm, Denmark, 257, 1995.
Leinonen, K., Leikola, M., Peltonen, A., and Räsänen, P.K., Natural regeneration of Norway spruce in Pirkka-Häme Forestry Board District, southern Finland, *Acta Forestalia Fennica*, No. 209, The society of Forestry in Finland and the Finnish Forest Research Institute, 1989.
LePage, P.T., Canham, C.D., Coates, K.D., and Bartemucci, P, Seed abundance versus substrate limitation of seedling recruitment in northern temperate forests of British Columbia, *Can. J. For. Res.*, 415, 2000.
Löf, M., Establishment and growth in seedlings of *Fagus sylvatica* and *Quercus robur*: influence of interference from herbaceous vegetation, *Can. J. For. Res.*, 30, 855, 2000.
Lüpke, B.v., Einflüsse von Altholzüberschirmung und Bodenvegetation auf das Wachstum junger Buchen und Traubeneichen, *Forstarchiv*, 18, 1987.
Lüpke, B.v., Silvicultural methods of oak regeneration with special respect to shade tolerant mixed species, *For. Ecol. Manage.*, 19, 1998.
Lundqvist, L., Simulation of sapling population dynamics in uneven-aged *Picea abies* forests, *Ann. Bot-London*, 76, 371, 1995.
Lundqvist, L. and Fridman, E., Influence of local stand basal area on density and growth of regeneration in uneven-aged *Picea abies* stands, *Scand. J. For. Res.*, 11, 364, 1996.
Lust, N., Regeneration patterns and problems in mixed forests, *Silva Gandavensis*, 59, 1, 1994.
Madsen, P., Effects of seedbed type on wintering of beech nuts (*Fagus sylvatica*) and deer impact on sprouting seedlings in natural regeneration, *For. Ecol. Manage.*, 37, 1995.
Malanson, G.P. and Armstrong, M.P., Dispersal probability and forest diversity in a fragmented landscape, *Ecol. Model.*, 103, 1996.
McClanahan, T.R., The effect of a seed source on primary succession in a forest ecosystem, *Vegetatio*, 175, 1986.
McWilliams, W.H., Stout, S., Bowersox, T.W., and McCormick, L.H., Adequacy of advance tree-seedling regeneration in Pennsylvania's forests, *North. J. Appl. For.*, 187, 1995.
Messier, C. and Puttonen, P., Spatial and temporal variation in the light environment of developing Scots pine stands: the basis for a quick and efficient method of characterizing light, *Can. J. For. Res.*, 343, 1995.
Morris, L.A., Moss, S.A., and Garbett, W.S., Competitive interference between selected herbaceous and woody plants and *Pinus taeda* L. during two growing seasons following planting, *Forest Sci.*, 39, 1, 166, 1993.
Mortzfeldt, Über horstweisen Vorverjüngungsbetrieb, *Z. Forst-Jagdwesen*, 2, 1896.
Mosandl, R. and El Kateb, H., Die Verjüngung gemischter Bergwälder — Praktische Konsequenzen aus 10jähriger Untersuchungsarbeit, *For. Centralbl.*, 2, 1988.

Mosandl, R. and Kleinert, A., Development of oaks emerged from bird-dispersed seeds under old-growth pine stands, *For. Ecol. Manage.*, 35, 1998.

Müller, K. and Wagner, S., Fine root dynamics in gaps of Norway spruce stands in the German Ore Mountains, *Forestry*, 76, 2, 149, 2003.

Muona, O. and Harju, A., Effective population sizes, genetic variability, and mating system in natural stands and seed orchards of *Pinus sylvestris*, *Silvae Genet.*, 38, 5-6, 221, 1989.

Nörr, R., Ganz, M. and Waechter, A., Wildlinge, *Allgemeine Forstzeitschrift/DerWald*, 5, 225, 2002.

Nyland, R.D., *Silviculture: Concepts and Applications*, McGraw-Hill Companies, New York, 1996.

Oliver, C.D. and Larson, B.C., *Forest Stand Dynamics*, John Wiley & Sons, New York 520 pp., 1996.

Otto, H.-J., Die Ausbreitung spontaner Verjüngung in den Wäldern des nordwestdeutschen Flachlandes während des letzten Vierteljahrhunderts, *Forstarchiv*, 236, 1996.

Pacala, S.W., Canham, C.D., Saponara, J., Silander, J.A., Kobe, R.K., and Ribbens, E., Forest models defined by field measurements: estimation, error analysis and dynamics, *Ecol. Monogr.*, 1, 1996.

Perala, D.A. and Alm, A.A., Reproductive ecology of birch: a review, *For. Ecol. Manage.*, 1, 1990a.

Perala, D.A. and Alm, A.A., Regeneration silviculture of birch: a review, *For. Ecol. Manage.*, 39, 1990b.

Ponder, F., Shoot and root growth of northern red oak planted in forest openings and protected by treeshelters, *North. J. Appl. For.*, 1, 36, 1995.

Preuhsler, T. and Pinto da Costa, M.E., Growth of mixed-species regeneration below *Pinus* shelter, in *Symposium of Working Groups of S4.01: Mensuration, Growth and Yield: Mixed Stands — Research Plots, Measurement and Results. Models*, IUFRO, 25 – 29 Apr. 1994, Lous/Coimbra, Portugal, 207, 1994.

Qinghong, L. and Hytteborn, H., Gap structure, disturbance and regeneration in a primeval *Picea abies* forest, *J. Veg. Sci.*, 391, 1991.

Reed, K.L., Shumway, J.S., Walker, R.B., and Bledsoe, C.S., Evaluation of the interaction of two environmental factors affecting Douglas-fir seedling growth: light and nitrogen, *Forest Sci.*, 29, 1, 193, 1983.

Ribbens, E., Silander, J.A., and Pacala, S.W., Seedling recruitment in forests: calibrating models to predict patterns of tree seedling dispersion, *Ecology*, 1794, 1994.

Richardson, D.M., *Ecology and biogeography of Pinus*, Cambridge University Press, Cambridge, 1998.

Röhrig, E., Gussone, H. A., and von Alfred Dengler, Waldbau auf ökologischer Grundlage, Paul Parey Hamburg u. Berlin, 6. Auflage, 1990.

Runkle, J.R., Patterns of disturbance in some old-growth mesic forests of eastern North America, *Ecology*, 1533, 1982.

Runkle, J.R., Synchrony of regeneration, gaps, and latitudinal differences, in tree species diversity, *Ecology*, 546, 1989.

Schütz, J.-P., *Der Plenterwald*, Parey, Berlin, 2001.

Schupp, E.W., Seed-seedling conflicts, habitat choice, and patterns of plant recruitment, *Am. J. Bot.*, 399, 1995.

Shibata, M. and Nakashizuka, T., Seed and seedling demography of four co-occurring *Carpinus* species in a temperate deciduous forest, *Ecology*, 1099, 1995.

Shimoda, K., The regeneration of pioneer tree species under browsing pressure of Sitka deer in an evergreen oak forest, *Ecol. Res.*, 85, 1994.

Skarpe, C., Hytteborn, H., and Cramer, W., Femtio års skogsutveckling på Granskär, Norra Uppland (Fifty years of forest development at Granskär, E Central Sweden), *Svensk Botanisk Tidskrift*, 83, 177, 1989.

Smith, D.M., Larson, B.C., Kelty, M.J., and Ashton, P.M.S., *The practice of silviculture: applied forest ecology*, 9th ed., John Wiley and Sons, New York, 1997.

Sork, V.L., Evolutionary ecology of mast-seeding in temperate and tropical oaks (*Quercus* spp.), *Vegetatio*, 107/108, 133, 1993.

Stimm, B. and Böswald, K., Die Häher im Visier — Zur Ökologie und waldbaulichen Bedeutung der Samenausbreitung durch Vögel, *For. Centralbl.*, 204, 1994.

Stöckli, B., Moderholz für die Naturverjüngung im Bergwald, *Wald Holz*, 8, 1995.

Suchant, R., Baritz, R. and Armbruster, F., Werden Wildlinge weniger verbissen? *Allgemeine Forstzeitschrift/DerWald*, 5, 251, 2000.

Sutton, R.F., *Soil Properties and Root Development in Forest Trees: A Review*, Forestry Canada, Ontario Region, Ontario Information Report, No. O-X-413, Sault Ste. Marie, 42 pp., 1991.

Tappeiner, J.C., II and Wagner, R.G., Principles of silvicultural prescriptions for vegetation management, in *Forest Vegetation Management for Conifer Production*, Walstad, J.D. and Kuch, P.J., Eds., John Wiley and Sons, Inc., New York, 399, 1987.

Tilman, D., *Resource competition and community structure*, Princeton University Press, Princeton, NJ, U.S.A., 1982.

Ulrich, B., Process hierarchy in forest ecosystems: an integrative ecosystem theory, in *Effects of Acid Rain on Forest Processes*, Godbold, D.L. and Hüttermann, A., Eds., Wiley-Liss Verlag, New York, 353, 1994.

Van Der Meer, P.J., Dignan, P., and Savaneh, A.G., Effect of gap size on seedling establishment, growth and survival at three years in mountain ash forest in Victoria, Australia, *For. Ecol. Manage.*, 33, 1999.

Van Lear, D.H. and Waldrop, T.A., Prescribed burning for regeneration, in *Forest Regeneration Manual*, Duryea, M.L. and Dougherty, P.M., Eds., Kluwer Academic Publishers, Dordrecht, 235, 1991.

Wagner, R.G., Research directions to advance forest vegetation management in North America, *Can. J. For. Res.*, 23, 2317, 1993.

Wagner, S., Height increment of single plants in mixed beech and ash regeneration beneath old growth, in *Modelling Regeneration Success and Early Growth of Forest Stands, Proceedings from IUFRO Conference*, Skovsgaard, J.P. and Johannsen, V.K., Eds., Copenhagen, Danish Forest and Landscape Research Institute, Hoersholm, 518, 1996.

Wagner, R.G., Competition and critical-period thresholds for vegetation management decisions in young conifer stands, *For. Chron.*, 76, 961, 2000.

Wagner, R.G., Mohammed, G.H., and Noland, T.L., Critical period of interspecific competition for northern conifers associated with herbaceous vegetation, *Can. J. For. Res.*, 29, 7, 890, 1999.

Walters, M.B. and Reich, P.B., Growth of *Acer saccarum* seedlings in deeply shaded understories of northern Wisconsin: effects of nitrogen and water availability, *Can. J. For. Res.*, 237, 1997.

Walters, M.B. and Reich, P.B., Trade-offs in low-light CO_2 exchange: a component of variation in shade tolerance among cold temperate tree seedlings, *Funct. Ecol.*, 14, 2, 155, 2000.

Wardle, P., The regeneration of *Fraxinus excelsior* in woods with a field layer of *Mercurialis perennis*, *J. Ecol.*, 483, 1959.

Watt, A.S., On the ecology of British beechwoods with special reference to their regeneration, Part I: the causes of failure of natural regeneration of the beech, *J. Ecol.*, 1, 1923.

Willson, M.F., Dispersal mode, seed shadows, and colonization patterns, in *Vegetatio*, No. 107/108; *Frugivory and Seed Dispersal: Ecological and Evolutionary Aspects*, Fleming and Estrada, Eds., 261, 1993.

Young, T.P., Restoration ecology and conservation biology, *Biol. Conserv.*, 73, 2000.

Zeibig, A. and Wälder, K., Untersuchungen zum Einzelbaumeinfluss auf die Walddynamik in Mischbeständen — Räumliche Variation der Fruchtausbreitung von Hänge-Birke (*Betula pendula* Roth.). *Verhandlungen der Gesellschaft für Ökologie*, 32, 165, Cottbus, 2003.

chapter eleven

Cultural practices for restoring and maintaining ecosystem function

David H. Van Lear
Clemson University
Tricia L. Wurtz
US Department of Agriculture Forest Service

Contents

11.1 Introduction

Forest restoration, in a general sense, suggests a transition from a degraded state to some "natural" condition, presumably devoid of human influence (Stanturf, this volume). Yet, because nearly all temperate and boreal forests have been influenced to varying and unknown degrees by aboriginal man, as well as being subject to continually changing climate and other natural disturbances, seeking to restore an ecosystem to a single assumed pre-human condition is neither a realistic nor relevant goal. Instead, managers must study the past disturbance history of the system, and develop a working cognitive model of the desired future condition for that ecosystem. Because a "desired condition" is, by definition, an artifact of society, the desire for restoration will often simply reflect a shift in societal land-use preferences. In other cases, restoration efforts will focus on forest ecosystems that have been altered or degraded by urbanization, agriculture, or silvicultural practices.

Various disturbances, both anthropogenic and unrelated to man, historically maintained a shifting mosaic of different successional stages, forest types, and structural conditions across forested landscapes. A variety of vegetative conditions made ecosystems and species resilient in the face of infrequent severe disturbances such as hurricanes or stand-replacing wildfires. But the increasing influence of humans on temperate and boreal forests has led to widespread changes in these systems. The primary impact of humans on some forests has come from intensive timber management, while other forests have been impacted by fire exclusion. In some systems, a variety of human-mediated forces have worked together. Many forest ecosystems today show reduced biodiversity, fewer structural components, less coarse woody debris, and different stand and landscape-level patterns than those that previously characterized them (Walker 1993; DeLong and Tanner 1996; White and Walker 1997; Linder and Östlund 1998; Siitonen 2001; Carroll et al. 2002). These current traits are believed to be related to lowered, overall resilience of forest systems (Franklin 1993; Franklin and Forman 1987; Hunter 1999).

Reliance on natural disturbances to restore forest ecosystems to their previous condition would not be feasible due to their stochastic nature, and would ignore the important role often played by man. Although silviculture was not historically used for restoration purposes, it can provide landowners and managers with the tools to culture forests for restoration. Indeed, silviculture has been described as a process of creating, maintaining, or restoring an appropriate balance of essential components, structures, and functions to ecosystems that ensure their long-term vitality, stability, and resiliency (Nyland 2002). Silviculture provides many of the tools to restore and maintain ecosystems, especially if they are used to approximate those disturbance agents that historically shaped the system. Restoration is a relatively new focus, but the time-tested responses of forest ecosystems to various silvicultural practices (prescribed burning, regeneration methods, herbicides, etc.) are already fairly predictable and well understood (Table 11.1). Once ecosystems have been restored to specific desired conditions, silvicultural practices can be used to maintain them.

In this chapter, we discuss the challenges inherent in determining restoration goals and fitting appropriate cultural practices to them. We then review the role of fire as a

Table 11.1 Restoration Uses of Some Traditional Silvicultural Practices

Silvicultural Practice	Restoration Uses
Prescribed burning	To achieve desired stand structures; to enhance plant community diversity; to improve wildlife habitat; to maintain fire-dependent communities; to restore function
Harvesting	To create environmental conditions necessary to regenerate even-aged or uneven-aged stands; to achieve desired stand and landscape structure; to provide income for landowner
Intermediate cuttings	To tend established stands; to provide desired stand density, composition, and structure; to reduce fuel loading; to provide income
Planting and seeding	To reestablish desired species on areas where they cannot be regenerated naturally
Herbicide application	To control or remove exotic species; to alter or maintain desired stand structure and composition

disturbance agent and the effects of nearly a century of fire exclusion on temperate and boreal ecosystems. We consider a number of different case studies in the restoration of temperate and boreal forests, with a special emphasis on the use of fire.

11.2 Determining restoration goals

Ecosystem restoration can have many goals. It may involve reintroducing particular species to the forest; in northern Europe, native broadleaved species such as beech (*Fagus* spp.), birch (*Betula* spp.), and oaks (*Quercus* spp.) are being planted on sites previously converted to nonnative conifers (Madsen et al., this volume; Spiecker and Hansen, this volume). On public lands in the U.S., a common objective is the restoration of old-forest conditions, a concept popular in the public mind, although often poorly defined (O'Hara and Baker, this volume). Restoration goals on private lands may be entirely different. Because most private landowners must generate income from their lands to pay taxes and other management costs, timber harvest may be a component of any restoration activity. Some landowners may factor in additional values of restored ecosystems (aesthetics, conservation, wildlife, biodiversity, etc.) to balance the benefits and costs of the restored ecosystem (Mitchell et al. 2000). Whatever the goal, a basic tenet is that restoration, to be successful in the long term, must benefit local people in some manner (Landers et al. 1995).

Aesthetics should be considered in the development of restoration goals. Although often not explicitly stated, the physical appearance of the forest is a critical aspect of the perception of a desired future condition, especially on public lands (Gobster 1996). Treatments that result in aesthetically attractive forest stands are likely to garner public acceptance, while manipulations producing results perceived as unattractive may be controversial. Managers should consider aesthetics when planning, and work to educate the public on the ecological value of intermediate forest conditions that may be perceived as aesthetically unattractive.

A basic requirement of any restoration plan is a clear definition of a desired future condition. Yet such objectives are moving targets. Ecosystems continue to evolve in response to changing climates, disturbance patterns, and landuses, and the desires of landowners and society will continue to shift as well. Information is often lacking on the range of variation, dynamics, and characteristics of the ecosystem being restored (Noss 1985; White and Walker 1997), making it impossible to know the exact condition of an ecosystem at a prior time, or the precise sequence of disturbances that shaped it. Because the degree to which any mix of cultural practices simulates the historical disturbance regime of a particular ecosystem is a matter of interpretation, restoration efforts are as much art as science.

Managers face the further challenge of recreating an ecosystem that probably varied widely over temporal and spatial scales. Issues such as fragmentation and the invasion of exotic species require landscape-level and even regional perspectives (Galley and Wilson 2001). Because of all these factors, restoration must be a long-term effort and goals are likely to be general rather than specific. Silviculture can provide the tools to create and sustain the desired future condition. However, successful ecosystem restoration efforts will require both perseverance and a flexible, adaptive management approach that combines knowledge and skills from both the biological and social sciences (Figure 11.1).

11.3 Fitting cultural practices to restoration goals

Resource managers and landowners who attempt to restore and maintain ecosystems must educate themselves about the nature of the disturbance regimes, both anthropogenic and natural, that historically shaped the system. Once managers have developed a cognitive model of a desired future condition, they can develop a plan to restore those conditions to the landscape. Silvicultural practices provide the appropriate tools.

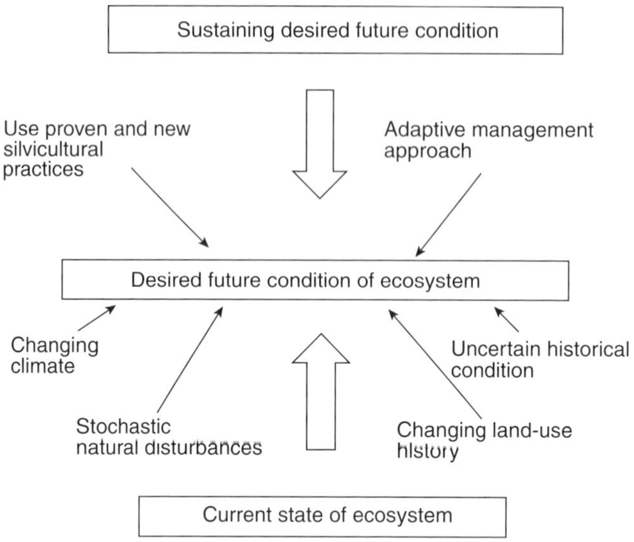

Figure 11.1 Factors that influence the desired future condition of ecosystems and the management approach to sustain that desired condition.

Cultural practices can mimic many features of natural disturbances. For example, clearcuts open a portion of a forest to greater solar radiation, much as a windstorm might, and soil moisture can increase soon after harvest, just as it would after many types of natural stand-replacing disturbance. Soil nutrients are more available after anthropogenic or natural disturbances, and early-successional species often invade a clearcut just as they might after a windstorm that blows down a portion of a forest. But silvicultural practices may differ from natural disturbance in ways that may affect the condition of the forest such that the goals of the landowner are compromised. The chief difference is that harvesting removes woody biomass while a natural disturbance does not. In natural disturbances, the dead trees remain in the system and are important to wildlife and as sources of coarse woody debris (Jonsson and Kruys 2001). Managers can mitigate the effects of woody biomass removal by designing harvest strategies that leave tops of cut trees on site, or that retain some of the overstory standing as either dead or live trees.

Similarly, group-selection harvesting can approximate small openings in a forest that might have resulted from a lightning strike or a small group of trees killed by insects or disease. Thinning can imitate the natural mortality that occurs as stands age, or the selective mortality that occurs in windstorms, disease or insect outbreaks, etc. Again, these harvesting disturbances differ from natural disturbance by the removal of trees. The significance of this biomass removal to the desired future condition of the stand, and whether the impact can be lessened through the use of partial overstory retention or other means, will depend on the ecosystem in question.

Access is another way that silvicultural practices differ from natural disturbances in their impact on ecosystems. Road access is necessary to carry out most stand-based treatments, to transport people and equipment to the stand or to remove merchantable fiber, or both. Because establishing a new road on a landscape is the equivalent of adding a new ecosystem to an existing one (Lugo and Gucinski 2000), road networks have the potential to greatly impact forest landscapes (Trombulak and Frissell 2000). Building, operating, and maintaining roads constitute new forms of disturbance; the earth movement during road building can disturb whole watersheds. Roads fragment the landscape and can allow access by invasive species. Yet roads can also help landowners respond to wildfire, manage insects or disease, and realize a profit on their forested land. The effects of roads may be thus viewed as positive as well as negative, depending on the ecosystem component in question (Trombulak and Frissell 2000; Lugo and Gucinski 2000). Opportunities to mitigate the adverse impacts of roads include careful choice of site, careful construction practices, use of temporary or seasonal roads, and abandonment and restoration of old roadbeds.

Fire has long been used to accomplish land management objectives (Hämet-Ahti 1983; Wade and Lunsford 1989; Pyne et al. 1996). Historically, fire acted as an ecological process on varying gradients of soil, topography, hydrology, and climate to shape vegetative patterns and character. Prescribed fire can be used to complement other agents of disturbance, both anthropogenic and natural, and to maintain a shifting mosaic of different seral stages of the restored ecosystem over the landscape. If properly used, prescribed fire can restore structure, composition, and function of certain ecosystems to more closely resemble desired conditions.

In many situations, however, the use of fire is not feasible. Fire carries with it risks to human habitation, concerns about liability, difficulties in obtaining burning permits, effects of smoke, limited burning days, and the costs of applying, controlling, and monitoring burns (Wigley et al. 2002). In some circumstances, selective herbicides can provide the disturbance functions of fire. Selective herbicides allow managers to manipulate species composition in the understory, control stand structure, and selectively remove undesirable exotic plants (Grilz and Romo 1995; Wigley et al. 2002). At the same time, herbicides do not have the same ecological impact of fire; they cannot scarify leguminous seeds to enhance germination or open serotinous cones. In some parts of the boreal zone, such as Scandinavia, the use of herbicides in forest management is prohibited.

Neither prescribed fire nor herbicide application can address every restoration objective; sometimes a combination of the two practices may be called for. Because public support is critical, efforts to educate the public about the environmental effects of both prescribed fire and herbicides, and the reasons for their use, may pay large dividends. In any case, managers must exercise considerable judgment when deciding to use either fire or herbicides or both for restoration purposes.

Most decisions regarding desired future conditions require a landscape-level approach (Boyce 1995). Although cultural treatments are applied to stands, a landscape-level perspective is needed for such issues as connectivity, fragmentation, wildlife habitats, sustained yield, and endangered and threatened species. At a minimum, treatments applied at the level of the stand should complement landscape-level objectives.

11.4 Fire as a disturbance agent in temperate and boreal forests

Historically, fire was the dominant disturbance that shaped the character of many temperate and boreal ecosystems (Pyne et al. 1996; Bonnicksen 2000; Brown and Smith 2000). It often complemented other types of disturbances, such as hurricanes, tornadoes, ice storms, insect infestations, disease outbreaks and, more recently, harvests, because these events created conditions that predisposed disturbed areas to burn (Myers and Van Lear 1997; Brown and Smith 2000).

Fire regimes vary widely among and within geographic regions, making it difficult to generalize about regional responses to fire (DeLong 1998; Bergeron et al. 2002). The historical fire cycle of many ecosystems is subject to continuing scientific study and debate. For example, how did the fire cycle change in response to the arrival of the first aboriginal humans in the system? How did it change with the arrival of European settlers? How did a changing climate affect it?

Despite these challenges, fire scientists have devised systems for characterizing fire regimes, using variables such as intensity, severity, frequency, and extent (Agee 1993; Brown and Smith 2000). In parts of North America, for example, in moist to wet systems in northern hardwood forests, fires were so rare as to be effectively absent from the system (Table 11.2). Other ecosystems were subject to very frequent fires that burned with low intensity (understory fire regime). Typically, such understory fires were less severe, had short flame lengths, and had relatively little direct effect on the soil and overstory vegetation. Other areas burned infrequently, but with great intensity, and essentially all overstory trees were killed (stand-replacement fire regime). In other cases, fire caused selective mortality in the dominant vegetation or the effects of fire varied spatially and/or temporally between understory and stand-replacement regimes (mixed severity fire-regime).

In the temperate forests of North America, fire has been integral to the processes that maintained the structure, composition, and function of most ecosystems (Agee 1993; Duchesne 1994; Whelan 1995; Brown and Smith 2000). The pine and oak forests of the eastern U.S., the sequoia forests of the west coast, the conifer forests of the Rocky Mountains and the Sierra Nevada, as well as grassland ecosystems all depended on frequent fire. Boreal forests, on the other hand, are slower growing and less productive than most temperate forest systems, with slower accumulation of woody fuels that persist longer on the forest floor. While normal fire regimes in much of the boreal forest zone include infrequent, large, and often severe fires (Pyne et al. 1996; Viereck 1973), in other parts of the boreal zone, fire regimes are significantly different (Bergeron et al. 2001). There is a wider range of variation in what constitutes a natural boreal fire cycle than has previously been believed (Bergeron et al. 2002).

In unraveling fire histories, the role of anthropogenic burning must be considered. Aboriginal humans were often the dominant ignition sources, using fire to manage ecosystems for their benefit (Pyne et al. 1996; Bonnicksen 2000; Brown and Smith 2000). As early people learned to use and control fire, they complemented the natural lightning regime, increasing the frequency of fire and changing its seasonality (Bonnicksen 2000; Carroll et al. 2002). The frequent and often extensive anthropogenic and lightning-ignited fires, along with the herbivory that fire encouraged, created and maintained open woodlands, savannas, and prairies throughout much of the temperate forest zone in the U.S. When Columbus arrived, most of the eastern U.S. was a managed landscape composed of a mosaic of dense forests, woodlands, savannas, and prairies in various stages of succession, all created and maintained by burning. After populations of Native Americans plummeted in the 16th and 17th centuries, the level of burning declined and these open and dispersed ecosystems gradually became closed, contiguous forests (Williams 1989; McCleery 1993; Mann 2002; Pyne et al. 1996; Carroll et al. 2002).

11.5 Fire exclusion and the decline of fire-dependent ecosystems

Not until the early 1900s were there serious efforts to exclude fire as an ecological process in North America. A series of large wildfires in Wisconsin, Minnesota, and Idaho around the

Table 11.2 Oak Restoration Options with the Shelterwood-Burn Method

Option	Silvicultural Treatment
Regeneration of new oak-dominated stand	Harvest shelterwood following prescribed burn to release oak regeneration
Culture a two-aged stand with oak dominating both age classes	Retain shelterwood and withhold additional burns
Develop oak woodland or savanna	Retain shelterwood and resume burning at 2–4-year intervals

From Brose et al. Using Shelterwood harvests and prescribed fire to regenerate oak stands on productive upland sites, *For. Ecol. Manag.*, 113. 125, 1998.

turn of the last century aroused public attention and concern. Cutover forests in the southeastern Coastal Plain were burning so frequently — often annually — that forest regeneration was impossible. Slash fires following logging in the Appalachian Mountains burned severely, causing devastating off-site effects, such as erosion, sedimentation, and air pollution from smoke. In the early decades of the 20th century, the public, led by the Forest Service and state forestry commissions, began to see fire as an enemy to be suppressed at all costs (Pyne et al. 1996; Johnson and Hale 2002). These actions resulted in a widespread policy of fire exclusion.

Fire exclusion in fire-dependent ecosystems can be considered a form of disturbance, that is, a state of disorder, which changes the very nature of that ecosystem. Exclusion of fire had many unexpected and undesirable consequences (Covington and Moore 1994; Brennan et al. 1998; Landers et al. 1995; Brose et al. 2001). In fact, the ecosystem restoration movement in the U.S. can be linked closely to the threatened or declining status of fire-dependent ecosystems and the species within them, such as longleaf pine in the southeast (Brockway et al., this volume) and ponderosa pine in the southwest (Kaufmann et al. this volume).

In Europe, a similar situation has evolved, but with earlier origins. The agricultural and forest practices of the British and French empires became the standard by which to measure forest ecosystem health (Pyne et al. 1996). Intensive agriculture supplanted fire and heavy grazing by domestic animals reduced fuel accumulation. A century of intensive forestry removed the threat of wildfire and prescribed burning was prohibited.

The situation is quite different in the boreal forest of Canada and Alaska. In some areas that have yet to be accessed by roads, the major human impact has been fire suppression. Other vast tracts receive no suppression effort at all and still experience the large wildfires that are part of the natural fire regime. For example, between 1976 and 1991, northwestern Ontario had at least one wildfire of greater than 100 km^2 per year (Racey et al. 1991), while in Alaska, at least one wildfire of more than 400 km^2 occurs in a typical year. While fire suppression efforts in parts of Canada and Alaska began only at the start of the 20th century, in other areas suppression was not instituted until the 1970s. Because of the vast areas involved, lack of access, and limited budgets, the effectiveness of organized suppression activities is debated. There are indications that, at least in western Canada, the number of acres burned increased in the second half of the 20th century (Kurz and Apps 1999; Van Wagner 1988). Indeed, Bergeron et al. (2002) make the point that for vast areas of Canada, maintaining existing biodiversity is a more appropriate management goal than seeking to restore it.

It is apparent that fire has been an important factor shaping many forest ecosystems around the world. Attempts to exclude it have resulted in many undesirable consequences. While it is unlikely that fire will ever return to its historical importance, certain fire-dependent ecosystems can be managed with fire to restore and maintain their historical character. Where it is not possible to use fire in a management context, other silvicultural tools may be appropriate to create and maintain forests similar to those created and maintained by fire.

11.6 Cultural practices for restoring and maintaining ecosystem function in temperate forests

11.6.1 Maintaining oaks on productive sites in the eastern U.S.

Fire played a major role historically in maintaining oak (*Quercus*)-dominated forests on productive sites in the eastern U.S. As a result of fire exclusion, oak forests on good sites are now declining. Because oaks are typically unable to regenerate in dense shade, they have been replaced by shade-tolerant species where small canopy-gap disturbance regimes dominate. When large canopy openings occur, such as after clearcut regeneration harvests, oaks are unable to compete with the fast-growing mesophytic species that colonize recently harvested sites (Abrams 1992; Loftis and McGee 1993; Lorimer 1993; Brose and Van Lear 1998). For decades, silviculturists puzzled over the problem of securing vigorous oak regeneration on good-quality sites in the eastern U.S. Prescribed burning in hardwood stands was not considered feasible because of the widely recognized damage to boles of overstory trees from wildfires, and more recent accounts of bole damage from prescribed fires (Wendell and Smith 1986).

Recently, we have come to understand better the role of fire in these forests. Fire often swept through hardwood forests of the eastern U.S. in the past, shaping the character of forests and maintaining oak as a dominant component in the landscape (Abrams 1992; Brose et al. 2001; Carroll et al. 2002). Recent studies have shown that understory burning in mature mixed-hardwood stands at 2- to 4-year intervals enhanced oak advance regeneration in the Piedmont region of South Carolina (Barnes and Van Lear 1998). These low-intensity burns produce little damage to boles of larger crop trees and reduce understory and midstory density, allowing more sunlight to reach the forest floor. However, a disadvantage of this technique is that multiple burns are required to reduce competition and increase the abundance and size of oak regeneration sufficiently to where harvest cuts would result in successful oak regeneration.

Brose and Van Lear (1998) and Brose et al. (1998) developed a more efficient and financially attractive technique using fire to regenerate oaks — a shelterwood-burn method that has produced excellent results in the Virginia Piedmont (Figure 11.2). The prescription calls for an initial shelterwood cut in oak-dominated hardwood stands followed in several years by a moderate to high-intensity fire (flame lengths of about 1 m) through the advance regeneration. This technique has successfully converted advance regeneration under mature mixed-hardwood shelterwood stands from yellow poplar (*Liriodendron tulipifera*) domination to predominately oak regeneration. In addition, the oak regeneration is of good form, competitively sized, and sufficiently free to grow so that it should be capable of forming a new oak-dominated stand when the overstory is removed (Brose et al. 1998).

A key ingredient of the prescription is achieving fires of sufficient intensity to significantly set back competing species. The silvicultural prescription simulates, to a degree, the combined events of overstory disturbance caused by wind or ice storms followed by fire, related disturbances that have shaped the composition of oak ecosystems for millennia (Brose et al. 2001; Van Lear and Brose 2002). This method assumes that there is advance regeneration of oak, albeit noncompetitive, prior to the initial cut. If not, it will be necessary to begin understory burning and delay the initial shelterwood cut until adequate oak advance regeneration is in place.

There are several restoration options following the burn in this technique (Figure 11.2). If regeneration of an even-aged oak-dominated stand is the goal, the shelterwood can be removed to allow the regeneration to develop in full sunlight. If, for aesthetic and other reasons, a continuous cover of high forest is desired, the shelterwood can be retained to allow the oak-dominated regeneration and the shelterwood to develop into a two-aged stand. Both of these options will maintain oaks after harvest and reverse the decline of oak forests on good-quality sites. A third option is to continue burning at frequent intervals (2- to 4-year intervals) to encourage the development of oak woodlands, or savannas if continued long enough. These open, grass-dominated ecosystems are now rare in the eastern U.S., but were once common when frequent fires controlled vegetative patterns (Carroll et al. 2002).

11.6.2 *Maintaining Table Mountain/pitch pine ecosystems in the southern Appalachians*

Table Mountain pine (*Pinus pungens*)/pitch pine (*Pinus rigida*) stands became rarer in the southern Appalachian mountains during the past century (Clinton et al. 1993). As with

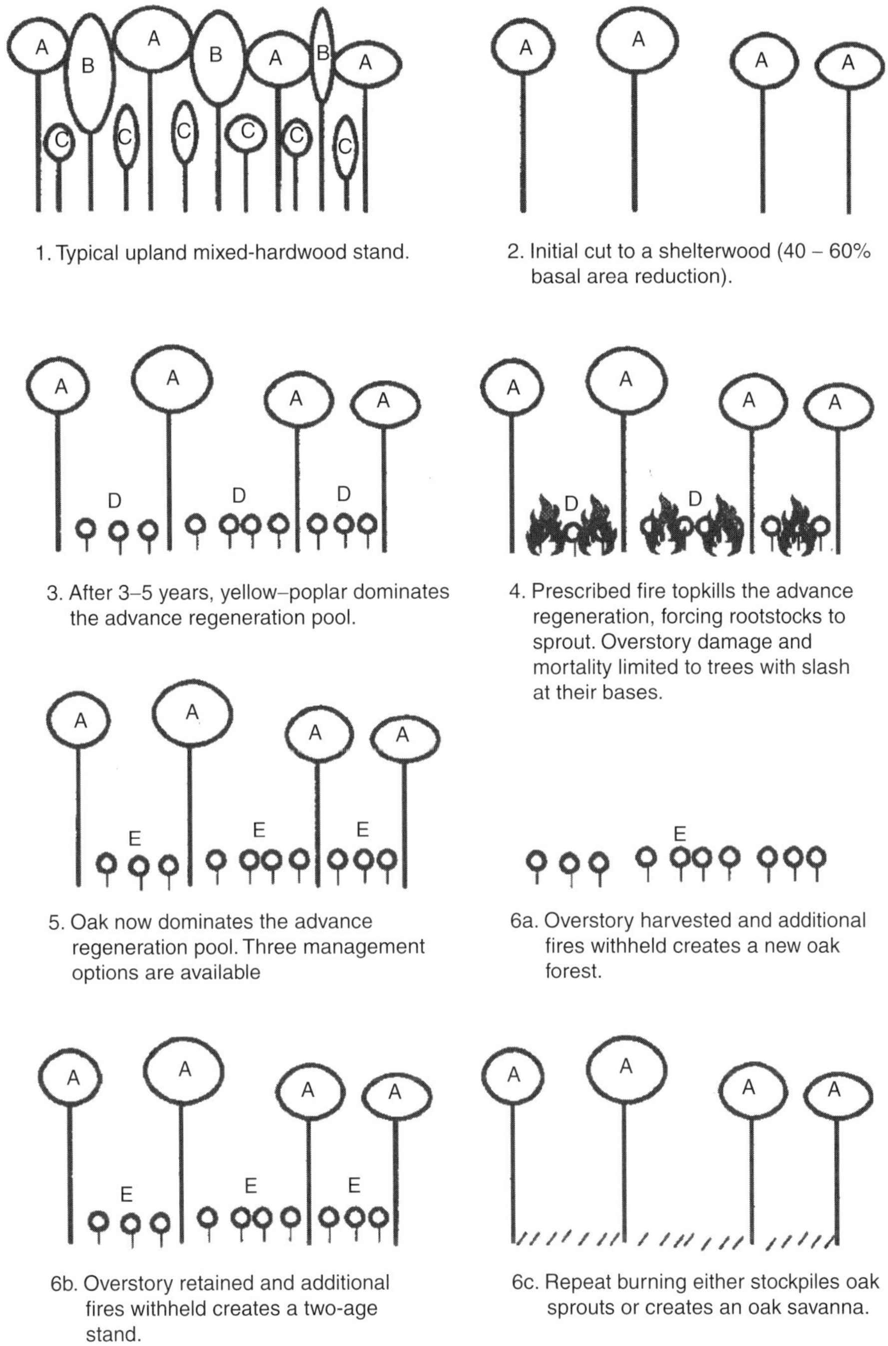

Figure 11.2 Shelterwood-burn technique for regenerating oak stands on productive upland sires in the Piedmont. A = high-quality dominant oaks; B = hickories, poor-quality oaks, and yellow poplar; C = American beech, flowering dogwood, and red maple; D = mixed hardwood regeneration dominated by yellow poplar; E = mixed hardwood regeneration dominated by oak. (From Brose and Van Lear, Responses of hardwood advance regeneration to seasonal prescribed fires in oak-dominated shelterwood stands, *Can. J. For. Res.*, 28, 331, 1998.)

(A)

(B)

Figure 11.3 (A) Open woodland conditions in a Table Mountain/pitch pine stand following four surface burns over a 12-year period. (Photo by Russell Randles.) (B) Stand replacement prescribed burn for fuel reduction and regeneration in Table Mountain/pitch pine stands. (Photo courtesy of Joint Fire Science Program and Tom Waldrop.)

other fire-dominated pine ecosystems (Brockway et al., this volume; Kaufmann et al., this volume), the Table Mountain/pitch pine ecosystem is in decline because of fire exclusion. Isolated stands of Table Mountain pine and pitch pine (these two species often coexist) occur on exposed, generally southwest-facing slopes, and grow on xeric sites. Because of inaccessibility, small size, and low quality of trees in this type, it is considered unmerchantable. Maintaining it as a component of the larger southern Appalachian ecosystem is primarily a matter of conservation and land stewardship rather than economics, and is of concern mostly on public lands.

Ecologists and fire scientists debate the type of fire regime that historically maintained this pine ecosystem. In the absence of fire, low-quality hardwood species and mountain laurel (*Kalmia latifolia*) replace these pine stands (Elliott et al. 1998; Waldrop and Brose 1999). Cyclic southern pine bark beetle (*Dendroctonus frontalis*) epidemics accelerate the rate of succession to hardwoods.

Single high-intensity prescribed burns have limited success in restoring Table Mountain/pitch pine stands (Elliott et al. 1998; Waldrop and Brose 1999). Thick layers of accumulated duff must be reduced to obtain seedling regeneration of these species (Mohr et al. 2002). Periodic burning at 3- or 4-year intervals with moderate- to high-intensity fires creates and maintains the type in open woodland conditions (Figure 11.3A) (Randles et al. 2002). Historically, a mixed-severity fire regime with both surface fires and stand-replacement fires (Figure 11.3B) was probably typical in this type. This fire regime kills part of the overstory, reduces the importance of ericaceous shrubs such as mountain laurel, prepares a favorable seedbed, and favors herbaceous cover and pine regeneration.

11.7 Cultural practices for restoring and maintaining ecosystem function in boreal forests

The level of human impact in the circumpolar boreal zone and the associated need for forest restoration varies greatly. Much of the forestland of Fennoscandia has been managed intensively for centuries, via widespread slash-and-burn conversion of forests to farmland, followed by intensive silviculture and the exclusion of wildfire (Hämet-Ahti 1983; Östlund et al. 1997). At the other end of the spectrum lie Alaska and portions of boreal

Canada, where large areas of forest have never been harvested or even accessed by roads (Sanderson et al. 2002). For some of boreal North America, the primary human impact comes in the form of fire suppression. Fire has shaped the boreal biota over centuries and has directed the biological structure of the landscape (Rowe and Scotter 1973; Zackrisson 1977; Pyne et al. 1996). The fire-return interval is long relative to temperate zones; fires can be large, intense, and smolder for a long time. While there is variation across the region, the typical boreal forest wildfire is stand-replacing (Figure 11.4). Trees regenerate from seed from nearby unburned patches of forest (e.g., *Picea glauca*), from seed released from serotinous cones (e.g., *Pinus banksiana* and *Pinus contorta*), or by sprouts from surviving belowground parts (e.g., *Betula* spp. and *Populus tremuloides*).

Boreal silvicultural strategies are being widely redesigned with the goal of maintaining or restoring specific attributes of forest ecosystems at the same time that they yield usable fiber. Harvest prescriptions can be designed for a variety of different purposes, including approximating some aspects of wildfire (Bunnell 1995; Delong and Tanner 1996), promoting the development of old-forest characteristics (Coates and Burton 1997; Singer and Lorimer 1997; Bergeron and Harvey 1997; Burton et al. 1999), or increasing the diversity of the ecosystem (Haila 1994; Lieffers et al 1996; Lämås 1996). In both Canada and Alaska, timber is commonly harvested during the winter, when frozen ground can improve access to merchantable stands. Ice bridges and temporary winter roads can be

Figure 11.4 Stand replacement fire in black spruce in Alaska. (Photo courtesy US Department of Agriculture, Forest Service.)

used to avoid some of the negative effects of permanent, all-season roads (Rhoads 1974; Blinn 1998).

11.7.1 *Restoring coarse woody debris and biodiversity to the managed forests of Sweden*

The boreal forests of Fennoscandia are adapted to centuries of anthropogenic- and lightning-caused fires (Zackrisson 1977; Kuusela 1992). Beginning early in the last century, however, intensive agriculture and plantation forestry gradually changed these forests (Östlund et al. 1997). The long history of intensive human use included slash-and-burn conversion to farm and grazing land, establishment of plantation forests of Scots pine (*Pinus sylvestris*) and Norway spruce (*Picea abies*), and exploitative harvesting for charcoal and sawlogs (Hämet-Ahti 1983). The amount of coarse woody debris and the number of standing dead trees declined precipitously, and with this loss of fuel, wildfires became rare (Lämås 1996; Pyne et al. 1996). Wildfires that did occur were suppressed. By the 1980s, fire was virtually eliminated from the forests of Fennoscandia.

In Sweden, the combination of slash-and-burn agriculture, intensive forestry, conversion to conifer plantations from native broadleaved forests, and ultimately fire exclusion resulted in a dramatic loss of biodiversity (Esseen et al. 1997). Nearly 200 species of vascular plants, 55 species of bryophytes, 100 species of lichens, and 10 species of mammals were considered endangered or regionally extinct by the year 2000 (Gardenfors 2000), and many of these species are dependent on old, dying, or dead trees (Berg et al. 1994). In 1993, the maintenance of biodiversity was made equally important with timber production in Swedish forest policy (Lämås and Fries 1995). A new strategy was adopted to restore biodiversity, which is based on a system of forest reserves and management that mimics natural processes. However, because only 0.5% of productive forestland in Sweden is in reserves, most of the restoration effort is occurring in concert with fiber production (Fries et al. 1997; Hansen et al. 1991).

Attention to coarse woody debris is a fundamental part in this approach. Coarse woody debris provides habitat for plants, animals, and fungi and is important in nutrient cycles and carbon budgets (Harmon et al. 1986; Sturtevant et al. 1997). New forest management strategies are being advanced that allow some trees, or some patches of trees, to become very large and old, or that include provisions for generating snags. In addition to acting as a bank of coarse woody debris, large, old trees are a source of future small-scale disturbance. In northern Sweden, uprooting of large trees during natural windthrow has been shown to lead to greater habitat heterogeneity and a significant increase in bryophyte diversity (Jonsson and Esseen 1990).

Fries et al. (1997) and Angelstam (1998) describe the three major site types in boreal Sweden and suggest management options. Under natural conditions, two of these ecosystems, Scots pine sites and mesic sites with either deciduous species or Norway spruce, are fire disturbed, and are viewed as well adapted to large-scale disturbances. On Scots pine sites, harvest strategies would leave 5 to 20 seed trees standing per ha, and low-intensity prescribed burning would be conducted beneath seed trees. Efforts would be made to conserve or reestablish native broadleaved tree species, either through leaving birch and other deciduous species uncut, or by planting. Such postfire, successional forests are considered a biotope of high conservation value (Lämås and Fries 1995); the Swedish government provides financial incentives for landowners to use prescribed burning rather than mechanical site preparation.

For mesic-type sites, clear felling of conifers with retention of deciduous species is recommended (Fries et al. 1997). Rotations would be prolonged, with some stands allowed to develop to domination by aspen and birch. The third type of site, an uneven-aged Norway spruce forest, is viewed as highly sensitive to severe disturbance. The traditional management approach for such forests has been to clearcut and drain (or "ditch") them, followed by replanting with Norway spruce. Partial overstory retention, using either selection harvest systems or dense shelterwood harvesting systems, is now recommended on these sites (Fries et al. 1997), along with allowing at least a third of the shelterwood trees to remain on the site as a source of large snags.

Concerns about species conservation are forcing a reexamination of attitudes toward fire in Fennoscandia (Kuuluvainen et al. 2002). As more is learned about the ecological role of fire in European boreal forests, it is likely that the use of prescribed fire will increase. Nevertheless, the beneficial role of fire in the perpetuation of nature preserves and historically valued landscapes has yet to be widely accepted in Europe (Pyne et al. 1996).

11.7.2 Restoring components of natural disturbance to managed forest landscapes in Canada

Although harvesting in the boreal forest of Canada began in the early 20th century, the vast majority has occurred since the 1970s, when rates of harvest accelerated substantially. Some of the interest in new silvicultural approaches in the boreal zone of Canada thus comes from a desire to maintain natural characteristics of newly-accessed land, rather than from a restoration need. In other situations, researchers are modifying the methods developed to maximize wood production and redesigning them to provide a wider variety of forest attributes.

The forest landscape that results from intensive forest management differs dramatically from forests generated by wildfire (DeLong and Tanner 1996; Bergeron et al. 2002). Wildfire creates a complex pattern of disturbance on the landscape, consisting of a mosaic of different-aged patches and irregular boundaries. Many patches exceed 500 ha in size, containing unburned areas within them in which trees sometimes grow very old (Eberhart and Woodward 1987). As it is currently practiced, intensive forest management results in a forest landscape composed of regularly dispersed clearcuts of less than 60 to 100 ha in size (DeLong and Tanner 1996). As this approach moves toward a fully regulated forest, trees older than the rotation age will gradually become underrepresented on the landscape until they disappear completely (Bergeron et al. 2002).

Different silvicultural approaches have been suggested to mimic natural disturbances. Increasing the size of some harvest units to greater than 500 ha would result in a landscape pattern of disturbance more similar to the pattern of wildfire (DeLong and Tanner 1996). Such occasional large harvest units could have economic advantages that would compensate for the costs of irregular unit boundaries and leaving patches of trees uncut. Because the vertebrate fauna of boreal and sub-boreal British Columbia are adapted to large-scale disturbance, the effects of clearcutting would more closely approximate those of wildfire if more woody debris, standing snags, and live trees were left (Bunnell 1995). Natural forest age structure could be maintained through the use of rotations of various lengths (Burton et al. 1999).

11.7.3 Mixed-wood management in Canada's southern boreal forest

The northern mixed-wood forest occurs on mesic sites in a swath extending from Alaska to eastern Canada. It is composed primarily of white spruce, balsam fir (*Abies balsamea*), and trembling aspen, and lesser amounts of black spruce (*Picea mariana*) and jack pine (*Pinus banksiana*) (Greene et al. 2002). The mixed-wood ecological region includes about 250,000 km^2 of northern Alberta (Cumming et al. 1996). Most of the area has never been harvested, and remains inaccessible by road, although it is crisscrossed by seismic investigation lines and natural gas pipelines. Fire is the main natural disturbance agent, and fire suppression is likely the most significant impact of European settlers (Cumming et al. 1996). Trembling aspen is usually the first species to dominate following fire, while white spruce gradually colonizes and develops in the understory. If undisturbed, uneven-aged stands of white spruce and balsam fir become dominant.

The traditional silvicultural methods used to manage these stands have focused on establishing relatively pure stands of white spruce (Greene et al. 2002). Regenerating white spruce is difficult; however, the species grows slowly and is easily overtopped by broadleaved trees, shrubs, and grass. Even when pure spruce stands can be established successfully, however, they have lower biodiversity than mixed species stands (Burton et al. 1992). At the same time, stands in which the majority of the overstory is trembling aspen are typically not managed for conifer production at all (Lieffers et al. 1996).

In an effort to capture the potential spruce yield on such primarily deciduous sites, as well as improve the biodiversity value of sites that have been managed for white spruce, a variety of mixed-wood silvicultural options have been proposed (Lieffers et al. 1996; Man and Lieffers 1999). For sites where a spruce seed source is lacking, clearcutting aspen followed by underplanting of white spruce early in the development of the regenerating aspen stand has been proposed. Another option is using heavy mechanical site preparation to control the shrub understory of an open, low-vigor stand of aspen, followed by underplanting with white spruce 10 or more years before harvesting the trembling aspen overstory. Sites would need to be carefully chosen so as to allow for the removal of the aspen overstory without damaging the young white spruce (DeLong 2000). On sites where spruce occurs in the overstory, shelterwood and seed tree harvesting options are recommended as a means of retaining a spruce seed source on the site. Such ecologically based, mixed-wood management may sustain more ecosystem components than the current management system, and the public support it garners may result in greater security of tenure for industrial forestry on public lands (Lieffers et al. 1996).

11.7.4 Landscape-scale burning in Alaska

The boreal forests of Alaska have been harvested very little. Impacts of historical as well as recent timber harvesting are limited to the areas surrounding towns, along the limited State Forest road system, near major rivers where wood was cut for stern-wheeled riverboats, and around mining areas where wood was used for fuel (Wurtz and Gasbarro 1996). The total area disturbed in this way is a small fraction of Alaska's boreal zone. Fire suppression has had a larger impact than timber harvesting; however, the natural fire cycle remains the subject of some debate. Between 1898 and 1939, an estimated 1 million acres burned annually. With the founding of permanent settlements in the early 1900s, concerns about wildfire arose (Jewkes 1999). Due primarily to limited access, early suppression efforts were limited in scope and effectiveness.

Today, most of the land in Alaska is classified into one of four fire management categories. Inhabited areas are designated "Critical" because of the immediate threat of wildland fire to human life and shelter. Areas having desired resources such as accessible timber or late-seral wildlife habitat are classified as "Full," where initial attack is again practiced to reduce the spread of fire. Many areas remote from human occupation, constituting the majority of the state, are designated "Limited," and fires in these areas are monitored but not suppressed unless they threaten to spread to areas where fire protection is desired. Thus, most of Alaska still experiences a natural or seminatural fire regime. The fourth class, "Modified," is intended to provide a higher level of protection when fire danger is high and a lower level of protection when fire danger decreases, thus providing

Figure 11.5 In interior Alaska, landscape-scale prescribed burns are conducted to enhance habitat for species such as moose. (Photo courtesy of Alaska Department of Fish and Game.)

increased flexibility in the selection of suppression strategies (Alaska Wildland Fire Coordinating Group 1998).

In interior Alaska, aggressive suppression in the immediate vicinity of settled areas has led to a disproportionate representation of conifers in older age classes. This pattern of succession in the absence of fire increases fire risk to nearby structures or communities and lowers the productivity of many wildlife species adapted to fire disturbance (Haggstrom and Kelleyhouse 1996). In such areas, managers are experimenting with stand-scale treatments such as willow crushing or aspen felling to maintain cover and browse on sites known to be productive for game species, such as moose (*Alces alces*) and grouse (*Bonasa* spp.). Stand-scale burning of aspen in spring, just before leaf-out, is also used on sites that are accessible by forest roads but away from communities.

In remote areas, large landscape-scale burns have been carried out using aerial ignition. These large burns are indistinguishable from natural fires and dramatically enhance habitat for a variety of wildlife species requiring early-seral vegetation, including moose, grouse, and a number of meso-carnivores (Figure 11.5).

11.8 Conclusions

Restoring ecosystems and maintaining them in a restored state is as much an art as a science. In most cases, managers are faced with the challenge of approximating some version of the ecosystem in question, with some uncertainty about the nature of this desired system. Managers can make use of a variety of tools to develop a cognitive model for the restored ecosystem, including fire histories, palynological reconstructions, comparative analysis of other systems, accounts of early explorers and settlers, and even old photographs. Once the desired future condition of the ecosystem has been determined, managers can use well-documented silvicultural practices to maintain the system in that general state. In keeping with the adaptive management approach, managers must continually reevaluate not only the target state of the ecosystem but also the methods used to maintain it there.

Long-established silvicultural practices can be used to culture restored ecosystems. Prescribed burning can simulate many of the effects of anthropogenic and lightning-ignited wildfire regimes and can be used to maintain historical stand structures and to direct composition and function of restored ecosystems. Harvesting systems can emulate, at least partially, those types of disturbances that created environmental conditions favoring reestablishment and maintenance of desired species and communities. Intermediate cuttings can control stand density and composition. Portions of the overstory can be retained as wildlife habitat, and as a source of future coarse woody debris. Herbicides can control exotic species and maintain desired stand structure. Species mixtures can be accepted and encouraged. Native species that have been lost from the system can be reintroduced. Seasonal roads can be used or permanent roads can be sited to minimize adverse roading impacts. Silviculture provides opportunities for landowners to realize a financial return from their forested land — an important consideration in the long-term maintenance of restored ecosystems.

Fire was the dominant factor that historically shaped the character of most temperate and boreal forests. However, a century-old policy of fire exclusion in North America and a longer one in Europe has erased the memory of fire's importance. Because the field of restoration ecology is rekindling the search for knowledge of the ecological processes that shaped ecosystems, interest in the ecology and use of fire will likely grow. Fire should be considered in the context of its role as an ecological process that shaped vegetative patterns in most ecosystems for thousands of years. Whether ignited by man or by lightning, fire has the potential to play an important role in forest restoration. Fire and other silvicultural practices can reestablish and maintain pattern and process in ecosystems around the world.

Acknowledgments

We thank Lars Edenius, Sybille Haeussler, Craig Delong, and Peter Kapeluck for helpful reviews of the manuscript.

References

Abrams, M.D., Fire and the development of oak forests, *Bioscience*, 42, 346, 1992.

Agee, J.K., *Fire Ecology of the Pacific Northwest Forests*, Washington, DC, Island Press, 493, 1993.

Alaska Wildland Fire Coordinating Group, *Alaska Interagency Wildland Fire Management Plan*, 61 pp., 1998.

Angelstam, P.K., Maintaining and restoring biodiversity in European boreal forests by developing natural disturbance regimes, *J. Veg. Sci.*, 9, 593, 1998.

Barnes, T.A. and Van Lear, D.H., Prescribed fire effects on advanced regeneration in mixed hardwood stands, *South. J. Appl. For.*, 22, 138, 1998.

Berg, Å., Ehnström, B., Gustafsson, L., Hallingbäck, R., Jonsell, M., and Weslien, J., Threatened plant, animal, and fungus species in Swedish forests: distribution and habitat associations, *Conserv. Biol.*, 8, 718, 1994.

Bergeron, Y., and Harvey, B., Basing silviculture on natural ecosystem dynamics: an approach applied to the southern boreal mixed wood forest of Quebec, *For. Ecol. Manage.*, 92, 235, 1997.

Bergeron, Y., Gauthier, S., Kafka, V., Lefort, P., and Lesieur, D., Natural fire frequency for the eastern Canadian boreal forest: consequences for sustainable forestry, *Can. J. For. Res.*, 31, 384, 2001.

Bergeron, Y., Leduc, A., Harvey, B.D., and Gauthier, S., Natural fire regime: a guide for sustainable management of the Canadian Boreal Forest. *Silva Fenn.*, 36, 81, 2002.

Blinn, C., *Ice bridges. Forest Management Practices Fact Sheet: Crossing Options*, Series #4, University of Minnesota Extension Service, FS-7004-GO, 1998.

Bonnicksen, T.M., *America's Ancient Forests: From the Ice Age to the Age of Discovery*, John Wiley & Sons, Inc., New York, 594, 2000.

Boyce, S.G., *Landscape Forestry*, John Wiley & Sons, Inc., New York, NY, 239, 1995.

Brennan, A., Engstrom, R.T., Palmer, W.E., Herman, S.M., Hurst, G.A., Burger, L.W., and Hardy, C.L., Whither wildlife without fire, in *Transactions of the 63rd North American Wildlife and Natural Resources Conference*, 1998.

Brose, P.T. and Van Lear, D.H., Responses of hardwood advance regeneration to seasonal prescribed fires in oak-dominated shelterwood stands, *Can. J. For. Res.*, 28, 331, 1998.

Brose, P.T., Van Lear, D.H., and Cooper, R., Using shelterwood harvests and prescribed fire to regenerate oak stands on productive upland sites, *For. Ecol. Manage.*, 113, 125, 1998.

Brose, P., Schuler, T., Van Lear, D., and Berst, J., Bringing fire back — the changing regimes of the Appalachian mixed-oak forests, *J. For.*, 99, 30, 2001.

Brown, J.K., and Smith, J.P., *Wildland Fire in Ecosystems: Effects of Fire on Flora*, U.S. Department of Agriculture, Forest Service, Rocky Mountain Research Station, General Technical Report RMRS-GTR-42, 2000.

Bunnell, F.L., Forest-dwelling vertebrate faunas and natural fire regimes in British Columbia: patterns and implications for conservation, *Conserv. Biol.*, 9, 636, 1995.

Burton, P.J., Ballisky, A.C., Coward, L.P., Cumming, S.G., and Kneeshaw, D.D., The value of managing for biodiversity, *For. Chron.*, 68, 225, 1992.

Burton, P.J., Kneeshaw, D.D., and Coates, K.D., Managing forest harvesting to maintain old growth in boreal and sub-boreal forests, *For. Chron.*, 75, 623, 1999.

Carroll, W.D., Kapeluck, P.R., Harper, R.A., and Van Lear, D.H., Historical overview of the southern forest landscape and associated resources, in *Southern Forest Resource Assessment*, U.S. Department of Agriculture, Forest Service, Southern Research Station General Technical Report SRS-53, 583, 2002.

Clinton, B.D., Vose, J.M., and Swank, W.T., Site preparation burning to improve southern Appalachian pine-hardwood stands: vegetation composition and diversity of 13-year-old stands, *Can. J. For. Res.*, 23, 227, 1993.

Coates, K.D. and Burton, P.J., A gap-based approach for development of silvicultural systems to address ecosystem management objectives, *For. Ecol. Manage.*, 99, 337, 1997.

Covington, W.W. and Moore, M.M., Southwestern ponderosa forest structure: changes since Euro-American settlement, *J. For.*, 92, 39, 1994.

Cumming, S.G., Burton, P.J., and Klinkenberg, B., Boreal mixed-wood forests may have no "representative" areas: some implications for reserve design, *Ecography*, 19, 162, 1996.

DeLong, S.C., Natural disturbance rate and patch size distribution of forests in northern British Columbia: implications for forest management, *Northwest Science Special Issue*, 72, 35, 1998.

DeLong, S.C., *Planting White Spruce under Trembling Aspen: 7-Year Results of Seedling Condition and Performance*. Research Branch, Ministry of Forests, Working Paper 54/2000, Victoria, B.C., 2000.

DeLong, S.C. and Tanner, D., Managing the pattern of forest harvest: lessons from wildfire. *Biodivers. Conserv.*, 5, 1191, 1996.

Duchesne, L.C., Fire and diversity in Canadian ecosystems, in *Biodiversity, Temperate Ecosystems, and Global Change*, T.J.B. Boyle and C.E.B. Boyle, Eds., Springer-Verlag, Berlin, 247, 1994.

Eberhart, K.E. and P.M. Woodward, Distribution of residual vegetation associated with large fires in Alberta, *Can. J. For. Res.*, 17, 1207, 1987.

Elliott, K.J, Hendrick, R.L., Major, A.E., Vose, J.M. and Swank, W.T., Vegetation dynamics after a prescribed fire in the southern Appalachians, *For. Ecol. Manage.*, 114, 199, 1998.

Esseen, P.-A., Ehnström, B., Ericson, L., and Sjöberg, K., Boreal forests, *Ecol. Bull.*, 46, 16, 1997.

Franklin, J.F., Preserving biodiversity: species, ecosystems or landscapes, *Ecol. Appl.*, 3, 202, 1993.

Franklin, J.F. and Forman, R.T.T., Creating landscape patterns by forest cutting: ecological consequences and principles, *Landscape Ecol.*, 1, 5, 1987.

Fries, C., Johansson, O., Pettersson, B., and Simonsson, P., Silvicultural models to maintain and restore natural stand structures in Swedish boreal forests, *For. Ecol. Manage.*, 94, 89, 1997.

Galley, E.M. and Wilson, T.P., Eds., *Proceedings of the Invasive Species Workshop: The Role of Fire in the Control and Spread of Invasive Species, Fire Conference 2000: The First National Congress on Fire Ecology, Prevention, and Management*, Miscellaneous Publication No. 11, Tall Timbers Research Station, Tallahassee, FL, 2001.

Gärdenfors, U., Ed., Rödlistade arter I Sverige — The red list of Swedish species, Art Databanken, SLU, Uppsala, 393, 2000.

Gobster, P.H., Forest aesthetics, biodiversity, and the perceived appropriateness of ecosystem management practices, in *Defining Social Acceptability in Ecosystem Management: A Workshop Proceedings*, M.W. Brunson, L.E. Kruger, C.B. Tyler, and S.A. Schroeder, Eds., U.S. Department of Agriculture, Forest Service, Pacific Northwest Station, General Technical Report PNW-369, 77, 1996.

Grilz, P.L. and Romo, J.T., Management considerations for controlling smooth brome in fescue prairie, *Nat. Areas J.*, 15, 148, 1995.

Greene, D.F., Kneeshaw, D.D., Messier, C., Lieffers, V., Cormier, D., Doucet, R., Coates, K.D., Groot, A., Grover, G., and Calogeropoulos, C., A biological and economic analysis of silvicultural alternatives to the conventional clearcut/plantation prescription in Boreal Mixed-wood Stands (aspen/white spruce/balsam fir), *For. Chron.*, 78, 281, 2002.

Haggstrom, D.A. and Kelleyhouse, D.G., Siliviculture and wildlife relationships in the boreal forest of Interior Alaska, *For. Chron.*, 72, 59, 1996.

Haila, Y., Preserving ecological diversity in boreal forests: ecological background, research, and management. *Ann. Zool. Fenn.*, 31, 203, 1994.

Hansen, A.J., Spies, T.A., Swanson, F.J., and Ohmann, J.L., Conserving biodiversity in managed forests, *Bioscience*, 41, 382, 1991.

Harmon, M.E., Franklin, J.F., Swanson, F.J., Sollins, P., Lattin, J.D., Anderson, N.H., Gregory, S.V., Cline, S.P., Aumen, N.G., Sedell, J.R., Lienkaemper, G.W., Cromack Jr. K., and Cummins, K.W., Ecology of coarse woody debris in temperate ecosystems, *Adv. Ecol. Res.*, 15, 133, 1986.

Hunter, M.L. Jr., Ed., *Maintaining biodiversity in forest ecosystems*, Cambridge University Press, Cambridge, U.K., 698 pp., 1999.

Hämet-Ahti, L., Human impact on closed boreal forest (taiga), in *Man's Impact on Vegetation*, W. Holzner, M.J.A. Werger, and I. Ikusima, Eds., Dr. W. Junk Publishers, The Hague, 201, 1983.

Jewkes, H.A., *History of fire control in Alaska*, University of Alaska Fairbanks, Fairbanks, AK, 153, 1999.

Johnson, A.S. and Hale, P.E., The historical foundations of prescribed burning for wildlife: a Southeastern perspective, in *The Role of Fire in Nongame Wildlife Management and Community Restoration: Traditional Uses and New Directions, Proceedings of a Special Workshop*, W.M. Ford, K.R. Russell, and C.E. Moorman, Eds., Nashville, TN, U.S. Department of Agriculture, Forest Service, Northeastern Station, General Technical Report NE-288, 11, 2002.

Jonsson, B.G. and Esseen, P., Treefall disturbance maintains high bryophyte diversity in a boreal spruce forest, *J. Ecol.*, 78, 924, 1990.

Jonsson, B.G. and Kruys, N., Eds., Ecology of woody debris in boreal forests, *Ecol. Bull.*, 49, 283, 2001.

Kurz, W.A. and Apps, M.J., A 70-year retrospective analysis of carbon fluxes in the Canadian forest sector, *Ecol. Appl.*, 9, 526, 1999.

Kuuluvainen, T., Aapala, K., Ahlroth, P., Kuusinen, M., Lindholm, T., Sallantaus, T., Siitonen J., and Tukia, H., Principles of ecological restoration of boreal forested ecosystems: Finland as an example, *Silva Fenn.*, 36, 409, 2002.

Kuusela, K., Boreal forestry in Finland: a fire ecology without fire, *Unasylva*, 43, 22, 1992.

Landers, J.L., Van Lear, D.H., and Boyer, W.D., The longleaf pine forests of the Southeast: Requiem or renaissance, *J. For.*, 93, 39, 1995.

Lieffers, V.J., Macmillan, R.B., MacPherson, D., Branter, K., and Stewart, J.D., Semi-natural and intensive silvicultural systems for the boreal mixedwood forest, *For. Chron.*, 72, 286, 1996.

Linder, P. and Östlund, L., Structural changes in three mid-boreal Swedish forest landscapes, 1885–1996, *Biol. Conserv.*, 85, 9, 1998.

Loftis, D. and McGee, C.E., Eds., *Oak Regeneration: Serious Problems, Practical Recommendations. Symposium Proceedings*, 8–10 Sept. 1992, Knoxville, TN, U.S. Department of Agriculture, Forest Service, Southeastern Research Station, General Technical Report SE-84, Asheville, NC, 319, 1993.

Lorimer, C.G., Causes of the oak regeneration problem, in *Oak Regeneration: Serious Problems, Practical Recommendations*, D.L. Loftis and C.E. McGee, Eds., 8–10 Sept. 1992, Knoxville, TN, U.S. Department of Agriculture, Forest Service, Southeastern Station, General Technical Report SE-84, Asheville, NC, 14, 1993.

Lugo, A.E. and Gucinski, H., Function, effects, and management of forest roads, *For. Ecol. Manage.*, 133, 249, 2000.
Lämås, T., *Forest management planning for biodiversity and timber production*, Swedish University of Agricultural Sciences, Umeå; Report 3, 56, 1996.
Lämås, T. and Fries, C., Emergence of a biodiversity concept in Swedish forest policy, *Water Air Soil Pollut.*, 82, 57, 1995.
Man, R. and V.J. Lieffers, Effects of shelterwood and site preparation on climate and establishment of white spruce seedlings in a boreal mixedwood forest, *For. Chron.* 75, 837, 1999.
Mann, C.C., *The Atlantic Monthly*, 1491, March 2002.
McCleery, D.W., *American Forests — A History of Resiliency and Recovery*, revised edition, U.S. Department of Agriculture, Forest Service, and the Forest History Society, Durham, NC, 58, 1993.
Mitchell, R.J., Neel, W.L., Hiers, J.K., Cole, F.T., and Atkinson Jr., J.B., A model management plan for conservation easements in longleaf pine-dominated landscapes, Joseph W. Jones Ecological Research Center, Newton, GA, 24, 2000.
Mohr, H.H., Waldrop, T.A., and Shelburne, V.B., Optimal seedbed requirements for regenerating Table Mountain pine, in *Proceedings, Eleventh Biennial Southern Silvicultural Research Conference*, 20–22 Mar. 2001, Knoxville, TN, U.S. Department of Agriculture, Forest Service, Southern Research Station, General Technical Report SRS-48, 306, 2002.
Myers, R.K. and Van Lear, D.H., Hurricane–fire interactions in coastal forests of the south: a review and hypothesis, *For. Ecol. Manage.*, 103, 265, 1997.
Nyland, R.D., *Silviculture: Concepts and Applications*, 2nd ed. McGraw Hill, New York, NY, 682, 2002.
Noss, R.F., On characterizing presettlement vegetation: how and why, *Nat. Areas J.*, 5, 5, 1985.
Östlund, L., Zackrisson, O., and Axelsson, A.-L., The history and transformation of a Scandinavian boreal forest landscape since the 19th century, *Can. J. For. Res.*, 27, 1198, 1997.
Pyne, S.J., Andrews, P.L., and Laven, R.D., *Introduction to Wildland Fire*, 2nd ed., John Wiley & Sons, Inc., New York, NY, 769, 1996.
Racey, G.D., Abraham, K., Darby, W.R., Timmerman, H.R. and Day, Q., Can woodland caribou and the forest industry coexist: the Ontario scene, *Rangifer*, 7, 108, 1991.
Randles, R., Van Lear, D., Waldrop, T., and Simon, D., Periodic burning in Table Mountain-pitch pine stands, in *Proceedings, Eleventh Biennial Southern Silvicultural Research Conference*, 20–22 Mar. 2001, Knoxville, TN, U.S. Department of Agriculture, Forest Service, Southern Research Station, General Technical Report SRS-48, 114, 2002.
Rhoads, E.M., Ice crossings, *North. Eng.*, 5, 19, 1974.
Rowe, J.S. and Scotter, G.W., Fire in the boreal forest, *Q. Res.*, 3, 444, 1973.
Sanderson, E.W., Jaiteh, M., Levy, M.A., Redford, K.H., Wannebo, A.V., and Woolmer, G., The human footprint and the last of the wild, *Bioscience*, 52, 891, 2002.
Siitonen, J., Forest management, coarse woody debris, and saproxylic organisms: Fennoscandian boreal forests as an example, *Ecol. Bull.*, 49, 11, 2001.
Singer, M.T. and Lorimer, C.G., Crown release as a potential old-growth restoration approach in northern hardwoods, *Can. J. For. Res.*, 27, 1222, 1997.
Sturtevant, B.R., Bissonette, J.A., Long, J.N., and Roberts, D.W., Coarse woody debris as function of age, stand structure, and disturbance in boreal Newfoundland, *Ecol. Appl.*, 7, 702, 1997.
Trombulak, C.T. and Frissell, C.A., Review of ecological effects of roads on terrestrial and aquatic communities, *Conserv. Biol.*, 14, 18, 2000.
Van Lear, D.H. and Brose, P.H., Fire and oak management, in *Oak Forest Ecosystems — Ecology and Management for Wildlife*, The Johns Hopkins University Press, Baltimore, MD, 269, 2002.
Van Wagner, C.E., The historical pattern of annual burned area in Canada, *For. Chron.*, 64, 182, 1988.
Viereck, L.A., Wildfire in the taiga of Alaska, *Q. Res.*, 3, 465, 1973.
Wade, D.D. and Lunsford, J.D., *A Guide for Prescribed Fire in Southern Forests*, U.S. Department of Agriculture, Forest Service, Region 8, Technical Publication R8-TP 11, 1989.
Walker, J., Rare vascular plant taxa associated with the longleaf pine ecosystems: patterns in taxonomy and ecology, in *Proceedings, 18th Tall Timbers Fire Ecology Conference*, 30 May–2 June 1991, Tallahassee, FL, Tall Timbers Research Station, 105, 1993.
Waldrop, T.E. and Brose, P.H., A comparison of fire intensity levels for stand replacement of Table Mountain pine (*Pinus pungens* Lamb.), *For. Ecol. Manage.*, 1999.

Wendell, G.W. and Smith, H.C., *Effects of Prescribed Fire in a Central Applachian Oak-Hickory Stand*, U.S. Department of Agriculture, Forest Service, NE-RP-594, 8, 1986.

Whelen, R.J., *The ecology of fire*, Cambridge University Press, Cambridge, U.K., 1995.

White, P.S. and Walker, J.L., Approximating nature's variation: selecting and using reference information in restoration ecology, *Restor. Ecol.*, 5, 338, 1997.

Wigley, T.B., Miller, K.V., de Calesta, D.S., and Thomas, M.W., Herbicides as an alternative to prescribed burning for achieving wildlife management objectives, in *The Role of Fire in Nongame Wildlife Management and Community Restoration: Traditional Uses and New Directions, Proceedings of a Special Workshop*, W.M. Ford, K.R. Russell, and C.E. Moorman, Eds., Nashville, TN, U.S. Department of Agriculture, Forest Service, Northeastern Station, General Technical Report NE-288, 124, 2002.

Williams, M., *Americans and their Forests — A Historical Geography*, Cambridge University Press, New York, NY, 1989.

Wurtz, T.L. and Gasbarro, A.F., A brief history of wood use and forest management in Alaska, *For. Chron.*, 72, 47, 1996.

Zackrisson, O., Influence of forest fires on the North Swedish boreal forest, *Oikos*, 29, 22, 1977.

Part 5

Case Studies: Afforestation

chapter twelve

Restoration of birch woodlands in Iceland

Ása L. Aradottir
Soil Conservation Service
Thröstur Eysteinsson
Iceland Forest Service

Contents

12.1 Introduction

Land degradation and soil erosion have drastically changed Icelandic ecosystems since human settlement: vegetation cover has decreased, most of the woodlands have disappeared, and the remaining vegetation is severely degraded. The result of the degradation processes is often sparsely vegetated land with shallow and poor soils and in other cases barren land, a desert (Arnalds and Kimble 2001). Once extensive, birch woodlands and shrublands now cover about 1% of the country. Downy birch (*Betula pubescens* Ehrh.) is the only species that has formed woodlands in Iceland during the Holocene (Hallsdottir 1995).

1-56670-635-1/05/$0.00+$1.50

Preservation of the remaining native woodlands and restoration of productive birch ecosystems on degraded landscapes are important conservation goals in Iceland. Traditional landuse, however, is a major obstacle to large-scale restoration efforts, particularly sheep grazing. Birch woodland restoration must also compete for funding and popularity with other forms of reclamation such as seeding with grasses to improve grazing or planting of exotic tree species.

In this chapter, we discuss the ecological and socioeconomic context of restoring birch woodlands and describe potential restoration strategies.

12.2 Natural birch woodlands

12.2.1 Physiography

Iceland is a 103,000 km^2 island in the North Atlantic Ocean, situated between 63°23′N and 66°32′N. The island is mountainous: only 24% of the land is below 200 m elevation, 58% is above 400 m elevation, 37% is above 600 m, and glaciers cover about 11% of the country (NLSI 2002). Iceland is where the Mid-Atlantic Ridge is elevated above sea level by a mantle plume or "hot spot." Volcanic activity is frequent, with one eruption every 3 to 4 years (Guðmundsson 1996).

The climate of Iceland is maritime, cool-temperate to sub-arctic. July mean temperatures are 9 to 11°C for most lowland weather stations, but January mean temperatures drop to −2 to +1°C (Einarsson 1976). Long periods in winter with temperature fluctuating around the freezing point and numerous freeze–thaw cycles contribute to intensive cryoturbation (movement caused by frost action) of Icelandic soils (Arnalds 2004). Annual precipitation ranges from 400 mm in the northeastern part to 4,000 mm in the southeast highlands (Einarsson 1976). Important characteristics of the climate in Iceland are a high frequency of strong winds and frequent changes in the weather.

Icelandic soils are formed in parent materials of tephra layers and aeolian sediments consisting mainly of volcanic glass (Arnalds 2004). Substantial soil has been removed by large-scale erosion, leaving desert areas with unstable surfaces covering 35 to 45% of the country (Arnalds and Kimble 2001). The Vitrisols of these barren deserts are sandy and shallow and have low (<1%) organic C content (Arnalds and Kimble 2001). Brown Andosols are the main soils of freely drained sites; wetlands are chiefly gleyic Andosols with a relatively high mineral content due to a steady influx of aeolian materials from erosion areas (Arnalds 2004).

Classified satellite images show that 14% of the country has a continuous vegetation cover, 13% is fairly well-vegetated, and 23% has poor vegetation cover (NLSI 1993). The most common vegetation communities on dryland soils include moss heath dominated by *Rhacometrium* mosses, dwarf shrub heath dominated by one or more of *Empetrum hermafroditum*, *Vaccinium uliginosum* and other dwarf shrub species, shrub heath dominated by *Salix* species and *Betula nana*, sedge or rush heaths characterized by *Carex bigelowii*, *Juncus* species or *Kobresia myosuroides*, and grasslands (Steindorsson 1980). Most of the wetlands are characterized by *Carex* species, but *Salix* shrubs and *B. nana* can also be important components in some bog types (Steindorsson 1980). All these communities are influenced to some extent by long-term grazing, mostly by sheep.

The desert areas have low vegetation cover (1 to 10%) characterized by mosses, lichens, and vascular plant species such as *Armeria maritima*, *Cardaminopsis petraea*, *Cerastium alpinum*, *Festuca richardsonii*, *Poa alpina*, *Poa glauca*, and *Silene uniflora* (Arnalds et al. 1987). Many of these surfaces were once covered with fertile, brown Andosols that were eroded away, mostly after human settlement. The deserts have been classified by surface geomorphology; sandy lag gravel is the most common, followed by lag gravel, sandy fields, and sandy lava surfaces

(Arnalds and Kimble 2001). Intensive cryoturbation, wind and water erosion, and sand accumulation make these surfaces unstable and hostile environments for plants.

12.2.2 *History of the birch woodlands*

During the Holocene, birch woodlands were common in lowland areas of Iceland, but their extent varied with fluctuations in the climate (Einarsson 1963; Hallsdottir 1995). Birch woodland and shrubland cover has been estimated as > 20% of the country at the time of settlement in the 9th century AD, perhaps over 30% (Figure 12.1A) (Bjarnason 1974; Sigurðsson 1977), based on historical records, pollen analyses, old place names, and the current distribution of the woodlands (Thorarinsson 1961; Einarsson 1963; Arnalds 1987; Hallsdottir 1995; Kristinsson 1995). Extensive clearing, burning, and cutting for fuel rapidly reduced woodland area after settlement, and grazing by domestic herbivores inhibited regeneration (Thorsteinsson 1986; Arnalds 1987).

Widespread ecosystem degradation and soil erosion likely started soon after settlement (Thorarinsson 1961; Sigbjarnarson 1969). Woodlands are more resilient following natural disturbances than the heathlands or grasslands that replaced them, and woodland deterioration probably contributed to accelerated soil erosion (Aradottir and Arnalds 2001). Birch woodlands can withstand substantial volcanic ash deposition, as demonstrated by woodland remnants near active volcanoes such as Mount Hekla. Woodland cover reduces wind speed at the soil surface, thus stabilizing ash falling on woodland and enabling vegetation to regenerate. Lower wind speed under woodland cover also reduces abrasion of the vegetation by blowing ash. Woodlands accumulate snow, which insulates the soil surface and reduces the intensity of cryogenic processes. Hillslope erosion has been common in the mountainous landscapes of Iceland (Arnalds et al. 2001). The woodlands contribute to slope stabilization not only through their effects on solifluction (soil creep) and other cryogenic processes, but also because woodlands have higher infiltration rates than other vegetation types (Orradottir 2002), which limits the risk of running water on the slopes.

Birch woodlands in Iceland probably reached a postglacial minimum at the beginning of the 20th century of about 1% cover of total land area. Legislation was approved in 1907 to protect the remaining woodlands and to create new forests. For the next few decades, emphasis was on enclosing some remaining birch woodlands to protect them from sheep grazing. The purpose was to prevent further degradation and promote restoration of the woodlands, and in many cases this led to new expansion (Blöndal and Gunnarsson 1999).

(A) (B)

Figure 12.1 (A) Estimated cover of birch woodlands in Iceland at the time of settlement, A.D. 870, based principally on elevation limits of birch distribution today (by Sherry Curl, Iceland Forest Service). (B) Cover of birch woodlands in 1990 based on a survey of birch woodlands conducted in 1987 to 1991. (By Daði Halldórsson, Iceland Forest Service. With permission).

Even though improvements in agriculture during the latter part of the 20th century reduced the need to use birch for fodder, most native birch woodlands are open for summer sheep grazing, which has continued to prevent natural expansion of woodlands outside of protected areas.

Planting exotic conifers started around 1935, and planting of conifers in birch woodlands was encouraged and continued until the 1980s. Although these plantations are not extensive, less than 2,000 ha in all, they have affected many birch woodlands. Conifers usually grow taller than birch and planting them in birch woodlands can result in replacement.

12.2.3 *Characteristics and distribution of the birch woodlands*

Information about the birch woodlands is based on a qualitative survey in 1972 to 1975 (Sigurõsson 1977) and a quantitative survey in 1987 to 1991 (Aradottir et al. 2001). Natural birch woodlands in Iceland occupy 1,165 km^2 (Figure 12.1B). Small, scattered woodland units form a large portion of the woodlands, which are unevenly distributed over the country. Most of the woodlands are in west Iceland and the western fjords (about 47%) and southwest and south Iceland (23%), but hardly any natural woodland can be found in northwest Iceland. Most of the woodlands are in the lowlands, below 300 m elevation, but vigorous woodlands are found at higher elevations. The highest elevation where birch forms continuous stands is at 500 to 550 m, although groups of scattered trees or shrubs may exist at higher elevations.

Low-growing trees (<4 m) and a shrubby, multistemmed growth form characterize most birch woodlands, although woodlands with taller trees and one to four stems are also found (Figure 12.2). The multistemmed shrub form of birch is especially prevalent under

(A) (B)

Figure 12.2 (A) Straight, single-stemmed birch tree in a North Iceland birch woodland. (B) Remnants of a shrubby, low, and multistemmed birch woodland adjacent to eroded landscape (front) near the volcano Hekla in South Iceland. (Photos: A.L. Aradottir.)

strong oceanic conditions and at higher elevations. The taller trees are more common inland at low to mid-elevations, especially in northern, eastern, and southeastern Iceland.

Other tree species include scattered individuals of *Sorbus acuparia*, which are found in many woodlands but never form pure stands. A limited distribution of clones of *Populus tremula* has been found in eastern and northern Iceland (Blöndal 2002).

Willows commonly form shrub layer in woodlands. *Salix phylicifolia* is the most frequent species, and may form shrublands on its own. The lower-growing *Salix lanata* and the prostrate *Salix arctica* are also common in the shrub layer of birch woodlands. Where the birch is especially low growing, *S. phylicifolia* and *S. lanata* can be codominant or even dominant over the birch. Dwarf shrubs, such as *V. uliginosum*, *V. myrtillus*, and *E. nigrum*, often dominate the understory of birch woodlands, but grasses, tall forbs, and *Equisetum* species can also occur in the understory (Steindorsson 1980; Aradottir et al. 2001).

12.2.4 Economic importance, past and present

Throughout history, birch woodlands have been valuable for fuel, building material, and animal fodder (Thorarinsson 1974). Charcoal from woodlands was used for iron working and they were extensively grazed (Thorarinsson 1974). These uses of birch woodlands declined in importance after the 1930s when they were replaced by geothermal heat, imported fossil fuels, and cultivated hayfields.

Presently, the wood products of birch are not economically important, although a small amount (less than 500 m^3) is felled each year and sold for firewood and handicrafts. On the other hand, birch woodland properties are in high demand for summerhouses and some of the woodlands are popular recreation areas. Birch woodlands are also valued as representing ecosystems that once covered large areas but today are limited in extent.

The most common use of birch woodlands is still grazing by sheep, although winter grazing and browsing has declined due to increased haymaking and housing of the animals during winter. Farmers used to have a negative attitude toward the woodlands, as wool, the most valuable farm product, got tangled in branches of the birch and was lost. Some sheep farmers now consider birch woodlands an important part of their grazing lands and they are trying to increase the woodland cover on their land through management actions.

12.2.5 Land ownership and land tenure

Land suitable for restoration of birch woodlands can be divided into farms, grazing commons, state-owned or managed land, and land owned by municipalities. Farms usually include cultivated areas and uncultivated home grazing areas. About 90% of Icelandic farms are privately owned and the state or the national church own most of the rest (Icelandic Agricultural Statistics 2000). Although the majority of land best suited for reestablishing birch woodlands is on farms, the greatest competition with other types of landuse, especially livestock grazing, is also here.

Farmers jointly manage communal grazing areas in each municipality or local community for sheep grazing and sometimes horse grazing. Ownership is often unclear and is sometimes contested. Large areas of the grazing commons are above the tree line, but there are also some areas at lower elevations that could support birch and willow scrublands or woodlands.

The Iceland Forest Service (IFS), the Environment Agency (EA), and the Soil Conservation Service (SCS) manage nonfarm land owned by the state, including about 15% of Icelandic birch woodlands as well as areas suitable for their restoration. Land managed by the IFS includes remnant birch woodlands and surrounding areas that are protected from grazing and land for afforestation. About 4% of the birch woodlands in Iceland are under the supervision of the IFS. The EA manages national parks and reserves that include over 10% of birch woodlands (NCC 1996). Most of the land managed by the SCS has been badly degraded or has active erosion and is under reclamation. Land owned by municipalities includes mostly towns and their surroundings. Because of public interest and availability of both volunteer and paid labor, afforestation is commonly practiced in these areas. However, restoration of the birch woodland ecosystem is rarely a specific goal of these afforestation efforts.

12.3 Afforestation and restoration of birch woodlands

12.3.1 Afforestation in Iceland

Organized forestry began in Iceland early in the 20th century and focused on protecting the birch forest remnants. The IFS acquired several forest areas for protection. From 1935 to 1951, native birch was the most planted species in Iceland, with planting ranging from a few thousand to over 150,000 seedlings per year.

After 1950, emphasis has been on afforestation through increased planting of exotic conifers, principally *Picea abies, Picea sitchensis, Pinus sylvestris, Pinus contorta,* and *Larix sibirica* (Blöndal and Gunnarsson 1999). Planting was at 500,000 to 1.5 million seedlings annually until 1989. During this entire 40-year period (1950 to 1990), native birch planted was 5 to 15% of the seedlings planted annually or 50,000 to 150,000 seedlings per year (Petursson 1999, unpublished data).

Afforestation has increased again since 1990 to about 5 million seedlings planted annually, which corresponds to a planted area of 1,200 to 1,500 ha per year. Planting of native birch has increased as a proportion of the total, comprising almost 30% of seedlings planted in 1999 to 2001. *Larix sibirica* is planted to almost the same extent as native birch (30%), followed by *Picea sitchensis* (12%), *Pinus contorta* (10%), and *Populus trichocarpa* (6%). Over 20 other species comprise the remaining 12% of trees planted in Iceland (Petursson 2002).

Today, the main actors in Icelandic afforestation are government-funded regional farm afforestation projects and forestry societies (NGOs). Since 1990, many forestry societies have participated in the Land Reclamation Forest (LRF) project, a government-funded program with the aim to afforest eroded or degraded land. Numerous private individuals also participate, but usually on a smaller scale. The IFS, once active in afforestation, now manages the National Forests and plays a leading role in research, planning, and policy.

The LRF project is the first large-scale program of planting tree seedlings for reclamation of eroded and degraded sites in Iceland. About 1 million seedlings are planted annually in this program, of which 40 to 75% are native birch (Figure 12.3). The survival and condition of the first plantations have been surveyed on several occasions in order to learn from mistakes and successes (Table 12.1). There was large variation in the survival of all species at individual sites reflecting differences in environmental conditions, vegetation cover and surface characteristics of the planted areas, and different fertilization treatments. The survival of birch was overall better than that of larch and pine, even though pine was consistently planted at more favorable sites than the other two species.

Multiple-use objectives underlie Icelandic afforestation planning and management of cultivated forests (Eysteinsson 1999a). Most farm afforestation plans emphasize timber production for areas where it is feasible, accounting for roughly 60% of the seedlings planted annually (Petursson 2002). In peripheral areas, emphasis is on protective functions, shelterbelts, and sometimes wildlife value, improved grazing for livestock, and outdoor recreation (Eysteinsson 1999b). Indirectly, afforestation projects promote substantial natural regeneration of birch in grazing exclosures.

Figure 12.3 A 10-year-old Land Reclamation Forestry plantation in South Iceland. Birch interplanted with Nootka lupin as a nurse crop. (Photo: A.L. Aradottir.)

Table 12.1 Survival in 1996 of the Most Commonly Planted Species in the LRF Project During 1991–1992 (Native Birch, Siberian Larch, and Lodgepole Pine)

Three Species	Planting Year	No. of Sites in Survery	Survival of Planted Seedlings		
			Mean of all Surveyed Sites	Lowest Site Average	Highest Site Average
Birch	1991	9	69	32	94
Birch	1992	4	72	55	89
Larch	1991	6	41	39	83
Larch	1992	4	48	11	89
Pine	1991	2	64	10	90
Pine	1992	3	63	60	68

The numbers are based on a survey of over 3,200 tree seedlings at nine sites, selected in a randomly stratified manner from more than 80 planting sites (Aradottir and Gretarsdottir, unpublished data). Not all species were represented at all sites.

12.3.2 *Restoration of birch woodlands*

12.3.2.1 *Goals*

Afforestation with birch can have several different objectives, including restoration of native woodlands, ecosystem functions, and biodiversity. Other objectives include carbon sequestration and improved options for future landuse. Although using exotic tree species can meet many of these objectives, only the use of native birch and willows can restore the native woodlands.

Birch is useful for the rehabilitation of degraded land. It is an early colonizer in both primary and secondary succession in Iceland (Persson 1964; Aradottir 1991; Gretarsdottir 2002), and can be a key species in ecosystem development. It performs relatively well in plantations on degraded and eroded sites (Table 12.1), where restoration of birch woodlands also means restoration of ecosystem structure and function and increased options for land-use. Restoration of birch woodlands on sites with low soil organic matter has the potential to sequester considerable carbon by increasing carbon in the soil, understory vegetation, and in the trees (Arnalds et al. 2002).

12.3.2.2 *Extent of restoration needs*

Somewhere on the order of 100,000 ha (about 1% of Iceland) are now available for birch woodlands restoration, or could be within a short period. However, active restoration of birch woodlands is currently limited in Iceland. Current opportunities for the restoration of birch woodlands may be the greatest on public land, which is largely managed by state agencies. These agencies have similar goals of protecting existing woodlands and increasing woodland area where appropriate.

Potentially, 1 to 2 million ha of additional land could be available for woodland restoration, but this depends on developments in landuse and farm economy, factors largely beyond the control of foresters or ecosystem restorationists. Most of this land is located within the home grazing areas of farms and some grazing commons include large areas at 200 to 500 m elevation. Many of these areas are badly eroded and in need of reclamation, but for most summer grazing currently limits birch and willow woodland or shrubland restoration.

Sheep numbers in Iceland declined by 50% from 1978 to the 1990s due to market conditions and government initiatives, but production still exceeds home market demand (Arnalds and Barkarson 2003). Farmers are, however, reluctant to change landuse and seek to maintain landuse rights for fear of losing certain traditions, such as the fall roundup of the commons that serves as an important social function in rural communities. Nevertheless, there should be opportunities for changes in landuse such that large areas could be managed or protected from grazing by sheep in order to enhance the restoration of birch ecosystems.

12.3.2.3 *Restoration techniques and strategies*

Restoration of birch woodlands can take a long time and will often require cultural inputs, even on sites previously covered by birch. Reintroduction of birch may be necessary in areas where birch has disappeared or has low abundance. Planting of seedlings is the most common method for restoration and direct seeding is only used on a small scale. Spread of birch can be effective through natural recruitment in the vicinity of birch woodland remnants or established birch stands.

Research on birch woodland restoration in Iceland has focused on methods for establishing birch as a key species in woodland development and the dynamics of birch colonization. Establishment can be especially difficult on eroded sites, where it is impeded by factors such as frost heave, low nutrient status, low water holding capacity of the soil, and active soil erosion (Arnalds et al. 1987; Magnusson 1997; Arnalds and Kimble 2001).

12.3.2.3.1 Planting The 1.4 million birch seedlings planted per year are almost all container seedlings grown in 50 or 100 ml multipot containers. The seedlings are usually 1 year old at planting and 10 to 25 cm in height. Most of the seed comes from seed orchards of select material or identified seed stands of birch originating in southeast Iceland. Seedlings are produced in private nurseries.

Frost heaving is one of the main causes of mortality during the first years after planting (Aradottir and Gretarsdottir 1995; Oddsdottir et al. 1998). The extent of cryoturbation depends on factors such as differences in soil texture, snow cover, and frequency of frost–thaw cycles (Bergsten et al. 2001; Matsuoka 2001), but is most serious where seedlings are planted in spots with little or no vegetation cover. This can be reduced by sowing annual grasses and fertilization at planting (Oddsdottir et al. 1998), by using nurse crops such as lupine (*Lupinus nootkatensis*) (Figure 12.3) that can improve growing conditions

Figure 12.4 A group of children assist with seeding of birch at a reclamation area in South Iceland. Individuals and groups have the opportunity to participate in excursions organized by SCS, IFS, or other agencies, where environmental education and recreation are combined with reclamation and afforestation. (Photo: O. Arnalds.)

at harsh sites (Aradottir 2004), or by mixing organic fertilizer with the soil before planting (Aradottir, unpublished data).

Fertilization increases birch seedling vigor and growth, but nitrogen fertilizers may increase mortality in the first year after planting due to salt stress and competition by other vegetation (Oskarsson et al. 1997; Oddsdottir et al. 1998). Fertilizer-induced mortality in young plantations can be avoided by using slow-release or low N–high P fertilizers (Oskarsson et al. 1997). The increased vigor and long-term survival of fertilized seedlings, as well as reduced mortality from frost heaving, may compensate for early mortality on poor and eroded soils (Oskarsson et al. 1997; Oddsdottir et al. 1998), at least if stocking density is less important than survival. Furthermore, birch is likely to seed into any gaps within 20 years.

Salix lanata and *Salix phylicifolia*, the native willows that form the shrub layer in many birch woodlands, can be propagated from fresh cuttings (Svavarsdottir and Aradottir 2004). Because these species are found in most parts of the country, local populations can be used in restoration programs.

12.3.2.3.2 Direct seeding Direct seeding of birch has been used on a small scale; a few small stands were established in the first half of the 20th century by this method (Magnusson and Magnusson 1990). During the last 10 to 15 years, there has been renewed interest in direct seeding of birch. Birch seed can be easily collected. In most years seed is readily available, although viability can vary greatly between stands and between years (Thorarinn Benedikz, unpublished data). Simple instructions on seed collection and seeding are available; also, a demonstration video has been made. Volunteer groups have collected seed for their own use or donated to the SCS or the IFS. Direct seeding is often used in small reclamation projects involving volunteer groups or school children (Figure 12.4). In many cases, the seed collection and seeding are done in the same field trip, providing an opportunity for educating volunteers about reclamation work, soil, and nature conservation.

Characterization of safe sites for birch seedling establishment is useful for targeting suitable areas for direct seeding. Seedling emergence of birch is generally much higher in open microsites where the seeds have good contact with the mineral soil than in microsites with dense vegetation cover (Kinnaird 1974). Greatest seedling emergence is on microsites of mineral soil, but lowest emergence is where the sward thickness exceeds 1 to 2 cm (Aradottir 1991; Magnusson and Magnusson 1990). Winter survival, however, is lower in open microsites due to cryoturbation (Magnusson and Magnusson 1990; Aradottir 1991, 2003), but this can be ameliorated by low-level fertilization (Aradottir 1991). Fertilization can also stimulate the formation of biological soil crusts (Elmarsdottir 2001; Gretarsdottir 2002) that stabilize the soil surface and are favorable for seedling establishment of many species (Aradottir 1991; Elmarsdottir 2001).

Direct seeding of birch is sometimes used with fertilization and grass seeding to revegetate severely degraded areas. A sufficient vegetation cover has to be established in order to stabilize the soil surface and reduce cryoturbation, but the vegetation should not be so dense that it will curtail birch seedling establishment. Conditions for birch establishment are often most favorable on older revegetation sites where the seeded grasses have declined (Aradottir 1991; Gretarsdottir 2002).

The main problems associated with direct seeding are limited seedling emergence and slow growth for the first several years. Using seeding shelters (cones and tubes) to establish birch can improve seedling emergence and growth (Aradottir 2003; Aradottir unpublished data), but it has not been evaluated whether these are practical to use on a large scale. Nevertheless, they may be useful for small projects to avoid discouragement from the lack of visible results.

12.3.2.3.3 Natural expansion Substantial natural expansion of birch can occur in the vicinity of established birch stands, especially in areas protected from sheep grazing. However, only a few examples have been well documented. For example, aerial photographs and field measurements documented extension on a disturbed site in southern Iceland. The birch was initially established by seeding several small plots. After 20 years, birch cover had extended to 9,000 m^2; during the following 24 years, birch cover expanded to 30,000 m^2 (Aradottir 1991; Aradottir and Arnalds 2001). Some of the extension was due to early transplants from the initial birch clusters to new sites, but most was by natural recruitment of birch around the edges of established clusters and the establishment of new clusters that subsequently served as foci for additional colonization (Aradottir 1991).

Two factors govern natural regeneration from seed: availability of safe sites and density of seed rain. Seed rain decreases in a log-linear manner with increased distance from the seed source (Hughes and Fahey 1988; Aradottir 1991); closer to the seed source, recruitment is primarily limited by the availability of safe sites. The direction of seed dispersal is correlated with strong prevailing winds (Aradottir et al. 1997).

12.3.2.3.4 Restoration strategies The potential area for restoring birch woodlands in Iceland is on the order of thousands of km^2. Planting can successfully establish birch, but it is costly and labor intensive and therefore not well suited for woodland restoration of extensive areas. Direct seeding is relatively inexpensive, but is more sensitive to environmental conditions and initial growth is slow; thus, results are not immediately visible. Natural expansion of birch is limited to areas close to seed sources, such as planted birch stands and woodland remnants. As existing birch woodlands are of limited distribution (Figure 12.1B), the lack of seed sources limits colonization of birch in large areas of the country. Furthermore, safe sites for seedling establishment may be limited even when seed sources are available.

Natural expansion can be encouraged by establishing islands of birch and willows by planting or seeding at strategic locations across landscapes, thus forming seed sources from which trees can gradually colonize the whole area (Aradottir 1991; Ludwig and Tongway 1996). Such a strategy would concentrate resources (plants, seeds, fertilizer, etc.) at selected

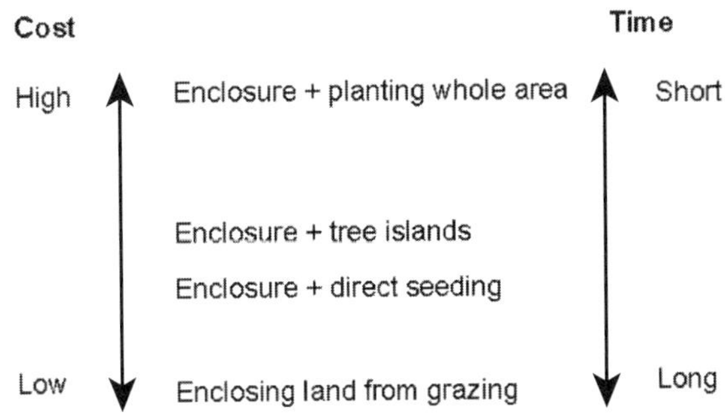

Figure 12.5 The cost of restoration strategies is probably inversely related to the time until woodland cover is restored, other factors being equal. The success of enclosing land from grazing without introduction of birch will depend on available seed sources. If they are plentiful, this method may give faster results than direct seeding or planting of tree islands.

locations to enhance the probability of local establishment. Stands serving as seed sources should be established close to areas that have a high density of potentially safe sites for seedling establishment, or where the availability of safe sites can be increased by strategic fertilizer applications after good seed years. This strategy will cost less than wholesale planting of large areas, but should give quicker results than unassisted natural expansion in areas where birch is absent or has a very limited distribution (Figure 12.5). Planting birch islands for seed sources is not likely to be effective at high elevation sites because availability of viable seed decreases with increasing altitude (Kullman 1984).

12.4 Economic, social, and political contexts

Although afforestation and reclamation activities are widespread in Iceland, restoration of birch ecosystems is not emphasized. The main obstacles are lack of interest among landowners and others who practice afforestation and lack of specific goals in reclamation and afforestation projects. Landowners do not generally consider birch woodlands as a potential economic resource in the same way as plantations of exotic species that grow faster and yield timber. Many landowners also prefer planting for variety, and they tend to use exotic species or species mixtures, even for plantations that are not particularly aimed at timber production. Extension service personnel in forestry have traditionally promoted the use of exotic species, although this is changing.

Many Icelanders agree with the goal of restoring land to birch woodlands. However, those individuals or organizations involved with afforestation generally plant a number of exotic tree and shrub species along with birch, most often in mixed stands. Vague objectives, for example, to bring back woodland, rather than to restore a certain type of woodland or ecosystem, or to just simply plant trees because it is fun to plant trees, are the main reason for this. For example, participants in a recently completed afforestation grants project (Eysteinsson, unpublished data) were given seedlings to plant, if they provided an equal number of seedlings. Participants were able to choose from among the seven most planted tree species in Iceland (native birch and six exotic species). Although most participants had nothing against planting birch, only 5 out of 127 participants chose to plant only native birch. Some wanted only one or another exotic species. Most participants wanted two or more species, indicating interest in variety when planting trees for amenity. The commonly slow growth and low, scrubby habit of Icelandic birch is perceived by many to be less desirable than forests of taller, straighter trees.

Soil erosion and land degradation are more active and widespread in Iceland than in other European countries. In a Western democracy with a high standard of living and a well-educated population, there should be strong incentives to reclaim some of what has been lost, including the woodlands. The SCS historically has emphasized soil conservation and reclamation of eroded land for various uses, including grazing, but restoration of birch woodlands has not been an important goal. This has changed in recent years, and success stories of establishment and spread of birch stands on reclamation sites (Aradottir 1991) show that restoration of birch woodlands on eroded sites can be a feasible option.

The Icelandic Forest Service has emphasized exotic tree species rather than native birch (Blöndal and Gunnarsson 1999), although this is changing somewhat. Recently, the IFS obtained areas with the aim of protecting relatively intact birch forest ecosystems as well as land with birch scrub remnants to promote natural invasion. Government subsidies for afforestation on farms through Regional Afforestation Projects have increased, mainly for timber production, but carbon sequestration and soil protection are also important objectives. Native birch is often a component of these plantations, either as a nurse species or in plantation edges, but is rarely planted for wood production due to slow growth and poor form.

The environmental sector, including both NGOs and state agencies, has not emphasized restoration of birch woodlands. To be sure, afforestation with native birch is less controversial than afforestation with exotic species, but there has been no organization that specifically promotes protection and restoration of birchwoods. The lack of interest by the general public and other stakeholders is reflected in the political arena. Despite increased government appropriations for afforestation in recent years, support has been limited for ecological restoration, except for a few projects to restore drained wetlands. However, grants schemes could be set up to encourage restoration of birch woodlands and other important ecosystems, just as farmers are subsidized for producing meat and dairy products or as a part of Regional Afforestation Projects. The need for afforestation projects with a primary goal of restoring birch woodland is increasingly recognized and there are currently several small restoration projects and at least one large-scale project under way (Land Reclamation Forest Project).

12.5 Conclusions

Despite a general lack of interest in Iceland in ecosystem restoration and hindrances resulting from conflicting landuses, there is reason for optimism. Native birch is increasingly used in afforestation, making it the most planted tree species in Iceland. Large-scale birch woodland restoration projects are under way, which will serve as demonstrations and may evoke interest in further restoration activities. Sheep farming seems likely to decline, which may alleviate some problems with conflicting landuse and make more land available for restoration. Birch is regenerating naturally within enclosures wherever a seed source is available. In some cases, birch woodland is expanding by natural recruitment as fast or faster than planting, resulting in mixed woodlands or even abandonment of plans for planting (Skúli Björnsson, pers. com.). Birchwoods being planted today will be seed sources for further expansion in the future.

Techniques for establishing birch under various conditions are fairly well known, and lack of knowledge should not be a limiting factor for woodland restoration. Nevertheless, further development of methods for large-scale operations is needed. The necessity to introduce other components of the birch ecosystem in order to ensure ecosystem recovery is also unknown.

In order to increase interest in birch woodland restoration, stakeholders for afforestation and reclamation must be better informed about the need to focus on this objective in individual projects. Governmental agencies, such as the IFS, SCS, and EA, can encourage others to work on birchwood restoration by increasing emphasis in their own projects on this objective. A change in policy such that some governmental funding for regional afforestation programs would be earmarked for restoration projects would greatly stimulate birchwood restoration in Iceland.

References

Aradottir, A.L., *Population Biology and Stand Development of Birch* (*Betula pubescens* Ehrh.) *on Disturbed Sites in Iceland*, Ph.D. dissertation, Department of Range Sci., Texas A&M University, College Station, TX, *Diss. Abs. International*, 52-06 B, 2869, 1991.

Aradottir, A.L., Does the Nootka lupin facilitate or impede colonization and growth of native birch in Iceland? in *Wild and Cultivated Lupins from the Tropics to the Poles, Proceedings of the 10th International Lupin Conference*, van Santen, E., Ed., Laugarvatn, Iceland, 19–24 June 2002, International Lupin Association, Canterbury, New Zealand, 2004.

Aradottir, A.L. and Arnalds, O., Ecosystem degradation and restoration of birch woodlands in Iceland, in *Nordic Mountain Birch Ecosystems*, Wielgolaski, F.E., Ed., UNESCO, Paris, and Parthenon Publishing, Carnforth, 2001, chap. 24.

Aradottir, A.L. and Gretarsdottir, J., Úttektir á gróðursetningum til landgræðsluskóga 1991 og 1992 (Surveys of reclamation forest plantations from 1991 and 1992), *Icel. For. Res. Stat. Rep.*, 9, 1995 (in Icelandic).

Aradottir, A.L., Robertson, A., and Moore, E., Circular statistical analysis of birch colonization and the directional growth response of birch and black cottonwood in south Iceland, *Agr. For. Meteorol.*, 84, 179, 1997.

Aradottir, A.L., Thorsteinsson, I., and Sigurðsson, S., Distribution and characteristics of birch woodlands in North Iceland, in *Nordic Mountain Birch Ecosystems*, Wielgolaski, F.E., Ed., UNESCO, Paris, and Parthenon Publishing, Carnforth, 2001, chap. 6.

Arnalds, A., Ecosystem disturbance in Iceland, *Arctic Alpine Res.*, 19, 508, 1987.

Arnalds, O., Volcanic soils of Iceland, *Catena*, 56, 3, 2004.

Arnalds, O. and Kimble, J., Andisols of deserts in Iceland, *Soil Sci. Soc. Am. J.*, 65, 1778, 2001.

Arnalds, O. and Barkarson, B.H., Soil erosion and land use policy in Iceland in relation to sheep grazing and government subsidies, *Environ. Sci. Policy*, 6, 105, 2003.

Arnalds, O., Aradottir, A.L., and Guðbergsson, G., Organic carbon sequestration by restoration of severely degraded areas in Iceland, in *Agricultural Practices and Policies for Carbon Sequestration in Soil*, Kimble, J.M., Lal, R., and Follett, R.F., Eds., CRC Press LLC, Boca Raton, FL, 2002, chap. 25.

Arnalds, O., Aradottir, A.L., and Thorsteinsson, I., The nature and restoration of denuded areas in Iceland, *Arctic Alpine Res.*, 19, 518, 1987.

Arnalds, O., Thorarinsdottir, E.F., Metusalemsson, S., Jonsson, A., Gretarsson, E., and Arnason, A., *Soil Erosion in Iceland*, The Soil Conservation Service and the Agricultural Research Institute, Reykjavík, 2001.

Bergsten, U., Goulet, F., Lundmark, T., and Lofvenius, M.O., Frost heaving in a boreal soil in relation to soil scarification and snow cover, *Can. J. For. Res.*, 31, 1084, 2001.

Bjarnason, H., Comments on the history of Icelanders in relation to woodland destruction, *Yearb. Icel. For. Soc.*, 30. 1974 (in Icelandic; English summary).

Blöndal, S., Íslensku skógartrén 2 (Icelandic forest trees 2), *Skógræktarritið 2002*, Icel. For. Soc., Reykjavík, 9, 2002 (in Icelandic).

Blöndal, S. and Gunnarsson, S.B., *Íslandsskógar (Iceland's Forests)*. Mál og Mynd, Reykjavík, 1999. (in Icelandic).

Einarsson, T., Pollen analytical studies on the vegetation and climate history of Iceland in Late and Post-Glacial times, in *North Atlantic Biota and their History*, Löve, A. and Löve, D., Eds. Pergamon Press, Oxford, 355 pp., 1963.

Einarsson, M., *Veðurfar á Íslandi (The Climate of Iceland)*, Iðunn, Reykjavik, 1976 (in Icelandic).

Elmarsdottir, A., *Colonization of Native Plants on Degraded and Reclaimed Sites in Iceland*, MS thesis, Colorado State University, Ft. Collins, CO, 2001.

Eysteinsson, T., Sjálfbær þróun skógræktar og hin sex þverevrópsku viðmið (Sustainable development in forestry and the six pan-European criteria), *Skógræktarritið 1999, 2*, Icel. For. Soc., Reykjavík, 73, 1999a (in Icelandic).

Eysteinsson, T., Forest planning in Iceland, in *Forest Planning Today in the Nordic Countries and Scotland*, Makinen, T., Ed., *TemaNord* 580, 19, Nordic Council of Ministers, 1999b.

Gretarsdottir, J., *Long Term Effects of Reclamation Treatments on Plant Succession at Two Localities in Iceland*, Can. Sci. Thesis, University of Bergen, 2002.

Guðmundsson, A.T., *Volcanoes in Iceland. 10,000 Years of Volcanic History*, Vaka-Helgafell, Reykjavík, 1996.

Hallsdottir, M., On the pre-settlement history of Icelandic vegetation, *Icel. Agr. Sci.*, 9, 17, 1995.

Hughes, J.W. and Fahey, T.J., Seed dispersal and colonization in a disturbed northern hardwood forest, *B. Torrey Bot. Club*, 115, 89, 1988.

Icelandic Agricultural Statistics, The Farmers Organization, Reykjavík, 2000.

Kinnaird, J.W., Effect of site conditions on the regeneration of birch (*Betula pendula* Roth and *B. pubescens* Ehrh.), *J. Ecol.*, 62, 467, 1974.

Kristinsson, H., Post-settlement history of Icelandic forests, *Icel. Agr. Sci*, 9, 31, 1995.

Kullman, L., *Germinability of Mountain Birch* (Betula pubescens *ssp.* tortuosa) *Along two Altitudinal Transects Down Slope from the Tree-Limit in Sweden*, Reports of the Kevo Subarctic Research Station, 19, 11, 1984.

Ludwig, J.A. and Tongway, D.J., Rehabilitation of semi-arid landscapes in Australia. II. Restoring vegetation patches, *Restor. Ecol.*, 4, 398, 1996.

Magnusson, S.H., Restoration of eroded areas in Iceland, in *Restoration Ecology and Sustainable Development*, Urbanska, K.M., Webb, N.R., and Edwards, P.J., Eds., Cambridge University Press, Cambridge, 188, 1997.

Magnússon, S.H. and Magnússon, B., Seeding of birch for land reclamation and woodland establishment, *Yearb. Icel. For. Soc.*, 9, 1990 (in Icelandic; English summary).

Matsuoka, N., Solifluction rates, processes and landforms: a global review, *Earth-Sci. Rev.*, 55, 107, 2001.

NCC, *Nature Conservation Register*, 7th ed., The Nature Conservation Council, Reykjavík, 1996 (in Icelandic; English summary).

NLSI, *Digital Vegetation Index Map of Iceland*, National Land Survey of Iceland, Akranes, 1993.

NLSI, *Iceland in Statistics*, National Land Survey of Iceland, Akranes, Retrieved on 15 April 2002 from http://www.lmi.is.

Oddsdottir, E.S., Halldorsson, G., Aradottir, A.L., and Guðmundsson, J., Methods to prevent frost heaving of planted seedlings, *Yearb. Icel. For. Soc.*, 72, 1998 (in Icelandic; English summary).

Orradottir, B., *The Influence of Vegetation on Frost Dynamics, Infiltration and Surface Stability in Icelandic Andisolic Rangelands*, MS thesis, Department Rangeland Ecology and Management, Texas A&M University, College Station, TX, 2002.

Oskarsson, H., Sigurgeirsson, A., and Helgason, B., Fertilization of afforestation plantations in poor soil. I. Results after two growing seasons, *Skógræktarritið 1997*, Icel. For. Soc., Reykjavík, 42, 1997 (in Icelandic; English summary).

Persson, Å., The vegetation at the margin of the receding glacier Skaftafellsjökull, southeastern Iceland, *Bot. Notiser*, 117, 323, 1964.

Petursson, J.G., Skógræktaröldin (The first 100 years of tree planting in Iceland), *Skógræktarritið 1999*, Icel. For. Soc., Reykjavík, 49, 1999 (in Icelandic).

Petursson, J.G., Skógræktarstarfið 2001: Tölulegar upplýsingar (Forestry figures 2001), *Skógræktarritið 2002:2*, Icel. For. Soc., Reykjavík, 107, 2002 (in Icelandic).

Sigbjarnarson, G., The loessial soil formation and soil erosion on Haukadalsheidi, *Náttúrufræðingurinn*, 39, 49, 1969 (in Icelandic; English summary).

Sigurðsson, S., Birki á Íslandi (Birch in Iceland), in *Skógarmál*, Guðmundsson, H., Ed., Edda Print, Reykjavík, 146, 1977 (in Icelandic).

Steindorsson, S., Vegetation classification in Iceland, *J. Agric. Res. Icel.*, 12, 11, 1980 (in Icelandic; English summary).

Svavarsdottir, K. and Aradottir, A.L., Test of methods for propagating native willows, in *Innlendar víðitegundir og notkun þeirra við landgræðslu*, Svavarsdottir, K. Ed., Soil Conservation Service, Gunnarsholt, 2004 (in press) (in Icelandic; English summary).

Thorarinsson. S., Wind erosion in Iceland. A tephrochronological study, *Yearb. Icel. For. Soc.*, 17, 1961 (in Icelandic; English summary).

Thorarinsson, T., The nation survived but the forest died, *Yearb. Icel. For. Soc.*, 16, 1974 (in Icelandic; English summary).

Thorsteinsson, I., The effect of grazing on stability and development of northern rangelands: a case study of Iceland, in *Grazing Research at Northern Latitudes*, Gudmundsson, O., Ed., Plenum Press, New York, 37, 1986.

chapter thirteen

Afforestation in Denmark

Palle Madsen
Danish Centre for Forest, Landscape and Planning, KVL
Finn A. Jensen
Hedeselskabet
Søren Fodgaard
Danish Forest Association

Contents

13.1 Landscape history

Denmark is a small (43,000 km^2) Scandinavian country characterized by its 7,000 km coastline and almost 500 islands. Approximately 11% (486,000 ha) of the country is forest land. Denmark shares its only land border (50 km) with Germany on the Jutland peninsula (near 55°N latitude), which is the largest part of the country. Other interesting characteristics of this nation, with a population of 5.3 million, are the lack of fully exposed bedrock in the landscape (only on the island of Bornholm in the Baltic Sea) and the low relief; the highest elevation is only 170 m above sea level. The last ice age (the Weichselian) lasted 110,000

1-56670-635-1/05/ $0.00+$1.50

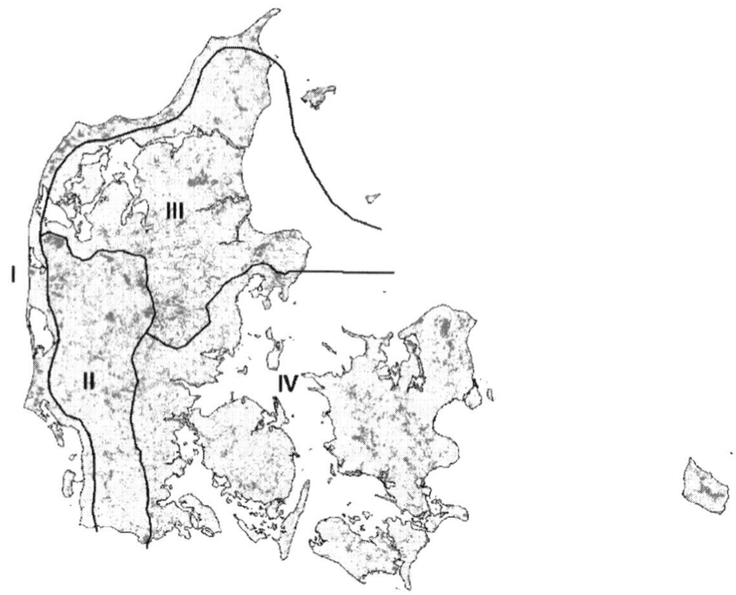

Figure 13.1 Map of Denmark showing Forest Regions (see text) and the present forest cover. Used with permission from Danish Forest and Nature Agency, March 2004.

years, and it ended 12,000 years before present (ybp). Ice advances and retreats considerably shaped and rearranged the landscape, coastline, and distribution of soils. The land surface was changed not only by the force of moving glaciers but also from melt water in front of the glacier running in rivers, floating in deltas, or passing though lakes. As the ice retreated, the windy periglacial climate also affected the landscape.

Glacial ice came from Norway and Sweden, covering northeastern Jutland and leaving sandy deposits of limited fertility (Figure 13.1, Forest Region III). Another ice lobe left more loamy deposits containing lime in the southeastern part of Jutland, as well as at the southeastern islands (Figure 13.1, Forest Region IV). The southwestern part of Jutland was never covered with ice (Figure 13.1, Forest Region II); it was a tundra landscape with sparse vegetation during the whole glacial period. Soils in the west were not "renewed" by materials left by the glaciers; deposits here were sand and gravel from glacier melt water that created sandy outwash plains between slightly more elevated structures, which are deposits from a previous ice age (the Saalian). The soils of western Jutland are generally sandy and poor, mostly classified as Spodosols or Entisols, whereas Alfisols and Inceptisols are the most common soil types in the rest of the country (Vejre et al. 2003). Sands moving along the west coast of Jutland carried by the near-coastal and sea currents in the sea formed dunes (Figure 13.1, Forest Region I). The glacial ice was probably more than 2 km thick, and when it melted, the land rebounded from the released weight. Simultaneously, sea level rose and the relative speed of the two processes has changed the coastlines several times. At times, sea level was lower than now and sometimes it was higher. This change is illustrated by Viking inlet harbors (fjords), which were in cities that today are impossible to reach by boat.

13.2 Climate

The mild and windy Atlantic climate supports annual mean temperatures between 7.5 and 8.5°C, with January and July means close to 0 and 16°C, respectively, and annual mean precipitation between 500 and 900 mm. Despite the overall mild climate, weather can be harsh for forests, in particular for regeneration. Atlantic weather is unpredictable and different from year to year. Late spring frost is a major problem for the regeneration of many species such as beech (*Fagus sylvatica* L.), Sitka spruce (*Picea sitchensis* [Bong.] Carr.), and European silver fir (*Abies alba* Mill.). Warm weather in early spring, from late March to early April, may stimulate an early flush of growth but the risk of late spring frost is high, depending on the site. Near-coastal areas are much less exposed than interior sites in Jutland or on Zealand. In fact, late July to early August may be the only time in central Jutland when there is no risk of frost. Site-specific features such as local terrain, grass cover, and shelterwood may influence the risk of late spring frost.

Denmark is a windy country; low-pressure fronts travel east from the Atlantic across the country, creating the predominant westerly or northwesterly winds. The almost perpetual west wind mainly influences the trees in the western, northern, and central regions of Jutland. Wind can stress trees in Denmark, not only when the low pressures develop into hurricanes and blow trees and forests down but also due to long-term effects. Young trees may be killed or severely damaged in the first several years after planting. A worst case occurs on exposed sites in winter, when there may be no snow cover, intense frost, and a dry and cold east wind.

13.3 Land-use history — loss and gain of forestland

13.3.1 From ice age to black death

Humans arrived shortly after the ice left, hunting reindeer in the open arctic landscape. As the temperature gradually increased, tree species such as aspen (*Populus tremula* L.) and birch (*Betula pendula* Roth. and *B. pubescens* Ehrh.) arrived, followed later by Scots pine (*Pinus sylvestris* L.). Hazel (*Corylus avelana* L.) was one of the first shade-tolerant species to invade these open forests and became dominant. Later, oak (*Quercus robur* L. and *Q. petraea* [Matt.] Liebl.), lime (*Tilia cordata* Mill. and *T. platyphyllos* Scop.), elm (*Ulmus glabra* Huds.), and ash (*Fraxinus excelsior* L.) dominated and formed a dense forest. Many large animals, now extinct in Denmark (e.g., wild boar, moose, aurochs, lynx, and wolf), inhabited the virgin forest. Our hunting and fishing ancestors mainly settled at the coasts. The average temperature reached a climax (2 to 3°C higher than now) about 5,000 ybp; the first steps toward agriculture about 6,000 ybp were the beginning of a long-lasting decline in forest cover. Beech did not arrive until 3,500 ybp in the Bronze Age, when the forests were already strongly influenced by human activity. Other chapters in this volume explore these aspects of the early Danish landscape and land-use history (Bradshaw, this volume; Hahn et al., this volume).

Gradually increasing population and developing technology put more pressure on the forests. Too-short rotations of shifting cultivation and heavy livestock grazing gave ericaceous plants such as heather (*Caluna vulgaris* [L.] Hull) the opportunity to establish on the poorest soils. These developments were sometimes interrupted by periods of war or epidemic diseases such as the Black Death, which arrived in Denmark in 1348 and killed about one third of the population. Forest vegetation invaded the abandoned fields, but when the Black Death was over and human population recovered, forest exploitation began anew. Wood was used for many purposes, mainly construction timbers for houses and ships, as well as firewood and charcoal to manufacture steel, glass, and bricks.

13.3.2 Cutting, grazing, and fire

Wood consumption was by itself a serious threat to the forest, but intensive grazing by domestic stock also significantly contributed to the rapid forest decline. Until the late 18th

Figure 13.2 Historic landscape of western Jutland, showing a peasant herding sheep on heathland. (Painted by F. Vermeren in 1855.)

century, peasants had rights to coppice the understory trees and to graze the forest with domestic stock. The trees of the overstory belonged to the landowner, often the local landlord. Consequently, trees were cut for timber and firewood, but the regeneration was removed or excluded by grazing. If the landowner had an interest in growing high forest for good-quality timber, there would be a conflict with the peasants over the rights to graze the forest. A closed canopy would not support good grazing and a fence would have been necessary during the regeneration phase. Stories of early attempts to plant trees are among the more amusing anecdotes of Danish forest history. The peasants, who did all the work of planting, made all kinds of "mistakes" that killed the trees and thereby made sure the planting failed.

Through the 18th century, the population faced increasingly severe shortages of wood and agricultural products. Almost complete deforestation was reached by 1800, when forest cover reached a low of only 4%. Jutland, except for the southeast, was largely covered by heathland (Figure 13.2), which had no trees and supported only sheep grazing. The heather was regularly burned so that it regenerated and retained its grazing value. In many locations, overgrazing or fires destroyed the vegetation and exposed the underlying sand to serious wind erosion. Dunes developed primarily on the west coast of Jutland but also in the interior, and in many places the drifting sand threatened the populace. Blowing and drifting sand covered villages, churches, farms, and fields.

13.3.3 The complete ecological disaster

Today, such natural disasters and their impact on communities are difficult for modern people to envision. The now revegetated dunes and remaining heathland are seen as interesting landscape elements or habitat for rare species. However, these past events and the landscapes they created were truly ecological disasters caused by exploitation of the forests and landscape. Old travel accounts from interior Jutland described the harsh conditions and the poverty of the poor people (Kousgaard 2000). Many locations with names that indicate a past forest cover were barren heathland in the mid-19th century. This level of forest and landscape degradation and accompanying sand drift and heathland development, however, did not develop in the eastern parts of Denmark to the same extent it did in western Jutland.

From the mid-18th century, the government tried to alter this unfortunate situation by various means. Reforms were launched over decades to change farmland ownership and tenure from large estates to independent peasant owners. Different governmental initiatives were launched over nearly 50 years, including assistance from German foresters who introduced basic forestry knowledge such as planning, testing species, regeneration methods, silvicultural systems, and forest education. In 1805, new forest legislation separated forest from farmland by fences in order to exclude grazing livestock from the forests. Fencing and the gradually changing ownership structure of farmland and forest, including a new social class of independent peasants/farmers, were important elements for the restoration of forests and landscapes. This became a crucial step forward for forestry, particularly management of high forests.

Late in the 18th and early 19th centuries, the state forest service established plantations at several sites on the heathland of western Jutland. Afforestation was slow; only a few thousand ha were established as plantations by the mid-19th century. Regeneration was difficult at these exposed sites, with infertile, acid soils containing a dense subsoil pan (B-horizon) that perched water and impeded root penetration. Severe competition from heather as well as the risk of severe frost added to the difficulty. There was hardly any silvicultural knowledge or experience to guide afforestation, and no rich private association or government agency to support the movement in the beginning. Other elements of the initial failures were choosing the wrong species and provenances, as well as inadequate cultivation methods (direct seeding).

13.3.4 An aid project of the 19th century

As a result of war, Denmark lost southern Jutland in 1864 to Prussian Germany, led by Bismarck. The northern third of this region became Danish again in 1920 by plebiscite, after Germany's defeat in World War I. The southern two thirds of the region in question had a clear German majority and remained Germany. A national movement arose, in response to this loss of southern Jutland in 1864, that boosted reclamation of bare heathland and other low-productivity areas in Denmark, particularly western Jutland. This afforestation movement had three main objectives: timber production, soil erosion control, and, in the 20th century, job creation in times of high unemployment. Additionally, tree shelterbelts have been regularly distributed in the landscape to create windbreaks to shelter agricultural crops.

In 1866, an independent organization (Hedeselskabet) was established to support reclamation by public relations, coordination, and knowledge. The goal was to persuade wealthy people to invest, government to subsidize, and local people to work, invest, and believe that reclamation would improve their future. The symbol of this land reclamation movement became E.M. Dalgas, an army engineer and officer. The timing was undoubtedly right for him and his vision, furthered by his organizational skills combined with his persuasiveness and speaking gifts. This seemed to be the necessary blend of elements to catalyze the whole process. Dalgas and his organization (Hedeselskabet) supported the reclamation by a range of activities, including:

- Find relevant sites
- Agitate at meetings for the reclamation movement
- Negotiate with the landowners, often poor peasants

- Find investors willing to invest in land, planning, amelioration, and planting
- Obtain government subsidies for planting
- Establish experiments to develop and demonstrate planting methods
- Transfer knowledge through an extension service, lectures, newspaper articles, and other publications
- Plan, manage, and control specific projects.

Hedeselskabet worked very much like a modern nongovernmental organization (NGO), but within the same country where it was formed. The approach today seems very modern: local people and cooperative organizations were freely involved, knowledge that was based on research and development was utilized, documented, and distributed, practical planning and implementation was properly guided, and (last but not least) proper financial support was obtained.

Getting the finances in place is an interesting story on its own. Rich landlords, nobility, merchants, and industrialists were willing to invest in reclamation (Kousgaard 2000; Jensen 2001). Fortunes developed due to the growth of industry and the economy. It became fashionable and popular for such people to invest in reclamation projects, perhaps for a good reputation and image in society as a patriotic individual, while becoming a kind of landlord on their own estate, where they could recreate with their family. Local people and communities also invested in reclamation and considerably contributed to afforestation. Because these people were generally not wealthy, their plantations were financially constructed like corporations. Even relatively poor people could participate by buying one or a few shares, and thereby contribute to the movement.

The reclamation of the barren heathland was almost completed by 1950; by then, there were 180,000 ha in plantations and the forestland cover in Jutland increased from 2.7% in 1860 to 10% in 1950 (Figure 13.3). Thus, much of the reclamation led to farmland. Hedeselskabet established 59% of the new plantations from 1866 to 1950 and the government was responsible for 41%. Local people and rich investors were the main actors in this nongovernmental afforestation (unpublished data, Hedeselskabet). The local people (45%) were mainly farmers; rich landlords and businessmen from other parts of the country

Figure 13.3 Maps showing the heathland distribution in 1800 and 1950. (Jensen et al. 2001 with permission from Hedeselskabet.)

accounted for 38% of the investment. Hedeselskabet itself (9%) and local authorities such as counties and communities (8%) were relatively minor investors. Today it may seem that there was complete agreement on the advantage of reclamation and on the distribution between afforestation and reclamation for agriculture, but such overall agreement never existed (Jensen 2001). Voices of disagreement have always been present, but only in recent decades have nature conservation, biodiversity, or environmental protection interests gained sufficient support to influence afforestation efforts.

This first wave of Danish afforestation in most of Jutland undoubtedly involved hard work and sometimes even hunger and disappointment for at least some of the workers and peasants. No doubt, the present farmland and conifer plantations appear more successful and productive than even the visionary pioneers and rich urban investors imagined some 70 to 140 years ago.

13.3.5 How was the reclamation done?

Generally, the best sites were reclaimed to farmland and the poor sites to forestry. When E.M. Dalgas and Hedeselskabet increased the speed of reclamation, there was still much to learn about it. However, the knowledge was soon greatly improved and reliable planting methods were developed by the late 1870s (Kousgaard 2000). From today's perspective, the approach was very intensive; planning might involve careful surveying, land leveling, and mapping site types. Often it was a practical question of how to divide the land between "poor" and "very poor" sites. Access to fertilizers, lime, and heavy machinery was not an option then, as it is today. Instead, labor was inexpensive and horses were, apart from the workers, the main source of power.

Dalgas soon realized that mountain pine (*Pinus mugo* Turra) was extremely tolerant of the harsh conditions of frost, drought, and infertile soil, and served as an effective nurse crop that could shade out heather and assist the establishment of the main species such as Norway spruce (*Picea abies* [L.] Karst.) (Kousgaard 2000). Mountain pine became the only suitable species for the first-generation plantation on the poorest sites. However, mountain pine does not tolerate shallow rooting, such as from a dense subsoil pan that blocks root penetration. A single-stem subspecies of mountain pine (ssp. *uncinata* (Ramond) Domin) that gained popularity in the 20th century has similar nurse tree effects as the regular mountain pine. Generally, the species choice was simple:

- The very poor sites were planted with approximately 3,500 mountain pine seedlings per ha.
- Medium sites were planted with 50% mountain pine and 50% Norway spruce (approximately 3,500 per ha in total).
- Good sites were planted with approximately 3,500 Norway spruce seedlings per ha.

Various soil preparation methods were employed, depending on the site; on very sandy sites, mountain pine was just planted in holes prepared by hand and spade. Plowing was intensively employed on sites where heather dominated because it was important to control heather and stimulate decomposition and humus mineralization. Such superficial but intensive soil preparation also overcame some of the problems created by pans. Deep plowing to approximately 45 cm depth was used to some extent, but it was expensive and slow. Deep plowing was mainly used to establish plantations on the most exposed sites in the west, where white spruce (*Picea glauca* Moench) was planted instead of Norway spruce.

Norway spruce became the main timber species because it also tolerated frost, drought, and infertile soil, although not as well as mountain pine. It was reliable, relatively inexpensive to plant, and produced high-quality timber, which until recently brought high stumpage prices. Norway spruce is not without problems; it suffers from root rots, windthrow, and poor stability (Hahn et al., this volume). Windthrow problems became apparent early on, and silviculturists soon attempted to create more diverse and stable forests. Japanese larch (*Larix kaempferi* [Lamp.] Carr.) is a fast-growing pioneer species, which serves well as a nurse crop except that it is sensitive to drought. Consequently, larch somewhat replaced mountain pine as a nurse crop on many sites. Microclimate and soil fertility improved as the plantations developed, and so did the conditions necessary to establish a number of other species such as Sitka spruce, European silver fir, Douglas fir (*Pseudotsuga menziesii* Mirb. Franco), and grand fir (*Abies grandis* Dougl. Lindley). Great effort was dedicated toward establishing European silver fir, often mixed with Norway spruce. European silver fir resists root rots and was thought to be more stable in older stands. However, it is difficult and expensive to establish, as it needs shelter and will not tolerate late spring frost. Additionally, it is very attractive to deer. Consequently, it is easily suppressed if mixed with Norway spruce, and many expensively planted mixtures have ended up today as pure Norway spruce stands. Improved microclimate and soil fertility led to shorter regeneration periods and faster growth. Norway spruce productivity has increased; the typical average annual production (100-year rotation) has increased from 7 to 11 m^3 ha^{-1}.

The leading role of Norway spruce is uncontested: 94% of the area was in conifer species and, on average, 67% of this was Norway spruce (Neckelmann 1986). No other conifer species exceeded 5% on average, although some species reached, exceeded, or came close to a 10% share as they were popular for a period (e.g., lodgepole pine [*Pinus contorta* Dougl.], European silver fir, and Sitka spruce). Broadleaves only covered 6% of the heathland afforestation, with oak as the most important species, covering 58% of the broadleaf area. Scots pine never attained the leading role expected at the beginning of the afforestation program (Neckelmann 1986). Although it was the main species planted during the first 50 to 60 years in the early 19th century, the German provenances used were poor choices and the species gained a bad reputation. Today we know that it is definitely a tolerant species and as such suitable as a nurse crop species, but not as highly productive as Norway spruce.

13.3.6 *Along the west coast*

Close to the west coast of Jutland, site conditions are quite different from the interior. The North Sea deposits high levels of salt, carried by the predominant westerly winds, which is not tolerated by Norway spruce. On the other hand, frost problems are less frequent along the coast. Besides the "blanket of mountain pine" that covered the Danish west coast, the relatively mild coastal climate called for species that are sensitive to frost but tolerant of the wind and salt deposition such as Sitka spruce and to a lesser extent European silver fir.

13.3.7 *Afforestation in eastern Denmark, 1800 to 1950*

In the rest of the country (southeastern Jutland and the islands of Funen and Zealand), there were still forest remnants in the early 19th century. Between one third and one half of the forestland was given up and cleared for farmland, and the land-use reforms initially reduced the forestland. Future land-use (farmland and forestry) was separated; the forests were fenced against livestock and hay harvest ended. It was generally easier to regenerate the forests and plant trees in eastern Denmark, and this part of the afforestation history appears less spectacular than reclamation of the heath plains. Afforestation in eastern

Denmark included expansion of existing forests and afforestation on sites unsuitable for agriculture because of steep terrain or poor soil. The forest land area gradually increased in this part of the country as well, and the better growing conditions made it possible to use a wider range of tree species of both broadleaves and conifers.

Growing conditions in the southeastern part of the country (Figure 13.1, Forest Region IV) were better than on the heath plains. The soils offered good nutrient and water supply, hilly terrain protected against late spring frost, and often regeneration was present, although browsed. In particular, beech regeneration sprouted in many places. Beech, oak, and ash were the main species planted, mixed with alder (*Alnus glutinosa* [L.] Gaertn.) and elm. New species were introduced, such as European silver fir, Norway spruce, larch, and sycamore maple (*Acer pseudoplatanus* L.). Interestingly, oaks were planted or sown by the state forest service (e.g., in the royal forests) in the early 19th century to meet the demands for naval timbers. The motivation for this planting increased after Denmark lost its navy ships in the Napoleonic wars to Great Britain in 1807. Nevertheless, conifers gradually became more frequently used for afforestation in eastern Denmark, particularly in the 20th century until the 1980s.

13.4 *Recent afforestation*

13.4.1 *Changing goals — like a moving target*

The first wave of afforestation (1800 to 1950) was very successful. Forest cover increased to the present 11% and two thirds are exotic conifers. Although 11% forest cover may seem a limited success from an international perspective, Danish culture is deeply rooted in agriculture, which has served its economy well and remains one of our largest exports. Nevertheless, the success of the early afforestation effort depends on the viewer, and today there are alternative views on the success or failure of what foresters accomplished decades ago. Even as we harvest what they planted, our modern affluent society questions whether the result of their efforts is optimal after all. Prices for softwood timber are at historically low levels, which has reduced or even eliminated the formerly important source of forestry income. Considerable windthrow risk in conifer stands before the trees reach target diameter further stresses the poor economics of spruce timber production. Such plantations often show a poor resilience and consequently both aspects of stability (resistance and resilience) (Larsen 1995) suffer in these plantations (Hahn et al., this volume).

Yet, afforestation is still under way. After two decades (1960s and 1970s) with limited activity, afforestation became an issue again from the mid-1980s. The goals differ now from those of 150 years ago, and even from the goals of 15 years ago. No doubt, they will probably change again. Subsidized production that produced a surplus of agricultural products fueled new governmental interest in afforestation, in order to reduce farm production and related subsidies. Another attractive goal was increased timber production, as the annual wood consumption substantially exceeded harvest (8 to 9 million m^3 wood consumed but only 2 to 3 million m^3 harvested). Additionally, the potential for more multifunctional forests came into focus. Pure conifer stands were still planted, but in 1989 the Danish government initiated a new afforestation program with the aim of doubling forest cover within one tree generation (80 to 100 years). Timber production was still an issue, but other values such as nature conservation, biodiversity, recreation, carbon sequestration, bioenergy, and protection of environment and groundwater increasingly gained support (Skov- og Naturstyrelsen 2000; Danish Forest and Nature Agency 2003). These are typical national goals, well known in many countries with afforestation programs.

The regeneration stages challenge foresters to anticipate management goals 50 to 120 years into the future, at least until the end of a rotation. Goals may well change because of the long production time, and the longer the rotation length, the greater the difficulty in anticipating preferences. The typical rotation age in Denmark is between 50 and 120 years so that stable and flexible forests, which may be subject to changing preferences, seems to be a reasonable strategy and should allow even for periods of no management. In Denmark, this includes using site-adapted and stable species, which on a certain site may be both native species and exotic species such as Douglas fir, Sitka spruce, or European silver fir. Norway spruce is still planted but not to the same extent as before.

13.4.2 *Public afforestation*

The central government and some communities may buy farmland from farmers and plant trees. Such afforestation is often done by the Danish Forest and Nature Agency (Skov- og Naturstyrelsen), in cooperation with local community authorities or water companies responsible for drinking water supply. Public afforestation usually creates new and relatively large forests (greater than 200 ha). These are located close to urban areas to protect important groundwater resources and provide recreational opportunities. Participatory planning and design involving local people encourages use of the forest from the earliest stages of development. Such young forests provide valuable recreation for people living nearby (walking the dog, jogging, etc.); the closer the forest is located to the neighborhood, the better (Jensen and Koch 1997). Moreover, proximity of residential housing to forests affects house prices positively, up to 500 m, where the effect fades out (Anthon and Thorsen 2002; Præstholm et al. 2002).

Afforestation was not planned only for government land; private landowners are subsidized to participate in the afforestation program. Subsidies seem necessary to reach afforestation goals, because the most common alternative landuse is agriculture, which is heavily subsidized. Private landowners require subsidies for establishment costs as well as additional compensation for the lost agricultural income. Currently, subsidies for afforestation on private land may be as high as US\$4,800 ha^{-1} (US\$1 = 7 DKR) plus an annual income compensation of US\$320 ha^{-1} year^{-1} for the first 10 years (Table 13.1).

Table 13.1 Subsidies for Afforestation on Private Land 2004

	Afforestation Areas (US\$ ha^{-1})	Outside Afforestation Areas (US\$ ha^{-1})
(A) Establishment		
Planting broadleaved forest or forest edge	3,333	2,133
Planting conifer forest	2,133	1,333
Direct seeding	2,133	1,333
(B) No pesticides used	400	400
(C) "Gentle" soil preparation — no deep preparation	400	400
(D) Fence	2 m^{-1}	2 m^{-1}
(E) Income compensation for 10 years	320 year^{-1}	0
(F) Preparatory investigations		
Site mapping	133 + 27 ha^{-1} [a]	133 + 27 ha^{-1} [a]
Mapping and survey	67 + 6.7 ha^{-1} [a]	67 + 6.7 ha^{-1} [a]

[a]Example: Subsidy for 5 ha site mapping: 133 + (5 × 27) = US\$ 268.

Source: Modified after Skov-og Naturstyrelsen, Tilskud til skovrejsning, http://www.sns.dk/skov/tilskud/skovrejs/, 2004.

13.4.3 *How is afforestation done today?*

The high subsidy rates reflect the high cost for regeneration under Danish conditions. Stock density is high (2,500 to 3,500 conifer seedlings ha^{-1} or 4,000 to 6,000 broadleaf seedlings ha^{-1}; Figure 13.4). Bare-root seedlings presently are the most common stock type, usually 2-0 or 2-1 seedlings (2 to 3 years in the nursery). Price negotiations for materials and planting make it difficult to obtain a clear picture of costs. The price per planted seedling, including machine-planting, ranges from US$0.30 to US$0.55, depending on stock-type, species, and area to be planted. The total regeneration cost usually ranges between US$2,600 and 5,000 ha^{-1}, depending on material and planting costs plus the need for fence, weeding, or other measures.

Soil preparation and weeding before planting, followed by weeding the first years after establishment, is common. Fencing, mainly against roe deer (*Capreolus capreolus* L.), is also common practice. Depending on species, from 2,500 to 4,000 saplings must reach an average height of 1 m within 8 years after planting in order to obtain the full subsidy. Because faster growth means a quicker subsidy payment, intensive regeneration methods including weeding, fencing, and deep plowing are favored.

Deep plowing (to 70 to 80 cm depth), in particular, has large and positive effects on survival and growth in poor sandy soils, due to several factors. The topsoil contains many weed seeds and has a relatively high water-holding capacity due to higher humus content than the subsoil. Deep plowing buries the topsoil, which eliminates most of the weed competition for the first growing season, protects the buried topsoil from evaporation as the hydraulic connectivity is broken, and plow pans, that may have developed in soils with higher clay content, are broken.

However, deep plowing is not popular among people who value cultural heritage, which may be severely disturbed or destroyed. Landowners must consult archeological experts before a site is deeply plowed. Further, establishment without deep plowing receives a higher subsidy (Table 13.1). Similarly, afforestation without using herbicides obtains additional subsidies. The Danish Forest and Nature Agency does not use pesticides or herbicides, as the agency has converted to close-to-nature silviculture. Afforestation by water companies is probably done without using herbicides for competition control; harrowing or rotary cultivation between the rows is a common alternative. These mechanical methods are more expensive, may be difficult in a rainy summer, and may increase nutrient leaching, which is undesirable for groundwater quality (Pedersen et al. 2000). Increased nutrient leakage may also result from deep plowing.

(A) (B)

Figure 13.4 Typical recent afforestation in eastern Denmark. (A) Beech with larch as nurse trees. The beech is approximately 2 m tall. An old broadleaved forest is in the background. (B) Ash planted at the same location but on a moister site, showing the dense stocking; the lowest branches are approximately 1 m above the forest floor.

Preparatory survey of soil conditions, mapping, and survey may also release additional subsidies (Table 13.1). Site classification is supposed to guide the choice of tree species, for example, planting ash and alder in moister areas and oaks primarily on sites with hard pans, pseudogley, or on relatively dry sites. The ultimate goals are better stability of the future forest and a higher probability of successful natural regeneration. Such attention to matching site and species was absent during most of the 20th century, when conifers were planted almost everywhere.

13.4.4 Restrictions, priorities, and landscape planning

The Danish landscape is intensively used and managed and the administration of afforestation subsidies is a good example of the intensity of landuse. Several restrictions and priorities seek to regulate how and where afforestation is carried out on private land in order to reflect the interests of society. Such restrictions include the following:

- The new forest must be more than 2 ha and forests between 2 and 5 ha are only subsidized close to urban areas, and the public must have access.
- Habitats such as lakes, meadows, heathland, and moors must not be drained or otherwise changed and planted with trees or shrubs; they may, however, become integrated open areas within a new forest.
- New forests need at least 20 m wide forest edges with a minimum 20% shrubs to the north and the west, and a minimum of 10 m forest edge is needed to the south and the east.
- Broadleaved stands must have minimum 75% broadleaved species.
- Both broadleaved and conifer stands must be intermixed with minimum 10% other species than the main species.

Budget limits prevent the government from supporting all afforestation projects on private land. The highest priority, and therefore the greatest probability to successfully gain subsidies, is reached under the following conditions:

- In areas classified as afforestation areas by the regional planning authority
- In areas where groundwater is extracted for drinking water
- When native species are more than 75% of the planted seedlings
- When no deep plowing or other deep soil preparation is used
- When the new forest area is more than 5 ha, better when it is more than 10 ha
- When the new forest is located close to an urban area or in an area with low forest cover
- When the afforestation project is coordinated with other landscape restoration and afforestation projects such as environmentally friendly farming, habitat restoration, or improved public access to forest and nature.

13.4.5 Regional planning — where is afforestation wanted?

These restrictions and priorities offer government detailed control of afforestation on private land, which can be changed as new needs are identified. Additionally, regional authorities are responsible for general land-use planning, which is reviewed and adjusted at regular intervals. The landscape is divided into three main categories: suitable, unsuitable, or neutral for afforestation. At present, 6% of the land area is classified as suitable for afforestation, whereas 25% is classified as unsuitable. The most recent classification guidelines stress the importance of groundwater and recreational interest near urban areas for areas suitable for afforestation. The guidelines also encourage regional authorities to favor "green corridors" in the landscape for biodiversity, connectivity, and to show consideration for agricultural interests.

Generally, afforestation areas should not be located in areas where it conflicts with interests of geology, cultural heritage (e.g., churches and castles), and landscape. New forests should not cover or hide buildings, geological formations, or landscape vistas. Such areas may instead be classified as unsuitable, along with areas where afforestation conflicts with future urban development or technical installations such as windmills.

13.4.6 *Why do private landowners enter afforestation?*

Private landowners undertaking afforestation are of two kinds. One group includes part-time farmers, who earn their main income elsewhere and own relatively small properties. Their average afforestation project includes 7 ha of new forest, primarily for recreation and nature conservation. The other group includes professional and full-time farmers who want to create a hunting property out of part of their farm, perhaps the least suitable part for farming.

Professional full-time farmers seem to need increasingly more farmland to stay in business. In addition to arable land, a farmer must have access to a certain amount of land per cow or pig to spread manure in order to avoid concentrated loss of nutrients to the groundwater or surface water. Consequently, farmland is a limited and expensive resource; a price for farmland of US\$15,000 ha^{-1} would be usual. Forestland is less subsidized, and thus less expensive, approximately US\$7,000 to 17,000 ha^{-1} including standing volume. Thus, subsidies for afforestation are in competition with subsidies for continued farming, which explains the importance of income compensation. By planting a minimum 35 ha without subsidies, the landowner acquires a property with a very attractive feature: it can be sold to anybody who can afford to buy it, without having to live there year-round, as with ordinary farms. This new status of forestland makes such properties very valuable and earns the farmer who planted the 35 ha a good return.

Wood production is not a driving force for afforestation of private land (Skov- og Naturstyrelsen 2000). The value of timber in the distant future is theoretical, not real, for most individuals. Private landowners may simply afforest land to enjoy ownership of forestland for recreation, wildlife watching, and hunting. Hunting rights on private property belong to the landowner and may generate recreation and pleasure for the owner or rental income of US\$50 to 100 ha^{-1} or more annually. Projects may call for tree planting on about half of the afforestation area, with the rest left open for livestock grazing or small lakes or ponds, which may be designed for duck hunting. The most important economical aspect of afforestation for the landowner may be its influence on overall property value, which typically includes the transformation of a traditional farm dominated by bare fields to an attractive property with amenity values. Consultants and extension services offer assistance to private landowners for designing, planning, and implementing afforestation, as well as applying for subsidies.

13.4.7 *The afforestation program is behind schedule*

Both public afforestation and subsidized private afforestation are expensive. Expenses for public afforestation include purchasing land and regeneration costs, whereas the subsidies for private landowners include compensation for agricultural income lost over 10 years. Costs are US\$17,000 ha^{-1} for public and US\$9,500 ha^{-1} for private afforestation (Skov- og Naturstyrelsen 2003). Not surprisingly, afforestation is behind schedule (Skov- og Naturstyrelsen 2000, 2003; Danmarks Statistik 2002). Over the period 1991 to 2001, the program has resulted on average in 1,650 ha of new forest annually, far behind the target set in 1989 of 4,000 to 5,000 ha annually. Public afforestation in particular is behind schedule; only 3,000 ha (11%) were established in the 10-year period, whereas 8,500 ha of subsidized private and 5,000 ha of private afforestation without subsidies were established.

Competition with subsidized farming for land is clearly an obstacle for the afforestation program. An additional subsidy to compensate for the loss of annual income, launched in 1998, boosted subsidized private afforestation in 1999 to 2001 to 1,300 to 3,300 ha annually (Skov- og Naturstyrelsen 2003). Private afforestation without subsidies was probably reduced. When first introduced, the income-compensation period was 20 years but private landowners are now supported for only 10 years. Subsidized private forests must not be cleared again. The future will show how agricultural subsidies and earnings will influence land prices and the need for income compensation.

The development of new and inexpensive regeneration methods may provide some help for the afforestation program. New methods need to be independent of intensive weeding, either by herbicides or machinery, and deep soil preparation in order to meet the requirements of public afforestation, and for private afforestation to receive the highest subsidies and priorities. Direct seeding and small container stock, in combination with cover crops and nurse trees, have shown promising but preliminary results. These regeneration techniques better match the principles of close-to-nature silviculture than traditional, more expensive and intensive means. However, the methods require further development and documentation before they are ready to be recommended. Natural colonization (invasion) is an even cheaper means of afforestation that will probably gain in importance. Although it is indeed very close to nature, landowners and the agencies administering subsidies will need more patience than is needed with other methods.

References

Anthon, S. and Thorsen, B.J., Værdisætning af statslig skovrejsning — en husprisanalyse. Dansk Skovbrugs Tidsskrift, 87, 73, 2002.

Danish Forest and Nature Agency, *The Danish National Forest Programme in an International Perspective*, Ministry of the Environment, Danish Forest and Nature Agency, Haraldsgade 53, DK-2100 Copenhagen Ø, http://www.sns.dk/internat/dnf-eng.pdf, 2003.

Danmarks Statistik, Skove og Plantager 2000, Danmarks Statistik, 171 pp., http://www.dst.dk, 2002.

Jensen, F.A., De jyske heders opdyrkning, in *Aktieselskabet Sønder Omme Plantage år 2001*, Jensen, F.A., Ed., Aktieselskabet Sønder Omme Plantage, 96 pp., 2001.

Jensen, F.S. and Koch, N.E., *Friluftsliv i skovene 1976/77–1993/94*, Forskningsserien nr. 20, Forskningscentret for Skov & Landskab, Hørsholm, 1997.

Kousgaard, P., *Hvor heden før var gold nu skov og læhegn findes*, St. Hjøllund Plantage A/S og Palle Kousgaard, 128 pp., 2000.

Larsen, J.B., Ecological stability of forests and sustainable silviculture, *For. Ecol. Manage.*, 73, 85, 1995.

Neckelmann, J., Træartsvalg og dyrkningserfaringer i hedeplantagerne. Vækst 107, 20 and Vækst 107, 12, 1986.

Pedersen, L.B., Riis-Nielsen, T., Ravn, H.P., Dreyer, T., Krag, M., Nielsen, A.O., Matkowski, A., and Sunde, P.B., Alternativer til pesticissprøjtning i skovkulturer, *Skoven*, 32, 355, 2000.

Præstholm, S., Jensen, F.S., Hasler, B., Damgaard, C., and Erichsen, E., Forest and afforestation in the neighbourhood: attractiveness and value of local areas in Denmark, in *Proceedings from the Conference "The Changing Role of Forestry in Europe; Between Urbanisation and Rural Development,"* 11–14 Nov. 2001, Proceedings 2002-02, Forest and Nature Conservation Policy Group, Wageningen University, the Netherlands, 2002.

Skov- og Naturstyrelsen, *Evaluering af den gennemførte skovrejsning 1989-1998*, Skov- og Naturstyrelsen, Haraldsgade 53, DK-2100 Copenhagen Ø, http://www.sns.dk/skov/net-pub/evaluering, 2000.

Skov- og Naturstyrelsen, Skov og Natur i tal 2003. Miljøministeriet, Skov- og Naturstyrelsen, http://www.skovognatur.dk, 2003.

Skov- og Naturstyrelsen, *Tilskud til skovrejsning*, http://www.sns.dk/skov/tilskud/skovrejs/, 2004.

Vejre, H., Callesen, I., Vesterdal, L., and Raulund-Rasmussen, K., Carbon and nitrogen in Danish forst soils — contents and distribution determined by soil order, *Soil Sci. Soc. Am. J.*, 67, 335, 2003.

chapter fourteen

Baltic afforestation

Kalev Jõgiste, Aivo Vares, Veiko Uri, and Hardi Tullus
Estonian Agricultural University

Contents

14.1 Introduction

The Baltic States of Estonia, Latvia, and Lithuania have undergone drastic upheaval in their political systems during the last century. The Baltic States were independent nations between the two world wars; otherwise, their recent history was as part of Tsarist Russia and later the Soviet Union. Changes in economic systems and land tenure, from the centrally planned economy of the Soviet Union to the free market and private ownership of today's newly independent states, significantly affected landuse, especially the balance between forestry and agriculture. In all the Baltic States, large areas of agricultural land were abandoned during the last decade and are available for afforestation. Although afforestation was practiced in the Baltic States for nearly a century, the past decade since independence has been the most active period. Mining operations also result in areas open to reclamation. Many abandoned areas are undergoing old-field succession following natural invasion by pioneer hardwood species such as birches, alders, aspens, and willows.

The main aim of restoration has been to establish commercially valuable trees, particularly conifers. Spruce and pine plantations on abandoned farmland, however, are often plagued with root rot, and timber quality is reduced by low stocking density that allows large branches to develop. Recently, new objectives for the afforestation have been set, based on concerns for environmental effects and the desire to enhance biodiversity. The protection of soils against erosion and pollution, and protection of water resources,

1-56670-635-1/05/$0.00+$1.50

increasingly drives programs to establish forests. Our objectives in this chapter are to provide a historical backdrop to the present effort to afforest public and private land in the Baltic States.

14.2 Natural conditions

The Baltic States of Estonia, Latvia, and Lithuania are situated in the transition zone between boreal and temporal forests (56 to 59°N; 24 to 26°E). The region is regarded as a hemiboreal zone; southern Lithuania belongs to the temperate zone (Ahti et al. 1968). Average temperatures range from +20C° in July to −6C° in February, and mean annual precipitation varies from 500 to 930 mm. Dominant soil types are podzols and gleyic podzols. In areas with calcareous parent material, podzolization processes in soil development are retarded.

Mixed-coniferous and deciduous woodlands are characteristic natural vegetation. Nearly half of the territory of Estonia and Latvia is covered by forest (Tullus and Uri 2002); Lithuania is approximately one third covered by forest (Gaizutis 1998). The main tree species in the Baltic region are Scots pine (*Pinus sylvestris* L.), Norway spruce (*Picea abies* [L.] Karst), birches (*Betula pendula* Roth and *B. pubescens* Ehrh. and hybrids), alders (*Alnus incana* [L.] Moench and *Alnus glutinosa* [L.] Gaertn.), common aspen (*Populus tremula* L.), pedunculate oak (*Quercus robur* L.), European ash (*Fraxinus excelsior* L.), and others. Oak is more common in Lithuania.

The variety of different forest habitats is high in the Baltic region. Dry sandy soils are predominantly occupied by Scots pine, with Norway spruce found on more fertile soils. Mixtures of conifers and broadleaved tree species are common. The proportion of seminatural communities such as grasslands, pasture, and abandoned agricultural land is considerable. Seminatural vegetation is typical for coastal areas (pastures covered with *Juniperus communis* L.) and also inland woody meadows. Over time, many of these communities will become woodlands.

During the last century, most attention has been devoted to cultivating conifers, primarily Scots pine and Norway spruce. Seedlings of these species have mostly been planted, but direct seeding has been used as well (mainly in the case of Scots pine). Few exotic tree species have been planted; nevertheless, hybrid aspen is a promising plantation species for pulpwood.

14.3 History

Forest and agricultural land uses are closely linked in the Baltic States. By the end of the 19th century, the forest resource was at its lowest ebb, exploited for construction timbers and fuelwood (Anon. 2000). Clearing for agriculture was intense early in the 20th century. Deforestation was also due to peat extraction, which left areas without vegetation. A considerable area of peatlands has been drained and afforested.

Agricultural land uses predominated between the two world wars in the independent Baltic States, and forest cover fell. At the end of the 1930s, for example, the forest area of Estonia was less than 30% land area (Figure 14.1). Nevertheless, during this same period, many of today's most valuable commercial forests were planted.

After World War II, the Communist government repressed farmers, causing abandonment of large amounts of agricultural land. Forest area increased in Estonia (Figure 14.1), as well as the other Baltic States. During the period 1950 to 1987 in Estonia, pine and spruce planted on former fields created about 65,000 ha of forest (Tullus 2000). In Lithuania, during the period 1951 to 1985, afforestation of nonforest area was considerably higher: 279,000 ha (Table 14.1), mainly with Scots pine and Norway spruce. Afforestation in Latvia is documented from the 19th century (Daugaviete 2000). Although farmland was

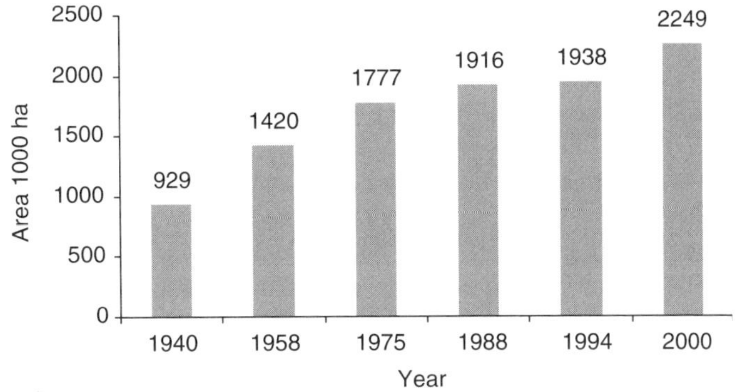

Figure 14.1 Forestland changes in Estonia during 1940 to 2000; the increase is mainly due to the abandonment of agricultural lands.

Table 14.1 Afforested Area Relative to Total Area Planted and Predominant Species in Lithuania, over the Period 1951 – 1985 (modified from Gabrilavicius 1999)

Period	Area (1,000 ha)		Predominant Tree Species as a Percentage of Total Area Planted			
	Total	On Non-forest	Scots Pine	Norway Spruce	Oak	Other
1951–1955	96.7	64.0	66	21	9	4
1956–1960	79.8	53.4	65	25	5	5
1961–1965	75.0	63.4	58	34	3	5
1966–1970	67.2	49.6	35	58	1	6
1971–1975	49.8	16.2	34	62	1	3
1976–1980	55.3	16.2	26	68	—	6
1981–1985	51.7	16.3	21	71	—	8

abandoned to natural invasion after wars, planting trees was easier on farmland compared to clear-felled forest areas. Between 1935 and 1983, forest area increased over 1 million ha through afforestation or natural invasion (Anon. 2002).

After regaining their independence, the Baltic States transitioned to market economies. Land reform measures and restitution processes are returning public land to private ownership. Nevertheless, forest establishment on abandoned agricultural land did not start instantly after regaining independence in Estonia. Massive economic restructuring often left the ownership status of land unclear and, in this unstable environment, financing for forest and agricultural management was unavailable. Consequently, natural invasion processes occurred on many fields, including the establishment of pioneer species such as gray alder, silver birch, and *Salix* sp. Even as recently as 2001, land abandonment in Latvia increased 150,000 ha over 1998 (Anon. 2002).

Today, approximately half of the forests in all Baltic States will be privatized, with the other half remaining in state ownership. In Latvia, for example, the state owned 51.1% of the forestland in 2001, private forest owners managed 45.0%, and municipalities 3.9% (Anon. 2002). The situation is similar in Lithuania: 50% of forest area belongs to the state, 23% was privatized in the course of the restitution process, and 27% was reserved for later privatization. The state forest proportion is lower in Estonia, approximately 40%.

These changes in forest ownership have greatly influenced management, especially the afforestation methods used in abandoned fields. Private forest management is oriented to more profitable landuse and intensive management. Timber prices have pushed afforestation toward softwoods; the market value of the timber in 50-year-old Norway spruce stands is 4950 EUR/ha; similar values for birch and gray alder are 3410 and 1695 EUR, respectively (Maamets 2003). The high cost of establishing plantations, however, has meant less mechanical and more manual work carried out in privately owned forests.

The national forest policies of all Baltic States include afforestation as a high priority. In Estonia, a Swedish–Estonian cooperation program initiated research and demonstration areas in 1999. The Forestry Development Plan 2002–2010 includes afforestation; legislation to offer subsidies is being developed. Preliminary calculations estimate a realistic support level at 960 EUR/ha. Today, the modest estimate of afforestation needed is 150,000 ha of farmlands, less than half of the 442,000 ha of abandoned land.

A similar Lithuanian afforestation project was launched in 1999 as a cooperative program of the Danish Ministry of Environment and Energy and the Ministry of Environment of the Republic of Lithuania (Gabrilavičius 1999). Afforestation in Lithuania has focused on planting coniferous tree species (Danusevicius, pers. comm.). At least 228,000 ha of abandoned agricultural land is considered to be afforested; most afforestation has been by natural invasion. Scarce funding has limited afforestation by planting to 2,500 ha in private and state forests between 1997 and 2001.

In 1994, the State Forest Service of the Latvian Ministry of Agriculture commissioned a research program on afforestation. Several legislative acts issued between 1998 and 2000 included exemptions for real estate tax on young afforestation stands. Afforestation subsidies in Latvia are set at 150 EUR/ha annually (Daugaviete 2000). Surplus farmland in Latvia available for afforestation exceeds 300,000 ha. Since 1994, 150 ha of experimental and demonstration stands were established (Daugaviete 2002). A rural development program began planting abandoned farmland in 1999; the goal is to plant 10,000 ha (2% of the total abandoned farmland) by 2006, mostly with birch (Anon. 2002).

Oil shale is an important raw material for energy and chemicals; 98% of the electric energy produced in Estonia is from oil shale. Exhausted opencast mines from oil shale extraction are afforestation targets (Figure 14.2). In 2001, 77% of the abandoned mined

Figure 14.2 Opencast oil shale mine area reclaimed with Scots pine (*Pinus sylvestris*). The plantation is 3 years-old and the planting stock was 2-1 seedlings. (Photo by Veiko Uri.)

Figure 14.3 Successful forest restoration: reclamation of an opencast oil shale mine with a 40-year-old Scots pine plantation. (Photo by Veiko Uri.)

land had been afforested (Kaar 2002). Afforestation of such areas is difficult because of high soil pH, low organic matter content, altered water regime, and stoniness. Scots pine is the primary tree species planted on these areas, accounting for 86% of the forest growing on abandoned opencast mines (Figure 14.3). Mostly 2-year-old seedlings have been planted at a density of 5,360 to 6,667 seedlings per ha (Kaar 2002). Norway spruce forms 4% of the plantations on opencast mines, but spruces are generally unsuitable because of their sensitivity to frost damage and infertile soil. Recent research suggests that the proportion of deciduous tree species should be higher on opencast mines (Kaar 2002).

14.4 *Afforestation practice*

14.4.1 *Species selection*

The Baltic States were similar to the Nordic countries in that initially the preference was to plant conifers due to high production value and low establishment costs. Foresters today understand that hardwood species were undervalued during the last century, when silvicultural recommendations aimed to totally clean young stands of hardwoods. Experimental hardwood plantations established in different locations of Estonia showed that silver birch was the best hardwood for afforestation of abandoned agricultural lands (Vares et al. 2001). Polyethylene mulch was the most effective cover material in birch plantings, reducing grass competition and conserving moisture. The cost for afforestation with deciduous species was 857 EUR/ha (Kaimre 2001). One of the most expensive operations was planting: planting stock cost 195 EUR/ha, protection fencing 253 EUR/ha, and soil preparation 5% of the total costs. Estimated thinning costs were 130 EUR/ha at age 15.

Latvian experiments on soil types and suitable species have looked at a number of species, including birch, aspen, black alder, wild cherry, oak, ash, beech, larch, spruce, and pine (Daugaviete 2000, 2002). In general, the ash, oak, and beech were slower growing and more susceptible to late-season frosts. Based on these experiments and observations in older abandoned farmland, most afforestation in Latvia will be with birch, which reaches veneer size in 30 years (Anon. 2002).

Biodiversity, or the lack of it, is a concern when afforestation is with single-species plantations over large areas. Thinning and pruning in young stands, which allows undergrowth to develop, improves the structure of monospecific plantations (Fujimori 2001). Under Baltic conditions, Norway spruce is the most common shade-tolerant species creating undergrowth. Planting several species in a region, according to site adaptation, is another approach to increase biodiversity. Thus, sites that are more fertile can be afforested with hardwood species. Sites with lower fertility can be afforested with pioneer hardwoods (mainly alders) and Scots pine. Mixtures are a third approach. Various methods are available to create mixtures, but performance varies depending on species characteristics and environmental conditions (Vares et al. 2001). For example, pedunculate oak suffers from late frost and game damage in Estonia; similar problems have been reported form Latvian afforestation sites (Daugaviete 2002).

Few exotic species have been used widely in the Baltic States. Several experimental stands were established after World War II, including exotic spruces, pines, larches, and Douglas fir (*Pseudotsuga menziesii* [Mirb. Franco]). Douglas fir and larches (*Larix decidua* Mill., *Larix sibirica* Ledeb., and *Larix kurilensis* Mayr) produced successful results, comparable to Scots pine. Larch plantations form 2% of the plantations growing on opencast oil shale mine areas (Kaar 2002). Hybrid aspen has become an important species during the last decade (Tamm 2000; Vares et al. 2003). Due to relatively short rotation, this species (*Populus* × *wettsteinii* Hämet-Ahti) is not susceptible to trunk rot caused by *Phellinus tremulae* that is common in European aspen. Hybrid aspen achieves maturity for pulp production at 25 to 30 years in northern Europe. Today, there are more than 600 ha of hybrid aspen plantations in Estonia (Vares et al. 2003).

14.4.2 Site preparation

Experience in Finnish afforestation (Hynönen 2000) indicates that after the choice of species for planting, the next most important factor is competition by ground vegetation. Site preparation reduces competition from weeds and improves soil conditions, but it is expensive. Moreover, polyethylene and paper mulch materials reduce grass competition (Vares et al. 2001). Mulch materials or herbicides can also increase seedling growth (Table 14.2). Usually, the most common method for site preparation is plowing in Estonia. If soil preparation is not used, afforestation success depends on former landuse and time since last tilling. In Latvia, fertilization is considered essential for afforestation, especially in nutrient-poor old fields. Fertilization enhanced the performance of the main species tested: pedunculate oak, ash, wild cherry (*Prunus avium* L.), Scots pine, and Norway spruce. Fertilizer prescriptions were estimated after observing deficiencies of N, P, and K; applying fertilizer in the ratio 5:8:11 as NH_4^+, P_2O_5, and K_2O targeted optimal fertility for conifers and hardwoods (Daugaviete 2000).

Table 14.2 Height of Young Trees from Different Experimental Treatments in Latvia, Expressed as Percentage of Growth of Untreated Trees (Control) (Modified from Daugaviete 2000)

Tree Species	Tending Applied				
	Hoeing	Mowing	Herbicide Application	Mulching	Control
Birch (*Betula pendula*)	108	100	156	132	100
Oak (*Quercus robur*)	108	134	125	100	100
Aspen (*Populus tremula*)	100	100	115	120	100
Pine (*Pinus sylvestris*)	125	130	144	100	100
Spruce (*Picea abies*)	100	171	133	100	100

14.4.3 *Spacing*

Spacing in timber plantations is debated in all Baltic States. Although low spacing increases individual stem growth, it is believed to reduce wood density. However, the wood density of deciduous tree species is not affected by spacing. Experience in Finland indicates that the low density of birch is beneficial; pruning was quite the same with 1,600 and 5,000 plants/ha (Niemistö 1995). Tests of hardwoods in Latvia have included densities of 1,111 stems/ha to 10,000 stems/ha (espacement of 3 by 3 m to 1 by 1 m) (Daugaviete 2000).

14.4.4 *Planting and direct seeding*

Although natural invasion has proven the effectiveness of seed dispersal (Suchockas 2002), planting is preferred in Estonia because it results in better stem quality. Planting bareroot seedlings is the most common method. Although root suckers of gray alder have been planted experimentally, growth is inferior compared to seedlings (Uri 2001). Nevertheless, direct seeding is an inexpensive method for afforestation. Establishment success of direct seeding depends very much on weather conditions not being too dry or too wet. Rodents and weevils are problems for direct seeding of large seeds or acorns. Lithuanian experience in afforestation points out critical questions in technology development and species choice (Table 14.3). Failures are apparently due to weed competition and poor quality of planting stock (Gabrilavičius 1999).

14.4.5 *Natural invasion*

Old-field succession on abandoned agricultural lands usually leads to the development of forests (Bazzaz 1998). Low-intensity afforestation can rely upon natural invasion, which has proven effective under the growing conditions of the Baltic region. Natural regeneration is a tool for regenerating harvested hardwood forests. The main problems with reliance on natural invasion are the long time period for forest establishment on harsh sites, uneven densities that vary from a few individuals of desirable species to thousands of trees/ha and lowered wood quality, and the time it takes for dense stands to differentiate through self-thinning, even for birch (Jõgiste et al. 2003). The main benefit is that it is essentially free and may be the only option for some private landowners.

14.4.6 *Protection and tending*

Browsing by wildlife significantly damages forest regeneration in all Baltic States. Moose (*Alces alces* L.) is a serious threat to both deciduous and coniferous species in plantations.

Table 14.3 Establishment Success of Several Tree Species Planted on Forestland and on Farmland in Lithuania; The Greatest Success was with Norway Spruce (modified from Gabrilavičius 1999)

Land Use	Total Area Planted (1,000 ha)	Establishment Success for Several Species, as a Percentage of the Total Area Planted				
		Scots Pine	Norway Spruce	Larch	Oak	Ash
Forest	522	62.9	71.5	11.6	9.8	29.7
Farmland	95	60.0	86.4	17.0	20.0	25.0

Even though fencing is important for establishment success, it is very expensive and can account for as much as 30% of costs (Kaimre 2001). Different individual seedling protectors have been tested and proven effective against small mammals (Daugavieate 2000; Vares et al. 2003); but long-term problems due to poor light quality (light absorption by tube walls), and instability of the sapling, if the tube is not removed, have put the method under question (Madsen and Löf 2000).

The planning of improvement measures (tending, thinning, and supplemental planting for diversity) is an important aspect of forest restoration today (Tullus 2001). Optimal tending of stands has been stressed in afforestation strategies (Tullus 2000; Daugavieate 2000). The EU programs in Latvia state clearly the need for tending the plantations after 3 years (Anon. 2002). Restoration for objectives other than timber, however, is just beginning to receive attention in the Baltic States.

14.5 Research needs

The greatest needs in the Baltic States for successful afforestation are to develop management capacity, methods for establishing forests with high biodiversity, and funding for landowners. Alternative species to the traditional conifers have produced highly stable and sustainable stands. Birch species, in particular, produce commercially valuable forests under the right conditions. Nevertheless, research is needed to increase the variety of suitable species and affordable afforestation methods. Species adaptations to sites, spacing, and stand development of both planted and direct-seeded stands need to be better understood. Stand density and age structure can be affected by silvicultural practice, requiring research on tending methods.

Natural invasion can be included in afforestation schemes as a low-cost alternative. Tending and thinning, however, may be needed to overcome poorly stocked or overly dense stands. Supplemental planting may be needed to increase diversity. Research is needed to describe conditions for safely relying on natural invasion and to estimate the stand dynamics and development pathways. However, the experimental work is quite a complex task because of a variety of conditions for stand establishment.

Monocultures, especially of conifers such as Norway spruce, have been recognized as problematic in all Baltic States and the need for diverse restored forest is understood. Nevertheless, there is little experience with converting conifers to hardwoods except by natural regeneration of birch. Perhaps the experience gained with afforestation can help us make more informed regeneration decisions in forest plantations.

Afforestation stands will increasingly be regarded as essential for meeting national obligations for biodiversity conservation, carbon sequestration, and providing other ecological services. Research is needed to better understand these processes and to quantify benefits. One of the primary tasks should be the collection of existing information into a database. The adequate documentation of afforestation results can improve the basis for decisions.

References

Ahti, T., Hämet-Ahti, L., and Jalas, J., Vegetation zones and their sections in Northwestern Europe, *Ann. Bot. Fennici*, 5, 169, 1968.

Anon., *Majanduslicult väheväärtuslike lehtpuualade hooldamise ja rekonstrueerimise vajaduse analüüs (Economic Analysis of Reconstruction of Hardwood Areas of Low Commercial Value)*, Project report, Estonian Forest Survey Centre, Tallinn, Manuscript, 2000 (in Estonian).

Anon., *Forest Sector in Latvia*, Ministry of Agriculture of the Republic of Latvia, http://www.zm.gov.lv/data/forest_sector_2002_web.pdf, 2002.

Bazzaz, F.A., *Plants in Changing Environments. Linking Physiological, Population, and Community Ecology*, Cambridge University Press, Cambridge, 1998.

Danusevicius, D., personal communication, 2003.

Daugaviete, M., Afforestation of agricultural lands in Latvia, in *NEWFOR — New Forest for Europe: Afforestation at the Turn of the Century*, Weber, N., Ed., EFI Proceedings No. 35, 175, 2000.

Daugaviete, M., *Research Results on the Afforestation of Surplus Farmland in Latvia*, Finnish Forest Research Institute, Research Paper 847, 96, 2002.

Fujimori, T., *Ecological and Silvicultural Strategies for Sustainable Forest Management*, Elsevier, Amsterdam, 2001.

Gabrilavičius, R., *Review and Assessment of Lithuanian Experience with Afforestation of Abandoned Agricultural Land*. Lithuanian Forest Research Institute, Project No. 055-17, Manuscript, http://miskai.gamta.lt/agro/reports.htm, 1999.

Gaizutis, A., The role of forestry in the economy of Lithuania, in *Social Sustainability of Forestry in the Baltic Sea Region*, Hytönen, M., Ed., The Finnish Forest Research Institute, Research Paper 704, 111, 1998.

Hynönen, T., *Pellometsitysten onnistuminen Itä-Suomess (Field Afforestation Results in East-Finland)*, The Finnish Forest Research Institute, Research Paper 765, 2000 (in Finnish).

Jõgiste, K., Vares, A., and Sendrós, M., Restoration of former agricultural fields in Estonia: comparative growth of planted and naturally regenerated birch, *Forestry*, 76, 209, 2003.

Kaar, E., Coniferous trees on exhausted oil shale opencast mines, *Metsanduslikud Uurimused (Forestry Studies)*, XXXVI, 125, 2002.

Kaimre, P., Väheväärtuslike põllumajandusmaade metsastamise ökonoomilised aspektid, Summary: Economic aspects of afforestation on abandoned agricultural lands, in *Proceedings of Estonian Academical Forestry Society*, 14, 68, 2001.

Maamets, L., *Metsastamistoetuste põhjendatuse finantsmajanduslik analüüs (The Financial and Economic Analysis as a Basis for Subsidies of Afforestation)*, Forest Inventory Bureau Ltd., Tallinn, Manuscript, 2003 (in Estonian).

Madsen, P. and Löf, M., Direct seeding of *Fagus sylvatica* L., *Quercus robur* L., *Prunus avium* L. and *Craetaegus monogyna* Jacq. on clearfelled sites and under coniferous shelterwoods, in *Forest Ecosystem Restoration. Proceedings of the International Conference*, Hasenauer, H., Ed., Vienna, Austria, 10 – 12 April, *2000*, 357, 2000.

Niemistö, P., Influence of initial spacing and row-to-row distance on the crown and branch properties and taper of silver birch (*Betula pendula*), *Scand. J. Forest Res.*, 10, 235, 1995.

Suchockas, V., Seed dispersal and distribution of silver birch (*Betula pendula*) naturally regenerating seedlings on abandoned agricultural land at forest edges, *Baltic Forestry*, 8, 71, 2002.

Tamm, Ü., *Haab Eestis (Aspen in Estonia)*, Eesti Loodusfoto, Tartu, 2000 (in Estonian; English summary).

Tullus, H., Põllumajandusmaade metsastamine ja metsastumine (Natural and artificial afforestation of abandoned agricultural lands), *Agraarteadus (J. Agric. Sci.,)*, 11, 22, 2000 (in Estonian; English summary).

Tullus, H., Lehtpuupuitute kasvatamine (Growing broadleaves), in *Proceedings of the Academical Forestry Society*, XIV, 5, 2001 (in Estonian; English summary).

Tullus, H. and Uri, V., Baltic afforestation, in *Proceedings of the IUFRO Conference on Restoration of Boreal and Temperate Forests — Documenting Forest Regeneration Knowledge and Practices in Boreal and Temperate Ecosystems*, Gardiner, E.S. and Breland, L.J., Eds., Danish Centre for Forest, Landscape and Planning, 48, 2002.

Uri, V., *The Dynamics of Biomass Production and Nutrient Status of Grey Alder and Hybrid Alder Plantations on Abandoned Agricultural Lands*, Dissertationes Scientarium Naturalium Universitatis Agriculturae Estoniae IX, 2001.

Vares, A., Jõgiste, K., and Kull, E., Early growth of some deciduous tree species on abandoned agricultural lands in Estonia, *Baltic Forestry*, 7, 52, 2001.

Vares, A., Tullus, A., and Raudoja, A., *Hübriidhaab. Ökoloogia ja majandamine. (Hybrid aspen. Ecology and Management)*, Estonian Agricultural University, Tartu, 2003 (in Estonian; English summary).

chapter fifteen

Restoration of bottomland hardwood forests in the Lower Mississippi Alluvial Valley, U.S.A.

Emile S. Gardiner
US Department of Agriculture Forest Service
James M. Oliver
Ducks Unlimited, Inc.

Contents

15.1 Introduction

The world's third largest river, the Mississippi, extends more than 3700 km through the central U.S. where it drains 41% of the conterminous U.S. before reaching the Gulf of Mexico. The river historically served as a primary travel corridor and trade route for Native American Indians, provided access for exploration and colonization to early European settlers, and today remains the backbone of commerce for much of the central U.S. The southernmost reach of the river is flanked by a 10 million ha physiographic region, the Lower Mississippi Alluvial Valley (LMAV). The rich alluvial soils in this floodplain historically supported vast expanses of mixed-species, deciduous forests. These bottomland hardwood forests are unique because their species composition and structure are linked to the past and current dynamics of the fluvial processes of the river. Inherently diverse in tree, shrub, and vine species, bottomland hardwood forests provide habitat for a rich fauna, function to abate floodwater and improve water quality, produce an exceptional timber commodity, and offer an abundance of recreational opportunities.

The aims of this chapter are to describe the LMAV and its forests, illustrate land-use history and current extent of forests, and summarize forest restoration activities in the region. Although many of the existing forests in the LMAV have been degraded through indiscriminate timber harvests, alterations to hydrologic regimes, or improper management, the focus of this chapter will be on restoration of deforested areas. Readers interested in management or rehabilitation of existing bottomland hardwood forests are referred to Hodges (1995, 1998) and Kellison et al. (1998).

15.2 *Physiography, climate, soils*

The LMAV is shared by seven states, which include Illinois, Missouri, Kentucky, and Tennessee, with the largest portion (over 85%) situated in the tri-state area of Arkansas, Mississippi, and Louisiana (Figure 15.1). The region roughly extends between 29° and 37°N latitude, and ranges between 89° and 92°W longitude. The LMAV is situated in a humid, subtropical region of the Northern Temperate Zone where annual precipitation ranges between 1000 and 1600 mm. January temperatures range from 3°C in the northern reach, 7.5°C in the central, and 11°C in the southern reach of the region, while temperatures in July average about 27°C across the region, with daily highs often exceeding 38°C (Gurley 1976; Matthews 1980; Scott and Carter 1962). Growing season length is typically 185 days in the northern portion of the LMAV, 229 days in the central, and as long as 245 days in the southern portion of the region (Gurley 1976; Matthews 1980; Scott et al. 1975).

Geological processes giving rise to the dissection of the LMAV began as early as 2 million years before present (ybp) during the late Tertiary to early Quaternary Periods (Saucier 1994). Advances and retreats of several glacial stages cycled during the Quaternary Period, although only the very most northern portion of the LMAV may have contacted glacial ice. Nevertheless, glacial advances generally resulted in dissection and widening of the LMAV, while glacial retreats typically led to alluviation of the LMAV (Saucier 1984). Alluvial deposits of the Holocene Epoch (beginning c. 10,000 ybp) characterize nearly 46% of the current surface

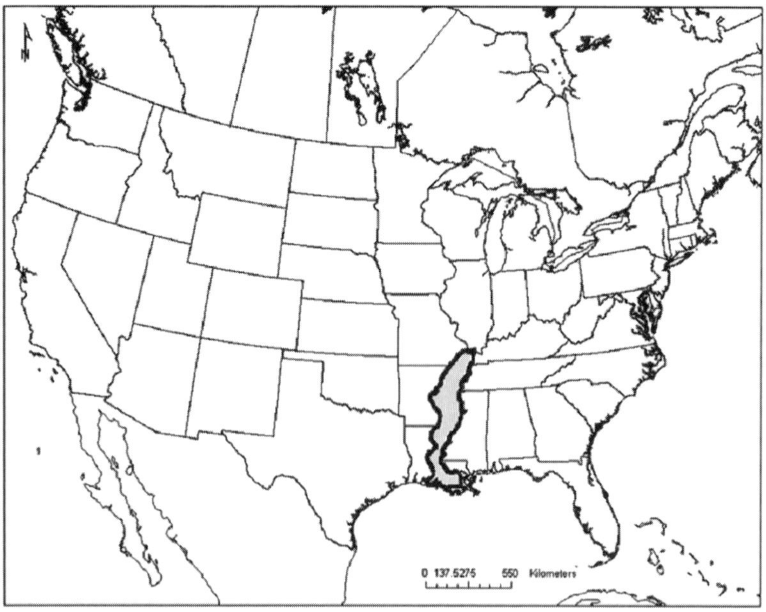

Figure 15.1 Location and extent of the LMAV in North America.

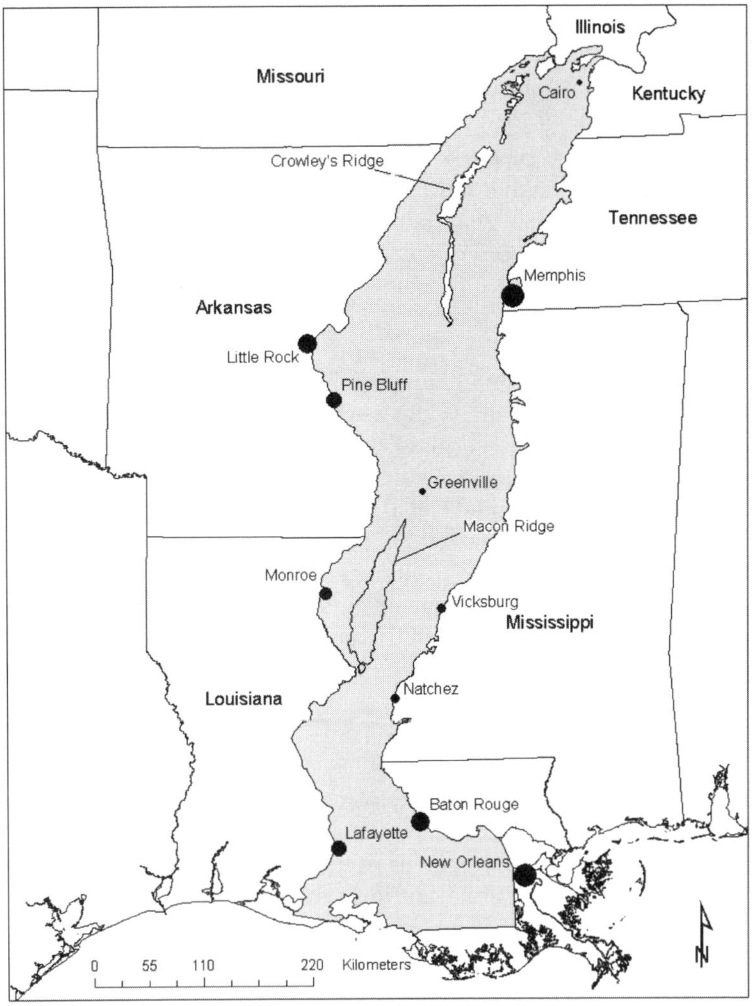

Figure 15.2 Prominent physiographic and urban areas in and adjacent to the LMAV.

geology of the LMAV, with other prominent terrace features such as Maçon Ridge (early Wisconsin Stage, c. 70,000 ybp) dating to much earlier geologic times (Figure 15.2). Crowley's Ridge, a prominent inclusion in the LMAV, is a significantly older upland remnant that is thought to have divided earlier valleys of the Mississippi and Ohio Rivers (Saucier 1994).

During the last 10,000 years, fluvial processes in the LMAV created geomorphologic features that are recognized today in the Holocene deposits of the floodplain. Characteristic features in former meander belts of the LMAV include ridge and swale topography, as well as abandoned channels or oxbow lakes that exhibit a range of alluviation (Saucier 1984). Alluvium from as far away as the Rocky Mountains, the northern Appalachian Mountains, and the upper Midwestern U.S. provides parent material for soil in the LMAV (Stanturf and Schoenholtz 1998). Fluvial processes of the river guide the distribution of these parent materials as it meanders through or inundates the floodplain. Due to the dynamics of fluvial processes, textures of surface soil horizons range from coarse sands to fine clays. Coarse soil particles (sands) are generally deposited near the channel, medium-textured particles (very fine sand and silt) are deposited away from the channel, and fine-textured soil particles (clays) are deposited in backwater areas where stream energy is lowest (Hodges 1997; Stanturf and Schoenholtz 1998). The activity of these alluvial soils ranges from alkaline to acid, with relatively young soils exhibiting the highest pH. Vertisols, Inceptisols, Entisols, and Alfisols are among the most prevalent soil orders developing from the alluvial substrates in the LMAV.

The importance of fluvial processes of the Mississippi River to the landscape of the LMAV was tempered by extensive anthropogenic modification during the 20th century. Engineering projects designed to maintain navigability of the Mississippi River and its tributaries, protect populated areas from flooding, and secure land for agriculture production altered the natural hydrologic regime and reduced the connectivity of the Mississippi River to its floodplain (Smith and Winkley 1996). Much of the alluvial valley no longer functions as an active floodplain since the river is largely contained within an artificial levee network. Documented alterations to the hydrologic regime and fluvial processes include an increase in the water surface slope and stream power of the main channel, reduced erosion and accretion adjacent to the main channel, alteration to river stage fluctuations, and a reduction in the magnitude and duration of high- and low-flow events (Biedenharn et al. 2000; Franklin et al. 2003). These and other engineered modifications to the channel and floodplain impact critical processes, such as lateral channel migration, flood pulses, and point bar and cutoff formation, which drive biogeochemical cycles in low-gradient alluvial floodplain ecosystems. Ecological implications of these engineering activities radiate from modification of the natural disturbance regime of the river and are potentially manifest throughout the LMAV ecosystem.

15.3 The Forest

Bottomland hardwood forest types have predominated throughout the LMAV since about 5000 ybp up through the present (Delcourt et al. 1980; King and Allen 1977). The recent rise in dominance of deciduous tree species in the LMAV followed the end of a 3500-year dry period (the Hypsithermal Period), during which grassland and marsh species were the predominant flora (King 1981; King and Allen 1977). By examining documentation of early naturalists and field notes of surveyors working in the LMAV, some authors have attempted to reconstruct the structure of pre-European settlement bottomland hardwood forests (Foti 2001; Ouchley et al. 2000). However, problems with note interpretation, sample methods, and species taxonomy limit the strength of these descriptions. Tanner (1986) published perhaps the most cited description of an ancient bottomland hardwood forest of the LMAV. His work was based on a vegetation sample collected five decades prior to the publication of his findings. Of the more than 60 tree species endemic to bottomland hardwood forests of the LMAV (Little 1971; Putnam et al. 1960), dominant species include sweetgum (*Liquidambar styraciflua* L.), green ash (*Fraxinus pennsylvanica* Marsh.), baldcypress (*Taxodium distichum* [L.] L.C. Rich.), sugarberry (*Celtis laevigata* Willd.), maples (*Acer* spp.), American sycamore (*Platanus occidentalis* L.), water tupelo (*Nyssa aquatica* L.), eastern cottonwood (*Populus deltoides* Bartr. ex Marsh.), black willow (*Salix nigra* Marsh.), elms (*Ulmus* spp.), hickories (*Carya* spp.), and at least nine species of oak (*Quercus* spp.) (Putnam et al. 1960). In addition to a highly diverse composition of overstory species, bottomland hardwood forests are characteristically rich in woody vines and shrubs, and the understory may feature large monocots such as switchcane (*Arundinaria gigantea* [Walter] Muhl.) and palmetto (*Sabal minor* [Jacquin] Persoon). The composition and density of midstory and understory layers, however, are generally determined by hydrologic regimes with the best-drained, less frequently flooded sites exhibiting the rankest vegetation.

Many tree species of bottomland hardwood forests are segregated in associations that are distinct to particular floodplain site types (Putnam 1951; Putnam and Bull 1932; Putnam et al. 1960; Shelford 1954; Tanner 1986) (Figure 15.3). This stratification results from species

Figure 15.3 Examples of species–site associations that can be found in bottomland hardwood forests of the LMAV: (A) black willow growing on newly accreted land on a bar adjacent to the river channel; (B) baldcypress and water tupelo growing in a swamp; (C) green ash, American elm, sugarberry, and Nuttall oak growing on a low flat; and (D) water oak and sweetgum growing on a ridge.

physiology interacting with the edaphic and hydrologic conditions inherent to each site type. As a result, the abundance of a particular species in a given forest is determined mainly by the availability of the particular site or sites on which it thrives, and the abundance and distribution on the floodplain of these sites are a function of fluvial processes of the river.

The work of Tanner (1986), Putnam and Bull (1932), Putnam (1951), and Putnam et al. (1960) provides an insight into species associations and distribution in natural bottomland hardwood forests, but they fail to fully illustrate the complexity of bottomland hardwood forest structure across the LMAV landscape prior to European settlement, particularly in estimating the extent of anthropogenic disturbance. Hamel and Buckner (1998) recognized that cultural activities of Native Americans impacted the structure and composition of the early forests, and argued that most historical descriptions were based on observing forests that developed after the decline of extensive Native American cultural activities due to population crashes as a result of introduced disease. The earliest evidence of humans in the LMAV dates back to nomadic hunters of the late Pleistocene (12,000 – 11,000 ybp) (Brain 1971). However, the period 9,000 – 8,500 ybp marked the expansion of grasslands during the Hypsithermal, and the apparent influx of a more localized bison-hunting, plains Indian culture into the LMAV.[1] A shift from grassland vegetation to forest, fostered by climate change subsequent to the Hypsithermal, accompanied more sedentary occupation of the LMAV by Native Americans. Archeological evidence indicates that apparently sedentary human societies were established in the LMAV prior to 5,000 ybp (Connaway 1977; Smith 1986). One of the most studied cultures, the extensive Poverty Point Culture,

[1] Presented in 1996 by S. O. Brookes and M H. Reams at the 61st Annual Meeting of the Society for American Archaeology, New Orleans, LA.

was the epicenter of economy and technology throughout the LMAV during the period 3,700 to 3,300 ybp (Gibson 1999). This culture comprised several societies that inhabited numerous permanent villages ranging in size from less than 1 ha to more than 40 ha, all linked by the Mississippi River and its tributaries (Gibson 1999). One might surmise that a culture as extensive as the Poverty Point Culture would have cleared extensive areas of the LMAV to sustain its societies with agriculture, but it is believed that societies based on agriculture did not inhabit the LMAV until more recent times (c. 900 ybp) (Fritz and Kidder 1993). Nevertheless, the vast network of these early villages undoubtedly had a significant influence on the structure of bottomland hardwood forests in the LMAV. Science has yet to reveal the actual extent and distribution of clearing, burning, and cultivation by prehistoric man in the LMAV. It appears that Native American disturbance to the forests of the LMAV rose dramatically and apparently peaked in the period between 3,200 and 1,350 ybp (Galloway 1994). Thus, the existing bottomland forests of the LMAV have apparently always received anthropogenic disturbance.

Land-use practices that removed forest cover accelerated in the LMAV with the onset of European settlement in the 1700s. The settlers found very fertile soils that could be cleared and drained for growing agricultural crops, particularly cotton (*Gossypium hirsutum* L.). Waves of settlement and the demand for farmland in this region were such that by the early 1900s only about half of the estimated original forest remained (MacDonald et al. 1979; Stanturf et al. 2000). Additional deforestation throughout the region was facilitated in the early 1900s by the expansion of rail transportation through the region and by the Flood Control Act of 1928, federal legislation that authorized construction of the mainline levee system (Kelley 1963; MacDonald et al. 1979). This 2500 km system, which supplanted earlier local and state attempts at levee construction and maintenance, straddles the river channel and prevents overland flooding of the Mississippi River onto much of its original floodplain. Deforestation continued throughout the 20th century as the levee system was constructed and clearing peaked in the 1960s and 1970s as the global market demand for soybean (*Glycine max* [L.] Merrill) escalated (MacDonald et al. 1979; Sternitzke 1976).

Drainage and land clearing, primarily for agricultural production, has reduced forest cover in the region to about 26% of the original extent (Figure 15.4). Settlement and deforestation began on the highest and driest sites, and progressed towards lower and wetter sites (Kelley 1963). Indeed, the well-drained riverfront soils were particularly favored for cotton production because of their arable qualities and proximity to water transportation (Winters et al. 1938). As with land impacted by cultural activities of Native Americans, land deforested for agriculture by European settlers sometimes reverted back to forest cover after abandonment for crop production (Hudson 1979). For example, significant areas of abandoned farmland reverted to bottomland forests in the northeastern Louisiana portion of the LMAV during the Civil War (1861–1865) and the Great Depression of the 1930s (Winters et al. 1938). The extent of land-use cycling over time in the LMAV has not been critically assessed. Yet, an example of the magnitude can be found in Winters et al. (1938), who during 1933–1934 documented that 6.5% of agricultural land in the northeastern Louisiana portion of the LMAV was either abandoned or idle.

Of the estimated 2.6 million ha of bottomland forest existing in the LMAV today, the largest contiguous tracts remain primarily where land is not protected from flooding, and most of the current bottomland forests exist on land too wet to sustain profitable agricultural production. Examples of large forested tracts remaining in the LMAV include the Atchafalaya Basin, Delta National Forest, Tensas National Wildlife Refuge, and White River National Wildlife Refuge. The Atchafalaya Basin (c. 240,900 ha) in south Louisiana contains the greatest amount of forested land remaining in the LMAV. The basin is largely privately owned, but portions are maintained by the U.S. Army Corps of Engineers as floodways to divert water from the main channel of the Mississippi River when it

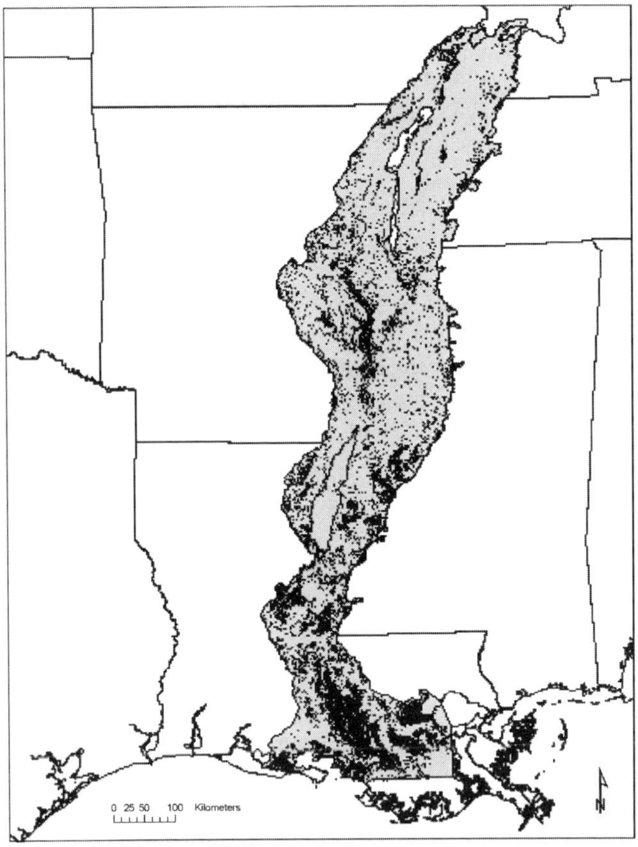

Figure 15.4 Status of deforestation and forest fragmentation in the LMAV.

approaches flood stage. The Delta National Forest (c. 25,000 ha), Tensas National Wildlife Refuge (c. 22,000 ha), and White River National Wildlife Refuge (c. 65,000 ha) represent the largest public holdings of bottomland hardwood forests in the LMAV.

Although deforestation in the LMAV has nearly halted and forest restoration predominates current land-use changes, several factors continue to contribute to degradation of functions and values associated with the remaining bottomland hardwood forests. Alterations to hydrologic regimes, deposition of agricultural runoff, invasion by exotic plant species, indiscriminate timber harvesting, and forest fragmentation are some of the factors contributing to continuing degradation of existing forests (Kellison et al. 1998; Rudis 1995). Afforestation of deforested land in the LMAV not only promises to restore forest functions and values directly but will also indirectly reduce the severity of degradation of existing forest tracts.

15.4 *Current land ownership and use in the LMAV*

A significant majority of the 10 million ha land base in the LMAV is under private ownership, with federal, state, or local governments retaining ownership of a much smaller percentage of the area. Major land-use categories in the LMAV include agriculture (> 65%), forests (26%), open water (4%), municipalities (2%), and aquaculture (1%). Cotton, corn (*Zea mays* L.), soybeans, rice (*Oryza sativa* L.), sorghum (*Sorghum vulgare* Persoon), and wheat (*Triticum aestivum* L.) are the primary agricultural crops raised in the LMAV. Of the forestland in the LMAV, 77% is privately owned, leaving 23% under public ownership.

Metropolitan areas in or adjacent to the LMAV include New Orleans, LA (population 1,305,479), Baton Rouge, LA (population 578,946), Lafayette, LA (population 377,238), Monroe, LA (population 146,627), Little Rock, AR (population 559,074), Pine Bluff, AR (population 80,785), and Memphis, TN (population 1,105,058). Additionally, a substantial population exists in rural areas and smaller towns and cities established throughout and on the fringes of the LMAV. A range of these population centers includes Cairo, IL (population 4,846), Natchez, MS (population 18,732), Vicksburg, MS (population 26,407), and Greenville, MS (population 46,000) (Figure 15.2). The proximity of metropolitan areas and the network of developed urban and rural communities within the LMAV have created a significant demand on forest and aquatic resources of the region. Fishing and hunting are primary forms of recreation provided on public property, and private landowners can readily draw a premium by leasing hunting and angling rights. Because of the high demand from urban residents, management of bottomland hardwood forests for quality wildlife and fish habitat often supersedes management for other objectives such as timber production.

15.5 Restoration policy, goals, and objectives

Beginning in the late 1960s, forest restoration efforts instituted on federal or state property were aimed to reestablish forest cover for wildlife habitat (Haynes and Moore 1988; Savage et al. 1989). The primary objective for most of these early plantations was to provide a hard mast, primarily oak, component on the site, because acorns are favored in the diet of many wildlife species found in bottomlands, and the seeds are not readily dispersed (Allen 1997; Newling 1990). Goals achieved by these initial plantings usually did not extend beyond the boundaries of public property. As the early plantations aged, and forest restoration gained political support, a greater array of restoration objectives and goals emerged from a growing awareness of the unique functions and values of bottomland hardwood forests of the LMAV. Current objectives are linked to the particular missions of the conservation agencies and private organizations implementing forest restoration programs, and established goals often encompass issues relevant to the entire LMAV or beyond.

Several key federal government actions during the 1980s led to the current policy in the LMAV emphasizing conservation of existing forested wetlands and facilitating forest restoration on private land (Haynes and Moore 1988). Included in this legislation was Section 906 of the Water Resources Development Act of 1986, which required mitigation of bottomland hardwood forest impacted by Federal water projects (Haynes and Moore 1988). However, widespread restoration of bottomland hardwood forests on private land began with the approval of federal legislation, including the Food Security Act of 1985 and the Food, Agriculture, Conservation, and Trade Act of 1990 (Loesch et al. 1995). These and subsequent "Farm Bills" have encouraged afforestation of agricultural land through the creation and funding of the Conservation Reserve Program and particularly the Wetland Reserve Program (Kennedy 1990; Loesch et al. 1995).

Numerous public agencies currently play key roles in promoting, facilitating, and implementing afforestation in the LMAV. The Wetland Reserve Program is administered through the U.S. Department of Agriculture, Natural Resource Conservation Service, with the objective of restoring and protecting farmed wetlands, and a primary goal of "achieving the greatest wetland functions and values, along with optimum wildlife habitat, on every acre enrolled in the program." Likewise, the Conservation Reserve Program, administered through the U.S. Department of Agriculture, Farm Service Agency, targets protecting "the Nation's long-term capability to produce food and fiber; to reduce soil erosion and sedimentation, improve water quality, and create a better habitat for wildlife." The U.S. Fish and Wildlife Service is another federal agency pivotal in forest restoration throughout the LMAV. Their primary mission is to provide for and maintain habitat for migratory birds. In addition to work conducted on the National Wildlife Refuge System, the Service promotes forest restoration for wildlife habitat by facilitating structured partnerships, including the Lower Mississippi Valley Joint Venture and the Southeast Regional Working Group of Partners in Flight (Loesch et al. 1995). The Southeast Regional Working Group of Partners in Flight, for example, has established a strategy to restore forested patches that would provide critical habitat for forest-breeding birds (Twedt et al. 1999). The U.S. Army Corps of Engineers has secured and reestablished forest cover on extensive tracts of agricultural land in the LMAV to mitigate impacts to bottomland hardwood forests sustained through various development projects, particularly flood control. Additionally, the U.S. Forest Service has begun promoting forest restoration in the LMAV by organizing a partnership program designed to integrate federal, state, and nongovernmental organizations as partners in afforestation (Young et al. 2002). The goal of this partnership program, which was organized in 2000, is to restore forest cover to 800,000 ha within the LMAV over a 20-year period. State agencies continue to play a fundamental role in forest restoration efforts throughout the LMAV. State conservation agencies in all states of the LMAV promote forest restoration on public land, and most are active members of conservation partnerships.

In addition to the forest restoration activities of public entities, private interests are proving essential in increasing forest restoration opportunities and management alternatives for landowners. Landowner assistance programs offered by forest industry in the LMAV often provide technical assistance with afforestation efforts, and provide landowners an opportunity to draw revenue from investments in forest plantation establishment. Private interest in forest restoration promises to increase in the LMAV as a market for carbon sequestration develops in the U.S. Several utility cooperatives, speculating that a market for carbon offsets will soon develop, have already invested in afforestation.

15.6 Previous restoration and extent of current restoration needs

Documented forest restoration efforts in the LMAV date back to the 1940s with the establishment of experimental hardwood plantations to "reclaim abandoned submarginal agricultural land" (Maisenhelder and McKnight 1962). USDA Forest Service researchers working at the Stoneville Experimental Forest, Washington County, Mississippi, initiated these early experiments on forest restoration techniques. Early plantations, such as those established on the Stoneville Experimental Forest, are probably scattered through the LMAV, but they were established infrequently, poorly documented, and were of minor significance in relation to the extent of deforestation occurring at the same time.

The earliest sustained forest restoration effort in the LMAV probably dates back to the late 1960s, with the establishment of oak plantations on former agricultural fields. The majority of this effort focused on public property recently acquired from the private sector by federal and state natural resource management agencies, such as the U.S. Fish and Wildlife Service and the Louisiana Department of Game and Fish. The U.S. Fish and Wildlife Service, for example, upon acquiring agricultural land in their National Refuge System, began a concerted effort to restore forest cover in order to improve wildlife habitat (Haynes and Moore 1988; Allen 1990). This afforestation effort continued primarily on public holdings until the 1990s, when governmental incentives were instituted to encourage private landowners to remove land from agricultural production (Figure 15.5). Over the past three decades, forest restoration in the tri-state LMAV portions of Louisiana, Mississippi, and Arkansas has increased dramatically, leading to a significant removal of land from agricultural production by establishing hardwood plantations. Current esti-

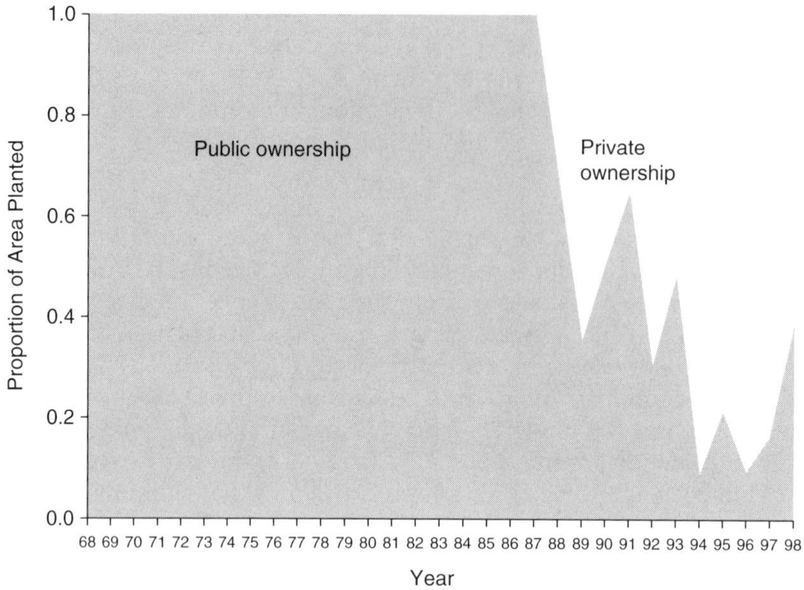

Figure 15.5 Relative proportion of afforestation supported by State or Federal appropriated dollars practiced on public vs. private land in the Louisiana, Mississippi, and Arkansas portions of the (LMAV) (unpublished data filed by S.H. Schoenholtz at the Department of Forestry, Mississippi State University, Mississippi State, MS, U.S.A.).

mates of afforested land in the LMAV approach 194,000 ha. With this increase, the effort has shifted from a predominance of afforestation on public land to predominantly private land (Schoenholtz et al. 2001) (Figure 15.5). In the tri-state area of Louisiana, Mississippi, and Arkansas, an estimated 75% of the acreage converted to forest cover as of 1999 is under private ownership (unpublished data filed by S.H. Schoenholtz at the Department of Forestry, Mississippi State University, Mississippi State, MS, U.S.A.).

Since federal conservation programs that provide the impetus for afforestation target economically marginal farmland, it is the wettest soils that are generally removed from agricultural production. If a proposed tract of land meets qualification requirements for a particular conservation program, participation is based on ranking of the land according to an environmental benefits index (see www.fsa.usda.gov for Conservation Reserve Program guidelines). Environmental benefit indices typically generate a ranking by factoring wildlife habitat benefits, water quality benefits, air quality benefits, and other conservation priorities associated with the land and proposed restoration practice.

In addition to the almost 200,000 ha already removed from agricultural production, much additional land appears available to be restored to forest cover. In 2002, Louisiana, Mississippi, and Arkansas had 95,000 ha that qualified for easements under the Wetland Reserve Program and were awaiting funding. Provisions of future Farm Bills will determine to a large degree the amount of land that will be enrolled in the future. Additionally, the development of a market for carbon sequestration could potentially influence future land-use trends involving forest cover in the LMAV.

15.7 Restoration techniques

Guidelines for restoring bottomland hardwood forests have been published by Allen and Kennedy (1989) and Allen et al. (2001), providing basic information to landowners interested in establishing forest cover on former agricultural land. These guides are useful starting points to gain a knowledge of recommended afforestation practices, commonly used materials, and general costs. Additional information on current incentive programs and enrollment procedures is provided in local field offices of federal and state conservation agencies, particularly field offices of the Natural Resource Conservation Service.

The general approach to restoring forest cover on former agricultural land in the LMAV has been extensive rather than intensive application of plantation establishment practices. This extensive approach of afforesting to most areas at the least cost has been fostered by the agencies that administer the governmental incentive programs that target the removal of land from agricultural production. The relatively large size of enrolled easements also contributes to this approach.

Site preparation practices employed for afforestation vary by conservation organization, but are generally minimal. If plantations are to be established on land immediately removed from agricultural production, site preparation is generally not practiced (Gardiner et al. 2002). One exception is that subsoil plowing is sometimes applied to break up plow pans and facilitate seedling planting. Herbicide applications for weed control and fertilizer applications to improve the nutrient status of soils are rarely practiced. Where fields have been removed from agricultural production for several years prior to plantation establishment, site preparation typically involves mowing, primarily as a method of facilitating the planting job. Disking is also used to break up heavy herbaceous vegetation, and provides the additional benefit of reducing rodent habitat (Gardiner et al. 2002).

Early plantations established in the LMAV were primarily bottomland oak plantings. The various oak species were deployed because most of the early research on plantation establishment in the LMAV was conducted on oak species, and oaks are favored for their timber and wildlife habitat value. Furthermore, it was believed that if heavy-seeded oaks were established, light-seeded species would naturally invade the site (Allen 1990). This has not proven true where mature stands of light-seeded species are not immediately adjacent to the restoration site (Stanturf et al. 2001). In current practice, species assignments are based on soil types, with mixtures of 2 to 3 overstory species established on each soil type. Thus, combinations of 8 to 10 species may be planted in a single afforestation project that encompasses several different soil types. Although current afforestation practices encourage establishment of a greater diversity of species than past practices, species mixtures are not established based on compatibility (Gardiner et al. 2002). Additionally, little effort has been directed toward developing reliable techniques for establishing the midstory and understory species endemic to bottomland hardwood forests.

Bottomland hardwood plantations can be established with various types of stock depending on the target species. Cuttings can be used for vegetative reproduction of fast-growing species such as black willow and eastern cottonwood (Stanturf et al. 1998). Direct seeding provides a cost-effective method of establishing hard-mast or heavy-seeded species such as the oaks and hickories (Bullard et al. 1992; Lockhart et al. 2003; Stanturf et al. 1998), while seedlings, bareroot or container, can be deployed to establish any bottomland hardwood species. The primary stock type used in afforestation in the LMAV is a 1-0 bareroot seedling. King and Keeland (1999) reported that over 60% of the plantations established before 1997 were established with bareroot seedlings, and this percentage has continued to increase as conservation agencies have greatly reduced the use of direct seeding (Schoenholtz et al. 2001). Bareroot seedlings are typically grown from unimproved seed, and the seedlings are produced and sold by state and private nurseries. Most agencies place a minimum standard on the size of bareroot seedlings suitable for outplanting; seedlings generally have to be greater than 45 cm tall with a root-collar diameter of 10 mm. These seedling quality standards are primarily based on ease of handling and are not reflective of an optimal seedling size for survival or growth (McKnight and Johnson 1975).

On the majority of afforestation sites in the LMAV, planting density is typically set at 746 seedlings per ha. The selection of this planting density is quite arbitrary, but it is expected to produce at least 309 surviving seedlings per ha at year three. Although harsh site factors such as excessive flooding, herbivory, and heavy competition can reduce survival, planting density is generally not altered to account for potential risks of these factors. Seedlings can be hand or machine planted equally as successful so long as a conscientious planting crew is employed (Gardiner et al. 2002; Schoenholtz et al. 2001) (Figure 15.6), but machine planting is generally limited to relatively drier soils that can sustain machine traffic. The density of sown seed is typically much greater than the density of planted seedlings and generally targets about 2,470 to 3,700 seed per ha. Direct seeding is generally accomplished with modified agricultural seeders.

Although postplanting cultural practices have demonstrated increased survival and growth of hardwood seedlings, current forest restoration activities generally do not incorporate such practices as weed control, seedling protection from herbivory, and fertilization (Gardiner et al. 2002). Silvicultural practices that increase seedling growth could reduce mortality risks on flood-prone bottomland sites and sites with high rodent populations.

With the strong emphasis of forest restoration in the LMAV, alternative afforestation practices have been developed that offer landowners the ability to meet several objectives not optimized by conventional afforestation practices. For example, it has been demonstrated that the establishment of a fast-growing pioneer species can be interplanted with slower-growing hardwoods such as oaks (Gardiner et al. 2001). This interplanting system enables the manager to establish a plantation that provides a quick economic return to recoup the costs of afforestation (Stanturf and Portwood 1999), provides a rapid establishment of vertical structure that advances the development of forest bird habitat (Hamel 2003; Twedt and Portwood 1997), and may be beneficial in rapid restoration of soil quality and other ecosystem processes. Another innovative approach can be found in the northern reaches of the LMAV. Some practitioners are deploying "Root Production

Figure 15.6 Hand planting bareroot hardwood seedlings on a former agricultural field in the LMAV.

Method" (RPM) oak seedlings; these are very large container seedlings that have undergone a root development process that speeds maturation and catalyzes the early production of hard mast on the restoration site. The current emphasis on forest restoration in the LMAV will certainly spawn the development of other viable techniques and practices.

15.8 Costs of restoration techniques and expected benefits

Costs of forest restoration practices in the LMAV are commensurate with the level of intensity of the assigned afforestation practices. Low-intensity plantation establishment practices are generally applied on an extensive basis, while high-intensity plantation establishment practices are applied less extensively (Stanturf et al. 2000). The vast majority of afforestation projects implemented in the LMAV, those established through the Conservation Reserve Program and the Wetland Reserve Program, would be considered low intensity, and as such have received minimal input per ha in terms of silvicultural practices and financial investment. Drawbacks associated with relatively low-input, extensively applied restoration programs include a relatively low confidence level of obtaining desired benefits, a potentially longer time needed to obtain desired benefits, or both. Conversely, establishment of intensively managed plantations usually increases the confidence of meeting desired objectives, but requires significant investment in time and money. It is incumbent upon the landowner to define restoration goals, management objectives, and expected outputs for the restoration project so that an optimal approach with minimal risk to obtaining the objectives can be developed.

Approximate costs of various silvicultural practices that may be performed on afforestation sites in the LMAV are listed in Table 15.1. Prices would typically vary for a particular practice based on the size of the job, experience of the contractor and crew, soil types, and condition of the afforestation site. Research has not provided basic cost–benefit

Table 15.1 Approximate Costs of Silvicultural Practices and Materials Used for Afforestation Purposes on Former Agricultural Land in the LMAV

Practice	Cost
Site Preparation	
Mowing	US$27 – 35 per ha
Disking	US$30 – 45 per ha
Subsoil plowing	US$24 – 32 per ha
Herbicide application	US$30 – 37 per ha
Burning	US$12 – 15 per ha
Plant Material[a]	
Seed	US$2 – 5 per kg
Seedlings	US$185 – 210 per thousand
Cuttings	US$200 per thousand
Establishment	
Direct seeding	US$85 – 100 per ha
Planting	US$85 – 110 per ha
Other Practices	
Fertilization[b]	US$37 – 40 per ha
Postplanting herbicide application	US$27 – 37 per ha
Postplanting disking	US$30 – 45 per ha

[a] Seed prices are based on general collections of uncertified oak seed; acorns are generally seeded at 2,470 to 3,700 per ha with 45 to 230 acorns per kg depending on species. Seedling prices are for 1-0 bareroot seedlings from unimproved stock. Cuttings are 45 cm long from superior clones.

[b] Fertilization cost is for a treatment of 90 kg per ha nitrogen injected 45 to 50 cm deep into the soil.

Table 15.2 Wetland Reserve Program Agreement Options and Benefits Available to Landowners in the Mississippi Portion of the (LMAV), 2003

Option	Landowner Obligation[a] (%)	Subsidy[b] (%)
10-year agreement	25	None
30-year agreement	25	75
Perpetual easement	0	100

[a] Landowner obligation is the percentage of the costs of approved afforestation practices that are paid by the landowner. Approved afforestation practices generally do not exceed a total cost of US$300 to 345 per ha.

[b] Subsidy to the landowner is based on the appraised value of the property and is not to exceed US$2,223 per ha.

analyses for these silvicultural practices, but practices that increase seedling vigor and growth may be beneficial in reducing the risk of plantation failure and the need to replant the site. A few researchers have investigated the economics of hardwood plantation establishment on former agricultural fields of the LMAV. Forest restoration on former agricultural land is generally not a viable economic decision, unless the landowner receives cost subsidies provided by federal or state conservation programs (Amacher et al. 1998; Stanturf and Portwood 1999). Subsidies are particularly important economically if the restoration site is predominated by soil of relatively low productivity (Amacher et al. 1998).

A majority of approved afforestation expenses may be defrayed through subsidies provided by governmental conservation programs. The Wetland Reserve Program, for example, provides three options for subsidizing landowners willing to convert agricultural land to forest cover (Table 15.2). Contract options provided through the Wetland Reserve Program range from 10-year agreements to perpetual easements, and reimburse the landowner for 75% to 100% of the costs of approved practices. Additionally, landowners who engage in relatively long-term agreements benefit by receiving cash subsidies based on agreement length and the appraised value of the enrolled land (Table 15.2).

Landowners throughout the LMAV have readily enrolled land in government conservation programs that remove land from agriculture production and establish forest cover. In the tri-state region of Louisiana, Mississippi, and Arkansas, nearly 900 easement contracts were established between 1991 and 2001 through the Wetland Reserve Program. Through this voluntary enrollment, forest cover was established on more than 139,000 ha of former agricultural fields, at a cost of over $230 million.

15.9 Research, management, and policy needs

There is no lack of research, management, or policy needs with respect to restoration of bottomland hardwood forests in the LMAV. Questions abound around topics such as the characterization of suitable planting stocks, sources of appropriate genetic material, development rates of restored ecological functions, development of alternative practices to achieve particular restoration objectives, the appropriate distribution and size of restored forests across the landscape, and proper methods for assessing restoration success. However, economic constraints are probably the primary limitations on the rate, extent, and intensity of current restoration practices in the LMAV. The development of ecologically viable restoration systems that are also economically sustainable would have the potential to shift land-use practices toward a greater balance of forest cover in the LMAV. The bottomland hardwood forest of the LMAV has continued to function and provide valuable resources from the time it migrated into the physiographic region with early man up through the 20th century. The demonstrated resilience to disturbance of this ecosystem over the centuries offers a promise that existing forest restoration practices will provide sustainability to the unique flora and fauna of this magnificent environment into the next century.

Acknowledgments

The GIS Laboratory of the Southern Regional Office, Ducks Unlimited, Inc. provided graphics for Figures 15.1, 15.2, and 15.4. We thank our peers, who provided anonymous reviews of this manuscript, and the editors John Stanturf and Palle Madsen, who provided us with an opportunity to present this material.

References

Allen, J.A., Establishment of bottomland oak plantations on the Yazoo National Wildlife Refuge Complex, *South. J. Appl. For.*, 14, 206, 1990.

Allen, J.A., Reforestation of bottomland hardwoods and the issue of woody species diversity, *Restor. Ecol.*, 5, 125, 1997.

Allen, J.A. and Kennedy, H.E. Jr., *Bottomland Hardwood Reforestation in the Lower Mississippi Valley*, U.S. Department of the Interior, Fish and Wildlife Service, National Wetlands Research Center; U.S. Department of Agriculture, Forest Service, Southern Forest Experiment Station, 28 pp., 1989.

Allen, J.A., Keeland, B.D., Stanturf, J.A., Clewell, A.F., and Kennedy, H.E. Jr., *A Guide to Bottomland Hardwood Restoration*, U.S. Geological Survey, Biological Resources Division Information and Technology Report USGS/BRD/ITR-2000-0011, U.S. Department of Agriculture, Forest Service, Southern Research Station, General Technical Report SRS-40, 132 pp., 2001.

Amacher, G.S., Sullivan, J., Shabman, Zepp, L., and Grebner, D., Reforestation of flooded farmland, *J. For.*, 96, 10, 1998.

Biedenharn, D.S., Thorne, C.R., and Watson, C.C., Recent morphological evolution of the Lower Mississippi River, *Geomorphology*, 34, 227, 2000.

Brain, J.P., *The Lower Mississippi Valley in North American prehistory*, Report submitted to the National Park Service, Southeastern Region, Tallahassee, FL, 104 pp., 1971.

Bullard, S., Hodges, J.D., Johnson, R.L., and Straka, T.J., Economics of direct-seeding and planting for establishing oak stands on old-field sites in the South, *South. J. Appl. For.*, 16, 34, 1992.

Connaway, J.M., *The Denton Site: A Middle Archaic Occupation in the Northern Yazoo Basin, Mississippi*, Mississippi Department of Archives and History, Archaeological Report No. 4, 146 pp., 1977.

Delcourt, P.A., Delcourt, H.R., Brister, R.C., and Lackey, L.E., Quaternary vegetation history of the Mississippi Embayment, *Q. Res.*, 13, 111, 1980.

Foti, T.L., Presettlement forests of the Black Swamp area, Cache River, Woodruff County, Arkansas, from notes of the first land survey, in *Bottomland Hardwoods of the Mississippi Alluvial Valley: Characteristics and Management of Natural Function, Structure, and Composition*, P.B. Hamel and T.L. Foti, Eds., U.S. Department of Agriculture, Forest Service, Southern Research Station, General Technical Report SRS-42, pp. 7–15, 2001.

Franklin, S.B., Wasklewicz, T., Grubaugh, J.W., and Gruelich, S., Hydrologic stage periodicity of the Mississippi River before and after systematic channel modifications, *J. Am. Water Resour. Assoc.*, 39, 637, 2003.

Fritz, G.J. and Kidder, T.R., Recent investigations into prehistoric agriculture in the Lower Mississippi Valley, *South. Archaeol.*, 12, 1, 1993.

Galloway, P., Prehistoric population of Mississippi: a first approximation, *Mississippi Archaeol.*, 29, 44, 1994.

Gardiner, E.S., Russell, D.R., Oliver, M., and Dorris, L.C. Jr., Bottomland hardwood afforestation: state of the art, in *Proceedings of a Conference on Sustainability of Wetlands and Water Resources: How Well can Riverine Wetlands Continue to Support Society into the 21st Century*? M.M. Holland, M.L. Warren, and J.A. Stanturf, Eds., U.S. Department of Agriculture, Forest Service, Southern Research Station, General Technical Report SRS-50, pp. 75–86, 2002.

Gardiner, E.S., Schweitzer, C.J., and Stanturf, J.A., Photosynthesis of Nuttall oak (*Quercus nuttallii* Palm.) seedlings interplanted beneath an eastern cottonwood (*Populus deltoides* Bartr. ex Marsh.) nurse crop, *For. Ecol. Manage.*, 149, 283, 2001.

Gibson, J.L., *Poverty Point, A terminal Archaic Culture of the Lower Mississippi Valley*, Louisiana Archaeological Survey and Antiquities Commission, Department of Culture, Recreation and Tourism, Anthropological Study Series No. 7, 36 pp., 1999.

Gurley, P.D., *Soil Survey of Dunklin County*, Missouri, U.S. Department of Agriculture, Soil Conservation Service, Washington, DC, 137 pp., 1976.

Hamel, P.B., Winter bird community differences among methods of bottomland hardwood forest restoration: results after seven growing seasons, *Forestry*, 76, 189, 2003.

Hamel, P.B. and Buckner, E.R., How far could a squirrel travel in the treetops? A prehistory of the southern forest, in *Transactions of 63rd North American Wildlife and Natural Resources Conference*, Orlando, FL, 20–24 Mar. 1998, pp. 309–315, 1998.

Haynes, R.J. and Moore, L., Reestablishment of bottomland hardwoods within national wildlife refuges in the southeast, in *Increasing Our Wetland Resources: Proceeding of the Conference on National Wildlife Federation — Corporate Conservation Council*, Washington, DC, pp. 95–103, 1988.

Hodges, J.D., The southern bottomland hardwood region and brown loam bluffs subregion, in *Regional Silviculture of the United States*, J.W. Barrett, Ed., 3rd ed., John Wiley & Sons, Inc., New York, pp. 227–269, 1995.

Hodges, J.D., Development and ecology of bottomland hardwood sites, *For. Ecol. Manage.*, 90, 117, 1997.

Hodges, J.D., Minor alluvial floodplains, in *Southern Forested Wetlands, Ecology and Management*, M.G. Messina and W.H. Conner, Eds., Lewis Publishers, Boca Raton, FL, pp. 325–341, 1998.

Hudson, J.C., The Yazoo–Mississippi Delta as plantation country, in *Proceedings Tall Timbers Ecology and Management Conference*, 16, 66, 1979.

Kelley, A., Levee building and the settlement of the Yazoo Basin, *South. Q.*, 1, 285, 1963.

Kellison, R.C., Young, M.J., Braham, R.R., and Jones, E.J., Major alluvial floodplains, in *Southern Forested Wetlands, Ecology and Management*, M.G. Messina and W.H. Conner, Eds., Lewis Publishers, Boca Raton, FL, pp. 291–323, 1998.

Kennedy, H.E. Jr., *Hardwood Reforestation in the South: Landowners can Benefit from Conservation Reserve Program Incentives*, U.S. Department of Agriculture, Forest Service, Southern Forest Experiment Station, Research Note SO-364, 6 pp., 1990.

King, J.E., Late Quaternary vegetational history of Illinois, *Ecol. Monogr.*, 51, 43, 1981.

King, J.E. and Allen, W.H., Jr., A Holocene vegetation record from the Mississippi River Valley, southeastern Missouri, *Q. Res.*, 8, 307, 1977.

King, S.L. and Keeland, B.D., Evaluation of reforestation in the Lower Mississippi River Alluvial Valley, *Restor. Ecol.*, 7, 348, 1999.

Little, E.L. Jr., *Atlas of United States Trees: Volume 1, Conifers and Important Hardwoods*, U.S. Department of Agriculture, Forest Service, Miscellaneous Publication No. 1146, 9 pp., 1971.

Lockhart, B.R., Keeland, B., McCoy, J., and Dean, T.J., Comparing regeneration techniques for afforesting previously farmed bottomland hardwood sites in the Lower Mississippi Alluvial Valley, USA, *Forestry*, 76, 169, 2003.

Loesch, C.R., Twedt, D.J., and Reinecke, K.J., Conservation partnerships in the lower Mississippi Alluvial Valley, *Wildl. Soc. Bull.*, 23, 791, 1995.

MacDonald, P.O., Frayer, W.E., and Clauser, J.K., *Documentation, chronology, and future projections of bottomland hardwood habitat loss in the Lower Mississippi Alluvial Plain.* Vol. 1, U.S. Fish and Wildlife Service, Washington, DC., 133 pp., 1979.

Matthews, D., *Soil survey of Jefferson Parish, Louisiana*, U.S. Department of Agriculture, Soil Conservation Service, Washington, DC, 95 pp., 1980.

McKnight, J.S. and Johnson, R.L., Growing hardwoods in southern lowlands, *For. Farmer*, 34, 38, 1975.

Maisenhelder, L.C. and McKnight, J.S., Southern hardwood research: past–present–future, *Southern Lumberman*, 15 December issue, 1962.

Newling, C.J., Restoration of the bottomland hardwood forest in the Lower Mississippi Valley, *Restor. Manage. Notes*, 8, 23, 1990.

Ouchley, K., Hamilton, R.B., and Barrow, W.C. Jr., Historic and present-day forest conditions: implications for bottomland hardwood forest restoration, *Ecol. Restor.*, 18, 21, 2000.

Putnam, J.A., *Management of Bottomland Hardwoods*, U.S. Department of Agriculture, Forest Service, Southern Forest Experiment Station, Occasional Paper 116, 60 pp., 1951.

Putnam, J.A. and Bull, H., *The Trees of the Bottomlands of the Mississippi River Delta Region*, U.S. Department of Agriculture, Forest Service, Southern Forest Experiment Station, Occasional Paper 27, 207 pp., 1932.

Putnam, J.A., Furnival, G.M., and McKnight, J.S., *Management and Inventory of Southern Hardwoods*, U.S. Department of Agriculture, Agriculture Handbook No. 181, 102 pp., 1960.

Rudis, V.A., Regional forest fragmentation effects on bottomland hardwood community types and resource values, *Landscape Ecol.*, 10, 291, 1995.

Savage, L., Pritchett, D.W., and DePoe, C.E., Reforestation of a cleared bottomland hardwood area in northeast Louisiana, *Restor. Manage. Notes*, 7, 88, 1989.

Scott, F.T. and Carter, R.C., *Soil survey of Sharkey County, Mississippi,* U.S. Department of Agriculture, Soil Conservation Service, Washington, DC, 36 pp., 1962.

Scott, F.T., Walton, L.B., Nail, E.E., and McGehee, V.H., *Soil survey of Yazoo County, Mississippi*, U.S. Department of Agriculture, Soil Conservation Service, Washington, DC, 51 pp., 1975.

Saucier, R.T., *Quaternary Geology of the Lower Mississippi Valley*, Arkansas Archaeological Survey Research Series No. 6, 26 pp., 1984.

Saucier, R.T., *Geomorphology and Quaternary geologic history of the Lower Mississippi Valley,* Vol. 1, U.S. Army Corps of Engineers, Waterways Experiment Station, Vicksburg, MS, 364 pp., 1994.

Schoenholtz, S.H., James, J.P., Kaminski, R.M., Leopold, B.D., and Ezell, A.W., Afforestation of bottomland hardwoods in the Lower Mississippi Alluvial Valley: status and trends, *Wetlands*, 21, 602, 2001.

Shelford, V.E., Some Lower Mississippi Valley Flood Plain biotic communities; their age and elevation, *Ecology*, 35, 126, 1954.

Smith, B.D., The archaeology of the southeastern United States: from the Dalton to de Soto, 10,500 B. P.–500 B. P, in *Advances in World Archaeology*, F. Wendorf and A.E. Close, Eds., Academic Press, Inc., Orlando, FL, pp. 1–92, 1986.

Smith, L.M. and Winkley, B.R., The response of the Lower Mississippi River to river engineering, *Eng. Geol.*, 45, 433, 1996.

Stanturf, J.A. and Portwood, C.J., Economics of afforestation with eastern cottonwood (*Populus deltoides*) on agricultural land in the Lower Mississippi Alluvial Valley, in *Proceedings of the Tenth Biennial Southern Silvicultural Research Conference*, J.D. Haywood, Ed., U.S. Department of Agriculture, Forest Service, Southern Research Station, General Technical Report SRS-30, pp. 66–72, 1999.

Stanturf, J.A. and Schoenholtz, S.H., Soils and landforms, in *Southern Forested Wetlands, Ecology and Management*, M.G. Messina and W.H. Conner, Eds., Lewis Publishers, Boca Raton, FL, pp. 123–147, 1998.

Stanturf, J.A, Schweitzer, C.J., and Gardiner, E.S., Afforestation of marginal agricultural land in the Lower Mississippi River Alluvial Valley, U.S.A., *Silva Fenn.*, 32, 281, 1998.

Stanturf, J.A., Gardiner, E.S., Hamel, P.B., Devall, M.S., Leininger, T.D., and Warren, M.E. Jr., Restoring bottomland hardwood ecosystems in the Lower Mississippi Alluvial Valley, *J. For.*, 98, 10, 2000.

Stanturf, J.A., Schoenholtz, S.H., Schweitzer, C.J., and Shepard, J.P., Achieving restoration success: myths in bottomland hardwood forests, *Restor. Ecol.*, 9, 189, 2001.

Sternitzke, H.S., Impact of changing land use on Delta hardwood forests, *J. For.*, 74, 25, 1976.

Tanner, J.T., Distribution of tree species in Louisiana bottomland forests, *Castanea*, 51, 168, 1986.

Twedt, D.J. and Portwood, J., Bottomland hardwood reforestation for Neotropical migratory birds: are we missing the forest for the trees? *Wildl. Soc. Bull.*, 25, 647, 1997.

Twedt, D.J., Hamel, P.B., Cooper, R.J., and Woodrey, M.S., An evaluation strategy for conservation goals in the Mississippi Alluvial Valley, in *Strategies for Bird Conservation: The Partners in Flight Planning Process*, R. Bonney, D.N. Pashley, R. Cooper, and L. Niles, Eds., http://www.ornith.cornell.edu/pifcapemay/twedt.html, 1999.

Winters, R.K., Putnam, J.A., and Eldredge, I.F., *Forest Resources of the North-Louisiana Delta*, U.S. Department of Agriculture, Miscellaneous Publication No. 309, 49 pp., 1938.

Young, G.L., Hopkins, C., and Rheinhardt, G., Restoring the Delta: a partnership approach to sustainable forest wetland restoration in the Lower Mississippi River Alluvial Valley, U.S.A., in *Proceedings of the IUFRO Conference on Restoration of Boreal and Temperate Forests*, E.S. Gardiner and L.J. Breland, Comps., Vejle, Denmark, 28 Apr.–2 May 2002, Skov and Landskab, Report 11, 238 pp., 2002.

chapter sixteen

Restoration of oak woodlands in Mediterranean ecosystems

Doug McCreary
University of California-Berkeley
Isabel Cañellas
Centre for Forest Research, INIA

Contents

16.1 Introduction

Mediterranean ecosystems are characterized by a lengthy period of hot, dry weather, and in California and southern Europe this dry period usually extends from April until

1-56670-635-1/05/$0.00+$1.50

October. The array of plants that evolved in this environment has adapted to survive and reproduce. Oak woodlands, dominated by an overstory of native *Quercus* species, are common landscape features in both California and southern Europe. These woodlands typically consist of one to several species of native oaks with an understory of predominantly annual plants, principally grasses. Shrubs may or may not be present, depending on past management, environmental conditions, and edaphic features.

The oak woodlands in the dehesa region of Spain and Portugal and the foothills of California are threatened by a lack of natural regeneration and share the need for some intervention to ensure that oaks remain a component of the ecosystem. Such concern has led to efforts to develop restoration practices to ensure that oak woodlands can be sustainably managed.

Nevertheless, significant differences exist between California and the dehesa regions in Spain and Portugal; in the latter, oak woodlands have been intensively managed for centuries. The dehesa system has, and continues to provide, highly valued products, including forage for domestic livestock, habitat favorable for game animals, acorns for fattening hogs, and cork from *Quercus suber* L. trees. To maintain the production of these goods, a well-established system has evolved with proven management guidelines. Still there are concerns, not only for inadequate regeneration but also recent oak removals. In response, there are aggressive efforts to increase woodland cover through government-funded planting programs. In California, on the other hand, the market products derived from oak woodlands are limited to extensive grazing and firewood. Although the oaks are ecologically and aesthetically valuable, active management has, until recently, been concerned more with increasing forage production than restoring degraded habitats.

16.2 California's oak woodlands

16.2.1 Distribution

The approximately 3 million ha of woodlands in California (Bolsinger 1988) are slow-growing hardwood forests unsuitable for timber production. Oak woodlands, also called hardwood rangelands, are located primarily in the foothills of the coastal, western Sierra Nevada, and Transverse mountain ranges (Figure 16.1). There are substantial differences in vegetation between these woodlands as well as the biotic and abiotic factors that influence them, but all have a Mediterranean climate with a dry period extending from early spring until early fall.

The 20 native species of oak in California (Nixon 2002) are from three subgenera, including white oaks (Section *Quercus*), black or red oaks (Section *Lobatae*), and intermediate oaks (Section *Protobalanous*). The average rainfall within California's oak woodlands varies from 30 cm in the southern interior regions to over 100 cm in northern coastal woodlands. The wetter portions of these woodlands have higher tree density, and in lower rainfall areas, oak woodlands grade into savannahs with tree density often less than 12 trees ha^{-1}.

16.2.2 Ownership

More than 70% of California's oak woodlands are privately owned (Bolsinger 1988), in contrast to higher elevation coniferous forests that are mostly publicly owned. Livestock grazing is the primary economic activity for these private woodlands. Understory forage plants are productive and nutritious; many of these lands are unsuitable for intensive agriculture due to uneven terrain, poor soils, and inadequate moisture. Livestock grazing has been the main economic use of California's oak woodlands since the introduction of domestic herd animals in the 1700s. Livestock have impacted the oak woodlands by

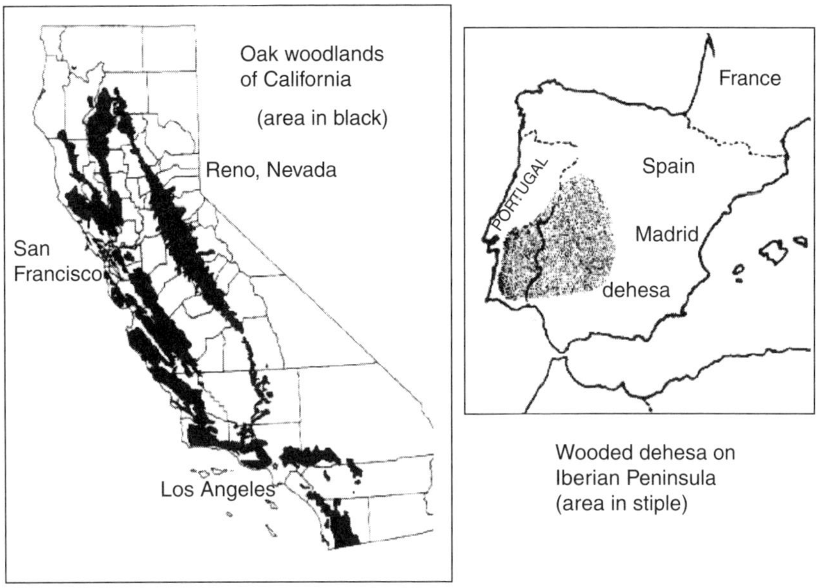

Figure 16.1 Oak woodlands in California and the dehesa region in the southwestern Iberian Peninsula. (From Campos-Palacín et al., Working woodlands: public demand, owner management, and government intervention in conserving Mediterranean ranches and dehesas, in *Proceedings of the Fifth Oak Symposium: Oak in California's Changing Environment*, Standiford, McCreary and Purcell, Tech. Coords., U.S. Department of Agriculture, Forest Service, Pacific Southwest Research Station, General Technical Report PSW-184, 511, 2002.)

reducing organic matter, compacting soil, and dispersing exotic annual plants from seeds brought from Europe in hay bales and grain sacks that supplied Spanish settlements (Pavlik et al. 1991). These introduced plants probably displaced many native perennials. Introduced annual grasses such as wild oats (*Avena fatua* L.), ripgut brome (*Bromus diandrus* Roth.), and Italian rye (*Lolium multiflorum* Lam.) dominate the understory of most oak woodlands in California today.

16.2.3 Historic uses and values

Indigenous peoples populated California's woodlands for approximately 10,000 years prior to the arrival of Europeans. These woodlands may have supported the highest population densities on the North American continent. Native Americans actively managed woodlands principally by their use of fire (Blackburn and Anderson 1993). Oaks were revered by California's native peoples and figured prominently in their culture. Acorns were a main food source for many tribal groups and are still consumed today.

Oaks continued to be important to later residents. The deep and endearing value of oaks in the psyche of early settlers is apparent in the many cities and landmarks that carry oak or the Spanish equivalents "encina" and "roble" in their names. There are even towns named "Dehesa" and "Bellota," the latter meaning acorn in Spanish. Today, golden brown hills dotted with gnarled oak trees epitomize the California landscape, and native oaks symbolize values that are held dear: strength, beauty, adaptability, and longevity.

The value of oaks in California is more than aesthetic; oaks and oak woodlands support a diverse fauna. More than half of the 600-plus species of terrestrial vertebrates in California utilize oak woodlands during the year for essential food and shelter. Oaks are critical in protecting watersheds and ensuring the quality of water resources. The majority

Table 16.1 General Regeneration Strategies for Oak Woodlands in California and the Dehesa Region of Spain and Portugal

	California Oak Woodlands	Dehesa Region of Spain and Portugal
Acorn collection	Local genotypes recommended. Acorns considered ripe when caps easily dislodge. Acorns should be collected directly from tree branches if possible.	Provenance regions established in Spain. Acorns collected when ripe, fat, light brown, and smooth. Generally collected from the ground from late October to December.
Acorn storage	Store in a refrigerator or cooler preferably near, but above, freezing, in plastic bags that are left loosely sealed at the top. White oaks cannot be stored for longer than 3 months.	Keep under controlled conditions at -1 to -3°C, and 40 to 45% humidity, in non-hermetic containers with adequate ventilation. Mixing acorns with hygrophilous material (peat) sometimes used to lengthen the storage period.
Acorn planting	Better performance for acorns sown in winter, rather than in spring.	Acorns sown in autumn should be treated with a rodent repellent. Sowing in uneven terrain should be done by hand. When sown in furrows, sow 1 to 2 acorns every 5 m.
Seedling production	Most seedlings are produced in containers of varying sizes. Limited numbers of bareroot seedlings are produced, but root pruning in nursery beds is critical.	Most seedlings are grown for 1 year in containers larger than 300 cm^3 that promote wide, well-developed root systems and prevent root circling.
Seedling storage	Seedlings generally kept in refrigerated units prior to planting. Blue oak seedlings can be stored for up to 3 months without degradation in quality.	Seedlings should be kept cool. Those taken to field should be placed in a sheltered place to ensure that the roots stay sufficiently cool and moist.
Seedling planting	Seedlings should be planted in winter. Augering planting spots can greatly reduce the soil bulk density and make it much easier for the oaks roots to grow.	Autumn is almost always the recommended time for planting seedlings. In areas that have had complete plowing, a furrow created by a ripper is often used for planting.
Weed control	Weed control in areas around individual seedlings using herbicides, mulch, or mechanical removal is essential for adequate field performance.	Preparing sites using plowing is necessary in areas that have shrub cover. In grassy areas, weeds should be controlled by hoeing, cultivation, or with herbicides.
Seedling protection	Treeshelters have proven very effective in protecting oak seedlings from a variety of damaging animals.	Both fencing and individual protectors are used to keep grazing animals out of planted pastures for up to 20 years to prevent damage.

of the state's water is stored as snow pack in high elevation mountains, flowing through oak woodlands on the way to rivers that support fisheries, farms, and cities. Oak trees anchor the soil, preventing erosion and sedimentation.

16.2.4 Oak woodland losses

Although there are few records to indicate the total area of oak woodlands before European settlement, the best estimate is that roughly half of the pre-settlement area has been converted to other uses (Burcham 1981 [1957]). The original settlers converted substantial areas of oak woodland to agriculture, especially in the Central Valley where deep, fertile, alluvial soils that supported valley oak (*Q. lobata* Née) forests were converted to orchards and other crops. During the Gold Rush, oaks and other trees were important for firewood and for mining timbers and harvested for fuel for steamships, railroads, and industrial purposes. A large charcoal business that utilized native oaks flourished in the 19th century.

Loss of oak woodlands continued into the 20th century. From the end of World War II until 1975, approximately 800,000 ha of woodlands and savannah were converted to treeless pastures in order to promote greater forage production for livestock (Bolsinger 1988), often supported by government grants. In the 1970s, the sharp increase in the cost of petroleum caused by the oil embargo increased the demand for oak firewood for home heating. Thousands of ha were cut, primarily in the upper Sacramento Valley in north-central California.

Today, pressures on oak woodlands continue. Approximately 3,000 ha were converted annually for residential and commercial development in the 1970s and 1980s (Bolsinger 1988), and development pressure continues. In the 1990s, the area in vineyards increased, largely at the expense of oak woodlands, driven by the demand for high-quality wines and the higher prices for premium grapes. In the same period, a new disease, Sudden Oak Death (*Phytophthora ramorum*), killed thousands of oak trees (Garbelotto et al. 2003).

16.2.5 Historical oak woodland restoration efforts

Until the 1980s, there was little interest in oak conservation or restoration of degraded woodlands in California. Because native oaks were widely distributed and had little economic value, they were often considered weeds and there was no incentive to learn how to propagate or establish them. The recognition in the mid-1980s that some oak species were not adequately regenerating naturally and that oak woodlands were being lost caused the state to support a research and education program on oak management (Passof 1987). In part, this program funded oak regeneration research addressing such fundamental questions as the best time to collect acorns, how they should be stored, how to grow seedlings in nurseries, and how to plant and maintain oak seedlings in the field. These efforts demonstrated that successful artificial regeneration of oaks is possible, but often requires considerable maintenance and follow-up (McCreary 2001).

16.3 Oak woodlands in the dehesa region of Spain and Portugal

16.3.1 What is the dehesa?

The term dehesa has many meanings and comes from the Latin word "deffesa," the name for a fenced or "defended" pasture or enclosed area that protected livestock from wild animals. The most widely accepted definition today is that of an agro-silvo-pastoral system consisting of a sparse overstory of trees sheltering herbaceous understory. Dehesa management usually includes cultivation of the understory every 2 to 5 years. Besides producing food or fodder, cultivation limits the invasion of grassland by aggressive shrubby vegetation.

The dehesa is a cultural ecosystem created and maintained by agricultural use, characterized by the diversity of taxonomic groups (flora and fauna) and structures and products derived from it (Margalef 1980). It is a versatile and stable system ecologically and with regard to productivity. The dehesa has worked efficiently over the centuries, and even with today's technology, so different from that available at the origin of the dehesa, it is still considered the most efficient way to use an area's natural resources (San Miguel 1994).

16.3.2 Distribution and characteristics

Because a precise definition of the dehesa is lacking, there are few reliable statistics on the area it occupies. Nevertheless, various authors (ICONA 1980; Abreu 1983; Díaz Pineda 1987; Joffre et al. 1998; San Miguel 1994) suggest that over 4 million ha of Spain and Portugal is occupied by dehesa. It is mainly found on infertile soils in the west and southwest of the Iberian Peninsula (Figure 16.1). The dehesas that occurred on fertile soils were replaced by cultivated crops. The dominant tree species of typical dehesas are evergreen oaks, holm oak (*Quercus ilex* L. subsp. *ballota*), cork oak (*Q. suber* L.), and even quejigo oak (*Q. faginea* Lam. subsp. *broteroi*) in areas of higher soil moisture. However, there are also dehesas where the primary tree species are *Q. ilex* L. subsp. *ilex*, *Q. faginea* Lam. subsp. *faginea*, *Q. faginea* Lam. subsp. *alpestris*, *Q. pyrenaica* Willd., other Mediterranean and sub-Mediterranean oaks (*Q. canariensis* Willd. and *Q. pubescens* Willd.), ash (*Fraxinus angustifolia* Vahl), and even pines (especially *Pinus pinea* L.). Other locally important tree species include wild olive (*Olea europaea* L.), juniper (*Juniperus oxycedrus* L. and *J. thurifera* L.), and carob (*Ceratonia siliqua* L.).

16.3.3 Ownership and historic uses

All of the dehesa woodlands in Spain and Portugal are privately owned. For centuries, they have been intensively managed to maximize the system's direct products in the form of grazing, browse, acorns, cereals, firewood, and charcoal. Human influence and transformation of the dehesa have been intense, and what we now see are merely partial remnants of the original, more densely forested, woodland systems. These systems have been drastically altered because of the importance of their products, including food (acorns, grazing, cereals, livestock), fuel (firewood and charcoal), and packing material (wood and cork). It appears that the woodlands we see today, most of which are in the southwestern quadrant of the peninsula, began to develop after the re-conquest from the Moors and these systems must be relatively young, about 900 years old (Manuel and Gil 1996).

During the first half of the 20th century, there was a major increase in the land under cereal cultivation for human food in the dehesa region. Thus, few dehesas remained untilled. Woodland in general was subject to the same efforts, and much open woodland was cut and cultivated in this period. The economic expansion of the 1960s and 1970s brought about intense development of cropland through deforestation and mechanization. Tractors and machinery revealed their tremendous power for altering the landscape, further hindering already low rates of tree regeneration.

16.3.4 Poor natural regeneration

The main problem affecting the dehesa today is insufficient natural oak regeneration. In most dehesas, livestock production overshadows trees in importance, and as more productive feeds have been introduced, herds have increased. The decline in traditional pastoral systems such as transhumance, where grazing animals are moved to higher-elevation pastures in the summer, has increased year-round grazing. Exhaustive utilization of acorns by feral hogs, indiscriminant periodic plowing, and shrub clearance has further restricted oak natural regeneration. Many stands with insufficient tree density are limited by harsh climatic conditions and compacted soils. Unless the practices causing poor seedling establishment are not stopped or reversed, large areas of dehesa will not regenerate. Unfortunately, most landowners show little interest in oak regeneration, seeing it as a long-term problem without immediate consequences. They ignore the potential for tree loss that threatens the short-term existence of the dehesa in some places, as well as the medium- and longer-term threat to the dehesa region (Montero and Cañellas 2001).

16.4 Oak restoration practices

16.4.1 Acorn collection, storage, and planting

Although there is little information about geographic variation in germination, establishment, and growth of oaks planted from acorns collected from different regions in Spain, Portugal, or California, it is generally agreed that local seed sources should be used, or at least seed from regions with similar growing conditions. To facilitate the choice of seed, provenance regions have been defined for the main Spanish Mediterranean *Quercus* species based on ecological uniformity. Presumably, there is a corresponding similarity in genetic structure. In California, there is also recognition that it is important to plant local genotypes, but provenance regions have not been established. Nurseries are encouraged to keep records of where seed was collected. No doubt, there have been instances where nonlocal stock has been planted, but the possible contribution to regeneration problems is unknown.

Acorn yield is highly variable in both California and southern Europe and typically there are no more than two or three abundant crops over a 10-year period. In the dehesa region, good crops are almost always preceded by a wet spring. Acorns cannot be stored longer than 3 or 4 months without greatly lowering the germination rate, and in some low-yield years there may not be enough acorns for direct seeding or nursery production of seedlings for outplanting the following year (Cañellas 1994; Montero and Cañellas 1999).

Acorns should be collected shortly after they are physiologically mature. While there are various indicators, such as moisture content, carbohydrate percent, or acorn color, that have been used to predict ripeness for oak species in other locales in the U.S. (Bonner and Vozzo 1987), the most useful indicator for many oak species in California is the ease of dislodging acorns from the cupule or cap. After collection, acorns are especially sensitive to drying and their ability to germinate can decrease rapidly with even small losses in moisture. For example, a 10% reduction in moisture content caused a 50% drop in germinating blue oak (*Q. douglasii* Hook. & Arn.) (McCreary and Koukoura 1990). Because acorns are so sensitive to drying and the weather conditions in California when they drop can be quite hot and dry, acorns should be collected directly from tree branches. Once acorns fall to the ground, their quality can decline quickly and they can be consumed by animals. Sometimes, collecting from branches is impossible because they are so high that they are beyond reach, even with long poles. In these cases, frequent collections from the ground minimize exposure.

In the dehesa, healthy acorns are generally fat, light brown, and smooth when ripe. Acorns are collected from the ground from late October to December. Acorns should be rejected if attacked by insects, have broken radicles, or have dried out. Damaged acorns rarely produce seedlings with the vigor and quality needed to adequately root and fully develop. Some parasites, especially the larvae of *Balaninus*, may be present inside the acorns, but infested acorns can be eliminated by float-testing in a bucket of water. After discarding the acorns that float, the healthy ones are collected from the bottom of the container and dried in the air to prevent fungal development during storage.

After collection, acorns in California should be stored in plastic bags that are loosely sealed at the top, in a refrigerator or cooler, just above freezing. Caps should be removed before storage. Because acorns continue to respire during storage, some gas exchange with the atmosphere is necessary and airtight containers should be avoided. Keeping acorns cool during storage slows down respiration, which utilizes energy and can deplete carbohydrate reserves. Cool temperatures also reduce the tendency for acorns to germinate, which is especially common for white oaks. Finally, lower temperatures reduce the damage from harmful microorganisms that can rapidly degrade acorn quality in storage.

In Spain and Portugal, it is not customary to sow acorns immediately after collection, although this would be desirable. Commonly, acorns collected from the dehesa are stored until February or March. They are then directly planted in the field (if this has not been carried out in the autumn, see below) or sown in the nursery for outplanting the following autumn. In the dehesa, transportation from the place of collection to the storage facility must be done rapidly, with the acorns in coarse-weave sacks that allow ventilation. At the storage facility, acorns should be emptied into trays or onto the floor in layers less than 10 cm deep and stirred frequently to allow ventilation (Montero et al. 2000). Acorns should be stored at cool temperatures, between −1 and −3°C but not below −5°C. They should be kept under controlled conditions at 40 to 45% humidity in non-hermetically sealed containers with adequate ventilation. In order to reduce damage during storage, respiration should be minimized by storage in an inert CO_2 or nitrogen atmosphere, in a low-pressure container, or by lowering storage temperature and humidity. Storage in an inert atmosphere is expensive and as yet unproven for cork oak. Because it is easier and cheaper, reducing temperature and humidity is the usual method. Mixing seeds with hygrophilous material such as peat can help lengthen the storage period.

The best time of year to directly sow acorns depends mainly on climate. In areas with very cold winters and moderately dry summers, it is advisable to sow at the end of winter. But in areas with mild winters and very hot, dry summers, autumn sowing is better. This will allow plants to enter the summer with a more developed root system and the plant will be better prepared to withstand drought. A disadvantage of autumn sowing is the seed usually takes longer to germinate and is exposed longer to attacks by mice and other seed predators. Acorns sown in autumn should be treated with a rodent repellent (Torres 1995; Cañellas et al. 2000).

In California, acorns generally ripen in late summer to mid-fall before the first heavy fall rains, and soils can still be extremely dry. Although even fairly dry soils can have relatively high humidity under the surface, the hard dry surface will restrict root penetration. Therefore, acorns should not be directly sown until there has been sufficient rainfall to soak the soil to a depth of at least several cm. Acorns should be planted as early as possible after the soil is sufficiently wet (McCreary 1990); as a rule of thumb, acorns should not be sown later than the end of January. Even this may be too late in areas with low rainfall and short winters.

In general, acorns should be planted horizontally and placed 1- to 3-cm deep; however, it may be better to plant them even deeper if rodents are present, which could dig them up. Even though acorns tend to naturally fall on the soil surface, and some are able to germinate and become established, exposed acorns are more likely than buried acorns to suffer from depredation and desiccation damage. Burying acorns may not eliminate rodent damage, but it will reduce losses (Griffin 1971). Germination of buried acorns may be twice that of surface-sown ones (Borchert et al. 1989).

16.4.2 Seedling propagation

The majority of native oaks produced by nurseries in California are grown in containers rather than as bareroot seedlings. Seedlings of most oak species invest tremendous energy into producing roots, which initially have a taproot or carrot-like configuration. Such rapid growth means that seedlings can quickly exceed the volume of the container root systems and become pot-bound. In general, better quality oak seedlings are produced in narrow, deep containers than in wide, shallow ones. Oak taproots will often grow to the bottom of a container before the shoots even emerge from the soil surface. At the bottom, roots will tend to circle around the container unless checked. Such root deformation can persist for decades after field planting and create problems of weakness, poor growth, and lack of stability years later. Many container production systems employ air pruning to thwart root circling, which produces a fibrous root system.

A limited number of bareroot oak seedlings are produced in California and are sold to the public by the California Department of Forestry and Fire Protection (CDF). Seedlings must be undercut (root-pruned) with an oscillating bar towed behind a tractor while they are in the nursery beds in order to produce acceptable plants (Krelle and McCreary 1991). The timing of this pruning is also critical; if done too early in the season before the roots have grown down at least 20 cm, pruning has little effect on root form. If pruning occurs after seedlings have produced thick, deep, roots, too much of the roots are lost during pruning and seedlings usually die.

Seedlings can be lifted from early December to early February without seriously affecting seedling quality. They can be left in cold storage for 2 months without damage, as long as roots do not dry out. However, the combination of late lifting (February or later) and long storage can result in poor field survival and growth, because seedlings are not able to initiate sufficient root growth before conditions become harsh (McCreary and Tecklin 1994).

Plant quality is one of the most important factors governing the success of seedling-planting projects in the dehesa. Genetic, morphological, and physiological characteristics of seedlings are critical, and must be within established limits so that a good-quality seedling roots well and grows quickly. Most afforestation stock is 1-year-old container seedlings with wide, well-developed root systems grown under as natural conditions as possible (Montero et al. 2000). Seedlings are moved outdoors from greenhouses as soon as possible, where they are exposed to higher light intensities and can become hardened- off before outplanting.

Both direct seeding and planting of young seedlings in the dehesa have advantages and disadvantages, but, in general, planting seedlings is preferred. Nevertheless, direct seeding may be advantageous as compared to seedlings, including the following (Montero and Cañellas 1999):

- Direct seeding is simpler and cheaper, especially transportation to the field.
- Successful sowing will produce more plants per ha, enabling selection of the most vigorous trees.
- Oaks growing from directly sown acorns from containers may be better adapted without root deformation.

The main advantages of planting seedlings over direct seeding are as follows:

- Seedlings have a better chance of success under difficult conditions, since the plants arrive at the site more developed and more weather-resistant.
- Seedlings are a year older when planted, potentially reducing the time it takes to develop into trees.
- The initial planting density for seedlings can be less, thereby reducing the costs of later operations such as thinning.
- Fewer seeds are needed for growing seedlings in the nursery than for direct seeding in the field.

16.4.3 Planting and maintenance

16.4.3.1 Planting date and site preparation

Sowing or planting date can influence subsequent field performance. The greatest problems in California arise from planting seedlings too late in the season. Since seedlings in the white oak subgenera are able to grow roots during winter, late fall or early winter planting allows them to develop well-established root systems while the soil is still cool and moist. Planting after March 1 can result in poor establishment, since soils generally become too dry.

The interval suitable for planting in the dehesa is not very long and is even shorter for direct seeding because the seed must first be obtained. Autumn is almost always the recommended time for planting seedlings. Subsequent rains firm the soil around the roots, creating favorable conditions for the root system to begin growing when soils warm. Dehesa sites usually have intense competition and should be prepared before planting or oak establishment will be difficult. The intensity of plowing (complete or in strips) largely depends on the characteristics of the site. Areas with high densities of competing scrub need intense tillage. Partial strip-plowing is suitable for mountain areas, and complete plowing is better for flatter, low-lying land. Complete plowing improves water storage by reducing plant competition (Montero and Cañellas 1999). Complete plowing is accompanied by the opening of furrows with a ripper to provide seedling roots access to moisture at depth, which enhances survival and growth through the driest months of the summer.

For direct seeding, 1 to 2 acorns should be sown every 5 m along the furrow opened by the ripper, for a density of 1,000 acorns ha^{-1}. Sowing more acorns in the furrows, to reduce the risk of failure due to seed predation, may require culling before age 3 if all the acorns germinate. High densities, nevertheless, have the advantage that poorly shaped trees or trees producing poor-quality cork may be removed, increasing yield and quality (Montero and Cañellas 2001). Most of the potential areas for afforestation are flat or slightly hilly, abandoned farmland where mechanized planting is possible. In uneven terrain, hand sowing is recommended as machine sowing runs the risk of burying the acorns too deep.

Many of the hardwood rangelands in California have been grazed almost continuously for the past century and a half, and soil compaction is commonplace. Augering planting spots can greatly reduce the bulk density of the soil and make it much easier for the oak's roots to grow downward, especially shallow-planted acorns. The depth of augering apparently makes little difference because most of the compaction is in the upper portion of the soil. As long as this area was broken up, augering as shallow as 30 cm is sufficient (McCreary 1995).

16.4.3.2 Transporting seedlings

As a general rule, transportation should not last more than 24 h and both roots and shoots must be protected. On long journeys, drying out can be prevented by the use of special packing or, preferably, by using refrigerated trucks. Typically, seedlings are not planted immediately after arrival, but rather over periods of days, or even weeks. Therefore, it is essential that seedlings be protected against cold, heat, and drying by holding them in a sheltered place to ensure that the roots stay sufficiently cool and moist, and away from domestic or wild animals.

16.4.3.3 Planting densities

The initial planting density of plantations in the dehesa will depend on management objectives. The minimal initial density should be around 625 plants ha^{-1} (Montero and Cañellas 1999). If cork is the main product, higher initial densities can be used. The disadvantage of high densities is that both holm and cork oaks sprout vigorously from stumps;

to prevent sprouting, stumps have to be removed. The disadvantage of wide spacing (5 m × 6 m or 6 m × 6 m) is that large unproductive areas result if several adjacent seedlings die, requiring replacement plantings.

16.4.3.4 Controlling competing vegetation

Competition from grasses and other ground vegetation at the planting site is a primary obstacle to successful artificial regeneration of oaks in California. Competing vegetation can greatly reduce soil moisture and may severely limit survival and growth. Controlling competing plants around planted acorns or seedlings is essential. These plants also compete with oak seedlings for nutrients and light, and provide cover for population pests such as voles (*Microtus californicus scirpensis*) and grasshoppers (*Melanoplus* spp.) that can seriously damage young seedlings. There are a variety of methods that can be used to eradicate competing plants, including herbicides, mulches, and physical removal. The actual procedure or technique chosen may depend on a host of variables, including soil, topography, equipment or materials available, cost, regulation, and the oak species planted (deciduous or evergreen).

In the dehesa, afforestation on abandoned farmland usually means a heavy grass cover that competes with the oaks for moisture. The grass cover can smother small plants, increases the risk of fires in summer, and provides refuge for acorn and seedling predators. It is best to remove grasses before they flower in order to prevent seed dissemination. This may be done mechanically, with a cultivator or a disc harrow, or chemically.

16.4.3.5 Protecting seedlings from animals

Animal damage is another common obstacle to successful regeneration in both California and the dehesa region. Without animal protection, oak plantings often stand little chance of survival. Tree shelters have proven particularly effective in California (Tecklin et al. 1997). They not only protect seedlings from a variety of animals but also stimulate aboveground growth by elevating temperature, humidity, and CO_2 concentrations (Tuley 1983, 1985).

In the dehesa, wire netting tree protectors (2 to 3 m high) are the most suitable way to protect plantations from domestic and wild animals. Seedlings inside these enclosures should also be protected from rodents with short (60 to 70 cm high) protectors. Although their unit cost is high, these devices allow livestock to graze after planting. Because of the cost, however, planting density should be lowered to 300 to 350 stems ha^{-1}.

Game fencing is desirable on estates with high deer populations. Livestock fencing is less expensive, but keeps all stock away during the first 5 years after planting. Sheep grazing may be allowed at that point; however, cattle, goats, and deer should be excluded from the planted area for at least 20 years.

16.5 Current restoration efforts

16.5.1 California

Local governments are usually responsible for requiring oak woodland conservation efforts following tree removal or loss. Programs and approaches vary, depending on local threats to the resource. Regulations usually focus on removals of individual trees, rarely addressing habitat conservation. Voluntary oak conservation guidelines seem to work in rural areas, where financial incentives to convert oak woodlands to other uses such as vineyards or housing are lacking. Recently, conservation easements have become fashionable. By agreeing to an easement, a woodland owner can sell the right to develop a property, while retaining other rights such as grazing.

Lacking the financial incentives common in Europe, the total area planted to oaks in California in the last 20 years has been relatively low, probably less than 1,000 ha. Such planting as has occurred results from local requirements to replace trees removed for development. These plantings rarely attempt to replace habitat and are often failures within 3 years.

The largest oak plantings in California have been undertaken by The Nature Conservancy, an international conservation organization that has focused on restoring riparian forests with valley oaks and other hardwoods (Ballard et al. 2002). They have planted several hundred ha in reserves along the Sacramento and Cosumnes Rivers in northern California. Some of their plantings are now 15 years old.

16.5.2 Dehesa region

The lack of regeneration and rejuvenation of the dehesa was apparent from the late 1970s. Although it was clear that something was required to prevent the gradual loss of production, it was not until 1993 that legislation was passed (decree number 378/93). Known as the Reforestation Decree, this legislation originated from the European Union land-use policy on the afforestation of abandoned farmland. It has led to the restoration of over 150,000 ha in the Iberian Peninsula, with typical Mediterranean species, evergreen and cork oak.

The high proportion of private ownership of the dehesas has been an obstacle to a general restoration program by central and regional administrations. Nevertheless, a variety of economic incentives are available to encourage landholders to afforest or regenerate their woodlands. Greater use of technological and other resources has contributed to successful restoration, and dehesa afforestation is becoming widespread in the southwest of the peninsula (Montero and Cañellas 1999).

References

Abreu, J.M., *La dehesa*, ICONA, Madrid (Documento interno no publicado), 1983.

Ballard, H., Kraetsch, R., and Huntsinger, L., Collaborative monitoring in Walnut Creek, California, in *Proceedings of the Fifth Oak Symposium: Oaks in California's Changing Environment*, Standiford, R.B., McCreary, D.D., and Purcell, K., Tech. Coords., U.S. Department of Agriculture, Forest Service, Pacific Southwest Research Station, General Technical Report PSW-184, 617, 2002.

Blackburn, T.C. and Anderson, K., Introduction: managing the domesticated environment, in *Before the Wilderness: Environmental Management by Native Californians*, Blackburn, T.C. and Anderson, K., Eds., Ballena Press, Menlo Park, CA, 15, 1993.

Bolsinger, C.L., *The Hardwoods of California's Timberlands, Woodlands and Savannas*, U.S. Department of Agriculture, Forest Service, Pacific Northwest Research Station, Research Bulletin PNW-148, 148, 1988.

Bonner, F.T. and Vozzo, J.A., *Seed Biology and Technology of* Quercus, U.S. Department of Agriculture, Forest Service, Southern Forest Experiment Station, New Orleans, LA, General Technical Report SO-66, 22, 1987.

Borchert, M.I., Davis, F.W., Michaelson, J., and Oyler, L.D., Interactions of factors affecting seedling recruitment of blue oak (*Quercus douglasii*) in California, *Ecology*, 70, 389, 1989.

Burcham, L.T., *California rangeland: a historic-ecological study of the range resources of California*, University of California, U.C. Davis, CA, 1981 [1957].

Campos-Palacín, P., Huntsinger, L., Standiford, R., Martin-Barroso, D., Mariscal-Lorente, P., and Starrs, P.F., Working woodlands: public demand, owner management, and government intervention in conserving Mediterranean ranches and *dehesas*, in *Proceedings of the Fifth Oak*

Symposium: Oaks in California's Changing Environment, Standiford, R.B., McCreary, D.D., and Purcell, K., Tech. Coords., U.S. Department of Agriculture, Forest Service, Pacific Southwest Research Station, General Technical Report PSW-184, 511, 2002.

Cañellas, I., Producción de bellotas en alcornocales, *Comunicación invitada presentada al Simposio Mediterráneo sobre Regeneración del Monte Alcornocal y Seminario de Política sobre el Alcornocal en el área Mediterránea*, Octubre, Mérida-Portugal-Sevilla, Editado: Instituto de Promoción del Corcho (Junta de Extremadura), 223, 1994.

Cañellas, I., Pardos, M., and Montero, G., El efecto de la sombra en la regeneración natural del alcornoque (*Quercus suber* L.), *Reunión del Grupo de Trabajo de Selvicultura sobre "Regeneración Natural,"* Lugo, Octubre, 2000.

Díaz Pineda, F., Interés científico de la dehesa: perspectivas de la investigación, *Seminario sobre Dehesas y Sistemas Agro-silvo-pastorales similares*, MaB. Madrid, 1987.

Garbelotto, M., Davidson, J.M., Ivors, K., Maloney, P., Huberli, D., Koike, S.T., and Rizzo, D. M., Non-native plants are main hosts for sudden oak death in California, *Calif. Agric.* 57, 18, 2003.

Griffin, J.R., Oak regeneration in the upper Carmel Valley, California, *Ecology*, 52, 862, 1971.

ICONA, *Las frondosas en el Primer Inventario Forestal Nacional*, Ministerio de Agricultura, Madrid, 1980.

Joffre, R., Rambal, S., and Ratle, J.P., The dehesa system of southern Spain and Portugal as a natural ecosystems mimic, *Agroforest. Systems*, 45, 57, 1998.

Krelle, B. and McCreary, D., Propagating California native oaks in bareroot nurseries, in *Proceedings of the Intermountain Forest Nursery Association*, Landis, T.D., Tech. Coord., U.S. Department of Agriculture, Forest Service, Rocky Mountain Research Station, General Technical Report RM-211, 112, 1991.

Manuel, C. and Gil, L., *La transformación histórica del paisaje forestal en España*, Ministerio de Medio Ambiente, Madrid, 1996.

Margalef, R., Ed., *Ecología*, Omega, Barcelona, 1980.

McCreary, D.D., Acorn sowing date affects field performance of blue and valley oaks, *Tree Planters Notes*, 41, 6, 1990.

McCreary, D.D., Augering and fertilization stimulate growth of blue oak seedlings planted from acorns, but not from containers, *West. J. Appl. For.*, 10, 133, 1995.

McCreary, D. D., *Regenerating Rangeland Oaks in California*, University of California Agriculture and Natural Resources Publication 21601, 62, 2001.

McCreary, D.D. and Koukoura, Z., The effects of collection date and pre-storage treatment of the germination of blue oak acorns, *New For.*, 3, 303, 1990.

McCreary, D.D. and Tecklin, J., Lifting and storing bareroot blue oak seedlings, *New For.*, 8, 89, 1994.

Montero, G. and Cañellas, I., *Manual de Reforestación y cultivo de alcornoque* (Quercus suber L.), Monografía INIA, 103, 1999.

Montero, G. and Cañellas, I., Silviculture and sustainable management of cork oak forest in Spain, in *IUFRO Meeting on Silviculture of Cedar* (Cedrus atlantica) *and Cork Oak* (Quercus suber), Rabat, Morocco, 2001.

Montero, G., San Miguel, A., and Cañellas, I., *Systems of Mediterranean Silviculture "La dehesa,"* 48, 2000.

Nixon, K.C., The oak (*Quercus*) biodiversity of California and adjacent regions, in *Proceedings of the Fifth Oak Symposium: Oaks in California's Changing Environment*, Standiford, R.B., McCreary, D.D., and Purcell, K., Tech. Coords., U.S. Department of Agriculture, Forest Service, Pacific Southwest Research Station, General Technical Report PSW-184, 3, 2002.

Passof, P.C., Developing an educational program to address the management of California's hardwood rangelands, in *Proceedings of the Symposium on Multiple-Use Management of California's Hardwood Resources*, Plumb, T.R., and Pillsbury, N.H., Tech. Coords., U.S. Department of Agriculture, Forest Service, Pacific Southwest Research Station, General Technical Report, PSW-100, 394, 1987.

Pavlik, B.M., Muick, P.C., Johnson, S., and Popper, M., *Oaks of California*, Cachuma Press, Inc., Los Olivos, CA, 122, 1991.

San Miguel, A., Ed., *La dehesa española, Origen, tipología, características y gestión*, Fundación Conde del Valle de Salaza, 1994.

Tecklin, J., Connor, J.M. and McCreary, D.D., Rehabilitation of a blue oak restoration project, in *Proceedings of the Symposium on Oak Woodlands: Ecology, Management and Urban Interface Issues*, Verner, J. and Tietje, W.D., Tech. Coords., U.S. Department of Agriculture, Forest Service, Pacific Southwest Research Station, General Technical Report PSW-160, 267, 1997.

Torres, E., *Estudio de los principales problemas selvícolas de los alcornocales del macizo de Aljibe (Cádiz y Málaga)*, Tesis Doctoral, ESTI Montes, 1995.

Tuley, G., Shelters improve the growth of young trees in the forest, *Q. J. For.*, 77, 77, 1983.

Tuley, G., The growth of young oak trees in shelters, *Forestry*, 58, 181, 1985.

Part 6

Case Studies: Rehabilitation

chapter seventeen

Data and tools for conservation, management, and restoration of northern forest ecosystems at multiple scales

Per Angelstam
Swedish University of Agricultural Sciences and Örebro University
Jean-Michel Roberge
Swedish University of Agricultural Sciences
Tommy Ek
County Administration Board of Östergötland
Lars Laestadius
World Resources Institute

Contents

17.1 Introduction

The transition from sustained yield of wood from forests to ecological sustainability based on ecosystem management requires expanded data collection to monitor outcomes and an

1-56670-635-1/05/$0.00+$1.50

improved toolbox of analytical techniques. Broadening the thematic range of data and spatially explicit analysis is crucial for determining the relative efficacy of protection, management, and restoration by rehabilitation and re-creation of forests. Sustainable forest management is being redefined both in policy and practice as industrial societies transition from sustained yield of timber toward multifunctional ecosystem management (Hunter 1999; Schlaepfer and Elliot 2000) and ecological sustainability (Goodland and Daly 1996; Kennedy et al. 2001). Depending on the country and region, this transition results in some concern for managing forests to provide ecological products and services, such as the maintenance of viable populations (Sjögren-Gulve and Ebenhard 2000), biodiversity (Heywood 1995; Larsson et al. 2001), and protective functions (Kräuchi et al. 2000).

The trend to sustainable management is particularly pertinent to boreal forest management, where forests have served mainly as sources of wood for densely populated regions. The boreal forest is the world's largest biome, and the one with the largest proportion of intact forests (Aksenov et al. 2002). The ecology of the boreal forest is reasonably well understood and threats are diverse, ranging from legal and illegal overexploitation of timber (Marchak 1995), severe impacts caused by extraction of oil and gas (Schneider 2002), pollution, to global climate change (Intergovernmental Panel on Climate Change 1998). In addition to maintaining remnants of forests of different kinds, there is the challenge of securing a representative suite of large intact natural areas that can be used as references for the restoration of highly altered boreal forests (Angelstam et al. 1997; Bryant et al. 1997; Yaroshenko et al. 2001).

Restoration of viable populations of all native species (Larsson and Danell 2001; Korpilahti and Kuuluvainen 2002), managing overabundant herbivores (Angelstam et al. 2000), and use of fire as an ecosystem process are important new aspects of forest management in Europe. For example, in Sweden and Finland, managing for viable populations of species that have suffered under intensive forest management (Gärdenfors 2000) started in the 1980s with leaving trees on clearcuts (stand-scale variable retention). This development was continued in the 1990s with a landscape perspective (Lämås and Fries 1995; Angelstam and Pettersson 1997; Angelstam 2003) and, today, all large forest companies have landscape plans that include an objective of maintaining biodiversity.

Both conservation and restoration are incremental processes where each significant improvement is important. Adaptive management that includes planning, action, monitoring, and evaluation should be the goal (Walters 1986). Certified forest industry companies that have promised to consider carefully the way in which wood is extracted from high conservation value forests (Higman et al. 1999) are developing such approaches. The "triad" approach to landscape conservation (Seymour and Hunter 1999) divides land into zones of intensive wood production near processing plants, extensively managed forests, and preservation areas. Two issues become paramount: minimizing loss of existing forests of high conservation value and emphasizing wood production in areas closest to forest industry. Maintenance of viable populations of specialized species is most feasible in the former (Scott et al. 2002). In the latter, where the landscape is already impoverished, rehabilitation would be costly and success would be uncertain (Tilman and Kareiva 1997; Hanski and Ovaskainen 2002).

Assessing ecological sustainability requires information on biodiversity at multiple scales. Biodiversity conservation is more complex than simply zoning forestland (Larsson et al. 2001) and there is a need to develop transparent, understandable, and robust methods for the rapid assessment and integration of data that managers can carry out. Initially, analytic tools should use the routine information collected about forests by companies and other organizations. Later, as better and more specific data become available, the tools should be applied with increasing detail and spatial resolution.

The aim of this chapter is to use the "near-to-nature silviculture" paradigm to identify data and analyses needed to measure progress toward ecological sustainability of boreal forests. First, we use biodiversity as a proxy for sustainability. We define it and give examples of new kinds of data that are needed at different spatial scales. Second, we review the emerging analytic tools for assessing forest ecosystem maintenance and provide an example of their use at an operational scale. Finally, we discuss the need for effective communication of the status and trends of forests.

17.2 Data on biodiversity at multiple spatial scales

Biological diversity has been defined and redefined and is now a buzzword (Kaennel 1998). Nevertheless, genes, species, habitats, and functions all remain parts of the biodiversity concept (Harper and Hawksworth 1995), although different interest groups may stress different aspects. In Europe, the use of biodiversity indicators in forest monitoring programs was formalized by the Ministerial Conference on the Protection of Forests in Europe (Liaison Unit in Lisbon 1998).

17.2.1 Elements of biodiversity

Assessment of biodiversity requires that its different components (genes, species, and habitats) be defined as clearly as possible, and that the factors underlying their loss be identified and communicated (Larsson et al. 2001). Moreover, a number of important natural processes that affect forest composition and structure (e.g., fire, flooding, browsing, fungal and insect infestation) should be considered (Noss 1990). For practical forest planning and management, the maintenance of forest biodiversity requires a knowledge of the viability of populations of different species, forest habitat quality, and important ecosystem processes (Noss 1990; Larsson et al. 2001).

17.2.2 Spatial scale

Forests can be examined at different spatial scales (Larsson et al. 2001); forest management usually focuses on the stand level, and social planners work at the scale of landscapes or regions. Patches of trees of different species, age, size, and spacing form stands of a particular stand condition; stands with different tree species composition, tree age structure, etc. form landscapes with different amounts and types of forest cover; landscapes with different forest composition form forest regions. As a system for evaluating biodiversity must include diverse data at multiple spatial scales (Table 17.1), more than one method of measuring biodiversity must be employed.

17.2.3 Composition

Species with different life-history traits have different levels of specialization and spatial demands (Angelstam 1996; Mykrä et al. 2000). Restricting consideration to species that have no landscape-scale requirements, such as those used to identify Woodland Key Habitats (Norén et al. 2002), is too limiting. Incorporating species with landscape-scale requirements, such as grouse (Swenson and Angelstam 1993), woodpeckers (Mikusinski and Angelstam 1998; Mikusinski et al. 2001), overwintering passerine birds (Jansson and Andrén, 2003), as well as some insects (Wikars, in press), will markedly increase our understanding of the status and trends of biodiversity composition in a region. Additionally, genes (provenance) are important for reintroductions of extirpated species or afforestation.

17.2.4 Structure

Compared with dynamic natural boreal forests, managed forests have altered structural variability at all spatial scales. Typical changes are truncated age class distribution,

Table 17.1 Examples of Elements of Biodiversity in the Boreal Forest Across Spatial Scales

Elements of Biodiversity	Spatial Scale			
	Tree	Stand	Landscape	Ecoregion
Composition (species, genes)	Plants, fungi	Lichens, birds,wood-living insects	Birds	Large mammals
Structure (habitats)	Coarse woody debris, very large trees, exposed root plates, tree cavities	Composition of the field, shrub, and tree layers, stand age class, foliage height diversity, tree density (or basal area/volume)	Distribution of patch size of stands with different properties, integrity of the local hydrological network	Ecosystem integrity
Function (processes)	Mycorrhizal symbiosis, single-tree gap dynamics, wood and litter decay	Flooding, large-scale windthrow, infestations of defoliating insects or of fungal pathogens	Moose browsing, predation,fire, dispersal of vertebrates among habitat patches, dispersal of invertebrates among subpopulations	Trophic interactions including top predators, pollution, dispersal of vertebrates among subpopulations

declines in the amount of large patches (Aksenov et al. 2002; Mykrä et al. 2000), altered tree species composition, increased homogeneity in tree spacing, truncated diameter distributions, favoring smaller trees (Nilsson et al. 2002), and reduced decadence, including standing live damaged trees, standing dead trees, and fallen trees with different decay stages (Siitonen 2001). Similarly, there has been a disproportionate loss of forests on rich sites (Angelstam et al. 2003b) to agriculture and other uses.

17.2.5 Function

Processes that maintain function are not distinct elements of biodiversity, but intactness of forests is clearly related to important processes, including disturbance regimes. Altered fire frequencies (Niklasson and Granström 2000) and hydrologic regimes (Bergquist 1999) are examples in boreal forests. Less obvious examples are disruption of predator–prey relationships, such as favoring generalist predators that have affected the breeding success of old forest species such as capercaillie *(Tetrao urogallus)* (Kurki et al. 2000), and browsing by superabundant wild herbivores on certain deciduous trees species (Angelstam et al. 2000) that has changed forest composition. Additionally, air pollution is causing leaching of nutrients such as nitrogen from sensitive soils and changing vegetation in some regions (Ellenberg 1996). Socioeconomic changes in rural communities followed by land abandonment constitute another example (Kennedy et al. 2001).

17.2.6 Data needs

Data on forests have long been collected on stands within forest management units. At the landscape scale, sampling for national or regional forest inventories has been aggregated to the biome level. However, sustainability necessitates collecting data on a variety of societal values. Traditionally collected information must become spatially explicit and needs to be complemented at both finer and coarser spatial scales, by data on a wider range of themes.

Work in northern forests provides several examples of successful efforts to collect data on high conservation value forests of different kinds, and can be viewed as building blocks both for conservation and restoration of forest ecosystems. At the global scale, mapping of large intact forest areas in the Russian Federation (Yaroshenko et al. 2001; Aksenov et al. 2002) provides an example of the kinds of data required to map whole forest ecosystems. There are many examples of regional compilations of areas suitable for nature reserves such as the Estonian Forest Conservation Area Network (Viilma et al. 2001). Similarly, the Baltic Forest Mapping Project is an initiative aimed at mapping areas that could host species with larger area requirements (Angelstam et al. In Press). This project used information in current forest management plans and presented it as thematic maps for different forest types, illustrating the suitability of habitats for several focal species (Angelstam 2002; Angelstam et al. 2003a, 2003b, In Press). In Sweden, private land was inventoried for Woodland Key Habitat; this approach has been refined and applied to the Baltic republics (Auzins et al. 2000). In addition, remote sensing has been used to produce spatially complete databases of current land cover. However, remote sensing products generally cannot resolve key characteristics of high conservation value forests, such as age class distribution, tree species composition, or stand structure (Angelstam et al. 2003b).

17.2.7 Combining data and tools

The differences in these efforts illustrate differences in perspectives of the stakeholders initiating the projects. These differences present a major problem for developing a unified framework for collecting data describing the status and trends in forest ecosystems. A geometric series for the minimum size of unit mapped (Table 17.2) further illustrates these differences.

Division of the boreal forest into different ecoregions (Laasimer et al. 1993) provides an initial stratification into different units. However, differences in species composition and natural disturbance regimes call for further subdivisions. Additionally, the human transformation from naturally dynamic to managed ecosystems is gradual. There are clear effects of past management on the level of ecological sustainability of forests, which calls for stratification according to the level of anthropogenic transformation of the forest

Table 17.2 Forest Data at Multiple Scales and How They are Considered in Different Countries

Pixel size (ha)	<0.5	0.5–5	5–50	50–500	500–5,000	5,000–50,000	>50,000	≥50,000
United Kingdom	Veteran trees	Ancient woodland	Ancient woodland					
Sweden	Variable retention	Riparian corridors	Woodland key habitats					
Latvia	Variable retention	Riparian corridors	Woodland key habitats	Baltic forest mapping	Baltic forest mapping			
NW Russia						Intact landscape residuals	Large intact forest areas	
Siberia and N Canada								Intact systems

ecosystem. Different sets of data are relevant, depending on the phase in the development of forest utilization (Table 17.2). This staircase approach to identifying forests of high conservation value takes into account global, national, and regional conditions and requirements. This same approach can be adopted to assess progress toward sustainability. Periodic inventory, monitoring, and analysis can achieve such evaluation.

17.3 Tools for multiple scales

Ecosystem management for protection and restoration requires planning at multiple scales. The approach used in most planning systems for large-scale forestry is hierarchical within a forest management unit (FMU) (Weintraub and Cholaky 1991; Jonsson et al. 1993; Holmgren 1995; Higman et al. 1999). The planning process is usually divided into three subprocesses: strategic planning to decide long-term goals covering an entire rotation, tactical planning to select among different alternatives within the strategic goals but on a shorter time horizon, and operational planning to manage actual operations within a year. This same logic can be applied to planning for conservation and restoration of biodiversity. Landscape management should be based on ecological targets to which measurements of biodiversity could be compared with increasing detail and at different spatial scales (Higman et al. 1999). To assess the ability of a forest landscape to maintain biodiversity, its departure from natural dynamics must be evaluated (Angelstam 2002). Construction of indices to ecological thresholds (Muradian 2001), such as nonlinear response of populations to habitat loss (Fahrig 2001; Angelstam and Breuss 2001), provides a possible starting point for the assessment of different elements of forest biodiversity. With such a quantitative and qualitative approach, strategic decisions could be made about where sustainability of the forest ecosystem could be achieved at the lowest cost.

17.3.1 Strategic planning

17.3.1.1 Quantitative gap analysis

Gap analysis identifies the most endangered and the most rare but ecologically valuable types of habitat within an ecoregion (Scott et al 1996; Perrera et al. 2000). At the strategic level, the focus is on the shortfall in the amount of the different forest types that are needed in the long term to maintain viable populations of the naturally occurring species that cannot survive in conventionally managed landscape. Questions to ask include what are the long-term needs for protected areas to maintain viable populations of species of concern in different forest types? How much of those forest types exist and how well are they protected? Is there a need for restoration or rehabilitation of habitats?

Angelstam and Andersson (2001) estimated the need for protected areas for several broad ecoregions in Sweden based on new information about the dynamics of different forest types, land-use history, habitat loss thresholds for forest specialists, and current management practices. They recommended setting aside all existing forests and restoring or re-creating naturalistic forests in an additional 3 to 11% of the landscapes, depending on ecoregion. For example, in Dalarna and Gävleborg in central Sweden (Angelstam et al. 2003b), the following analytical procedure was used:

- Estimate the amounts of the various types of potential forest vegetation based on modeling the distribution of different natural forest disturbance regimes, and on a knowledge about age distribution within these different disturbance regimes
- Estimate the current amount of those various forest types using remote sensing calibrated with forest stand data

- Estimate the amounts of various forest types needed to maintain viable populations of the most demanding species based on new knowledge about nonlinear responses to habitat loss
- Estimate the difference between current and needed habitat; a negative value implies a shortfall in habitat area and a need for habitat rehabilitation, re-creation, or both
- Estimate how much of the existing forest types that are in short supply are protected as well as how much is planned for future protection.

17.3.2 *Tactical planning*

17.3.2.1 *Habitat modeling*

Gap analysis will identify the forest types in short supply within an ecoregion, but leaves open the question of their spatial distribution and the extent to which they provide functional networks of habitat for dependent species (Angelstam et al. 2003a, 2003b, In Press). Habitat modeling combines spatially explicit land cover data with the requirements of specialized species to build maps describing the probability that a species is found in a landscape (Scott et al. 2002). A habitat model for a given species thus requires data for land cover type(s) constituting habitat, habitat patch size, landscape-scale proportion of suitable habitat, and habitat duration. With adequate habitat models for several carefully selected focal species, a picture of landscape functionality can be obtained. This requires quantitative information on the habitat requirements at many scales of the several species (Angelstam et al. 2003a). Vegetation cover must be mapped in units that are relevant for the focal species, at a scale that matches the home range of individuals. Habitat for a species is often composed of a combination of several land cover types. The necessary size and number of patches of each suitable land cover type must be defined for the species, as well as their spatial distribution and connectivity (Forman 1995). In most cases, the proportion of sufficiently large patches in a landscape can be used as a surrogate for connectivity (Fahrig 2001; Scott et al. 2002). Patch duration is also important; if patches are ephemeral, the landscape must be large enough to contain a stable proportion of patches of this type (Pickett and White 1985).

Focal species should be chosen from among the most demanding species for a range of landscape attributes (Lambeck 1997; Roberge and Angelstam 2004). Because the most demanding species vary among habitats and scales, representatives from several taxa should be included (e.g., Angelstam 1998; Nilsson et al. 2001). Each model should be validated in order to test how reliably one can predict occurrences of the focal species in real-world landscapes, even if such validation efforts are costly. Many specialized species appear to have indicator value, for example, some birds (Mikusinski et al. 2001) and lichens (Uliczka and Angelstam 1999). Moreover, there is growing interest in using umbrella or keystone species for conservation planning (Roberge and Angelstam 2004). Protecting these species, for example, some woodpeckers and grouse, is expected to provide protection to numerous co-occurring species.

Habitat modeling provides a general basis for the planning and assessment of habitat networks on a species-by-species basis. However, such models are subject to some uncertainty depending on a knowledge about the parameters in the models. For example, the knowledge available for a suite of specialized forest birds in northern Europe was evaluated, and although the requirements of most species at the patch scale were known, the knowledge was lacking of requirements for viable populations in landscapes and regions (Angelstam et al. 2003a). The accuracy of models also depends on the quality of land cover data. For example, assessing the habitat of species dependent on deadwood (e.g., many species of woodpeckers, beetles, and wood-decay fungi) requires spatially explicit data on the occurrence of this resource across the landscape. Such data are not currently available from routine forest management maps or classified satellite images, and therefore additional data need to be collected in the field or surrogates obtained from modeling.

17.3.3 Operational planning

In regions with long histories of using land primarily to produce tangible products useful to people, the identification of high conservation forest usually focuses on small remnants of natural forest. In Sweden (Hansson 2001) and the Baltic countries (Auzins et al. 2000; Andersson et al. 1999), small forest stands (usually <10 ha) called Woodland Key Habitats (WKH) are mapped. The WKH are identified based on a combination of field judgments of the naturalness of the stand's structure, the presence of indicator species with small area requirements, signs of historical use, and cultural heritage values. Although often wrongly considered as conceptually homogenous, WKH come in many different kinds having different ecological properties. WKH is an example of an operational approach that can be used for ecosystem restoration (Fries et al. 1997; Angelstam 2002). WKHs differ considerably in their composition and structure, and their preservation requires that different ecosystem processes be maintained over the long term. Management of WKH is one of the most important conservation efforts in the Baltic Sea Region, an area lacking large intact forests (Aksenov et al. 2002).

Maintaining forest biodiversity requires an understanding of the range of natural disturbance regimes, the resulting habitats to which species have adapted, and the management regimes that must be used to mimic natural processes (Hunter 1999; Angelstam 2002). Broadly speaking, disturbance regimes in the boreal forest range from large-scale dynamics, such as fire and wind, to small-scale dynamics associated with tree fall gaps in the canopy (Angelstam 1998; Gromtsev 2002). The relationship in the boreal forest between fire, the main disturbance, and site conditions is described in the ASIO model (Angelstam 1998), based on the occurrence and behavior of wildfire. Four groups of average fire frequencies are assigned (Absent or Almost never occurring, Seldom, Infrequent, or Often — ASIO) that are inversely related to average fire intensity. The interaction between fire and local and regional site conditions was used to deduce three main stand dynamics in European boreal forest: succession from young to old-growth mixed deciduous/coniferous forest following large-scale natural disturbance, multicohort Scots pine (*Pinus sylvestris* L.) after repeated low-intensity disturbance, and gap-phase Norway spruce (*Picea abies* (L.) Karst.) caused by mortality of trees in small patches followed by regeneration (Angelstam 1998, 2002).

Forest management potentially can be used to emulate natural disturbance regimes. Silvicultural systems were divided into three groups (even-age, multi-age, and uneven-age)

Table 17.3 Matrix Ecological and Management Dimensions

Management Dimension	Disturbance Regime		
	Gap Phase (Shade-Tolerant Tree Species)	Cohort Dynamics (e.g., Pinus spp. on Dry Soils)	Successional (Fire and Strong Wind)
Single-tree and group selection systems	+++	+	
Shelterwood systems	+	+++	+
Stand-replacing systems		+	+++

Source: Simplified from Angelstam, Reconciling the linkages of land management with natural disturbance regimes to maintain forest biodiversity in Europe, in *Landscape Ecology and Resource Management: Linking Theory with Practise*, Bissonette, J.A. and Storch, I., Eds., Island Press, Covelo, CA and Washington, DC, 193, 2002.

and arrayed against ecological dimensions (Table 17.3) to summarize the extent to which different combinations will maintain species that have evolved under different disturbance regimes. Even-age management for biodiversity conservation and restoration must maintain more successional stages and tree species combinations than is typical when applied to wood production only. Additionally, most traditional management practices are poor at maintaining coarse woody debris in all decay classes, very large and old trees, and other components of naturally dynamic forests. Consequently, conservation areas with both laissez-faire and active management strategies will usually be a necessary part of a complete approach to maintain forest biodiversity.

To expedite their evaluation for biodiversity conservation, WKHs can be grouped into natural forest disturbance regimes (Angelstam 1998, 2002). We present an example using the forest types and WKH found in Latvia (Auzins et al. 2000). Latvia was chosen because it contains boreal coniferous and nemoral broadleaf forest types (Laasimer et al. 1993) and its land-use history has favored the survival of many natural forest types (Boruks 2001; Laivins 1997; Dzintara 1999). It should be noted that the WKH concept, as it has been applied in Sweden, is limited to securing the occurrence of plants, lichens, and fungi, all species without landscape-scale requirements. By contrast, in Latvia also invertebrates were used (Mollusca and Insecta).

In Latvia, 15 undrained and 8 drained site types form the basis for silviculture (Buss 1997; Avis 1997). The 15 undrained site types can be classified according to nutrient status (Table 17.4) and disturbance regime (Table 17.5). Dry pine forests (upper left corner of Table 17.4) and pine forests on peat with the shrub *Ledum palustre* (lower left corner of Table 17.4) typically have cohort dynamics due to fire. Even on wet peats, fires may occur during hot and dry summers when the peat has dried out. Another source of mortality is physiological drought, which can occur when the peat is frozen in early spring and the daytime air temperature is above freezing. These disturbances would generally result in cohorts, but sometimes also succession, depending on the intensity of the disturbance. In the central part of the matrix, succession after intensive stand-replacing fire or windfall should be

Table 17.4 The 15 Undrained Site Types in the Latvian Classification According to Buss (1981, 1997) With Their Nutrient Status Based on Laivins (1997)

	Oligotrophic		Mesotrophic		Eutrophic	
Dry	Sl	Mr	Ln	Dm	Vr	Gr
Wet mineral	Gs	Mrs	Dms		Vrs	Grs
Wet peat	Pv		Nd	Db	Lk	

Sl (Cladinoso-callunosa), Mr (Vaccionosa), Ln (Myrtillosa), Dm (Hylocomiosa), Vr (Oxalidosa), Gr (Aegopodiosa), Gs (Callunoso-sphagnosa), Mrs (Vaccinio-sphagosa), Dms (Myrtilloso-sphagnosa), Vrs (Myrtilloso-polytrichosa), Grs (Dryopteriosa), Pv (Sphagosa), Nd (Caricoso-phragmitosa), Db (Dryopteriosa-caricosa), and Lk (Filipendulosa).

Table 17.5 Occurrence of Different Forest Disturbance Regimes in the Site and Forest Types Listed in Table 17.4

	Oligotrophic		Mesotrophic		Eutrophic
Dry	Cohort	Cohort	Cohort	Succession	Gap phase (Broadleaf)
Wet mineral	Cohort	Cohort	Succession		Gap phase (*Picea*)
Wet peat	Cohort (Succession)		Succession	Gap phase (*Picea*)	Gap Phase (*Alnus, Fraxinus*)

The cells within Table 17.5 correspond to the forest types at a similar location within Table 17.4.

expected. In the lower right-hand part of the matrix, different kinds of gap-phase dynamic should be expected because of the general absence of stand-replacing disturbance by fire or wind. Depending on the nutrient status and moisture, these forests would be dominated by *Picea*/*Betula* or *Alnus glutinosa*/*Fraxinus*, the latter type on more eutrophic and slightly wetter soil. On dry eutrophic soil, lime (*Tilia* spp.) and elm (*Ulmus* spp.) dominate and gap-phase dynamic is the typical natural disturbance regime. Semi-open *Quercus* forests once developed on dry eutrophic soil, due to mowing or cattle, and may have been semi-open *Quercus* forests in pre-human times because of browsing and grazing of now extinct large herbivores (Vera 2000). Today, such forests are mostly *Populus tremula* (L.) stands.

The matrices in Tables 17.3 and 17.4 were simplified into Figure 17.1, which shows the different WKHs. Some WKHs were excluded from Figure 17.1, including those dependent on topography and geology such as ravines, steep slopes, riparian forest, springs, calcareous forest, fen/meadow, and bog-forest mosaic. The differences among different types of WKH clearly affect the extent to which they require intervention. Table 17.6 gives examples of different kinds of management activities at different spatial scales. Minimum intervention is generally the best in hydrologically intact wet forest stands or other intact forest stands with gap-phase dynamics. No intervention is needed in large intact boreal forest areas with unaltered large-scale disturbance dynamics such as fire regime.

Buffer zones are needed when harvesting near small, wet forest stands or other forest stands with a moist microclimate, in order to improve the probability of survival for the associated habitat specialists. In drained wet forest stands, it is necessary to restore upslope hydrology if the wet forest specialist species in the stands are to remain. Similarly, filling the ditches could restore nonforested peatland at least to some extent.

In areas where fire has been suppressed, prescribed burning may be used to mimic natural disturbance regimes. To some extent, removing invading fire-sensitive Norway spruce from fire-adapted Scots pine forest stands with high conservation value will have the same effect. In regions where the fires are suppressed on a larger landscape scale, it may be necessary to encourage the recruitment of new aspen stands in order to maintain a stable patch dynamic originally created by regular fire episodes. However, overabundant populations of wild browsers, such as moose (*Alces alces*), may hinder the recruitment of aspen (Angelstam et al. 2000); intense hunting or the introduction of natural predators would then be needed.

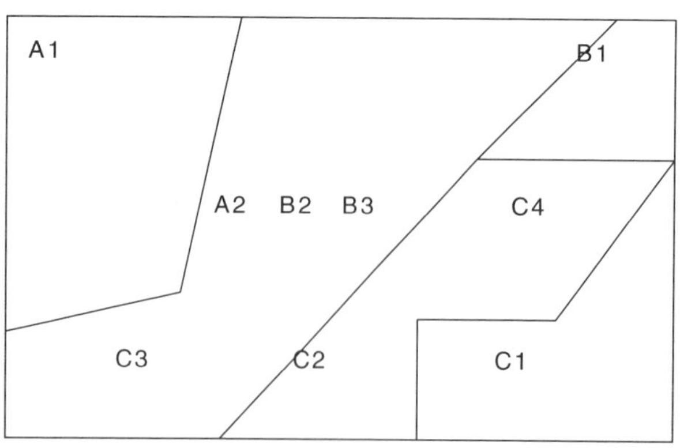

Figure 17.1 Moisture–nutrient matrix and types of woodland key habitats: coniferous forest (A1), mixed coniferous–deciduous forest (A2), broadleaf forest (B1), aspen forest (B2), other deciduous forest (B3), black alder wetland forest (C1), spruce and mixed spruce wetland forest (C2), and pine and birch wetland forest (C3). See Tables 17.5 for a description of the associated natural disturbance regimes. The axes are the same as in Tables 17.4 and 17.5.

Table 17.6 Matrix Illustrating the Ecological Differences Among Different Types of Woodland Key Habitats, and How Management Needs to Consider These Differences

Management within WKH?	
No	Yes
• Spruce forest in ravine with *Usnea longissima* or other "old-growth" lichens • Wet black alder forest in hydrologically intact catchments • Large area of old-growth coniferous or mixed coniferous/deciduous forest • Leave 50–100 m buffer around old-growth forest with dead wood and need for moist microclimate	• Grazing, mowing, and/or removing invading spruce in semi-open oak wooded areas • Cut undergrowth of spruce and/or burn in pine forest not subject to low-intensity fire • Peatland restoration by filling ditches
• Ditto by restoring the hydrology upslope	• Maintain patch dynamics by allowing recruitment of aspen forest in the landscape • Reduce density of overabundant wild large browsers

Reintroduced or increased grazing or mowing is needed for the maintenance or restoration of semi-open oak pastures and oak meadows. The removal of invading Norway spruce could, to some extent, be used as a substitute for grazing or mowing, thereby increasing the amount of light reaching the ground.

It should be evident that not all forest types lend themselves to landscape-scale planning. Ephemeral successional stages, such as recently burned or windblown areas or the deciduous phase in the boreal forest succession, represent a kind of habitat patch dynamic. By contrast, wet sites, wells, ravines, etc. may provide local microclimates that are linked to the specific site conditions, and cannot be re-created.

17.3.4 *Communication of status and trends of forest ecosystems*

Effective communication and mutual learning can mitigate the frequently occurring gap between policy and practice. Maps are efficient in this respect; they can delineate and describe different properties, allow integration of complex information, show data gaps, and are transcultural and have heuristic value. A simple way of communicating biodiversity status and trends is to discuss the fate of particular species. Many people better understand the consequences for species than effects on processes such as litter decomposition and nutrient cycling. The loss of species richness associated with systematic habitat loss over time can be viewed as a dramatic journey where species with different life-history traits pass through a series of thresholds representing the levels of communities, populations, and individuals. Presenting results in relation to thresholds is an effective tool for communicating biodiversity status in terms that the general public finds easy to understand.

The status and trends of biodiversity in sustainable forest ecosystems must be assessed for aggregated conservation efforts across spatial scales and, if possible, compared with threshold values for habitat loss at different spatial scales (Angelstam et al. In Press). Thresholds for occupancy, viability, function, and resilience can be identified. However, there is considerable variation in the range of spatial domains of different actors in land-use management. To understand the total consequences of management for forest biodiversity, the efforts of all actors across spatial scales need to be aggregated for each forest type. Data should be provided on the effects of silvicultural treatments on quantities of remaining trees, snags, and downed wood. For the landscape, critical information would be the degree of naturalness (or intactness) of the forests, the proportions of forested and nonforested land-use, and how connected is the system of conservation areas within and outside reserves. The aggregated results can be compared with a threshold value based on the most demanding species in each main forest ecosystem; any deficit would require habitat restoration.

Research is needed to test whether we can assess the functionality of habitat networks using loss thresholds at multiple scales (Angelstam et al. 2003a). Such research would require selecting a suite of species representing multiple land cover types and identifying quantitative targets based on the requirements of these focal species at multiple scales. Habitat modeling would be used to build spatially explicit maps describing the probability that existing habitat patches really contribute to the functional connectivity in the landscape. Habitat modeling would augment regional gap analysis for the different land cover types, since gap analyses alone neglect quality, size, duration, and configuration of land cover patches, and therefore overestimate the amount of functional habitats.

Acknowledgments

Writing this chapter was triggered both by the invitation from Palle Madsen to attend the IUFRO meeting on "Restoration in boreal and temperate forests," and the senior author's recently commenced cooperation with Sveaskog, the national forest company of Sweden. Funding from WWF, Mistra, Naturvårdsverket, FORMAS, and SLU made this work possible. We thank the participants at the Woodland Key Habitat workshop in Bialowieza (Asiya Zagidullina, Petri Heinonen), at the World Resources Institute workshop on rapid mapping of forests with high conservation value (Dmitri Aksenov, Hans Djurberg, Steve Jennings, Peter Lee, Bryan Lee, Fiona Schmiegelow, Alexey Yaroshenko), and at seminars in Latvia (Janis Donis, Maris Laivins, Imants Liepa, Inga Straupe) for stimulating discussions on the topic of this chapter. Stefan Leiner, John Stanturf, Petteri Tolvanen, and two referees provided valuable comments to the manuscript and Kjell Andersson assisted with the technical editing.

References

Aksenov, D., Dobrynin, D., Dubinin, M., Egorov, A., Isaev, A., Karpachevskiy, M., Laestadius, L., Potapov, P., Purekhovskiy, A., Turubanova, S., and Yaroshenko, A., Atlas of Russia's intact forest landscapes, *Global Forest Watch*, Russia, Moscow, 184 pp., 2002.

Andersson, L., Ek, T., and Martverk, R., *Inventory of Woodland Key Habitats*, Final Report, National Forestry Board, Estonia, Tallinn, Estonia, 38 pp., 1999.

Angelstam, P., Ghost of forest past — natural disturbance regimes as a basis for reconstruction of biologically diverse forests in Europe, in *Conservation of Faunal Diversity in Forested Landscapes*, DeGraaf, R. and Miller, R.I., Eds., Chapman & Hall, London, 287, 1996.

Angelstam, P., Maintaining and restoring biodiversity by simulating natural disturbance regimes in European boreal forest, *J. Veg. Sci.*, 9, 593, 1998.

Angelstam, P., Reconciling the linkages of land management with natural disturbance regimes to maintain forest biodiversity in Europe, in *Landscape Ecology and Resource Management: Linking Theory with Practice*, Bissonette, J.A. and Storch, I., Eds., Island Press, Covelo, CA, 193, 2002.

Angelstam, P., Forest biodiversity management — the Swedish model, in *Towards Forest Sustainability*, Lindenmayer, D.B. and Franklin, J.F., Eds., CSIRO Publishing, Canberra and Island Press, Washington, DC, 143, 2003.

Angelstam, P. and Andersson, L., Estimates of the needs for nature reserves in Sweden, *Scand. J. For. Res.*, Supplement 3, 38, 2001.

Angelstam, P. and Pettersson, B., Principles of present Swedish forest biodiversity management, *Ecol. Bull.*, 46, 191, 1997.

Angelstam, P., Anufriev, V., Balciauskas, L., Blagovidov, A., Borgegård, S-O., Hodge, S., Majewski, P., Ponomarenko, S., Shvarts, E., Tishkov, A., Tomialojc, L., and Wesolowski, T., Biodiversity and sustainable forestry in European forests — how west and east can learn from each other, *Wildl. Soc. Bull.*, 25, 38, 1997.

Angelstam, P. and Breuss, M., Eds., *Critical Habitat Thresholds and Monitoring Tools for the Practical Assessment of Forest Biodiversity in Boreal Forest.*, Report to MISTRA, Downloadable at http://iufro.boku.ac.at/iufro/iufronet/d8/hp80206.htm, 2001.

Angelstam, P., Bütler, R., Lazdinis, M., Mikusinski, G., and Roberge, J.-M., Habitat thresholds for focal species at multiple scales and forest biodiversity conservation — deadwood as an example, *Ann. Zool. Fenn.*, 40, 473, 2003a.

Angelstam, P, Eriksson, J.A., Jaxgård, P., Kellner, O., Koffman, A, Mikusinski, G., Ranneby, B., Roberge, J.-M., Rosengren, M., Rystedt, S., and Seibert J., *Gap Analysis and Planning of Habitat Networks for the Maintenance of Boreal Forest Biodiversity — A Technical Report from the wRESEx Case Study in Sweden*, Department of Natural Sciences, Örebro University, 2003b.

Angelstam, P. and Roberge, J.-M., Lõhmus, A., Bergmanis, M., Brazaitis, G., Breuss, M., Edenius, L., Kosinski, Z., Kurlavicius, P., Lärmanis, V., Lükins, M., Mikusinski, G., Račinskis, E., Strazds, M., and Tryjanowski, P., Habitat suitability index modelling as a tool for landscape-scale conservation — a review of parameters for focal forest birds, *Ecol. Bull.*, 51, in press.

Angelstam, P., Wikberg, P.E., Danilov, P., Faber, W.E., and Nygrén, K., Effects of moose density on timber quality and biodiversity restoration in Sweden, Finland and Russian Karelia, *Alces*, 36, 133, 2000.

Auzins, R., Bermanis, R., Ek, T., and Thor, B., *Inventory of Woodland Key Habitats — Report After 2 Years of Full-Scale Inventory*, State Forest Service, Riga, Latvia, 49 pp., 2000.

Avis, P.G., The forest typology of Latvia: an overview and comparison, in *Proceedings of the Latvian Academy of Sciences*, Section B, 51, 5/6 (592/593), 195, 1997.

Bergquist, B., Påverkan och skyddszoner vid vattendrag i skogs- och jordbrukslandskapet, *En litteraturöversikt, Fiskeriverket Rapport*, 3, 1999.

Boruks, A., Ed., Zemes izmantosana un kadastrs Latvija, in *Latvijas Republikas valsts zemes dienensts*, Riga, Latvia, 408 pp., 2001.

Bryant, D., Nielsen, D., and Tangley, L., The last frontier forests, in *Ecosystems and Economies on the Edge*, World Resources Institute, Washington, DC, 42 pp., 1997.

Buss, K., Forest ecosystem classification in Latvia, in *Proceedings of the Latvian Academy of Sciences, Section B*, Vol. 51, No. 5/6 (592/593), 204, 1997.

Dzintara, A., Ed., Latvijas mezu vesture lidz 1940 gadam, in *WWF — Pasaules Dabas Funds*, Riga, Latvia, 364 p., 1999.

Ellenberg, H., *Vegetation Mitteleuropas mit den Alpen*, 5. Auflage, Verlag Eugen Ulmer, Stuttgart, 1096 pp., 1996 (in German).

Fahrig, L., How much is enough? *Biol. Conserv.*, 100, 65, 2001.

Forman, R.T.T., *Land Mosaics — The Ecology of Landscapes and Regions*, Cambridge University Press, Cambridge, 632 pp., 1995.

Fries, C., Johansson, O., Pettersson, B., and Simonsson, P., Silvicultural models to maintain and restore natural stand structures in Swedish boreal forests, *For. Ecol. Manage.*, 94, 89, 1997.

Goodland, R. and Daly, H., Environmental sustainability: universal and non-negotiable, *Ecol. Appl.*, 6, 1002, 1996.

Gromtsev, A., Natural disturbance dynamics in the boreal forests of European Russia: a review, *Silva Fenn.*, 36, 41, 2002.

Gärdenfors, U., Ed., *The 2000 Red List of Swedish Species*, Artdatabanken, Uppsala, 397 pp., 2000.

Hanski, I. and Ovaskainen, O., Extinction debt at extinction threshold, *Conserv. Biol.*, 16, 666, 2002.

Hansson, L., Key habitats in Swedish managed forests, *Scand. J. For. Res.*, Supplement 3, 52, 2001.

Harper, J.L. and Hawksworth, D.L., Preface, in *Biodiversity — Measurement and Estimation*, Hawksworth, D.L., Ed., Chapman & Hall, London, 5, 1995.

Heywood, V.H., Ed., *Global Biodiversity Assessment*, Cambridge University Press, Cambridge, 1140 pp., 1995.

Higman, S., Bass, S., Judd, N., Mayers, J., and Nussbaum, R., *The Sustainable Forestry Handbook*, Earthscan Publications Ltd, London, 289 pp., 1999.

Holmgren, P., *Geographic Information for Forestry Planning*, Reports in forest ecology and forest soils, Swedish University of Agricultural Sciences Report 68, Uppsala, 1995.

Hunter, M.L., Ed., *Maintaining Biodiversity in Forest Ecosystems*, Cambridge University Press, Cambridge, 698 pp., 1999.

Intergovernmental Panel on Climate Change, *The Regional Impacts of Climate Change: An Assessment of Vulnerability: A Special Report of IPCC Working Group II for the Intergovernmental Panel of Climate Change*, Watson, R.T., Zinoyowera, M.C., and Moss, R.H., Eds., Cambridge University Press, Cambridge, 1998.

Jansson, G. and Andrén, H., Habitat composition and bird diversity in managed boreal forests, *Scand. J. For. Res.*, 18, 225, 2003.

Jonsson, B., Jacobsson, J., and Kallur, H., The forest management planning package, Theory and application, *Studia Forestalia Suecica*, 189, 56, 1993.

Kaennel, M., Biodiversity: a diversity in definition, in *Assessment of Biodiversity for Improved Forest Planning*, Bachmann, P., Köhl, M., and Päivinen, R., Eds., Kluwer Academic Publishers, Dordrecht, 71, 1998.

Kennedy, J.J., Thomas, J.W., and Glueck, P., Evolving forestry and rural development beliefs at midpoint and close to the 20th century, *For. Policy Econ.*, 3, 81, 2001.

Korpilahti, E. and Kuuluvainen, T., Eds., Disturbance dynamics in boreal forests: defining the ecological basis of restoration and management of biodiversity, *Silva Fenn.*, 36, 447, 2002.

Kräuchi, N., Brang, P., and Schönenberger, W., Forests of mountain regions: gaps in knowledge and research needs, *For. Ecol. Manage.*, 132, 73, 2000.

Kurki, S., Nikula, A., Helle, P., and Lindén, H., Landscape fragmentation and forest composition effects on breeding success in boreal forests, *Ecology*, 81, 1985, 2000.

Laasimer, L., Kuusk, V., Tabaka, L., and Lekavicius, A., Editors, *Flora of the Baltic Countries 1. Compendium of Vascular Plants*, Estonian Academy of Sciences, Tartu, 1993.

Laivins, M., Latvijas mezu regionala analize, Latvijas valsts mezzinatnes instituts "Silava," Salaspils, *Mezzinatne* 7, 40, 1997.

Lambeck, R.J., Focal species: a multi-species umbrella for nature conservation, *Conserv. Biol.* 11, 849, 1997.

Larsson, S. and Danell, K., Science and management of boreal forest biodiversity, *Scand. J. For. Res.*, Supplement 3, 123, 2001.

Larsson, T.-B., Angelstam, P., Balent, G., Barbati, A., Bijlsma, R.-J., Boncina, A., Bradshaw, R., Bücking, W., Ciancio, O., Corona, P., Diaci, J., Dias, S., Ellenberg, H., Manuel Fernandes, F, Fernandez-Gonzalez, F., Ferris, R., Frank, G., Friis Møller, P., Giller, P.S., Gustafsson, L., Halbritter, K., Hall, S., Hansson, L., Innes, J., Jactel, H., Keannel Dobbertin, M., Klein, M., Marchetti, M., Mohren, F., Niemelä, P., O'Halloran, J., Rametsteiner, E., Rego, F., Scheidegger, C., Scotti, R., Sjöberg, K., Spanos, I., Spanos, K., Standovar, T., Svensson, L., Tømmerås, B. Å., Trakolis, D., Uuttera, J., VanDenMeerschaut, D., Vanderkerkhove, K., Walsh, P.M., and Watt, A.D., Biodiversity evaluation tools for European forest, *Ecol. Bull.*, 50, 236, 2001.

Liaison Unit in Lisbon, *Third Ministerial Conference on the Protection of Forests in Europe*, General declarations and resolutions adopted, Ministry of Agriculture, 1998.

Lämås, T. and Fries, C., Emergence of a biodiversity concept in Swedish forest policy, *Water Air Soil Pollut.*, 82, 57, 1995.

Marchak, M.P., *Logging the Globe*, McGill-Queen's University Press, Montreal, 404 pp., 1995.

Mikusinski, G. and Angelstam, P., Economic geography, forest distribution and woodpecker diversity in Central Europe, *Conserv. Biol.*, 12, 200, 1998.

Mikusinski, G., Gromadzki, M., and Chylarecki, P., Woodpeckers as indicators of forest bird diversity, *Conserv. Biol.*, 15, 208, 2001.

Muradian, R., Ecological thresholds: a survey, *Ecol. Econ.*, 38, 7, 2001.

Mykrä, S., Kurki, S., and Nikula, A., The spacing of mature forest habitat in relation to species-specific scales in managed boreal forests in NE Finland, *Ann. Zool. Fenn.*, 37, 79, 2000.

Niklasson, M. and A. Granström, Numbers and sizes of fires: long-term spatially explicit fire history in a Swedish boreal landscape, *Ecology*, 81, 1484, 2000.

Nilsson, S.G., Niklasson, M., Hedin, J., Aronsson, G., Gutowski, J.M., Linder, P., Ljungberg, H., Mikusinski, and G., Ranius, T., Densities of large living and dead trees in old-growth temperate and boreal forests, *For. Ecol. Manage.*, 161, 189, 2002.

Nilsson, S.G., Hedin, J., and Niklasson, M., Biodiversity and its assessment in boreal and nemoral forest, *Scand. J. For. Res.*, Supplement 3, 10, 2001.

Norén, M., Nitare, J., Larsson, A., Hultgren, B., and Bergengren, I., *Handbok för Inventering av Nyckelbiotoper*, Skogsstyrelsen, Jönköping, 105 pp., 2002 (in Swedish).

Noss, R.F., Indicators for monitoring biodiversity: a hierarchical approach, *Conserv. Biol.*, 4, 355, 1990.

Perrera, A.H., Euler, D.L., and Thompson, I.D., Editors, *Ecology of a Managed Terrestrial Landscape: Patterns and Processes of Forest Landscapes in Ontario*, UBC Press, Vancouver, 336 pp., 2000.

Picket, S.T.A. and White, P.S., *The Ecology of Natural Disturbance and Patch Dynamics*, Academic Press, Inc., New York, 472 pp., 1985.

Roberge, J.-M. and Angelstam, P., Usefulness of the umbrella species concept as a conservation tool, *Conserv. Biol.*, 18, 76, 2004.

Schneider, R.R., Alternative futures, in *Alberta's Boreal Forest at the Crossroads*, The Federation of Alberta Naturalists, Edmonton, 152 pp., 2002.

Schlaepfer, R. and Elliot, C., Ecological and landscape considerations in forest management: the end of forestry? in *Sustainable Forest Management*, von Gadow, K., Pukkala, T., and Tomé, M., Eds., Kluwer Academic Publishers, Dordrecht, 1–67, 2000.

Scott J.M., Tear T.H., and Davis, F.W., Eds., *Gap Analysis: A Landscape Approach to Biodiversity Planning*, American Society for Photogrammetry and Remote Sensing, Bethesda, MD, U.S.A., 320 pp., 1996.

Scott, J.M., P.J. Heglund, M. Morrison, M., Haufler, J.B., Raphael, M.G., and Samson, F.B., Eds., *Predicting Species Occurrences: Issues of Scale and Accuracy*, Island Press, Covelo, CA, U.S.A., 868 pp., 2002.

Seymour, R.S. and Hunter, M.L., Principles of ecological forestry, in *Maintaining Biodiversity in Forest Ecosystems*, Hunter, M.L., Ed., Cambridge University Press, Cambridge, 22, 1999.

Siitonen, J., Forest management, coarse woody debris and saproxylic organisms: Fennoscandian boreal forests as an example, *Ecol. Bull.*, 49, 11, 2001.

Sjögren-Gulve, P. and Ebenhard, T., Eds., The use of population viability analyses in conservation planning, *Ecol. Bull.*, 48, 2000.

Swenson, J.E. and P. Angelstam, Habitat separation by sympatric forest grouse in FennoScandia in relation to forest succession, *Can. J. Zool.*, 71, 1303, 1993.

Tilman, D. and Kareiva, P., The role of space in population dynamics and interspecific interactions, *Monographs in Population Biology 30*, Princeton University Press, Princeton, 368 pp., 1997.

Uliczka, H. and Angelstam, P., Occurrence of epiphytic lichens in relation to tree species and age in managed boreal forest, *Ecography*, 22, 396, 1999.

Vera, F.W.M., *Grazing Ecology and Forest History*, CABI Publishing, Wallingford, 2000.

Viilma, K., Öövel, J., Tamm, U., Tomson, P., Amos, T., Ostonen, I., Sorensen, P., and Kuuba, R., *Estonian Forest Conservation Area Network*, Ministry of the Environment of the Republic of Estonia, 2001.

Walters, C., *Adaptive Management of Renewable Resources*, Macmillan, New York, 1986.

Weintraub, A. and Cholaky, A., A hierarchical approach to forestry planning, *For. Sci.*, 37, 439, 1991.

Wikars, L.O., Habitat requirements of the pine wood-living beetle *Tragosoma depsarium* (Coleoptera: Cerambycidae) at log, stand, and landscape scale, *Ecol. Bull.*, 51, in press.

Yaroshenko, A. Yu., Potapov, P.V., and Turubanova, S.A., *The Intact Forest Landscapes of Northern European Russia*, Greenpeace Russia and the Global Forest Watch, Moscow, 2001.

chapter eighteen

Ecological restoration of forested ecosystems in Finland

Timo Kuuluvainen
University of Helsinki
Harri Tukia and Kaisu Aapala
Finnish Environment Institute

Contents

18.1 Ecological characteristics of forested ecosystems in Finland

Twenty thousand years ago, ice up to thousands of meters thick covered Finland. About 11,900 years ago, the front of the melting continental ice sheet was positioned on the south coast of Finland (Nenonen 1995). Thus, in evolutionary terms, all species arrived in Finland relatively recently. Deciduous tree species such as *Betula pendula* Roth, *Betula pubescens* Ehrh., *Populus tremula* L., and various species belonging to genus *Salix* L. first colonized the open land that was exposed under the retreating ice. *Pinus sylvestris* L. and *Alnus glutinosa* (L.) Gaertn. came shortly afterward. From 5,000 to 10,000 years ago, the forests of Finland were dominated by *Pinus sylvestris* and deciduous species. The shade-tolerant *Picea abies* (L.) Karst. colonized from the east 5,000 years ago (Bradshaw, this volume) and reached its current northern limit approximately 3,000 years ago.

Finnish forests are generally poor in tree species; the only forest-forming coniferous species are *Pinus sylvestris* and *Picea abies*. Deciduous tree species such as *B. pendula* and *B. pubescens*, *Populus tremula*, and *Salix caprea* L. occur in almost all forest types but rarely grow as single-species stands. However, deciduous trees often dominate the early successional stage after severe fire or wind disturbance, especially on fertile sites. In the southernmost part of Finland (boreonemoral vegetation zone) (Figure 18.1), deciduous hardwood species

1-56670-635-1/05/$0.00+$1.50

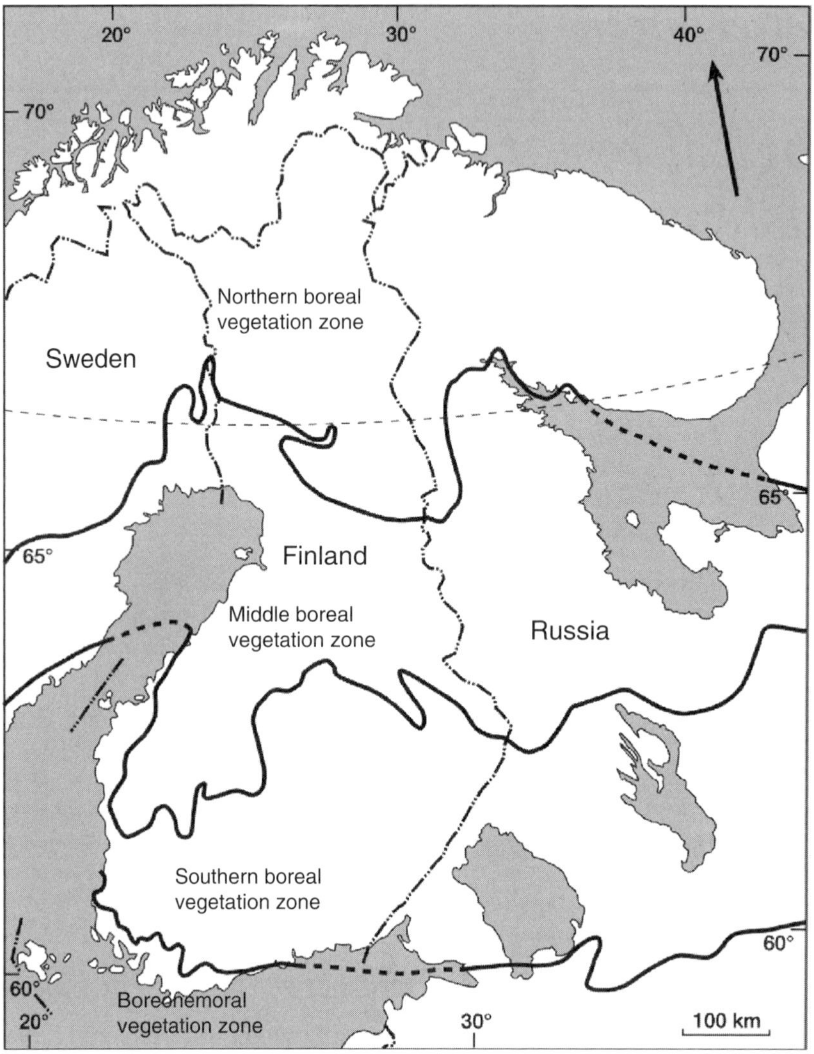

Figure 18.1 Finland can be divided into three main vegetation zones: northern, middle, and southern boreal zones (Kalela 1961; Ahti et al. 1968). In addition, the boreonemoral zone covers the southwestern coastal area of the country.

that are characteristic of nemoral (temperate) forests of central Europe, such as *Acer platanoides* L., *Tilia cordata* Mill., *Fraxinus excelsior* L., and *Quercus robur* L., grow naturally.

The biological communities in Finland are mixed combinations of southern temperate, northern, and eastern boreal (or taiga) elements. Many species are of continental European origin. These often generalist species colonized Finland mainly from the east, southeast, and southwest soon after the ice retreated. Eastern specialists are common, especially in eastern and northern parts of the country. Finland is the westernmost part of the range of many taiga species.

The climate in Finland is affected by high latitude, the vicinity of the Atlantic Ocean, and the continental influence of Eurasia. Because of the maritime influence, winters are milder and summers are cooler than in more continental boreal areas. The mean annual temperature is several degrees warmer than would be typical of boreal regions at comparable latitudes. The oceanic influence, together with a northern location, makes the climate of

Finland a unique combination of annual variability of light and temperature. Low topographic variation and a cool and humid climate give rise to extensive peatlands, which originally covered a third of the land area. For the same reasons, numerous lakes occur in many landscapes, especially in the east-central part of the country.

Oceanic climate and landscape affect natural disturbance regime and ecology of Finnish forests. Although the fire regime without human influence is difficult to define, the natural fire regime would be affected by cooler and moister summers than in continental areas of Russia and Canada. We know from dendrochronological studies that *Pinus sylvestris*-dominated forests in the 17th to 19th centuries burned frequently, typically on 30 to 60-year intervals (Lehtonen 1997). During this period, human activity, such as slash-and-burn cultivation, significantly increased fire frequencies. Prior to significant human influence, fire return intervals were longer, in many cases hundreds of years (Pitkänen et al. 2002a, 2002b; Wallenius 2002). Regional and geographical variation in natural fire regimes occurred; lightning ignitions, for example, are much more frequent in southern Finland than in the northern part (Larjavaara et al. 2004).

Mixed-severity fires characterized the natural fire regime in *Pinus* forests (Pennanen 2002; Kuuluvainen 2002). According to one estimate, before settlement in eastern Finland about half of the fires were stand-replacing (Pitkänen 1999). Typically, a significant proportion of large fire-resistant *Pinus* trees would survive fires and form a persistent, mostly continuous cover of overstory trees (Pennanen 2002; Kuuluvainen et al. 2002b). Reliable estimates of fire size distributions in natural or near-natural conditions are lacking. The average size of fires is probably significantly smaller than fires in boreal forests of continental Canada and Siberia because of Finland's humid climate and landscapes fragmented with peatlands and lakes.

Natural *Picea*-dominated forests are associated with moist and fertile lowland areas. Disturbances in these forests are typically small-scale gap dynamics due to autogenic agents (Kuuluvainen et al. 1998; Engelmark 1999). Fires or severe storms, however, periodically cause severe and widespread destruction (Sirén 1955; Syrjänen et al. 1994).

18.2 History of utilization and current status of forests in Finland

Human influences on forests in Finland can be divided into four phases that overlap in time (Kuuluvainen et al. 2004): the earliest phase of hunting, gathering, and small-scale slash-and-burn cultivation; slash-and-burn cultivation and tar burning (1700 to 1900); selective logging (1870 to 1950); and the current phase of industrial forest utilization, which began in about 1950.

Humans have affected forests in southern Finland since the retreat of continental ice, approximately 11,000 years ago. Hunter–gatherers used timber for construction and for firewood. It is likely that they also ignited fires unintentionally, or intentionally to keep the forest more open to enhance berry production and the amount of plants favored by game animals. This may have increased fire frequency from its natural level, despite the low human population density. The first traces of agriculture are from the Bronze Age, 3,500 years ago (Tolonen 1978).

The human population increased rapidly, beginning in the mid-18th century. As humans increased, prey decreased and the importance of hunting decreased, especially in southern Finland. Grass–herb forests on the most fertile soils were converted to agriculture. Slash-and-burn cultivation was most common in the lake district in the southeastern boreal forest zone. With increasing population pressure, coniferous forests on poorer soils were burned for cultivation (Figure 18.1). This type of forest use was very extensive; in the eastern part of the southern boreal zone, 50 to 75% of forest area has been under slash-and-burn cultivation. Large forest areas never subject to slash-and-burn cultivation can only be found in northern Finland (Heikinheimo 1915).

The increased fire frequency associated with slash-and-burn cultivation favored fire-tolerant *Pinus sylvestris* over fire-sensitive *Picea abies* (Pitkänen 1999). Early successional deciduous trees such as *Betula* spp. and *Populus tremula* (L.) also benefited from increased fire disturbance. In core areas, slash-and-burn cultivation was practiced for centuries and ceased only in the early 20th century (Heikinheimo 1915) with the development of industrial forest utilization.

Naval store production was an important use of Finnish forests. In tar production, *Pinus* trees, the main raw material, were first wounded to increase resin production. Some years later the wounded trees were cut and piled in a pit that was covered by soil, and the woodpile was lit to be combusted into tar. Tar burning was not as widespread as slash-and-burn cultivation, but was locally very influential. Tar burning was practiced since the 16th century, especially in the western and east-central parts of Finland. The time of large-scale production was in the 19th century, when tar became the main export of the country.

Forests were used to provide firewood in the cold climate of Finland, and extraction of construction timber was always an important local use. The production of sawn timber increased rapidly in the late 19th century, which led to large-scale selective logging of the largest and best-quality stems. Selective logging was practiced particularly in areas that were not used for slash-and-burn cultivation. The effect of selective logging on forest structure depended on the intensity of logging, which was highest in southern Finland where most of the sawmills were located. Selective logging probably led to the decline of large-sized living and dead trees over extensive areas (Kuuluvainen et al. 2004). However, at present the forests selectively logged c. 100 years ago are often comparable to natural stands in terms of deadwood and stand structural characteristics (Rouvinen et al. 2002a; Lilja and Kuuluvainen 2004).

Forests were used to pasture domestic animals. Forest grazing was widespread in the late 19th and early 20th centuries. In the 1930s, almost half of the forest area in privately owned forestlands was affected by grazing. Grazing was detrimental for tree regeneration and the forests grazed for a long time changed to open woodlands.

Since World War II, the structure and development of forests have been shaped by industrial utilization, which means intensive management over large areas. At present, forest management and silviculture are based on compartments of 0.5 to 10 ha as basic operational units. The dominant harvesting method is clearcutting, covering about two-thirds of the harvested area, with retention trees left in groups of 10 to 20 trees (Working Group 2000). In *Pinus sylvestris* forests, seed tree cuttings are common (one-third of the harvested area). In regeneration, whether by natural seeding or by planting, soil preparation continues to be used extensively. First plowing was common, but lighter scarification methods are now favored.

Silvicultural treatments, such as thinning from below, are widely used to create homogeneous even-aged stands, to fully utilize the sites' wood production potential. The management ideal has been a fully regulated even-aged forest, where each stand-age class covers an equal area. Although industrial forest management has profoundly changed the structure of forests over the past decades, the regeneration material has always been local; exotic species have not been used in Finnish forestry.

Drainage of peatlands for forestry has been very extensive in Finland. Drainage started in the early 20th century and was most intensive in the 1960s and 1970s, when up to 300,000 ha of peatland were ditched annually; drainage was also commonly combined with fertilization. In southern Finland, for example, 78% of *Picea* mires and 72% of *Pinus* mires and treeless mires have been drained (Virkkala et al. 2000). The structure and functioning of forested peatland ecosystems was changed by drainage in many ways; due to lowering of the water table, the accumulation of peat ceases, the number and size of trees increase, and the original fauna and flora decline or disappear (Laine et al. 1995).

A major change in the disturbance dynamics of forests has been the decline of fire as an ecological factor in forests. Modern fire prevention techniques, combined with a very dense forest road network, have practically eliminated wildfires. Prescribed burning, which was commonly used for forest regeneration in the 1950s and 1960s, has decreased substantially (Parviainen 1996). The absence of forest fires has affected the characteristics of small fragments of protected forests in southern Finland. For example, Lilja and Kuuluvainen (2004) found that protected areas in southern Finland had higher wood volumes and *Picea abies* dominance, and more scarce regeneration of *Pinus sylvestris* and deciduous tree species, compared with a near-natural landscape in Russian Karelia. These differences were mainly due to more recent occurrence of fires in Russian Karelia.

The long history of forest utilization, recent intensive forest management, and suppression of natural disturbance and successional processes, such as forest fires and gap dynamics, have led to a reduction in forest structural complexity, volume of deadwood, and overall biodiversity of forests (Working Group 2000; Siitonen 2001; Kuuluvainen 2002; Rouvinen et al. 2002a). A mosaic of more or less even-aged forest patches has replaced the historical landscape structure dominated by heterogeneous old-growth forests (Pennanen 2002; Kuuluvainen et al. 2004).

Deadwood provides an example of the effect of intensive human utilization on forests and its threat to biodiversity. Dead trees provide habitat for about 5,000 species in Finland (Siitonen 2001). In southern Finland's managed forests, there is only 2 to 10 m^3ha^{-1} of coarse woody debris in managed forests, while in natural forests there is an average 60 to 90 m^3ha^{-1} of deadwood. The average volume of coarse woody debris has decreased by over 90%. Based on general species-area models, over 50% of the species dependent on deadwood (*Saproxylic* species) could disappear (Siitonen 2001).

It is evident that extensive restoration of both managed and protected forest ecosystems will be needed to sustain their natural level of biological diversity (Working Group 2000; Kuuluvainen et al. 2002a). Restoration is urgently needed, in particular, to accelerate the formation of deadwood and other structural features resembling those of natural forests, in order to enhance the conservation function of both protected and managed forests. In southern Finland especially, many protected areas have a long history of utilization before receiving protection. Their value as a habitat for native species has been reduced by past utilization. In addition, they cover only a small area, about 1% of the forests in southern Finland. The need for restoration is particularly urgent in *Picea* mires in boreonemoral and southern boreal zones, where less than 1% of them have been protected and almost half of these protected *Picea* mires have been drained (Virkkala et al. 2000).

18.3 Principles and methods of forest restoration

In Finland, restoration of forested ecosystems aims at reestablishing structures and processes that are typical of natural forest ecosystems (Table 18.1) (Tukia 2000; Tukia et al. 2001; Kuuluvainen et al. 2002a; Working Group 2003). Restoration actions aim at initiating long-term ecological processes, such as tree and dead tree succession, and restoring wetland hydrology (paludification). In the long run, landscape restoration should create and maintain mosaics of habitat types and successional stages that maintain viable populations of species that would naturally occur in the area (Working Group 2003).

Recent advances in understanding the structure, dynamics, and species diversity of natural or near-natural forest ecosystems provide a more firm basis for defining how much we have changed the forests and for setting restoration goals (Korpilahti and Kuuluvainen 2002; Working Group 2002, 2003). However, defining the characteristics of natural forests is not a simple task (Kuuluvainen 2002). A natural forest may be so variable

Table 18.1 Stand- and Landscape-Level Structures and Processes Typical of Natural Boreal Forests that have Diminished Because of Forest Utilization and that Restoration Attempts to Rehabilitate

Structural features to be restored
At Stand Level • Old trees, especially deciduous trees (above all, *Populus tremula* and *Salix caprea*) • Broken, leaning, damaged, and cavity trees • Trees with abundant epiphytic lichen flora • Various kinds of fallen deadwood, especially large logs • Standing dead trees (snags) • Burned living and dead trees • Mounds and pits caused by uprooting • Mixtures of coniferous and deciduous tree species • Multi-aged and -sized tree stand structures • Structurally and compositionally diverse understory canopies • Diverse microhabitat mosaics in relation to water table in peatland forests
At Landscape Level • Natural variability of postfire and other successional stages • Natural-like landscape connectivity • Natural-like ecotone structures
Processes to be restored
At Stand Level • Small-scale gap disturbances • Fine-scale soil disturbances • Postfire successions • Successions following other disturbances besides fire • Natural tree stand succession and self-thinning • Multiple pathways of wood decay successions • Natural successions of peatland forests
At Landscape Level • Natural variability of fire regime • Natural variability and distribution and spatial pattern of young deciduous successional stages • Natural variability and distribution and spatial pattern of old successional stages • Natural variability and distribution and spatial pattern of fire-free areas (fire refugia) • Natural variability of dynamics and spatial distribution of deadwood • Natural-like hydrology of peatland forests

Structural features describe the state of forest ecosystems at a given point in time, while processes refer to the dynamics of disturbance and successional processes in time. Although listed here separately, structures and processes are closely interlinked, since structures are created processes, while processes are affected by structures (modified from Kuuluvainen et al., Principles of ecological resotoration of boreal forested ecosystems: Finland as an example, *Silva Fenn.*, 36, 409, 2002a).

over time that it does not provide any static targets for restoration (White and Walker 1997; Landres et al. 1999). In frequently disturbed ecosystems such as the boreal forest, we often lack knowledge of the natural range of variability of forest structure in areas to be restored. Because forested ecosystems are complex, hierarchical, and dynamic systems (Pickett et al. 1997; Kuuluvainen 2002), hierarchical and holistic approaches to restoration are needed (Pickett and Parker 1994; Christensen et al. 1996; Kuuluvainen 2002). Therefore, the goals of restoration can vary within the framework set by the natural range of variability (Landres et al. 1999; Kuuluvainen 2002; Pennanen 2002).

18.3.1 Methods of forest restoration on mineral soils

Fire is an important component of the ecology of boreal forests (Zackrisson 1977; Pitkänen 1999). Many fire-dependent species have declined in Finland in recent decades, due to suppression of wildfire and lack of prescribed burning. There are 34 species depending on fire that are threatened or have become regionally extinct, and 11 species are near-threatened (Rassi et al. 2001). Restoring boreal forests requires the use of fire. By using fire in a controlled manner, it is possible to increase the complexity of stand structures, increase deadwood, and create open and warm habitats with young successional stages dominated by deciduous trees (Figure 18.2). On the landscape or regional levels, it is necessary to ensure that fire continues to be used in order to provide habitat for fire-dependent species (Working Group 2003).

The selective effect of fire on different tree species increases the structural complexity of the postfire stand and enhances conditions for tree regeneration (Vanha-Majamaa et al. 1996). When a forest fire occurs, some trees die immediately, some within a few years, and some survive the fire. The ecological effects of fire are not limited to overstory trees. Fire consumes some of the soil organic layer, enhancing tree regeneration and activating the dormant soil seed bank of herbaceous species. The resulting increase in deadwood, open sunny conditions, and competition-free substrates created by fire provide important habitat for many decomposer fungi and saproxylic insects (Penttilä and Kotiranta 1996; Wikars 1997).

Creation of a gap-phase structure, which in natural boreal forests is perhaps most typical of old-growth *Picea abies*-dominated forests, may be relevant as part of a restoration project, because natural gap dynamics only emerge late in the successional sequence (Kuuluvainen et al. 1998). In natural forests, small gaps are created when single trees or groups of trees die. The size of such gaps can vary between 0.001 and 0.1 ha. The creation of small gaps diversifies forest structure and tree species composition and produces more deadwood, while still maintaining the moist and shady microclimate typical of old-growth forests. When making gaps artificially, it is important to ensure that fine-scale soil disturbances occur, because pits and mounds are important for both species diversity (Jonsson and Esseen 1990) and tree regeneration (Kuuluvainen 1994).

Figure 18.2 Controlled use of fire is an important method of boreal forest restoration in Finland. Fire is used to increase the complexity of stand structures, to increase deadwood and to create open and warm habitats and young successional stages dominated by deciduous trees (photograph by Saara Lilja).

Damaging and felling of individual trees and groups of trees are used to imitate fine-scale gap disturbances found in natural forests (Tukia et al. 2001; Rouvinen et al. 2002b). The intent is to accelerate the formation of forest structures such as deadwood and canopy gaps that naturally require a long period of time to appear. Single treefalls can be emulated by cutting trees, or by pushing and pulling them down with an excavator to create the pit-mound structures typical of wind-thrown trees. Mechanical harvesters can be used to cut trees at a height of several meters to imitate stem breakage in natural forests. The cambial zone of living trees can be damaged to create slowly dying trees, which are common in natural forests and form an important habitat for many species. It is preferable to use different methods to damage trees in order to imitate the multiple pathways of tree death and wood decay succession found in natural forests (Tukia et al. 2001; Siitonen 2001; Karjalainen and Kuuluvainen 2002).

18.3.2 Methods of restoring forested peatlands

Water is the most important element in restoring forested mires. In pristine mires, natural variability in water flow, such as high water due to exceptionally wet weather, can cause disturbances. Flow patterns may also change, causing waterlogging and tree death in some areas, while enhancing tree growth and seedling establishment in areas where the water level drops. Restoration of drained, forested peatlands requires restoration of both hydrological and biological features and processes. Drainage always changes the natural flow pattern of waters. Restoration of mires begins by filling the drainage ditches in order to restore hydroperiod.

In drained peatlands that were originally forested, restoration of the structure and dynamics of the forest stand is also needed. Pristine spruce mires characteristically had long continuity (mean interval between fires over 400 years), trees of all sizes and ages, large volumes of deadwood, and gap-phase regeneration dynamics (Hörnberg et al. 1998; Siitonen and Saaristo 2000). Peatland forests maintained their uneven-sized and -aged structure for quite a long time after drainage (Hökkä and Laine 1988; Uuttera et al. 1996), but thinning removed larger trees and evened stand structure (Uuttera et al. 1997; Päivänen 1999).

When restoring peatlands that were sparsely forested mires, such as *Pinus* mires, reducing stem density of the present stand by partial harvesting is often plausible for both ecological and economical reasons. If the original mire type was open, trees grown after drainage should be cut and removed. Restoring the natural uneven-aged structure of *Picea* mires requires attention to regeneration conditions. Raised microhabitats, such as logs, are crucial for the regeneration of spruce (Hörnberg et al. 1997) but, due to previous management, logs are not necessarily available in drained peatlands. Some trees may die due to waterlogging, but if this is not expected, logs should be created by felling trees. Restoring the hardwood component of the stand is usually not a problem since at least birch readily colonizes restored mire habitats. Birch may become a problem if it forms a dense thicket that prevents regeneration of other tree species and recolonization by the original mire plant species.

18.4 Forest restoration in Finland

18.4.1 Restoration in protected forests

Protected forests are distributed unevenly in Finland. Most protected areas are in northern and northeastern Finland, and very little forest area is protected in the south of the country. The greatest percentage of protected forest area is in the northern boreal zone (16.9%).

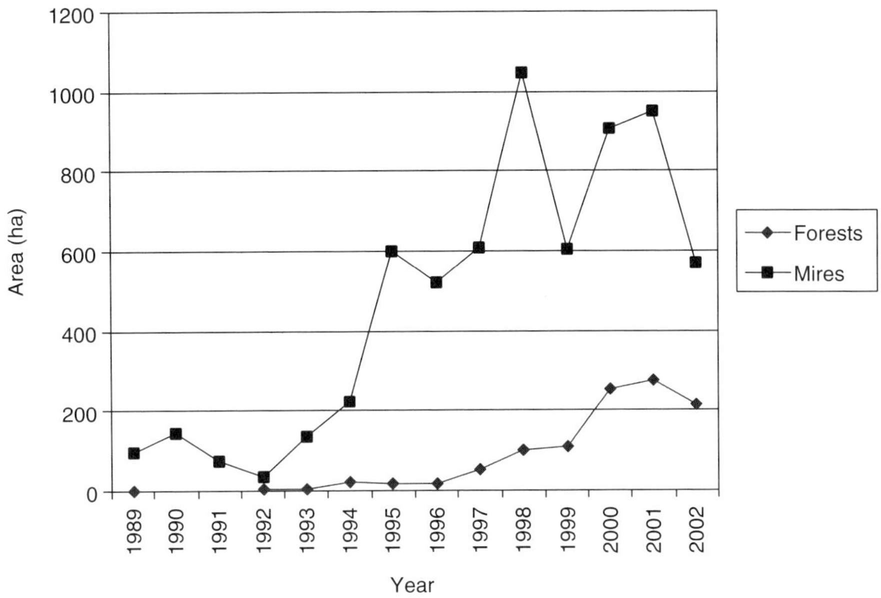

Figure 18.3 Restoration of forests and mires in protected areas in Finland from 1989 to 2002 (Working Group 2003).

The proportion of protected forests drops as one moves south: only 2.4% of the middle boreal, 0.7% of the southern boreal, and 1.6% of the boreonemoral zones are protected (Figure 18.1).

The first restoration experiment in Finland was carried out in 1989, when two small forested mineral soil islands were burned in the middle of a large mire in the Patvinsuo National Park. Not until the end of the 1990s was more than 100 ha of forests restored annually (Figure 18.3). Until 2003, approximately 1,300 ha of forests have been restored in protected areas (Working Group 2003). Until 1998, burning was the main method of restoration on mineral soils. Since then, creating gaps and deadwood has increased in importance as methods of forest restoration.

Today, the most commonly used restoration methods in protected forests have been forming deadwood by tree girdling or felling (46% of all restored forests), burning of stands (23%), and imitating gap dynamics by creating small openings in even-aged stands (16%). Restoration of all mires have included filling in the ditches. In 2002, the total area of restored mires in protected areas was 7,300 ha (Working Group 2003).

Using restoration to create functional, connected, conservation networks is a major challenge because even the largest protected areas in Finland are quite small; only 7 of 35 National Parks are larger than 100 km^2. An important goal is to plan restoration actions to better connect the small, separated, protected areas in the network and to buffer them against the unwanted effects from the surrounding managed forests.

18.4.2 Restoration in managed forests

Finnish forestry law was reformed in 1997 and management has been modified to meet the equally important goals of ecological and social sustainability, preservation of biodiversity, and sustainable yield of forests. Governmental, industrial, and private forestry organizations have all modified their management guidelines during the 1990s (Monimuotoisuus UPM-Kymmenen 1998; Korhonen and Savonmäki 1997).

New management practices that are already widely applied include setting aside habitats of special importance for forest biodiversity (so-called key biotopes), retention of living and dead trees in harvesting, and favoring prescribed burning and deciduous admixture. Landscape ecological planning is practiced on all state forests managed by Metsähallitus (Forest and Park Service) (Karvonen 1999). The ASIO model, which aims at demonstrating the natural fire dynamics and landscape structures (Angelstam and Rosenberg 1993; Angelstam 1998), is used as one of the starting points in landscape planning (Karvonen 1999). All these measures aim at adding structural features of natural forests into managed forests; hence, the measures can be regarded as restoration in its widest sense (rehabilitation) (Stanturf, this volume). However, at present, we have only a limited understanding of the ecological efficiency of these new management practices.

18.5 *Future prospects and challenges of restoration*

In 1999, the Ministry of Environment commissioned an expert working group to evaluate the current state of forest protection in southern Finland. The commission report (Working Group 2000) lists a number of measures that should be taken to improve protection, including: additional conservation of poorly represented forest types, such as herb–grass forests and spruce mires; restoration of forests both within protected areas and managed forests surrounding them; formation of larger conservation networks around the existing core areas; and enhancing biodiversity-oriented silvicultural methods in managed forests.

Restoration tools will be needed to implement all these recommendations. Restoration can be used in nature conservation to shift forest structure within protected areas closer to their natural state in order to enhance habitat quantity and quality for focal species; to enlarge and complement small and fragmented areas in the vicinity of protected areas to form larger and better connected units that can maintain landscape-level successional and disturbance dynamics; to create buffer zones between managed and protected areas; and to improve connectivity of protected areas (Kuuluvainen et al. 2002a).

Restoration of existing protection areas and adjacent managed forests may be a better, more effective ecologically, and more acceptable economically, strategy than setting aside new protection areas in previously managed forests. Simulation models suggest that restoration of managed forests located close to existing source areas of sensitive species should be most efficient in terms of species conservation (Tilman et al. 1997; Huxell and Hastings 1999; Hanski 2000). Restored habitats are more likely to be colonized if the restored patches are close to existing sources and within the colonization abilities of potential species. Thus, in the future, restoration principles could be used in managed forests to complement the protection function of reserves.

Effective restoration considers the roles of both protection area networks and managed forests. The most urgent task, however, is the restoration of protection areas where at least 39,000 ha of forests and 20,000 ha of mires are in need of restoration. Deadwood in protected forests in southern Finland must be increased up to 30 m^3 ha^{-1}. A network of fire continuum areas covering the whole country should be established that would enhance the viability of fire-dependent species (Working Group 2003).

Only a small portion of the protected forests in southern Finland is close to its natural state (Kuuluvainen et al. 2002a). Restoration in these protected areas could increase their capacity to support populations of many threatened species (Working Group 2003). However, these protected areas may also be too small and fragmented to allow necessary natural forest dynamics for maintenance of these species (Hanski 2000). Restoration can be used to enlarge and combine smaller areas into larger landscape units that better meets the goals of ecosystem and species protection.

In general, restoration can be regarded as the most important means of maintaining and complementing networks of protected forests in areas where the forest landscape has been strongly altered by human impacts, such as in southern Finland.

Maintenance of structural and species diversity of forests has not been a goal in traditional silvicultural systems, which have concentrated on wood production. Although recent management guidelines aim at adding or maintaining some structural components important for biodiversity (leaving key biotopes and retention trees), current silvicultural systems will not restore the essential multiscale characteristics of the natural forest (Kuuluvainen 2002). Natural forest can be used as a reference when developing forest management methods aiming at biodiversity restoration and management (Haila et al. 1994; Angelstam 1998; Lähde et al. 1999; Kuuluvainen 2002). Cutting treatments must be diversified to create more variability in disturbance severity, quality, extent, and repeatability. In practice, this means that there should be a shift from harvesting dominated by clearcutting to management using a range of partial cutting methods, inspired by tree mortality patterns in natural forests (Bergeron et al. 2002). Additional forest management methods and strategies must be developed that are based on natural disturbances or natural variability of forests and that are aimed at restoration and management of structural complexity and biodiversity (Kuuluvainen 2002).

One of the future tasks is to link restoration actions with monitoring and research (Young 2000; Ehrenfeld 2000; Block et al. 2001). Monitoring is a prerequisite to evaluation of the outcome of restoration so that practices can be changed, if needed (the principle of adaptive management) (Walters 1986; Walters and Holling 1990). Incorporation of research into management would also create synergy benefits, such as facilitating experiments on scales that are both ecologically and managerially relevant. Research and monitoring have been part of restoration since the beginning in Finland (Heikkilä and Lindholm 1995, 1997; Tukia 2000) but sufficient funding for monitoring must be ensured.

References

Angelstam, P., Maintaining and restoring biodiversity in European boreal forests by developing natural disturbance regimes, *J. Veg. Sci.*, 9, 593, 1998.

Angelstam, P. and Rosenberg, P., Aldrig Sällan Ibland Ofta, *Skog och Forskning*, 1, 34, 1993 (in Swedish).

Bergeron, Y., Leduc, A., Harvey, B., and Gauthier, S., Natural fire regime: a guide for sustainable management of the Canadian boreal forest, *Silva Fenn.*, 36, 81, 2002.

Block, W., Franklin, A., Ward, J., Caney, J., and White, G., Design and implementation of monitoring studies to evaluate the success of ecological restoration on wildlife, *Restor. Ecol.*, 9, 293, 2001.

Christensen, N.L., Bartuska, A.M., Brown, J.H., Carpenter, S., D'Antonio, C., Francis, R., Franklin, J.F., MacMahon, J.A., Noss, R.J., Parsons, D.J., Peterson, C.H., Turner, M.G., and Woodmansee, R.G., The report of The Ecological Society of America Committee on the scientific basis for ecosystem management, *Ecol. Appl.*, 6, 665, 1996.

Engelmark, O., Boreal forest disturbances, in *Ecosystems of Disturbed Grounds; Ecosystems of the World 16*, Walker, L., Ed., Elsevier, Amsterdam, 161, 1999.

Ehrenfeld, J., Defining the limits of restoration: the need for realistic goals, *Restor. Ecol.*, 8, 2, 2000.

Haila, Y., Hanski, I.K., Niemelä, J., Punttila, P., Raivio, S., and Tukia, H., Forestry and the boreal fauna: matching management with natural forest dynamics, *Ann. Zool. Fenn.*, 31, 187, 1994.

Hanski, I., Extinction debt and species credit in boreal forests modeling the consequences of different patterns of forest restoration, *Ann. Zool. Fenn.*, 37, 271, 2000.

Heikinheimo, O., Kaskiviljelyksen vaikutus suomalaisiin metsiin (The effects of slash-and-burn cultivation to forests in Finland), *Acta Forestalia Fennica 4*, 1915 (in Finnish; German summary).

Heikkilä, H. and Lindholm, T., The basis of mire restoration in Finland, in *Restoration of Temperate Wetlands*, Wheeler, S.C., Shaw, S.C., Fojt, W.J., and Robertson, R.A., Eds., John Wiley & Sons Ltd., New York, 549, 1995.

Heikkilä, H. and Lindholm, T., Soiden ennallistamistutkimus vuosina 1987-1996 (The mire restoration study from 1987 to 1996), *Metsähllituksen luonnonsuojelujulkaisuja, Sarja A* 81, 1, 1997 (in Finnish).

Huxell, G.R. and Hastings, A., Habitat loss, fragmentation and restoration, *Restor. Ecol.*, 7, 309, 1999.

Hökkä, H. and Laine, J., Suopuustojen rakenteen kehitys ojituksen jälkeen (Post-drainage development of structural characteristics in peatland forest stands), *Silva Fenn.*, 22, 45, 1988.

Hörnberg, G., Ohlson, M., and Zackrisson, O., Influence of bryophytes and microrelief conditions on *Picea abies* seed regeneration patterns in boreal old-growth swamp forests, *Can. J. For. Res.*, 27, 1015, 1997.

Hörnberg, G., Zackrisson, O., Segeström, U., Svensson, B.W., Ohlson, M., and Bradshaw, R., Boreal swamp forests. Biodiversity "hotspots" in an impoverished forest landscape, *BioScience*, 48, 795, 1998.

Jonsson, B.G. and Esseen, P.-A., Treefall disturbance maintains high bryophyte diversity in a boreal spruce forest, *J. Ecol.*, 78, 924, 1990.

Karjalainen, L. and Kuuluvainen, T., Amount and diversity of coarse woody debris within a boreal forest landscape dominated by *Pinus sylvestris* in Vienansalo wilderness, eastern Fennoscandia, *Silva Fenn.*, 36, 147, 2002.

Karvonen, L., *Alue-ekologisen suunnittelun ohje (Guide to Landscape Ecological Planning)*, Finnish Forest and Park Service, Helsinki, 23 pp., 1999 (in Finnish).

Korhonen, K.-M. and Savonmäki, S., Eds., *Metsätalouden ympäristöopas (Environmental Guide of Forestry)*, Metsähallitus, Helsinki, 130 pp., 1997 (in Finnish).

Korpilahti, E. and Kuuluvainen, T., Eds., Disturbance dynamics in boreal forests: defining the ecological basis of restoration and management of biodiversity, *Silva Fenn.*, 36, 2002.

Kuuluvainen, T., Gap disturbance, ground microtopography, and the regeneration dynamics of boreal coniferous forests in Finland, a review, *Ann. Zool. Fenn.*, 31, 35, 1994.

Kuuluvainen, T., Natural variability of forests as a reference for restoring and managing biological diversity in boreal Fennoscandia, *Silva Fenn.*, 36, 97, 2002.

Kuuluvainen, T., Syrjänen, K. and Kalliola, R., Structure of a pristine *Picea abies* forest in northeastern Europe, *J. Veg. Sci.*, 9, 563, 1998.

Kuuluvainen, T., Aapala, K., Ahlroth, P., Kuusinen, M., Lindholm, T., Sallantaus, T., Siitonen, J., and Tukia, H., Principles of ecological restoration of boreal forested ecosystems: Finland as an example, *Silva Fenn.*, 36, 409, 2002a.

Kuuluvainen, T., Mäki, J., Karjalainen, L., and Lehtonen, H., Tree age distributions in old-growth forest sites in Vienansalo wilderness, eastern Fennoscandia, *Silva Fenn.*, 36, 169, 2002b.

Kuuluvainen, T., Jäppinen, J.-P., Kivimaa, T., Rassi, P., Salpakivi-Salomaa, P., and Siitonen, J., Ihmisen vaikutus Suomen metsiin (Human impact on forests in Finland), in *Metsän kätköissä — Suomen metsäluonnon monimuotoisuus (Hidden in the Forest — Biodiversity of Forests in Finland)*, Kuuluvainen, T., Saaristo, L., Keto-Tokoi, P., Kostamo, J., Kuuluvainen, J., Kuusinen, M., Ollikainen, M., and Salpakivi-Salomaa, P., Eds., *Edita*, Helsinki, 2004 (in Finnish).

Lähde, E., Laiho, O., and Norokorpi, Y., Diversity-oriented silviculture in the boreal zone of Europe, *For. Ecol. Manage.*, 118, 223, 1999.

Laine, J., Vasander, H., and Sallantaus, T., Ecological effects of peatland drainage for forestry, *Environ. Rev.*, 3, 286, 1995.

Landres, P.B., Morgan, P., and Swanson, F.J., Overview of the use of natural variability concepts in managing ecological systems, *Ecol. Appl.*, 9, 1179, 1999.

Larjavaara, M., Kuuluvainen, T., Rita, H., and Venäläinen, A., Spatial distribution of lightning ignited forest fires in Finland, *Silva Fenn.*, In Press.

Lehtonen, H., Forest fire history in north Karelia: dendroecological approach, University of Joensuu, Joensuu, *Research Notes*, 59, 1, 1997.

Lilja, S. and Kuuluvainen, T., Structural changes in old *Pinus sylvestris* dominated forests along a geographic and human impact gradient in boreal Fennoscandia: implications for restoration and conservation, (submitted manuscript).

Monimuotoisuus UPM-Kymmenen metsissä, Tausta ja toimenpiteet (*Biodiversity in the Forests of UPM-Kymmene. Background and Action Plan*), UPM-Kymmene Metsä, Valkeakoski, 64 pp., 1998 (in Finnish).

Nenonen, K., Pleistocene stratigraphy of southern Finland, in *Glacial Deposits in North-East Europe*, Ehlers, J., Kozarski, S., and Gibbard, P. L., Eds., A.A. Balkema, Rotterdam, 11, 1995.
Parviainen, J., Impact of fire on Finnish forests in the past and today, *Silva Fenn.*, 30, 273, 1996.
Päivänen, J., Tree stand structure on pristine peatlands and changes after forest drainage, *Int. Peat J.*, 9, 66, 1999.
Pennanen, J., Forest age distribution under mixed-severity fire regimes — a simulation-based analysis for middle boreal Fennoscandia, *Silva Fenn.*, 36, 213, 2002.
Penttilä, R. and Kotiranta, H., Short-term effects of prescribed burning on wood-rotting fungi, *Silva Fenn.*, 30, 399, 1996.
Pickett, S.T.A. and Parker, V.T., Avoiding the old pitfalls: opportunities in a new discipline, *Restor. Ecol.*, 2, 75, 1994.
Pickett, S.T.A., Ostfeld, R.S., Shachak, M., and Likens, G.E., *The Ecological Basis of Conservation: Heterogeneity, Ecosystems, and Biodiversity*, Chapham & Hall, New York, 466 pp., 1997.
Pitkänen, A., Paleoecological study of the history of forest fires in eastern Finland, University of Joensuu, *Publications in Sciences*, 58, 1999.
Pitkänen, A., Huttunen, P., Jungner, H., and Tolonen, K., A 10,000-year local fire history in a dry heath forest in eastern Finland, *Can. J. For. Res.*, 32, 1875, 2002a.
Pitkänen, A., Huttunen, P., Tolonen, K., and Jungner, H., Long-term fire frequency in the spruce-dominated forests of the Ulvinsalo strict nature reserve, Finland, *For. Ecol. Manage.*, 176, 305, 2002b.
Rassi, P., Alanen, A., and Kanerva, T., Eds., *The 2000 Red List of Finnish Species*, Ministry of the Environment and Finnish Environment Institute, Helsinki, 432 pp., 2001.
Rouvinen, S., Kuuluvainen, T., and Karjalainen, L., Coarse woody debris in old *Pinus sylvestris*-dominated forests along a geographic and human impact gradient in boreal Fennoscandia, *Can. J. For. Res.*, 32, 2184, 2002a.
Rouvinen, S., Kuuluvainen, T., and Siitonen, J., Tree mortality in a *Pinus sylvestris*-dominated boreal forest landscape in Vienansalo wilderness, eastern Fennoscandia, *Silva Fenn.*, 36, 127, 2002b.
Siitonen, J., Forest management, coarse woody debris and saproxylic organisms: Fennoscandian boreal forests as an example, *Ecol. Bull.*, 49, 11, 2001.
Siitonen, J. and Saaristo, L., Habitat requirements and conservation of *Pytho kolwensis*, a beetle species of old-growth boreal forest, *Biol. Conserv.*, 94, 211, 2000.
Sirén, G., The development of spruce forest on raw humus sites and its ecology, *Acta For. Fenn.*, 62, 1, 1955.
Syrjänen, K., Kalliola, R., Puolasmaa, A., and Mattson, J., Landscape structure and forest dynamics in subcontinental Russian European taiga, *Ann. Zool. Fenn.*, 31, 19, 1994.
Tilman, D., Lehman, C.L., and Kareiva, P., Population dynamics in spatial habitats, in *Spatial Ecology. The Role of Space in Population Dynamics and Interspecific Interactions*, Tilman D. and Kareiva, P., Eds., Monographs in Population Biology, 30, 3, 1997.
Tolonen, K., Paleoecology of annual laminated sediments in Lake Ahvenainen, S. Finland. I. Pollen and charcoal analyses, and their relation to human impact, *Ann. Bot. Fenn.*, 26, 53, 1978.
Tukia, H., Metsien ennallistaminen suojelualueilla — lähtötilanne 1995 (Restoration of forests in protected areas — initial state in 1995), *Metsähallituksen luonnonsuojelujulkaisuja, Sarja A* 118, 1, 2000 (in Finnish).
Tukia, H., Hokkanen, M., Jaakkola, S., Kallonen, S., Kurikka, T., Leivo, A., Lindholm, T., Suikki, A. and Virolainen, E., Metsien ennallistamisopas (The handbook of ecological forest restoration), *Metsähallituksen luonnonsuojelujulkaisuja, Sarja B*, 58, 87 pp., 2001. (in Finnish).
Uuttera, J., Maltamo, M., and Hotanen, J.-P., Stand structure of undrained and drained peatland forests in central Finland, *Suo*, 47, 125, 1996.
Uuttera, J., Maltamo, M., and Hotanen, J.-P., The structure of forest stands in virgin and managed peatlands: a comparison between Finnish and Russian Karelia, *For. Ecol. Manage.*, 96, 125, 1997.
Vanha-Majamaa, I., Tuittila, E.-S., Tonteri, T., and Suominen, R., Seedling establishment after prescribed burning of a clearcut and partially cut mesic boreal forest in southern Finland, *Silva Fenn.*, 30, 31, 1996.

Virkkala, R., Korhonen, K.T., Haapanen, R., and Aapala, K., Metsien ja soiden suojelutilanne metsä- ja suokasvillisuusvyöhykkeittäin valtakunnan metsien 8. inventoinnin perusteella. (Protected forests and mires in forest and mire vegetation zones in Finland based on the 8th National Forest Inventory) *Suomen ympäristö*, 395, 1, 2000 (in Finnish; English summary).

Wallenius, T., Forest age distribution and traces of past fires in a natural landscape dominated by *Picea abies*, *Silva Fenn.*, 36, 201, 2002.

Walters, C.J., *Adaptive Management of Renewable Resources*, McGraw-Hill, New York, 1986.

Walters, C.J. and Holling, C.S., Large-scale management experiments and learning by doing, *Ecology*, 71, 2060, 1990.

White, P.S. and Walker, J.L., Approximating nature's variation: selecting and using reference information in restoration ecology, *Restor. Ecol.*, 5, 338, 1997.

Wikars, L., Effects of forest fire and the ecology of fire-adapted insects, Comprehensive summaries of Uppsala dissertations, *Fac. Sci. Technol.*, 272, 1, 1997.

Working Group on the need for forest protection in southern Finland and Ostrobothnia, Forest protection in southern Finland and Ostrobothnia, *Finnish Environ* 437, 284, 2000 (in Finnish; English summary).

Working Group on restoration, Restoration in protected areas: report by the working group on restoration, *The Finnish Environment 618*, 220, 2003 (in Finnish; English summary).

Young, T., Restoration ecology and conservation biology, *Biol. Conserv.*, 92, 73, 2000.

Zackrisson, O., Influence of forest fires on the North Swedish boreal forest, *Oikos*, 29, 22, 1977.

chapter nineteen

Forest rehabilitation in Denmark using nature-based forestry

Katrine Hahn, Jens Emborg, J. Bo Larsen, and Palle Madsen
Danish Centre for Forest, Landscape and Planning, KVL

Contents

19.1 Denmark: A case of forest rehabilitation

Forestry in Denmark is presently experiencing an economic crisis, which coincides with society's increasing demands for sustainable, multipurpose forestry. From a practical as well as a political perspective, nature-based forestry has been proposed as a tool to rehabilitate our forests to enable them to fulfill our future expectations. In this chapter, we describe and discuss the background for rehabilitation, the practical implications, and the possible outcomes, well aware that practical as well as scientific experiences with nature-based forestry are still rather limited.

Although, or perhaps because, Denmark is a small country with a very limited forest resource, the debate about the future needs for rehabilitation (Stanturf and Madsen 2002)

1-56670-635-1/05/$0.00+$1.50

of Danish forests has been rather intensive (Forfang et al. 1995; Heding et al. 2002; Larsen and Emborg 2002). The debate has included the whole range of interests tied to our woodlands, from the economic perspectives of the forest owners, to the public seeking recreation, and the hunter, or horseback rider paying the landowner for access. Additionally, the debate has focused on the forests' potential to serve society in general by, for example, protecting groundwater, hosting flora, fauna and wildlife, accumulating CO_2, and producing wood. Our existing monoculture plantation-like forests are indeed expected to be multitasking in the future!

Traditional Danish forest management with homogeneous single species, single-storied stands established by intensive and expensive plantings does not seem to meet the present needs. Therefore, forest owners and managers are paying increasing attention to alternative silvicultural strategies as a means to rehabilitate the forests to meet the present and future challenges. The so-called nature-based forestry is now considered as an alternative silvicultural approach, which seems to draw more and more attention among scientists and practitioners. The overall vision of nature-based forestry is to make use of the natural processes as regards production and organization of the ecosystem for economic as well as ecological optimal control of the production processes (Larsen 2000). This new trend in Denmark is a modern modification of an old silvicultural phrase: "*... to follow and support Nature in her performance*" (The Royal Forest Decree 1781).

The Ministry of Environment and society in general find that nature-based forestry is a promising means to fulfill the goals of sustainable forest management including the integration of nature conservation and environmental care. From the point of view of a private landowner, nature-based forestry is attractive due to much less intensive and more inexpensive regeneration and stand tending efforts. Moreover, lessons have been learned about the importance of forest stability, including both resistance and resilience (Larsen 1995).

Here we analyze and discuss the recent changes toward nature-based forestry within a historical, political, and economic context. Examples of unstable conifer plantations and plantation-like beech forest are used to illustrate the challenges as well as the needs for more practical experience and scientifically based knowledge that can successfully guide the rehabilitation process.

19.1.1 Ecological–historical background

Denmark is situated in the Atlantic lowland region of northern Europe, where mixed deciduous forest is the predominant potential natural vegetation type (Ellenberg 1996; Svenning 2002). The last ice age, which ended approximately 12,000 years ago, left large outwash plains of poor sandy soils in the western part, and rich clayey till in the eastern part of the country. After the retreat of the ice, the land gradually became covered by woodland. Closed forest was the predominant vegetation type, but longer-lasting openings and light-demanding vegetation occurred in wetland areas and on dry, infertile soils (Svenning 2000). The original woodland composition included many deciduous species—beech (*Fagus sylvatica*), ash (*Fraxinus excelsior*), oak (*Quercus robur* and *Q. petraea*), lime (*Tilia cordata* and *T. platyphyllos*), elm (*Ulmus glabra*), hornbeam (*Carpinus betulus*), birch (*Betula pendula* and *B. pubescens*), —but few coniferous species—the only ones being Scots pine (*Pinus sylvestris*), yew (*Taxus baccata*), and juniper (*Juniperus communis*). Studies of natural forest remnants show that the natural forests were characterized by high structural diversity, shifting mosaic cycles (up to 300 years) driven by small-scale disturbances such as wind, low-intensity fires, and insects, and occasional large-scale windstorms (Emborg et al. 2000; Hannon et al. 2000).

Parallel to many other European countries, Denmark came close to complete deforestation around AD 1800, reaching a low of 2 to 3% forest cover (Fritzbøger 1994). By then, western Denmark was dominated by vast, impoverished heathland (*Calluna vulgaris*), whereas the eastern region was turned into highly productive agricultural land. Due to large afforestation projects within the last 250 years, forest cover has now increased to 11% (approximately. 486,000 ha) (Larsen and Johannsen 2002), and continues to increase at a rate of approximately 2,000 ha per year (Kirkebæk and Thormann 2000).

The early afforestation projects (AD 1850 to 1900) were dominated by the establishment of plantations of exotic conifers, mainly Norway spruce (*Picea abies*), primarily of the infertile, western heathland soils. These sites were not able to support first-generation hardwoods with a reasonable growth rate. The difficulty of cultivating the heathland historically was also a major obstacle to practicing hardwood silviculture. Later, conifer plantations were also established on both marginal farmland and as a replacement for clearfelled deciduous woodlands in the eastern part of the country (Figure 19.1). Deciduous afforestation did not become common until the last decade, when government policy, including subsidies, generated a larger interest among forest owners (Madsen et al., this volume). Most plantations were established without regard to indigenous plant associations and little attention to ecological site conditions, and the actual tree species composition is therefore rather different from the potential, natural vegetation cover. Presently, two-thirds of the Danish forests are plantations with exotic conifers, the main species being Norway spruce. The remaining one-third is structurally homogeneous forest stands of native broadleaved tree species, primarily beech (Figure 19.2). The long-term decrease in the proportion of deciduous tree species combined with a recent expansion of Norway spruce is by no means unique for Denmark, but can be recognized in most of northern Europe, for example, Germany (Zerbe 2002) and Sweden (Björse and Bradshaw 1998).

19.1.2 Economical, social, and political background

Approximately one half of the Danish forest area is private forests, one-fourth is public forests, and organizations, trusts, and companies own one fourth (Table 19.1). Very few, large forest estates (>500 ha) account for over two-thirds of the total forest area, whereas the many, small ownerships contribute little to the overall forest area (Figure 19.3). Typically, owners of private forests combine forestry with agricultural activities. Public forests often include other types of open spaces such as lakes, heath, and grasslands.

Throughout most of the 20th century, economic calculations have shown much better returns from exotic conifers than from the native hardwood species. Norway spruce was planted outside or on the border of its ecophysiological range due to its rapid production of quality timber and low regeneration costs. However, forest economists did not properly include the risk of windthrow or the poor health and stability of Norway spruce and other exotic conifers, nor were they able to foresee the continuously decreasing prices of softwood timber over the last 20 years. Currently, earnings from timber are very small, stand

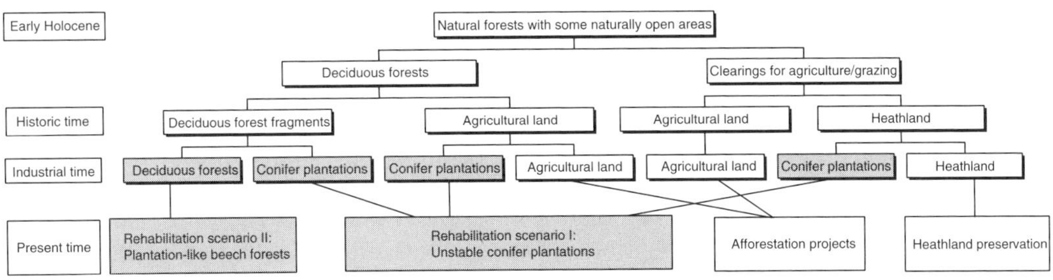

Figure 19.1 Changing land cover in Denmark, from the early Holocene (10,000 BC) to the present day. The gray boxes highlight the two scenarios of forest rehabilitation that are presented in this chapter.

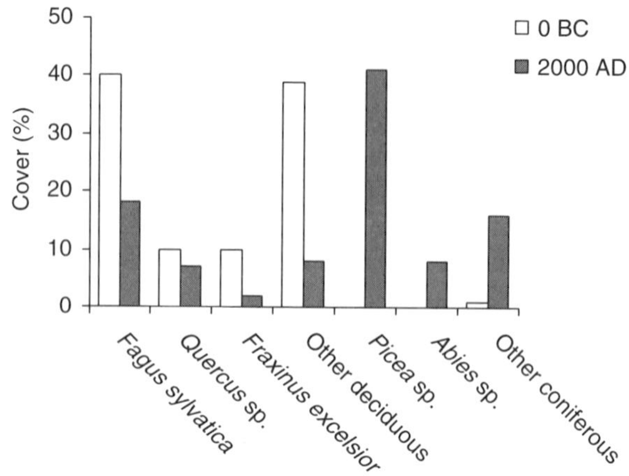

Figure 19.2 Change in species composition in Danish forests over the past 2,000 years (data from Bradshaw et al. [1999] and Larsen and Johannsen [2002]).

Table 19.1 Ownership Structure of Danish Forestland (data from Larsen and Johannsen 2002)

Ownership Type	Forest Area (ha)
Private	224,000
Public, state	113,000
Public, other	25,000
Organizations, trusts, and companies	124,000

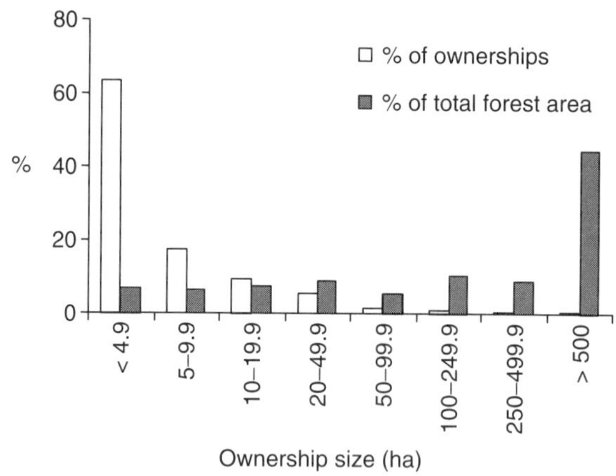

Figure 19.3 Size class distribution of forest ownerships in Denmark (data from Larsen and Johannsen 2002).

establishment costs are high, and the proportion of unstable conifer stands is high. Future economic returns from forestry seem very uncertain.

Presently, the annual economic balance for private forest districts varies between a surplus of US\$ 10 ha^{-1} and a deficit of US\$ 190 ha^{-1}. Hunting licenses in particular, but also Christmas trees and greenery, positively contribute to profitability (the export value of Christmas trees and greenery was approximately US\$ 100 million in 1999); timber production contributes little or even negatively (Dansk Skovforening 2000; Larsen and

Johannsen 2002). Recreational hunting, especially for deer, is an important feature provided by the forests, with an annual harvest of 100,000 to 120,000 deer (Larsen and Johannsen 2002). The forest owner typically obtains an annual rent of 25 to 60 US$ ha^{-1} from forestland for hunting privileges (Thorsen and Strange 2000). However, from a national economic perspective, forestry makes a small contribution to gross national income and has declined within the last decade from 0.16% in 1990 to 0.11% in 1999 (Larsen and Johannsen 2002). The combination of decreasing timber prices and increasing wages and taxes has initiated an economic crisis in Danish forestry, forcing private, as well as public, forest owners to minimize the expenditure for management operations.

Compared to the world's forests in general, Danish forests hold a story of intensive monoculture management. Nevertheless, many Danes consider their forests as "genuine nature," since forest management is low-intensive landuse compared to agriculture, which dominates the Danish landscape. Similar to many other countries throughout Europe, the regard for nature and the environment has grown considerably during the past 20 years in Denmark and has become an important political issue. The recreational use of Danish forests is high, due to a general law giving public access to most forests, including those privately owned (Hansen et al. 1997). With about 50 million visits to the Danish forests every year (Larsen and Johannsen 2002), there is no doubt that the forests are more frequented than the most popular museums and entertainment parks.

From a social perspective, the changes from well-known monoculture plantations to heterogeneous, multifunctional forests may change the recreational appearance of the forest more than professional foresters can imagine. Although the general reaction of the public to the change from monoculture spruce plantations to mixed stands is positive, the public perceives forests as a stable and unchanging element in a constantly changing world (Jensen 2002). It has also been postulated that a heterogeneous forest structure with a dense understory will remove the desirable transparent appearance of the column-like beech stands, and that structures such as creeks, rolling hills, and burial mounds would possibly become blurred by a dense green carpet (Stahlschmidt 2002).

The large contrasts in economic, biological, and social aims are one of the greatest challenges for forestry today (Table 19.2). It is clear that we must integrate an increasing number of goals and functions into our forest management. The diverse and multifunctional aims of forestry have emphasized the need for the forests to be flexible with respect to future outputs of both wood and nonwood products. The importance of flexibility is magnified by the relatively long production periods in Danish forestry, ranging between 50 and 150 years.

19.1.3 Forest rehabilitation goals

In 2002, the Danish Ministry of Environment decided on a new national forest program (Skov- og Naturstyrelsen 2002a, 2000b) based on international guidelines as well as national reports and recommendations (Thorsen and Strange 2000; Emborg et al. 2001; Larsen and Madsen 2001; Larsen et al. 2001; Rune 2001; The Wilhjelm Committee 2001). In this program, nature-based forest management is seen as an efficient and economically desirable tool for forest rehabilitation. Nature-based forestry, in combination with other incentives, will be used as a rehabilitation tool for changing our forests fundamentally from monoculture plantations, largely based on nonnative species, to sustainable natural forest ecosystems with more site-adapted, often native, species. Similar ideas are being launched by private forest owners, often in association with Pro-Silva, the European organization for sustainable forestry (Hahn and Fanta 2001).

The goal is a gradual readjustment of all public forests (133,000 ha) within 80 to 100 years plus additional voluntary conversion in private forests. It has been estimated that

Table 19.2 Present Attributes of Typical Danish Forests, as Compared to Desired Attributes of Future Forests.

Danish forests are typically...
1. Plantations established since 1800
2. Dominated by exotic conifers
3. Established in regions with low population densities
4. Established primarily with the aim of wood production
5. Showing considerable problems with stability
6. Often deficit-making

Our future forests are expected to...
1. Be multifunctional forests
2. Protect biological diversity
3. Provide areas for recreation near cities and in nature parks
4. Provide wood and wood products of high quality
5. Be stable and flexible forests
6. Strengthen landscape values and amenities, provide environmental protection, groundwater protection, and CO_2 storage

Source: Skov- og Naturstyrelsen, *Danmarks mationale skovprogram*, Miljømisteriet, 2002a.

nature-based forest management can be implemented on the first 25% of the area within 10 years (The Wilhjelm Committee 2001; Skov- og Naturstyrelsen 2002a, 2002b). Moreover, it is suggested that 10% of the forests should be protected as larger woodland reserves in combination with other landscape elements such as heathland, wetlands, and coastal areas, possibly in the form of national parks (Skov- og Naturstyrelsen 2002a, 2002b).

19.2 Nature-based forestry — how?

The future management guidelines should be sufficiently flexible to accommodate changing conditions in climate, sites, economy, and social needs. When looking to our Scandinavian and Baltic neighbors, two approaches can be observed: integration of functions by development of silvicultural models to rehabilitate natural stand structure and functions, that is, by application of nature-based forest management, and separation of functions so that intensive forestry in some areas is compensated by strict forest reserves, with high biological values in other areas. Nature-based forestry, which represents the integrating approach, is discussed intensively all over Europe (Bradshaw et al. 1994; Fries et al. 1997; Führer 2000; Hüttl et al. 2000) and has been applied in all public forests in Germany within the last decade, for example, the LÖWE program from Lower Saxony (Niedersachisches Ministerium für Ernährung, Landwirtschaft und Forsten 1991). In Denmark, the decision has been made to combine the two approaches, integration and separation, using them where they fit best, as the areas of existing and potential forest reserves in Denmark are very small. Therefore, we need to manage most forestland for both production and biodiversity. Additionally, this also reflects the Danish political tradition of seeking consensus.

19.2.1 Silviculture of nature-based forestry

The idea behind nature-based forestry is to take advantage of and mimic the natural processes and structures of the forest ecosystem rather than try to control the system throughout the whole rotation (Figure 19.4). This approach is supposed to work as a

Figure 19.4 Old beech stand, approximately 110 years old, in the regeneration phase. In nature-based silviculture, the stand is allowed to develop a more heterogeneous structure with beech, sycamore (*Acer pseudoplatanoides*), and perhaps ash (*Fraxinus excelsior*) in the stand. Traditional management would thin and regenerate to achieve a uniform, plantation-like structure.

general guideline for an economically and ecologically sound management of the forest. Originally, the economic rationale of nature-based forest management was to reduce the expenses using natural processes, such as natural regeneration and self-thinning in the early stages of stand development. This is still a very important argument for nature-based forest management. However, the high priority given to the use of site-adapted species and natural stand and regeneration dynamics are expected to develop forest structures that are more stable relative to storms, climatic changes, and biotic damaging agents. Moreover, the forests would be more flexible for use by future generations due to the ability for functional integration (Björse and Bradshaw 1998; Larsen 1995, 1997). The Ministry of Environment, the Danish Forest Association, and NGOs have agreed to operational guidelines for nature-based forestry in Denmark (Table 19.3).

Nature-based forestry includes planting or direct seeding only if natural regeneration is insufficient with respect to stock density or tree species composition. Planting and direct seeding may aim for high densities (3,000 to 5,000 saplings per ha) in order to support high-quality wood production. Alternatively, a so-called enrichment planting, which typically includes the planting of small groups of trees (100 trees per group), is used, with the purpose of introducing a certain tree species at a site. An optimal production value for each tree during the later part of the rotation period is obtained by harvest of single trees as they reach their specific target diameter. The target-diameter harvesting regime should not be misinterpreted as "high grading" known from overexploitation of natural stands, but rather a system where a sustained yield is provided by making thinnings among the various age classes to assure that these are retained in their desired proportions, and that a suitable mixture of species is maintained (Matthews 1989). It should also be stressed that

Table 19.3 Operational Guidelines for Nature-Based Forest Management in Denmark, where Guidelines 1 to 10 are Core Guidelines and 11 to 13 Focus on Biodiversity, Recreation, Landscape Values, and Intensive Christmas Tree and Greenery Production

	Guidelines for Nature-Based Forest Management	Specific Actions
1.	Secure the production and the basis for production	Reduce the use of clearcutting, use productive and stable tree species, which secure soil fertility
2.	Secure a broad and continuous supply of various forest products	Use the local variation to secure a varied supply of species
3.	Use site-adapted tree species and provenances	Use native species where possible, use only exotic species where these can fulfill the requirements to site adaptation and biological integration
4.	Promote stands with more species and age classes	Use natural succession, protection of rare native species
5.	Promote and use natural regeneration	Promote a favorable regeneration climate, use harvesting systems that favor natural regeneration
6.	Improve the forest structure including edges	Stand management that supports the development of species and structural variation and promotes the use of native, site-adapted species along interior and exterior edges, and along streams, rivers, and lakes
7.	Promote target-diameter harvesting and increase the standing volume	Use of long regeneration periods, where single trees are harvested when they reach their economically optimal target diameter
8.	Use biological plant protection instead of chemical pesticides	Use species, whose survival and growth do not depend on pesticides. Use biotechnical and mechanical methods to control biotic and abiotic damaging agents
9.	Apply ecosystem-based wildlife regulation	Wildlife management that secures a balance between the game populations and the desire to make successful natural regeneration without substantial use of fencing
10.	Apply ecosystem-based operations techniques	Use of harvesting and regeneration techniques that are gentle to the stand and the site, including historical artifacts and soil biodiversity
11.	Make specific actions to secure biological diversity	Forest reserves, old silvicultural systems, deadwood in managed forests, habitat trees, key-biotopes registrations, natural wetlands
12.	Make specific actions to secure and promote recreation	Securing and developing the landscape functions, esthetical values, historical sites, and recreational facilities
13.	Apply intensive Christmas tree and greenery production only on small sites	These sites should be chosen with care and the use of pesticides and fertilizers should be minimized

Source: Larsen and Madsen, Eds., Naturnær skovdrift — erfaringer, status for forskningen og muligheder i Damark, *Skovbrugsserien*, 29, 2001 and Bjerg et al., Danske retningslinjer for bæredygtig skovdrift på ejendomsniveau, *Skoven*, 5, 206, 2001.

the heterogeneity of the stands is not the goal itself, but a way of allocating the species to various soil conditions and creating good forest floor conditions for natural regeneration (Madsen et al. 2002).

Nature-based forestry, as such, does not fulfill all the requirements for sustainable forestry (Table 19.3, guidelines 11 and 12). Several structural components, spatial patterns and processes, which are important for the maintenance of biodiversity, will not necessarily be taken care of in such managed forests. Structural components such as decaying trees, deadwood, open water, natural drainage, and open areas are underrepresented or not present in managed forests (Table 19.4). These components and patterns would naturally be present in nonintervention forest reserves (Christensen and Emborg 1996). Also, ecological processes, which typically affect large-scale areas (natural disturbances like storm and flooding), and the biological needs of large wildlife stress the need for forest and nature areas of a much larger size than the average forest estates in Denmark are able to provide. Finally, recreational needs are not automatically fulfilled in forests managed by nature-based principles. The silvicultural approach may need adjustment in urban forests close to larger towns and cities.

19.2.2 Economy of nature-based forestry

The profitability of nature-based forest management is primarily obtained through low establishment costs, postponement of the thinning expenses, and an optimization of the

Table 19.4 Requirements for Maintenance of Biodiversity, the Fulfillment Rate by Nature-Based Forestry, and Additional Biological Considerations;

Requirements for Maintenance of Biodiversity	Fulfilled by Nature-Based Forestry	Additional Biological Considerations
Structural components		
Old broadleaved trees	(+)	+
Trees with holes and cavities		+
Dead standing trees		+
Dead logs and branches		+
Burned wood		+
Stumps with uneven-aged surfaces		+
Natural water levels		+
Spatial patterns		
Developed understory of saplings and shrubs	+	
Mixed stands	+	
Patchy distribution of species and stems	+	
Multilayered tree canopies	+	
Uneven-aged stand structure	+	
Open areas		+
Processes		
Post windthrow succession	(+)	+
Succession with tree species replacement	+	
Self-thinning	+	
Gap formation	+	
Deadwood formation	(+)	+

+ = complete fulfillment, (+) = partial fulfilment.

Source: Bradshaw et al. (1994) and Bengtsson et al. (2000).

production value of single trees. Low establishment costs are obtained by using natural regeneration (Table 19.5), and the expenses for precommercial thinning may be avoided by postponing the first thinning until positive net revenue can be obtained (Nord-Larsen et al., 2003). Thus, when comparing nature-based beech stand management with traditional beech management regimes such as clearfelling and cyclic shelterwoods, it may provide an economically sound alternative. In addition, the prolonged retention of the holdovers created an accumulation of capital, which provides the manager with greater economic flexibility (Thorsen 1999; Nord-Larsen et al., 2003). Conversely, by prolonging rotation age, there is a potential risk of sudden quality losses of the wood due to red heart formation, which lowers the price significantly.

The economic yield and stumpage value show less fluctuation for forests managed after sustainability guidelines than traditionally managed forests (Boon et al. 1997). Considering pesticide bans, risk of windthrow, provenance variation, and other factors, conversion to nature-based forestry would be economically favorable at most sites (Boon et al. 1997; Thorsen and Strange 2000). However, there are still many unanswered questions regarding the conversion from traditional monospecific silviculture to nature-based forestry. Depending upon the actual stand condition, the conversion process could be expensive, especially when it is done too rapidly.

19.2.3 *Implementation of nature-based forestry*

The implementation of nature-based forestry may require political–economical, scientifically based practical instruments as well as education at several levels. As the political pressures for the public forests are a conversion to nature-based forestry within a period of 80 to 100 years, with 25% converted within 10 years, some rather drastic changes are awaiting the public forest managers and workers (Skov- og Naturstyrelsen 2002a, 2002b). The goal of the Danish Forest and Nature Agency is that public forests show the way, acting as guiding examples for private forest districts to follow, if desired. For private forests, the conversion period is not fixed and the motivation factor is planned to be incentives rather than legislation. However, it should be mentioned that some private forest districts already have changed their management plans to nature-based forestry, thus providing case studies and practical experiences of high interest for other forest owners regarding the conversion phase (Hahn and Fanta 2001).

Another tool of more substantial character is an upcoming revision of the Danish forest legislation of 1996, which will probably allow unthinned stands and the use of natural

Table 19.5 Approximate Costs for Traditional (planting) and Nature-Based Rejuvenation (Natural Regeneration) in Denmark

	Traditional Rejuvenation (Planting)	Nature-Based Regeneration (Natural Seeding)
Site preparation	0–700	0–400
Transplants	1,200–2,400	0–700 (supplementary planting)
Planting	400–1,000	0–500
Fence	0–1,300	
Total	1,600–5,400	0–1,600
Precommercial thinning	500–2,200	0–700

All prices are in US\$ ha^{-1}.

Source: Thorsen and Strange (2000), Madsen et al. (2002), and Madsen and Petersen (2002).

succession (Skov- og Naturstyrelsen 2002a, 2002b). Such changes are in sharp contrast to the present restrictive law, strictly aimed at securing efficient reforestation and wood production. In addition, changes in the present subsidies are expected to be a steering instrument of large importance for the conversion phase. Finally, the chance of higher prices on certified forest products might also act as an incentive for conversion.

Practical instruments include extension service, case studies, experimental sites, excursions, discussions, and, most importantly, education of forest workers and managers. Experience shows that the educational aspect is invaluable to a successful conversion (Hahn and Fanta 2001). Changing silvicultural practices involves all employees at the forest district. A straightforward method is to arrange short-term courses at the forest districts, where the ideas and application of nature-based forestry are discussed on site. It is considered important that both forest workers and staff are involved in the process of identifying the goals and methods suitable at their forest district. As the theories behind and the practical applicability are primarily brought in from Germany (Jakobsen 1995), many forest districts find it natural to turn toward German forest districts for inspiration, information, educational study-tours, etc. Finally, the discussion is often motivated in forest practice networks, such as Pro-Silva.

19.3 Nature-based rehabilitation — two common stand types

In order to give a more detailed picture of the goals and problems with rehabilitation, two common stand types are discussed. They illustrate the typical situations of unstable conifer plantations and plantation-like beech forests (Figure 19.1), which have a number of ecological and economic problems highlighting the need for forest rehabilitation. In each example, the aims and visions are described followed by an overview of the problems and challenges connected with rehabilitation. Finally, the possible means and strategies to fulfill the goals are presented and discussed.

19.3.1 Unstable conifer plantations

19.3.1.1 Aims and visions

The conifer plantations as they now exist are not considered the final goal, but provided they are not clearcut or windthrown (Figure 19.5), they represent a fairly good starting point for further forest development, far better than bare land (Emborg and Larsen 1999). The vision is to rehabilitate these plantations into stable, self-regenerating, mixed stands with site-adapted species and a high degree of stability and flexibility (Figure 19.6). Because expenses for silvicultural activities presently exceed the likely income from timber sales, an economic aim would be to promote stands that require little silvicultural input. Presently, the potential for regeneration and conversion in the sheltered and favorable forest microclimate is hardly utilized in traditional forest management. Only 10% of the coniferous plantations are regenerated by underplanting shelterwoods or natural regeneration (Larsen and Johannsen 2002). This emphasizes the need for the development of new and cheaper methods for reliable establishment of the species that we need to rehabilitate the plantations.

19.3.1.2 Problems

A major problem in the rehabilitation of conifer plantations is their poor wind stability, especially with shelterwood systems. The two most recent significant wind events in Denmark (1981 and 1999) have shown that conifers are extremely prone to windthrow. The 1999 windstorm, which hit the southern part of the country, caused the total destruction of some coniferous plantations. Three million m^3 of softwood timber (equal to 1.5 times the national annual harvest) but only 0.3 million m^3 of hardwood timber forest was

Figure 19.5 A typical Norway spruce stand, approximately 45 years old, on a fertile site in the eastern part of Denmark. This stand exhibits fast growth but poor stability; it probably will be stormfelled soon and replaced with broadleaves.

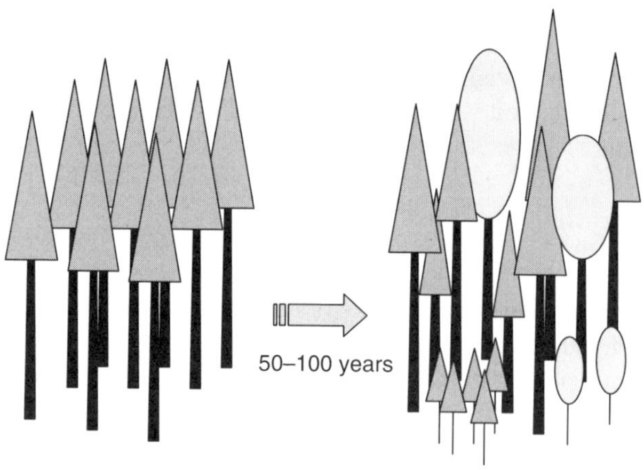

Figure 19.6 Change in conifer plantations from rehabilitation showing the introduction of deciduous species and structural diversification.

blown down (Larsen and Johannsen 2002). Such catastrophes destroy the trees and worsen the economic situation for landowners. They also cause considerable problems with regeneration, due to the exposed climatic conditions that increase the risk of late-spring frosts, weed competition, and soil degradation. The homogeneous, monoculture plantations are also susceptible to large-scale destruction by biotic disturbance agents like pathogens and insects. Norway spruce in particular has shown poor health and problems with the long-term stability because of insects (*Ips typographus*) and fungi (*Heterobasidium annosum*) (Larsen and Wellendorf 1997).

The coniferous plantations, contrary to the deciduous forests, often have no advance regeneration or understory that can rapidly reestablish the forest climate and improve the

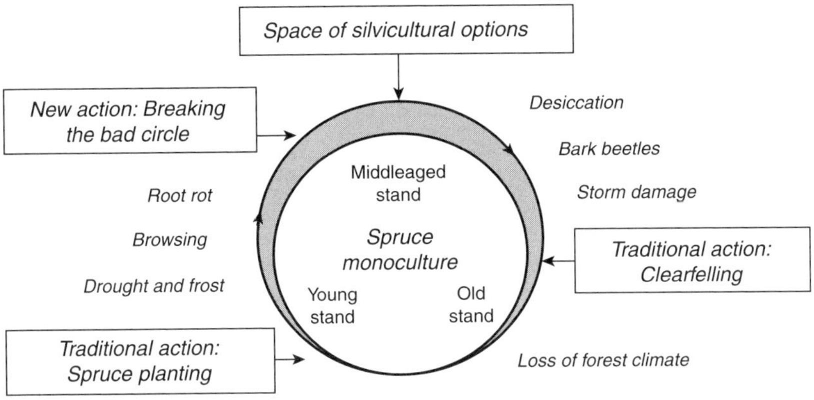

Figure 19.7 Breaking the negative cycle of spruce monoculture. The gray band indicates the best opportunities for successful rehabilitation. (Modified from Larsen, from plantation management to nature-based silviculture — a Danish perspective, *Third International Congress on Sustainability in Time and Space,* Pro Silva Europe, 187, 2000.

conditions for the regeneration after such disturbances. The possible choice of species is reduced to only a few, typically Norway spruce or Scots pine, which are tolerant of the harsh conditions on the clearfelled sites. Using the terminology of Larsen (1995), both resilience and resistance are low in such plantations. Thus, by traditional silvicultural means, the forester is caught in a trap in Denmark, known as the plantation syndrome (Clausen 1995), making it difficult to improve species diversity, stability, and flexibility of the forest within the next rotation (Figure 19.7).

The only factors keeping estate prices up are hunting access and the intrinsic value of forest ownership. Hunting conflicts with timber management, however, because roe deer and red deer cause browsing damage to regeneration, especially of broadleaved species (Madsen 1995). In many cases, failure is certain without expensive fencing to keep out the deer during the regeneration period, as few species are able to reach a height of >1.5 m at unfenced sites (Zerbe 2002).

19.3.1.3 Means and strategies

The choice of rehabilitation strategy may depend on the starting point, the objective, and the time available for the conversion (Emborg and Larsen 1999). Here, we will describe some possible strategies for the rehabilitation of young and mid-aged stands, mature stands, and clearfelled sites.

Rehabilitation of mature stands can be applied where the present stand has a high degree of stability or where wood production has lower priority. For conifers, this is only the case on protected sites and areas where soils allow deep rooting by trees. In this case, it is important to use the stand production potential, aiming at economic optimal rotation ages for single trees by target diameter harvesting. Harvesting will create small gaps, in combination with small-scale calamities that naturally occur in the latter part of the rotation such as root rot, bark beetle attack, and windthrow of single trees. These will be the starting points for the natural regeneration of species present in and around the stand, supplemented by enrichment plantings of desirable native broadleaved species (Emborg and Larsen 1999; Zerbe 2002). Structural diversity can be achieved partly by using species with different growth rhythms, but especially by group management in combination with long regeneration phases. This emphasizes the need for starting as early as possible with the first groups of regeneration and waiting as long as possible with the initiation of regeneration in the last spots. A boreal model of almost pure Norway spruce plantations rejuvenated by gap regeneration may be an alternative for those forest owners who are unwilling to bear the high costs of planting hardwood seedlings, or who experience very high browsing pressures. This method places a priority on structural diversity higher than species diversity (Fries et al. 1997). In any case, the aim is to minimize the regeneration costs while maintaining the highest possible production value during the transformation phase, even though the desired increase in biological and recreational values might be delayed.

Rehabilitating young or mid-aged stands is often the only possibility on high-risk sites where the stability of the stand is low. Many spruce stands are extremely susceptible to windthrow and the instability increases with stand development, which limits the silvicultural options with time (Figure 19.7). Thus, the need for an early action discards the principle of economic rotation age in order to ascertain transformation into mixed forests (Emborg and Larsen 1999; Larsen 2000). The traditional way of stand transformation is with underplanting of shelterwood systems (Neckelman 1995). Regeneration has to be initiated early enough to take advantage of the stable canopy needed to provide forest climate required for successful regeneration, but as late as possible to minimize the inevitable production losses in the overstory (Emborg and Larsen 1999). On highly unstable sites, the transformation should take place at stand height 12 to 15 m (Nielsen et al. 2002), but can possibly wait until the trees are up to 18 m on protected sites with lower risk (Clausen 1995; Emborg and Larsen 1999). The choice of machinery used for soil preparation should be balanced with the stand conditions, as the stability of the overstory trees decreases with increased intensity of machinery. Point cultivation appears to be the most cost-efficient and least damaging way of carrying out soil preparation in spruce plantations (Scuadicani 2002). To avoid repetition of the even-aged monoculture of the previous stand, structural heterogeneity should be promoted by the application of a mixed-regeneration approach including natural regeneration, direct seeding, and enrichment planting to establish different species with contrasting growth habits in groups.

In clearfelled or windthrown conifer stands, there is great public interest in a shift toward deciduous stands. This is illustrated by the high subsidies that were given after the 1999 windstorm, with an average of US\$ 5,600 ha^{-1} for mixed broadleaved afforestation (Fodgaard 2002). However, on these sites the environmental conditions for regeneration, particularly for broadleaves, is far more difficult than under shelterwood. There is a great need for new silvicultural methods that can meet the challenges of reduced cultivation costs, game browsing, and harsh regeneration conditions. Possible solutions include using natural regeneration, direct seeding, or planting. Approaches will depend on local conditions and site-specific judgment of what is necessary to obtain satisfactory regeneration comprised of the desired tree species. This is clearly an approach different from the traditional, technically intensive regeneration methods based on planting alone. Improved shelterwood methods, as well as nurse crops (alder and poplar), are being tested for their ability to preserve or restore forest conditions and support the establishment of small and inexpensive seedlings regardless of whether they are naturally regenerated, direct seeded, or planted. Direct seeding may offer low cultivation costs (30 to 50% of planting costs), provided the method can produce successful establishment. Promising preliminary results indicate that it may be possible to inexpensively establish broadleaves and other species by means that make the regeneration more tolerant of deer browsing. First, it is hypothesized that higher stock densities of the main species along with the occurrence of cover crops (legumes) that are attractive to deer will reduce the browsing pressure on the main species. Second, sown seedlings are expected to be less attractive to the deer compared to planted seedlings.

19.3.2 Plantation-like beech forests

19.3.2.1 Aims and visions

The plantation-like deciduous forests are primarily even-aged beech (Figure 19.8), where a successful rehabilitation strategy should be to attain forests that are more mixed. The aim is a greater variety of age classes and stand structures, including an improved distribution of tree species in accordance with their demands for the site.

19.3.2.2 Problems

In contrast to the possibly radical conversions of spruce plantations, rehabilitating pure even-aged plantation-like stands of beech is more a question of adjustments into more natural, uneven-aged, and mixed broadleaved stands. Although the plantation-like beech forests are characterized by structural homogeneity, species monotony, lack of late-successional stages, and lack of coarse woody debris (Christensen and Emborg 1996), they have the advantage of consisting of native, site-adapted, late-successional species, which are far less wind-prone than those in the coniferous plantations. The main challenges are the single-story overmature stands with an unfavorable forest climate and critical forest floor conditions, caused by a combination of low decomposition rates and soil compaction due to skidding. Moreover, beech has previously been planted on many sites with poor natural drainage (clayey soils or pseudogley) or on sites that were improved by artificial drainage. These drains are no longer intensively maintained and this generally leads to difficulties with red heart formation and low regeneration success. A high level of deer browsing often limits regeneration and contributes to the continuation of poor species diversity. As in the case for conifer plantations, the economic situation is rather unreliable and has worsened considerably within

Figure 19.8 Plantation-like beech stand, middle-aged (approximately 50 years old), on a fertile site in eastern Denmark.

the last decade. At present, more than three-fourths of the deciduous forests are regenerated by clearcutting with subsequent planting of shelterwood with underplanting, whereas natural regeneration takes place on less than one-fourth of the area (Larsen and Johannsen 2002).

19.3.2.3 Means and strategies

In Denmark, the dynamics of most natural forests is characterized by single-tree gap formation causing a mosaic structure and cycle (Emborg et al. 2000). Hence, a model that mimics the shifting mosaic steady state could be used to develop applied silvicultural methods for converting plantation-like deciduous forest (Figure 19.9). Using group-selection cutting with groups of various sizes can be a solution. By varying the group sizes within a stand, a mixture of species with different light demands can be accomplished. Where a shelterwood thinning has already taken place, a combined approach using both shelterwood cutting and group-selection cutting could be a feasible solution. A third solution would be to apply single-tree selection to even-aged stands where individual trees are harvested when the optimal target diameter is reached. Simulations show that it potentially takes more than 100 years to convert even-aged stands into sustainable selection stands (Morsing 2001), making it economically desirable but a rather slow rehabilitation tool.

The use of heavy machinery needs to be reduced in order to avoid soil compaction and improve the conditions for natural regeneration. The use of permanently marked skidding rows, a strict ban of machinery within the stands, and the use of more flexible machinery achieve this. It is especially important where the frequency of harvesting activities increases compared to traditional clearcutting or shelterwood systems.

Expected effects of the rehabilitation are an improvement of the forest microclimate, the reduced soil compaction by prevention of heavy machinery within the stands, and possibly less intensive deer browsing pressure due to more abundant and patchy regeneration patterns. However, just as with the coniferous plantations, we have a lot to learn and new methods to develop. Due to the stable shelterwood, regeneration can be established by natural regeneration. However, direct seeding or planting is strongly needed where seed sources other than beech are either too few or completely absent. Direct seeding may be more difficult than in the coniferous plantations on the former heathland due to a higher pressure of seed predators (voles and mice) and weeds. Planting has considerable problems with deer browsing on many sites, which makes group plantings by traditional methods very expensive since the small regeneration patches usually need

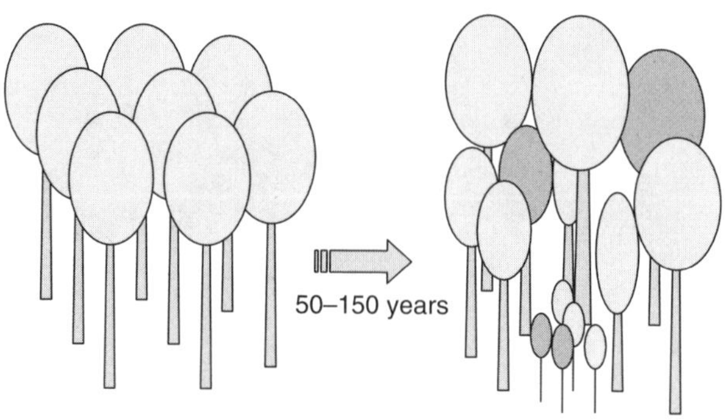

Figure 19.9 Rehabilitation of plantation-like beech forests by introducing other deciduous species, natural regeneration, and structural diversification.

protective fences. New methods to solve the deer browsing problems are, therefore, strongly needed. Like in the coniferous plantations, a range of complementary means including population management and alternative food supplies may offer solutions to balance the conflicts between wildlife and wood production.

For all rehabilitation strategies, it is important to be aware that the relevant knowledge for Danish conditions is limited and scientific research on nature-based systems has just started. The scientific basis for the relationship between game population densities and browsing damage is not fully developed. Moreover, if the present trend of negative incomes from timber continues, strategies less active than presented here will be more attractive.

19.4 Conclusion

The economic and ecological problems, which presently challenge Danish forestry, coincide with society's increasing demands for sustainable, multipurpose forestry. Nature-based forestry is favored as a major approach to rehabilitate our plantations and forests. In 2002, a political decision was made that nature-based forest management should be used as a rehabilitation tool for changing our forests fundamentally from monoculture plantations to stable, flexible, and self-rejuvenating forests of site-adapted species. Private forest owners are encouraged to convert by a range of economic incentives. The economic argument for choosing nature-based forest management is based on stable and flexible forests with low establishment costs, postponement of the thinning expenses, and an optimization of the production value of single trees. However, the rehabilitation phase that awaits our forests and plantations may take several decades and contains several challenges.

If the present trend of negative incomes on timber continues, strategies more passive than presented here will be more appropriate. This means that advanced silvicultural models are not an issue. However, we should not preclude the possibility for future economically feasible harvests of good-quality timber. The issue of wildlife as income (hunting) vs. the substantial problem with natural regeneration (browsing) must be solved. Therefore, rather than saying what to do, we can summarize what the forest manager should not do:

- One should not select tree species for sites where they are prone to windthrow, even if projected growth rates are high compared with more stable species.
- One should not strive for technically perfect and homogeneous stands, but rather concentrate on the utilization of natural regeneration and regeneration dynamics, focusing on what is good enough as a basis for future stand quality.
- One should not jeopardize the forest climate through large clearcuts, since a good forest climate is a prerequisite for natural regeneration.

Finally, it is important to be aware that nature-based forestry is not an excuse for poor silviculture. A repetition of former exploitative forestry without care for the future forest and its wood quality must be avoided.

Acknowledgments

We thank an anonymous reviewer for constructive comments and suggestions for improvements of an earlier version of the chapter.

References

Bengtsson, J., Nilsson, S.G., Franc, A., and Menozzi, P., Biodiversity, disturbances, ecosystem function and management of European forests, *For. Ecol. Manage.*, 132, 39, 2000.

Bjerg, A.-M., Søndergaard, J., Simonsen, J.P., and Stoltze, M., Danske retningslinjer for bæredygtig skovdrift på ejendomsniveau, *Skoven*, 5, 206, 2001.

Björse, G. and Bradshaw, R., 2000 years of forest dynamics in southern Sweden: suggestions for forest management, *For. Ecol. Manage.*, 104, 15, 1998.

Boon, T.E., Holten-Andersen, P., Madsen, K., and Feilberg, P., *Privat og samfundsøkonomisk vurdering af bæredygtig skovdrift på baggrund af 4 skove*, Projekt Bæredygtig Skov, Skov- og Naturstyrelsen, 63, 1997.

Bradshaw, R., Gemmel, P., and Björkman, L., Development of nature-based silvicultural models in southern Sweden: the scientific background, *For. Landscape Res.*, 1, 95, 1994.

Bradshaw, R., Hansen, J.M., and Møller, P. F., Om begrebet natur – natur versus menneske i kvartærtiden, *Skoven*, 3, 117, 1999.

Christensen, M. and Emborg, J., Biodiversity in natural versus managed forests in Denmark, *For. Ecol. Manage.*, 85, 47, 1996.

Clausen, J.T., Fra hedeplantage til skov, in *Skovbrugets grønne alternativ*, Forfang, A. S., Sørensen, P., and Feilberg, P., Eds., Nepenthes forlag, Århus, 37, 1995.

Dansk skovforening, Regnskabsoversigter for dansk privatskovbrug, *Beretning nr. 55, Generel del*, 2000.

Ellenberg, H., *Vegetation Mitteleuropas mit den Alpen in ökologischer, dynamischer und historischer Sicht*, Ulmer, 1095, 1996.

Emborg, J. and Larsen, J.B., How to develop plantations into forests in order to achieve stability and functional flexibility? A North European perspective, in *Proceedings of International Expert Meeting on the Role of Planted Forests in Sustainable Forest Management*, Barros, S. and Campodónico, M. I., Eds., 135, 1999.

Emborg, J., Christensen, M., and Heilmann-Clausen, J, The structural dynamics of Suserup Skov, a near-natural temperate deciduous forest in Denmark, *For. Ecol. Manage.*, 126, 173, 2000.

Emborg, J., Hahn, K., and Christensen, M., Eds., Urørt skov i Danmark — status for forskning og forvaltning, *Skovbrugsserien*, 28, 2001.

Fodgaard, S., 429 milllioner kr efter stormfaldet, *Skoven*, 10, 454, 2002.

Forfang, A.S., Sørensen, P., and Feilberg, P., Eds., *Skovbrugets grønne alternativ*, Nepenthes forlag, Århus, 37, 1995.

Fries, C., Johansson, O., Pettersson, B., and Simonsen, P., Silvicultural models to maintain and restore natural stand structures in Swedish boreal forests, *For. Ecol. Manage.*, 94, 89, 1997.

Fritzbøger, B., *Kulturskoven: dansk skovbrug fra oldtid til nutid*, Gyldendal, 439, 1994.

Führer, E., Forest functions, ecosystem stability and management, *For. Ecol. Manage.*, 132, 29, 2000.

Hahn, K. and Fanta, J., Eds., *Contemporary Beech Forest Management in Europe*, Deliverable 4 of the NatMan project (Nature-based management of beech in Europe), 2001.

Hannon, G., Bradshaw, R., and Emborg, J., 6000 years of forest dynamics in Suserup Skov, a seminatural Danish woodland, *Global Ecol. Biogeogr.*, 9, 101, 2000.

Hansen, R., Emborg, J., Neergaard, M., and Dalsgaard, S., The public nature of private forestry, *J. For.*, 95, 2, 20, 1997.

Heding, N., Ed., Bæredygtighed og flersidig træproduktion i hede- og klitplantager, *Dansk Skovbrugs Tidsskrift*, 87, 1–2, 2002.

Hüttl, R., Schneider, B.U., and Farrell, E.P., Forests of the temperate region: gaps in knowledge and research methods, *For. Ecol. Manage.*, 132, 83, 2000.

Jakobsen, M.K., Naturnær skovdyrkning i med- og modgang, in *Skovbrugets grønne alternativ*, Forfang, A.S., Sørensen, P., and Feilberg, P., Eds., Nepenthes forlag, Århus, 108, 1995.

Jensen, F.S., Fra ensartethed til variation — hvordan vil befolkningen se på omstillingen af hedeskovbruget fra ensaldrende rødgranplantager til blandskov? *Dansk Skovbrugs Tidsskrift*, 87, 1, 5, 2002.

Kirkebæk, M. and Thormann, A., *Evaluering af den gennemførte skovrejsning 1989–1998*, Skov og Naturstyrelsen (http://www.sns.dk/skov/netpub/evaluering/kolofon.htm.), 2000.

Larsen, J.B., Ecological stability of forests and sustainable silviculture, *For. Ecol. Manage.*, 73, 85, 1995.

Larsen, J.B., Skovbruget ved en skillevej — teknisk rationalisering eller biologisk optimering, *Dansk Skovforenings Tidsskrift*, 82, 277, 1997.

Larsen, J.B., From plantation management to nature-based silviculture — a Danish perspective, *Third International Congress on Sustainability in Time and Space*, Pro Silva Europe, 187, 2000.

Larsen, J.B. and Wellendorf, H., Rødgran — proveniensvariation, forædling og frøkildevalg, in *Træarts- og proveniensvalg i et bæredygtigt skovbrug*, Larsen, J. B., Ed., Dansk Skovbrugs Tidsskrift, København, 144–157,1997.

Larsen, J.B. and Madsen, P., Eds., Naturnær skovdrift — erfaringer, status for forskningen og muligheder i Danmark, *Skovbrugsserien*, 29, 1–57, 2001.

Larsen, J.B., Emborg, J., Rune, F., and Madsen, P., Skov og biodiversitet — bidrag til en handlingsplan for biologisk mangfoldighed og det nationale skovprogram 2001, *Skovbrugsserien*, 30, 1–55, 2001.

Larsen, J.B. and Emborg, J., Fremtidens skovbrug i Danmark, in *Visioner for fremtidens jordbrug*, Jensen, E.S., Vejre, H., Bügel, S.H., and Emaluelsson, J., Eds., Gads Forlag, København, 279, 2002.

Larsen, P.H. and Johannsen, V.K., Eds., *Skove og plantager 2000*, Danmarks Statistik, Skov & Landskab, Skov- og Naturstyrelsen, 2002.

Madsen. P., Effects of seedbed type on wintering of beech nuts (*Fagus sylvatica*) and deer impact on sprouting seedlings in natural regeneration, *For. Ecol. Manage.*, 73, 13, 1995.

Madsen, P., Aradottir, A.L., Gardiner, E., Gemmel, P., Lund Høie, K., Löf, M., Stanturf, J., Tigerstedt, P., Tullus, H., Valkonen, S., and Uri, V., Forest Restoration in the Nordic Countries, in *Proceedings of a Conference on Sustainability of Wetlands and Water Resources: How Well Can Riverine Wetlands Continue to Support Society into the 21st Century?*, Holland, M.M., Warren, M.L., and Stanturf, J.A., Eds., U.S. Department of Agriculture, Forest Service, Southern Research Station, Asheville, NC, General Technical Report SRS-50, 2002.

Madsen, E.M. and Petersen, H.C., Udrensning i meget planterige bøgeforyngelser, *Dansk Skovbrugs Tidsskrift*, 4, 109, 2002.

Matthews, J.D., *Silvicultural Systems*, Oxford Science Publications, Oxford, 284 pp., 1989.

Morsing, M., *Simulating Selection System Management of European Beech (Fagus sylvatica L.)*, Ph.D. thesis, The Royal Veterinary and Agricultural University, Denmark, 176 pp., 2001.

Neckelmann, J., To foryngelsesforsøg i rødgran på midtjysk heleflade, *Forskningsserien, Forskningscentret for Skov & Landskab*, 16, 1995.

Niedersachisches Ministerium für Ernährung, Landwirtschaft und Forsten, *Langfristige ökologische Waldentwicklung in den Landesforsten*, Program der Landseregierung Niedersachsen, 1991.

Nielsen, C.N., Glent-Madsen, B., Gaarde, M, and Nord-Larsen, T., Wind stability in semi-mature Norway spruce trees after extensive release in a shelterwood, *Dansk Skovbrugs Tidsskrift*, 87, 2, 41, 2002.

Nord-Larsen, T., Bechsgaard, A., Holm, M., and Holten-Andersen, P., Economic analysis of near-natural beech stand management in Northern Germany, *For. Ecol. Manage.*, 184, 149, 2003.

Rune, F., Ed., Biodiversitet i dyrket skov, *Skovbrugsserien*, 27, 2001.

Scuadicani, K., Fra plantagedrift til naturnær skovdrift — om skovning og kulturanlæg, *Dansk Skovbrugs Tidsskrift*, 87, 1, 15, 2002.

Skov- og Naturstyrelsen, *Danmarks nationale skovprogram*, Miljøministeriet, 2002a.

Skov- og Naturstyrelsen, *The Danish National Forest Programme in an International Perspective*, Miljøministeriet, 2002b.

Stahlschmidt, P., Spændende skove har også brug for den klassiske skovdrift, *Skoven*, 8, 349, 2002.

Stanturf, J. and Madsen, P., Restoration concepts for temperate and boreal forests of North America and Western Europe, *Plant Biosyst.*, 136, 2, 143, 2002.

Svenning, J. -C., A review of natural vegetation openness in north-western Europe, *Biol. Conserv.*, 104, 133, 2002.

The Royal Forest Decree, *Forordning angaaende de kongelige Skove og Tørvemoser udi Danmark*, 5, §63, 1781.

The Wilhjelm Committee, *Danish Nature — status, trends and recommendations for future biodiversity policies*, Skov- og Naturstyrelsen, 2001.

Thorsen, B.J., *Studies in Stochastic Decision Analysis in Forest Management and the Behaviour of Roundwood Prices*, Ph.D. dissertation, The Royal Veterinary and Agricultural University, 241 p., 1999.

Thorsen, B.J. and N. Strange, *En økonomisk vurdering af konvertering til naturnær skovbrug*, Rapport udarbejdet for Skov- og Naturstyrelsen, 2000, 84.

Zerbe, S., Restoration of natural broadleaved woodland in Central Europe on sites with coniferous forest plantations, *For. Ecol. Manage.*, 167, 27, 2002.

chapter twenty

Great Britain — conifers to broadleaves

Ralph Harmer
Forest Research, Alice Holt Lodge
Richard Thompson and Jonathan Humphrey
Forest Research, Northern Research Station

Contents

1-56670-635-1/05/$0.00+$1.50

20.1 Introduction

During the last few thousand years, Great Britain has been denuded of most of the forest cover that developed after the last ice age. Much of this loss was due to clearance for agriculture, and there are no remnants of the original natural woodland that have not been managed in some way. However, some woodlands have a very long history and those sites that have been continuously wooded since AD 1600 are classified as ancient woodland sites. Those sites that also clearly comprise native species are classified as ancient semi-natural woodland. The rationale and development of the terminology can be read in Peterken (1977, 1993, 1996). Although many ancient semi-natural woodlands have been regularly managed, they are regarded as very important sites for conservation and biodiversity; an inventory found that they cover only about 2% of Great Britain's land area (Spencer and Kirby 1992). Another category of ancient woodland sites comprises those that have been converted to woodlands of species that are not site-native: these are termed Plantations on Ancient Woodland Sites (PAWS) and often consist of exotic conifers.

In the 20th century, considerable effort was expended in establishing tree cover on unwooded sites, primarily using conifers, but during the last decade there has been a growing interest in using native species to create new woodlands (Rodwell and Patterson 1994; Pryor et al. 2002), and in restoring PAWS to semi-natural woodland (Humphrey and Nixon 1999). Restoration in this latter sense concerns the re-establishment of a functioning native woodland ecosystem.

This chapter, which gives a brief overview of forestry in Great Britain, concentrates on some of the factors that influence the restoration of native broadleaved woodland from plantations of conifers on ancient woodland sites.

20.2 A brief history of forestry in Great Britain

The history of forestry and woodlands in Great Britain has been described in detail by Anderson (1967), Fernow (1913), James (1981), and Linnard (2000). Here, we provide a very brief summary with an emphasis on the practices during the 20th century that have resulted in the need for restoration.

It is generally accepted that without human influence the natural vegetation of Great Britain would be dominated by woodland of some description, but woodland cover has declined for much of recorded history, and in Great Britain forest cover is one of the world's lowest. The Roman geographer Strabo described Great Britain as a densely wooded country, but following clearance for agricultural purposes, woodland cover had declined to about 5% of land area by the early years of the 20th century. Although there is evidence to suggest that coppice woodlands were often well managed on short rotations to produce a regular supply of small round wood products (Rackham 1980), the management of trees and woodlands to produce timber has, from the viewpoint of timber production, a long and unsatisfactory history. However, timber supply is not the sole outcome of forest management, and other features such as deer hunting, pheasant shooting, and shelter for farm animals were, and continue to be, important.

The series of invasions that occurred until the middle of the 11th century brought with them rulers who enjoyed hunting, and they established large areas of Royal Forest for the pursuit of their sport. These forests had special laws, often had little tree cover, and were managed primarily for hunting rather than the promotion of tree growth. Concerns about the supply of timber were frequent, and despite the introduction of several laws to promote forestry, uncontrolled felling often occurred during periods of need. Worries about the supplies of timber for shipbuilding during the 17th century stimulated the production of John Evelyn's (1664) classic book *Sylva, or a Discourse on Forest Trees*, but neither its publication nor the periodic, irregular grumbling about the state of British forestry, or the frequent investigative committees, had much overall effect on the quality of forest management for the next 250 years. At the beginning of the 19th century, the Royal Forests were placed under the control of a board of management, but their quality continued to decline and 100 years later there were only about 30,000 ha that could be regarded as true forest, and this was more or less mismanaged. Forest management and silvicultural practice were virtually unknown, and the use of working plans was restricted to a few estates, the majority of which preferred to use their woodlands primarily for shooting.

Shortages of timber during World War I again revealed the poor state of Great Britain's forests and home-grown timber supply, but on this occasion the government acted. In 1919, the Forestry Commission, a newly established government department, was given the task of developing a strategic reserve of timber (Pringle 1994). Large areas of land were purchased or leased, and during the remainder of the 20th century many new plantations were established and managed by the state. A variety of grant schemes have also encouraged private owners to follow government policy by both new planting and improving the management of existing woodlands. This program to increase the amount of woodland cover has been very successful: in 1905, the total woodland cover for Great Britain was about 1.1 million ha (4.9% of land area), but by 1998 this had increased to 2.7 million ha (11.7% of land area) (Figure 20.1). A large part of the increased area resulted from the afforestation of upland grassland, moors, and bogs, which form a larger proportion of the land area of Scotland and Wales. Consequently, the distribution of forest cover has changed; for much of the 20th century most of the woodland was found in England, but by 1998 the greatest amount occurred in Scotland (Figure 20.1). The proportion of land area covered by woodland increased by 300 to 400% in Wales and Scotland, but by only 60% in England. The expansion of forest area has led to a change in the types of woodland found, with a general increase in the amount of conifer relative to broadleaves (Figure 20.2). In addition, as *Pinus sylvestris* (L.) is the only native conifer suitable for use in large-

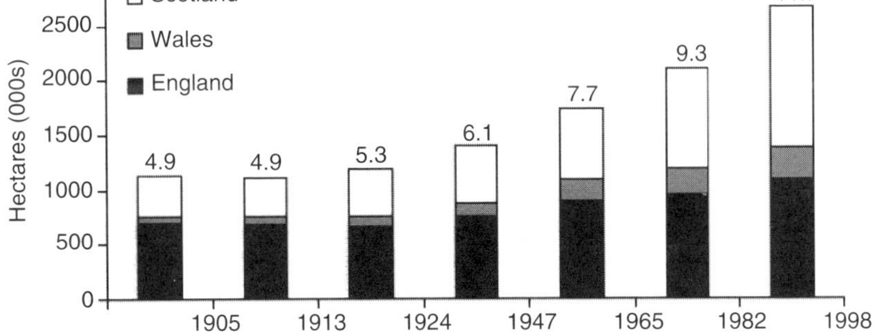

Figure 20.1 Areas of woodland recorded for each country in censuses carried out during the 20th century (Anon. 1928, 1952, 2001a; Locke 1079, 1987). The areas include all types of woodland recorded at each date including scrub, felled, devastated, and open areas. The figures at the top of each bar are the amount of woodland as a percentage of the total land area of Great Britain.

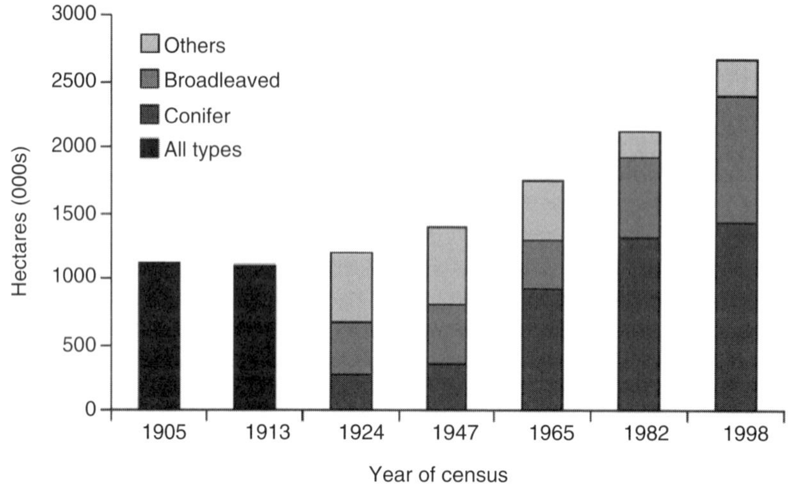

Figure 20.2 Areas of different types of woodland in Great Britain in the censuses carried out during the 20th century (Anon. 1928, 1952, 2001a; Locke 1979, 1987). In 1905 and 1913, conifers and broadleaves were not distinguished: broadleaved woodland includes all types of coppice. Others include scrub, felled, devastated, and open areas.

Table 20.1 Amount of Each Coniferous Species Present as a Percentage of Total Conifer Area in Each Country

Species	England	Wales	Scotland	Great Britain
Scots pine (*Pinus sylvestris*)[a]	24.0	3.1	15.0	15.9
Corsican pine (*Pinus nigra spp.laricio*)	12.1	2.3	0.2	3.3
Lodgepole pine (*Pinus contorta*)	2.1	4.2	13.4	9.7
Sitka spruce (*Picea sitchensis*)	24.1	55.9	57.8	49.6
Norway spruce (*Picea abies*)	9.2	7.6	3.8	5.5
European larch (*Larix decidua*)	4.0	0.4	1.0	1.6
Japanese larch (*Larix kaempferi*)[b]	9.4	14.8	6.1	7.8
Douglas-fir (*Pseudotsuga menziesii*)	7.3	7.4	1.1	3.3
Other conifers	5.4	4.0	0.6	2.1
Mixed conifers	2.4	0.3	0.9	1.2
Total area (ha)	329,832	145,523	904,155	1,379,510

[a] Native to Scotland.

[b] Includes hybrid larch (*Larix × eurolepis*).

Figures are for woodlands >2 ha in area.

scale commercial plantations, much of the afforestation that has occurred during the last century has relied on the use of exotic species (Table 20.1).

In their drive to create fast-growing, productive plantations, foresters attempted to replace existing uneconomic, slow-growing, broadleaved woodland with conifers or more valuable broadleaved trees (frequently beech, *Fagus sylvatica* L.). Although these existing woodlands had been managed, in some cases for many centuries, many were ancient and semi-natural in origin and of considerably more biological interest than the conifer stands that often replaced them. About 40% of the ancient woodland area that existed in the 1930s has been converted to PAWS (Table 20.2), mostly by planting conifers. About 60% of the total area of PAWS in Great Britain is on land managed by Forest Enterprise (part of the government's forest service), covering an area of about 130,000 ha, which is about 15% of their estate (Pryor 2003).

Table 20.2 Areas of Ancient Semi-natural Woodland and Plantations on Ancient Woodland Sites

Species	England	Wales	Scotland	Great Britain
ASNW	193	27	65	285
Conifer PAWS	80	13	46	139
Other PAWS	60	7	8	75
Total	333	47	119	499
Area of country	13,000	2,100	7,700	22,800

Figures are 1,000's ha; these data are currently under review, but any changes will be relatively small. Conifer PAWS includes mixtures with broadleaves.

20.3 *Forest and woodland types in Britain*

20.3.1 *Ecological background*

On a global scale, much of Great Britain falls within Zonobiome VI (Walter 1985). The climate is generally temperate with short periods of frost, but the west of the country is more oceanic. Cold, temperate, boreal conditions (Zonobiome VIII) prevail in the Scottish mountains. Forest brown earths and podzols, which are typical soils of these biomes, are widespread in Great Britain. The zonal vegetation type is broadleaf deciduous forest. However, at the local scale, Great Britain is a country of great ecological contrasts and it has a wide range of different climates and soils. Ecological Site Classification provides an objective framework for characterizing the variability of site and climatic conditions across Great Britain, and can be used to guide the establishment of trees and woodland (Pyatt et al. 2001). The majority of the south and east of Great Britain has a warm dry climate where precipitation is relatively low (< 800 mm year^{-1}). For forestry purposes, the north and west of Great Britain are termed the uplands (Figure 20.3). In these areas, the climate is cooler and more temperate with high precipitation (up to 3,000 mm year^{-1}), and a greater risk of damage due to strong winds.

20.3.2 *Commercial forestry and PAWS*

Commercial afforestation in the lowlands is dominated by Corsican pine (*Pinus nigra* spp. *laricio* Maire), Norway spruce (*Picea abies* (L.) Karst), and Scots pine (*Pinus sylvestris* L.) (Figure 20.4). Although some forests were established on more fertile, base-rich soils, the most extensive plantings were on acid heaths in the south and east of England. The distribution of Corsican pine is limited by winter frosts (Pyatt et al. 2001); hence, there are few extensive strands north of the English Midlands.

In the uplands, the altitudinal limit to the development of high forest varies from about 400 m in the northwest of Scotland and 750 m in the eastern Scottish Highlands. Paleozoic base-poor metamorphic and igneous rocks and glacial deposits dominate the geology of the uplands, giving rise to acid, nutrient-poor soils. The dominant conifer species in the uplands is Sitka spruce (*Picea sitchensis* (Bong.) Carr,) (Figure 20.4), which outperforms most other species on the predominantly wet, nutrient-poor sites. However, both Sitka spruce and the two most common larch species (*Larix* × *eurolepis* Henry and *L. kaempferi* (Lam.) Carr.) are restricted in their distribution by the occurrence of severe winter frosts. Thus, in the drier, upland areas with more continental climates (e.g., the eastern Scottish Highlands), Norway spruce and Scots pine are the preferred species (Figure 20.4).

Windthrow is a significant threat to forestry in the uplands (Quine et al. 1995) and restricts the range of management options. Normal management practice is to leave stands unthinned and then clearfell them before they reach full economic maturity, to reduce the risk of catastrophic damage. However, as part of recent sustainable forestry

Figure 20.3 Map of Great Britain showing the borders between England, Scotland, and Wales (dotted lines), the areas regarded as uplands for forestry purposes (shaded gray), and the locations of the case studies. The Shetlands and other smaller islands are not shown.

initiatives (Anon. 2001b), there has been a move to increase the irregularity of forest structure at both stand and landscape scales. This may, in time, reduce the threat of windthrow and provide a greater permanency of forest structure, offering both ecological and aesthetic benefits.

PAWS are found throughout Great Britain. They occur at locations that have different soils and climates, which influenced the range of trees planted and how well they grew. The trees and woodlands now growing on PAWS vary in age, precise composition, structure, intensity of management, and other quality characteristics, such as how well the planted crop established and how many components of the original woodland remain. This wide variability in PAWS will have an important influence on the management needed to restore the woodland.

20.3.3 Classification of semi-natural woodlands

Several methods are available for the classification of semi-natural woodlands in Great Britain, depending on the requirements of the user. At the end of the 20th century, ecologists and other specialists adopted the National Vegetation Classification (NVC), which is a phytosociological classification that relies entirely on the plant species present. Within this system, there are 18 woodland and 7 scrub communities that are subdivided into 73

Figure 20.4 Distribution of important commercial conifer species in Great Britain derived by permission from the Forestry Commission's subcompartment database. Each pixel represents a 1 km^2 containing a conifer stand, and therefore does not give an accurate reflection of the actual area covered by each species.

subcommunities (Rodwell 1991). However, for many practical purposes, broader groupings may be sufficient. In 1994, the Forestry Commission issued a series of eight guides giving recommendations for the management of semi-natural woodlands. The guides describe wet woodlands, three classes of lowland wood and four classes of upland wood. In general, classes are identified by the species present and comprise at least two NVC woodland communities that are usually exclusive to each class. A broadly similar classification has been developed for woodlands with Habitat Action Plans, which have similar names, but there are some differences in the NVC communities included within each. The relationship between these classifications and other European types is shown in Table 20.3. The broadleaved woodlands with Habitat Action Plans are briefly described below; for more information, see Hall and Kirby (1998). There are about 500,000 ha of woodland

Table 20.3 Relationships Between Different Types of Woodland Classification

HAP Type	Forestry Commission Guide Type[a]	NVC Type[b]	CORINE	Habitats Directive Annex 1 Type
Lowland beech and yew	1. Lowland acid beech and oak woods	W13	42.A71	*Taxus baccata woods*
	2. Lowland beech-ash woods	W12	41.13	*Asperulo-Fagetum* beech forests
		W14, W15	41.12	Beech forest with *Ilex* and *Taxus*, rich in epiphytes
			(41.16)	
Lowland mixed broadleaved[c]	1. Lowland acid beech and oak woods	W8a-d (e-g)	41.23, 41.32	(*Tilio-Acerion* ravine forests)
	3. Lowland mixed broadleaved woods	W10a-d (e), W16	(41.24)	(*Stellario-Carpinetum* oak-hornbeam forests)
			(41.51)	(Old acidophilous oak woods with *Quercus robur* on sandy plains)
			41.52	
Upland mixed ash	4. Upland mixed ashwoods	(W7c) W8(a-c) d-g, W9	41.31, 41.32, 41.41	
		W13	42.A71	*Tilio-Acerion* ravine forests
			(62.3)	*Taxus baccata* woods
				limestone pavement
Upland oak	5. Upland oakwoods	W10e, W11, W16b, W17	41.53, 41.52	Old oak woods with *Ilex* and *Blechnum* in the British Isles
Upland birchwoods[c]	6. Upland birchwoods	W10e, W11, W17	41.53, 41.52	Old oak woods with *Ilex* and *Blechnum* in the British Isles
		W4a, b		
Native pine woodlands	7. Native pinewoods	W18, (W19)	42.51	Caledonian forest
		W4(a), b, c	44.A2	Bog woodland
Wet woodlands	8. Wet woodland	W4(a, b), c	44.A1	Bog woodland
		W5 – W7 (W8)	44.31	Resdual alluvial forests
		W1, W2, W3	44.13	
			44.92	

HAP type = name of woodland type with a Habitat Action Plan. Where a type is thought to be only a minor component of the HAP type, it has been placed in brackets. Adapted from Hall and Kirby, *The Relationship Between Biodiversity Action Plan Priority and Broad Woodland Habitat Types, and Other Woodland Classification*, JNCC, Report No. 288, JNCC, Peterborough, 1998, with permission.

[a] There are eight Forestry Practice Guides (Anon. 1994a) that give guidance on the management of different woodland types.

[b] Community described by the National Vegetation Classification (Rodwell 1991).

[c] Do not yet have approved Habitat Action Plans, but included because of their importance as semi-natural woodland types.

included within the Habitat Action Plans, about half of which is lowland mixed broadleaved (Table 20.4). The types are not evenly distributed and consequently the target area for restoration of each type varies for each country.

20.3.3.1 *Lowland mixed broadleaved*

This is a diverse category of woodland that occurs on a wide range of soils and includes most of the semi-natural woodland that occurs in south and east England; it is also found in some lowland areas of Wales and Scotland. The frequency and abundance of different species within these woodlands are variable, but trees such as ash (*Fraxinus excelsior* L.), oak (*Quercus robur* L.), silver birch (*Betula pendula* Roth), and hazel (*Corylus avellana* L.) are widespread.

20.3.3.2 *Lowland beech and yew*

Beech (*Fagus sylvatica*) is generally considered to be native only in England and South Wales, and restoration will be concentrated within this area. These woodlands occur on both calcareous and acidic soils. Woodlands with a substantial amount of yew (*Taxus baccata* L.) tend to occur on more calcareous sites where other species, such as ash and whitebeam (*Sorbus aria* (L.) Crantz.), may also be found within the overstory.

20.3.3.3 *Upland birch woods*

These are dominated by downy birch (*B. pubescens* Ehrh.) and/or silver birch mixed with varying amounts of rowan (*Sorbus aucuparia* L.), aspen (*Populus tremula* L.), and juniper (*Juniperus communis* L.). Grasses and dwarf ericaceous shrubs often dominate the ground flora.

20.3.3.4 *Upland mixed ash*

These woodlands are found in the north and west of Great Britain, typically on free-draining, base-rich soils. Ash is usually a major component of the overstory, but other species such as oak, silver birch, elm (*Ulmus glabra* Huds.), and small-leaved lime (*Tilia cordata* P. Mill)

Table 20.4 Amounts of Each Woodland Type Included in a Habitat Action Plan and the Target Areas for Restoration

Woodland Type	GB Area	England		Wales		Scotland	
	(1,000's ha)	%	Target	%	Target	%	Target
Lowland beech and yew	30	90	1.35	10	0.15	—	0
Lowland mixed[b] broadleaved	250	88	15[a]	4	?	8	?
Upland mixed ash	68	55	1.5	11	0.25	33	0.8
Upland oak	70–100	25	1.75	31	2.2	42	2.95
Upland birch[b]	30–40	5[c]	—	5[c]	—	90	3.2[a]
Native pine	16	—	0	—	0	100	0[d]
Wet	50–70	33	1.05	16	0.5	50	1.6

Note: Percentage and target are likely to be revised as progress is made.

GB area = area of woodland type in Great Britain; Woodland type = woodland defined by Habitat Action Plan; % = estimated percentage of total GB area of each woodland type in each country; Target = target Area (1,000's ha) for restoration.

[a] Proposed total areas for restoration, to be allocated when Habitat Action Plans are published.

[b] Habitat Action Plans for these woodlands not yet published.

[c] These small areas will be treated within upland mixed ash and upland oak plans.

[d] No target for restoration.

may be locally abundant. The shrub understory and the ground flora are often well-developed and species-rich.

20.3.3.5 Upland oakwoods

These generally occur in the north and west of Great Britain, and while they are usually found in upland areas they may occur at low-level, particularly along the coast of Scotland, Wales, and southwest England. They are found on neutral to acidic soils and canopy dominants are typically sessile oak (*Q. petraea* (Mattuschka) Liebl.) and birches, with hazel, rowan, and holly (*Ilex aquifolium* L.) in the understory.

20.3.3.6 Wet woodlands

These occur throughout Great Britain on poorly drained or seasonally wet soils, typically on floodplains, successional habitats on fens, mires and bogs, and hillside flushes. They are found on a range of soil types including nutrient-rich mineral soils, and those that are acidic, organic, and nutrient-poor. Alder (*Alnus glutinosa* (L.) Gaertn.), birches, and willows (*Salix* spp.) dominate the tree canopy.

20.4 Reasons for restoration

20.4.1 Conservation and biodiversity

Toward the end of the 20th century, the extensive use of exotic species became increasingly unpopular, and concern was expressed about the impact of production-oriented government policy and management objectives leading to the loss of semi-natural habitats. Plantations were composed largely of non-native conifers with little room for non-productive, native, broadleaved species. Pressure from conservation bodies led to changes in legislation and government policy. Forestry practice changed, with more emphasis being placed on the use of native species and natural processes to re-stock woodlands. The international commitments made by the government regarding biodiversity, and the resulting U.K. Forestry Standard (Anon. 1998a) and U.K. Biodiversity Action Plan (Anon. 1994b, 1998b) have encouraged the restoration of PAWS, with woodland Habitat Action Plans giving targets for the area that should be restored.

20.4.2 Certification schemes

The public's increased environmental awareness, reflected in policy changes toward native woodlands, has also led to demands for wood products produced from sustainably grown timber. In addition, large retailers are keen to pursue "green" marketing opportunities. State-owned forests and many private woodlands are now certified under the U.K. Woodland Assurance Scheme (Anon. 2000), an accredited body under the Forestry Stewardship Council.

The scheme requires owners to consider options for managing PAWS. These options range from the maintenance of biodiversity within an otherwise conventionally managed plantation, to full restoration to site native species. In broad terms, the scheme requires restoration of at least 10% of plantations on ancient woodland sites within the first 10 to 15 years following certification.

20.5 Aims of restoration

The short-term aim of restoration is to restore sites to woodland, comprising site- or locally-native tree and shrub species. This is probably all that is achievable within the 10 to 15-year time-scale of the Habitat Action Plans, which also indicate that 10% of the area of PAWS should be restored. In the longer term, it is hoped that the restored woodlands will develop more varied structures, undergo more natural ecological processes, and contain a wider variety of native flora and fauna than plantations. However, future development will depend on how the woodland is managed, and even under the most favorable conditions such a full restoration will take many decades to achieve.

20.6 Costs and expected benefits of restoration

The issue of restoring ancient woodland sites raises strong feelings with two broad sectors of opinion evident. Traditional timber producers often see restoration as a threat to their livelihood, and there is some resistance to the Habitat Action Plan process and any new initiatives to increase the area restored. In contrast, conservation organizations are keen to see the process accelerated with alternative options to restoration as the exception. However, many pursue a balanced approach (e.g., in state forests where an ambitious yet realistic program of restoration is being undertaken). A Forest Practice Guide (Thompson et al. 2003) supports the developing Forestry Commission policy on the restoration of PAWS.

20.6.1 Biodiversity

There is a tacit acceptance that restoring PAWS to native woodland will enhance woodland habitats for a range of plant and animal species. This is undoubtedly true in most cases, especially where the plantation is growing very quickly and shading out ground flora communities (Humphrey and Nixon 1999). However, non-native conifer species can have habitat value in their own right — providing refugia for key species of conservation concern such as threatened fungi (Humphrey et al. 2000), red squirrel (Thompson et al. 2003), and dormice (R. Trout, personal Communication). In the future, there will be a need for careful surveys of existing PAWS before embarking on restoration, to ensure that biodiversity is safeguarded, at least in the short to medium term.

20.6.2 Cost of restoration

The British government will fund much of the cost of restoration. This will be by direct management of state-owned PAWS sites managed by Forest Enterprise, or indirectly by the payment of grant-aid to private owners via the Woodland Grant Scheme (Table 20.5). Some money to pay for the restoration of PAWS may be obtained from the European Union (Case Study 3). The precise costs of restoring PAWS are unknown and will vary with site, but indicative costs for 2002 of individual elements of the process are shown in Table 20.6.

20.6.3 Economics

Although restoration may have an effect on income, in many of the woodlands where there would be a high ecological benefit from restoration, the continued use of non-native trees is not economically viable. This could be due to site type (where competition from weeds or coppice regrowth of native species would make re-establishment of non-native trees difficult) or due to difficult access (in some areas of the Scottish Highlands where extensive networks of new forest roads would be required to extract timber). In contrast, some ancient woodland sites can support productive stands of non-native trees (Douglas-fir [*Pseudotsuga menziesii* (Mirb.) Franco] in the southwest of Great Britain) and owners may forego revenue from future timber production if they change species to less productive broad-leaves. In a broad analysis, Pryor (2003) concluded that although the long-term

Table 20.5 Types of Grant Payable under the Forestry Commission's Woodland Grant Scheme (Anon., 2001c) that are Available for Funding the Restoration of PAWS

Type of Grant	Amount Paid	Notes
Restocking	£525 ha^{-1}	Minimum of 1,100 stems ha^{-1}. Established by year 10. Use of approved species and%'s in mixtures.
Natural Regeneration	Initially 50% of agreed costs	Discretionary payment for work necessary to facilitate natural regeneration (e.g., scarifying, fencing).
	Additional fixed payment of £525 ha^{-1}	Paid when regeneration reaches 30–45 cm height.
Annual Management Grant	£35 ha^{-1}	To fund any operation (e.g., thinning) that would aid restoration.
Woodland Improvement Grant	50% of agreed costs	Maximum of £10,000 for each application. Funds work that will improve biodiversity.

These grants are currently under revision.

Table 20.6 Indicative Costs in 2003 of Materials and Operations Necessary for the Restoration of PAWS with Broadleaved Trees

	Cost	Notes
Fencing		These represent a selection of commonly used types.
Stock	£3.00 m^{-1}	
Rabbit	£4.50 m^{-1}	
Permanent deer	£6.50 m^{-1}	
Permanent deer and rabbit	£7.00 m^{-1}	
Plants		
Supply		
Bare-rooted	£220	For 1,000 plants.
Container	£320	
Planting	£100	1000 plants, by hand.
Planting with tree shelter	£1.70	For each tree with a 1.2 m shelter.
Forest operations		
Thinning	£20 m^{-3}	Fell, sned (lop), and extract.
Felling	£8.70 t^{-1}	By harvester, average trees size 0.6 m^{-3}, and extract to roadside.
Scarification	£130 ha^{-1}	
Chemical weed control	£115 ha^{-1}	Spot spraying 2,500 trees.

impact of restoration on timber production would be significant, the costs for individual owners over the next 30 years are modest.

Native woodlands may offer opportunities to generate revenue (production of quality hardwood timber) and improve the capital value of landholdings (improve conditions within watercourses for game fishing).

20.6.4 Cultural aspects

Many people value woodlands as much for their visual and cultural qualities as their ecological, conservation, and biodiversity interests. Woodlands are often embedded in the culture of a locality and any changes to their existing state, even by normal forest management, can be opposed by the public. Factors including visual appearance at the local or landscape level, the archeological history of the site, and the role of the woodland in local history, folklore, and recreation may all constrain the decision on whether to restore a site. For example, non-native trees, however recently planted, can become accepted as part of the normal countryside, and local communities may not welcome their conversion to native species. In contrast, there are many examples where ownership boundaries divide ancient woodland sites; moreover, where plantations exist on one side, with native woodland on the other, restoration can be undertaken to improve the visual appearance of the whole woodland.

20.7 Research

The 20th-century expansion of plantation forestry was supported by a significant program of applied research into subjects such as establishment, nutrition, and growth. In contrast, there has been little direct research into the restoration of PAWS. Some desk studies and field observations have determined the areas, locations, and types of PAWS (Pryor and Smith 2002; Pryor et al. 2002), and there has been some work on the influence of non-native species on biodiversity (Humphrey et al. 2002, 2003). However, there have been no systematic studies of the procedures used during restoration that have influenced current silvicultural practice. Recently established experimental work includes monitoring the recovery of sites after clearfelling Norway and Sitka spruce, the influence of thinning intensity on the restoration of Corsican pine stands, and the influence of brash on tree seedlings. Given the long-term nature of field experiments in forestry, the small number of researchers working on restoration, and the targets for restoration, it is almost inevitable that large areas of PAWS will be treated before the final results of these studies are available.

20.8 Choosing target woodland type

Achievement of the ambition to restore PAWS to more naturally functioning mixtures of site-native species will require the identification of the appropriate type of native woodland for the site. Choice of the correct type should, given appropriate management, maximize the likelihood of achieving the biological aims of restoration. Thompson et al. (2003) also stress that the identification of the potential woodland type will help with decisions regarding:

- Selection of sites in relation to regional targets — for example, identification of local and regional priority woodland types, uncommon stand types, etc.
- Choice of silvicultural regime — clearfelling vs. selection or groupfelling
- Planting vs. natural regeneration
- Suitability of the adjoining semi-natural woodland as a source of colonizing species; for example, does it contain species that are needed to colonize the restored woodland?

Identification of the potential native woodland type for the site may be achieved by a survey of the flora and soils of the least disturbed parts of the existing stand, or in undisturbed rides and the surrounding woodland. Records of site history, combined with local knowledge, can also be informative. A number of decision support tools are available to help link current vegetation and soils to future woodland potential. Rodwell and Patterson (1994) provide a qualitative method where "pre-cursor" ground flora species are used to give an indication of future woodland community type. Ecological Site Classification provides a quantitative method, where field-derived measures of the frequency and abundance of "indicator" plant species are used to determine the woodland types most suited to the site (Ray 2001). At the landscape or regional scale, strategic planning tools such as the Scottish Natural Heritage/Macaulay Land-Use Research Institute native woodland model link broad soil and land cover types to potential woodland types (Gray and Stone 2003).

20.9 Selecting sites to restore

A method for selecting sites, which maximizes the potential benefits for biodiversity without placing unnecessary burdens on woodland owners, is described in Thompson et al. (2003). The approach combines an assessment of the ecological potential of a site or number of sites with an assessment of whether restoration is realistic in relation to practical opportunities and constraints. The ecological criteria listed below are separated into two groups: importance and ecological development.

1. *Importance*: This is the extent to which the restored site will contribute to local, wider landscape, and regional and national biodiversity objectives (e.g., will restoration achieve aims of priority species and habitat action plans?, does the site contain species or habitats that have some other form of conservation status?, how will restoration improve the connectivity of semi-natural habitats within the wider landscape?).

Indicators of high importance include:

- The site or an adjoining area is designated due to the quality of semi-natural woodland, and restoration will contribute significantly to safeguarding and enhancing this site.
- The potential woodland type will contribute to the local area's Biodiversity Action Plan targets for restoration.
- The restoration of the site will contribute significantly to the forest habitat network, and if the site is not restored, non-native trees will pose a threat to the ecological functioning of the network.
- Restoration will secure or benefit populations of nationally scarce or threatened species (those with Species Action Plans or Red Data Book species).
- The woodland is of a nationally or regionally rare type.
- A range of site types and habitat features are covered.

2. *Ecological development*: This relates to how successfully a native woodland habitat might be restored (i.e., are the building blocks for restoration present within the site or are they able to colonize from adjoining semi-natural habitats?).
   Features indicating high ecological value include:
   - The size of the woodland; larger sites are better.
   - A high density of mature trees with some veterans (very old trees that are often of great biological interest [Read 2000]). A good range of different native tree and shrub species and remnants of epiphytic communities.
   - A high diversity of specialist woodland ground flora species in robust patches.
   - The restoration site is adjacent to or surrounded by semi-natural woodland.
   - If the adjacent wood is of the same ecological type as that of the potentially restored wood and has high structural diversity, old large trees, and deadwood.

20.10 Methods of restoration and case studies

The preferred method of re-stocking is by natural regeneration, and most restoration has occurred in the uplands of Scotland where conditions for natural regeneration are generally more favorable than in the lowlands. When natural regeneration is used, any site-native species are accepted. If trees are planted then the species mixture will depend on management objectives, but the choice will be made from those species typical of the target woodland type. Where deer, rabbits, or other browsing animals cause problems, suitable methods of protection are recommended to promote success. On some sites, excessive weed growth can inhibit the survival, growth, and establishment of seedlings and transplants. However, as PAWS sites are generally being restored for ecological reasons, there is a reluctance to control weeds by using herbicides: at present, careful control of the existing canopy is used to try and restrict weed growth. Practical examples of how these methods are used are described in the following five case studies. As much of the planted area that is suitable for restoration is managed by Forest Enterprise, three of the studies are of their sites: one each from England, Wales, and Scotland.

20.10.1 Forest enterprise, southeast England forest district

This district is part of the Forestry Commission and is responsible for managing 186 woodlands, which have a total area of about 23,000 ha. Although some restoration has already begun, the process is largely at the planning stage. A thorough survey to classify all woodlands was carried out to identify how natural they were; about 12,000 ha of PAWS were identified, which includes about 2,000 ha of ancient woodland planted with beech and about 1,000 ha with sweet chestnut (*Castanea sativa* L.). The outcome of the survey was discussed with a wide range of interested organizations and a broad strategy for the restoration of PAWS was agreed. Initially, about 6,000 ha will be restored, the majority of which are planted with conifers.

Targets for restoration have been set for each of the 13 natural areas (Reid 1999) that occur within the forest district. The use of natural areas, which divide regions into areas with similar geology, soils, and landuse, is thought to be more appropriate than the use of administrative boundaries. Within each natural area, the PAWS identified for restoration have been selected on a variety of criteria, including their proximity to other ancient semi-natural woodland, the underlying soil type and probable difficulty of restoration, the potential NVC woodland community and the current state of transition to that community, and the need to retain conifers for both local industry and important species of animals that are dependent on the coniferous trees present.

Most restoration will be achieved by manipulation of the existing tree canopy, and clearfelling will only take place if the site is fully stocked with invasive conifers such as *Tsuga heterophylla* [(Raf.) Sarg.], open space is wanted, or existing production targets need fulfilling. Natural regeneration is the preferred method of restocking. Restoration is defined simply. Woodland will be considered restored when the percentage of canopy cover of trees that are not site-native falls below an acceptable threshold, which will vary with species but may be about 20% cover. Predicting the time course for restoration is difficult, but as management plans are revised, experience is gained, and the results of early work are assessed, a detailed timetable will evolve. However, given the simple criterion for restoration, it seems likely that many sites will be restored within a decade.

20.10.1.1 Example — Ampfield Wood

This wood is near Winchester in the south of England; it has an area of about 400 ha and is leased from two private estates. The natural broadleaved woodland was heavily cut during World War II and in the 1950s/1960s was generally planted with mixtures of conifers and broadleaves, using ground preparation and herbicides to kill the original broadleaved trees. The current 40 to 50-year-old crops comprise a wide variety of exotic conifers that are often in line-mixtures with oak and a few other broadleaved trees. To date, much of the restoration that has occurred within this wood has relied on the simple process of removing all conifers from the line-mixtures. As the stands of broadleaves remaining are 40 to 50 years old and adequately stocked, there has been no requirement to exclude damaging animals or control weeds. Access is good and operations have been in profit.

20.10.2 Forest enterprise, Coed y Brenin forest district

This 40,000 ha woodland district is located on the western and northern coasts of Wales. The majority of forests are upland plantations composed principally of Sitka spruce and other non-native species such as Japanese larch and Douglas-fir. However, there are substantial remnants of ancient woodland sites, including 98 upland oakwoods, many of which are designated at a European or Great Britain level for their nature conservation value. A survey of ancient woodland sites has recently been undertaken throughout Wales, with the primary objective of assessing their condition and restoration/enhancement potential, and identifying their woodland type. Secondary objectives were to assess the amount of deadwood, the extent of exotic scrub invasion (*Rhododendron ponticum* L.), and to note features of interest such as veteran trees (Read 2000).

20.10.2.1 Example — Coed y Brenin

This forest is approximately 2,500 ha and it is located in the Snowdonia National Park. Twelve oakwoods designated as Special Areas of Conservation are within the main woodland, with a further ten located in the surrounding area. Many of the remaining areas of upland oakwood are fragmented and the forest planning process has ensured that there will be good linkages between these patches by carefully targeting and prioritizing PAWS for restoration. Other criteria for deciding where to restore have included topography, geology, the needs of the Snowdonia National Park in which the forest is situated, existing vegetation communities (and, therefore, potential woodland type), plantation species, landscape impacts, access, and plans for adjoining land. A strategy has been developed to identify how restoration is to be integrated with other objectives for the forest.

Ancient woodland sites have played a major role in the new forest design plan. The aim is to retain some non-native trees, particularly stands of 80-year-old Douglas-fir that are valued as a landscape and recreation resource. Semi-natural woodland vegetation communities survive well under these stands and they will be managed using continuous-cover silviculture. The main long-term aim is to manage the woodland in a more traditional manner. This will include harvesting products such as charcoal, willow hurdles, and firewood, as well as low impact management for dormice. The timing of operations is difficult as there are a large number of visitors and many protected species of fauna breed within the forests.

Within Coed y Brenin, PAWS occupy approximately 1,000 ha. Restoration of these has been under way for 10 years, initially by clearfelling; natural regeneration of native broadleaves has been good. The move to more continuous-cover methods has been adopted, with clearfells only occurring where the planted species is very invasive (*Tsuga heterophylla*) or *Rhododendron ponticum* is present. External funding has been sought to remove exotic conifers and *Rhododendron*.

20.10.3 Forest Enterprise, Loch Lomond, and Trossachs National Park

In common with its restoration activities in England, Forest Enterprise has, since the early 1990s, also been involved in an extensive program of restoration in Scotland. In 1997, funding was obtained through the European "LIFE" program for a package of habitat restoration and enhancement activities within what is now the Loch Lomond and Trossachs National Park (an area of approximately 2,000 km^2 to the north and west of Glasgow). Two specific sites (Ross and Ptarmigan Woods, covering 425 ha in total) on the eastern shores of Loch Lomond were identified as priority areas for restoration. The primary motivation for the restoration work was to re-establish upland Atlantic oak–birch woodland (which is a variant of upland oakwood). Atlantic oakwoods have considerable value from natural, cultural, and heritage perspectives. They support a rich diversity of native flora and fauna, including many rare and threatened species whose continued survival depends on the stable ecological conditions created by a moist climate, continuity of woodland cover, and by the presence of native tree and shrub species. Originally, these woodlands covered much of the western seaboard of Scotland, Wales, and England, but centuries of conversion to other landuses have dramatically reduced their area.

Before any restoration was carried out, an extensive ecological survey was undertaken, identifying soil types and recording any surviving habitat features. Ecological Site Classification based on soils and remnant vegetation generated maps of potential semi-natural woodland types. These woodlands ranged from the expected oak–birch types, through to wet woodlands and, surprisingly, native pinewood. Scots pine originally occurred naturally in the Loch Lomond area, but died out 4,000 years ago for reasons that are unclear. Subsequent to the survey, a decision has been made to re-introduce pine to some suitable sites within the National Park.

Restoration began first in Ross Wood, largely because this site is the more accessible of the two and is located on flatter terrain. Conifers (mainly Sitka and Norway spruce) were clearfelled, taking care not to damage areas of wet ground and remnant mature broadleaved trees during extraction. To date, regeneration has been slow and hampered by weed growth on the richer sites. Deer and goats are controlled by culling, but are still having an impact on the pace of regeneration. Restoration has not yet begun on a large scale in the Ptarmigan Wood, as there are a number of logistical problems to address. Firstly, the wood is located on very steep terrain and the construction of access roads for timber extraction would have an adverse visual impact in an area that is highly visible to large numbers of visitors and tourists, and is valued for its "natural" landscape appearance. Secondly, large-scale timber extraction, by whatever means, would be visually intrusive. Alternatives to clearfelling are being considered, using low impact systems such as cable crane or skylines, and "felling to waste" (leaving the timber on site).

20.10.4 The Woodland Trust

The Woodland Trust was established in 1972 and is a charity dedicated entirely to the protection of Britain's woodlands; it has about 100,000 members. It currently owns more than 1,100 woods that are distributed throughout Great Britain and have a total area of greater than 18,000 ha. Of these woodlands, 113 are classified as PAWS, with an area of 2,377 ha, which is approximately 1% of Great Britain's total; 1,715 ha of this area have been identified as requiring restoration and 303 ha had already been completed by 2002. There is a commitment to restore all of its PAWS comprising conifers, but as a general principle it believes that more caution should be exercised on sites where non-native broadleaves have been planted. It is accepted that the woodlands that develop after removal of conifers may not always reflect the former natural composition, some sites are not suitable for restoration, and retention of some conifers may have cultural and wildlife value.

A project plan has been developed to facilitate and track progress. Sites have been prioritized on the basis of the threat to their long-term integrity from continued management as plantations; a four-point scale is used, which is essentially an indication of the amount of shade cast by the trees on site. It varies from no threat (sites with non-native broadleaves or few scattered exotics) to high threat where there is generally heavy shading from conifers. Approximately one-third of the work is at the planning stage, one-third has had an initial thinning, and the remainder has ongoing work. Practical experience has shown that clearfelling should be avoided, if possible, as it can often result in sites becoming overgrown with competitive weeds. In future, most restoration will take place by gradual thinning to promote established broadleaved trees and encourage the growth of naturally regenerating seedlings. The major problems encountered have been poor access to sites, poor markets for timber, and the small amounts harvested during thinning operations. The methods in use have been developed empirically, and in recognition of the need for research the Woodland Trust commissioned a study to investigate a number of characteristics of PAWS, including their nature, extent, and practical methods for restoration (Pryor and Smith 2002; Pryor et al. 2002).

20.10.4.1 Example — Grimshill and Church Woods

These two woods lie within Blean Wood, near Canterbury, in the southeast of England. They are ancient woodlands that were felled and replanted with conifers: either pure, or as line-mixtures with beech. Restocking will be by natural regeneration, and as there are many mature broadleaved trees adjacent to these sites, it is thought that this will not be constrained by a lack of seed. Access is generally good at this site, but soil conditions can restrict extraction during very wet periods. Restoration work began in 1996/1997 and the methods used have varied between compartments, depending on the structure and mixture of species present. Pure crops of Corsican pine and Norway spruce that established successfully have been line-thinned with additional selective felling to favor established broadleaves. Areas where Corsican pine failed to establish successfully and significant amounts of broadleaved trees grew were cleared of conifer that was cut to waste if insufficient timber could be harvested. The beech established poorly in the line-mixtures with Corsican pine. The first thinning of restoration in these areas removed every fourth row of pine, with additional selective felling around any good beech or other broadleaves. Further selective thinning is planned at 5-year intervals, with final restoration planned for 2018. These sites have not been protected from browsing animals as damage, if any, is slight. Weed control is by maintenance of canopy cover by slow, careful thinning.

20.10.5 The Nanteos Woodlands Group

This is a partnership of private woodland owners, centered around a historic estate in Ceredigion, Mid Wales. The group is composed of silviculturists, ecologists, and wood craftsmen who manage their own woodlands and run courses such as "Enhancing Biodiversity in Woodlands" and "Restoration of Broadleaved Woodland" for other practitioners.

20.10.5.1 Example — Coed Tamsin

This woodland, owned and managed by Felicia Jervis and Bob Shaw, is 11 ha in area, half of which is designated as an ancient woodland site. Historically, it was used for hunting game, but it was sold to a syndicate who wished to manage the site as a commercial forest. Clearfelling took place in 1962 and most of the area was replanted with Douglas-fir and Norway spruce. The present owners purchased the woodland in 1987. The present objectives are to manage the woodland for timber production, nature conservation, training in woodland skills/crafts, and health activities (e.g., retreats).

This woodland has well-drained, base-rich soils and is not typical of other site types in the region, which are dominated by poor, acidic soils. Ash would be the dominant native species. Stands are selectively thinned under a continuous-cover system to produce timber and small roundwood as needed, and to gradually increase the percentage of broadleaves. Timber is extracted using horses, a tractor and winch, or by hand. Restocking is mostly by natural regeneration with a small amount of planting, mostly of oak, alder, and sweet chestnut (the latter two planted in windblown gaps). All planted trees are grown from locally collected seed. The long-term vision for the balance between broadleaves and conifers is about 90/10%, respectively, with natural regeneration of Douglas-fir encouraged to maintain the conifer element.

Materials from Coed Tamsin are used to make a range of wood products, including fence hurdles, turnery, tool handles, chairs, tables, birch besoms, ash rakes, and artists' charcoal. Income is from educational courses and sale of woodcraft products. As this area is remote from large populations, a major constraint is distance from potential customers. However, the woodland is now self-financing, grant-aid having ceased in 1998.

Acknowledgments

Thanks are due to: Tim Hodges and John Tucker of the Woodland Trust; Wendy Joss, Stan Abbott, and Harry Oram of Forest Enterprise; Justin Gilbert of the Forestry Commission's Woodland Survey Branch; and Felicia Jervis and Bob Shaw of the Nanteos Woodlands Group.

References

Anderson, M.L., *A History of Scottish Forestry*, Volumes 1 and 2, Nelson, 1967.

Anon., *Report on Census of Woodlands and Census of Production of Home-grown Timber 1924*, HMSO, London, 1928.

Anon., *Census of Woodlands 1947–1949*, HMSO, London, 1952.

Anon., *Forestry Practice Guides* 1–8, Forestry Commission, Edinburgh, 1994a.

Anon., *Biodiversity: The UK Action Plan*, HMSO, London, 1994b.

Anon., *The UK Forestry Standard*, Forestry Commission, Edinburgh, 1998a.

Anon., *UK Biodiversity Group Tranche 2 Action Plans Volume II — Terrestrial and Freshwater Habitats*, HMSO, London, 1998b.

Anon., *Certification Standard for the UK Woodland Assurance Scheme*, UKWAS, Edinburgh, 2000.

Anon., *National Inventory of Woodland and Trees*, Separate publications for England, Wales, Scotland and Great Britain, Forestry Commission, Edinburgh, 2001a.

Anon., *Woodlands for Wales: The National Assembly for Wales Strategy for Trees and Woodlands*, Forestry Commission, Aberystwyth, 2001b.

Anon., *Woodland Grant Scheme*, Forestry Commission, Edinburgh, 2001c.

Evelyn, J., Silva, or a discourse of forest trees in *The Writings of John Evelyn*, G. de la Bédoyère (1995), Ed., The Boydell Press, Woodbridge, 1664.

Fernow, B.E., *History of Forestry*, Toronto University Press, Toronto, 1913.

Gray, H. and Stone, D., Applications of spatial data in strategic woodland decisions in Scotland: an example from the Isle of Mull, in *The Restoration of Wooded Landscapes*, J. Humphrey, A. Newton, J. Latham, H. Gray, K. Kirby, E. Poulsom, and C. Quine, Eds., Forestry Commission, Edinburgh, 2003.

Hall, J.E. and Kirby, K., *The Relationship Between Biodiversity Action Plan Priority and Broad Woodland Habitat Types, and Other Woodland Classification*, JNCC Report No. 288, JNCC, Peterborough, 1998.

Humphrey, J.W. and Nixon, C.J., The restoration of upland oakwoods following the removal of conifers: general principles, *Scottish For.*, 53, 68, 1999.

Humphrey, J.W., Newton, A.C., Peace, A.J., and Holden, E., The importance of conifer plantations in northern Britain as a habitat for native fungi, *Biol. Conserv.*, 96, 241, 2000.

Humphrey, J.W., Ferris, R., Jukes, M.R., and Peace, A.J., The potential contribution of conifer plantations to the UK Biodiversity Action Plan, *Bot. J. Scotland*, 54, 49, 2002.

Humphrey, J.W., Ferris, R., and Quine, C.P., Eds., *Biodiversity in Britain's Forests: Results from the Forestry Commission's Biodiversity Assessment Project*, Edinburgh, 2003.

James, N.D.G., *A History of English Forestry*, Basil Blackwell, Oxford, 1981.

Linnard, W., *Welsh Woods and Forests*, Gomer, Llandysul, 2000.

Locke, G.M.L., *Census of Woodlands 1965–67*, HMSO, London, 1979.

Locke,G.M.L., *Census of Woodlands and Trees 1979–82*, Forestry Commission Bulletin 63, HMSO, London, 1987.

Peterken, G.F., Habitat conservation priorities in British and European woodlands, *Biol. Conserv.*, 11, 223, 1977.

Peterken, G.F., *Woodland Conservation and Management* 2nd ed., Chapman & Hall, London, 1993.

Peterken, G.F., *Natural Woodland*, Cambridge University Press, Cambridge, 1996.

Pringle, D., *The Forestry Commission — The First 75 years*, Forestry Commission, Edinburgh, 1994.

Pryor, S., The costs and benefits of restoring plantations versus creating new native woodland, in *The Restoration of Wooded Landscapes*, J. Humphrey, A. Newton, J. Latham, H. Gray, K. Kirby, E. Poulsom, and C. Quine, Eds., Forestry Commission, Edinburgh, 115, 2003.

Pryor, S.N. and Smith, S., *The Area & Composition of Plantations on Ancient Woodland Sites*, The Woodland Trust, Grantham, 2002.

Pryor, S.N., Curtis, T.A., and Peterken, G.F., *Restoring Plantations on Ancient Woodland Sites*, The Woodland Trust, Grantham, 2002.

Pyatt, D.G., Ray, D., and Fletcher, J., *An Ecological Site Classification for Forestry in Great Britain*, Forestry Commission Bulletin 124, Forestry Commission, Edinburgh, 2001.

Quine, C.P., Coutts, M.P., Gardiner, B.A., and Pyatt, D.G., *Forests and Wind; Management to Minimise Damage*, Forestry Commission Bulletin 114, HMSO, London, 1995.

Rackham, O., *Ancient Woodland*, Edward Arnold, London, 1980.

Ray, D., *Ecological Site Classification Decision Support System*, Forestry Commission, Edinburgh, 2001.

Read, H., *Veteran Trees: A Guide To Good Management*, English Nature, Peterborough, 2000.

Reid, C.M., Using natural areas to promote woodland conservation in England, *Q. J. For.*, 93, 21, 1999.

Rodwell, J.S., *British Plant Communities Vol. 1 Woodlands and Scrub*, Cambridge University Press, Cambridge, 1991.

Rodwell, J.S. and Patterson, G. S., *Creating New Native Woodlands*, Forestry Commission Bulletin 112, Forestry Commission, Edinburgh, 1994.

Spencer, J.W. and Kirby, K. J., An ancient woodland inventory for England and Wales, *Biol. Conserv.*, 62, 77, 1992.

Thompson, R.N., Humphrey, J. W., Harmer, R., and Ferris, R., *Restoration of Native Woodland on Ancient Woodland Sites*, Forest Practice Guide, Forestry Commission, Edinburgh, 2003.

Walter, H., *Vegetation of the earth*, 3rd ed., Springer-Verlag, Berlin, 1985.

chapter twenty-one

*Conversion of Norway spruce (*Picea abies *[L.] Karst.) forests in Europe*

Jörg Hansen and Heinrich Spiecker
Albert-Ludwigs University Freiburg

Contents

21.1 Introduction

Traditional forestry in much of northern and central Europe has concentrated on growing conifers in plantations for timber and pulp. Increased concern over ecological stability, nature conservation, possible climate change (Ulrich and Puhe 1994), and damage from major storms, with the resulting periods of reduced prices for softwood timber, however, has increased interest in alternative management approaches such as close-to-nature forestry (Dafis 2001). As a consequence, the widespread reliance on conifer monocultures has come into question and more stable alternatives, such as site-adapted mixtures of species, often dominated by broadleaves, are under consideration or already employed (Hansen et al. 2004; Spiecker et al. 2004).

Norway spruce (*Picea abies* [L.] Karst.) is found in almost every European country. It is affected by severe calamities caused by storms, snow, or ice (Ebert 2002; Kuhn 1995), and the corresponding secondary damage, such as decline of damaged trees or bark beetle attacks. As there are potentially unstable Norway spruce forests in many European countries, this species is the primary candidate for conversion to structured or mixed broadleaf-dominated forests (Teuffel et al. 2004; Gardiner and Breland 2002; Hasenauer 2000; Klimo et al. 2000). The stability of forests can be expressed in terms of the resistance and resilience of the forest ecosystem (Larsen 1995). Poor resistance can be manifested as a higher susceptibility than would be expected of a natural forest to damage or destruction by strong winds, drought, fire, or a complex of factors. Poor resilience may entail considerable

1-56670-635-1/05/$0.00+$1.50

difficulty regaining a closed-forest condition after a catastrophe, because of regeneration problems.

Norway spruce is an example of a species that can regenerate successfully on sites where it is not adapted in the long run, or where browsing or management intervention has favored it over other more site-adapted species (Johann et al. 2004). A large proportion of potentially unstable Norway spruce stands exist today that were artificially established by direct seeding or planting. A critical question facing managers is which Norway spruce stands should be converted. Selection of conversion stands cannot be based solely on the natural range of Norway spruce, but also needs to be compatible with the objectives of forest management.

21.2 Distribution of Norway spruce

Norway spruce is one of the most profitable tree species in Europe and has been widely planted outside its natural range. The highest relative proportion of Norway spruce, as compared to the total national territory, is found in Sweden (~34%) followed by Austria with over 22% and the Czech Republic with around 17%. Germany, Luxemburg, Slovakia, and Switzerland have Norway spruce proportions of approximately 10% and therefore correspond to the European average (Teuffel et al. 2004). For Finland and Norway, there have been reports of Norway spruce being between 10 and 25% (Spiecker 2000). Although these countries are located mostly within the natural distribution of the species, Norway spruce was introduced to new sites (Heinzel and Peters 2001; Koch 2000) where it had been absent or, at most, was codominant or admixed.

Norway spruce's natural range is somewhat contentious. Because it was little affected by human interference until the 3rd or 4th century AD (Hasel 1985), its range at that time provides a starting point for what can be considered "natural." At the end of the last glacial maximum, Norway spruce probably spread into central Europe from the east at the same time as other species, such as European silver fir (*Abies alba* Mill.), returned to central Europe from southwestern refugia. With its wide ecological amplitude and ability to compete against other species (Mayer 1984), Norway spruce might have migrated further into the northwest of continental Europe even without human intervention. Thus, the determination of which stands are not site-adapted is not a simple matter of referring to a map of its natural range (Figure 21.1).

Based on its 3rd–4th century AD distribution, Norway spruce can be described as a species of a cool and humid climate. The focus of its natural distribution in Europe lies in the boreal coniferous forest of Nordic countries and Russia, as well as at higher elevations in the temperate zone forests of Switzerland, Austria, the Czech Republic, Poland, Byelorussia, and the Baltic countries of Lithuania, Latvia, and Estonia (Figure 21.1). Water availability in years with low precipitation levels is thought to limit the distribution of Norway spruce in the south. In the west and southwest of central Europe, natural, pure Norway spruce stands occur in mountainous areas where short growing seasons, low annual mean temperatures (≤5°C), and high annual precipitation (≥800 mm) are typical (Schmidt-Vogt 1977). With increasing competitiveness of European silver fir or European beech (*Fagus sylvatica* L.), Norway spruce is usually a codominant or admixed tree species.

Forest management has not only extended the geographical range of Norway spruce, but even within its natural range, Norway spruce has been planted on sites naturally dominated by broadleaves. This change in growing stock took place from the Middle Ages up until the 20th century and was mostly motivated by widespread overexploitation and devastation of forests, as well as the fear of timber shortage (Hasel 1985). From the 18th century onward, large-scale litter raking, which removed nutrients and diminished the

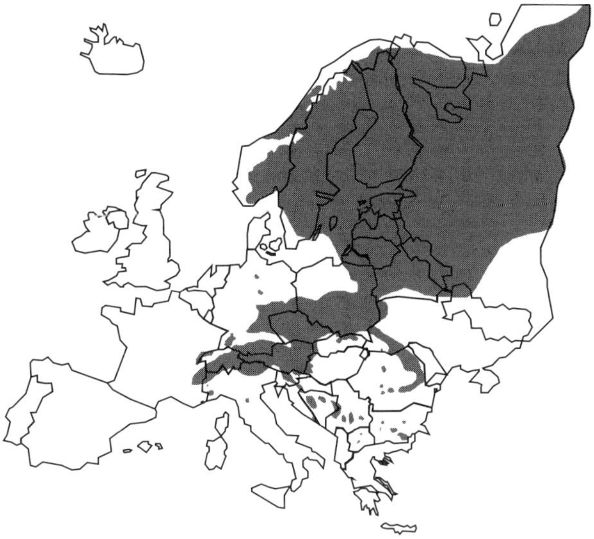

Figure 21.1 Natural distribution of Norway spruce (*Picea abies* [L.] Karst.) in Europe (modified from Schmidt-Vogt 1977).

humus layers, causing regeneration problems and lowering yield, degraded forests. Removals for charcoal and potash production, plus intensive timber consumption for shipbuilding and mining, resulted in extensive clearfelled areas. Cattle grazing and browsing by large populations of game animals added to the difficulty of obtaining adequate regeneration. Increasingly, managers turned to coniferous trees, notably Norway spruce and Scots pine (*Pinus silvestris* L.), for growing stock. Norway spruce is better adapted to the microclimate on clearcut sites than European beech, silver fir, or oak (*Quercus* spp.), which are more susceptible to damage by late frosts. Furthermore, Norway spruce exhibits high initial growth, an advantage in competing with other ground vegetation. The most recent wave of reforestation that increased the proportion of Norway spruce took place after World Wars I and II.

21.3 *Management of Norway spruce*

Industrialization in the 19th century dramatically increased the demand for timber and special timber assortments, including small dimensions. Economic models for organizing forest management and evaluating silvicultural decisions were developed at the same time. Norway spruce was ideal for these developments; it was easier and less expensive to regenerate than broadleaves, grew faster, and reached an economic size quicker. These economic decision models postulated profit maximization as the main goal and favored dense planting, heavy thinning, and relatively short rotations. As the focus of analysis shifted to the stand or harvesting unit level, forests were organized into even-aged cutting units. This form of profitable and efficient forest management became widespread in Europe and provided industry with a continuous and even volume of timber in typical and uniform assortments of consistent quality. The demand for small-dimension spruce timber was met by "Dunkelwirtschaft" (German for dark management), which is characterized by high density and low thinning to produce uniform material with little or no taper. Although appreciated by timber processors, such management increases the susceptibility of stands to storm and snow damage. The introduction of alternative materials such as steel and plastic in the 20th century reduced the demand for small-dimension

Figure 21.2 Storm-damaged Norway spruce stand in southern Germany.

timber. Without the demand for these timber assortments, needed thinning became precommercial and was often postponed, increasing the instability of dense spruce stands (Figure 21.2).

At present, forests in Europe are managed to fulfill a range of functions: economic (timber and employment), nonconsumptive recreation, and ecological and societal services (soil, air, and water protection, nature conservation, noise absorption, and avalanche prevention). In central Europe, Norway spruce is faced with the risk of various biotic and abiotic damages (Kulikova 2002; Nopp et al. 2001; Tomiczek 2000). Storm and snow are the most detrimental (Spiecker 2000). Further negative factors include bark beetles (*Ips typographus* L., *Pityogenes chalcographus* L.), fire, ice, fungal diseases (*Heterobasidion annosum* [Fr.] Bref.= *Fomes annosus* [Fr.] P. Karst., *Armillariella* spp.) or, in exceptional cases, damage by wildlife. The risk of damage to a stand increases if two or more of these detrimental factors coincide, or if they combine with other predisposing site or stand conditions such as poorly drained soils, few admixed species or non-site-adapted provenances of spruce, or great tree heights.

In some parts of Europe, high volumes of Norway spruce are harvested as incidental exploitations (Hanewinkel 2002), which are entirely unintended or not performed for economic motives (Kusché 2000; Spiecker 2000). Although implying an impaired stability and lack of resilience, general conclusions should not be drawn from such harvests without first analyzing the underlying causes. Damage review after the storms of the previous decades has shown that Norway spruce growing on physiologically shallow soil, such as on dual-layered soil profiles or poorly drained soils, develops a very shallow root system and bears an extreme risk for windthrow. This risk increases with high height-to-diameter ratios of individual trees, great tree height, and high stocking. Additional factors such as stand age and size, the proportion of other species, or trees impaired by fungi or cambial damage from wildlife also contribute to the risk of windthrow or windbreak.

21.4 The question of conversion

In order to achieve or safeguard the long-term stability of forests, foresters in many European countries are confronted with questions of whether to convert Norway spruce stands, which stands will benefit from conversion to other species or mixtures with Norway spruce, and how to accomplish conversion. A multifunctional approach implies that such decisions will be based on site factors, location, and potential effects on ecological processes. In order to answer all questions related to the option of conversion, further aspects also need to be taken into account, including the conservation status of a stand, the form of ownership, and the level of local infrastructure development. All these factors affect the approach and intensity of silvicultural measures to be taken. With increasing labor costs and stagnant timber prices, new models of forest management are required. Close-to-nature forestry, with species mixtures and uneven-aged structure, as opposed to pure stands and forest management in even-aged cutting units, is commonly regarded as the most effective option.

21.4.1 *Is conversion needed?*

Each forest owner determines the most dominant forest function(s) pursued in accordance with regulatory background, economic opportunity, and personal preference. In most instances, the objective will not be to achieve maximum fulfillment of all forest functions, but to have an optimized combination for implementing the forest owner's target system. The need for conversion will be manifest if a substantial disruption in one or several of the targeted forest functions occurs. Thus, the criteria to decide on conversion needs are always dependent on the objectives of forest management.

Management objectives differ greatly between forest ownerships. In several European countries, the conversion of Norway spruce stands in public ownership has progressed further than in privately owned forests (Schmid 1998). Timber from Norway spruce forests has been easily marketed due to its many uses and uniform assortments of standard quality. Unless excess supply, such as is experienced after calamities due to storm, heavy snow, or bark beetle outbreaks, causes a slump in timber prices, there is little short-term incentive for private owners to convert. Timber certification and its influence on marketability of forest products may be an incentive for conversion in the approximately 40% of the European forest area in private ownership. However, it is difficult to make general statements for private forest owners, as this group has many different attitudes toward management (Schraml and Härdter 2002; Schraml and Volz 2004). Nevertheless, reduced forestry subsidies in many European countries may enhance the appeal of low-input techniques such as close-to-nature silviculture.

21.4.2 *Which stands should be converted?*

A decision to convert a stand should only be taken if superior options for achieving desired forest functions can be identified. The timing of conversion will be determined in a second step, where the conversion priority for each stand is established on the basis of the overall situation in the forest enterprise. The strategy chosen for conversion, including the prescribed silvicultural measures, needs to be selected in accordance with the urgency for conversion. Large areas of Norway spruce blown down by a storm require immediate conversion. Adequate measures should be taken in the short-to-medium term in order to prevent replicating identical stand structures and species composition, for example, by introducing other species to the regenerating stand. Neighboring stands unaffected by blowdown should also be converted if similar site and stand conditions apply. If a Norway spruce stand has repeatedly suffered small-scale damage due to abiotic or biotic factors, such warnings of potential catastrophe should motivate stand conversion measures in the medium to long term. The conversion process can be delayed, however, if the risk is deemed low or if conversion will be undertaken to fulfill a new policy direction.

21.4.3 *How should conversion be accomplished?*

For each stand with an established need for conversion, the selection of a conversion strategy should be guided by the widest possible use of economic principles, both for conversion and for management of the target stand. The objectives for conversion are to optimize stability in the target forests, which is achieved by a high resilience against detrimental factors, and the greatest flexibility in the management system, so as to allow easy adaptation to future shifts in objectives or changed ecological conditions. As a practical matter, conversion strategies requiring a minimum of long-term intervention are preferable. Nevertheless, the long-term nature of some processes in forests (e.g., soil degradation) hinders or completely disables a direct and immediate analysis of the ecological impact after human intervention. The time lag between cause and effect, or between management measures and their long-term consequences, may be as long as a rotation length and therefore also as long as several generations of foresters.

There are two basic strategies for conversion: introduction of more complex structure within the same species, such as by creation of multiaged stands of Norway spruce, and expansion of species variety, as achieved by insertion of broadleaves (beech or maple [*Acer* spp.]) or other conifers (silver fir). To address a specific conversion task, these two strategies can be combined in varying degrees, which provide the forest manager with a broad range of conversion options. The silvicultural measures for conversion of pure secondary Norway spruce stands to mixed-species stands range from clearcutting Norway spruce and planting broadleaves that correspond to the potentially natural vegetation, to the less-intensive measure of tending broadleaves that naturally enter or are artificially inserted into pure Norway spruce stands. This range of measures reflects the large number of alternatives that can be chosen to fit site conditions, stand characteristics, management intensity, timing preference of the forest owner, and applicable laws and regulations. Each approach, however, will be based on a silvicultural strategy that must be implemented consistently over the long term. Some temporary destabilization of the stand may accompany intensive measures taken to convert Norway spruce.

Managing the light environment in the regeneration layer is the silvicultural challenge in conversion. Controlling light transmittance in the stand prevents suppression of broadleaves from dense Norway spruce regeneration or grass proliferation. Shade-tolerant broadleaves, which can be introduced into Norway spruce stands by underplanting or seeding, can be favored by these measures over Norway spruce regeneration (Lüpke et al. 2004). Knowledge and utilization of the shade tolerance of young broadleaves, such as European beech or oak, as well as utilization of gap development in Norway spruce forests, is the key component of a successful conversion strategy (Figure 21.3). Manipulating light levels in the stand requires more entries per planning period, which increases the risk of damage to residual trees and soil compaction from machinery traffic. Target-oriented (diameter limit) felling of individual trees or small groups to create small gaps usually constitutes the first and most important silvicultural measure in the conversion process. Harvesting will be spread across a larger area, as compared to conventional even-aged management. The initial phase of conversion presents severe economic challenges, including lower revenues and higher costs. As the conversion process may extend over several rotation lengths, the most effective strategies will include the fewest interventions with the greatest potential for revenue.

In addition to choosing an effective silvicultural technique for stand and site conditions, conversion decisions will be constrained by the socioeconomic differences within Europe and between forms of forest ownership. These differences not only cause a huge disparity in available infrastructure and equipment, as well as in available means for purchasing new technology, but are also manifest as large differences between countries in labor costs and supply. As a consequence, low-tech and mostly manual implementation may be highly superior to the use of machinery and high-tech methods in some countries (Nordfjell et al. 2004). Seasonal investment of the forest owner's labor in small privately owned forests, common in many farming families, may be a practical scenario to achieve low labor costs.

Figure 21.3 Seedlings of broadleaves and spruce following shelterwood treatments in Norway spruce stands.

21.5 Future research needs

A research initiative was begun in late 2000 as a Regional Project Centre (RPC) under the auspices of the European Forest Institute (EFI) to combine pan-European research efforts on CONversion of Norway spruce FORESTs (RPC-CONFOREST). The RPC-CONFOREST compiles available knowledge on the local or national level to determine the European dimension of conversion, to scale up findings, and to provide a platform for coordination and promotion of the corresponding research efforts (Hansen and Spiecker 2002; Spiecker et al. 2004).

Conversion is a highly complex process, as it exerts an impact on many sectors of forestry and the timber industry, as well as on political life. In order to encompass this complexity, the RPC-CONFOREST has eight focal areas, which not only examine the phase of conversion itself but also compare the management of pure Norway spruce stands to their mixed or broadleaf counterparts. These thematic areas include developing a common terminology, compiling and analyzing geo-referenced data, investigating land-use history, ecological, and economic aspects of conversion, and developing silvicultural strategies and operational options for implementation, as well as analyzing the political environment. The RPC-CONFOREST emphasizes equal intensity of effort in all thematic areas, as well as interdisciplinary and transnational cooperation within and across the thematic areas to achieve pan-European validity.

The long-term scientific goal of this effort is to compile validated conversion guidelines for various scenarios, and to forecast the impact of the conversion of Norway spruce stands on the forestry sector. Conversion research regarding other coniferous species is also envisaged. Improved and scientifically based management strategies in European forestry will contribute to increased profitability in forest enterprises and related industries.

References

Dafis, S., Ecological impacts of close-to-nature forestry on biodiversity and genetic diversity, in *Ecological and Socio-Economic Impacts of Close-to-Nature Forestry and Plantation Forestry: A Comparative Analysis*, Proceedings of the Scientific Seminar of the 7th Annual EFI Conference, Green, T., Ed., 3 Sept. 2000, EFI Proceedings 37, 21–25, 2001.

Ebert, H.P., Schäden durch 'Lothar' im Lehrrevier Rottenburg, *Allgemeine Forstzeitschrift/Der Wald*, 24, 1261, 2002.

Gardiner, E.S. and Breland, L.J., Eds., *Proceedings of the IUFRO Conference on Restoration of Boreal and Temperate Forests - Documenting Forest Restoration*, Vejle, Denmark, 28 Apr.–2 May 2002, Skov ob Landskab, Report 11, 238 pp., 2002.

Hanewinkel, M., Climatic hazards and their consequences for forest management - an analysis of traditional methodological approaches of risk assessment and alternatives towards the development of a risk control system, in *Risk Management and Sustainable Forestry*, Arbez, M., Birot, Y., and Carnus, J.M., Eds., EFI Proceedings, 45, 21, 2002.

Hansen, J. and Spiecker, H., Nucleus Network RPC-CONFOREST: The question of conversion of secondary coniferous forests in Europe, in *Forest Research and the 6th Framework Programme - Challenges and Opportunities*, Joensuu, Birot Y., Päivinen, R., and Roihuvuo, L., Eds., Report of the Open Seminar held 25 November 2002, Paris, France, European Forest Institute, 99, 2002.

Hansen, J., Spiecker, H., and Teuffel, K. von, Eds., *The Question of Conversion of Coniferous Forests*, Abstracts of the International Conference, 27 Sept.–2 Oct. 2003, Freiburg im Breisgau, Germany, Freiburger Forstliche Forschung, Berichte, 47, 85, 2004.

Hasel, K., *Forstgeschichte. Ein Grundriß für Studium und Praxis*, Paul Parey, Hamburg, Berlin, *Pareys Studientexte*, 48, 258 pp., 1985.

Hasenauer, H., Ed., Ecological and economic impacts of restoration processes in secondary coniferous forests, in *Proceedings of the International Conference on Forest Ecosystem Restoration*, Vienna, Austria, 10–12 April 2000, 418 pp., 2000.

Heinzel, K.U. and Peters, S., Umbau von Fichten- und Kiefernreinbeständen zu naturnahen Mischwäldern, *Allgemeine Forstzeitschrift/Der Wald*, 56 , 467, 2001.

Johann, E., Agnoletti, M., Axelsson, A.L., Bürgi, M., Östlund, L., Rochel, X., Schmidt, U. E., Schuler, A., Skovsgaard, J.P., and Winiwarter, V., History of secondary Norway spruce forests in Europe, in *Norway Spruce Conversion - Options and Consequences*, Spiecker, H., Hansen, J., Klimo, E., Skovsgaard, J.P., Sterba, H., and Teuffel, K. von, Eds., EFI Research Report 18, S. Brill Academic Publishers, Leiden, 25, 2004.

Klimo, E., Hager, H., and Kulhavy, J., Eds., *Spruce Monocultures in Central Europe: Problems and Prospects*, International Workshop held in Brno, Czech Republic, 22–25 June 1998, EFI Proceedings, 33, 208 pp., 2000.

Koch, G., Vergleich potenzieller natürlicher und aktueller Baumartenverteilung in forstlichen Problemgebieten, in *Mariabrunner - Waldbautage 1999: Umbau sekundärer Nadelwälder*, Müller, F., Ed., Berichte der Forstlichen Bundesversuchsanstalt in Wien, 111, 31, 2000.

Kuhn, N., Die standörtliche Abhängigkeit der Vivian-Windwürfe, *Informationsblatt des Forschungsbereichs Landschaftsökologie*, 28, 5, 1995.

Kulikova, E.G., Phytosanitary risks, in *Risk Management and Sustainable Forestry*, Arbez, M., Birot, Y., and Carnus, J. M., Eds., EFI Proceedings, 45, 43, 2002.

Kusché, W., Geschichtliche Entwicklung, Umfang und Verteilung sekundärer Fichtenwälder im Alpenvorland, in *Mariabrunner - Waldbautage 1999: Umbau sekundärer Nadelwälder*, Müller, F., Ed., Berichte der Forstlichen Bundesversuchsanstalt in Wien, 111, 45, 2000.

Larsen, J.B., Ecological stability of forests and sustainable silviculture, *For. Ecol. Manage.*, 73, 85, 1995.

Lüpke, B. von, Ammer, C., Bruciamacchie, M., Brunner, A., Ceitel, J., Collet, C., Deleuze, C., Di Placido, J., Huss, J., Jankovic, J., Kantor, P., Larsen, J.B., Lexer, M., Löf, M., Longauer, R., Madsen, P., Modrzynski, J., Mosandl, R., Pampe, A., Pommerening, A., Stefancik, I., Tesar, V., Thompson, R., and Zientarski, J., Silvicultural strategies for conversion, in *Norway Spruce Conversion - Options and Consequences*, Spiecker, H., Hansen, J., Klimo, E., Skovsgaard, J.P., Sterba, H., and Teuffel, K. von, Eds., EFI Research Report 18, S. Brill Academic Publishers, Leiden, 121, 2004.

Mayer, H., *Waldbau auf soziologisch-ökologischer Grundlage*, 3rd ed., Gustav Fischer, Stuttgart, 514 pp., 1984.

Nopp, U., Netherer, S., Eckmüller, O., and Führer, E., Parameters for the assessment of the predisposition of spruce-dominated forests to various disturbing factors with special regard to the 8-toothed Spruce Bark Beetle (*Ips typographus* L.), in *Criteria and Indicators for Sustainable Management at the Forest Management Unit Level*, Franc, A., Laroussinie, O., and Karjalainen, T., Eds., EFI Proceedings, 38, 99, 2001.

Nordfjell, T., Bacher, M., Eriksson, L., Kadlec, J., Stampfer, K., Suadicani, K., Suwala, M., and Talbot, B., Operational factors influencing the efficiency in conversion, in *Norway Spruce Conversion - Options and Consequences*, Spiecker, H., Hansen, J., Klimo, E., Skovsgaard, J.P., Sterba, H., and Teuffel, K. von, Eds., EFI Research Report 18, S. Brill Academic Publishers, Leiden, 197, 2004.

Schmid, S., Struktur und Entwicklung des Kleinprivatwaldes in Baden-Württemberg, *Forst und Holz*, 7, 199, 1998.

Schmidt-Vogt, H., *Die Fichte. Ein Handbuch in zwei Bänden, Taxonomie, Verbreitung, Morphologie, Ökologie, Waldgesellschaften*, Vol.I, Paul Parey, Hamburg, Berlin, 647 pp., 1977.

Schraml, U. and Härdter, U., Urbanität von Waldbesitzern und von Personen ohne Waldeigentum - Folgerungen aus einer Bevölkerungsbefragung in Deutschland, *Allgemeine Forst- und Jagdzeitung*, 7–8, 140, 2002.

Schraml, U. and Volz, K.-R., Conversion of coniferous forests - social and political perspectives. Findings from selected countries with special focus on Germany, in *Norway Spruce Conversion - Options and Consequences*, Spiecker, H., Hansen, J., Klimo, E., Skovsgaard, J.P., Sterba, H., and Teuffel, K. von, Eds., EFI Research Report 18, S. Brill Academic Publishers, Leiden, 97, 2004.

Spiecker, H., Growth of Norway spruce (*Picea abies* [L.] Karst.) under changing environmental conditions in Europe, in *Spruce Monocultures in Central Europe: Problems and Prospects*, Klimo, E., Hager, H., and Kulhavy, J., Eds., International Workshop held in Brno, Czech Republic, 22–25 June 1998, EFI Proceedings, 33, 11, 2000.

Spiecker, H., Hansen, J., Klimo, E., Skovsgaard, J.P., Sterba, H., and Teuffel, K. von, Eds., *Norway Spruce Conversion - Options and Consequences*, EFI Research Report 18, S. Brill Academic Publishers, Leiden, 269 pp., 2004.

Teuffel, K. von, Heinrich, B., and Baumgarten, M., Present distribution of secondary Norway spruce in Europe, in *Norway Spruce Conversion - Options and Consequences*, Spiecker, H., Hansen, J., Klimo, E., Skovsgaard, J.P., Sterba, H., and Teuffel, K. von, Eds., EFI Research Report 18, S. Brill Academic Publishers, Leiden, 63, 2004.

Tomiczek, C., Überblick über die Forstschutzprobleme in sekundären Nadelwäldern, in *Mariabrunner - Waldbautage 1999: Umbau sekundärer Nadelwälder*, Müller, F., Ed., Berichte der Forstlichen Bundesversuchsanstalt in Wien, 111, 81, 2000.

Ulrich, B. and Puhe, J., Auswirkungen der zukünftigen Klimaveränderung auf mitteleuropäische Waldökosysteme und deren Rückkopplungen auf den Treibhauseffekt, in *Enquete-Kommission 'Schutz der Erdatmosphäre' des Deutschen Bundestages*, Studienprogramm, Volume 2, Wälder, Deutscher Bundestag, Ed., Economica Verlag, Bonn., III + 208 pp., 1994.

chapter twenty-two

Regeneration by direct seeding — a way to reduce costs of conversion

Helmuth Baumhauer
Forstamt Weissenhorn
Palle Madsen
Danish Centre for Forest, Landscape and Planning, KVL
John A. Stanturf
US Department of Agriculture Forest Service

Contents

22.1 Introduction

Many western European countries need less expensive regeneration methods for their Norway spruce (*Picea abies* [L.] Karst.) forests. Financial returns from timber production have declined severely over the past 15 years and levels of future returns are impossible to predict. Additionally, the increasing frequency of severe windthrow from winter storms has focused more attention on stand stability and has prompted efforts to convert pure Norway spruce forests to broadleaf or mixed stands. Foresters of Forstamt Weissenhorn in Bavaria, Germany implemented a technique for direct seeding beech (*Fagus sylvatica* L.) and European silver fir (*Abies alba* Mill.) in Norway spruce stands that costs about 20% of the cost for traditional planting methods and results in a mixed-species stand (Baumhauer 1996). The direct seeding approach seems to solve an important challenge to restoring off-site and pure Norway spruce plantations in Europe and provides low costs and high stocking density, thereby ensuring high wood quality in the future.

Forstamt Weissenhorn is a state forest in the Bavarian region of southern Germany. Soils are mostly poorly drained due to high clay contents. Norway spruce grows well, up to 40 $m^3\ ha^{-1}\ year^{-1}$, but develops a shallow root system that increases the risk of windthrow. In 1984, Norway spruce plantations accounted for 89% of the area and dominated the older age classes (80 to 120 years). About one third (2,000 ha) of the spruce plantations have been damaged by wind within the past 15 years. Predictions of future climate change suggest that Norway spruce will be eliminated as a commercial species in the region within the next 100 years due to summer drought. In response to this grim future, a decision was taken to quickly increase the regeneration of shade-tolerant and wind-stable species such as beech and European silver fir in order to take advantage of the favorable regeneration conditions under the remaining spruce stands. The goal is mixed stands with beech, fir, and spruce in equal amounts. Norway spruce regenerates naturally and vigorously; for the other species, direct seeding seemed a promising and cost-effective alternative to planting.

22.2 Silvicultural treatments

Traditional silvicultural approaches to manage spruce stands restrict regeneration treatments to older, often unstable stands that leave limited flexibility for regeneration treatments. Blowdown, followed by mortality from bark beetle attacks in old spruce stands, causes severe regeneration problems. Mortality opens the canopy, releasing natural spruce regeneration before sown fir and beech regeneration is able to establish. Until recently, fir and beech were sown in many old stands that are now too open due to the bark beetle attacks. Instead, they are underplanted with fir seedlings and beech wildings that can compete with the flush of spruce regeneration. Planting wildlings is cheaper than planting beech nursery stock. In younger spruce stands, direct seeding of fir and beech is now used to avoid future regeneration problems in the old stands.

Control of the light environment for regeneration by manipulating the canopy cover is essential for successfully establishing mixed stands. Thinning a spruce stand too heavily or too early will favor the spruce in the entire stand. This is particularly the case with directly sown fir and beech. Consequently, direct seeding generally requires lower and better-controlled light conditions than planting. Fir is a slow starter, but it can establish under a rather dense shelterwood; therefore, it is sown several years before the beech. Beech is intermediate between fir and spruce in shade tolerance, but it also needs to gain a foothold before the natural regeneration of spruce is released by canopy opening.

Direct seeding starts after crown thinning to open the canopy in spruce stands as young as 35 years old. Fir is sown 3 years later and beech is sown 10 to 18 years later. There are two major advantages of this strategy. First, the young spruce stands produce less seed than older stands, reducing the competition with beech and fir. Second, young spruce stands have a lower risk of blowdown or bark beetle attack and are likely to keep a shelterwood longer. The light shade of the shelterwood may facilitate individual tree differentiation in the late regeneration stages, reducing the need for precommercial thinning in the future.

Table 22.1 A Typical Timeline for the Regeneration of Spruce Stands

Regeneration Timeline (year)	Treatment
0	Heavy thinning to open the closed canopy. Canopy cover is reduced to 80% in stands beginning as young as 35 years old
3	Fir sowing
7	Shelterwood thinning where fir has established. Canopy cover is reduced to 70%
10–18	Beech sowing or planting
20–35	Two further thinnings to release the fir and beech regeneration and naturally regenerate spruce
Beyond 35	A final cut when the regeneration is complete

We can only provide an approximate schedule for the sowing and shelterwood thinning (Table 22.1). The timing and selection of the silvicultural treatments are decided by the forester, based on experience and site and stand characteristics. Additionally, seed years are important for the precise timing. Generally, fir is sown 3 years after the stand has been thinned to 80% canopy cover. Beech is sown or planted after the fir regeneration reaches a height of 15 to 20 cm (7 to 15 years later). Before sowing the beech, the shelterwood is thinned again to 70% canopy cover. After the beech regeneration is about 2 m tall, two more crown thinnings open the canopy. These thinnings gradually release the fir and beech regeneration and the naturally regenerated spruce. The final harvest of the stand is done about 35 years after the regeneration was initiated.

In the Forstamt Weissenhorn, hunting regulates the deer population to a level where fencing to protect the regeneration is unnecessary. If browsing on the new germinants or planted seedlings is a problem, jars with a commercial deer repellent are placed half-buried in the topsoil in a 20 × 20 m grid for 2 months. Direct seeding is best following major seed years, when seed predation by rodents, birds, and squirrels seems to be less and the germination percentage better than in years of lower seed production. Moreover, seed years offer inexpensive seed if they are collected and handled in large amounts.

22.3 Seed storage and handling

The seed stock requires no pretreatment if sown in the fall, whereas spring seeding necessitates storage and pretreatment to break seed dormancy. Nevertheless, spring seeding is recommended for beech in order to shorten the time beechnuts are on the forest floor and vulnerable to seed-eating mammals and birds. Controlled moisture content (28 to 32%) and 4°C is needed to break the dormancy of beech and prevent germination before seeding. Normally, the pretreatment takes 12 weeks, after which the beechnuts can be sown or stored at about −2°C until seeding. Stored seed should be mixed every week to prevent fungal attack. A local storage facility for agricultural crops was modified for storing beechnuts. Seeds are blown into the storage facility by airflow, and a built-in cooler controls the temperature.

Fir cones are collected in September from both standing and felled trees and placed on a wooden floor to dry, in a layer of up to 15 cm in thickness. In about 2 weeks, the cones

Figure 22.1 Locally developed sowing machine (Sämagrub) with two flexible seeding units. The machine, which is pulled by a 4WD farm tractor, can scarify a mineral soil seedbed without injuring the roots of the overstory trees.

Table 22.2 Costs for Direct Seeding

Beechnuts — 40 kg ha^{-1}, \$10 kg^{-1}	\$400 ha^{-1}
Fir seed — 3.5 kg ha^{-1}, \$50 kg^{-1}	\$175 ha^{-1}
Sowing with machine, 2 × \$125	\$250 ha^{-1}
Deer repellent protection	\$75 ha^{-1}
Total	\$900 ha^{-1}

Figure 22.2 Direct seeding can support high stock density.

fall apart. The seeds are sorted from the scales and sown during the following fall and winter, which makes seed pretreatment unnecessary.

22.4 Sowing technique

Basically, all spruce stands are accessible for our locally developed sowing machine (Figure 22.1). Sowing by hand is also possible, but is more expensive. A 1.3-m wide 4-wheel-drive tractor pulls the sowing machine. The seeding density is controlled from the driver's position. This combination can climb slopes up to 30%, and can sow 1 ha h^{-1}. The sowing machine has two flexible seeding units, which scarify a mineral soil seedbed without injuring the roots of the overstory trees. The capacity to sow 8 to 10 ha daily considerably reduces costs; large regeneration targets are reached in a short time, thereby avoiding seeding in less favorable years. Fir seeds are sown on the mineral soil seedbed surface and beechnuts are placed deeper in the mineral soil.

Generally, the amount of seed is 3.5 kg fir and 40 kg beechnuts, per ha provided the germination rate is high, which typically means 40% for fir and 80% for beech. Seed densities are increased if germination rates are lower. Sometimes, sycamore maple (*Acer pseudoplatanus* L.) is sown; this species is broadcast sown without seedbed preparation at 4 kg of seed per ha. Sycamore maple is sown less often than fir and beech because the target for this species is only 3 to 4% of the forest district. Largely used for supplementary planting if some of the main species fail to regenerate, sycamore maple can be sown on larger areas in a mixture with ash (*Fraxinus excelsior* L.) along small streams and rivers.

22.5 Costs

To establish a mix of species in pure spruce stands, fir and beech are traditionally planted in 30 m × 60 m groups, at 2 m × 2 m spacing for fir and 1 m × 1 m for beech. This corresponds to 833 fir and 3,333 beech seedling per ha. Fir and beech seedlings from nurseries cost \$1.87 and \$1.5 per seedling including planting, totaling \$6,600 ha^{-1}. At Forstamt Weissenhorn, foresters use beech wildings (seedling from natural regenerations) for

Figure 22.3 Direct-seeded beech develops good root systems.

planting, which cost less than nursery stock. Beech wildings cost \$0.31 per seedling and a further \$0.62 to plant. The total cost to establish planted fir (nursery stock), beech (wildings), and Norway spruce (natural regeneration) is typically \$4,660 ha^{-1} at Forstamt Weissenhorn.

This technique of direct seeding costs \$900 ha^{-1} (Table 22.2), which is approximately 20% of the cost of the least expensive planting method using beech wildings. Sowing by hand is more expensive, as the seedbed preparation is done by hand on terrain that is inaccessible for machinery. It takes about 15 h of labor per ha (\$50 h^{-1} including benefits) to sow fir and an additional 15 h ha^{-1} to sow the beech. Hand sowing has been discontinued at Forstamt Weissenhorn for fir and beech. Because no seedbed preparation is needed, sycamore maple can be inexpensively sown by hand; the only costs are for 4 kg of maple seed (\$90 kg^{-1}) and 1 h ha^{-1}, a total of \$410 ha^{-1}.

22.6 Conclusion

Direct seeding offers a very cost-effective alternative to planting. The main disadvantages are: the higher risk due to the many predators of seeds and newly germinated seedlings, as well as the need for more intensive management to select suitable stands, properly treat and store seed, and apply deer repellents, if necessary. Among the advantages: direct seeding fits well with the principles of nature-based silviculture, and it supports high future stand quality by the high stock density (Figure 22.2) and by the natural development of root systems (Figure 22.3). Inventories have shown up to 160,000 sown beech saplings per ha (50 to 80 cm average height) 8 years after establishment. Moreover, it is a highly mechanized method that can overcome the combined problems of large regeneration areas to be treated in a short time with limited manpower. Direct seeding is an important silvicultural method in modern forestry as it reduces regeneration cost.

Reference

Baumhauer, H., Verjüngung durch Saat — ein Beitrag zur Kostensenkung, *Allgemeine Forst Zeitschrift/Der Wald*, 51, 1192, 1996.

chapter twenty-three

Forest restoration in the French Massif Central Mountains

Philippe Balandier and Bernard Prevosto
CEMAGREF, Clermont-Ferrand Regional Centre
Jean-Luc Guitton
CRPF Auvergne (Auvergne Regional Centre of Private Forest)

Contents

23.1 Introduction

23.1.1 Background

The Massif Central is an upland area of igneous origin in southern France (Figure 23.1) with a mean elevation of 800 m, from 300 m in the Limagne plain to 1,886 m in the Sancy Mountains and covers about 70,000 km^2 (12.7% of France). Only the higher elevations will be considered here, as the low fertile plains are farmed (Figure 23.2). The climate is semi-continental, with low winter temperatures (≥ -20°C), late spring frosts (until June), and high summer temperatures (≤ 30°C). A striking feature of the climate is the high amplitude of daily temperature variation, sometimes more than 25°C. Temperatures vary with elevation, about -0.6°C per $+100$ m elevation (Guitton 1986). Rainfall is about 800 mm per

1-56670-635-1/05/$0.00+$1.50

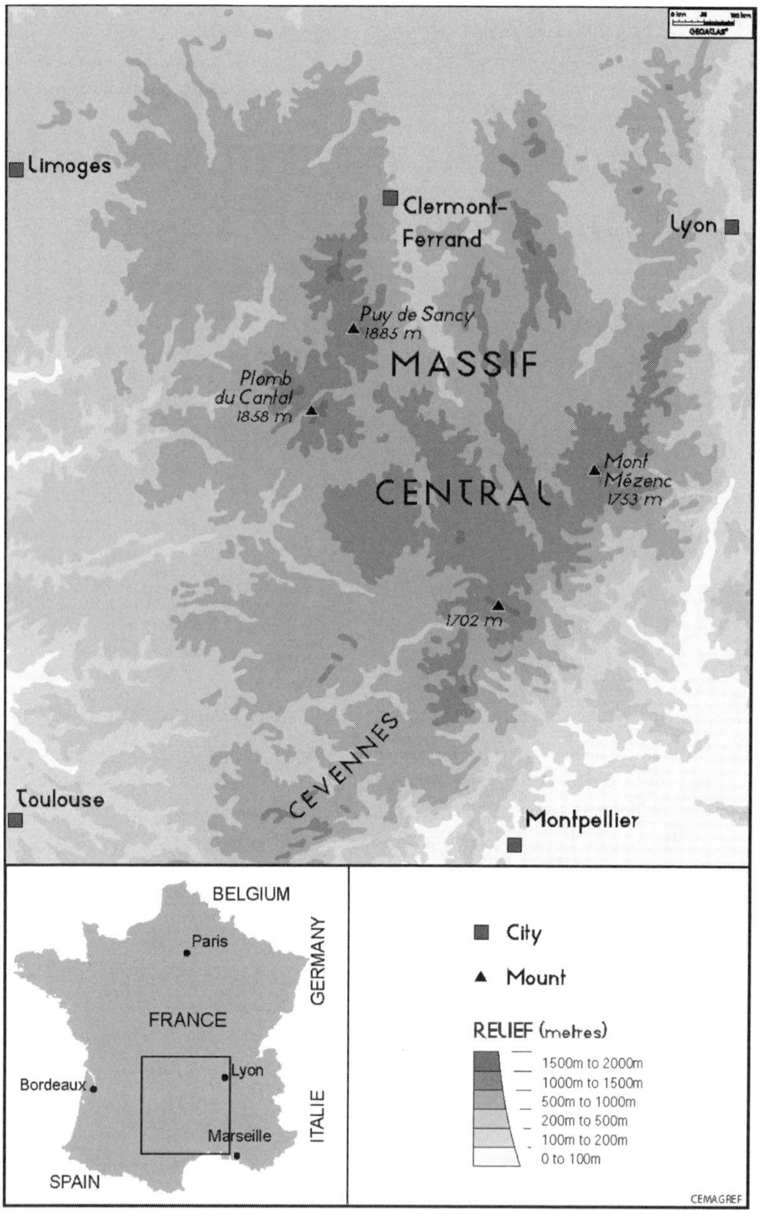

Figure 23.1 Geographical location, relief, and elevation of the Massif Central in France.

year but varies widely with exposure, from less than 600 mm in the rain shadow of hills to more than 1,500 mm in the west slopes of the mountains. Rainfall is fairly evenly distributed throughout the year (Figure 23.3), especially in areas benefiting from oceanic influences, but summer droughts occur about one year in every five (Balandier et al. 2003a).

Soils are derived from granite, basaltic flows, and ash-fall deposits. The most frequently occurring soils are acid brown soils, podzolic podzols, and andosols on volcanic material. Soils derived from granite are often infertile: low pH (5 to 5.5), low cation exchange capacity (15 to 20 m-equiv 100 g^{-1}), and low organic matter content (5 to 10%). In contrast, soils in volcanic areas are more favorable to forest production due to higher

Figure 23.2 Massif Central landscape, with forested higher elevations and farmed lowlands. Sancy Mountain is in the background. (Photo: P. Balandier.)

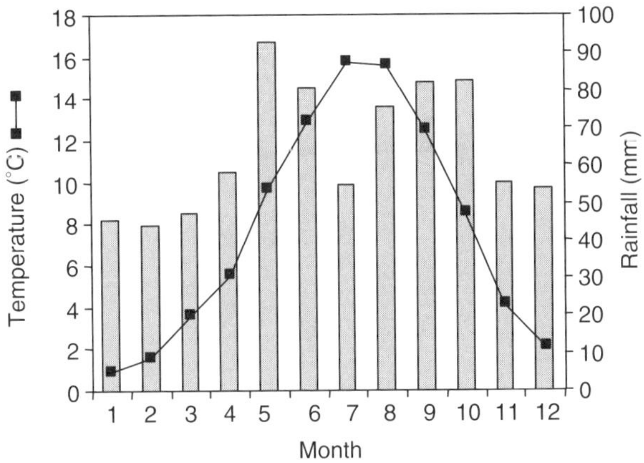

Figure 23.3 Mean monthly temperature and mean monthly rainfall for the period 1971 to 1994 at Saint-Genès-Champanelle (890 m above sea level).

pH (about 6), higher cation exchange capacity (30 to 35 m-equiv 100 g^{-1}), and higher organic matter content (10 to 15%).

Vegetation history of the Massif Central is well documented due to numerous pollen analyses of peat bogs, swamps, and lakes (De Beaulieu et al. 1988). During the postglacial period, the area was successively dominated by *Betula* and *Pinus* forests (9,900 to 9,500 ybp), *Coryllus* (9,500 to 8,000 ybp), and *Quercus* forests (8,000 to 4,200 ybp). From 4,200 ybp, *Abies alba* and *Fagus sylvatica* forests were the dominant type until human disturbances occurred on a large scale. At the beginning of the 20th century, forest cover was minimal and beech and fir forests remained only in some mountainous areas.

At the beginning of the 18th century, most of the land was in subsistence farms producing cereal and typically possessing a few cattle and a horse. In 1800, less than 10% of the total area was wooded (Michelin 1995). Agricultural specialization came with the arrival of the railway at the end of the 19th century, and with it began a slow and continuous decline of agriculture. Production systems are now mostly restricted to forage crops and ruminant livestock (cattle and sheep) for dairy products and meat. After World War II, further specialization and intensification of agriculture accelerated rural depopulation. The onset of common European agricultural policies (which favored the most productive farming systems) and natural hindrances (such as harsh climate and steep slopes) led to the abandonment of the most difficult upland areas (Balandier et al. 2002b). Thus, old grasslands, heathlands, and alpine pastures gradually have been taken over, first by brushwood and later, or directly, by pioneering tree species, among which *Pinus sylvestris* and *Betula* sp. have played a major role (Prévosto et al. 1999). Alongside this natural colonization of former upland agricultural areas by trees, national aid programs have favored planting conifer stands.

23.1.2 Forests today

Natural forests are very rare, as most disappeared or were drastically modified by human activities. Natural forests at the highest elevations were dominated by *Fagus Sylvatica – Abies alba* communities; oak (generally *Quercus pubescens*) takes the place of fir with decreasing altitude or decreasing latitude and becomes the dominant species in the south of the Massif Central. Often mixed with those dominant species, trees or shrubs such as *Prunus avium*, *Acer pseudoplatanus*, *Fraxinus excelsior*, and *Sorbus torminalis* can enrich these forests. Plantations, mainly conifers (*Pinus sylvestris*, *Picea abies*, and *Pseudotsuga douglasii*), were the main species of afforestation and, consequently, today are the most abundant forests. Upland areas today have greater than 60% forest cover, but with wide differences between regions (Balandier et al. 1997). In the Auvergne Region, coniferous plantations account for 33% of private woods; for the entire Massif Central, conifer plantations account for about 25% of the forested area, that is, 500,000 ha. Afforestation initially responded to a shortage of wood for industrial purposes, such as mining and coal, beginning in the 19th century and continuing into the 20th century (RTM: restoration of mountainous forests, FFN: national forest fund, Bianco 1998). Coniferous species (successively *Pinus sylvestris*, *Picea abies*, and until 1980 *Pseudotsuga douglasii*) were the main species used in these plantations because of their fast growth and high productivity.

Forests stemming from spontaneous regeneration by pioneer trees such as Scots pine are also widespread, comprising as much as 45% of the forested area in some parts of the Massif Central (Lifran et al. 1997). This woodland is generally very fragmented, mostly privately owned (80%) with a mean holding of 2.2 to 2.5 ha per landowner in Auvergne (ONF 1993; Orientations Régionales Forestières 1999).

Utilization developed steadily as postwar plantations matured. Timber production is expected to double in the next 10 years from 3 million m^3 at present to 6 million m^3 (AFOMAC 2001). Salvage logging after the last catastrophic windstorm in 1999 changed the face of the wood industry toward mechanization; harvesting machines have replaced individual chain saws, except in areas with large trees on steep slopes. Small sawmills have closed and larger mills are expanding and modernizing. The wood pulp industry, however, is small and logs have to be transported to remote factories. Pulpwood demand has fallen owing to waste paper recycling and the direct supply of paper mills with sawmills waste. This low demand for pulpwood reduces income from first thinnings and stands consequently lose their stability to wind and heavy snow (De Champs 1997).

23.1.3 Forest restoration

Timber production is no longer the sole function expected from forests; management has to provide other goods and services demanded by the public, such as recreation, biodiversity restoration and conservation, and landscape value. For instance, coniferous afforestation is often disliked by local people (Balandier et al. 1997), because of its adverse effect on the landscape (screening, dark-colored vegetation) or because it symbolizes the abandonment of earlier agriculture (Bouvarel and Larrere 1981). In addition, management of these generally small-sized plantations is often desultory and wood of low economic value accumulates (Lifran et al. 1997). The severe windstorm of December 1999 destroyed more than 20% of the standing volume in some parts of the Massif Central, representing seven times the annual harvest (140 million m^3 for the whole of France) (Rérat 2000) and coniferous stands were the most frequently injured (mainly *Picea abies* and *Pinus sylvestris,* although *Pseudostuga douglasii* and *Larix* sp. were less affected).

The public also favors sound and sustainable management practices and dislikes mechanical and chemical competition control in plantations (Frochot et al. 2002). It welcomes the use of local tree species and a return to mixed woodland with broad-leaved species (Collective Report 1998; Bianco 1998).

The aim of forest restoration in the Massif Central in France is to find sustainable solutions for forest management that are environmentally friendly and responsive to public demands, while remaining economically profitable (eco-efficient), especially for the small owners who manage most of the forested area.

In light of these considerations, we shall look at three main issues for forest restoration in the Massif Central:

1. How can we treat competing ground vegetation in the most environmentally friendly way (i.e., with limited use of herbicides or mechanical operations) in afforestation and rehabilitation?
2. How can we accelerate stand development in spontaneous forests of pine or birch toward near-climax forests (such as beech and silver fir)?
3. Is it possible to convert monospecific coniferous stands into plurispecific mixed stands combining both softwood and broadleaved species?

23.2 Sound management of ground vegetation in afforestation or regeneration of forest stands

Unless ground vegetation (herbs or shrubs) that competes with young seedlings is controlled, tree establishment fails in plantations on former agricultural land or in forest regeneration (Nambiar and Sands 1993; Collet et al. 1996; Balandier and De Montard 1999; De Montard et al. 1999). Cutting back competing vegetation has only a temporary effect; it grows again, often more vigorously (Reinecke et al. 2002). Mechanical and chemical applications are effective in controlling this ground vegetation (Davies 1987) but may harm the environment and are sometimes dangerous for human health.

In a search for more ecologically sound approaches to help reduce the use of herbicides, two techniques are currently being tested: using accompanying woody species to control the herbaceous stratum, and using a cover of herbaceous plants that competes only weakly with tree seedlings.

23.2.1 Woody accompanying species

Ground vegetation has long been considered unfavorable to forest tree establishment. Nevertheless, the herbaceous stratum is often more damaging than the shrubby stratum (Perrin 1963) and microclimate of a large open area is often unfavorable to forest tree establishment (Aussenac 1986; Friedrich and Dawson 1984). Certain woody species may be more beneficial than detrimental by providing protection against wind, late spring frost, and other hazards. Lateral shelter may even favor young tree growth (Collet et al. 1992). These accompanying woody species, also termed nurse trees, nurse vegetation, or nurse crops, also improve forest tree shape (Hubert 1992). Other benefits of accompanying species include limiting the development of competing herbaceous species, mainly grasses (Frochot et al. 1986; Davies 1987; Hubert 1992), financial return from harvesting intermediate wood products from the accompanying species (e.g., pulpwood, firewood), and improving tree growth by adding nitrogen by using N-fixing species such as *Alnus* sp.

Managing accompanying species is difficult in practice. Decisions include choice of species, plantation density, and plantation design. The accompanying vegetation is beneficial only if it is sufficiently close to the crop trees and has a comparable height (Hubert 1992). However, increasing the density of the accompanying species around crop trees also intensifies the competition among species. Although the highest densities of the woody accompanying species accelerate height growth of crop trees, diameter growth tends to decrease correspondingly (Figure 23.4). At the highest densities (more than 2,000 stem ha^{1}), crop tree stems are slender, with a height/diameter ratio greater than 120, and thus vulnerable to strong winds (Balandier and Marquier 1998). The windstorm in France in December 1999 showed that this effect cannot safely be ignored.

Using woody accompanying species generally benefits tree stem shape (straightness), but these effects may only be appreciable at very high densities of accompanying species. For example, in a plantation of ash (*Fraxinus excelsior*) with alder (*Alnus glutinosa*) as an accompanying species at densities up to 1,665 alder ha^{-1}, the ash trees showed no difference in the number of forks or trunk deformations (Balandier and Marquier 1998). However, the size and number of lateral shoots were strongly influenced by the density of the accompanying species (Table 23.1), increasing the density producing fewer, thinner lateral shoots. Despite the reduction in lateral shoot vigor with increasing alder density, the natural pruning of lateral shoots is only obtained at very high densities (more than 5,000 stem ha^{-1}). Therefore, artificial pruning is necessary, in most cases, to obtain a knot-free bole.

Plantations with accompanying species are difficult to proportion (Balandier and Marquier 1998, and unpublished data). Desirable effects on tree shape, size, and number of lateral shoots and control of the ground herbaceous species are only obtained at the highest densities, but at high densities competition increases among species and leads to

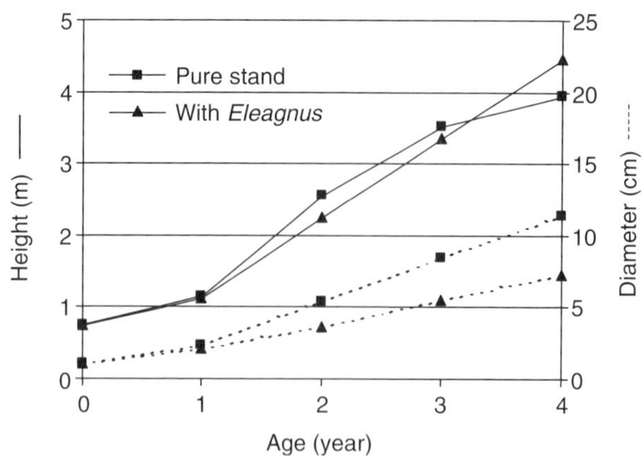

Figure 23.4 Stem height and basal diameter growth of hybrid walnut (*Juglans regia* × *nigra*) planted at 160 stem ha^{-1} near Clermont-Ferrand in the Massif Central in a pure stand or with *Eleagnus angustifolia* as an accompanying species (three *Eleagnus* around each walnut at a distance of 50 cm from the walnut trunk). Only the growth of the walnut is shown.

Table 23.1 Numbers of Forks and Trunk Deformations, and Lateral Shoot Number and Diameter on Ash Planted in 1991 with Different Densities of Alder as an Accompanying Species (see Balandier and Marquier 1998)

Density (stem ha^{-1})	Mean Number of Forks and Trunk Deformations per Meter of Pruned Bole	Mean Diameter (mm) of the Pruned Shoot	Mean Number of Pruned Shoots per Meter of Pruned Bole
555 ash seedlings	0.11 a [a]	14.6 a	12.8 a
555 ash + 555 alder	0.11 a	13.6 ab	11.9 a
555 ash + 1,110 alder	0.09 a	13.1 b	11.0 a
555 ash + 1,665 alder	0.10 a	11.0 c	8.0 b

[a] Densities in the same column followed by different letters are significantly different at $\alpha = 5\%$.

Trees were surveyed in 1995 for forks and trunk deformations; cumulative measurements of lateral shoot number and diameter are for the period 1997 to 2001.

decreased trunk diameters and results in slender trees, easily broken by wind. Harvesting the accompanying species may cause incidental damage to crop trees because of their proximity. Improved tree growth can be simply achieved using low densities of woody accompanying species as windbreaks, planted in lines regularly distributed in the stand. Windbreaks have no effect on tree shape and nor do they provide control of ground herbaceous vegetation. Therefore, forest managers must compromise between different effects according to management objectives and should probably avoid dense plantings of accompanying species in areas vulnerable to windthrow.

23.2.2 Cover plant mixtures to control competition

Using herbaceous vegetation that competes weakly against young seedlings in order to control highly competitive invasive species (e.g., *Agropyrum repens, Molinia caerulea, Deschampsia flexuosa, Agrostis stolonifera*, etc.) through competitive exclusion is not a new idea. Cereals, especially rye (*Secale cereale* L.), have been used in France in direct seeding mixes with oak and pine (Cotta 1822 in Dimkic 1997); blue lupin (*Lupinus perenis*) was tested in forest plantations in Germany in the 1950s.

One recent innovation is to use a mixture rather than a single species to control the spontaneous competitive species (mainly grasses) and thus favor young tree establishment (Reinecke 1985, 2000; Reinecke et al. 2002). The choice of the plants composing the mixture must have the following characteristics (Balandier et al. 2003b):

- Not be invasive species and must have low water and nutrient requirements to limit competition with tree seedlings
- Have different phenological developments and be capable of replacing one another in order to keep the soil continuously covered in space and time until the trees establish
- Disappear rapidly to give way to the natural vegetation
- Be available in seed form, because the cover vegetation is sown before trees are planted.

The sown vegetative cover, with its multilayered structure, is expected to exclude the most competitive species such as *Agropyrum repens,* which can be destroyed before sowing by a single herbicide application, and creates a favorable microclimate that protects young trees from wind and late-spring or early-autumn frosts. In summer, the cover creates favorable shade. Using a mixture of cover plants rather than a single species also favors biodiversity because fauna and insects (notably butterflies) proliferate (Reinecke et al. 2002).

The technique is still regarded as experimental, although more than 1,000 ha of forests have been established in Germany over the last 20 years with apparent success (Frochot et al. 2002). Tree mortality has been limited, tree growth seems better, and the sown vegetation disappeared in 3 to 5 years (Balandier et al. 2003b). In the Massif Central, however, only a few experimental plantations have been established and it is too early to conclude that the technique is effective here. The composition of the plant mixture used by Reinecke will probably need adjusting to fit local soil and climate, which is more favorable to the development of dense, herbaceous vegetation than the poor soils of northern Germany. Two experiments, one at high elevation (about 800 m, Charensat) and one in the plain (about 300 m, Montoldre), were established to assess different plant mixtures and compare their efficiency to woody accompanying species in terms of tree establishment (Provendier and Balandier 2002). If successful in trials, the cover plant technique may offer a real alternative to the intensive use of herbicides.

23.3 *Converting fallow land into more sustainable forest stands*

The forest cover of France has doubled over the last 150 years (DERF 1995) mainly by natural invasion of abandoned cropland or pasture (Figure 23.5). Natural invasion in the Massif Central has been primarily pine (*Pinus sylvestris*, *Pinus nigra*) and birch (*Betula* sp.) (Koerner et al. 2000). The resulting woodlands are usually of low economic value because of poor stem form and low wood quality. Over time, these natural secondary woodlands give way to forests of more shade-tolerant species such as oak (*Quercus* sp.) in lowlands and beech (*Fagus sylvatica*) and silver fir (*Abies alba*) at higher altitudes. Forest managers increasingly favor late-successional species, because they have more commercial value, are well-suited to site conditions, are better adapted than conifers to major disturbances such as windstorms, and native forests are valuable reservoirs of biodiversity (Hermy et al. 1999; Singleton et al. 2001).

Late-successional species will establish naturally under pioneer species; for example, *Fagus sylvatica* under *Pinus sylvestris* or *Pinus nigra* overstory. This apparent facilitation model (Connell and Slatyer 1977) is a slow process with variable success controlled by numerous factors, such as distance to seed sources, abundance and persistence of the seed bank, site conditions, competition, and ungulate herbivory. Managers can accelerate these processes by direct seeding or planting.

Figure 23.5 Progressive colonization of old pastures by natural invasion of woody species. (Photo A. Marquier, with permission.)

23.3.1 Direct seeding of late-successional species

Direct seeding can be an economical method for establishing deciduous species below conifer canopies, provided some prerequisites are met: high germination capacity of seeds, appropriate sowing technique, and absence of excessively competitive vegetation (Ammer et al. 2002).

Little is known about the environmental conditions that influence germination and survival of direct seeded species, but secondary woodlands certainly present unfavorable conditions because ground vegetation is abundant due to good light penetration. Ground vegetation competes with seedlings and provides favorable habitat for small mammals, leading to higher predation. We tried direct seeding of beech in pine stands and meadows in the Chaîne des Puys, a volcanic mid-elevation mountain of the Massif Central. We used three Scots pine stands of a range of density (4,000, 1,500, and 500 stem ha^{-1}) that created a light gradient from 9 to 46% of full light and an open meadow with full light. Conditions in the stands were comparable: same altitude (900 m) and topography (flat area), and similar volcanic soils. Annual rainfall of 820 mm is evenly spread over the year and soil is well watered in the upper horizon. Nevertheless, there is a risk of rapid summer dryness on these soils (Coll et al. 2003). A total of 1,200 seeds, with an 85% germination capacity (Suska et al. 1994), were sown in the different stands and meadow, either directly on the ground without modifying the existent vegetation, or on bare soil after removing vegetation and litter. Seedling emergence (cotyledons visible) was lower in open habitats (meadow and pine at 500 stem ha^{-1}) even in the absence of ground vegetation (Figure 23.6). After 20 weeks, survival was 23% in the high- and medium-density stands, 15% in the less crowded stand, and 0% in the meadow. One possible explanation is that small mammal predation was higher in open habitats owing to a more developed herb cover (Gill and Marks 1991). Another explanation could be that microclimatic conditions and variations in soil moisture may have affected seed germination and early survival (Ammer et al. 2002). Another explanation was that the ground vegetation, which followed the light gradient, may have negatively affected germination and young seedling survival.

23.3.2 Planting late-successional species

Planting beneath the canopy of pioneer species can favor late-successional species. Studies with planting 2-year-old beech seedlings were carried out using the same experimental

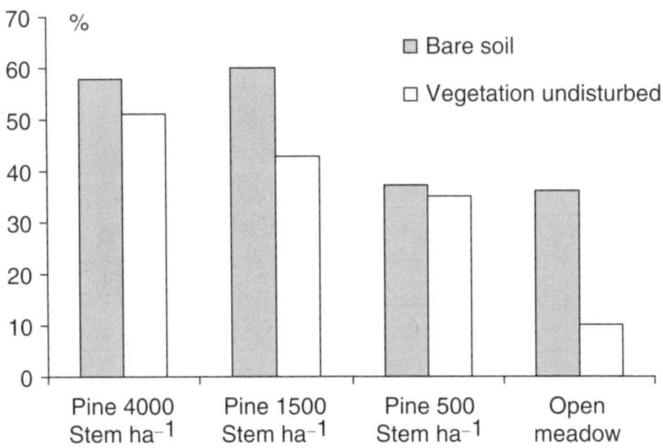

Figure 23.6 Cumulative percentage of beech seedling emergence (cotyledons visible) 20 weeks after seeding in Scots pine stands of different densities and in an open meadow. Two seeding modalities were compared: bare soil and undisturbed vegetation. Plants that died during this period were counted.

design as described above, where a light gradient was developed by thinning Scots pine stands and using an open meadow. This study focused particularly on the effect of light level on beech seedling growth and on the competitive interactions between beech and the ground vegetation. Beech seedling growth was quantified by computing relative growth rates (RGR) for the whole growing season (i.e., from May to September), which allows comparison of seedlings of different initial dimensions. Beech seedling growth was highest under the most heavily thinned Scots pine stand, which exceeded growth of seedlings in the open meadow on bare soil with full light (Figure 23.7). In contrast, beech growth was severely reduced in the denser pine plot (9% of relative irradiance) and in the unweeded, open meadow.

Leaf water potential in the beech seedlings was comparable in all treatments, except for that of the open, unweeded meadow, which exhibited a significantly lower value (Figure 23.8), indicating a higher water demand for beech seedlings in this treatment. The ground vegetation of mainly grasses in the meadow exerted a strong competition for water (Coll et al. 2003). Ground vegetation was absent in the most dense Scots pine plots

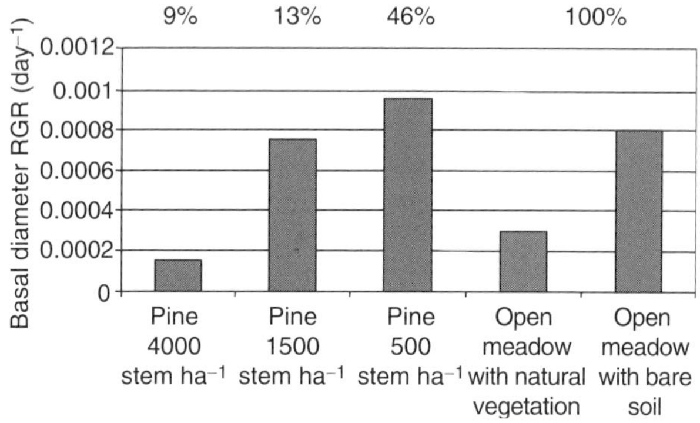

Figure 23.7 Annual relative growth rate of stem basal diameter of 2-year-old beech seedlings planted in Scots pine stands of different densities and in an open meadow, of which one part was weeded (data from Védrine 2000). The number at the top of the figure indicates the mean relative irradiance under canopies of Scots pine for the different densities.

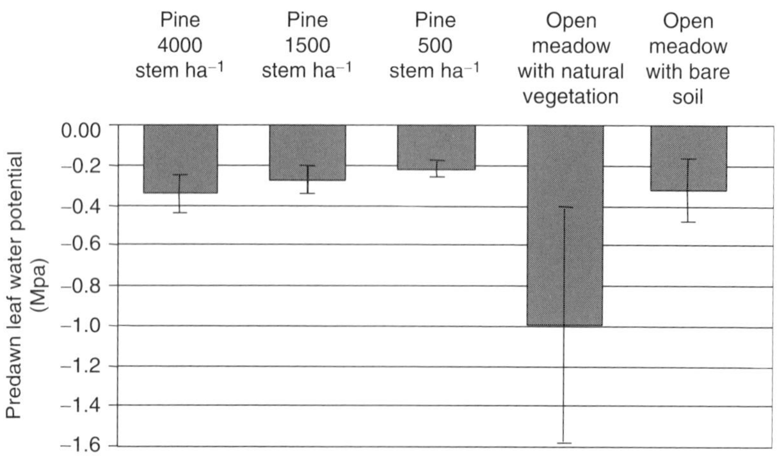

Figure 23.8 Predawn leaf water potential on 29 June 2000 of 2-year-old beech seedlings planted in Scots pine stands of different densities and in an open meadow, of which one part was weeded (data from Védrine 2000).

(4,000 stem ha^{-1}), weakly developed in the moderate density treatment, or well developed in the low-density treatment but composed of dicotyledonous species, which compete less for water (Coll et al. 2003) than the open meadow grasses.

Beech, the major late-seral species in this area, in the process of old field succession, establishes itself naturally under pioneer species but seldom on open sites. The dominant early-successional tree species develop both positive and negative interactions with young beech seedlings and saplings: direct competition for light, reduced water stress by control of the ground vegetation, and further modification of microclimatic conditions. Control of the shelterwood density of the established secondary woodlands (Scots pine or birch) can thus be a powerful tool to enhance beech establishment and growth.

23.4 *From monospecific coniferous to mixed coniferous–broadleaved stands*

In certain parts of the Massif Central, plantations of a single conifer species (*Picea abies* or *Pseudotsuga douglasii*) cover a large area. Forest managers are seeking to replace these stands with mixed forests (Figure 23.9) in order to reach a new equilibrium that combines the economic value of exotic species with the ecological value of native trees to ensure sustainable, highly productive woodlands. They are attempting to mix exotic species such as Douglas-fir, larch (*Larix* sp.), and sometimes red oak (*Q. rubra*), which are very productive, with native species (*Fagus sylvatica*, *Pinus sylvestris*, *Abies alba*).

Conifer plantations have always had high densities, with more than 3,000 stems ha^{-1} with pines, 2,500 with spruce, and 1,500 with Douglas-fir. High density was intended to ensure rapid occupancy of the site to forestall weed competition, allow a high selection ratio, and produce maximum wood volume. Before the first thinnings at age 40, the understory was poorly lit and nearly devoid of ground vegetation. Relative irradiance in these conifer plantations was very low, less than 2.5% of full sunlight. Height levels of 5% of available sunlight are necessary for understory plants to develop, and maximum plant biodiversity is obtained at about 20% (Balandier et al. 2002). An average tree density of about 800 stems ha^{-1} will produce adequate light levels for understory vegetation to develop, but this density is rarely obtained after the first thinning. Plant and overall biodiversity are generally reduced and only a few sources of biodiversity can be found in

Figure 23.9 Autumn view of a mixture of broadleaf and coniferous species. (Photo P. Balandier.)

such large plantation areas; mainly in gaps due to rocky or wet areas, deer damage, or uncleared spots. Even in areas where land ownership is more fragmentary, biodiversity inside woods is low.

Two approaches have been used to restore biodiversity in these plantations: natural or artificial regeneration.

23.4.1 Natural regeneration

Biodiversity can be increased in plantations by changing to natural regeneration. Natural regeneration relies on existing adult trees to produce seeds and obtain new tree populations. This means applying careful silviculture in young plantations to gradually reduce tree density to 150 to 200 stems ha^{-1} and selecting the best-formed trees as seed sources. According to the plantation's vitality (a function of soil, climate, and species adaptation), active management can obtain a final stand with a mean of 45 cm in 45 years for Douglas-fir and 60 years for Norway spruce. Longer rotations are required if thinnings are delayed; ideally, stands should be thinned as soon as the live crown height is less than half the total tree height. Stem conicity remains high and trees are resistant to wind damage.

Gaps in the canopy allow native species (beech, fir) to establish and be present when the stand is regenerated. Natural regeneration of the conifers is obtained by opening the stand and harvesting the less valuable timber; thus, the residual trees will produce seed in abundance, and seedlings will find enough light to develop. Fir and beech produce abundant and fertile seed crops every 2 to 3 years in the Massif Central and natural regeneration occurs readily if the soil is weed-free. Seeds from secondary species can also establish, adding to diversity. Presently, beech is establishing in many stands throughout the Massif Central, adding color and diversity to the landscape.

23.4.2 Artificial regeneration

Introducing biodiversity into new plantations is possible by planting at fairly low densities. There are few opportunities for afforestation in some areas due to agriculture subsidies, whereas in other areas with more than 70% high forest cover, there is pressure to preserve farmland and forbid new plantations. Nevertheless, planting is always used for reforestation after gale damage or clearcuts. Clearcuts are necessary in plantations where species were planted offsite or where upkeep such as thinnings were delayed and the stand cannot reach full maturity because of lack of wind firmness. About a quarter of the spruce plantations established in the 1960s at densities greater than 2,500, and unthinned, will not tolerate thinning. The only solution is to clearcut these plantations as soon as they reach an average of 30 cm, before they are fallen by wind or snowstorms.

Mixed-species plantations have rarely been used because of the difficulty of managing species with different growth rates. Attempts in the 1950s and 1960s with "half-and-half" mixtures always failed dismally as one of the two species took over and displaced the second one (Bernard 1987). The French Forest Administration (DERF 2000) does not allow step-by-step mixtures and recommends planting single species in bunches, clumps, or screens, where the second species covers less than 20% of the area and is scattered inside the main stand. When mature, the stand will include natural seed sources for future cohorts developing following regeneration (for instance, fir or beech planted with spruce).

In monospecific plantations, densities around 1,000 stem ha^{-1} are commonly used with rectangular spacing such as 4 m $\times$ 2.5 m. These patterns allow the natural establishment of other tree species inside the plantation. The forester must safeguard these "volunteers" when clearing the plantation and should release these stems when carrying out the first commercial or precommercial thinning. Because of the proliferation of roe deer, clearing is often done only in alternate rows to limit accessibility for deer and allow natural vegetation to grow in these areas. Because of low pulpwood demand, precommercial thinning (cut stems are left on the ground) is used when stands are about 5 to 6 m high to reduce tree density. Precommercial thinning gives indigenous tree species a greater chance to establish and become quality crop trees.

The enrichment of existing stands, whether established by natural or artificial regeneration, with a planting of a third species inside gaps under the tree canopy can give good results. In young stages, it is advisable to use individual tree tubes that distinguish the planted seedlings from the general vegetation and so that cleaning and weeding the stand can promote tree growth. Fast-growing species can be used, such as larch or maple in fertile soils, mixed for instance with spruce. In older forest stands with taller trees, shade-tolerant species must be used for supplemental planting. In the center of the Massif Central, fir has often been planted under Scots pine stands and there is transition toward fir stands.

23.5 Conclusions

Changing economic, environmental, and social conditions have challenged foresters to modify their silvicultural practices. Forests in the Massif Central begin the new century with new management strategies, including restoring diversity and transforming existing stands and plantations into diversified stands that are storm- and pest-resistant, using environmentally friendly soil preparation and weed control, and taking advantage of natural processes to favor mixtures of species. Experience to date suggests that more research should be devoted to fundamental exploration of four areas: competition between trees and ground vegetation, changes of biodiversity in relation to the main environmental variables, especially light level, ecology and uses of exotic species, and natural dynamics of forest stands in diverse contexts.

Acknowledgments

Many thanks are due to André Marquier, Fabrice Landré, Gilles Agrech, and René Jouvie for their field assistance. Experiments described in this chapter were partly funded by the Auvergne Region, the FNADT (National Fund for Territorial Development), and the IFB (French Institute for Biodiversity).

References

AFOMAC, Organiser l'offre de bois résineux dans l'espace central, compte-rendu de mission, France, 164 pp., 2001.

Ammer C., Mosandl R., and Kateb H.E., Direct seeding of beech (*Fagus sylvatica* L.) in Norway spruce (*Picea abies* [L.] Karst.) stands — effects of canopy density and fine root biomass on seed germination, *For. Ecol. Manag.*, 159, 59, 2002.

Aussenac, G., La maîtrise du microclimat en plantation, *Rev. For. Fr.*, 38, 285, 1986.

Balandier, P., Bergez, J.E., and Etienne, M., Use of the management-oriented silvopastoral model ALWAYS: calibration and evaluation, *Agroforest. Syst.* 2002a, 57, 159, 2003a.

Balandier, P. and De Montard, F.X., Tree and grass root growth in an agroforestry system of wild cherries in pasture: space sharing and competition for water, in *Proceedings of the Dynamics of Physiological Processes in Woody Roots*, Université H. Poincaré, Nancy, France, 48, 1999.

Balandier, P., Frochot, H., Charnet, F., Reinecke, H., Ningre, F., Koerner, W., Landré, F., Lefèvre, Y., and Boulet-Gercourt, B., *Restauration de la Biodiversité Floristique lors des Operations de Boidements*, Cemagref, INRA, IDF, Reinecke Forstingenieurbüro, IFB, Cemagref, Paris, France, 2003b.

Balandier, P., Landré, F., Lampin, N., and Ruchaud, F., Understorey vegetation diversity, composition and cover according to light level under canopies of larch stands thinned with different intensities, in *Popular Summaries of the Fourth International Conference on Forest Vegetation Management*, Frochot, H., Collet, C., and Balandier, P., Eds., INRA, Nancy, France, 157, 2002a.

Balandier, P. and Marquier, A., Vers une remise en question des avantages d'une plantation frêne — Aulne, *Rev. For. Fr.*, 50, 231, 1998.

Balandier, P., Rapey, H., and Guitton, J. L., Improvement and sustainable development of moderate altitude areas through agroforestry: tree-grass-animal association, in *Proceedings of the XI World Forestry Congress*, Antalya, Turkey, Vol. 1, topic 2, 1997.

Balandier, P., Rapey, H., and Ruchaud, F., Agroforesterie en Europe de l'Ouest: pratiques et experimentations sylvopastorales des montagnes de la zone tempérée, *Cahiers Agricultures*, 11, 103, 2002b.

Bernard C., *Etude sur les Potentialités de Production de Plantations Mélangées dans le Département du Puy-de-Dôme*, Cemagref, Clermont-Ferrand, France, 47 pp., 1987.

Bianco, J.L., *La forêt: une Chance pour la France, Rapport de Propositions D'orientation d'un Projet de loi de Modernisation Forestière*, Paris, France, 1998.

Bouvarel, P. and Larrere, G., Les forêts de montagne, in, *L'INRA et la montagne*, INRA, Versailles, France, 59, 1981.

Coll, Ll., Balandier, P., Picon-Cochard, C., Prévosto, B., and Curt, T., Competition for water between beech seedlings and surrounding vegetation in stands differing in light availability and vegetation composition, *Ann. For. Sci.*, 60, 593, 2003.

Collective report, *Prospective: La Forêt, sa Filière et Leurs Liens au Territoire*, INRA, Délégation permanente à l'agriculture, au développement et à la prospective, Paris, France, 1998.

Collet, C., Frochot, H., Pitsch, M., and Wehrlen, L., Effet d'un abri lateral artificial sur le développement de jeunes merisiers (*Prunus avium* L.) installées en pépinière, *Rev. For. Fr.*, 44, 85, 1992.

Collet, C., Guehl, J.M., Frochot, H., and Ferhi, A., Effect of two grasses differing in their growth dynamics on the water relations and growth of Quercus petraea seedlings, *Can. J. Bot.*, 74, 1562, 1996.

Connell, J.H., and Slatyer, R.O., Mechanisms of succession in natural communities and their role in community stability and organization, *Am. Nat.*, 111, 1119, 1977.

Davies, R.J., *Trees and Weeds: Weed Control for Successful Tree Establishment, Forestry Commission Handbook 2*, HMSO Publications, London, England, 1987.

De Beaulieu, J.L., Pons, A., and Reille, M., Histoire de la flore et de la végétation du Massif Central (France) depuis la fin de la dernière glaciation, *Cahiers de Micropaléontologie*, 3 , 5, 1988.

De Champs, J., Eclaircies et stabilité des peuplements, in *Le Douglas*, De Champ, J., Coord., AFOCEL, France, 1997, chap. 4.25.

De Montard, F.X., Balandier, P., Rapey, H., Lucot, F., and Drouot, J. P., Compétition pour l'azote entre arbre et herbe dans des plantations de noisetier (*Corylus avellana* L.) et de merisier (*Prunus avium* L.), in *Proc. Bois et Forêt des agriculteurs*, Cemagref, Ed., Antony, France, 73, 1999.

DERF (French Directorate of Rural Areas and Forests), *Les Indicateurs de Gestion Durable des Forêts Françaises*, Ministère de l'Agriculture et de la Pêche, Direction de l'Espace Rural et de la Forêt, Paris, France, 49 pp., 1995.

DERF (French Directorate of Rural Areas and Forests), *Actualisation des Conditions de Financement par le Budget de l'Etat des Projets de Boisement-Reboisement*, circulaire DERF/SDF/C2000-3021 du 18 août 2000, Ministère de l'Agriculture et de la Pêche, Paris, France, 2000.

Dimkic, C., *L'enherbement des Plantations Forestières*, IDF, Paris, France, 1997.

Friedrich, J.M. and Dawson, J.O., Soil nitrogen concentration and *Juglans nigra* growth in mixed plots with nitrogen-fixing *Alnus*, *Eleagnus*, *Lespedeza* and *Robinia* species, *Can. J. For. Res.*, 14, 864, 1984.

Frochot, H., Balandier, P., Reinecke, H., Boulet-Gercourt, B., Ningre, F., Lefèvre, Y., Landré, F., and Koerner, W., Using cover plants mixtures to favour tree establishment in afforestation: an alternative to repeated herbicides or mechanical vegetation controls? in *Popular Summaries of the Fourth International Conference on Forest Vegetation Management*, Frochot, H., Collet, C., and Balandier, P., Eds., INRA, Nancy, France, 233, 2002.

Frochot, H., Picard, J.F., and Dreyfus, P., La végétation herbacée obstacle aux plantations, *Rev. For. Fr.*, 37, 271, 1986.

Gill, D.S. and Marks, P. L., Tree and shrub seedling colonization of old fields in Central New York, *Ecol. Monogr.*, 61, 183, 1991.

Guitton, J. L., *Problèmes des Peuplements Forestiers de Montagne en Auvergne*, Report of Cemagref, Riom, France, 1986.

Hermy, M., Honnay, O., Firbank, L., Grashof-Bodkam, C., and Lawesson, J.E., An ecological comparison between ancient and other forest plant species of Europe, and the implications for forest conservation, *Biol. Conserv.*, 91, 9, 1999.

Hubert, M., La vegetation d'accompagnement: un auxilliaire sylvicole à surveiller de près, *Forêt-Entreprise*, 82, 39, 1992.

Koerner, W., Cinotti, B., Jussy, J.-H., and Benoît, M., Evolution des surfaces boisées en France depuis le début du XXe siècle: identification et localisation des territoires abandonnés, *Rev. For. Fr.*, 52, 249, 2000.

Lifran, R., Rapey, H., and Valadier, A., Nouveau regard sur la tradition sylvo-pastorale en Lozère, *Agreste — les cahiers*, 21, 17, 1997.

Michelin, Y., *Les jardins de Vulcain*, Maison des sciences et de l'homme, France, 180 pp., 1995.

Nambiar, E K.S. and Sands, R., Competition for water and nutrients in forests, *Can J. For. Res.*, 23, 1955, 1993.

ONF, l'Auvergne, Pays de hautes terres au coeur du Massif Central, *Arborescences*, 43, 25, 1993.

Orientations Régionales Forestières, Report of DRAF Auvergne, Clermont-Ferrand, France, 1999.

Perrin, H., *Sylviculture*, 2nd ed., Ecole Nationale des Eaux et des Forêts, Nancy, France, 174, 202, 1963.

Prévosto, B., Curt, T., Gueugnot, J., and Coquillard, P., Colonization by Scots pine (*Pinus sylvestris* L.) after field abandonment at mountain level on volcanic soils, in *Grasslands and Woody Plants in Europe*, Papanastasis, V.P., et al., Eds., Thessaloniki, Greece, 105, 1999.

Provendier, D., and Balandier, P., *Comparaison de Différentes Modalités de Reconstitution de la Forêt après Tempête en Auvergne dans une Perspective de Gestion Durable. Gestion de la Vegetation et Alternatives à L'utilisation D'herbicides en Plantation Forestière,* Cemagref, Clermont-Ferrand, France, 2002.

Reinecke, H., *Begleitwuchsregulierung*, Göttingen, Germany, 1985.

Reinecke, H., *Begleitwuchsregulierung, bestandesbegründung, Schutzpflanzendecken*, Gottingen, Germany, 2000.

Reinecke, H., Koerner, W., Frochot, H., Balandier, P., Boulet-Gercourt, B., and Ningre, F., Sowing cover plants mixtures as a mean to control competitive vegetation in new forest plantations, in *Popular Summaries of the Fourth International Conference on Forest Vegetation Management*, Frochot, H., Collet, C., and Balandier, P., Eds., INRA, Nancy, France, 421, 2002.

Rérat, B., Les dégâts des ouragans Lothar et Martin — France 1999, *Forêt-entreprise*, 131, 8, 2000.

Singleton, R., Gardescu, S., Marks, P.L., and Geber, M. A., Forest herb colonization of postagricultural forests in central New York State, USA, *J. Ecol.*, 89, 325, 2001.

Suska, B., Muller C., and Bonnet-Masimbert, M., *Graines des feuillus forestiers: de la récolte au semis*, INRA, Paris, France, 292 pp., 1994.

Védrine, V., *Evaluation de la Competition pour L'eau et la Lumière entre de Jeunes Hêtres et la Vegetation Environnante dans les Accrues de Pin Sylvestre,* Master report, Blaise Pascal University, Clermont-Ferrand, France, 2000.

chapter twenty-four

Rehabilitation for species enrichment in abandoned coppice forests in Japan

Takuo Nagaike
Yamanashi Forest Research Institute
Toshiya Yoshida
Hokkaido University
Hideo Miguchi and Tomohiko Kamitani
Niigata University
Tohru Nakashizuka
Research Institute for Humanity and Nature

Contents

24.1 *Introduction*

Japan is surprisingly forested for such a populous country, with 64% of the land area in forests (24.1 million ha in 2000). Forest plantations comprise 43% of the forested area (10.4 million ha); most other forests are naturally regenerated second growth, as the area of old forest decreased substantially during and after World War II (Figure 24.1 and Figure 24.2) (Nagaike and Kamitani 1997, 1999). The climate of most of Japan is suitable for forest development, but by the end of the 19th century, exploitive logging for timber, firewood, and mine props had degraded large areas of forest (Chiba 1973, 1991; Ogura 1992, 1996). For example, forests in Asio, 110 km northeast of Tokyo, were logged intensively to

1-56670-635-1/05/$0.00+$1.50

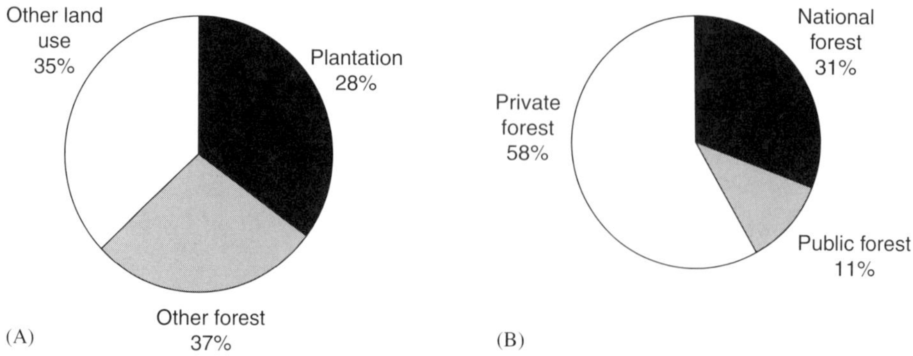

Figure 24.1 Percentage of (A) forest type and (B) ownership of Japanese forests in 2000.

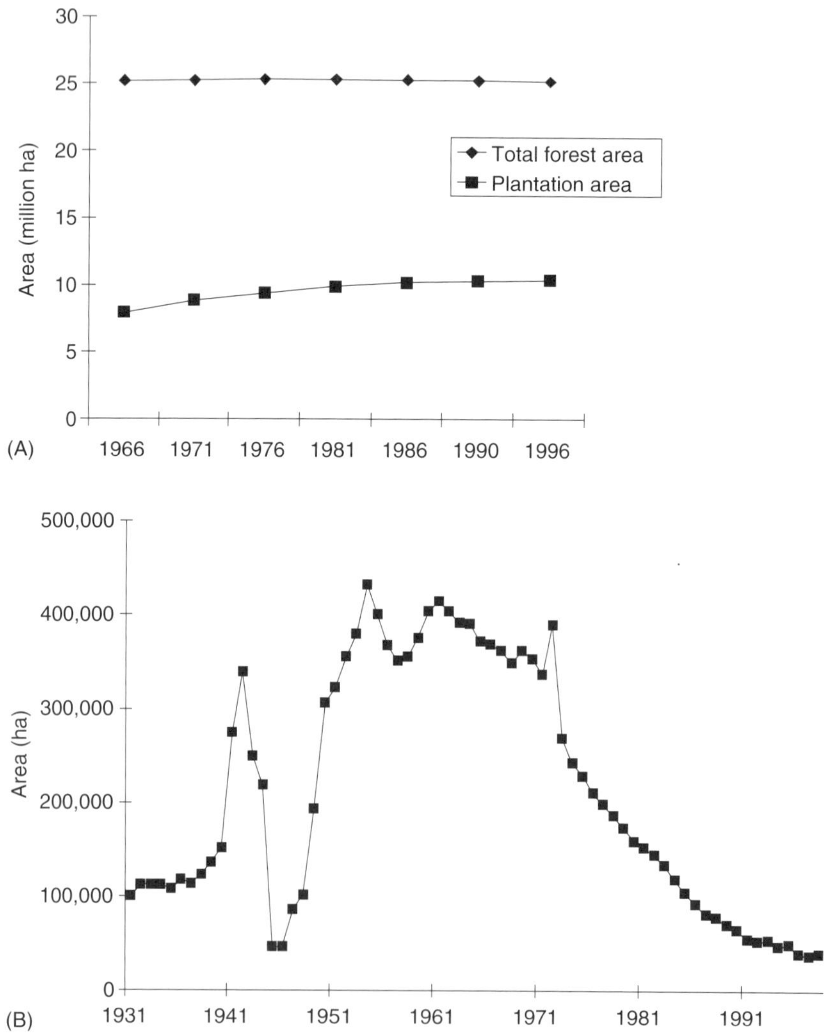

Figure 24.2 Change in (A) total forest area and plantation area and (B) the area of artificial regeneration in Japan. (*Source*: Shinrinbunkakyoukai at http://shinrinbunka.com)

provide fuel for copper smelting, which generated sulfurous acid gas that caused severe withering of trees in the vicinity of the mine (Tanimoto 1990). Other examples abound: a large area in the Kitakami Mountains on Honshu Island was deforested during the 1900s as a result of logging and grazing (Ohsumi, unpublished); the forest of Tanakamiyama, near Kyoto, was intensively logged to establish capital cities following repeated domestic battles in the region, and to build large temples and Buddhist statues during the Nara era, about 1,200 years ago. These degraded or deforested areas were reforested during the 20th century, primarily with conifers.

Two major needs for forest restoration have emerged in Japan, related to changing economic conditions. We briefly describe the problems associated with the widespread establishment of conifer plantations and the need to restore abandoned coppice forests, which provides a context for our primary focus. Our objectives in this chapter are to explore some ecological approaches to rehabilitating coppice forests and to test these ideas by applying them to abandoned coppice forests in central Japan, where heavy snowfall is a factor. Our test case is from our research in Kamikawa Village, in the cool-temperate region of central Japan.

24.2 Conifer plantations

After World War II, government policy affected a substantial increase in conifer plantations. The Forest Agency subsidized landowners to convert broadleaved forests into more commercially important and productive conifers (Figure 24.2B). This plantation movement lasted until the late 1980s and resulted in conifer plantations occupying more than 40% of the forest area. This rapid increase in the area of conifer plantations set the stage for the emergence of two major problems. First, conifers were planted in unsuitable areas. For example, after the primary forests of *Fagus crenata* (Bl.) were logged, *Cryptomeria japonica* (D.Don) was planted extensively on the western side of Honshu Island, which faces the Sea of Japan and experiences heavy snowfall. Because of the damage caused by heavy snowfall, the productivity of these plantations was below expectations (Hasegawa 1991; Yokoi and Yamaguchi 1998). Although many disturbance-dependent broadleaved tree species, shrub species, and dwarf bamboo species invaded these stands, *F. crenata* failed to regenerate successfully (Masaki et al. 2003).

The second problem with conifer plantations has been lack of management. Due to current economic conditions, forest owners have little incentive to invest in management because of high costs. Timber produced in Japan has become more expensive than imported timber because of the high labor costs in tending operations (e.g., weeding and thinning) and high trading costs. Moreover, the complex and steep terrain in Japan adds considerably to the harvesting costs. Although private forests in Japan occupy 58% of the total forest area (Figure 24.1), individual tracts are very small, often less than 1 ha. Another factor is the shortage of technical advice; the average age of current foresters is increasing and recruitment of young foresters is limited. Consequently, lack of thinning and subsequent overcrowding substantially shades out the understory vegetation, which can lead to soil erosion (Miura et al. 2003) and decreased species diversity, including lowered wildlife diversity.

24.3 Coppice forests

Coppice forests provided charcoal and fuel wood until the 1960s, when industry changed from using charcoal and firewood to natural and propane gas and oil. Most coppice forests were abandoned after this energy revolution, resulting in large areas of abandoned forests. Oaks (e.g., *Quercus serrata* Bl., *Q. crispula* [*Q. mongolica grosseserrata* (Bl.) Rehd.&Wils.],

Q. acutissima Carruthers, *Q. phillyreoides*, A. Gray, *Q. variabilis* Blume) were the major coppice species, with 10- to 30-year rotations (Tanimoto 1990). Once stems reach a size suitable for fuel wood (about 10 cm diameter), these species can produce many sprouts after cutting. Because coppice forests were kept in the stem exclusion phase (Oliver and Larson 1996), they usually had a less complex overstory structure and simpler composition than natural high forest. Although several decades have passed since active management ceased, derelict coppices still have less structural heterogeneity than natural forests.

Coppice forests have become important recreational areas for urban and suburban populations (Ishii et al. 1993; Takeuchi et al. 2003). Coppice forests also contribute to biodiversity in rural areas, where monospecific conifer plantations have increased in area (Fukamachi et al. 2002; Nagaike et al. 2003). Moreover, many endangered plant species have been recorded in coppice forests (Environment Agency of Japan 2000), and traditional coppice practices would maintain these species (Iida and Nakashizuka 1995). Thus, the management of coppice forests is a critical issue that warrants immediate attention.

24.4 Rehabilitation of abandoned coppice forests

In theory, there are three alternatives for managing abandoned coppice forests: reestablishment of the traditional management system of coppicing, continued abandonment, and restoration or improvement of their ecological quality as old forests. Near urban populations, it may be feasible to reestablish and maintain traditional coppicing to provide recreational opportunities. However, such areas are limited. For the remaining coppice forests, the question is whether intervention is needed. Since *Quercus* spp., which are the most important species in coppice forests, have the potential to grow larger than 1 m in diameter (Masaki et al. 1992; Nakashizuka et al. 1992), coppice could develop old-forest characteristics (complex structure and diverse composition) and augment the remaining small area of primary forest. Altering stand structure by creating gaps in the homogeneous canopy of abandoned coppice forests would be one way to restore plant species diversity from the understory to the canopy layer (Coates and Burton 1997; Kobayashi and Kamitani 2000). Although many studies have investigated the gap dynamics of natural forests, applications of gap dynamics to coppice forests are limited (Coates and Burton 1997; Yamamoto 2000).

Canopy gaps can create and facilitate vertical stratification in a forest, and enhance the coexistence of tree species through stand development (Nakashizuka 2001). Gap size is one of the most important factors determining seedling and sapling composition in gaps, since niche differentiation of species occurs on a gap-size gradient (Abe et al. 1995; Gray and Spies 1996; 1997; Van der Meer et al. 1999). It may be possible to diversify species composition and stand structure further by the spatial arrangement of patches (Hart 1995; Kohm and Franklin 1997; Smith et al. 1997). Given a homogeneous canopy structure, it is important to artificially create canopy gaps that resemble natural disturbance regimes in order to mimic ecological processes (Bawa and Seidler 1998; Emborg et al. 2000; Lindenmayer et al. 2000). Moreover, canopy gaps not only improve light conditions in the understory but also release neighboring canopy trees from competition, allowing the development of a canopy structure that resembles that of old-forest stages (Oliver and Larson 1996; Frelich and Puettmann 1999).

24.5 Test case — Kamikawa village

We applied these ecological concepts to abandoned coppice forests in central Japan. Our specific research objectives were to compare tree species diversity in the abandoned coppice forest to primary forest, to clarify the early effects of canopy gap creation on plant

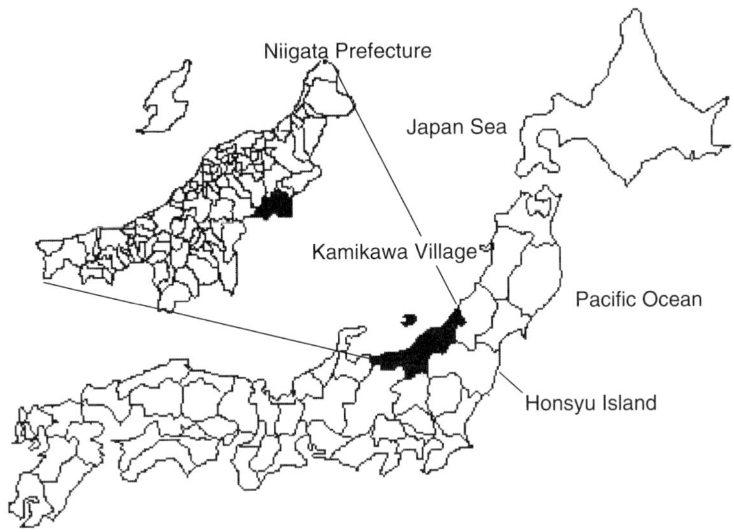

Figure 24.3 Location of Kamikawa Village, Niigata Prefecture in central Japan.

species diversity and tree regeneration in the understory, and to evaluate the effects of gap creation on competitive relationships and responses to disturbance among canopy tree species. This study was conducted in Kamikawa Village, Niigata Prefecture (37°N, 139°E) (Figure 24.3). The mean annual precipitation is approximately 2,200 mm, and the mean annual temperate is 11.7°C. Snow usually covers the ground from late November until mid-April, with a maximum depth of 2 to 3 m. The original vegetation in the area was *Sasa kurilensis – F. crenata* forest (Miyawaki 1985); *F. crenata* was the dominant canopy species. Management of most coppice forests in the region ended in the 1960s, and since then the stands have continued to develop (Nagaike and Kamitani 1997, 1999). Young coppice forests have changed markedly to tall coppice forests or have been converted to conifer plantations. Primary forest has decreased substantially. Consequently, even-aged secondary stands (30 to 50 years old) dominate the region. Near residential areas, abandoned coppice forests were converted to other uses.

24.6 Diversity in abandoned coppice

It is widely reported that repeated coppicing eliminates or diminishes the dominance of *F. crenata* (Kamitani 1993). In our study area, we inferred that abandoned coppice forests had experienced different intensities of coppicing (i.e., the number of times they had been harvested) based on their proportion of *F. crenata*; high proportions of *F. crenata* indicated less intensive management. In order to study the effects of management intensity on stand structure and species composition in the tree layer, thirty abandoned coppice forests and five primary *F. crenata* forests were studied. All coppice forests had closed canopies, but differed in intensity of management. Although no detailed management records for the stands were available, they appeared to be at least 40-years-old and to have been left unmanaged for at least 10 years.

The coefficient of variation (CV) in diameter at breast height (dbh) in the abandoned coppice forests (57%) was significantly lower than in primary forests (104%), suggesting that the structure of abandoned coppice forests is more homogeneous than primary forests. Understory species composition and species richness were closer to the primary forest understory in less intensively managed coppice than in intensively managed coppice. To restore the species composition of abandoned coppice forests to resemble that of primary forests, some management is required (Nagaike et al. 2003) and a high proportion of *F. crenata* remaining in the abandoned coppice stand makes success more likely.

24.7 Understory response to gap creation

One approach to rehabilitating abandoned coppice forests is to create artificial gaps in order to emulate natural disturbance regimes. Gaps cover 12% of the total land area in primary *F. crenata* forest and the mean gap size is 92 m^2 (Yamamoto 1989); gaps larger than 400 m^2 are infrequent (Ida 2000). Primary *F. crenata* forests form a regeneration complex composed of small patches at different stages of regeneration following canopy gap creation (Nakashizuka 1987; Yamamoto et al. 1995). The forest floor of *F. crenata* forests is usually dominated by dwarf bamboo species, which impede the regeneration of trees while also responding to canopy gaps. Moreover, soil disturbance is important in maintaining the coexistence of species that have different seedling establishment strategies. Thus, canopy gaps and other factors interactively affect regeneration (Nakashizuka 1989; Abe et al. 2001, 2002).

In 1995, we created 11 artificial canopy gaps by felling one or several trees in a cluster to test the effects of artificial gaps of different sizes on plant species diversity and tree regeneration in an abandoned coppice forest dominated by *F. crenata* and *Q. crispula*. Gaps ranged from 32 to 380 m^2 in area. The range of gap sizes matched that of natural canopy gaps in primary *F. crenata* forests (Yamamoto 1989). The 11 artificial gaps were classified into three size classes: small, 30 to 50 m^2 ($n = 5$); medium, 51 to 150 m^2 ($n=4$); and large, 151 to 400 m^2 ($n=2$). From 1996 to 1998, we systematically sampled vegetation under different canopy conditions. We estimated light levels using hemispherical photographs analyzed with the software HEMIPHOT (ter Steege 1993) that provided relative photosynthetic photon flux density (rPPFD). We estimated litter coverage ocularly.

Environmental factors varied significantly with gap size (Table 24.1). As gap size increased, the proportion of the surface covered by litter decreased from 99% under closed canopy to 76% in the larger gaps. Similarly, relative light intensity increased as gap size increased (Table 24.1). The soil surface was disturbed when canopy trees were cut to create the gaps. Presumably, these differences in environmental conditions affect germination and establishment of forest plant species (Kobayashi and Kamitani 2000). In the first year following gap formation, plant species diversity increased and the numbers of seedlings of *Acer mono* (Maxim.), *Betula grossa* (Sieb. & Zucc.), and *C. japonica* increased, but there was no significant effect on numbers of *F. crenata*. Although gap size was positively correlated with the number of seedlings of these four tree species that occurred in the first year, there was no consistent relationship of gap size with survival (Figure 24.4).

Table 24.1 Effect of Gap Size on Light Environment and Litter Coverage

Gap Size	rPPFD (%)		Litter coverage (%)	
	Mean	Standard deviation	Mean	Standard deviation
Small	8.0 a	3.6	93.1 a	15.3
Medium	18.0 b	5.4	85.6 ab	17.4
Large	33.9 c	13.3	76.3 b	20.8
Closed	6.7 a	3.9	98.8 a	2.1

rPPFD is the Relative Photosynthetic Photon Flux Density.

Means in a column followed by different letters are significantly different ($p<.017$), using Bonferroni-corrected multiple comparisons.

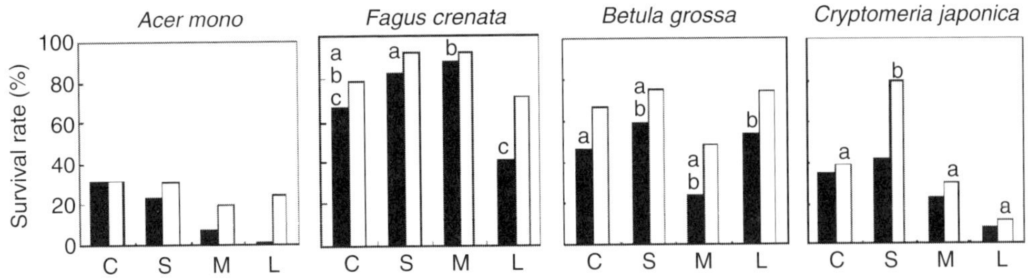

Figure 24.4 Average survival of tree seedlings in different sized gaps two and three years after gap creation. Letters above the bars denote significant differences among gap size in the same year. C = closed canopy, S = gap less than 100 m², M = gap between 100 and 200 m², and L = gap larger than 200 m².

Table 24.2 Competitive Characteristics of Canopy Tree Species in Secondary Stands, Based on Reference Studies (Yoshida and Kamitani 1999[1], 2000[2], 1998[3])

Species	Shade tolerance	Sensitivity to shading [1]	(b) Intensity of shading [2]	(c) Response to crown release [3]
Phellodendron amurense	Intolerant	Very high	—	—
Magnolia obovata	Intermediate	High	Very low	High
Quercus crispula	Intermediate	High	High	High
Fagus crenata	Tolerant	Low	Very high	Low
Acer mono	Tolerant	—	—	Low

The absence of a reference study is indicated by "—."

24.8 Canopy tree response to gap creation

The species composition of abandoned coppice forests is strongly affected by the intensity of past management, specifically the interval between coppice harvests (Kamitani 1993). To clarify the dynamics of canopy trees in abandoned coppice forests, we described the height, crown depth, and dbh of canopy trees in the study stands. We analyzed growth by extracting increment cores from the canopy trees. This analysis considered interspecific competition, incorporating asymmetric competitive effects (Nakashizuka and Kohyama 1995), and the effects of the 11 canopy gaps on recent diameter growth. The study stands consisted of combinations of deciduous broadleaved tree species (*F. crenata*, *Q. crispula*, *Q. serrata*, *Magnolia obovata* Sieb. & Zucc., and *A. mono* with several associated species). Stand structure was typical of secondary forests, in that there was little vertical stratification and height class distribution was bell shaped. Dead trees were found only in or beneath the subcanopy layer; hence, there were few marked openings in the canopy (Yoshida and Kamitani 1997).

Interspecific interactions were determined by differences in the height and crown depths of component species. *F. crenata*, which is shade tolerant, had the most pronounced suppressive effects on its neighbors (Table 24.2), because of its ability to maintain a deep crown (Yoshida and Kamitani 1997, 2000). *F. crenata* was also the most resistant to suppression by its neighbors (Yoshida and Kamitani 2000). In contrast, *Phellodendron amurense* (Rupr.), a shade-intolerant species, seemed to be the most sensitive to competition; it could grow well in the absence of suppression, but actual growth in the forest was significantly less than that of its neighbors (Yoshida and Kamitani 1999). *Q. crispula* and *M. obovata*, intermediate-tolerant species, showed greater sensitivity to competition than *F. crenata*. Their shallower crowns should have less shading effect on neighbors (Yoshida and Kamitani 2000).

24.9 Rehabilitation by creating canopy gaps

Shade-intolerant canopy species that occur in primary forests are likely to decline markedly as these forests mature. The dominance of shade-tolerant species such as *F. crenata* over less tolerant component species gradually increases as a stand matures (Yoshida and Kamitani 1997). Coppice forests contain fewer shade-intolerant species than primary forests as a result of repetitive cutting during the early stages of coppice development (Kamitani 1993). Considering the predominance in the region of even-aged stands in the stem exclusion phase (Nagaike and Kamitani 1997), the diversity of canopy tree species is likely to decrease under the current trend of abandonment of coppice forests. If rehabilitation is intended to develop stand structure and species composition that is characteristic of primary forests in the region, continued abandonment of coppice forests is inappropriate.

The relative proportion based on basal area of *F. crenata* in abandoned coppice forests should be considered a determining factor in deciding on appropriate management. Stands that contain a relatively high proportion of *F. crenata* are proposed as substitutes for primary forests, and are easily and quickly rehabilitated. However, management will be required to prevent shade-tolerant *F. crenata* dominating less tolerant species in the overstory. The creation of canopy gaps increases the heterogeneity of canopy structure by releasing less tolerant species from competition. In contrast, in stands with few *F. crenata*, gap creation should be intended to regenerate the whole forest. We found that in these coppice forests, gaps increased seedling establishment and species diversity.

Artificial control of the competitive relationships among canopy tree species is necessary to accelerate the development of old forest characteristics in coppice forests. Crown release by the creation of canopy gaps enhances the growth of surrounding canopy trees (Marquis and Ernst 1991; West and Osler 1995; Kerr 1996); however, the effects of intensive crown release differ among species. The stem growth of intermediate shade-tolerant species, such as *Q. crispula* and *M. obovata*, increases with increased cutting intensity (Table 24.2), whereas that of shade-tolerant species, *F. crenata* and *A. mono*, is not related to cutting intensity (Yoshida and Kamitani 1998). Canopy stratification, in which less tolerant species form an upper canopy above more tolerant species, is essential for the coexistence of species with different ecological niches (Wierman and Oliver 1979; Evans 1982; Kelty 1989). Shade-tolerant species adjacent to less tolerant target species should be removed. This will enhance the light environment for the less tolerant species, allowing them to stratify the crown by forming an upper canopy and prevent more tolerant species from dominating (Yoshida and Kamitani 1997, 2000). Low thinning, which is frequently practiced in Japan, is less useful to meet these objectives (Marquis and Ernst 1991; Miguchi 1996).

24.10 Research needs

Based on our limited experiments so far, we have concluded that artificial canopy gaps in abandoned coppice forests can play a significant role in the development of complex structure and composition. This appears to be true both for stands with a high proportion of shade-tolerant *F. crenata*, where the goal of canopy gap creation would be to maintain shade-intolerant species, and for stands with low amounts of *F. crenata*, where the goal would be to obtain a greater diversity of species composition. Additional research is needed to more quantitatively define the critical proportion of *F. crenata* in the stand, and the size and spatial arrangement of gaps needed to meet the species composition and structure objectives. Coppice stands with a high proportion of *F. crenata*, for example, probably require relatively large gaps to secure the establishment of desired shade-intolerant species. Planting and direct seeding into gaps, sometimes with soil preparation, have been effective elsewhere when adequate seed sources or dispersal mechanisms are lacking. A comprehensive research program would include monitoring abandoned coppice stands to determine how they develop without intervention. Such a monitoring program would have to include stands with a range of species composition, particularly the proportion of *F. crenata* and other shade-intolerant species, soil and site conditions, and risk of snow damage.

References

Abe, M., Miguchi, H., and Nakashizuka, T., An interactive effect of simultaneous death of dwarf bamboo, canopy gap, and predatory rodents on beech regeneration, *Oecologia*, 127, 281, 2001.

Abe, M., Izaki, J., Miguchi, H., Masaki, T., and Nakashizuka, T., The effects of *Sasa* and canopy gap formation on tree regeneration in an old beech forest, *J. Veg. Sci.*, 13, 565, 2002.

Abe, S., Masaki, T., and Nakashizuka, T., Factors influencing sapling composition in canopy gaps of a temperate deciduous forest, *Vegetatio*, 120, 21, 1995.

Bawa, K.S. and Seidler, R., Natural forest management and conservation of biodiversity in tropical forests, *Conserve Biol.*, 12, 46, 1998.

Chiba, T., *Culture of Bare Mountains*, Gakuseisya, Tokyo, 1973 (*in Japanese).

Chiba, T., *Research for Bare Mountains*, Soshiete, Tokyo, 1991 (*in Japanese).

Coates, K.D. and Burton, P.J., A gap-based approach for development of silvicultural systems to address ecosystem management objectives, *For. Ecol. Manage.*, 99, 337, 1997.

Emborg, J., Christensen, M., and Heilmann-Clausen, J., The structural dynamics of Suserup Skov, a near-natural temperate deciduous forest in Denmark, *For. Ecol. Manage.*, 126, 173, 2000.

Environment Agency of Japan, *Threatened Wildlife of Japan — Red Data Book 2nd ed. — Vascular Plants*, Japan Wildlife Research Center, Tokyo, 2000.

Evans, J., Silviculture of oak and beech in northern France: observations and current trends, *Q. J. For.*, 76, 75, 1982.

Frelich, L.E. and Puettmann, K.J., Restoration ecology, in *Maintaining Biodiversity in Forest Ecosystems*, Hunter, M.L., Jr., Ed., Cambridge University Press, Cambridge, 499, 1999.

Fukamachi, K., Oku, H., and Nakashizuka, T., The change of a satoyama landscape and its causality in Kamiseya, Kyoto Prefecture, Japan between 1970 and 1995, *Landscape Ecol.*, 16, 703, 2002.

Gray, A.N. and Spies, T.A., Gap size, within-gap position and canopy structure effects on conifer seedling establishment, *J. Ecol.*, 84, 635, 1996.

Gray, A.N. and Spies, T.A., Microsite controls on tree seedling establishment in conifer forest canopy gaps, *Ecology*, 78, 2458, 1997.

Hart, C., *Alternative Silvicultural Systems to Clear Cutting in Britain: A Review*, HMSO, London, 1995.

Hasegawa, M., The effects of weeding and improvement cutting on the establishment of hardwoods in an unsuccessful *Cryptomeria japonica* plantation, *J. Jpn. For. Soc.*, 73, 375, 1991 (in Japanese).

Ida, H., Treefall gap disturbance in an old-growth beech forest in southwestern Japan by a catastrophic typhoon, *J. Veg. Sci.*, 11, 825, 2000.

Iida S. and Nakashizuka, T., Forest fragmentation and its effect on species diversity in sub-urban coppice forests in Japan, *For. Ecol. Manage.*, 73, 197, 1995.

Ishii, M., Ueda K., and Shigematsu T., *Conserving the Satoyama Landscape in Japan*, Tsukijisyokan, Tokyo, 1993 (*in Japanese).

Kamitani, T., Ecological studies on regeneration of Beech (*Fagus crenata* Blume) coppice forests in heavy snowfall region, *Mem. Fac. Agric. Niigata Univ.*, 30, 1, 1993 (in Japanese with English summary).

Kelty, M.J., Productivity of New England hemlock/hardwood stands as affected by species composition and canopy structure, *For. Ecol. Manage.*, 28, 237, 1989.

Kerr, G., The effect of heavy or 'free growth' thinning on oak (*Quercus petraea* and *Q. robur*), *Forestry*, 69, 304, 1996.

Kobayashi, M. and Kamitani, T., Effects of surface disturbance and light level on seedling emergence in a Japanese secondary deciduous forest, *J. Veg. Sci.*, 11, 93, 2000.

Kohm, K.A. and Franklin, J.F., *Creating a Forestry for the 21st Century*, Island Press, Washington, DC, 1997.

Lindenmayer, D.B., Cunningham, R. B., Donnelly, C.F., and Franklin, J.F., Structural features of old-growth Australian montane ash forests, *For. Ecol. Manage.*, 134, 189, 2000.

Marquis, D.A. and Ernst, R.L., The effects of stand structure after thinning on the growth of an Allegheny hardwood stand, *For. Sci.*, 37, 1182, 1991.

Masaki, T., Suzuki, W., Niiyama, K., Iida, S., Tanaka, H., and Nakashizuka, T., Community structure of a species-rich temperate forest, Ogawa Forest Reserve, central Japan, *Vegetatio*, 98, 97, 1992.

Masaki, T., Nakamura, S., Ota. T., Otani, T., Oohara, H., Sugita, H., Saitoh, M., Kanbayashi, T., and Nagaike, T., Structure and 5-year change of tree communities of planted conifers and colonizing native trees in northern Japan, *Tohoku J. For. Sci.*, 8, 75, 2003 (in Japanese).

Miguchi, H., Thinning effects in a secondary beech forest, *Bull. Niigata For. Exp. Stn.*, 38, 17, 1996 (*in Japanese).

Miura, S., Yoshinaga, S., and Yamada, T., Protective effect of floor cover against soil erosion on steep slopes forested with *Chamaecyparis obtusa* (hinoki) and other species, *J. For. Res.*, 8, 27, 2003.

Miyawaki, A., *Vegetation of Japan, Chubu*, Shibundo, Tokyo, 1985 (in Japanese).

Nagaike, T. and Kamitani, T., Factors affecting changes in a landscape structure dominated by both primary and coppice forests in the *Fagus crenata* forest region of central Japan, *J. For. Res.*, 2, 193, 1997.

Nagaike, T. and Kamitani, T., Factors affecting changes in landscape diversity in rural areas of the *Fagus crenata* forest region of central Japan, *Landscape Urban Plan.*, 43, 209, 1999.

Nagaike, T., Kamitani, T., and Nakashizuka, T., Plant species diversity in abandoned coppice forests in a temperate deciduous forest area of central Japan, *Plant Ecol.*, 166, 145, 2003.

Nakashizuka, T., Regeneration dynamics of beech forests in Japan, *Vegetatio*, 69, 169, 1987.

Nakashizuka, T., Role of uprooting in composition and dynamics of an old-growth forest in Japan, *Ecology*, 70, 1273, 1989.

Nakashizuka, T., Species coexistence in temperate, mixed deciduous forests, *Trends Ecol. Evol.*, 16, 205, 2001.

Nakashizuka, T., Iida, S., Tanaka, H., Shibata, M., Masaki, T., and Niiyama, K., Community dynamics of Ogawa Forest Reserve, a species rich deciduous forest, central Japan, *Vegetatio*, 103, 105, 1992.

Nakashizuka, T. and Kohyama, T., The significance of the asymmetric effect of crowding for coexistence in a mixed temperate forest, *J. Veg., Sci.*, 6, 509, 1995.

Ogura, J., *History of Humans and Landscape in Japan, With Special Reference to Historical Drawing*, Yuzankaku, Tokyo, 1992 (*in Japanese).

Ogura, J., *Life Styles of Japan, with Special Reference to Vegetation*, Yuzankaku, Tokyo, 1996 (*in Japanese).

Oliver, C.D. and Larson, B.C., *Forest Stand Dynamics*, John Wiley and Sons, New York, 1996.

Smith, D.M., Larson, B.C., Kelty, M.J., and Ashton, P.M., *The Practice of Silviculture: Applied Forest Ecology*, 9th ed., John Wiley and Sons, New York, 1997.

Takeuchi, K., Brown, R.D., Washitani, I., Tsunekawa, A., and Yokohari, M., *Satoyama, the Traditional Rural Landscape of Japan*, Springer, Tokyo, 2003.

Tanimoto, T., *Ecology of the Management of Broadleaved Forests*, Sobun, Tokyo, 1990 (*in Japanese).

ter Steege, H., *HEMIPHOTO, A Programme to Analyze Vegetation Indices, Light Quality from Hemispherical Photographs*, The Tropenbos Foundation, Wageningen, 1993.

Van Der Meer, P.J., Dignan, P., and Saveneh, A.G., Effect of gap size on seedling establishment, growth and survival at three years in mountain ash (*Eucalyptus regnans* F. Muell.) forest in Victoria, Australia, *For. Ecol. Manage.*, 117, 33, 1999.

West, P.W. and Osler, G.H.R., Growth response to thinning and its relation to site resources in *Eucalyptus regnans, Can. J. For. Res.*, 25, 69, 1995.

Yamamoto, S., Gap dynamics in climax *Fagus crenata* forests, *Botanical Magazine Tokyo*, 102, 93, 1989.

Yamamoto, S., Forest gap dynamics and tree regeneration, *J. For. Res.*, 5, 223, 2000.

Yamamoto, S., Nishimura, N., and Matsui, K., Natural disturbance and tree species coexistence in an old growth beech-dwarf bamboo forest, southwestern Japan. *J. Veg. Sci.*, 6, 875, 1995.

Yokoi, S. and Yamagichi, K., Growth condition of planted trees and naturally regenerated broad-leaved trees in an unsuccessful *Cryptomeria japonica* plantation in a heavy snowy region in Japan, *Jpn. J. For. Environ.*, 40, 91, 1998 (in Japanese).

Yoshida, T. and Kamitani, T., The stand dynamics of a mixed coppice forest of shade-tolerant and intermediate species, *For. Ecol. Manage.*, 95, 35, 1997.

Yoshida, T. and Kamitani, T., Effects of crown release on basal area growth rates of some broad-leaved tree species with different shade-tolerance, *J. For. Res.*, 3, 181, 1998.

Yoshida, T. and Kamitani, T., Growth of a shade-intolerant tree species, *Phellodendron amurense*, as a component of a mixed-species coppice forest of central Japan, *For. Ecol. Manage.*, 113, 57, 1999.

Yoshida, T. and Kamitani, T., Interspecific competition among three canopy-tree species in a mixed-species even-aged forest of central Japan, *For. Ecol. Manage.*, 137, 221, 2000.

* The title is tentatively translated into English by the authors.

chapter twenty-five

Forest restoration and rehabilitation in Republic of Korea

Don Koo Lee
Seoul National University (SNU)
Seung Jin Suh
Korea Forest Research Institute (KFRI)

Contents

1-56670-635-1/05/$0.00+$1.50

25.1 Biogeography of the Korean Peninsula

The Korean Peninsula (124°11′–131°53′E, 33°7′– 43°01′N) shares borders with China and Far East Russia. The peninsula is divided into two nations: the Republic of Korea (ROK) in the south and the Democratic People's Republic of Korea (DPRK) in the north (Figure 25.1). The climate of Korea is primarily continental, except for August when it is oceanic. The long, hot summers are wet; monsoon rains account for 60% of the total annual rainfall. Winters are dry and freezing cold due to Siberian air masses that sweep in from the north. The annual mean temperature ranges from 3 to 14°C depending on altitude; annual precipitation ranges from 600 to 1,500 mm. In Seoul, the mean monthly temperature ranges from −2.5 to 25.4°C in the coldest and hottest months (Korea Meteorological Administration 2002).

Korea is a mountainous peninsula with complex geology. About 60% of the peninsula lies between 200 and 1,000 m above sea level (asl) with an average elevation of 482 m (Park 2002). The eastern regions are more mountainous and steeply sloped than the western regions. The peninsula is composed largely of Pre-Cambrian rocks such as granite and gneiss (70%) and other metamorphic strata topped by later sediments and granitic intrusions. Soils are susceptible to erosion owing to the changeable climate and localized torrential downpours in summer (Korea Forest Service 2002). Soils are either light brown, acidic, sandy soils or red-brown clays; black volcanic soils are common in Jeju and Ulleung Islands as well as in northern Gangwon-do. Podzolic soils and ash-gray forest soils are found in the northern mountain areas (Korea Forest Service 2000).

Figure 25.1 Location of the Korean Peninsula and the case studies.

The ROK falls within the temperate zone, although broadleaved evergreen plants and bamboo grow in the south. The number of species is unusually large due to climatic and topographic diversity within a small territory (99,600 km^2). The total number of animals is 18,029, and the plant diversity is recorded as 8,271 species: 4,662 for higher plants and 3,609 for lower plants in the ROK (Korea Ministry of Environment 1997).

The forest area in ROK occupies 6.4 million ha, about 65% of the entire land area. The dominant tree species are *Pinus densiflora* S. et Z. and *Quercus* species, which form mixed forests at lower elevations. *Pinus densiflora* naturally grows on relatively dry poor sites and can successfully regenerate with human intervention. During the last 10,000 years, this pine was the only tree available in quantities large enough to supply logs to build palaces, temples, public buildings, and farm cottages (Lee 2002). Its dominance has been much reduced; today more than 50% of the forest area is deciduous forests (Forestry Research Institute 1996).

25.1.1 Forest types

The three major vegetation types on the Korean Peninsula are warm temperate forest, cool temperate forest (southern, central and northern temperate forest subzones), and frigid forest (Yim 1985). Cool temperate forest accounts for 85% of the total forest area. The warm temperate forest covers the area south of 35°N, a part of the southern coastal regions, including Jeju Island and many smaller islands where the annual mean temperature is above 14°C. Evergreen broadleaved forests were once typical in this zone, but most natural forests were destroyed by overexploitation and fire. This area has become deciduous broadleaved, pine, or mixed forests. Dominant tree species in this zone are *Castanopsis sieboldii* (Makino) Hatusima, *Quercus acuta* Thunb., *Cinammomum japonicum* Siebold ex Nees, *Persea thunbergii* S. et Z., *Ternstroemia gymnanthera* (Wright et Arnold) Spargue, *Ilex integra* Thunb., *Camellia japonica* L., and various bamboo species.

Except for mountainous highlands, land between 35 and 43°N belongs to the cool temperate forest. The annual mean temperature ranges from 6 to 13°C. Deciduous broadleaved forests are representative in this zone, but most of the forests have been disturbed by human activities and changed into pine forests. Today, deciduous broadleaved tree species can be found in the lower strata of some pine forests where human-induced disturbance has not occurred. Dominant tree species are deciduous oaks (*Q. mongolica* Fischer ex Ledebour, *Q. acutissima* Carruther, *Q. variabilis* Blume), *Acer pictum* Thunb. var. *mono* (Maxim.) Maxim. ex Franch., *Carpinus*, *Betula*, and *Pinus densiflora*.

The frigid forest covers the northern part of the peninsula and the mountainous highlands where the annual mean temperature is below 5°C. Coniferous forests are representative of this zone. Forests disturbed by overexploitation or forest fire are usually replaced by deciduous broadleaved and coniferous forest. Dominant tree species are *Abies nephrolepis* (Traut.) Maxim., *Picea jezoensis* (S. et Z.) Carrier, *Juglans mandshurica* Maxim., and *Betula platyphylla* Sukatschew.

25.1.2 Forests of the ROK

Forestland in the ROK is classified into national (1.42 million ha, 22% of total forest land), public (0.49 million ha, 8%), and private (4.53 million ha, 70%) forests by ownership. National forests are managed for land conservation, forest management, scientific research, development of forestry techniques, preservation of both historical relics and cultural heritage, and other public benefits. Local governments and public organizations such as educational institutions own public forests. Individuals and groups such as corporations, nonprofit organizations, and temples own private forests. The number of private forest owners is about 2 million and 96% of the owners hold less than 10 ha of forestland. For these small forest owners, planting and tending have been subsidized since 1984. Forestland in the future is designed by the ROK government to be 3.5 million ha production forest, 1.5 million ha public forest including areas for environmental protection, and 1.4 million ha for other purposes including industrial and multiple uses.

The total value of forest products in 2000 was about US$1,600 million, representing 0.2% of the national GDP. The economic value of timber production did not exceed US$90.8 million, resulting in most of the economic value coming from mushrooms, nuts, and other nontimber products (Korea Forest Service 2001). Nontimber forest products such as nuts, mushrooms, wild medicinal or edible herbs, and tree sap are increasingly important to the people in both rural and urban areas (Korea Forest Service 2002). Currently, imported timber is needed to meet about 94% of domestic consumption.

25.1.3 Historical vegetation change

Glacial activity is not evident in Korea except in the highest mountains (Nelson 1993); however, periglacial climate affected the distribution of forest communities. Pollen analyses (Yamazaki 1940; Matsushima 1941; Oh 1971; Jo 1979; Yasuda et al. 1980; Choi 2001) indicate that the end of the last glacial period in Korea was approximately 10,000 BC. Between 10,000 and 6,000 BC, deciduous broadleaved forest (*Quercus* spp.) dominated and the number of subarctic conifers decreased rapidly (Choi 2002). Around 6,000 BC, forests occupied the lowlands of the eastern and western coasts. The western region changed from *Quercus* species to *Alnus* species followed by *Pinus* species. The eastern region changed from *Quercus* spp. to *Pinus* spp. (Choi 2002). Around 2,000 BC, the distribution of deciduous broadleaved forests dominated by *Quercus* and *Alnus* species, and of evergreen broadleaved forests dominated by *Cyclobalanopsis* spp. narrowed. The range of coniferous forests dominated by *Pinus* spp. (*Diploxylon*) increased.

Pine forests were protected for a long time, particularly during the Chosun Dynasty (1392 to 1910). The Chosun Dynasty initiated a forest control system that removed the provisions of the "Shiji" (lands for collecting fuelwood) and "Gumsan system," which regulated forests for specific national uses, such as providing coffins for the royal family (Park 1998). In the 19th century, the forests of the ROK were extensive, with much old-forest character. These forests were totally destroyed, however, by overharvesting and illegal cutting for fuel and building material throughout the chaotic periods of the Japanese occupation (1910 to 1945) and the Korean War (1950 to 1953). During this time, the average growing stock volume decreased from about 100 m^3 ha^{-1} in 1900 to only 10 m^3 ha^{-1} in 1960 (Table 25.1) (Korea Forest Service 2001).

The shortage of fuelwood needed for domestic cooking and heating continued even after this wartime turmoil. In 1960, there were about 2.4 million households in the country, each requiring about 0.5 ha of forest to meet fuel needs. One of the first priorities of the ROK government was to establish fuelwood plantations, which could meet these demands as well as start the rehabilitation of devastated forestland. During the last 40

Table 25.1 Forestland Area and Growing Stock by Year

Year	Area (10^3 ha)	Growing stock (10^3 m^3)	Growing stock (m^3 ha^{-1})
1960	6,700	63,995	9.55
1970	6,611	68,772	10.40
1980	6,567	145,694	22.18
1990	6,476	248,426	38.36
2000	6,430	387,758	60.30

Source: Korea Forest Service, *Statistical Yearbook of Forestry*, 468 pp., 2001.

years of rapid economic development and urbanization of the ROK, forests have been cleared for the construction of social infrastructure, industrial estates, and new towns. As a result of both anthropogenic activities and biological agents (pine gall midges), *Pinus densiflora* has declined and *Q. mongolica*, *Robinia pseudoacacia* L., and *Styrax japonicum* S. et Z. have increased in density (Lee 2002).

25.2 *Reforestation activities*

Since the end of the Korean War (1953), the ROK government has encouraged replanting forests and has promoted the search for alternative fuels, especially for domestic use. To rehabilitate deforested areas, the ROK government initiated successive reforestation plans that concentrated initially on fast-growing trees such as *Larix kaempferi* (Lamb.) Carr., *Pinus rigida* Miller, and *Populus* species. Nurse trees such as *R. pseudoacacia* and *Alnus* species were included from 1959, in order to improve soil conditions and control erosion. By 1999, about 97% of the deforested area was rehabilitated (Korea Forest Service 2000), making ROK one of the most successful countries for reforestation in the world. The widespread planting of *Pinus koraiensis* and *Larix kaempferi* without a detailed consideration of site characteristics, however, has resulted in failures and plantations of little economic value.

25.2.1 *National forest development plans*

25.2.1.1 *The first 10-year plan (1973 to 1978) — rehabilitation and fuelwood production*

The first National Forest Development Plan focused on reforestation of denuded forestlands and was a watershed event for Korean forestry. A national tree planting movement began and 1 million ha was planted (rehabilitated); new economic forest zones for land conservation and income enlargement were developed. Fast-growing tree species rapidly reforested denuded land (Korea Forest Service 1997). The Korea Forest Service (KFS) declared March 21 to April 20 as the "National Tree Planting Period," which is the best time for planting in the ROK. Groups from local communities, families, and schools participated in the reforestation program. In addition, November was declared the "Tree Tending Period" to promote thinning, pruning, fertilizing, weeding, and preventing insects and diseases. The KFS consolidated efforts to protect and minimize damage to forests by humans through restricting access to the mountains and developing fuelwood forests for rural people. The reforestation target of 1 million ha, planned to be completed in 1982, was accomplished in 1978 with the reforestation of 1.08 million ha.

25.2.1.2 *The second 10-year plan (1979 to 1987) — timber production*

The primary objective of the second 10-year plan was building large-scale commercial forest zones in order to develop a long-term timber resource. The KFS increased development funds for private forest management and expanded the national forests (Korea Forest Service 1997). Over 1 million ha was reforested and the KFS established 80 large-scale commercial forest zones that included 375,000 ha of reforestation. Major tree species used from 1973 to 1987 for reforestation were Italian hybrid poplar (25%), larch (18%), Korean pine (12%), hybrid aspen (8%), chestnut (7%), and hybrid pine (1%).

25.2.1.3 *Third 10-year plan (1988 to 1997) — establishment of multifunctional forests*

The objective of the third plan was to sustain efficient use of forest resources by increasing their economic value and improving public benefits from the forests. The plan focused on using forestland rationally, improving timber resources, establishing forest management infrastructure, maintaining distribution channels for forest products, and improving living standards (Korea Forest Service 1997). The KFS established commercial forest zones of 320,000 ha, conducted tending on over 3 million ha, and trained forest technicians. Establishing natural forest resorts, increasing water supply, and creating pleasant environment of forests with wildlife enhanced forest recreation opportunities. To meet future industrial wood needs, the government initiated overseas plantation projects in 1992 that attained the level of 42,000 ha in 2002, with planned expansion to 1 million ha by 2050.

25.2.1.4 The fourth 10-year plan (1998 to 2007) — sustainable forest management

The ROK government plans to complete its reforestation program and move to a forest management program based on self-regulation and promotion. To meet diverse social demands from forests and recent international forest policy objectives, the primary aim of the fourth plan is to establish and develop sustainable forest management. Recently, the KFS recommended 78 tree species for planting in Korea. These recommended tree species are comprised of 23 timber tree species, including *Pinus densiflora*, *Pinus koraiensis* S. et Z., and *Larix kaempferi*; four nut tree species including *Castanea crenata* S. et Z. and *Juglans sinensis* Dode; and 51 special-purpose tree species with specific properties for specialized uses including medicinal, pollution resistant, shade tolerant, and fire resistant.

25.2.2 Forest rehabilitation

The successful reforestation of the ROK's denuded mountains, begun during the first 10-year plan, marked the beginning of modern Korean forestry. The national tree planting movement and development of new economic forest zones not only reforested denuded forestlands and restored the beauty of the mountains but also provided employment for many people. The KFS contributed to the success of reforestation in the country by encouraging various local and civic groups to participate in tree planting and tending programs. Residents planted and tended trees near mountain villages. Forest fraternities (mutual aid association) were organized and leaders were trained to reinforce reforestation activities. The government paid local people to run village nurseries, implement erosion control, and plant trees. The cooperation among government agencies, nongovernmental organizations (NGOs), and villagers was infused with the spirit of Saemaul Undong (SU), meaning "New Village Campaign." It can be said that SU represents community involvement in rural development (Lee 1984). Recognizing the role of forestry in rural development, the KFS was moved from the Ministry of Agriculture and Forestry into the Ministry of Home Affairs in 1973.

Forestry in the ROK faces new challenges as reforestation policies have changed, transforming the goal from greening of the country into forest resources enhancement and sustainable management. Poor performance of exotic species in plantations has resulted in replacement by natural regeneration of native species.

25.3 Case studies

25.3.1 Rehabilitation of eroded land at Yeongil district

25.3.1.1 Background

The Yeongil district in southeast ROK is in the temperate forest zone. Original tree species included *Carpinus tschonoskii* Maxim. var. *tschonoskii*, *Zanthoxylum schinifolium* S. et Z., *Lindera glauca* (S. et Z.) Blume, *Celtis sinensis* Persoon, *Platycarpa strobilacea* S. et Z., *Pinus thunbergii*, and *Pinus densiflora*. Overexploitation of forests began with the invasion by

Japan and other countries in 1800 and continued during the 20th century. Political and social disorder, fuel shortages, continuous illegal harvest, and economical difficulty contributed to the devastation of forests in the district. The native species have disappeared due to illegal logging and the Korean War, with only dwarfed pine trees remaining. In spite of 50 small-scale erosion control projects since the 1950s, large areas of devastated lands in Yeongil district were ignored. Pushed by the central government following a presidential visit in 1971, erosion control activities were implemented beginning in 1973 (Gyeongsangbuk-do 1977).

The areas for erosion control project in Yeongil extend from 129°5′ to 129°26′ E and from 35°54′ to 36°16′ N (Figure 25.1), which includes Yeongil, Weolseong-gun (county) and Pohang-shi (city). The annual average temperature is 13°C, which ranged from a maximum of 37°C and a minimum of −14°C. Annual rainfall is around 1,100 mm, but 500 to 600 mm is concentrated during the months of June to August. Shale and mudstone parent rocks with low infiltration and water-holding capacity leads to much overland flow, erosion, and thin topsoil. Growing conditions are difficult; in addition to low soil moisture, dry westerly winds cause a substantial water deficit (560 mm annually) (Gyeongsangbuk-do 1977).

25.3.1.2 *Erosion control techniques*

Planning for the erosion control project at Yeongil began in 1971 and the 5-year project began in 1973. By 1977, the total completed area was 4,538 ha; general erosion control was implemented on 3,291 ha, special erosion control on 479 ha, and reforestation of 768 ha of bare land. Four erosion control techniques were used on gentle slopes, including: sod-patching channels to stabilize small valleys and safely discharge rain water; terrace-sodding structures to control rill erosion on slopes (Figure 25.2); contour-trench to improve water-holding capacity and promote vegetation growth; and underground structures (Gyeongsangbuk-do 1977).

Steep slope areas required more use of stone due to high water flow velocities. The five techniques used included stone-patching channels; stone soil-arresting structures to protect slopes from rill erosion; stone terracing structures to control overland flow and provide a stable surface to establish trees; latticed block works to control slope failure; and mulching structures to cover eroded surfaces. Valleys in uplands required the use of revetments to stabilize foot slopes in order to protect steep slopes; erosion check dams to stabilize foot slopes and control horizontal and vertical erosion; and erosion control dams to decrease stream run distances.

Figure 25.2 Contour terracing and transporting sod for erosion control planting.

Trees and grasses were seeded and planted to stabilize soil surfaces. Direct seeding used over 20 kg of seed per ha of *R. pseudoacacia*, *Lespedeza bicolour* Turcz., and *Cyperaceae* spp. The mixture was half grass and a quarter each of *Lespedeza bicolour* and *R. pseudoacacia*. Height growth in 1 year after seeding was 150 cm for *R. pseudoacacia*, 70 to 100 cm for *Lespedeza bicolor*, and 80 to 90 cm for grass. Two years after seeding, *R. pseudoacacia* occupied the overstory (Gyeongsangbuk-do 2002) (Figure 25.3). Other techniques include planting nurse trees (*Alnus* spp. and *R. pseudoacacia*) to supplement fertilization on strongly acid soils and pioneer species such as *Pinus thunbergii* and *Pinus rigida* for coastal areas and grassland. The project cost US$3.19 million, mostly for labor (US$2.7 million).

25.3.1.3 Ecological effects of rehabilitation and erosion control

The Yeongil area was surveyed in 1995, and the soil depth (A and B horizons) of the hillside had increased since 1971, from 1 or 2 cm to 11 cm. Soil depths were only 4 cm near the top of the hills and 112 cm at the foot of the hills, which are more than at the beginning of the program 24 years ago. Although the surface soil would be considered loose (penetrometer measurements of 0.7 kg/m^{-2}), the surface soil remains so shallow that root system development might be impeded, except at the base of the hill. Root system development, rainfall interception, and the windbreak effect of crown closure contributed to stabilizing the surface soils. Soil texture was changed from sandy-loam or silt-loam to fine sandy silt-loam as the ratio of clay increased. Soil acidity was lessened as pH changed from 4.1 to 4.5 to pH 4.6 to 5.5, but available calcium and cation exchange capacities were still less than typical of forest soils.

Natural species, *Pinus thunbergii* and *Pinus densiflora*, were in the overstory from natural regeneration in low density, 5 to 20 trees ha^{-1}. Middle- and lower-story vegetation was still not sufficiently dense, only 500 to 2,000 trees ha^{-1}, which were mostly planted and direct seeded. Nevertheless, the understory trees have grown well during the last 5 years due to the improved microclimate created by the faster growing trees (Lee 2001). Biodiversity has been improved: wildlife that was almost extinct for the last half century is found again, including *Parus majore* (great tit), *Phasianus colchicus karpowi* (pheasant),

Figure 25.3 Denuded forest lands at Hapchon, Gyongsangnam-do (A) and 2 years after reclamation (B).

Streptopelia spp. (dove), *Sus scrofa creanus* (wild boar), *Lepus brachyurus* (hare), and *Capreolus capreolus* (roe deer). Income from nontimber forest products for local people has increased; *R. pseudoacacia* has become a good source of honey.

25.3.2 Ecological restoration at Gyeonggi-do and Gangwon-do

25.3.2.1 Background

Because the ROK needed to quickly revegetate deforested land to control erosion and supply fuelwood, early forest planning used fast-growing trees, often of exotic species. As much as 70% of the total forest area was devoted to plantations. The major species planted were *R. pseudoacacia*, *Pinus rigida*, some *Alnus*, *Quercus*, and *Lespedeza* species. The choice of species shifted in the 1970s and 1980s to *Larix kaempferi* and *Pinus koraiensis* in order to increase the commercial value of the forest resource (Table 25.2). This approach produced

Table 25.2 List of Native Species, Exotic Species, Hybrid Pine, Poplars and Aspens, and Nut Trees in Reforestation During 1962–1992

	1962–1972		1973–1992	
	Trees		Trees	
Species/Hybrids	Planted (1,000)	Area (ha)	Planted (1,000)	Area (ha)
Native Species				
Pinus densiflora	42,275	14,902	51,150	17,050
Pinus thunbergii	204,364	68,121	90,761	30,254
Pinus koraiensis	159,998	53,333	831,093	277,031
Paulownia coreana	940	2,350	12,746	31,865
Rhus verniciflua	4,404	4,404	141	141
Quercus acutissima	78,232	26,077	18,819	6,273
Exotic Species and Hybrids				
Pinus rigida	1,252,387	417,462	634,483	211,494
Pinus taeda	19,052	6,351	12,787	4,262
Pinus rigida × *Pinus taeda*	31,458	10,486	83,124	27,708
Larix kaempferi	797,320	265,773	1,179,918	393,306
Cryptomeria japonica	106,262	35,421	160,459	53,486
Chamaecyparis obtusa	98,234	32,745	243,679	81,226
Robinia pseudoacacia	1,244,070	414,690	333,250	111,083
Alnus species	372,081	124,027	481,564	160,521
Populus euramericana	61,484	153,710	220,626	551,565
Populus alba × *Populus glandulosa*	—	—	142,438	142,438
Populus nigra × *Populus maximowiczii*	—	—	75,725	151,450
Nut Trees				
Castanea creanata	94,359	235,898	60,588	151,470
Juglans sinensis	1,956	4,890	1,830	4,575
Ginkgo biloba	667	1,668	3,857	9,643
Zizyphus jujuba var. *inermis*	122	305	1,024	2,560
Diospyros kaki	1,325	3,313	1,230	3,075
Diospyros lotus	1,353	3,383	170	425
Others	69,161	23,054	803,841	267,947
Total	4,641,504	1,902,363	5,445,303	2,690,848

Source: Korea Forest Service (1993).

forest ecosystems of simple structure and low biodiversity, which in turn has resulted in low stability and productivity. Recently, interest has turned to restoration that emphasizes native species (Urbanska et al. 1997). In the ROK, native species such as *Pinus densiflora* and *Quercus* species regenerate naturally in plantations (Lee et al. 2001) and future restoration will likely rely more on native species. Two studies, in Gyeonggi-do and Gangwon-do, illustrate this natural invasion process.

25.3.2.2 General description

Two study sites were established, in Seongnam-Shi (37°70′N, 127°20′E), Gyeonggi-do and in Yangyang-gun (38°05′N, 128°35′E), Gangwon-do (Figure 25.1). The Gyeonggi-do site is at Seongnam-shi near Seoul; it was planted in the 1970s with *Pinus rigida, R. pseudoacacia,* and *Alnus* species. This site has been much disturbed by human activities and has a simple forest structure dominated by *Pinus rigida* and *Quercus* species. The study site in Gangwon-do was located in the experimental forests at the Forest Works Training Centre in Yangyang-gun, close to the East Sea. This district is well known for producing mushrooms in the pine forests. *Pinus densiflora* was originally the dominant species at this site.

25.3.2.3 Ecological restoration at Gyeonggi-do

This area was dominated originally by *Pinus densiflora* and *Quercus* species before the early 1960s. Exotic species such as *Pinus rigida, R. pseudoacacia,* and *Alnus* were planted for rehabilitation after destruction of this area about 30 years ago. Mostly *Pinus rigida* was planted; however, a few stands were mixed with *R. pseudoacacia* and *Alnus* species for improving soil quality (Lee and Park 1999). Apparently well adapted, with good survival, *Pinus rigida* grew quickly at first but increasing competition from naturally invading *Quercus* species stunted *Pinus rigida*. With no active management of the plantation over the last 30 years, *Pinus rigida* has been replaced by the native *Quercus acutissima* Carr. The planted *R. pseudoacacia* have maintained a position in the stand (Table 25.3).

25.3.2.4 Ecological restoration at Gangwon-do

The *Pinus koraiensis* plantations planted in 1973 at Gangwon-do were established after harvest of the native *Pinus densiflora* forest. After 30 years, the *Pinus densiflora* has regenerated naturally in these plantations and has grown faster than the planted *Pinus koraiensis* (Figure 25.4). Based on these and similar results elsewhere, *Pinus densiflora* is recommended for future plantings for ecological and economical reasons (Cho and Lee 1995; Koo and Bilek 1998). An additional factor is the significant economic contribution of the mushroom production in *Pinus densiflora* forests. *Larix kaempferi* is another of the ten species recommended for timber production in the second 10-year plan. Also an exotic, it grows well on moist and fertile sites in valley bottoms. Because it performs poorly on slopes and ridges (Figure 25.5) that are the typical sites for reforestation, current recommendations are to confine it to valley bottom sites.

Table 25.3 Importance Value of the Major Tree Species in the *Pinus rigida* plantation at Gyeonggi-do

Species	Trees (no. ha^{-1})	Basal area/ha (m^2 ha^{-1})	Relative Density (%)	Relative Coverage (%)	Importance Value (%)
Quercus species	482	8.31	44.2	63.4	53.8
Robinia pseudoacacia	309	2.41	28.3	18.4	23.4
Pinus rigida	59	1.52	5.4	11.6	8.5
Alnus species	14	0.19	1.3	1.4	1.3
Others	227	0.69	20.8	5.2	13.0

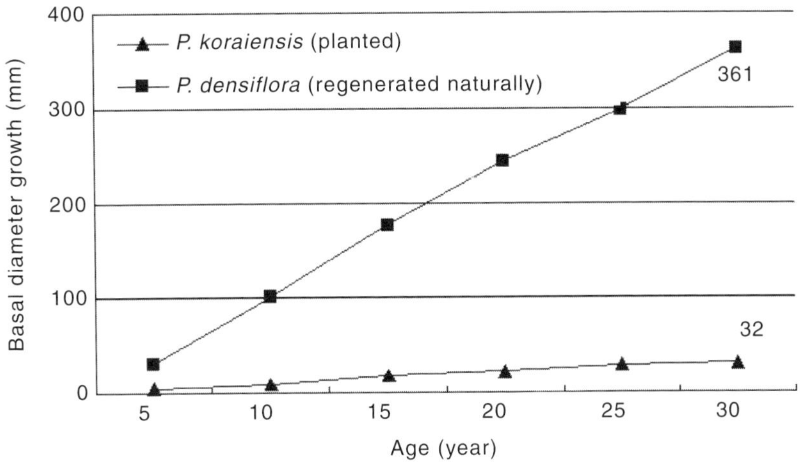

Figure 25.4 Basal diameter growth of *Pinus koraiensis* and *Pinus densiflora* in a *Pinus koraiensis* plantation. The naturally regenerated *Pinus densiflora* grows faster than the planted *Pinus koraiensis*.

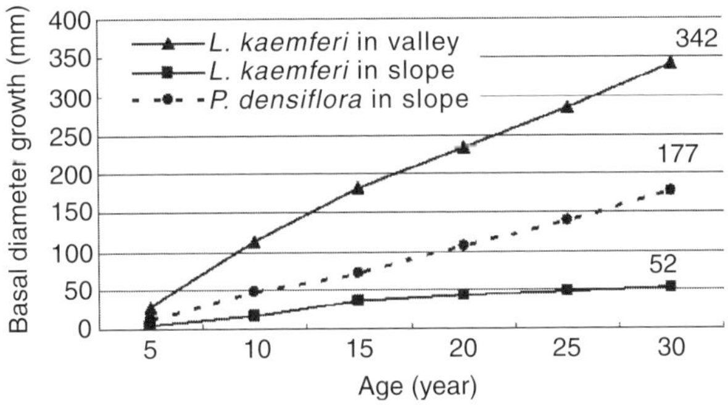

Figure 25.5 Basal diameter growth of *Larix kaempferi* and *Pinus densiflora* in *Larix kaempferi* plantation in the valley and on the slope.

25.4 Restoration challenges in northeast Asia

The northeast Asian countries of China, Mongolia, Japan, and Korea host about one fourth of the world population and consume 13% of world timber production. The region has many environmental problems that transcend national boundaries. In Mongolia, a large portion of the land is semiarid and has become denuded or disturbed by human activities. In the DPRK, 20% of the forests were destroyed, causing flooding problems and contributing to the current severe famine. Deforestation is also causing wind erosion in central China and Mongolia; this "yellow dust" passes over Korea and Japan every spring, reducing agricultural production. In the northeast Asia region, the major emphasis of international collaboration will be combating desertification and restoring degraded forest ecosystems. Greater public participation, as well as the increased presence of nongovernmental organizations (NGOs) in environmental decisions, will be the norm. NGOs in Korea are following the global trend and have, since the 1980s, become involved in both the political and the more practical aspects of supporting sustainable development, which includes forest restoration. The following descriptions of NGOs initiated by citizens groups in Korea illustrate these trends.

25.4.1 Forest for Life

The Forest for Life (FFL) campaign was born in the economic crisis of 1997 when millions of Koreans were unemployed. The goals of FFL are to help grow and tend the forests so that the resource could enhance employment. Nature conservation, a sense of community, harmony with nature, and respect for life were also goals of the FFL campaign. From 1998 to 2001, FFL programs have provided US$400 million to tend 383,000 ha of plantations. In 1999, about 20,000 people per day were involved in forest tending programs, and a total number of 4.83 million people have been involved. Selected employees were trained in forest tending technique, resulting in 10,000 trained as forest technicians as of 2001. Raw materials produced by forest tending of 120,000 m^3 per year were used for woodcarving, feed for livestock, and fuelwood.

25.4.2 Northeast Asian Forest Forum

One of the most crucial issues in the 21st century is the conservation of forests and environment. Despite global and national efforts to address environmental problems, deforestation progressed during the 20th century. As forests disappear, the efforts to save and restore them are also growing. The Northeast Asian Forest Forum (NEAFF) is a network of forestry academics in Korea, China, Mongolia, and Far East Russia that supports cooperation on sustainable and stable forest ecosystems. The forum is directed toward forest restoration and reforestation, increasing communication and environmental education for sustainable development, exchanging information for sustainable forestry, and forest conservation in northeast Asia. Particular interests include DPRK's Agricultural Rehabilitation and Environmental Protection Project and tree planting to combat desertification in China and Mongolia.

25.4.3 Forest for Peace

Forest for Peace (FFP), a sister NGO of NEAFF, is a national campaign to restore degraded forest ecosystems of the DPRK. Since its formation in 1999, FFP has provided planting materials and equipment to the DPRK, such as tree seeds, spray machines, branching shears, plastic sheets, and fertilizers. One aim of FFP is to establish about 200 forest nurseries (at least one for each of the counties in the DPRK) to produce tree seedlings for reforesting about 2 million ha of degraded forests. The total area of forested land within the

Figure 25.6 Degraded forest of DPRK in 1999.

DPRK is reported to be approximately 7.5 million ha, of which 6.3 million ha are natural and 1.2 million ha are plantations. Severe hailstorms in 1994, followed by floods in 1995 and 1996, and drought in 1997, devastated the agricultural productivity of the DPRK and caused massive damage to buildings, roads, and other infrastructure. One result has been indiscriminate clearing of forests and exposure to soil erosion and degradation. About 450,000 ha are currently recorded as urgently needing reforestation (Figure 25.6) (United Nations Development Programme 1998).

25.5 Summary and conclusion

After more than 50 years of social and political upheaval, the ROK faced a massive environmental problem of forest degradation and severe soil erosion. In addition, rural populations looked to the remaining forests to supply fuelwood for domestic heating and cooking needs. The government responded with a series of 10-year Forest Development Plans that attempted to meet the most pressing immediate needs and begin the long-term restoration of the country's forests. Beginning with the first and second Forest Development Plans, fuelwood plantations were successfully established and most of the degraded forests rehabilitated. Although driven by government programs, forest restoration was truly a national undertaking expressed as the spirit of Saemaul Undong. The cooperative effort of government organizations, NGOs, individuals, foresters, and the general public has carried over into successive plans that continue the rehabilitation effort in the context of sustainable management of planted and natural forests.

The first functional improvements of our environment have already appeared, such as erosion control, improvement of soil quality, and increasing biodiversity. Nevertheless, the full effects of the forest restoration will not be seen for at least another 50 years. Increasingly, rehabilitation has shifted from an emphasis on plantations and exotic species toward native species and natural regeneration. Two cases studies of successful rehabilitation were described to illustrate the restoration challenges in the ROK. In the Yeongil district with its severe and harsh conditions, much emphasis was placed on stabilizing erosion so that trees could be planted. In the Gyeonggi-do and Gangwon-do areas, the importance of matching species to site was illustrated as well as the promise of native species. The challenges faced in the ROK are similar to those facing other nations in the region and it is hoped that the experience gained in the ROK will be shared through international cooperation such as the NEAFF and FFP.

References

Choi, K.R., Vegetation and climate history of the lowland on the Korean peninsula, *Journal of the Korean Physical Society*, 39, 762, 2001.

Choi, K.R., Vegetational and climatic history of the lowland in Korea, in *Ecology of Korea*, Lee, D.W., Ed., Bumwoo Publ. Co., Seoul, 58, 2003.

Cho, D.H. and Lee, K.J., A relationship between climatic factors and matsutake productions in 29 sites during a 10-year period in Korea, *Journal of Korean Forestry Society*, 84, 277, 1995 (in Korean).

Forestry Research Institute, *Hardwood Forest Inventory Report No. 122*, Forestry Research Institute, 509 pp., 1996 (in Korean).

Gyeongsangbuk-do, *History of Yeongil District Erosion Control Project*, Gyeongsangbuk-do, 286 pp., 1977 (in Korean).

Gyeongsangbuk-do, *Erosion Control of Gyeongsangbuk-do Province*, Gyeongsangbuk-do, 140 pp., 2002 (in Korean).

Jo, W.R., Palynological studies on postglacial region, Korea peninsula. *Annuals of the Tohoku Geographical Association*, 31, 23, 1979 (in Japanese).

Koo, C.D. and Bilek E.M., Financial analysis of vegetation control for sustainable production of Songyi (*Tricholoma matsutake*) in Korea, *Journal of Korean Forestry Society*, 87, 519, 1998.

Korea Forest Service, *Statistical Yearbook of Forestry*, 528 pp., 1993.

Korea Forest Service, *Korea Forest Policy During Last 50 Years*, 1009 pp., 1997 (in Korean).

Korea Forest Service, *Forest and Forestry Technique: Chapter I, Forest*, Korea Forest Service, 2000 (in Korean).

Korea Forest Service, *Statistical Yearbook of Forestry*, 468 pp., 2001.

Korea Forest Service, *Green Korea*, 2002.

Korea Meteorological Administration, *Yearbook of Meteorology*, 2002 (in Korean).

Korea Ministry of Environment, *National Biodiversity Strategy*, 73 pp., 1997.

Lee, C.Y., *Traditional Knowledge for Soil Erosion Control in the Republic of Korea*, Korea Forest Research Institute, No. 184, 110 pp., 2001.

Lee, D.K., Kang, H.S., and Park, Y.D., Natural restoration of deforested woodlots in South Korea, in *Proceeding of International Seminar on Restoration Research on Degraded Forest Ecosystems*, 26, 2001.

Lee, D.K. and Park, Y.D., Nitrogen fixing woody plants in Korea, in *Proceeding of Korea–China International Joint Seminar on the Nutrient Cycling and Biomass Productivity in Forest Ecosystem*, 6, 1999.

Lee, J.H., The Korean model of community development — a case of Saemaul Undong, *Journal of SNU Saemaul Studies*, 4, 1, 1984.

Lee, K.J., Major factors causing disturbance in forests and its effects on forest ecosystem in Korea, in *Ecology of Korea*, Lee, D.W., Ed., Bumwoo Publ. Co., Seoul, 105, 2002.

Matsushima, S., Betrachtung zur Waldentwicklung in Korea auf grund von pollenststistik, *Journal of the Japanese Forestry Society*, 23, 441, 1941 (in Japanese).

Nelson, S.M., *The Archaeology of Korea*, Cambridge University Press, Cambridge, 1993.

Oh, C.Y., A pollen analysis in the peat sediments from Pyungtack country, Korea, *Korean Journal of Botany*, 14, 126, 1971 (in Korean).

Park, K.S., Change and establishment of Gumsan, Bongsan system in the Chosun dynasty, Korea Forest Research Institute, *Journal of Forest Science*, 57, 86, 1998 (in Korean).

Park, W.K, Tree-ring patterns of subalpine conifers in Korea: growth-trend analysis and climate reconstruction, in *Ecology of Korea*, Lee, D. W., Ed., Bumwoo Publ. Co., Seoul, 66, 2002.

Urbanska, M.K., Webb, N.R., and Edwards, P.J., *Restoration Ecology and Sustainable Development*, Cambridge University Press, Cambridge, 397 pp., 1997.

United Nations Development Programme, *Agricultural Recovery and Environmental Protection (AREP) Programme*, Working Papers, Vol. 3, 1998.

Yamazaki, T., Beitrage zur verwandlung der baunarten im sudlichen teile von Korea durch die pollenanalyse, *Journal of the Japanese Forestry Society*, 22, 73, 1940 (in Japanese).

Yasuda, Y., Tsukada, M., Kim, C.M., Lee, S.T., and Yim, Y.J., *History of Environmental Changes and Origin of Agricultural in Korea: History of Environmental Changes in Korea.* Nippon Monbushyo Kaigai Gakujutsu Report No. 404332, 1, 1980 (in Japanese).

Yim, K.B., *Principles of Silviculture*, Hyangmun Sa, 492 pp., 1985.

chapter twenty-six

Rehabilitation of temperate forests in Iran

Khosro Sagheb-Talebi
Research Institute of Forests and Rangelands

Contents

26.1 Introduction

Iran is a Middle Eastern nation that covers an area of 1,648,000 km^2 between the latitudes of 25°N and 40°N. Iran lies between the Caspian Sea, the Persian Gulf, and the Sea of Oman. Its geological past and geographical location resulted in open landscapes with vast mountain ranges, forests, wide plains, uncultivatable deserts, fertile plains, and large swamps (Figure 26.1). Deciduous forests that were destroyed by glacial advances in Europe and northern Asia survived in Iran. The Caspian forests are representative of forests from the Cenozoic period. The uplift that created the Zagros Mountains in the south and west of Iran increased the elevation of the Alborz and Zagros ranges and blocked rain-bearing clouds from the Mediterranean from reaching the interior plateau, creating the central deserts (Darreshouri and Kasraian 1998). Elevations vary from below sea level on the Caspian plain to mountain peaks over 5,000 m. Iran's climate is generally Mediterranean with some continental influence. Rainfall generally decreases from north to south and from west to east. As a result of these major topographical, geological, and climatic variations, Iran is rich in plant diversity. The country supports a total of around 8,000 plant species, and is one of the major centers of endemism in this part of the world (Jalili and Jamzad 1999). The Iranian plateau is a contact point of five phytogeographical regions: the Touranian, European, Indian, Mediterranean, and African (Sabeti 1976).

1-56670-635-1/05/$0.00+$1.50

Figure 26.1 Map of Iran showing the Alborz and Zagros Mountains, as well as the central desert region.

The total forest area of Iran is approximately 12 million ha, which covers only 7% of the total land area. Under UN and FAO definitions (Ball 1999), Iran is a Low Forest Cover Country (LFCC). Forests are divided into three ecological zones: western (the Zagros) with sub-Mediterranean characteristics and semiarid climate; central and southern parts with steppic arid and subtropical climate; and northern (the Caspian) with humid climate. In the western ecological zone, forest vegetation covers some 5 million ha of scattered stands of mainly oak (*Quercus brantii* Lindl., *Q. infectoria* Oliv., *Q. libani* Oliv.), wild pistachio (*Pistacia atlantica* Desf. subsp. *mutica* (F.&M.), *P. khinjuk* Stocks), and almond (*Amygdalus* spp.). There are also 5 million ha of forests in the central and southern regions of the country, mainly of drought-adapted species such as *Tamarix* spp., *Haloxylon* spp., *Pistacia* spp., and *Ziziphus* spp. The temperate commercial deciduous forests, with an area of almost 2 million ha, are located in the Caspian region of northern Iran. This humid area ranges from the southern coastal plain of the Caspian Sea to the northern slopes of the Alborz Mountains, at elevations ranging from 20 m below to 2,500 m above sea level.

26.2 Historical background

Civilization in Iran is more than 4,000 years old and has greatly impacted Iranian forests. Wood has been used for fuel and construction. Forests were cleared for agriculture, grazing, and urban and industrial areas. It is estimated that forests once covered 18 million ha, a loss of one third of the total forest area. Nearly all the natural vegetation of the valleys and cultivatable plains has been cleared for farming and fruit production. These fertile and easily irrigated areas are traditionally used for growing rice, tea, and tree crops such as orange and kiwi in the Caspian region, as well as wheat, barley, peach, apricot, almond, and walnut in the Zagros region. Until 1962, forest ownership was private and landowners and village inhabitants were the main forest owners. The government nationalized all forests in that year. Nationalization has caused problems of forest protection and governmental management of the forests because of a weakened sense of responsibility among the people and resistance by landowners. Nevertheless, the public is generally concerned by the decrease in woodland area and of trees in and around rural settlements, and supports and encourages forest restoration (Sagheb-Talebi et al. 1999).

The general objectives of forest restoration in Iran are to replace the loss of natural forests and planted forest cover, and to introduce forest to sites that never supported them, or that have been deforested for a long period. Besides improving degraded natural forest ecosystems, forest restoration focuses on afforestation to restore soil fertility, control erosion, and provide services and goods no longer available from natural forests, including fuelwood, fodder, nonwood products, and industrial wood. Ideally, these new plantations will be achieved with a clear vision of their economic, ecological, environmental, and landscape roles. They are intended to ease human and animal pressure on natural forest areas and to contribute to global ecological benefits such as gene conservation and biological diversity, as well as carbon sequestration. Except for the semidry and dry-steppic areas and deserts in the central part of the country, most of the recent plantations are concentrated in the northern and western parts of Iran. In the rest of the chapter, the Caspian and Zagros regions will be treated in more detail.

26.3 Natural forest communities

26.3.1 The Caspian region

The Alborz Mountains begin with the Azerbaijani frontier ranges (Caucasus) in the northwest and extend northeast, not far from the border of Turkmenistan (Figure 26.1). This high mountain chain forms more or less an unbroken wall, with elevations over 5,000 m, and receives most of the precipitation from the Caspian and Black Seas. Rainfall is evenly distributed over the year. The climate in the western part of the Caspian region is very humid, with cold winters and without a dry period; the annual rainfall is 2,000 mm and the mean annual temperature is 15°C. The eastern part is humid with mild winters and a short dry period (Figure 26.2); the annual rainfall is 600 mm and the mean annual temperature is 18°C. The growing season lasts between 7 and 9 months (Sagheb-Talebi and Dastmalchi 1997).

Massive sedimentary limestone and other calcareous rocks form the Alborz range, except for the small Asalem area in the northwest, with acidic bedrock. Soil types vary from rendzina and calcareous brown forest soils in good drainage sites to hydromorphic soils in very humid sites.

Mesophilic (subhumid) forest vegetation, originally of the Tertiary, called the Hyrcanian zone, characterizes the Alborz Mountains on the Caspian coast (Mobayen and Tregubov 1970). The Hyrcanian forests are one of the last remnants of natural deciduous forests in the world. Native and endangered tree species include honey locust (*Gleditsia caspica* Desf.), false walnut (*Pterocarya fraxinifolia* Lam. [Spach.]), ironwood (*Parrotia persica* C.A.Meyer), Siberian elm (*Zelkova carpinifolia* [Pall.] Dipp.), box tree (*Buxus hyrcana* Pojark), Caspian poplar (*Populus caspica* Bornm.), and a few conifer species such as yew (*Taxus baccata* L.) and *Thuja orientalis* L. (Sagheb-Talebi 2000). The natural forest vegetation is temperate deciduous forests containing broadleaved species such as beech

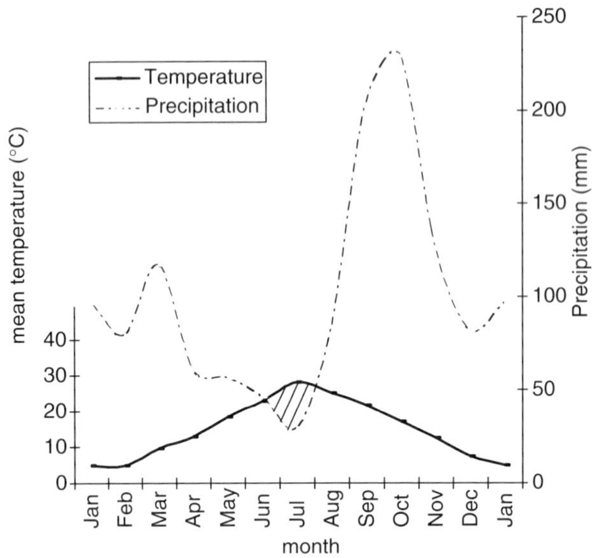

Figure 26.2 Climogram of the central parts of the Caspian region (based on Gaussen system).

(*Fagus orientalis* Lipsky), oak (*Q. castaneifolia* C.A.M, *Q. macranthera* F. & M.), hornbeam (*Carpinus betulus* L.), maple (*Acer velutinum* Boiss., *Acer cappadocicum* Gled.), ash (*Fraxinus excelsior* L.), alder (*Alnus subcordata* C.A.M., *Alnus glutinosa* Gaertn.), elm (*Ulmus glabra* Huds.), wild cherry (*Prunus avium* L.), wild service tree (*Sorbus torminalis* Crantz), and lime tree (*Tilia platyphyllus* Scop.). The main feature of the region is the lack of conifers; only relics of coniferous forests are present, which include *Taxus baccata* L., *Juniperus* spp., *Cupressus simpervirens* var. *horizontalis* L., and *Thuja orientalis* L.

On the plain, oak (*Q. castaneifolia*) is mixed with box tree, which forms the *Querco–Buxetum* community. On the lower slopes below 700 m, the oak and hornbeam are increasingly mixed with ironwood (*Parrotia persica*) and form the two communities of *Querco–Carpinetum and Parrotio–Carpinetum*. Between 700 and 1,500 m, beech is the dominant tree species and forms the *Fagetum hyrcanum* in pure and mixed stands with other noble hardwoods over a vast area in this cloudy zone. From its floristic composition, these beech forests are linked with European forests and with affinities to the beech forests of the Balkans. However, local conditions of aspect and edaphic factors, such as soil moisture and depth, are all of importance in determining the composition of the vegetation, which leads to the establishment of different beech subcommunities. Above the beech belt, oak (*Q. macranthera*) and hornbeam (*Carpinus orientalis*) build the *Querco macranthere–Carpinetum orientalis*, the forest community of higher elevations to the forest limit with short trees or shrubs.

26.3.2 The Zagros region

The Zagros Region consists of the western and southern slopes of the Zagros Mountains, which run from the northwest to southeast, from the Turkish border to the Persian Gulf (Figure 26.1). Elevations range from 200 to 2,000 m and the mean annual precipitation varies between 250 and 800 mm. Rainfall occurs only during the winter and spring, which causes severe summer droughts of 4 to 8 months' duration. Summers are hot and winters are cold. The mean annual temperature varies from 11°C in the northwest to 25°C in the southwest. Soils are of limestone origin with a pH of 7.5 to 8.5. Soil conservation problems in the region are serious and the local population has largely destroyed the natural forests for firewood and conversion to agriculture. The remaining natural forest is dominated by open stands of deciduous oak with other indigenous species; it grows slowly and has hardly any natural regeneration (Webb 1973).

The climax vegetation is an open xerophytic cold-resistant deciduous forest of oak (*Q. brantii* = *Q. persica*), the *Quercetum persicum* community, which dominates between 1,000 and 2,000 m elevation. Natural oak on good sites reaches a height of 12 to 15 m with a canopy cover of 50%. Stands in accessible areas, however, have been reduced to very open stands of deteriorating coppice stools, only 2 or 3 m in height. At higher elevations, scattered trees of *Juniperus excelsa* occur, and on the lower slopes a mixture of oak with almond, pistachio, and maple (*Acer monspessulanum* L.) appears.

26.4 Restoration activities

Forestry in Iran is in its infancy; before the Forest and Range Organization (FRO) initiated forest management in 1960, forests were exploited and there was very limited regeneration. In 1965, FRO initiated a considerable program of nonirrigated plantations in rural areas, and irrigated plantations around large towns and cities. A good example is the greenbelt around Tehran (Figure 26.3) using drought-resistant conifers and deciduous species (*Pinus eldarica* Medw., *Cuppressus arizonica* Greene, *Ailanthus altissima* [Mill.] Swingle and *Fraxinus rotundifolia* Mill.). The plantations were initially based on traditional knowledge, but beginning in the 1970s forestry research has supported plantings with spacing experiments using native species and fast-growing exotic species, mainly pines, poplars, and eucalyptus. Species selection for reforestation has been based on indigenous broadleaves; exotic species are used for afforestation. The system of afforestation trials with exotic species consists of three phases beginning with short-term elimination trials with 100 trees. Long-term growth trials consist of 500 trees and the results are used to select species to be introduced in large-scale plantations. All levels of trial are replicated to allow statistical analysis (Webb 1973). Higher priority is given to rare and noble native broadleaved species than exotics.

Not all experiments with exotic species prove satisfactory. Establishment of single- or limited-species mixtures (monocultures) over extensive areas may intensify some local

Figure 26.3 Afforestation with *Pinus eldarica* around Tehran.

pest problems, primarily if pest populations build up rapidly to high levels (Nyland 1996). For example, outbreaks of *Lymantria dispar* totally damaged large-scale monocultures of exotic pine species (*Pinus elliottii* Engelm. and *Pinus radiata* D. Don) in the western Caspian plain. Foresters use caution in selecting exotic species and avoid large plantations of those not previously tested in the region.

26.4.1 Nursery and plantation techniques

Degraded forests in the Caspian region potentially available for reforestation or afforestation total 600,000 ha, and about half has been restored, including reforestation on 240,000 ha. Afforestation of 50,000 ha includes 30,000 ha planted with poplar species in the Caspian plain. The 35 permanent and temporary, private and governmental nurseries produce 40 million seedlings annually. Depending on tree species, both bareroot and container stock are used of different sizes and ages. Usually, containers have 1 or 2-year-old seedlings; containers are 10 × 30 cm or 20 × 40 cm. Vegetative propagation of fast-growing species such as poplars has introduced new species or clones for widespread planting on suitable sites.

Some of the species produced in the nurseries and planted in the Caspian region include *Acer velutinum, Acer cappadocicum, Alnus subcordata, Alnus glutinosa, Eucalyptus* spp., *Fraxinus excelsior, Juglans regia, Populus deltoides* Marsh. (several clones), *Populus* × *euramericana* Dode (Guinier) (several clones), *Q. castaneifolia, Abies* spp., *Cedrus deodara* G. Don, *Cryptomeria japonica* D. Don, *Cupressus arizonica, Cupressus simpervirens, Juniperus excelsa, Juniperus virginiana* L., *Larix decidua* Mill., *Picea abies* Karst., *Pinus eldarica, Pinus elliottii, Populus nigra* (several provenances), *Pinus ponderosa* Douglas ex Lawson, *Pinus radiata, Pinus sylvestris* L., *Pinus taeda* L., and *Taxodium distichum* L.C.Rich. In the Zagros region, the main species are *Ailanthus altissima, Amygdalus* spp., *Cupressus arizonica, Cupressus simpervirens, Eucalyptus* spp., *Juglans regia, Pinus brutia* Ten., *Pinus eldarica, Pinus nigra* (several provenances), *Platanus orientalis* L., *Populus alba* L., *Populus nigra* L., *Q. brantii*, and *Robinia pseudoacacia* L.

Sometimes, seedlings are transplanted directly from the nursery bed into the plantation (1-0 or 2-0 stock). More commonly, seedlings are moved from the seedbed to a transplant bed for one or two growing seasons, after which they are lifted and planted in early to late winter. Direct sowing in forest sites may be used to fill gaps or to stabilize soil against landslides (using maple or alder). Soil preparation, consisting of weeding and digging pits, may be done manually or by machine (Shrivasta 1997). The general surface topography is usually improved by contouring, terracing, and leveling in steep terrain. Soil structure is improved in some cases in order to promote drainage or improve potential for root penetration, as well as enabling sufficient water supply in subhumid or subdry sites. On flat sites with a high water table in the Caspian plain, pioneer planting with indigenous alder or exotic baldcypress is used to improve soil drainage within the rooting zone. This alters many important hydrologic characteristics of a site and changes the vegetation community to species that grow in better-drained soil (Nyland 1996). Spacing depends on species, growth rate, and light demand (shade tolerance). Planting pattern is varied, as line, square, and triangle patterns are all used. Plantations are both pure and mixed, in line and block mixtures. Usually, fencing encloses the plantation area to exclude cattle and game and to protect the seedlings and young plants from grazing.

26.4.2 Replacement with exotic species

In 1972, about 1,000 ha of degraded hardwood forest in the Pilambra district of Gilan Province in the northwestern Caspian plain was replaced with an exotic conifer plantation. Elimination (adaptation) trails in the 1960s indicated that loblolly pine (*Pinus taeda* L.) was a promising species among the exotic conifers for the northwestern Caspian plain (Dastmalchi and Sagheb-Talebi 1997). The area is located at sea level; the soil is deep to the rock and heavy textured with poor internal drainage. The mean annual precipitation and mean annual temperature is 1257 mm and 18°C, respectively. The frost period lasts only 18 days, and pronounced dry periods are rare. The degraded oak and hornbeam forest was clearcut, and in late winter 1972 (February and March) 1-year-old seedlings of loblolly pine were planted at 2 × 3 m spacing and protected from browsing by fencing.

At age 17, survival was 80% (Figure 26.4). Mortality was due to competition as well as pest, snow, and wind damage (Gorji-Bahri 1993). The mean breast height diameter was 20.7 cm, the height was 14.5 m, and the total volume was 279 m^3 ha^{-1}. The loblolly pine has begun to regenerate naturally in this artificial stand, which suggests that wood production in the region may be increased by sustainable management of exotic forest stands.

26.4.3 *Afforestation with exotic species*

Poplars are fast-growing species suitable for afforestation on nonforest areas and former farmlands. For more than 30 years, exotic poplar species and clones (*Populus deltoides* and *Populus* × *euramericana*) were imported and tested together with native poplar species (*Populus caspica* Bornm., *P. alba*, and *P. nigra*). Research in different parts of the country has examined adaptation, spacing, and yield of poplars and has supported successful introduction of poplar clones to landowners in the Caspian region. Large privately and

Figure 26.4 A 17-year-old loblolly pine (*Pinus taeda* L.) stand established in the northwestern Caspian region.

publicly owned plantations have been established, which provide pulpwood for paper and roundwood for carpentry and other industrial purposes.

An experimental planting of ten different hybrid poplar clones illustrates the poplar plantations in this region (Figure 26.5). Soil was a fine-structured, alluvial, sandy-clay with a basic reaction (pH 7 to 8). The mean annual precipitation was 1,186 mm and the mean annual temperature was 17.5°C. The region has 108 rainy days and a dry period lasting two months. Frost occurs over a 14-day period. Hybrid poplar cuttings were grown for one year in a nursery before they were planted at 2 × 2 m and 4 × 4 m spacing in three replications. Preliminary results after 15 years showed that two clones of *Poplus × euramericana* (I-214 and I-488) were the most promising, with the lowest mortality and highest yield (Table 26.1). The highest standing volume was obtained by *Populus × euramericana* I-214 (408 m^3 ha^{-1}) at the 2 × 2 m spacing, and the mean annual volume increment after 15 years was 27 m^3 ha^{-1} $year^{-1}$ (Ziaie-Ziabary et al. 1988).

Moreover, the results showed that the wood produced by *Populus × euramericana* I-214 could be used in industrial applications. At the narrower spacing (2 × 2 m), 13% of the produced wood was suitable for carpentry and 87% for paper pulp and particle board. Larger diameter material was produced at the wider spacing (4 × 4 m), from which 26% was veneer wood and also suitable for match production, 62% for carpentry, and 12% for paper pulp. This has encouraged public and private managers to plant poplars and harvest them in short or long rotations.

Figure 26.5 Poplar plantation in the Caspian region.

Table 26.1 Quantitative Characteristics of Two Hybrid Poplars Clones (*Populus* × *euramericana* I-214 and I-488) 15 Years After Planting in the Caspian Region

Clone	Spacing (m)	N ha^{-1} (including mortality)	Mean DBH (cm)	Mean height (m)	Volume of mean tree (m^3)	Yield (m^3 ha^{-1})	Volume increment (m^3 ha^{-1} $year^{-1}$)
I-214	2 × 2	2,000	17.8	20.1	0.204	407.6	27.2
	4 × 4	500	25.6	25.6	0.552	276.1	18.4
I-488	2 × 2	2,000	15.2	19.8	0.168	336.0	22.4
	4 × 4	500	22.6	23.5	0.441	220.5	14.7

Source: Ziaie-Ziabary et al., Impact of scaling in production of Poplar plantation, Kimia, *Journal of Iranian Chemical on Cellulosical Industries*, 1, 31, 1988 (in Farsi; English Summary).

26.4.4 Mixed plantings on alluvial sites

Degraded alluvial forests around the cities in the Caspian region are suitable for plantations, including the Mashlak area in the central part of the Caspian plain. The mean annual precipitation and mean annual temperature is 1345 mm and 15.8°C, respectively. There is a short dry period. Soils are shallow alluvial, including skeletal soils with 39 to 59% calcareous stone fragments. During the rainy season, the site is waterlogged and thus unsuitable for agriculture (Sagheb-Talebi 1990). Both pure and mixed plantations of poplar (*Populus* × *euramericana* I-214), alder (*Alnus glutinosa*), and cypress (*Cupressus simpervirens* var. *horizontalis*) were established at 2 × 4 m spacing. Poplar cuttings and seedlings of alder and cypress were planted; all material was 1-year-old. Stock of the different species was planted alternately in lines.

Twenty years after planting, the highest rate of mortality was observed among the poplars, which varied between 6 and 10% in pure and mixed plantations, respectively. Alder had the lowest mortality. The mixed plantations formed two-storied stands, with the poplars in the upper layer and alder or cypress in the understory. Poplar and alder in mixture grew best; poplar had a maximum diameter of 18 cm and a height of 21.6 m. The mean diameter and height of alder was 16.5 cm and 16.2 m, respectively. These mixed plantations of alder and poplar were the most productive with a standing volume of 238 m^3 ha^{-1} and an annual volume increment of 12 m^3 $ha^{-1}year^{-1}$ (Sagheb-Talebi 1996). Compared to other sites, the production of these plantations is three times more than similar natural vegetation covers. This example shows that very high productivity rates are possible using appropriate tree species. The ability of alder to fix nitrogen can improve the physical and chemical properties of the soil and seems to improve the growth of the poplar.

26.4.5 Floodwater spreading and afforestation with exotic species

Iran is a pioneer among the countries of the region in controlling desertification through floodwater spreading. Uneven distribution of precipitation during the year and unexpected heavy rainfalls in the Zagros region lead to heavy floods, which bring millions of cubic meters of water onto the land in a short period. This huge amount of water usually destroys villages and agricultural lands along with other damage. Floodwater spreading is an inexpensive way to reduce the destructive effects of flashfloods and to produce other benefits from reclaiming eroded land (Kowsar 1991). A successful example of floodwater spreading in Iran is in the Gareh-Bygone plain, located in the Fars province in the southeastern Zagros region. The area floods three to seven times annually, with the diverted volume of water ranging from 0.1 to 15 million m^3 $year^{-1}$. Over a 6-year period (1983 to 1988), a cumulative water volume of about 38 million m^3 was diverted. A channel network is used to spread the flowing water onto the land plain. Channels are spaced 140 to 250 m apart. When floodwater reaches the end of a floodwater spreading system, it has lost most of its sediment load and is now suitable for artificial recharge or storage in surface reservoirs (Kowsar 1991).

Plantations of fast-growing Australian species have been established on the Gareh-Bygone plain (Mortazavi 1994). Seedlings of *Eucalyptus camaldulensis* (Dehnh.), *E. microtheca* (F. Muell.), *Acacia cyanophylla* (Lindl.), *Acacia salicina* (Lindl.), and *Acacia victoriae* (Benth.) were planted adjacent to the channels. The planting lines were ripped to a depth of 35 cm and the seedlings were planted at 3 m spacing. Perforated polyethylene bags (10 × 30 cm) with 9-month-old seedlings were used. A total of 30,000 seedlings were planted and protected from browsing for 9 months. Nine-month-old seedlings of quail bush (*Atriplex lentiformis* [Torr.], [Wats]) were planted at 3 m spacing in contour furrows, with a distance of 4 m between the furrows. Each seedling was irrigated with 2 l of water immediately following planting, and the plantations were irrigated by every flood event (Kowsar 1991). Eight years after establishment, the survival of *E. camaldulensis* was above 85%. All mortality occurred within the first year. The average height and diameter at breast height was 10.7 m and 12.2 cm, respectively. The total average yield was 10 m^3 ha^{-1} $year^{-1}$. Stem wood was approximately 75% and fuelwood was 25% (Kowsar et al. 1996).

Floodwater spreading and tree and shrub planting have transformed a wasteland and stabilized drifting sands. Introduction of quail bush (*Atriplex lentiformis*) and invasion of *Atriplex leucoclada* (Boiss.), gray-leaved sagebrush (*Artemisia sieberi* [Besser]), and other forage species have added more palatable fodder to a heavily grazed area. Growth of quail bush, which performs poorly on sands, was spectacular in sedimentation basins; 3-year-old bushes have grown to a height of 2 m and a crown diameter of 2.5 m. The return of a few hubara bustards (*Chlamydotis undulate*) and gazelles (*Gazella subgutturosa*) to their former haunts is an auspicious sign for the total recovery of the Gareh-Bygone plain. Human needs have been met with provision of irrigation water for 1,000 ha of cropland through 40 new and 16 old wells. The project has substantially raised the income of the four farming communities in the plain and the yield of 650 ha of flood-irrigated farmland has doubled (Kowsar 1991).

26.5 Conclusion

Iran is a nation of inherently low forest cover due to climatic limitations. Nevertheless, in several regions there are important forest resources. Thousands of years of grazing, fuelwood cutting, and conversion to agriculture have reduced the forest cover further and degraded the natural forests. Restoration involves a variety of activities, often utilizing exotic species that can better tolerate the rigorous growing conditions than native species. Replacement and afforestation with exotic species are restoring forest cover to this dry, open landscape. Reclamation of waste areas to control flash flooding using floodwater spreading is better suited to this environment than structural flood control measures, with the added benefits of irrigating both tree crops and agricultural crops.

References

Ball, J., Overview of low forest cover countries in developing region, in *Proceedings of the International Meeting on Special Needs and Requirements of Developing Countries with Low Forest Cover and Unique Types of Forests*, Mirsadeghi, S. M., Eds., Forest and Range Organization, Tehran, 190, 1999.

Dastmalchi, M. and Sagheb-Talebi, Kh., *Results of Elimination Trials (Needle-Leaved Species) in Gilan Province*, Part II, Technical Publication No. 168, Research Institute of Forests and Rangelands, Tehran, 136, 1997 (in Farsi; English abstract).

Dareshoori, F. and Kasraian, N., *Nature of Iran*, Rouzaneh Kar, Tehran, 198, 1998.

Gorji-Bahri, Y., Study of growth of Loblolly pine (*Pinus taeda* L.) in the Gilan province, *Research and Reconstruction*, 20, 34, 1993 (in Farsi; English abstract).

Jalili, A. and Jamzad, Z., *Red Data Book of Iran*, Publication No. 215, Research Institute of Forests and Rangelands, Tehran, 747, 1999.

Kowsar, A., *Floodwater Spreading for Desertification Control: an Integrated Approach, An Iranian Contribution to the Implementation of the Plan of Action to Combat Desertification*, Desertification Control Bulletin (UNEP), 19, 3, 1991.

Kowsar, A., Mortazavi, S.M., and Niazi Ardakani, M., *Eight-Year Performance of* Eucalyptus camaldulensis *in Waterspreaders in a Sandy Desert*, Desertification Control Bulletin (UNEP), 29, 35, 1996.

Mobayen, S. and Tregubov, V., *Guide Pour la Carte de la Vegetation Naturelle de l'Iran*, Bulletin No. 14, Universite de Tehran, Projet UNDP/FAO, IRA 7, 20, 1970.

Mortazavi, S.M., *Results of Eucalyptus Species Elimination Trials in Fars Province (Southern Iran)*, Technical Publication No. 99, Research Institute of Forests and Rangelands, Tehran, 71, 1994 (in Farsi; English abstract).

Nyland, R.D., *Silviculture; Concepts and Applications*, 1st ed., McGraw-Hill, Singapore, 633 pp., 1996.

Sabeti, H., *Forests, Trees and Shrubs of Iran*, Ministry of Agriculture and Natural Resources of Iran, Research Organization of Agriculture and Natural Resources, 810, 1976 (in Farsi; English summary).

Sagheb-Talebi, Kh., Climatological Condition of Noshahr District, *Pazhoohesh, Research in Science and Technology*, 9, 17, 1990 (in Farsi; English abstract).

Sagheb-Talebi, Kh., Study of pure and mixed plantations of Alder, Cypress and Poplar on the alluvial soils of Mashlak river (Noshahr) in Mazandaran province, *Research and Reconstruction*, 30, 100, 1996 (in Farsi; English abstract).

Sagheb-Talebi, Kh., Hyrcanian Forests (North of Iran), the unique ecosystem in Near East Region, in *Proceedings, XXI World Congress — Forests and Society: The Role of Research*, Kuala Lumpur, 8, 2000.

Sagheb-Talebi, Kh. and Dastmalchi, M., *Results of Elimination Trials (Broadleaved species) in Gilan Province*, Part I, Technical Publication No. 168, Research Institute of Forests and Rangelands, Tehran, 136, 1997 (in Farsi; English abstract).

Sagheb-Talebi, Kh., Sardabi, H., and Ball, J., Role of planted trees and forest plantations in Low Forest Cover Countries (LFCC), in *Proceedings of the International Meeting on Special Needs and Requirements of Developing Countries with Low Forest Cover and Unique Types of Forests*, Mirsadeghi, S.M., Ed., Forest and Range Organization, Tehran, 190, 1999.

Shrivasta, M.B., *Introduction to Forestry*, 1st ed., Vikas Publishing House, New Delhi, 262 pp., 1997.

Webb, D.B., *The Introduction and Trial of Exotic Tree Species in the Semi-Arid Zone of Iran*, Internal report, Research Institute of Forests and Rangelands, Tehran, 180 pp., 1973.

Ziaie-Ziabary, S.F., Gorji-Bahri, Y., and Sagheb-Talebi, Kh., Impact of spacing in production of poplar plantation, *Kimia, Journal of Iranian Chemical and Cellulosical Industries*, 1, 31, 1988 (in Farsi; English abstract).

chapter twenty-seven

Restoration practices in Brazil's Atlantic Rain Forest

Jorge Paladino Corrêa de Lima
Universidade Federal Rural do Rio de Janeiro and US Forest Service
Maísa dos Santos Guapyassú
Fundação O Boticário de Proteção à Natureza
Gunnars Hauff Platais
World Bank

Contents

27.1 Introduction

The Atlantic Rain Forest (Mâta Atlantica) extends along the southern coast of Brazil and inland into Argentina and Paraguay. Originally covering 15% of the land area of Brazil, it was a region of an estimated 1.3 million km^2 (MMA 2000). Today, remnants of the Atlantic Forest represent about 8% of the original area, or some 94,000 km^2 in Brazil

(Desmatamentozero 2001). Isolated from other major forest ecosystems in South America, the Atlantic Forest is a diverse mix of vegetation and forest types with a high rate of endemism. The two main ecoregions of the Atlantic Forest include the coastal forest, extending in a narrow strip only 50 to 100 km wide in northern Brazil. The interior Atlantic Forest stretches across the foothills of the Serra do Mar and related mountain ranges. These inland forests reach as far as 600 km from the coast and range in elevation as high as 2,000 m. The coastal forest has been the most disturbed, with only about 3% of its original forest cover remaining. The best-preserved areas of the Atlantic Forest in Brazil survive in relatively inaccessible areas of the states of São Paulo, Paraná, and central Rio de Janeiro. The largest contiguous stretch of Atlantic Forest, from Serra da Juréia in Iguape/SP to Ilha do Mel in Paranaguá/PR, was declared a Biosphere Reserve by UNESCO in 1991 and a World Heritage Site in 1999. This stretch covers an area of about 1.7 million ha with elevations ranging from sea level to 1,400 m. Half of Misiones Province in Argentina remains in forest, and 13% of the original interior Atlantic Forest survives in Paraguay.

The Atlantic Forest has been disturbed for 500 years; it has been logged and cleared for sugarcane and coffee cultivation, livestock breeding, and to provide industrial fuel. The development of sugarcane, coffee, and cacao for export was key to Brazil's economic and population growth from early colonial times until the mid-20th century. Early in the 19th century, large areas of the forest were burned and converted to coffee plantations. A steel industry developed early in the 20th century, expanding the demand for charcoal. Half the forests of Minas Gerais state were destroyed between 1920 and 1940, replaced today by extensive plantations of exotic *Eucalyptus* spp. Expanding human population and conversion to urban areas has been most intensive since 1960. Most timber production in the Atlantic Forest is exploitive and has fragmented and degraded the remaining forest (Dean 1995). Nevertheless, some forest remnants retain great diversity with as many as 450 different tree species per ha.

Despite legal protection aimed at conserving the forest, anthropogenic pressure continues to impact the Atlantic Forest. The relevant legislation (Decree 750/93) allows tree felling and the suppression and exploitation of Atlantic vegetation only under certain circumstances. Nevertheless, the most pressing need is to protect the remaining areas of forest from human encroachment. Restoration goals are to restore and expand the area of remnant forests and provide ecological corridors between blocks of remnant forest. Most restoration activity in Brazil is small, associated with remediation of construction projects or as partnerships on private land between conservation NGOs, government agencies, and private foundations. In the 1980s, hydroelectric companies attempted small-scale restoration to compensate for natural vegetation destroyed by construction of dams and reservoirs (Souza and Batista 2004). In general, restoration relies on natural regeneration and successional processes, augmented by planting. Small nurseries are often developed to provide planting material from local seed sources.

27.1.1 Ecological summary

The Atlantic Forest is located in the escarpment of the Serra do Mar spanning over 23° of latitude along the coast of South America. Over the years, there has been no common definition of the Atlantic Forest. A broad definition of the Atlantic Forest Domain in Brazil was developed in the early 1990s and approved by the National Council of the Environment in 1992, based on botanical and geological considerations together with environmental conservation issues (Pinto et al. 1996). Codified through Federal Decree 750/93, the Atlantic Forest Domain includes coastal forest, araucarian forest, deciduous forest and semideciduous inland forest, together with associated ecosystems, such as restinga, mangrove swamps, high plateaus, grassland enclaves, high-altitude swamps, and cerrado (MMA 2000). Climate in the region is bimodal, with two seasons characterized by the amount of rainfall. In the northeast, temperatures average 24°C. Temperatures in the south and southeast are lower and occasionally dip to −6°C. Brief ecological descriptions follow. In this chapter, however, we limit our consideration to the highly disturbed coastal forest.

- Dense ombrophilous forest, or hydrophilous evergreen coastal forest, of dense, tall stands with a typically tropical character, with many vines, epiphytes, tree-sized ferns, and palms. Composition and structure vary with elevation. Soils are sandy and infertile, with a superficial layer of humus in the southern coastal belt, and developed on tertiary or pre-Cambrian rocks in the north.
- Semideciduous seasonal, or subcadocifolia forest, occurring mainly on the high plateaus and inland mountains of São Paulo, Paraná, and Minas Gerais states, and characterized by a dry season. Soils vary from the relatively fertile *terra roxa* (red soils derived from limestone parent materials) to low-fertility sands. Characteristic vegetation is semideciduous arboreal and scrub. In the northeast, a transition from humid marshy forests to the caatinga vegetation includes forest enclaves in subcoastal and high arid regions.
- Mixed ombrophilous or subtropical subcadocifolia forest with Araucarias has a composition similar to the hydrophilous evergreen coastal forest but occurs on deeper, more fertile soils. It has an upper tree stratum of Brazilian pine *(Araucaria angustifolia)*, which grows to an average height of 30 m. Below the pine is another arboreal stratum, about 20 m tall, formed of *Lauraceae*. This forest is distributed continuously, or in isolated clumps in the midst of natural grassland.
- High-altitude grassland (Campos de Altitude) occurs above 1,800 m elevation with characteristic communities of tall mesophylous grasses, in some places xerophilous, interrupted by small heathland areas.
- High-elevation swamps are dense ombrophilous forests, with two arboreal strata and one scrub stratum found on the upper half of eastern slopes in the northeast, where rainfall is high due to orographic effects.
- Restinga is a vegetation formation typical of sandy coastal areas, comprised of widely varying vegetation communities such as sparse grasses, dense forests up to 12 m tall, or swamps with a dense aquatic vegetation.
- Mangrove swamps are located in brackish waters along estuaries and are an important source of nutrients to marine food chains.

27.1.2 *Environmental and economic importance*

The Atlantic Forest is the third most extensive biome in Brazil, after Amazonia and the Cerrado. It makes up 13% of the land base and even though the forest is restricted in area, it has great biodiversity. High rates of endemism remain; the Atlantic Forest has twice the diversity of Amazonia and 45 times the diversity of the North American hemlock forests (Lowe 1993). More than 6,000 of the estimated 20,000 plant species are endemic, 72 of 620 birds, 60 of 2,000 reptiles, and 253 of 280 amphibians (Paglia 2000). The rate of endemism rises to over 70% among palm trees, bromeliads, and other epiphytes. Some 60% of mammals are endemic (160 of 261). For comparison, there are 353 mammal species in the Amazon with five times the area. There are 250 known species of freshwater fish, of which about 56% are endemic (Paglia 2000). Brazil has greater primate diversity than any other country, and 6 of 16 genera and 25 of 77 species are found in the Atlantic Forest. Indeed, international conservation interest in the Atlantic Forest grew out of concern for primates. The black lion tamarin (*Leopardus pardalis*) is one of the most endangered primates in the world.

About 80% of the Brazilian population, distributed over four of the country's five regions, depend in some way on the Atlantic Forest for their livelihood, including ecological services such as maintenance of soil fertility and provision of clean water. The Atlantic Forest contains water sources that supply the large cities and inland communities; it also regulates the climate and provides shelter to hundreds of traditional communities, including indigenous peoples. As much as 70% of Brazil's GDP comes from the area of Atlantic Forest domain (MMA 2000). About 40% of the population relies on medicines derived from forest plants and 20% still use firewood for cooking and heating. The expansion of large cities such as São Paulo is a major cause of deforestation. Nevertheless, there are still relatively large stretches of forest remnant even in the area between São Paulo and Rio de Janeiro (Figure 27.1). The most affected areas are of most concern to conservation groups, the states of South Bahia and Espirito Santo.

Protected areas of Atlantic Forest are limited in scope, encompassing only 2.5% of the original forest, but approximately 35% of the remaining area. There are 14 national parks, 14 federal biological reserves, and 6 federal ecological stations. In combination with state and municipal parks, the area of protected forest is over 28,000 km^2. These areas are in danger of encroachment by landowners who aim to enlarge their areas, by landless peasants who see them as some of the last land in the region not privately owned, and by urban expansion.

27.2 Restoration practices

The Atlantic Forest in several regions of Brazil was reduced to isolated fragments incapable of accommodating fauna that require large territories. Such fragments are often

Figure 27.1 Forest enrichment in the Prainha Ecological Park, Rio de Janeiro, using *Ficus clusiaefolia* (Miq) Schott ex Spreng,

unable to maintain basic ecological processes. Although the top priority should be to halt the processes of deforestation and degradation in forest remnants, it is also important to restore forest landscapes by expanding and rehabilitating remnants, connecting them, and restoring degraded areas. Constraints on restoration include fragmentation of the remaining forest, high costs for restoration, and the need for long-term tending (Ferreira and Bechara 2000).

Until 1989, restoration in Brazil followed two mutually exclusive restoration strategies. The more popular strategy was to establish a "green carpet" of aggressive rapidly growing species, such as the herbaceous plants *Melinis multiflora* and *Brachiaria decumbens*, or trees such as *Eucalyptus* sp. (Griffith et al. 2000). This strategy provided rapid coverage and soil protection and was attractive to government environmental bodies (São Paulo 2000). Not surprisingly, problems arose, such as the need for frequent fertilizer applications, the vulnerability to pests and fire, erosion caused by lack of vegetation coverage, and visual impoverishment (Millar and Libby 1989; Einloft 1997). The second strategy was based on ecological succession or natural invasion (Klein 1994; Willians et al. 1999). This process culminates in vegetation assemblages that are more structured, diverse, and complex than the green carpet of exotic plants (Macedo 1993). This strategy, however, is slow and can leave the surface exposed to erosion for a time.

The current model has two steps: first, a degraded site or area is treated to create conditions favorable to rapid growth and, second, create conditions that enable ecological succession to proceed and attain equilibrium conditions in harmony with the regional landscape in order to guarantee self-sustainability with less need for future interventions. Brazilian legislation imposes conditions on the techniques and species that can be used to restore degraded areas. Permanent Preservation Areas are defined in the Brazilian Forestry Code (Law 4771/65) and only native species can be used for restoration in these areas. They include a 50 m strip around the margins of rivers, lakes, lagoons, reservoirs, and springs; the summits of hills, mountains, and mountain ranges; hill slopes; and restinga (Brasil 1965). Other degraded areas can be restored using exotic or native species. Multiple restoration systems can be used in a given area, depending on the presence and characteristics of existing vegetation. Where forests are present but degraded, combinations of natural and artificial regeneration can be used to rehabilitate the forest. If absent or highly degraded, several systems of planting can be used to restore the Atlantic Forest (Almeida 2000). Restoration designs that incorporate natural regeneration, such as in islands (e.g., discrete patches or clumps of pioneer trees and shrubs scattered across a pasture area), may be more cost-effective than conventional approaches where whole areas are planted (Kolb 1993).

27.2.1 Natural regeneration

Secondary forests result from natural succession following total or partial suppression of primary vegetation. The secondary forest may retain some of the structure and species found in the primary forest (IBGE 1992). Spontaneous natural regeneration of areas near remnant forests is an important part of the restoration of degraded areas. Remnant forest is an invaluable source of germ plasma, indispensable for genetic improvement (Simoes and Lino 2002). The efficiency of spontaneous regeneration depends on the availability of seeds, effective dispersal by wind, birds, and rodents, and the environmental conditions in which the new plant develops. Every plant species has its own optimum niche, which needs to be identified to ensure success (Kageyama et al. 1990). The process can be accelerated by practices to induce natural regeneration and is most useful in areas of slight to moderate disturbance that retain the biotic and abiotic characteristics of the natural forest formations. Areas to be regenerated are isolated from continuing disturbances (Kageyama et al. 1994). Additional practices include control of liana populations (climbing vines) and aggressive species such as *Graminea* spp. (Rodrigues and Gandolfi 1996).

27.2.1.1 Measures to support natural regeneration

Often, the main effort to prevent continuing degradation of a forest remnant is simply to isolate it. The main degradation factor of forest remnants today is fire; repeated fires destroy forest remnants. Surrounding the forest area with pastureland, or building and maintaining firebreaks, are relatively simple but very important measures (Moraes 1998). Clearly, these have a cost that needs to be considered when planning a forest restoration program. Cattle are another factor (Armour et al. 1991) and must be kept out of forest remnants by fencing.

Isolated forest remnants are frequently invaded by mostly exotic *Graminea* spp. The persistence of these plants along the border further encourages wildfires, especially in dry periods. Large masses of woody creepers are common, spreading over and suffocating border trees (Rodrigues and Gandolfi 1996). Controlling or eliminating the *Graminea* by periodic spraying not only prevents fire but also prevents them from competing with new seedlings. Simply attempting to eliminate creepers from disturbed forest fragments could mean a loss of local biodiversity (Morellato 1991). A specific management program can be introduced for excessive growth of some species of creeper (Willians et al. 1999).

Before planting begins, the potential of the seed bank to restore an area needs to be assessed (Seitz and Corvello 1984). Sometimes, plowing or disking can bring viable seeds to the surface from depths of 15 to 20 cm, which, after germinating, will provide partial recovery of the area (Leck et al. 1989). Another possibility is to transfer seed-laden soil from a recently deforested area to the restoration site (Skoglund 1992; Rodrigues and Gandolfi 1996). This method is used in mining areas where the seed bank is removed from an area about to be mined and this soil is used to recover stabilized areas that need to be restored (Campelo 1998; Pederson and Van Der Valk 1989; Seitz 1996).

Seed dispersal by animals is commonplace in tropical forests and an inexpensive way to accelerate restoration is to introduce food sources that attract birds and bats from neighboring forest remnants. This can be achieved by a correct choice of pioneer species, including those that provide a varied diet of fruit and areas for roosting (Guevara et al. 1986; Uhl et al. 1991). Plants known as *bagueiras*, which have berries that are extremely palatable to animals, play a key role in maintaining the dynamic equilibrium of forests and also in restoring degraded areas. Use of such plants can rapidly increase the number of species in a recovery area. Many birds characteristic of open spaces, such as the streaked flycatcher (*Myiodinastes maculatus*, Tyrannidae), the tropical kingbird (*Tyrannus melancholicus*, Tyrannidae), and thrushes (*Turdus* sp., Turdidae), prefer to roost in dry branches that are very common in the region. Distributing dry sticks over degraded areas provides birds with artificial roosting sites, and they drop the seeds ingested during feeding onto the soil (Guedes et al. 1997).

27.2.2 Enrichment

Forests that have been slightly disturbed can be rehabilitated using enrichment plantings. The technique is used in areas currently occupied by short vegetation dominated by early successional species (Rodrigues and Gandolfi 1996). Enrichment is intended to enhance the floristic composition of an area by introducing species of floristic or commercial importance through seeding or planting (Souza and Jardim 1993). Enrichment can increase the density of existing species or introduce new species. Secondary or climax species are usually planted under the canopy of pioneer species (Figure 27.2). To undertake enrichment planting in degraded forests, it is important to understand the ecophysiological demands of the species, particularly with regard to their tolerance of shade. Natural regeneration

Figure 27.2 Tijuca National Park, Rio de Janeiro, is the largest (34.5 km^2) urban park in the world and a reserve of Atlantic forest. In 1861, Pedro II, Emperor of Brazil, expropriated farmland for the park. Created mainly by artificial regeneration using direct seeding, in the beginning there were only 16,075 trees in the park. Manoel Gomes Archer, a major in the National Guard, planted more than 80,000 seeds of species from remnant Atlantic forest with the help of six slaves and paid workers.

can be combined with enrichment plantings where late successional species are introduced into previously disturbed zones, or throughout an area if desirable species are absent as a result of anthropogenic disturbance (Gandolfi and Rodrigues 1996).

27.2.3 Ecological corridors

Ecological corridors are strips of vegetation between forest remnants formed either through natural regeneration or through afforestation. Remnants may be of primary vegetation or secondary vegetation sufficiently advanced in development and height, capable of providing habitat or serving as a transit area for fauna living in the forest fragments (CONAMA 1993). Studies of biodiversity in forest islands reveal a correlation between the fragment size and the number of species that survive. The relationship varies from one locality to another; but, in a 100 ha forest island, reducing the area by 10% will cause 13% of the species to be lost in 50 years, while a 15% loss of area causes 40% loss of species (Iracambi 2001). Iracambi Recursos Naturais Ltda undertook one example of an ecological corridor project. They set up the Iracambi Atlantic Rainforest Research and Conservation Center, situated on a working farm near the village of Rosario da Limeira (lat. 21°00′S, long. 42°30′W) in the southeast of the state of Minas Gerais. The main objective is to link the farm's forest island with neighboring islands to create a continuous stretch of forest. To reduce the effect of fragmentation, corridors of about 20 m wide were established on either side of farm boundaries to connect neighboring fragments. These contain a variety of species known to have flowers and fruits that attract insects and seed dispersal agents (Iracambi 2001).

27.2.4 Artificial regeneration

Afforestation is used in highly disturbed areas that do not preserve any of the biotic characteristics of the original forest formations, typically where the original forest was replaced by agriculture and livestock grazing. Species are introduced in the following chronological sequence: pioneer species, secondary species, and climax species. In most cases, species are planted as seedlings in accordance with several restoration models that are variations on the plantation model (Gandolfi and Rodrigues 1996; Jesus 1997; Barbosa et al. 1999), depending on which species are planted, their spatial arrangement, and density.

27.2.4.1 Species selection

The choice of species for planting depends greatly on the purpose of the plantation. Apart from adaptation to the ecological conditions of the site, the selection should be based on the objective of the restoration activity: that is, whether the forest is intended to reconstitute permanent protection areas (APPs), or merely for the restoration of degraded areas (ADs), or even restoration associated with the production of timber and other forest products (Kageyama et al. 1990). The appropriate goal for disturbed or degraded APPs is a mixed-species planting with the maximum possible diversity of native species. Plantations intended to restore ADs that are not located in APPs can be made with less diversity, even including monoculture of an exotic species (Jesus 1997).

Priority should be given to species typical of the specific environments being recovered (Almeida 2000). This requires a knowledge of the successional process in the region and the autecology of the species to be planted. Certain species can catalyze the secondary succession process more efficiently than others, creating conditions that favor the emergence of spontaneously regenerated species (Parrota and Engel 2001). The availability of silviculture technology for native species is important, including seed production and preparation, production of seedlings, and planting of the forest (CESP 2000). Interaction between basic silvics, characteristics of the area to be restored, and available technology are factors that will decide which restoration models are suitable (Kageyama and Gandara 2000). Restoration costs vary significantly, depending on the method used.

27.2.4.2 Random planting model

The random planting model, or mixed planting of species with no predetermined order or arrangement for the different species in the plantation, is premised on the observation that the propagules of different species fall, germinate, and grow randomly in nature (Nogueira 1977). The key characteristic of this model is that it does not give importance to pioneer or climax species, concentrating more on the so-called "noble" forest species, which are basically the intermediate ones in succession (Kageyama and Gandara 2000). It is therefore a model that ignores differences between the groups of species in ecological succession; it considers all species to be equal when competing with each other, and that there are no differential demands between the species in terms of shade tolerance. One advantage of random planting is that it can proceed even when knowledge and technology are lacking, by planting many species and accepting that the survivors are the best-adapted species (Balistieri and Aumond 1997). One disadvantage of the random planting model is delay in achieving canopy closure. This delay may necessitate tending and further treatments to remove invasive vegetation (Barbosa et al. 1999; Kageyama and Gandara 2000).

27.2.4.3 Succession model

Use of ecological succession principles in mixed-forest plantations is an attempt to simulate natural regeneration (Piña-Rodrigues et al. 1990; Kageyama et al. 1990). The succession model separates species into ecological groups and places them in proximity such that the earliest species in the succession provide enough shade for species in the later stages (Kageyama et al. 1994). The basic concept is that pioneer species give more closed-shade conditions to climax species, while initial secondary species provide partial shade to the later secondaries (Kageyama et al. 1990). Plants in plantations can be arranged in modules or lines. In modules, a central plant of a late-successional species is surrounded and shaded by four or more plants of early-successional species. Line planting can be in alternating lines, with a line of early-successional species next to a line of late successional or climax species. Line planting is recommended for large-scale plantations (tens or hundreds of ha); module planting (which is more precise) is better in small-scale and experimental plantations (Rodrigues et al. 1992; Santarelli 1996). The original recommendations were for pioneer and nonpioneer trees (secondary and late-climax) to be planted in a ratio of 1:1 (Kageyama et al. 1994). Actual practice, however, has been to plant more pioneer trees (de Souza and Batista 2004).

27.2.4.4 Dense plantation

A system of very dense plantations used in the restoration of degraded areas in Japan was introduced to Brazil to restore degraded hillsides in the state of Rio de Janeiro (Piña-Rodrigues et al. 1997). The spacing used is 1 m × 1 m, which is equivalent to 10,000 seedlings per ha. Continuous lines of pioneer species are planted, together with lines containing mixed pioneer and nonpioneer species. As a result, pioneer plants, whose function is to provide shade, surround each nonpioneer. It is also important to ensure homogeneous distribution of species dispersed by biotic and abiotic means. Pinã-Rodrigues et al. (1997) maintained a proportion of 1:6 between nonpioneer and pioneer species, thereby simulating conditions existing in secondary areas, where colonizing species from more open areas predominate.

27.2.4.5 Direct seeding

Direct seeding is an alternative to planting seedlings and can be used when seeds are available in large quantities. It is advantageous when planting seedlings is difficult, such as limited access or lack of nurseries. This method is preferred in mountainous areas subject to soil erosion (Pompéia et al. 1990; Pompéia et al. 1992). Direct seeding can be used to introduce pioneer species into areas with no forest cover, and to introduce nonpioneer species to enrich secondary forests (Kageyama and Gandara 2000). Aerial sowing or broadcast seeding gives good results in areas with abundant and well-distributed rainfall and loosely compacted topsoil (Almeida 2000). In one project on 70 ha of the Serra do Mar, in Cubatão, São Paulo, the operating cost was approximately US$ 386 ha^{-1}. Most of the cost (74%) was labor to collect and prepare the seeds. The total cost per ha was about US$ 451, which included an investment in infrastructure for seed preparation and pelletizing. Other research has shown the need to test the suitability of a species for direct seeding and, most important, the need to use high-quality seed (Parrota and Engel 2001).

27.2.4.6 Cuttings

Cuttings of some forest tree and shrub species can be planted successfully. Several native Atlantic Forest species have given good results from this type of planting, for example, the cedars (*Cedrella fissilis* and *Cedrella odorata*), and various species of gameleira (*Ficus* spp.) (Almeida 2000). Limitations include the need for constant rainfall in the initial stage and irrigation after planting until the cutting produces buds and becomes established (Teixeira 1983).

27.2.5 Restoration in vegetation islands

The implantation of seedlings produced in forest nurseries is a way to generate nuclei capable of attracting greater biological diversity to damaged areas (Reis et al. 2003). The planting of an entire damaged area is especially difficult. One way to reduce the costs of restoration activities is to plant in restricted patches or islands. Such vegetation islands are areas with a few tree species surrounded by bare soil or nonforest vegetation (Eletronorte 1998). Small forest fragments or even isolated trees attract fauna that disperse seeds, thereby accelerating succession in the surrounding area (Guevara et al. 1986). These islands are centers of high diversity and sources of propagules for dispersion to the surrounding area (Willians et al. 1999). Vegetation islands should contain plant species from various stages of succession and include adaptations to pollinating and dispersion processes distributed throughout the year. Vegetation islands can be established by planting pioneer and nonpioneer species together in the islands or by planting nonpioneer species in the islands and pioneer species in the area as a whole (Kageyama and Gandara 2000). The first method is less costly, but the expansion of the island to occupy the entire area will be slower. The second method is more expensive but spreads faster and costs less than completely planting the area, because large numbers of seedlings of pioneer species are used, which costs less to produce.

27.3 Restoration projects

Restoration projects in the Atlantic Forest are small and fragmented, localized efforts. For example, a project by the Instituto Terra near Baixo Guandu is restoring native forest cover to 630 ha of private property (Instituto Terra 2003). The area was recognized as a Private Reserve of Natural Heritage in 1998. Activities are appropriate to the four main vegetation types on the property. In peppertree plantations (*Schinus terebinthifolius* Raddi), trees were thinned and seedlings of native species were planted in the understory. Removing cattle, which allowed herbaceous species to establish and eventually tree seedlings to invade from forest remnants, restored heavily grazed pasture. Secondary forests and shrubs were managed to promote natural regeneration and some native species were planted. A nursery was constructed as part of the project, producing 55,000 seedlings annually. Seeds are collected from forest fragments on the property. Excess seedlings are grown and provided to local landowners and other restoration projects. Since 1998, 309,000 seedlings have been planted on 125 ha (2,472 seedlings ha^{-1}) of abandoned pastureland. Twenty-two ha of secondary forest were cleared of liana and 20 ha of gallery forest have been enrichment planted with 22,000 seedlings (1,100 seedlings ha^{-1}).

Few projects provide details about the cost of the work. Public agencies provide total project costs, with few disaggregating costs by activity. Enrichment planting restored an Atlantic Forest fragment of approximately 50 ha, located in Mata do Convento, in Vila Velha, Espírito Santo, in the southeast of Brazil. The initial planting cost US\$1,890 ha^{-1} and an additional US\$1,503 ha^{-1} was spent on maintenance. Further tending cost US\$990 ha^{-1} on secondary maintenance and US\$660 ha^{-1} on further maintenance (Jesus 1997). Restoration of 10.81 ha following construction of the port of Sepetiba in Rio de Janeiro involved both physical stabilization and vegetation establishment. The total cost over the 13-year period was US\$103,594 or US\$737 ha^{-1} $year^{-1}$ (Valcarcel 1994). The Companhia

Energética de São Paulo (CESP) reported estimated costs of reclaiming areas of soil extraction used in the construction of hydroelectric plants. Costs for physical stabilization and revegetation were estimated to be US\$4,000 ha^{-1}, assuming 5 to 7 years for establishment of the plantation (CESP 2000). Research by the Escola Superior de Agricultura Luiz de Queiroz, under contract to CESP, enabled them to reduce to 2 years the time required for the plantations to become effectively established without the need for subsequent interventions. Costs were reduced to US\$1,500 ha^{-1} using the successional model (Kageyama and Gandara 2000).

27.4 Socioeconomic and policy factors

The importance of restoring and conserving Atlantic Forest remnants is not exclusively a question of maintaining biodiversity. Public support is more likely to come from the awareness of the environmental services indirectly and directly provided by natural systems. Among other things, these relate to the production of water and the maintenance of regularity in the supply, cycling and replacement of nutrients, moderation of climate and regulation of rainfall regimes, containment of erosion processes, and equilibrium in food chains. Nevertheless, there is immense political pressure to maintain the current status quo, especially from large landowners. Further, development, landuse, and natural resource policies are fragmented by sector and Brazil lacks an environmental policy that integrates these policies with urban policies. Agriculture, grazing, urbanization, industrialization, and transportation sectors are viewed and managed in mutual isolation.

Brazilian legislation is very complete on the removal of vegetation and related issues. On the other hand, the responsible are ill-equipped to carry out their functions because of inadequate funding and staffing. Inspection and control of environmental issues was decentralized because of the size of the country and the need for greater flexibility. Although this yields benefits by allowing greater agility and specificity of action, many environmental bodies on subnational levels are not structured to carry out these functions. Political pressures are often more intense at the local level. Still, restoration may generate employment and income from the production of seedlings, planting and tending, and products such as firewood, palmetto, and herbal medicines. Even in urban areas, such as on the hill slopes of the city of Rio de Janeiro, activities to restore these areas to prevent landslips at times of heavy rainfall employ labor from the same deprived communities that inhabit those hillsides and are the first to be affected by environmental imbalance.

27.5 Research needs

Actions to protect and restore the Atlantic Forest can be considered some of the most urgent conservation needs in Brazil. Experience and research show that natural regeneration strategies have much to contribute to the success of restoration activities. Often, there is a need to combine these with physical or physical–biological measures in an attempt to slow down intensive degradation processes and to physically stabilize an area before restoration measures are successful. Knowledge is lacking on the autecology of native species, as well as information about their economic potential in areas where sustainable forest management is not legally restricted.

Little is known about the true costs of restoration activities. Too often, the experiments that underpin recommendations made by technical experts do not consider cost effectiveness such that accurate economic analysis can be carried out and decisions taken with an understanding of cost–benefit ratios. One exception is a study comparing natural invasion (do nothing), direct seeding pioneer species, mixed-species line-planting, multipurpose commercial species, and planting native species (Parrota and Engel 2001). Establishment and maintenance costs for direct seeding compared favorably with planting seedlings, ranging from \$747 to \$912 ha^{-1} for direct seeding and \$1,200 to \$2,500 ha^{-1} for planting seedlings (Parrota and Engel 2001). The dissemination of information on restoration methods and techniques also leaves much to be desired. Many public bodies working with rural extension processes do not have access to these procedures, which makes it difficult to popularize them.

References

Almeida, D.S., *Recuperação Ambiental da Mata Atlantica*, Editus/Editora da UESC, Universidade Estadual de Santa Cruz, Ilhéus, Bahia/Brasil, 2000.

Armour, C.L., Duff, D.A., and Elmore, W., The effects of livestock grazing on riparian and stream ecosystems, *Fisheries*, 16, 7, 1991.

Balistieri, P.R.M.N. and Aumond, J.J., Recuperação Ambiental em Minas de Argila — Doutor Pedrinho, Santa Catarina, in *Simpósio Nacional de Recuperação de Áreas Degradadas, III*, Ouro Preto, Minas Gerais, Brasil, 42, 1997.

Barbosa, L.M., Barbosa, J.M., and Asperti, L.M., Estudos Ecofisiológicos visando o repovoamento vegetal em áreas degradadas, *2ª_Reunião de Pesquisa Ambiental na SMA — (SP)*, 147, 1999.

Brasil, *Lei 4771 de 15 de setembro de 1965*, institui o novo Código Florestal, 1965.

Campelo, E.F.C., Sucessão vegetal na recuperação de áreas degradas, in *Recuperação de Áreas Degradadas*, Dias, L.E.D. and Mello, J.W.V., Eds., SOBRADE, UFV, Editora folha de Viçosa, Viçosa-MG, 183, 1998.

CESP, *Manual de Produção de Mudas de Essências Florestais Nativas*, Série Divulgação e Informação, São Paulo, 55 pp., 2000.

CONAMA, *Resolução 10/93*, Estabelece os estágios sucessionais da Mata Atlântica, Brasília: Conselho Nacional do Meio Ambiente, Brasil, 1993.

Dean, Warren, *With Broadax and Firebrand: The Destruction of The Brazilian Atlantic Forest*, University of California Press, Berkeley, CA, 1995.

Desmatamentozero, *Rede de ONGs Mata Atlântica*. Desmatamento Zero, Brasil, http://www.desmatamentozero.ig.com.br, 2001.

Einloft, R., *Crescimento de Gramíneas e Leguminosas em Substrato Rico em Mangânes Proveniente de Área de Empréstimo*, Dissertação, Mestrado em Ciências Florestal, Viçosa, MG: UFV, Brasil, 67 pp., 1997.

Eletronorte, *Programa de Recuperação de Áreas Degradadas*, UHE Tucurui-PA/Brasil, 1998.

Ferreira, C.L. and Bechara, E., *Estratégias de Conservação para a Mata Atlântica e os Campos Sulinos*, MMA/Fundação SOS Mata Atlântica, Fundação Biodiversitas, Ipecológicas, SMA-SP, IEF-MG, Relatório Técnico, Brasil, 2000.

Gandolfi, S. and Rodrigues, R.R., Recomposição de florestas nativas: Algumas perspectivas metodológicas para o estado de São Paulo, in *Recuperação de Áreas Degradadas, III Curso de Atualização*, Curitiba-PR, 83, 1996.

Griffith, J.J., Dias, L.E., and De Marco, R. Jr., A recuperação ambiental, *Ação Ambiental*, 10, 8, 2000.

Guedes, M.C., Melo, V.A., and Griffith, J.J., Uso de poleiros artificiais e ilhas de vegetação por aves potencialmente dispersoras de sementes, *Ararajuba*, 5, 2, 1997.

Guevara, S., Purata, S.E, and Van Der Maarel, E., The role of remnant forest trees in tropical secondary succession, *Vegetation*, 66, 77, 1986.

IBGE, *Manual Técnico da Vegetação Brasileira*, Série Manuais Técnicos em Geociências, Vol.1. IBGE, DRN, Rio de Janeiro, Brasil, 92, 1992.

Instituto Terra, *Conservando a Mata Atlantica Brasileira*, http://www.institutoterra.org, 2003.

Iracambi, *Atlantic Rainforest Research and Conservation Project*, http://www.iracambi.com/english/research_center.htm, 2001.

Jesus, R.M., Restauração Florestal na Mata Atlântica, in *Simpósio Nacional de Recuperação de Áreas Degradadas*, Ouro preto-MG, Brasil, 544, 1997.

Kageyama, P.Y. and Gandara, F.P., Recuperação de áreas ciliares, in *Conservação e Recuperação*, Rodrigues, R. R. and Leitão Filho, H.F., Eds., Matas ciliares, São Paulo, Brasil, EDUSP, FAPESP, 249, 2000.

Kageyama, P.Y, Biella, L.C., and Palermo, J.R., A plantações mistas com espécies nativas com fins de proteção a reservatórios, in *Congresso Florestal Brasileiro, 6°*, Anais, Campos do Jordão-SP, Brasil, 1, 109, 1990.

Kageyama, P.Y., Santarelli, E., Gandara, F.B., Gonçalves, J.C., Simiovato, J.L., Antiqueira, L.R., and Geres, W.L., Restauração de áreas degradadas: Modelos de consorsiação com alta diversidade, in *Simpósio Nacional de Recuperação de Áreas Degradadas, II*, Foz do Iguaçu, Parana, Brasil, Anais, 569, 1994.

Klein, R.M., Síntese ecológica da floresta estacional daBacia do jacuí e importância do reflorestamento com essências nativas, RS, in *Congresso Florestal Florestal Do Rio Grande do Sul*, Brasil, Anais, 2, 265, 1994.

Kolb, S.R., *Island of Secondary Vegetation in Degraded Pastures of Brazil: Their Role in Reestablishing Atlantic Coastal Forest*, Ph.D. dissertation, University of Georgia, Athens, Georgia, U.S.A., 1993.

Leck, M.A., Parker, V.T., and Simpson, R., *Ecology of Soil Seed Banks*, Academic Press, San Diego, CA, 1989.

Lowe, J., Brazil's Atlantic Forest: A paradise pulped for profit, *Earth Island J.*, Fall, 8, 4, 1993.

Macedo, A.C., *Restauração Matas Ciliares e de Proteção Ambiental*, São Paulo, Fundação Florestal, SMA, 27 pp., 1993.

Millar, C.I. and Libby, W.J., Restoration: Disneyland or a native ecosystem: a question of genetics, *Restor. Manage. Notes*, 3, 1989.

MMA, *Programa Piloto Para a Proteção das Florestas Tropicais Brasileiras*, Sub-Programa Mata Atlântica-PPG7, Ministério do Meio Ambiente, Brasília, DF, Brasil, 2000.

Moraes, D.A.A., *Princípios Básicos para a Formação e Recuperação de Florestas Nativas*, 2nd ed., MA/SDR/PNFC, Brasília, Brasil, 55 pp., 1998.

Morellato, P.L.C., *Estudo da Fenologia de Arvores, Aarbustos e Lianas de Uuma Floresta Semidecidua no Sudeste do Brasil*, Ph.D. thesis, Universidade Estadual de Campinas, SP, 1991.

Nogueira, J.C., Reflorestamento Heterogêneo com Espécies Indígenas, *Boletim Técnico Institucional Florestal*, 24, 1977.

Paglia, A., *Avaliações e Ações Prioritárias para Conservação da Biodiversidade da Mata Atlântica e Campos Sulinos*, MMA, Fundação SOS Mata Atlântica, Fundação Biodiversitas, Ipecológicas, SMA-SP, IEF-MG, Relatório Técnico, Brasil, 2000.

Parrota, J.A. and Engel, V.L., An evaluation of direct seeding for reforestation of degraded lands in central São Paulo state, Brazil, *For. Ecol. Manage.*, 152, 169, 2001.

Pederson, R.L. and Van der Valk, A.G., Seed banks and the management and restoration of natural vegetation, in *Ecology of Soil Seed Banks*, Leck, M.A., Parker, V.T., and Simpson, R., Eds, Academic Press, San Diego, 329, 1989.

Piña-Rodrigues, F.C.M., Costa, L.G., and Reis, A Estratégias de estabelecimento de espécies arbóreas e o manejo de florestas tropicais, in *Congresso Florestal Brasileiro, 6°*, Campos do Jordão-SP, Brasil, 676, 1990.

Piña-Rodrigues, F.C.M., Lopes, L., and Bloomfield, V.K., Análise do desenvolvimento de espécies arbóreas da Mata Atlântica em sistemas de plantio adensado para revegetação de áreas degradadas em encostas, no entorno do Parque Estadual do Desengano, in *Simpósio Nacional de Recuperação de Áreas Degradadas, III*, Ouro Preto — MG, Brasil, 283, 1997.

Pinto, L.P.S., Costa, J.P.D., Fonseca, G.A.B., and Costa, C.M., *Mata Atlântica: Ciência e Conservação e Políticas*, Secretaria do Meio Ambiente, Governo do Estado de São Paulo, SP, Brasil, 1996.

Pompèia, S.L., Prandella, D.Z.A., Diniz, K.M., and Santos, R.P., Recuperação do ecossistema Mata Atlântica de encosta, in *Congresso Florestal Brasileiro, 6°*, Anais, Campos do Jordão-SP, 1, 146, 1990.

Pompèia, S.L., Prandella, D.Z.A., Diniz, K.M., and Santos, R.P., Comportamento dos manacás-da-serra (*Tibouchina sp*) semeados por via aérea em Cubatão, in *Congresso Florestal Brasileiro, 6°*, Campos do Jordão, Anais, São Paulo, Brasil, Sociedade Brasileira de Silvicultura, 1, 3, 1992.

Reis, A., Bechara, F.C., Espindola, M.B., Vieira, N.K., and Souza, L.L., Restoration of damaged land areas: using nucleation to improve successional processes, *Natureza & Conservação*, Fundação O Boticário de Proteção à Natureza, Curitiba, Paraná, Brasil, 1, 85, 2003.

Rodrigues, R.R. and Gandolfi, S., Recomposição de Florestas Nativas: Princípios Gerais e Subsídios para uma definição metodológica, *Ver. Bras. Hort Orn.*, Campinas, SP, Brasil, 2, 4, 1996.

Rodrigues, R.R., Gandolfi, S., and Ribeiro, C.A., Revegetação de áreas degradadas da bacia do Ceveiro, Piracicaba-SP, in *Simpósio Nacional de Recuperação de Áreas Degradadas*, Anais, Curitiba-PR, Brasil, 178, 1992.

Santarelli, e.g., Recuperação de Mata Ciliar/Seleção de Espécies e Técnicas de Implantação, in *Recuperação de Áreas Degradadas, Curso de Atualização, III*, Curitiba-PR, Brasil, 101, 1996.

São Paulo, *Secretaria de Meio Ambiente do Estado de Sao Paulo/Coordenadoria de Informações Técnicas*, Documentação e Pesquisa Ambiental, Manual sobre Princípios de Recuperação Vegetal de Áreas Degradadas, Elaborado por Luiz Mauro Barbosa, São Paulo, Brasil, SMA/CEAM/CINP, 76 pp., 2000.

Seitz, R.A. and Corvello, W.B.V., Qualidade de mudas obtidas na regeneração natural, in *Método de Produção e Controle de Qualidade de Sementes e Mudas Florestais*, Curitiba, Brasil, 392, 1984.

Seitz, R.A., Potencialidades da regeneração natural na recuperação de áreas degradadas, in *Recuperação de Áreas Degradadas, Curso de Atualização, III*, Curitiba-PR, Brasil, 45, 1996.

Simoes, L.L. and Lino, C.F., *Sustentável Mata Atlantica: A Exploração de Seus Recursos Florestais*, Editora Senac, São Paulo, Brasil, 2002.

Skoglund, J., The role of seed banks in vegetation dynamics in the restoration of dry tropical ecosystems, *J. Veg. Sci.*, 21, 357, 1992.

Souza, A.L. and Jardim, F.C., *Sistemas Silviculturais Aplicados às Florestas Tropicais*, Viçosa, Brasil, Documento SIF, 008, 125 pp., 1993.

Souza, F.M. and Batista, J.L.F., Restoration of seasonal semideciduous forests in Brazil: influence of age and restoration design on forest structure, *For. Ecol. Manage.*, 191, 185, 2004.

Texeira, S.L., Técnica de culturas de tecidos aplicáveis às espécies florestais, *Anais Simpósio IUFRO "Florestas Plantadas nos Neotrópicos como Fonte de Energia," Viçosa, 6–13 fevereiro, Viçosa*, 69, 1983.

Uhl, C., Nespstad, D., Silva, J.M.C., and Vieira, I., Restauração da floresta em pastagens degradadas, Universidade Estadual de Santa Cruz (UESC), USP, UNICAMP, MMA/PROBIO/FUNPAB, Brasil, *Ciência Hoje*, 13, 22, 1991.

Valcarcel, R., *Plano de Recuperação Ambiental*, Serviço de Engenharia, Rodoférrea, S. A., 64 pp., 1994.

Willians, D.D., Bugin, A. and Reis, J.L.B.C., *Manual de Recuperação de Áreas Degradadas Pela Mineração: Técnicas de Revegetação*, MINTER/IBAMA, Brasília; 96 pp., 1999.

chapter twenty-eight

Integrating ecosystem restoration into forest management in British Columbia, Canada

Tanis L. Douglas
Fernhill Consulting
Philip J. Burton
Symbios Research & Restoration

Contents

1-56670-635-1/05/$0.00+$1.50

28.1 Introduction

The Canadian province of British Columbia (B.C.) is an ecologically diverse jurisdiction occupying an area of 950,000 km^2, greater than the area of France and Germany combined. Its ecosystems range from offshore areas along its sinuous coastline with the Pacific Ocean to alpine habitats >4,000 m above sea level, and stretching from 48.3° to 60.0°N latitude. This diversity has been classified into 14 biogeoclimatic zones that identify potential climax vegetation and prevailing climatic constraints, and serve as the basis for most natural resources planning in the province (Pojar et al. 1987). Zones dominated by coastal influences include wet rainforests of the Coastal Western Hemlock (CWH) zone, the drier and sometimes more open forests of the Coastal Douglas-fir (CDF) zone (situated in the rain shadow of Washington state's Olympic Mountains and the Vancouver Island Mountains), and the high-elevation Mountain Hemlock (MH) zone in the coastal mountains. The southern interior of the province, much of which is in the rain shadow of the Coast and Cascade Mountains, includes Bunchgrass (BG) grasslands at the lowest elevations, above which are found Ponderosa Pine (PP) savannahs, and both uneven-aged and even-aged forests of the Interior Douglas-fir (IDF) zone. The Montane Spruce (MS) zone is found on high plateaus and mountain slopes above the IDF where the climate is dry, while the windward slopes of the Cariboo, Selkirk, and Rocky Mountains support an interior rainforest or "wet belt" known as the Interior Cedar–Hemlock (ICH) zone. Subalpine forests in both the southern interior and northern interior are classified as the Engelmann Spruce – Subalpine Fir (ESSF) zone. The Sub-Boreal Spruce (SBS) zone and the drier, colder Sub-Boreal Pine–Spruce (SBPS) zone dominate the central plateaus of the northern interior. Colder and transitional subzones of the ICH zone are also found in northern B.C. The extensive plains of northeastern B.C. (on the leeward side of the Rocky Mountains) and the northwestern valleys are part of Canada's boreal ecoregion, and are referred to as the Boreal White and Black Spruce (BWBS) zone. In the northern one third of the province, the subalpine tree cover is discontinuous and often dwarfed; therefore, these high-elevation areas are classified as the Spruce–Willow-Birch (SWB) zone. Alpine Tundra (AT) is found at the highest elevations throughout the province, encompassing alpine meadows as well as large areas of bare rock and permanent snow and ice fields. Summary descriptions of the ecosystems characteristic of each biogeoclimatic zone are provided in four-page brochures available at http://www.for.gov.bc.ca/hfd/pubs/Bro.htm, while Meidinger and Pojar (1991) provide more detailed descriptions of each zone.

Most land in B.C. has never been settled or cleared, and 95% of it remains in public hands. People come from around the world to visit its fog-enshrouded coastal rainforests, to hunt big game in its northern wilderness, and to fish for steelhead (sea-run) trout in its glacier-fed rivers. These facts and the impressions they engender would suggest that there is little need for ecological restoration in such a natural corner of the world. Yet, 4.1 million people and 150 years of agricultural expansion have replaced or degraded most of the province's rare southern, low-elevation ecosystems. A highly mechanized forest products industry logs an average of 190,000 ha of wild (previously uncut) forest every year, and domestic livestock roam over hundreds of thousands of ha of public land. A recent assessment documented examples of ecosystem degradation in every part of the province (Holt 2001), but institutional support for ecological restoration has been insufficient to meet the (generally agreed upon) needs identified by biologists and environmentalists.

Ecological restoration is the process of assisting the recovery of an ecosystem that has been degraded, damaged, or destroyed (SER 2002). As indicated in previous chapters, ecosystem restoration is a rapidly growing endeavor worldwide, and represents the widespread recognition that ecological systems have been negatively impacted by human actions in the past, and that well-conceived human actions can subsequently reverse some of that damage. In many ways, the roots of modern forestry and silviculture owe their origins to the same admissions in the 18th and 19th centuries, when controls on the rate of harvest and the planned regeneration of forests started to be instituted in Europe and subsequently in North America (Fernow 1913; Apsey et al. 2000). However, in its emphasis on the sustained yield of wood, the large-scale industrialization of forestry leaves much to be desired in terms of its ability to protect and regenerate all forest values. The concept of sustainable forest management has subsequently evolved from the practices of sustained yield planning and forest renewal, extending from an emphasis on wood fiber to encompass the sustainability of a wider basket of forest values (Adamowicz and Burton 2003).

28.2 *An overview of B.C. forestry*

Commercial management for timber and fiber is practiced on 42% of the area covered by B.C.'s public forests, the remainder being protected in parks and reserves (10%) or currently considered uneconomical for sustainable forest management (48%). The B.C. Ministry of Forests manages or administers approximately 82% of the land in B.C. Plans for road development, logging, and silviculture are prepared by government foresters on part of the forest land base, while most commercial forest lands are directly managed by company foresters responsible for Tree Farm Licenses (area-based tenures) or for particular operating areas within broader Timber Supply Areas (under a timber volume-based tenure), but still under government supervision.

Forest products industries, especially those associated with logging, silviculture, the milling of dimensional lumber, panel board manufacturing, and the production of pulp and paper, employ 90,000 B.C. residents directly. Together with associated economic spin-offs, this sector accounts for 14% of the employment and 13% of the gross domestic product of the province. Forest management policy is a hotly debated topic throughout the province, and is the dominant driver of landscape manipulation. Chambers of Commerce and resource-dependent workers support corporate officers and their investors in their effort to manage forests primarily for economic gain, while environmentalists, recreationists, and community activists (both urban and rural) promote a more holistic view of forest management.

Through decades of protest, land-use negotiations, and changes of government, the forest products industry has remained the primary agent of landscape change, and is a key player in the economy and politics of the province. Consequently, any serious effort at reforming forest management practices or introducing new paradigms such as ecosystem restoration must effectively engage land managers who come from a professional culture that has been driven by commercial imperatives and a fiber-based emphasis. There has been a growing receptivity to the concepts of ecosystem management, sustainable forest management, and ecological restoration as younger foresters make their way up the management hierarchy. World markets demand greater ecological integrity in the production of wood products, but the ability to practice ecological stewardship is limited by global competition and the desire to maintain profit margins.

From a global perspective, many of the forest management policies and silvicultural practices prevalent in B.C. are sustainable, environmentally sound, and represent implementation of many of the principles of conservation biology and restoration ecology (Burton 1998). For example, B.C.'s forests remain almost exclusively composed of indigenous species (unlike the widespread practice of establishing exotic tree crops in the British Isles, Chile, New Zealand, and elsewhere), and all cutover forests are likewise regenerated to native tree species of suitable genetic origins. These practices reflect B.C.'s good fortune in being home to some of the world's most productive temperate tree species, and in being characterized by sharp differences in climate where the necessity of genetic adaptation is readily apparent.

The basic elements of reforestation and the establishment of native species, key to the restoration of many other temperate forest ecosystems (Stanturf and Madsen 2002), have been a responsibility of forest products companies operating in B.C. since 1987. Ecologically based site stratification, diagnosis, and regeneration planning are now standard components of all silvicultural prescriptions. A mixture of tree species is frequently planted, and the tolerance for noncrop hardwood species is increasing, although the economic and regulatory bias favoring conifers remains. Recent trends in landscape planning and the designation of riparian buffers, wildlife tree patches, forested ecosystem networks, and old-growth management areas are serving to retain more diversity on the landscape than when trees were managed under the pretense that they were an agricultural crop. Interest in intensive short-rotation fiber production remains, but these practices are being channeled onto a smaller and more precisely defined (zoned) land base than in the past.

Although partial cutting and variable retention harvesting (Franklin et al. 1997) are on the increase (Beese et al. 2001), clearcut logging and even-aged stand management remain the norm. This may be appropriate in the northern and interior forests dominated by even-aged stands of lodgepole pine (*Pinus contorta*), where historic fire return intervals may have been 80 to 120 years, but clearcut harvesting and even-aged management are more dubious practices in coastal rain forests (and a number of interior forest types) that have experienced wildfire only rarely, and are historically multi-aged. As a matter of provincial forest policy, old-growth forests are being liquidated in order to establish more productive second-growth stands, and the rate of cut (the biggest environmental issue in both coastal and interior forests) is greater than the long-run sustained yield. Consequently, the structures, functions, and composition associated with older and naturally disturbed forests are not sufficiently protected or generated. Therefore, despite low human populations and wide-open spaces, much of the "managed" forestland in B.C. is in serious need of restoration. The sheer vastness of the province, with its diversity of forest types, ecosystems, and human pressures, means that we find many examples of healthy and degraded forests in the same political jurisdiction. By and large, ecological restoration is being undertaken on lands where damage (resulting from the more obviously short-sighted or single-purpose forest practices of the past) is being repaired to meet today's expectations for ecological stewardship.

Forest planning (by government and industry foresters operating on public land) in B.C. is currently moving from a prescriptive, rules-based approach, to a broader, results-based approach focusing on the preparation of sustainable forest management (SFM) plans and third-party certification. As a result, interest in incorporating some ecosystem restoration practices into day-to-day forest management in B.C. is increasing, in order to protect biodiversity, rebuild soil productivity, aid in the recovery of salmonid populations, or bolster the credibility of SFM plans and corporate applications for forest product certification. It has been argued that ecosystem restoration can make good business sense for forest products companies, aiding in certification and market access, minimizing costs associated with the liability for untreated ecological damage, and as a component of a risk management strategy (Douglas 2003). Examples of forest restoration activities in B.C. include the reintroduction of ground fires in dry-forest ecosystems, incorporating snags, coarse woody debris (CWD), or canopy gaps in homogeneous secondary forests, the rehabilitation of compacted landings and access roads, and the control of invasive species such as knapweed (*Centaurea* spp.), thistles (*Cirsium* spp.), and broom (*Cytisus scoparius*).

We here review some of the priorities for ecological restoration in different forest zones of the province, and then showcase a number of recurrent ecological issues and some recent cases of on-the-ground forest restoration activities that have addressed them. Eight broad ecological issues are identified, with general statements of their impact and options for remediation and prevention; five case studies provide a more detailed description of some on-the-ground restoration activities. This review reflects the priorities identified in a series of six regional workshops (Holt 2001), and borrows extensively from previous overviews and guidelines prepared by Gayton (2001) and Douglas (2003). The authors of this chapter participated in the evaluation of the restoration activities described below primarily through their advisory role in the Terrestrial Ecosystem Restoration Program of Forest Renewal B.C. (FRBC), but did not design or implement most of them, nor monitor their effectiveness. The approaches described below are not endorsed by the authors, but are simply being reported as being representative of restoration activities under way in B. C. at the time of writing.

28.3 *Forest restoration priorities in B.C.*

The most degraded ecosystems in B.C. are those southernmost habitats in which European settlement has been most pervasive over the last 150 years. Most of the land formerly occupied by Garry oak (*Quercus garrayana*) savannahs and meadows is now dominated by privately held residential and commercial land in the greater Victoria area on southern Vancouver Island and on the Gulf Islands. Likewise, much of the land formerly occupied by antelope brush (*Purshia tridentata*) scrub and bluebunch wheatgrass (*Pseudoroegneria spicata*) grasslands in the southern Okanagan Valley is now occupied by homes, orchards, vineyards, and recreational property. Restoration activities in these ecosystems have required a cooperative effort spanning all levels of governments and a large degree of public education and volunteer involvement. It is fortunate, yet ironic, that the greatest resources for ecological restoration (high public awareness, large pools of volunteers) are found in the very population centers that were responsible for the degradation of rare ecosystems in the first place. Recent symposia sponsored by the B.C. Chapter of the Society for Ecological Restoration have focused on the progress of ecological restoration in Garry oak (Burton 2002) and antelope brush (Seaton 2003) ecosystems, and they are not discussed further here.

In the other forested regions of the province, ecological degradation is mostly related to the impacts of industrial forestry and associated road development, rather than as a direct consequence of human population pressures and land-use conversion. The forested biogeoclimatic zones described in Table 28.1 have high-priority restoration needs in certain components, subzones, or areas. A brief description of the most pressing ecological issues is given for each zone, based on a process that determined the degree of ecological degradation at a regional scale,[1] with an overview provided by Holt (2001). Four sources of degradation are noted throughout many of B.C.'s public forestlands:

1. The loss of structural and compositional diversity, in terms of
   (a) physical habitat elements such as large trees, snags and fallen logs, canopy layering, and mixed tree species composition at the stand level and
   (b) natural variation in age- and size-class distributions at the landscape level
2. Liquidation of old-growth forests and stands of the largest and most productive trees as a result of commercial logging
3. Invasive species, especially shrubby and herbaceous plants of Eurasian origins, which tend to establish and persist in open habitats
4. The loss of wildfire (whether burning at ground level or through tree crowns) as a generator of biocomplexity and an agent of stand renewal and compositional diversity.

[1]For further information, "Strategic Ecological Restoration Assessments" are available for each of six administrative Forest Regions in the province at http://wlapwww.gov.bc.ca/wld/fia/ecosystem_restoration.html.

Table 28.1 Causes of Ecological Degradation and Associated Restoration Needs in the Forested Ecosystems of British Columbia

Ecosystem Type	Ecosystem Characteristics	Restoration Challenges			
		Loss of Biodiversity Features	Liquidation of Old Growth	Invasive Species	Loss of Wildfire
Interior Cedar-Hemlock	Interior wet belt, highly productive and diverse forests, some very old forests	***	***	**	*
Interior Douglas-fir and Ponderosa Pine	Low-elevation open forests of dry interior; historical regime of frequent surface fires	***	*	***	***
Montane Spruce	Interior transition between high and low elevations; lodgepole pine, stand-replacing wildfire, and insect outbreaks prevail	***	**		***
Engelmann Spruce – Subalpine Fir	High-elevation interior forests; whitebark pine decimated by blister rust	***	***	**	
Sub-Boreal Pine Spruce	Interior dry, cold, low-productivity zone dominated by lodgepole pine and wetlands; important habitat and lichen forage for caribou	**	**		**
Sub–Boreal Spruce	Productive interior forests with distinct tree species mix and extensive wetlands	***	***	**	***
Coastal Douglas-fir	Dry coastal forests of limited extent; high species diversity and high urban encroachment	***	***	***	*
Coastal Western Hemlock	Extremely productive, complex, and old coastal forests with very high biodiversity values.	***	***	*	
Mountain Hemlock	High-elevation coastal forests, now being logged as low-elevation forests have been depleted	*	***		
Boreal White and Black Spruce	Northern forest and muskeg east of the Rocky Mountains; extensive oil and gas exploration	**	**	*	**
Spruce–Willow–Birch	Sparse northern forests and shrublands west of the Rocky Mountains; primary land uses are mining and big-game hunting				**

The number of stars indicates the degree to which the restoration challenge applies to the ecosystem type, with three stars the highest degree.

B.C. shares many of the restoration needs and approaches found in the biogeographically similar U.S. Pacific Northwest and Inland Empire (O'Hara and Waring, this volume; Jain and Graham, this volume). Likewise, the dominance of industrial forestry and efforts to restore its ecological damage in Fennoscandia (Kuuluvainen et al., this volume; Angelstam et al., this volume) offer lessons applicable in B.C.

Despite the site- and problem-specific nature of all restoration projects, some general themes and approaches have emerged in most forest restoration efforts conducted recently in B.C. The components of these approaches are discussed below. These themes can also be found in the theory and practice of ecological restoration planning as developed in other parts of the world (Nuzzo and Howell 1990).

28.3.1 *General themes*

Prioritization and triage identifies those ecosystems (especially rare ones) and sites where intervention is urgently needed to save species at risk, or where restoration is expected to be comparatively effective for the least amount of effort. Triage implies the necessary postponement of work on more common ecosystems, those suffering from less severe problems, and those in which even heroic efforts may be unable to repair damage. Prevention identifies ongoing threats to ecological integrity so that the agents of ecological degradation at the project site and elsewhere can be stopped before undertaking restoration work. This avoids the need for even more restoration. A goal-setting approach focuses on measurable, meaningful, and achievable objectives. Stating goals clearly helps to recruit participation from the necessary agencies, corporations, and members of the public, and to keep projects "on track." Continuity is important, as most restoration efforts are multistep processes that typically need adjustment and repeat treatment in order to achieve their objectives. This approach follows through with effectiveness monitoring and maintenance activities, which create a record that can inform future projects.

28.3.2 *Programs*

In recognition of the restoration need, several government-sponsored programs were implemented in B.C. over the last decade. An important push for forest restoration was made under the auspices of Forest Renewal B.C. (FRBC), a crown (public) corporation established in 1994 and financed by timber cutting royalties imposed to forestall U.S. lumber tariffs. One of its major thrusts was the Watershed Restoration Program (WRP), which emphasized the rehabilitation of in-stream salmon habitat, the riparian zones influencing those streams, and upland sources of sediment. Prior to the demise of this corporation in early 2002, it restored more than 750 km of stream reach and riparian habitat, and stabilized sediment sources (landslides, unmaintained logging roads, and landings) in 20% of the 725 watersheds identified as a "high priority" for restoration (Underhill 2002). To balance the WRP emphasis on the restoration of habitat needed by anadromous salmon for spawning and rearing, FRBC started a Terrestrial Ecosystem Restoration Program (TERP) in 1998. This program specifically addressed many of the issues of terrestrial ecosystem degradation identified above, and provided much of the financial support for the projects described in the next section. Over its short lifespan, TERP supported the restoration of hundreds of ha of forests in B.C., primarily those fire-maintained ecosystems that had undergone serious ingrowth, and homogeneous second-growth forests lacking structural diversity.

The Habitat Conservation Trust Fund (HCTF) was started in 1996, and is a more modest program funded by hunting and fishing license fees, administered by the B.C. Ministry of Water, Land and Air Protection. This fund continues to aid small nonprofit groups and researchers in conserving and restoring habitat for rare and endangered species and ecosystems (especially fish and wildlife) in both forested and nonforested environments. A number of other programs supporting restoration activities also exist, such as B.C. Hydro's Columbia Basin Fish and Wildlife Habitat Compensation Fund. Private funding for ecological restoration work has also been available through various nonprofit foundations, and is commonly associated with local, volunteer-based projects. Solutions usually require site-specific assessments and customized prescriptions for human intervention to facilitate and accelerate ecosystem recovery. A growing sector of private restoration consultants and contractors is emerging in B.C. to meet these professional and technical needs, with assistance from specialized educational initiatives such as the "Restoration of Natural Systems" program at the University of Victoria (see http://www.uvcs.uvic.ca/restore/).

28.4 Recent experience in addressing key ecological issues

Integrating ecosystem approaches into management, such as mimicking natural disturbance patterns during forest harvesting (Seymour and Hunter 1999), should minimize the need for forest restoration in the future. However, most forest exploitation and management conducted in B.C. to date was completed to standards that were developed without today's ecological knowledge and societal expectations. As a result, there is an extensive backlog of land suffering from some degree of ecological degradation, calling for the application of a variety of restoration techniques. For instance, extensive areas of dense, young, closed-canopy stands provide little habitat for many forest species. Carefully designed commercial and precommercial thinning can increase the habitat value of these areas, while also accelerating the attainment of operable piece size for future timber harvests. In this section, we explore some of the more widespread issues and priorities described above, with emphasis on those being addressed through recent restoration activities.

Ecological restoration addresses these issues through a very broad range of activities, from reconstructing a local plant community such as a riparian fringe, to restoring ecological processes such as fire in a large area of forest. Impacts and implications are outlined for each issue, followed by suggested preventative and restorative measures. For some issues, case studies are presented to detail how government agencies, forest products companies, and community groups are currently approaching these types of restoration projects.

28.4.1 Ecological issue: lack of natural structural elements or species in second-growth stands

Natural stand structure or species is lost through highly uniform harvesting practices, particularly clearcutting without appropriate retention of wildlife trees, live and dead large wood pieces in riparian areas, and wildlife tree patches. Silvicultural practices can cause high stand uniformity, dense monocultures, highly uniform tree spacing, and very high levels of crown closure. Short-rotation forestry also results in lack of recruitment of important stand structural attributes. Plantations are of limited habitat value due to high uniformity and lack of stand structural elements such as large wildlife trees and CWD, virtual elimination of shrubs and herbs due to high crown closure, and high risk of pest or disease damage to highly uniform stands. Prevention methods include proactive design of harvest activities to retain structural elements, maintenance of species and genetic diversity during reforestation, and design of reforestation to produce higher diversity in densities through management of stocking levels and minimum spacings.

Restoration approaches could be patchy; precommercial; or commercial thinning to reduce crown closure and stand uniformity; or introducing stand structure into selected areas through techniques such as wildlife tree and CWD addition. Restoring the abundance of structural elements (such as wildlife trees, snags, CWD, and canopy gaps) is a common feature of many forest restoration activities. Such activities are also used for the fine-filter management of species dependent on old-growth forests (such as the northern spotted owl, *Strix occidentalis caurina*) in areas where most old-growth forest has already been logged, but second-growth forest is well developed. Northern spotted owls require relatively open habitats with large-diameter trees and snags, a multilayered canopy, and relatively high amounts of shrub and CWD cover, conditions typical in old forests (Tesky 1992). Commercial thinning is being used in southwestern B.C. to improve habitat for spotted owls in 60-year-old Douglas-fir (*Pseudotsuga menziesii* var. *menziesii*) stands, enhancing their future value from Type C ("suitable") to Type B ("moderate") spotted owl habitat, while at the same time making a profit.

28.4.1.1 Case study 1: fungal inoculation to create wildlife trees

Fungal inoculation is a relatively new and promising technique used to create wildlife trees. There methods are still under development, but appear to be a highly efficient and effective means for recruiting one of the most valuable structural elements of wildlife habitat, a tree that contains heart rot. Trees in this condition are excavated for use by wildlife and eventually become hollow trees, further increasing their habitat value. Eventually, they break apart to become fallen woody debris and hollow logs.

Fungal inoculation is used on live, healthy trees to initiate heart rot. The fungus does not usually kill the tree; instead, a compartmentalized decay column is produced in the live tree within 3 to 6 years. The inoculated tree is able to maintain its foliage and growth form, and continues to put on new incremental growth. Wildlife trees in this condition (as opposed to a dead snag) will provide habitat for a longer period of time, and will provide few worker safety or operational concerns. They will also continue to function as a seed source, and are less likely to be felled by firewood cutters.

Two inoculation procedures are currently under investigation in B.C. to determine which method produces the best results. In one method, spores of a native heart rot fungus (*Phellinus pini*) are injected into the tree by climbing it, drilling a hole, and inserting a wooden dowel that is cultured with a locally collected strain of the fungus (Figure 28.1). The second technique uses a rifle to shoot the tree trunk with a bullet that contains a smaller wooden dowel cultured with the same fungus. Both techniques result in fungal decay spreading within the tree above and below the point of inoculation. The decayed (and eventually hollow) bole center that results can provide cavity nesting and feeding habitats that are critical for many forest species. There is virtually no risk of unwanted spread to nontarget trees using these techniques because of the natural reproductive history of the fungi.

Data from the U.S. Pacific Northwest show that fungal inoculation is less expensive and faster than techniques such as topping trees with a chainsaw or explosives, top girdling, or cavity creation using chainsaws. However, all these techniques have shown some success at providing habitat for cavity- and snag-using wildlife (Lewis 1998). These wildlife tree creation techniques have the advantage of creating or enhancing habitat in a relatively short period of time (5 to 10 years), as opposed to recruiting similar stand structure through natural cycles. In contrast, it usually requires ≥100 years to naturally recruit trees of sufficient size and condition to function as useful wildlife trees.

In B.C., two operational trials using fungal inoculation are currently under way on Vancouver Island, sponsored by Weyerhaeuser Coastal Group and Canadian Forest Products Ltd (Manning 2004). These are both in second-growth Douglas-fir stands that have relatively few characteristics of older forest (i.e., wildlife trees, CWD). Partial cutting prescriptions are being implemented at these sites, with enhanced biodiversity value as one of the management objectives. The ecological and operational feasibility of fungal inoculation as a habitat enhancement tool will be evaluated over the next 5 years.

Fungal inoculation can be used in a variety of wooded habitats that require restoration or enhancement. For example, the method can be used to:

- Create more diverse stand structure and suitable wildlife habitat in riparian management areas
- Accelerate the production of mature forest elements (e.g., wildlife trees, snags, and CWD) in large areas of relatively homogeneous immature forest

(A) (B)

Figure 28.1 Individual trees in young second-growth or homogeneous forest stands can be inoculated with heart rot fungus, either by (A) climbing and drilling, or (B) using a rifle, in order to accelerate the development of suitable habitat for cavity-nesting birds (photos by Todd Manning).

- Manage for the habitat of particular species that require mature forest elements, either directly or indirectly (e.g., identified wildlife species, including woodpeckers, owls, and goshawks).

28.4.2 *Ecological issue: invasive species replacing native species*

Management activities and human disturbance have favored invasive species, leading to declines in native species. Impacts include loss of native ecosystems and habitat for native species, reduced production of economically valuable native species, or increased difficulty in reestablishing native species. This could be prevented by minimizing site disturbance, prompt revegetation and reforestation, and careful selection and control of plant materials for erosion control. Restoration approaches could include proactive removal of invasives and seeding or planting bare ground with appropriate native species.

28.4.2.1 *Case study 2: Scotch broom control*

Numerous groups and volunteers use manual pulling to reduce the populations of the invasive leguminous shrub Scotch broom (*Cytisus scoparius*) throughout the CDF zone on Vancouver Island. Like most invasive species, broom is exceptionally difficult to control due to its lack of natural enemies, its ability to out-compete native plants, and its profuse production of seeds that persist in seed banks. Broom can significantly reduce conifer regeneration in clearcuts, while in Garry oak meadows it threatens endangered plant communities

(Burton 2002). New invasions of broom have recently been found on the Queen Charlotte Islands, on the Lower Mainland, and around Castlegar and Kootenay Lake (Prasad 2003).

The most effective method of broom removal is by manual cutting or pulling, with repeated cutting or pulling usually required at any given site. Care needs to be taken to minimize soil disturbance in order to ensure that conditions are not re-created that are suitable for additional inseeding by broom or other exotic weeds. Preventing the spread of broom through these manual methods must continue until an acceptable biological control agent is developed. The Canadian Forest Service recommends the following actions to forest managers (Prasad 2000, 2003):

- Carefully inspect road ballast and materials brought from other areas for broom seed
- Bring new invasions to the attention of researchers, forest managers, and the local B.C. Ministry of Forests District office
- Cut broom stems as close to the ground as possible before the seed matures, taking care not to disturb the surrounding soil; hand pulling is preferable to prevent resprouting, but this is not always practical
- Remove broom before it has a chance to flower
- Inspect conifer plantations on a regular basis
- Support urban efforts to remove Scotch broom in parks
- Where broom invasion is likely, selective timber extraction should be considered over clearcutting
- On warm, well-drained sites, avoid excessive soil disturbance and exposed mineral soil.

28.4.3 Ecological issue: reduction and changes to open habitat types

Fire control and forest regeneration at densities greater than historic levels have reduced the extent and quality of open habitat types (grasslands, savannahs, and open fire-maintained forests). Impacts include loss of habitat for open-habitat and fire-dependent species, loss of economic values associated with open habitats (rangeland and vigorous trees), and increased risk of catastrophic fire, including increased risk of economic loss (timber, infrastructure). As prevention measures, open stand structures should be maintained, and surface fires should be used regularly as a stand management tool. Restoration approaches should include commercial and precommercial thinning to reduce stand densities in historically open stand types, clearing of trees from historic grassland areas, and controlled reintroduction of ecosystem-maintaining fire.

Extensive areas with densely ingrown stands (such as those shown in Figure 28.2) occur throughout the IDF, PP, and BG biogeoclimatic zones (Table 28.1) as a result of decreased fire frequencies since European settlement. With the exclusion of frequent surface fires, these forest stands now prevail on a landscape that was once dominated by more open-grown stands with greater species richness and diversity, greater tree vigor and growth rates, and greater forage values than found today. These ingrown stands also have a much-increased risk of catastrophic wildfire. This is a common problem in many of the western pine and Douglas-fir forests of the inland U.S. as well, and the use of thinning and prescribed fire to restore these ecosystems is described in other chapters of this book (O'Hara and Waring, this volume; Jain and Graham, this volume).

28.4.4 Ecological issue: extensive uniform stands

Uniform forest stand ages and characteristics over large areas are of concern and are the result of fire suppression, which increases the dominance of single-age classes, and

Figure 28.2 Treatment of densely ingrown Douglas-fir stands in the IDFxm subzone of the Cariboo Forest Region, B.C.: (A) prior to any treatment; (B) after logging (commercial thinning); (C) after logging slashing (noncommercial thinning); and (D) during a light underburn treatment (photos by Ordell Steen).

the harvest and regeneration of large areas over a short time. Uniform stands over large areas increase the risk of large-scale losses due to pests, disease or fire, and loss of critical habitat for species dependent on certain forest structures and stand ages. Prevention should include the proactive design of management activities to mimic natural disturbance patterns, in order to avoid lopsided stand age or structure distributions. Where logging now replaces wildfire as the dominant agent of ecological disturbance, timber harvests should be conducted in ways that create natural stand structures and mosaic patterns. Managing for mixed species and ages should increase stand heterogeneity. Restoration approaches could include increasing stand heterogeneity through precommercial and commercial thinning that produces a variety of stand densities and within-canopy openings. Stand structure can be enhanced in selected areas of the landscape through wildlife tree and CWD addition techniques. Major salvage activities should follow ecological best management practices.

28.4.4.1 Case study 3: restoring structural diversity to the Kitimat Valley

The Kitimat Valley in northwestern B.C. has been extensively logged, with less than 3% of its original CWH forest remaining. Since harvest, intensive silvicultural practices have been used to maximize the economic value of the stands. Even-density spacing and intensive pruning have created second-growth habitat with little structural diversity in horizontal or vertical dimensions. This extensive uniformity of secondary forest habitat limits opportunities for mammals, birds, and other elements of biodiversity dependent on snags, CWD, or canopy variability associated with old forests. The Kitimat Valley has approximately

19,400 ha of these heavily managed pole-sapling stands on a productive, but heavily roaded valley bottom. Coarse-filter biodiversity conservation measures such as streamside buffers and wildlife tree patches are no longer options in this landscape because little mature forest remains. The only option to improve biodiversity in this landscape is restoration. In 2001, a pilot project was initiated to reintroduce old-growth stand characteristics through stand management techniques (Pollard and Prins 2001). The long-term goal is to encourage the reintroduction of species temporarily extirpated from the area.

West Fraser Timber Co. Ltd. (Skeena Sawmills Division) and the B.C. Ministry of Forests worked together to complete Year 1 of the project. Variable spacing, girdling, and installation of snags and CWD piles were used to create islands of old-growth characteristics within the mosaic of homogeneous (young, second-growth, conifer-dominated) commercial forest. An important part of this project was determining the feasibility of incorporating these techniques into mainstream, volume-oriented stand management. Various baseline data were also collected to allow future monitoring of the success of this project.

The treated stands were primarily planted spruce (*Picea sitchensis* and its natural hybrids with *Picea engelmannii* and *Picea glauca*) affected by spruce leader weevil (*Pissodes strobi*), along with lower densities of naturally established western hemlock (*Tsuga heterophylla*), Pacific fir (*Abies amabilis*), and western redcedar (*Thuja plicata*). These stands were chosen because of their high densities (greater than 4,000 stems ha^{-1}), their lack of previous stand treatments, and because they were relatively young (20 to 30 years old). Spacing brought stand densities down to 800 stems ha^{-1} in many areas, 400 stems ha^{-1} in others, with no treatment of control areas. A proportion of trees were girdled as part of the treatments. Once spacing was complete, CWD piles were created, and imported snags were installed. The snags were logs purchased from Skeena Sawmills and were planted in upright positions using two backhoes. The Terrace Rod and Gun Club provided machine time and qualified operators at reduced cost.

Adding CWD and snags to second-growth stands to the levels naturally found in old-growth forests (e.g., average levels of 300 to 500 m^3 ha^{-1} in the CWH zone) (Stevens 1997) would be extremely costly. Instead, these structural elements were obtained and placed at lower densities on the basis of the availability of suitable logs. As a result, the snags and CWD piles were concentrated in two areas. However, it is expected that their addition will significantly improve habitat values over existing conditions.

Forest Renewal B.C. funded this project, and costs were tracked to determine operational feasibility. If one-time costs are removed and anticipated efficiencies are achieved, costs would be CAD[2] \$2,700 ha^{-1} for the package treatment of spacing, and snag and CWD addition. This compares very favorably to restoration silviculture projects elsewhere in the province, but is approximately twice the cost of standard precommercial thinning treatments at this stage of stand development. As this is a pilot project, its effectiveness remains to be determined. Breeding bird surveys and other monitoring activities conducted by volunteers will help assess its success over time. The obvious challenge in projects of this type is to treat enough area to have a landscape-scale effect. Nevertheless, the individual treated sites are expected to attract species that are scarce in the rest of the Kitimat Valley.

28.4.5 Ecological issue: under-represented stand types in the landscape

Some landscapes do not contain the full natural range of stand types. This lack may be due to preferential harvesting of particular species, age classes, or stand structures at

[2]1.00 Canadian dollar (CAD) = 0.76 United States dollar (USD) at the time of writing.

unsustainable levels, fire control that reduced the rate of replacement of specific stand types, and tree mortality due to exotic pests, diseases, and parasites. Impacts include loss of habitat for species dependent on certain stand structures, tree species, or age classes; low (and sometimes at-risk) population levels of species that require the under-represented stand type; reduction in the range of future economic opportunity; and increased risk of catastrophic damage in over-represented age or species types. Prevention can be achieved by designing harvesting to retain representative amounts of all stand types (based on compositional and structural criteria). Restoration approaches should include: silvicultural management (thinning, planting) to increase the future supply of under-represented stand types; use of slashing and prescribed burning to increase the supply of open forest stand types in the IDF and PP zones; and breeding and nursery research to support reintroduction or recovery of tree species devastated by exotic pests or overharvesting.

Stand types can be defined compositionally (those dominated by western white pine, Pinus monticola), structurally (uneven-aged, multistoried stands), or through combinations of composition, structure, and site type (stands of large, old white spruce, *Picea glauca*, on boreal floodplains). Western five-needle pines have been devastated by the arrival from Asia a century ago of white pine blister rust, *Cronartium ribicola*. Efforts to restore western white pine stands in B.C. and neighboring U.S. states have consisted of a two-pronged approach: reducing mortality in regenerating stands by pruning lower branches to a height of 3 m; and selecting rust-resistant lines in tree breeding programs. Earlier efforts to control the spread of the disease through eradication of alternate hosts (shrubs of the genus *Ribes*) were largely ineffectual (Anon. 1996; Maloy 1997). As the range of blister rust continues to expand northward and to higher elevations, the next challenge will be to save stands of whitebark pine (*Pinus albicaulus*), a high-elevation species with many trees less than 3 m in height and for which there has yet been little genetic research.

B.C. parks recently undertook restoration efforts after removing a campground from a rare old-growth Sitka Spruce–Salmonberry ecosystem on the Exchamsiks River in northwestern B.C. (Burton and Burton 2002). "Old-growth restoration" is certainly a long-term prospect, but it is expected that various structural and compositional goals can be accelerated through the placement of large CWD, and the planting of salmonberry (*Rubus spectabilis*) as well as Sitka spruce (*Picea sitchensis*) and other conifer seedlings (Figure 28.3). While this example is not situated in a commercially managed forest, old-growth restoration is likely to be increasingly needed in the restoration of landscape-level goals for forest seral-class distributions and connectivity. Many restoration efforts targeting particular stand types are often coupled with other objectives, such as maintaining critical wildlife habitat or riparian zone integrity, as discussed next.

28.4.6 Ecological issue: riparian habitat integrity

Riparian habitat has been degraded by historical timber harvesting in riparian areas without sufficient buffers and by current stand structures that do not sustain habitat integrity. Impacts include: loss of stand structural elements (wildlife trees, CWD) and older age classes in areas highly important for biodiversity conservation; limited habitat value in formerly biodiverse areas; lack of large woody debris input to stream channels, and subsequent impacts to fish habitat; conversion to deciduous stands or brush in some areas; loss or reduction of animal movement corridors; and possible loss of sufficient stream shading and filtering capacity. Prevention measures should include proactive design of harvest to include sufficient buffer width, taking windthrow probability into account. Silvicultural standards and techniques designed to maximize riparian habitat value should be used; for instance, use of more diverse species and spacing as well as retaining

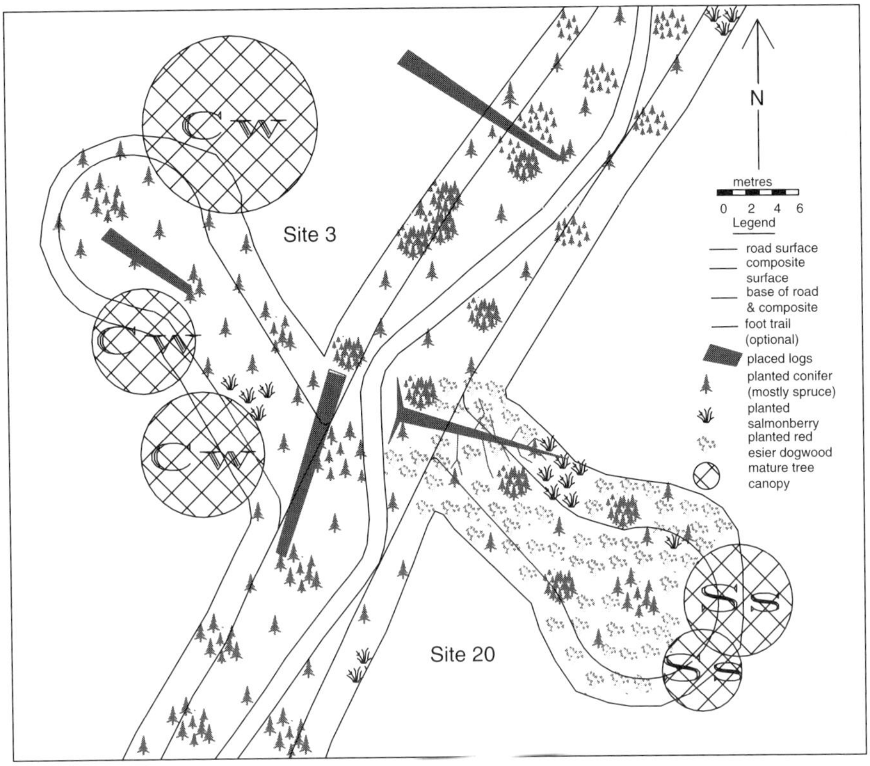

Figure 28.3 The first steps in restoring a rare old-growth Sitka spruce – Salmonberry ecosystem after the closing of a campground require not only the planting of trees but also the placement of very large pieces of coarse woody debris, and the planting of shrubs and ferns (from Burton and Burton 2002).

some structural diversity. Restoration approaches should include silvicultural approaches to restore stand composition (species and stocking levels); reintroducing stand structural elements such as wildlife trees and CWD; adding large woody debris to streams through restoration thinning; and bioengineering methods to rapidly establish cuttings of cottonwoods and willows for bank stabilization and cover production.

28.4.6.1 *Case study 4: Western Forest Products and Weyerhaeuser riparian silviculture*

Historic logging practices around streams have contributed to a loss of fish and riparian habitat in some important fisheries rivers in the chart area of Western Forest Products Limited on Vancouver Island. To deal with these issues, Western Forest Products (WFP) surveyed candidate areas to determine priorities for restoration silviculture. Three watersheds were selected for restoration on northern Vancouver Island, and a total of 171 ha were treated in high-priority riparian areas (Poulin and Simmons 2002a,b). The creation of stands of larger, well-spaced conifer trees and wildlife trees was an objective of the treatments. This work will speed the recovery of fish habitat and channel stability at the same time that wildlife habitat is improved.

Red alder (*Alnus rubra*) dominates many coastal riparian zones after disturbance, but these fast-growing deciduous trees cannot attain the structural size and durability of streamside conifers. Consequently, the objective of many riparian silviculture projects is to enhance the establishment and growth of long-lived conifers. Other typical objectives of riparian restoration projects include restoring riparian zone ecosystem functioning by managing for the provision of (Peters 1999; Bancroft and Zielke 2002):

- Large future wildlife trees
- Large future coarse woody debris on the ground and as large organic debris instream
- Open areas and canopy gaps for maintenance of shrub communities
- Ungulate and bear forage
- Habitat for birds, amphibians, small mammals, insects, plants, and other organisms
- Shade and nutrients to the stream
- Filtering of overland flow (run-off).

All the rivers in the WFP license area on northern Vancouver Island are in long-term forestry licenses. Most riparian stands were logged 25 to 45 years ago, with the exception of the retained mature or old-growth timber that occupies 18 to 52% of the total riparian area. Red alder, with an understory of scattered and suppressed conifers, often dominates the harvested stands. Restoration treatments were designed to increase and promote the conifer components of these stands, with a secondary objective of managing for large alder trees. Conifer growth was improved by removing all alder within a radius of 3 to 10 m of each conifer tree. This resulted in the removal of approximately 80% of the alder stems in the stand, the minimum necessary to achieve a target of 40% full sunlight to the understory trees. Post-treatment conditions depended on the site; alder stands with conifer understory were thinned to 100 to 300 stems ha^{-1} of overstory alder. Pockets of pure alder were also thinned, and up to 600 stems ha^{-1} were retained, depending on the age and height of the stand.

The drier riparian sites were generally overstocked with conifers, particularly western hemlock. These sites were also thinned with the goal of improving conifer growth and establishment. Both uniform and variable density thinning was used to take advantage of productive growing sites, while allowing natural gaps and clusters of trees to prevail. Thinning took densities down to 300 to 600 stems ha^{-1}, with preferential retention of western red cedar and Sitka spruce where these species were present. Cluster planting was also done in selected areas of both the alder- and conifer-dominated stands. Spruce and cedar were companion-planted with black cottonwood (*Populus balsamifera* ssp. *trichocarpa*) to increase the species and structural diversity of the riparian forests.

Thinning provided excellent opportunities for further enhancing the biodiversity value of these stands. Many of the cull trees were topped or top-girdled to create snags, while others were damaged to initiate heart rot. Habitat features like bat slits and cavity starts were also created in some trees. Over 900 such features have been introduced into Western Forest Product's riparian stands.

Thinning also provided the opportunity to add large woody debris to the stream. Approximately 80 instream structures were created using thinnings from both alder and conifer stands. The structures typically consisted of 2 to 10 trees, depending on the availability of thinnings and site suitability. Directional felling allowed the creation of self-locking jams that mimicked natural windthrow. Jams such as these are more likely to persist during floods.

Costs to complete the WFP work described above ranged from CAD $2,500 to $3,500 ha^{-1}. These costs included thinning, girdling, planting, and modifications for biodiversity and in-stream structures. Forest Renewal B.C. provided project funding. These treatments are based in part on experience garnered from similar ecosystem issues in the Pacific Northwest (Poulin et al. 2000), although project effectiveness in restoring fish habitat is as yet undetermined. The alder logjams are experimental and will need to be monitored over time for their durability.

In the last 4 years, Weyerhaeuser Company Limited has likewise treated 550 ha (approximately 55 linear km) in riparian reserve zones, and were in the process of treating at least another 180 ha in the fiscal year 2003/2004. This work has been part of a larger policy of variable-density spacing to meet a number of management goals in its coastal B.C. Tree Farm Licenses. In addition to providing for long-term stream stability and fish habitat, these density reduction treatments are undertaken with the objectives of creating uneven stocking and distribution, and horizontal and vertical diversity; retaining or promoting shrub and herb layers (particularly for use as ungulate browse); and promoting some conifer growth (through release) on hardwood-dominated areas while retaining a hardwood component on conifer-dominated areas (Perry and Muller 2002). The goal is spatial variability, often achieved by creating open space around key ecological attributes in order to enhance horizontal and vertical complexity.

Innovative features of the Weyerhaeuser program have included the interspersion of 5 m wide untreated leave strips within the treatment blocks, in addition to 3 m "no touch zones" left along all stream banks. Up to 50 stems ha^{-1} of western hemlock are partly screefed or girdled to develop trunks with a rotten core and a live hard shell for cavity-nesting birds and small mammals. Red elderberry (*Sambucus racemosa*) was randomly cut back to a height of 1 m to stimulate growth for browsing by black-tailed deer (*Odocoileus hemionus columbianus*) and Roosevelt elk (*Cervus canadensis roosevelti*).

Weyerhaeuser's riparian restoration treatments have ranged in cost from CAD $2,200 to $2,800 ha^{-1} with an experienced crew. The long-term success of these projects will not be clear for decades, and any future monitoring will be on an ad hoc basis. Short-term indications of success include the fact that stand diversity has increased through the promotion of previously suppressed species such as cedar, and that shrub growth (including browse for elk) has increased considerably in some areas.

28.4.7 Ecological issue: access impacts to sensitive ecosystems

Road and access development have negatively impacted sensitive ecosystems and populations. Impacts include: road-induced mortality of sensitive species; easier access by humans that leads to disturbance of sensitive species (e.g., grizzly bear and bull trout [*Salmo confluentus*]); fragmentation of important habitats; reduced value and use of retained habitat; spread of invasive species; and sediment generation and slope failures as a result of poor road construction or maintenance. Prevention measures should include minimizing roaded area through long-term operational planning and limiting public access through pre-planned road closure (physical barriers) and signage. Restoration approaches could include: road decommissioning to deter recreational use; restricted access or road closures to sensitive areas; signage and other extension methods to educate road users about closures and invasive species control measures; and prompt revegetation of disturbed and deactivated areas using native plant seed to preclude establishment of invasive species.

Roads created to facilitate timber harvesting in B.C. have also increased public access to the backcountry, mainly for recreation. This increased public use has the potential to negatively affect wildlife and plant species as well as their habitat. Control of human access and recreational activities in key habitats will reduce the level of disturbance; however, regulation and access management have their limits. Road closures that are instituted for conservation purposes are often ignored because the general public has not been adequately informed about the reasons for such access control (Byman and Hawkins 2003). Educating the public about sensitive areas and species will be a crucial part of any successful access management plan. Although access management and public education are not usually considered elements of forest ecosystem restoration, these activities are crucial to the recovery and health of forest wildlife populations. In addition, the decommissioning and restoration of old forest access roads, as financed for several years under the FRBC Watershed Restoration Program, is an important element of landscape restoration. Various degrees of road deactivation (from access and sediment control to full deconstruction and reforestation) can help restore aquatic values and populations of hunted wildlife.

28.4.8 Ecological issue: landscape fragmentation

Landscapes have been excessively fragmented by failing to plan for connectivity between patches of important habitat and by harvesting at scales and using methods that do not mimic natural disturbance. Impacts include loss of habitat for some large animal species and species dependent on forest interior habitat and connected habitats. There is also the potential for increased losses of timber due to blowdown, and increased pest and disease risk in regenerated stands due to the reduction of buffers between stands of similar age and species mix. Prevention could include design of harvesting activities to mimic the scale and frequency of natural disturbance and maintenance of connectivity between habitat types. Restoration approaches should restore connective corridors through silvicultural techniques that accelerate the recovery of mature forest attributes in areas designated as landscape corridors for old-growth recruitment.

28.4.8.1 Case study 5: planning for landscape connectivity in the Morice and Lakes IFPA

Restoring landscape connectivity once it is lost is a difficult, costly, and long-term endeavor. As a result, most examples of addressing landscape fragmentation are planning examples in areas where options still exist to connect important habitats on a landscape scale. One example of planning for connectivity is under way in the Morice and Lakes IFPA (Innovative Forest Practices Agreement; see http://www.moricelakes-ifpa.com), where program partners are working cooperatively on pilot projects to minimize landscape fragmentation. Both fine-and coarse-filter biodiversity conservation approaches are applied to maximize connectivity between important habitats.

Ecosystems that are particularly rare in the region or are important to the conservation of particular species are identified on the landscape prior to harvest. Often, these are ecosystems that form critical habitat for one or more rare species. Using grizzly bears (*Ursus arctos horribilis*) as an example, critical habitats would include avalanche chutes, seasonal foraging areas, denning areas, and traditional trails. Critical habitats are then mapped and an analysis is done to determine how they can be connected using coarse-filter biodiversity measures.

Coarse-filter conservation measures generally entail managing a proportion of the landscape in keeping with the natural disturbance regime. In areas with infrequent natural disturbance, this will require keeping a certain amount of the landscape in a condition that is less disturbed, consisting primarily of mature and older seral stages, under the assumption that this will provide sufficient habitat to sustain the majority of forest-dependent species that use the area. For the grizzly bear, coarse-filter measures that pertain to connectivity include unharvested riparian corridors and old-growth management areas. In the Morice and Lakes IFPA area, important grizzly feeding, denning, and trail areas can be connected using riparian corridors and old-growth patches to provide for all the bear's habitat needs on the landscape. Where inadequate continuity of riparian and old growth habitats can be identified, restoration activities (as described in the previous sections) may be prescribed to accelerate the attainment of large tree and log sizes and complex canopy structures.

28.5 Conclusions

There is a growing acceptance of the need to take nontimber values into account in the management of B.C.'s forests. This awareness, often linked to market demand for more environmentally sensitive practices, has resulted in the initiation of various restoration projects throughout the province. To date, forest restoration has been conducted most commonly in interior dry forests and coastal riparian forests, and often uses public funds intended for this purpose. The deterioration of salmon stocks and spawning beds in many of B.C.'s rivers and streams prompted a major effort at upland, riparian, and in-stream restoration work in the 1990s. The scope of restoration work has gradually expanded to include a variety of broader terrestrial ecosystem restoration and forest productivity maintenance goals as well, although available funding for all restoration work has declined in recent years.

Many of the issues of ecological degradation only emerge at the landscape scale. Homogeneous stand structures and compositions are not without natural precedent, and do not constitute a problem in themselves, but are associated with reduced biodiversity and less ecological resilience when they dominate large areas. Nevertheless, the remedial solutions to landscape-scale problems can only be implemented on a stand-by-stand basis, or even in pockets within stands. This bottom-up approach consumes considerable time and resources, and its effectiveness remains to be demonstrated. Top-down approaches of preventing the drastic alteration of landscape pattern and disturbance processes are clearly more practical and are likely more effective.

There is a diversity of restoration needs and solutions in B.C., and a high degree of public support and professional expertise to carry them out (Ritchlin 2001). Many innovative and effective techniques have been developed to restore ecological features and processes to B.C.'s forests. However, this growing awareness and capacity is limited by economic and political constraints. Political and corporate leaders see restoration activities as optional, and most actions are not carried out at the scale and with the continuity in funding required to truly address the ecological issues in B.C.'s forests. The high cost of implementing forest restoration work is helping to demonstrate the importance and cost-effectiveness of preventing ecological degradation in the first place. Yet, many processes of ecological degradation continue, driven by commercial expediency and issues of global competitiveness in the development of the province's forest and mineral resources.

Future challenges to forest restoration in B.C. will include:

- The growing threat of invasive species, including vascular weeds, exotic insects, and fungal pathogens
- The backdrop of climate change that makes historical or undisturbed benchmarks less relevant
- The ongoing human-caused degradation that makes the restoration of certain forest ecosystems and species ever more expensive and difficult
- The need to monitor and learn from past restoration efforts, usually with little institutional support for monitoring and follow-up work
- The degree of socioeconomic latitude and support for repairing ecosystems when there may be no immediate economic return on such an investment.

Forest restoration activities are gradually being integrated with a wide range of commercial forestry operations and public land management agency mandates. For example, road deactivation is now considered a standard component of forest planning and operations. Interagency cooperation, public participation through volunteer organizations, and the identification of mutually supportive objectives have characterized successful restoration efforts to date. Work on the ground is evolving, and ecological restoration work is becoming increasingly professionalized. The background work of planning, funding, and developing partnerships in support of forest restoration is typically more time-consuming and challenging than the physical work being undertaken. Yet, the politics of resource management and the culture of land stewardship in B.C. have not evolved to such a level that an ongoing commitment to ecological restoration is assured.

Acknowledgments

We are indebted to the contributions of the many original researchers and practitioners who provided the information given here, including the case study contributors Glen Dunsworth, Reinhard Muller, Eric Gagné, Vince Poulin, Todd Manning, Brad Pollard, Art Moi, Patience Byman, Morice and Lakes IFPA participants, Jan Jonker, and Raj Prasad. We thank Mike Fenger, Colene Wood, Robert Seaton, Rachel Holt, and the other members of the Technical Advisory Committee to FRBC's Terrestrial Ecosystem Restoration Program for identifying and structuring ecological issues in the manner reported above. The constructive comments of John Stanturf, Carla Burton, Chris Hawkins, and an anonymous reviewer are gratefully appreciated. The authors can be contacted by e-mail at tanis.douglas@shaw.ca <mailto:anis.douglas@shaw.ca> and pburton@pfc.cfs.nrcan.gc.ca.

References

Adamowicz, W.L., and Burton, P.J., Sustainability and sustainable forest management, in *Towards Sustainable Management of the Boreal Forest*, Burton, P.J., Messier, C., Smith, D.W., and Adamowicz, W.L., Eds., NRC Research Press, Ottawa, Ontario, 41, 2003.

Anon., *Forest Practices Code of British Columbia Pine Stem Rust Management Guidebook*, B.C. Ministry of Forests and B.C. Environment, Victoria, B.C., Available online at http://www.for.gov.bc.ca/tasb/legsregs/fpc/fpcguide/PINESTEM/pine-toc.htm [viewed 23 October 2003]. 30 pp., 1996.

Apsey, M., Laishley, D., Nordin, V., and Paillé, G., The perpetual forest: using lessons from the past to sustain Canada's forests in the future, *For. Chron.*, 76, 29, 2000.

Bancroft, B. and Zielke, K., *Guidelines for Riparian Restoration in British Columbia: Recommended Riparian Silviculture Treatments, 1st Approximation*, B.C. Ministry of Forests, Victoria, B.C., Available online at http://www.for.gov.bc.ca/hfp/pubs/riparian/Riparian%20Guidelines%20Oct%2011up%20locked.pdf [viewed 22 May 2003]. 71 pp., 2002.

Beese, W.J., Dunsworth, G., and Perry, J., The Forest Project: three-year review and update, *Ecoforestry*, 16, 10, 2001.

Burton, P.J., An assessment of silvicultural practices and forest policy in British Columbia from the perspective of restoration ecology, in *Helping the Land Heal: Ecological Restoration in British Columbia Conference Proceedings*, Egan, B., Comp. and Ed., B.C. Environmental Network Educational Foundation, Vancouver, B.C., 173, 1998.

Burton, P.J., Ed., Garry oak ecosystem restoration: progress and prognosis, in *Proceedings of the Third Annual Meeting of the B.C. Chapter of the Society for Ecological Restoration*, B.C. Chapter of the Society for Ecological Restoration, Victoria, British Columbia, 109 pp., 2002.

Burton, P.J. and Burton, C.M., *Phase II Campground Restoration Prescriptions for Exchamsiks River Provincial Park*, Prepared for B.C. Parks, Terrace Field Office, B.C. Ministry of Water, Land and Air Protection, Symbios Research and Restoration, Smithers, B.C., 40 pp., 2002.

Byman, P. and Hawkins, C.D.B., *Strategic Plan for Public Education — Impacts of Road Use on Species at Risk in the Prince George Timber Supply Area (PG TSA)*, Prepared for Canadian Forest Products Ltd., Prince George Woodlands Division, Mixedwood Ecology and Management Program, University of Northern British Columbia, Prince George, British Columbia, Available online at http://mixedwood.unbc.ca/pdf/Educating%20the%20Public%20Strategic%20Plan%20March%202003.pdf [viewed 23 October, 2003]. 27 p., 2003.

Douglas, T.L., *Integrating Ecosystem Restoration Into Forest Management — Practical Examples for Foresters*, B.C. Chapter of the Society for Ecological Restoration, Victoria, B.C., 35 pp., 2003.

Fernow, B.E., *A Brief History of Forestry in Europe, The United States and Other Countries*, 3rd revised edition, University of Toronto Press, Toronto, Ontario, 506 pp., 1913.

Franklin, J.F., Berg, D.R., Thornburgh, D.A., and Tappeiner, J.C., Alternative silvicultural approaches to timber harvesting: variable retention harvest systems, in *Creating a Forestry for the 21st Century: The Science of Ecosystem Management*, Kohm, K.A., and Franklin, J.F., Eds., Island Press, Washington, D.C., 111, 1997.

Gayton, D., *Ground Work: Basic Concepts of Ecological Restoration in British Columbia*, FORREX Series No. 3., Southern Interior Forest Extension and Research Partnership, Kamloops, B.C., Available online at http://www.forrex.org/publications/forrexseries/ss3.pdf [viewed 21 May, 2003]. 25 pp., 2001.

Holt, R.F., *Strategic Ecological Restoration Assessment: Provincial Overview*, Prepared for Forest Renewal B.C. and B.C. Environment, Nelson, B.C., 2001.

Lewis, J.C., Creating snags and wildlife trees in commercial forest landscapes, *West. J. Appl. For.*, 13, 97, 1998.

Maloy, O.C., White pine blister rust control in North America: a case history, *Annu. Rev. Phytopathol.*, 35, 87, 1997.

Manning, T., *Fungal inoculation of trees as a habitat enhancement tool in second-growth forests.* TFL 44 Operational Trial - 2003 (Year 2) progress report. Report prep. for Weyerhaeuser Coastal Group, Nanaimo, B.C. March, 2004.

Meidinger, D. and Pojar, J., Eds., *Ecosystems of British Columbia*, Special Report Series 6, B.C. Ministry of Forests, Victoria, B.C., Available online at http://www.for.gov.bc.ca/hfd/pubs/Docs/Srs/SRseries.htm [viewed 21 May 2003]. 330 pp., 1991.

Nuzzo, V.A. and Howell, E.A., Natural area restoration planning, *Nat. Areas J.*, 10, 201, 1990.

Perry, J. and Muller, R., *Forest Project — Technical Project Summary*, Report #3, July 2002, Riparian Restoration Program, BC Coastal Group, Weyerhauser Inc., Nanaimo, B.C., 2002.

Peters, C.M., *Riparian Zone Habitat Rehabilitation of the Eve River, Vancouver Island, BC — A Preliminary Report*, Work term report for Weyerhaeuser, Nanaimo Woodlands, University of Victoria, Dept. of Biology, Victoria, BC., 1999.

Pojar, J., Klinka, K., and Meidinger, D.V., Biogeoclimatic ecosystem classification in British Columbia, *For. Ecol. Manage.*, 22, 119, 1987.

Pollard, B.T. and Prins, M.C., *Onion Lake terrestrial ecosystem restoration project year one: progress report.* Contracted report for Ministry of Forests - Kalum District. Funded by FRBC, 2001.

Poulin, V.A., Simmons, B., and Harris, C., *Riparian Silviculture: An Annotated Bibliography for Practitioners of Riparian Restoration*, Prepared for the B.C. Ministry of Forests, Victoria, B.C., V.A. Poulin & Associates Ltd., Vancouver, B.C., Available online at http://www.clayoquot.org/restoration.html [viewed 23 October 2003]. 32 pp., 2000.

Poulin, V.A. and Simmons, B., *Keogh River Riparian Restoration Phase 3 - November 2001.* Activity Report. Riparian Treatment Activity Project No.:722300. Habitat Enhancement Riparian Activity No.:722566. FRBC Work Plan No.:PAM02604. Prepared for: Western Forest Products Limited, Campbell River, BC. 51 pp, 2002a.

Poulin, V.A. and Simmons, B., *Goodspeed River Riparian Restoration: 2001.* Activity Report. Riparian Treatment Activity Project No.: 722300. Habitat Enhancement Riparian Activity No.:722566. FRBC Work Plan No.:PAM02604. Prepared for: Western Forest Products Limited, Campbell River, BC. 51 pp, 2002b.

Prasad, R., Some aspects of the impact and management of the exotic weed, Scotch broom (*Cytisus scoparius* [L.] Link) in British Columbia, Canada, *J. Sustain. For.*, 10, 341, 2000.

Prasad, R., *Scotch Broom, Cytisus scoparius L. in British Columbia*, Pacific Forestry Centre, Canadian Forest Service, Victoria, B.C., Available online at http://www.pfc.forestry.ca/biodiversity/broom_e.html. 6 pp., 2003.

Ritchlin, J., *Healing the Land … Healing Ourselves: A Guide to Ecological Restoration Resources for British Columbia*, B.C. Environmental Network Foundation, Vancouver, B.C., 2001.

Society for Ecological Restoration Science and Policy Working Group (SER). 2002. The SER Primer on ecological restoration. Available online at http://www.ser.org/Primer.pdf [viewed February 24, 2004].

Seaton, R., Ed., Ecosystem at risk: Antelope brush restoration, in *Proceedings of a Workshop* March 28–30, 2003, Osoyoos, *British Columbia*, B.C. Chapter of the Society for Ecological Restoration, Victoria, B.C., 76 pp., 2003.

Seymour, R.S. and Hunter, M.L., Principles of ecological forestry, in *Maintaining Biodiversity in Forest Ecosystems*, Hunter, M.L., Ed., Cambridge University Press, Cambridge, U.K., 22, 1999.

Stanturf, J.A. and Madsen, P., Restoration concepts for temperate and boreal forests of North America and Western Europe, *Plant Biosyst.*, 136, 143, 2002.

Stevens, V., *The Ecological Role of Coarse Woody Debris: An Overview of the Ecological Importance of CWD in BC Forests*, Working Paper 30, Research Branch, B.C. Ministry of Forests, Victoria, B.C., Available online at http://www.for.gov.bc.ca/hfd/pubs/docs/Wp/Wp30.pdf [viewed 22 May 2003]. 26 pp., 1997.

Tesky, J.L., *Strix occidentalis*: Biological Data and Habitat Requirements, in *Fire Effects Information System (FEIS)*, U.S. Department of Agriculture, Forest Service, Rocky Mountain Research Station, Fire Sciences Laboratory, Missoula, MT. Available online at http://www.fs.fed.us/database/feis/animals/bird/stoc/biological_data_and_habitat_requirements.html [viewed 23 May 2003]. 1992.

Underhill, D.J., The legacy of WRP, *Streamline*, 6, 8, 2002., Available online at http://srmwww.gov.bc.ca/frco/bookshop/docs/vol6_no4.pdf.

chapter twenty-nine

Forest restoration practices in the Pacific Northwest and California

Kevin L. O'Hara and **Kristen M. Waring**
University of California – Berkeley

Contents

29.1 The region

The Pacific Northwest of North America in this chapter includes the area west of the Cascade Mountain range, the interior Columbia River basin, and northern California. Forests in Canada are not included, although they share many of the restoration issues

1-56670-635-1/05/$0.00+$1.50

presented. The forests in this region include the temperate rainforests of the Pacific coast, the mixed broadleaf/conifer forests and redwood forests of the California coast, the subalpine forests of the Cascades, Blue Mountains, and northern Rocky Mountains, and the dry interior forests that occur east of the Cascades and in the northern Rocky Mountains. Also within this region are the mixed-conifer forests of the Sierra Nevada and southern Cascades, and the mixed-conifer forests of northern Idaho and western Montana. More information on the forest types and their management is available in Eyre (1980) and Barrett (1995).

This chapter focuses on treatments that address important forest ecosystem restoration issues for this region. Although the objective of ecosystem restoration is relatively new to the Pacific Northwest and California, many of the tools needed are well developed, such as prescribed fire and silvicultural tools for enhancing structural diversity.

29.2 Forest issues

Forest ecosystem management in the Pacific Northwest and California is presently dominated by controversy over preservation of old forests, endangered species, and the management of public lands. These complex issues involve conflicting social and ecological objectives, often resulting in judicial intervention. In the case of old forests, there are conflicting pressures to preserve and restore old forests and maintain employment in rural communities in areas dependent on timber production.

Old forests (also known as old-growth, ancient, late-successional, or late-seral forests) are the most contentious forest management issue in the Pacific Northwest and California. Debates over the actual dependence of many old-forest species and over the amount of old-forest necessary are prolonged by disagreement over old-forest definitions. The objectives of restoration are to establish old-forest structures from younger stands and to a lesser extent grow new forests to eventually form part of the old-forest base. Restoration includes treatments to increase growth rates, enhance stand structural variability, and increase the number of snags (standing dead trees) and downed coarse woody debris. Merely having large trees and other structural features of old-forests, however, does not provide all the features of old-forests such as some of the processes associated with old forests. These restored stands will require many decades and even centuries to develop all the features associated with old forests. Additionally, the amount of old-forest structure, the size of old-forest patches, and the arrangement and interconnectivity of these patches represent important issues for landscape-level restoration.

Related to the old-forest issue, but broader, is the protection of wildlife species of concern. The 1973 Endangered Species Act required that all plants and animals be protected in their native habitat. This law gave special status to all species deemed endangered and has been the source of controversy, particularly in the Pacific Northwest and California. Endangered bird species such as the northern spotted owl (*Strix occidentalis caurina*) and the marbled murrelet (*Brachyramphus marmoratus*), and several anadromous fish species have become major considerations in forest management and in developing ecosystem management policy. The need for wildlife habitat at stand and landscape scales drives many forest restoration programs.

Controversy over old-forest preservation and endangered species protection has caused public dissatisfaction with current land management and the state of ecosystems in the region. Public land managers are responding by de-emphasizing strictly utilitarian goals and moving toward restoration and protection goals that will satisfy societal demands. Nevertheless, pressure to maintain rural communities dependent on using forest resources to provide employment often runs counter to preservation values associated with biodiversity and restoring forests to some form of presettlement condition.

29.3 Restoration issues

Of the many forest restoration issues of importance in the Pacific Northwest and California, four of the most visible and controversial are addressed in this chapter: restoration of old forests, stand structural diversity, forests affected by exotic pests, and fire-resistant stand structures. These issues are related in many ways, but for all, stand structure is key. For example, the scarcity of old-forest structures in the region contributes to the need for greater stand structural diversity. Organization of structures to form defensible fire protection zones has implications for the connectivity of wildlife habitat or spread of exotic pests.

29.3.1 Restoration of old forests

Old forests in the Pacific Northwest and California range from the complex multistoried, coastal temperate rainforests (Figure 29.1) to relatively simple single-storied savannah forests of the inland west (Figure 29.2), showing striking differences in stand structure (O'Hara et al. 1996). These varied ecosystems share a common feature; they include some old trees. Disturbances occur infrequently in the coastal forests, with long time periods between events. In contrast, the dry interior forests of ponderosa pine and western larch often experience very frequent low-intensity fires that maintain a relatively open understory. At one extreme, the old forests along the coast exist by becoming old during the relatively long intervals between disturbances or by surviving disturbance. At the other extreme, the frequent disturbance maintains the structure of the dry forest. Between these extremes are other stand structures that add to landscape complexity.

29.3.1.1 Replacement old forests

In many places the matrix of forest structures is lacking old forests, or the spatial relationship of old forests with other structures is inadequate. To overcome these deficiencies, managers may encourage the transformation of younger, even-aged forests into old-forest structures, or they may grow new forests that will become old forests. In either case, the solution will not be rapid. Management actions to increase structural diversity and efforts to enhance old-forest characteristics are often similar; treatments to enhance structural diversity are discussed in Section 29.3.2.

Figure 29.1 Coast redwood old forest in California.

Figure 29.2 Savannah structure of an old forest ponderosa pine stand on the Flathead National Forest in Montana.

29.3.1.1.1 Enhancing old-forest characteristics in younger even-aged stands For most forest types in the region, published ranges of old-forest characteristics are available that describe numbers of large trees, canopy strata, snags per unit area, amount of coarse woody debris (CWD), and other characteristics (Spies and Franklin 1991; Fites et al. 1992; National Research Council 2000). Generally, old forests have low density and relatively old trees. In comparison, younger even-aged forests generally have higher numbers of trees and lower average tree size. In many forest types, the structure of younger forests is often simple: one canopy stratum, primarily one species, high stocking, few snags, and scarce CWD. Encouraging old-forest characteristics in younger even-aged forests can take a variety of forms including thinning to produce an open stand, thinning to promote the development of additional canopy strata, or killing standing trees to develop snags (Oliver et al. 1994; Bailey and Tappeiner 1998).

It is also important to restore processes as well as structural features. For example, thinning in dry ponderosa pine forests may produce structures resembling presettlement forests formed by low-intensity surface fires, but thinning does not reduce litter, control shrubs, or raise crown heights. All ecological effects of low-intensity surface fires cannot be simulated with mechanical treatments and this may be true generally; silvicultural surrogates for natural disturbances are incomplete substitutes.

Likely treatments for enhancing old-forest structures include those to produce snag and cavity nesting trees. These trees may be purposely killed by girdling, herbicide injection, topping below the lowest live branch, using pheromones to attract bark beetle, or with fire (Ross and Niwa 1997). Living trees can also become habitat for some wildlife species by topping within the live crown to provide nest sites, carving nest holes, and injecting stem decay fungi to create cavities and attract insects. Felling trees to the forest floor or into streams can produce coarse woody debris.

29.3.1.1.2 Growing 'new' old forests Although a lengthy process, growing old forests from new stands is needed to ensure the future availability of old-forest structures. All forests in the region are susceptible to catastrophic disturbance; replacement stands can be grown in landscapes that are intended to include old forests indefinitely. Growing replacement old forests will require considerable foresight, planning, and, most importantly, consistent management objectives over time.

Historically, old forests in the Douglas-fir portion of the region often originated at low densities and individual trees grew rapidly when young (Tappeiner et al. 1997). Young Douglas-fir stands attained many old-forest characteristics within 100 years (Acker et al. 1998) and early thinnings may be critical to hastening the development of old-forest features. Whether old forests in other parts of the region began early stand development under similar conditions is unknown. It is also unclear how forests develop on sites of low productivity where trees grow slowly, regardless of spacing. However, the rapid attainment of size is a key feature of old-forest restoration, and wide spacings can provide sustained levels of fast growth (Oliver and Larson 1996).

Replacement old-forest stands may be established by planting at wide spacings or by heavy thinning of young stands. Artificial regeneration technologies are highly developed in the region and the potential for success is high (Hobbs et al. 1992). Natural regeneration is also effective, provided it can be applied within local forest practice regulations. Either method can produce an uneven spatial distribution of trees in the new stand. Once the trees emerge above shrubs and other competitors, height growth rates should be rapid.

In many parts of the region, aggressive shrub and broadleaf tree species can suppress conifers for decades. For example, red alder (*Alnus rubra*), manzanita (*Arctostaphylos* spp.), ceanothus (*Ceanothus* spp.), and various *Rubus* species are often treated manually or with herbicide. Whether sufficient conifers could emerge through this competition without such assistance and grow suitable old-forest structures is unknown, but these competitors possibly are partially responsible for the low stand densities reported in reconstructions of old forest development.

29.3.1.2 Landscape issues and old forests

Organizing types of stand structures over landscapes to meet diverse and often-conflicting land management objectives is greatly challenging. Because old forests are being restored for their value for specific wildlife species and biodiversity in general, additional spatial considerations of connectivity, edge, and patch size are important and add complexity to land management decisions. Understanding how present landscapes developed and reconstructing presettlement landscape patterns are critical to setting management objectives.

The management of many forested landscapes in the coastal areas of the Pacific Northwest and California began with the extraction of merchantable timber along river valleys by railroad. In many valleys, nearly all timber was removed, while in other valleys harvesting was uneven. Many areas burned during or near the time of the harvest. The resulting cutover and burned landscapes were the equivalent of massive clearings. In other areas, the landscape was a patchwork of stand structures that included unmerchantable old forests and younger stands, as well as cutovers. Later harvesting was staggered; relatively small clearcut harvest units were spread across the landscape (Figure 29.3). These coastal landscapes have therefore gone through periods of homogenization and fragmentation (Harris 1984).

Interior forests often received similar treatment, although there was a concerted effort to manage some dryer forests with single-tree selection (Meyer 1934). Land ownership patterns also contributed to the current mosaic as alternate 260 ha sections are often

Figure 29.3 Small clearcuts on the Gifford Pinchot National Forest in Washington.

owned by different organizations with different management objectives. The present inventory of old forests in the region includes isolated patches that were inaccessible or unmerchantable at the time of harvest, stands that were partially harvested but still maintain old-forest features, and stands on public ownerships that were not harvested. These old forests are arranged across landscapes in patterns that range from very small pockets to extensive areas of continuous old-forest structure.

Restoration begins with reconstructing historical landscape patterns (Morgan et al. 1994) and identifying desired future conditions. Dendrochronology and fire-scar analysis can be used to reconstruct landscape patterns of historical disturbance in areas with fire-dominated disturbance regimes, provided there are sufficient residual trees or stumps to provide evidence of fire history (Morrison and Swanson 1990; Arno et al. 1993; Cissel et al. 1994; Wallin et al. 1996). In coastal areas where wind disturbance dominates, or in areas where insects, pathogens, or other disturbances dominate, landscape reconstruction is more difficult. Modeling has been used to project spatial and temporal landscape change (Cissel et al. 1998) under different scenarios. These models are highly variable in their modeling approaches and ability to model landscape change. Examples include CRBSUM (Keane et al. 1996), LMS (McCarter et al. 1998), and SIMPPLLE (Chew 1995).

Once the desired future landscape condition is identified, restoration may consist of actively directing vegetation development, or passively waiting for desired stand structures to develop. Stand-level treatments are the primary means of directing these landscape-level changes, emphasizing the importance of silviculture for landscape management (O'Hara et al. 1994).

29.3.2 *Enhancing stand structural diversity*

The agricultural paradigm that favored high-yield single-species plantations through the 1980s minimized structural diversity. Although plantation forestry is still common in many areas, the demand for greater structural diversity in forests of the region is being met by the development of new techniques, including stands of mixed-species composition, variable spacing, multiaged, multistrata, and various combinations of these features. Ultimately, most restoration is a compromise between achieving a desired condition and the constraints created by previous human use and current resource demands.

29.3.2.1 Overstocked stands

Many overstocked stands in the Pacific Northwest and California are the result of unthinned plantations and normal development following stand replacement disturbance. Options for management can be limited in overstocked stands if they have stagnated for a long time (O'Hara and Oliver 1999). A reduction in density by thinning often results in massive windthrow or storm damage to residual trees (Wilson and Oliver 2000; Wonn and O'Hara 2001). In these cases, clearcutting may be the only viable option.

29.3.2.2 Reforestation

Increasing species diversity is one way to enhance structural diversity in natural stands and plantations. In natural regeneration, seed sources from several species can be encouraged instead of just one. In plantations, species can be randomly mixed at the planting stage, or planted in alternate rows to provide better access for later harvesting of select species. Seedlings can also be planted in species groups to minimize interspecific competition and assure that each species has some trees in the upper canopy.

29.3.2.3 Spacing, tree size, and structural diversity

Wide spacings that promote rapid tree growth also promote greater understory growth, thereby accelerating the development of understory reinitiation, with more or less continuous regeneration over time as tree size and canopy diversity increase. Unfortunately, forest practice regulations in many western states have minimum restocking requirements that may preclude reforestation at low densities. For example, Washington State requires about 470 trees ha^{-1} west of the Cascades and about 370 trees ha^{-1} east of the Cascades. The requirement in California is approximately 740 trees ha^{-1} on better sites and 370 ha^{-1} on poorer sites. These reforestation requirements were intended to ensure adequate restocking of cutover stands. However, there is generally more flexibility to enhance diversity as these stands age.

Precommercial thinning traditionally has been used to improve quality in even-aged stands. For example, trees are removed to leave more consistent spacing, and suppressed trees are removed. The residual stands have less structural diversity than before treatment. With different priorities, however, precommercial thinning can become a tool for enhancing structural diversity. Variable-density thinning is one approach that deliberately leaves trees on variable spacings so that some residuals are nearly in open-grown conditions whereas others are in heavy competition (Figure 29.4). The result is a variety of tree sizes in an even-aged stand (Thornburgh et al. 2000).

Variable-density thinning has been used in older coastal Douglas-fir stands to encourage the development of old-forest characteristics (Carey et al. 1999; Carey 2001; Thysell and Carey 2001). Other work indicates that traditional thinning in coastal Douglas-fir is effective for enhancing structural diversity by developing new age classes, encouraging the growth of advance regeneration and multistory stands (Newton and Cole 1987; Curtis and Marshall 1993; Tappeiner et al. 1997; Bailey and Tappeiner 1998). Less information is available on forest types other than coastal Douglas-fir; however, similar trends may be expected to occur based on spacing trials. Wider spacings with ponderosa pine result in greater understory growth (Barrett 1982; Cochran and Barrett 1999).

29.3.2.4 Stratified mixtures

The height stratification of mixed-species stands can potentially contribute to stand-level diversity (Oliver and Larson 1996). These stand development patterns offer more structural diversity than single-species stands. For the land manager, generally all that is

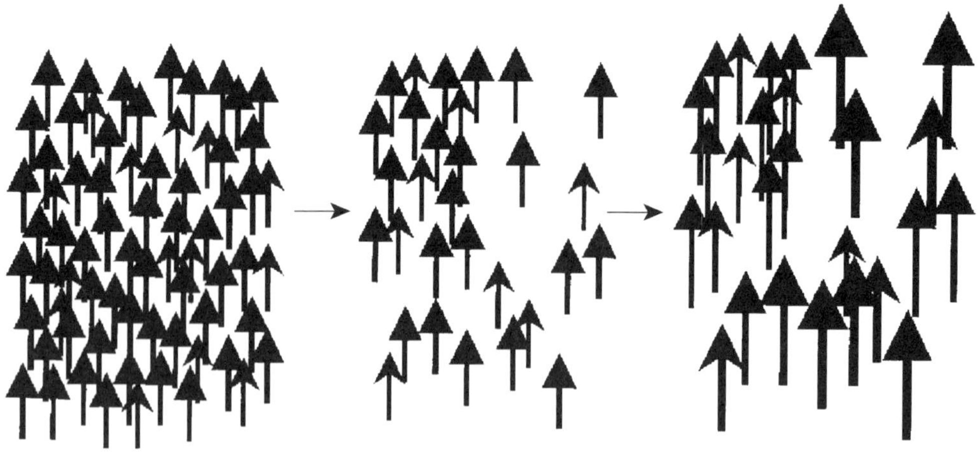

Figure 29.4 Variable-density thinning. Residual trees are thinned to have different amounts of growing space (middle) and grow to form more variable structures (right).

required is to establish the different species and give individual trees enough space for long-term development.

29.3.2.5 *Extended rotations*

Stands generally acquire more structural diversity as they age (Oliver and Larson 1996), so that a simple means of enhancing diversity in even-aged stands is to extend rotations beyond the economic and volume-based limits commonly used for timber objectives. Extended rotations in thinned coast Douglas-fir stands often do not reduce long-term volume production and probably increase timber value production (Curtis 1997). The effects of extended rotations have not been examined for other species, but other conifers in the Pacific Northwest and California may behave in a manner similar to Douglas-fir. The positive effects of extended rotations on aesthetics, wildlife habitat, and biodiversity are most evident in areas where shorter rotation, timber production is the norm (Curtis 1997; Curtis and Carey 1996).

29.3.2.6 *Options for retention*

Regeneration treatments present opportunities to enhance structure by leaving reserve trees, which are pole sized, or larger trees retained after the regeneration period (Helms 1998). Reserve trees have traditionally served three purposes: seed sources for subsequent regeneration; shelter for regeneration; and structural features to provide a second age class for a two-aged stand. The former two purposes are part of traditional regeneration methods such as seed tree or shelterwood methods. The latter is a relatively new purpose that falls within a number of restoration objectives (Franklin et al. 1997).

The spatial arrangement of reserve trees provides considerable opportunity to vary the resultant structural arrangement, from aggregated to dispersed patterns (Figure 29.5). The amount of reserve trees is important as well: too few reserve trees increase windthrow risk, and too many trees reduce understory growth and inhibit regeneration.

Gaps and patches are structural options related to dispersed or aggregated retention trees, but generally involve less thinning of uncut matrix areas. Creating canopy gaps to promote regeneration is analogous to group selection, where gaps are wide enough that some of the gap receives full sunlight. Compared to closed forests, seedlings in gaps are larger and increase in size toward the center of the gap (Gray and Spies 1996; York et al. 2003). Patches are clumps of trees left in conjunction with streamside protection zones and wildlife corridors to provide connectivity between forested areas. Patch size and location

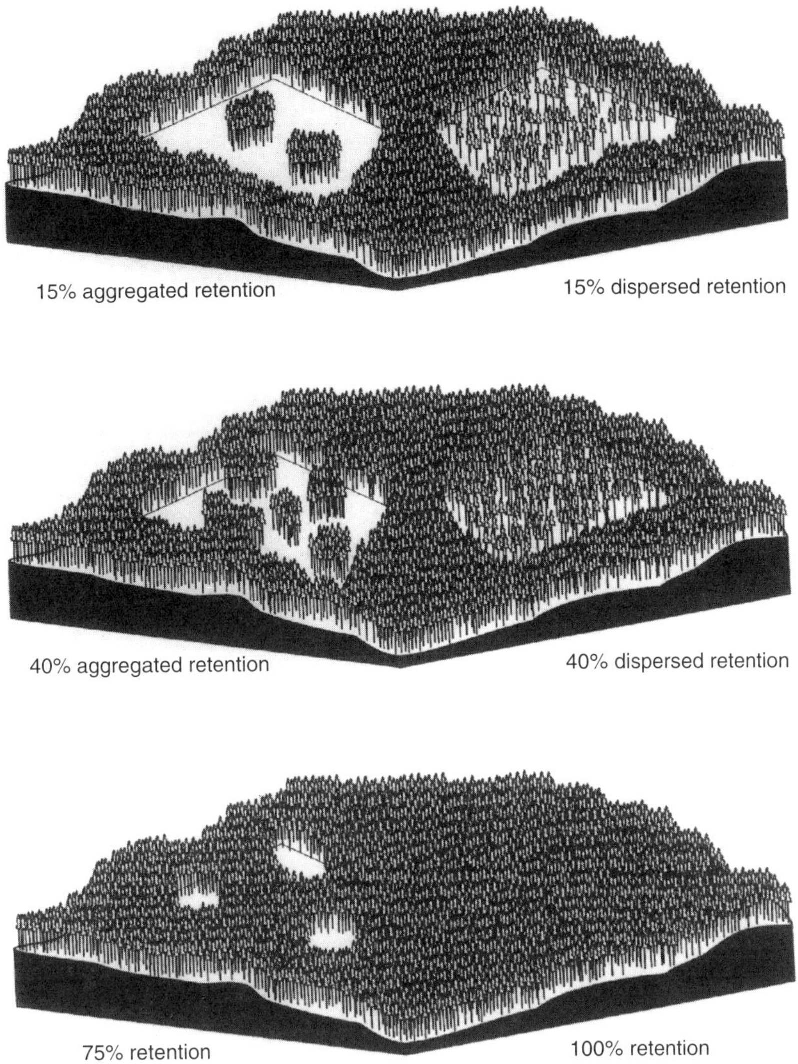

Figure 29.5 Diagram of variable retention options (from Aubry, K.B., et al. *Northwest Sci.*, 73, special issue, 12, 1999. Reprinted with permission). Copyright by Northwest Scientific Association

sometimes are a function of harvesting efficiency, particularly in areas with steep terrain or mixed property ownerships.

29.3.2.7 Multiaged stands

Multiaged stands are another means to enhance structural diversity. Techniques of multiaged silviculture that incorporate natural disturbance frequencies should provide more natural structures than either even-aged stands or traditional uneven-aged structures (O'Hara 1998). The even-to multiaged conversion (or transformation) process involves new age classes or cohorts to develop by encouraging understory reinitiation (Oliver 1980). In understory reinitiation, growing space is made available through canopy decline, mortality, or treatments that reduce stocking. Before any stand can be converted to multiaged, one must assess the potential for residual trees to remain stable and avoid storm damage. Schütz (2001) developed a flowchart (Figure 29.6) for the conversion process with three key decision nodes: the present level of differentiation or irregularity in the stand; stability of residual trees; and potential longevity of the trees. An even-aged stand can be continued as an even-aged stand, converted to multiaged, or regenerated (Figure 29.6).

Traditional structures characterized with reverse-J diameter frequency distributions are often the target for uneven-aged stands, but simpler structures may better resemble presettlement conditions and are often easier to implement (O'Hara 1998). Ponderosa pine is among the forest types easiest to manage in multiaged stands. Open canopies typically provide adequate light for regeneration and even- and multiaged stands are fairly common in all parts of the species range. Many of the mixed-species forest types, on the other hand, will tend to be dominated by shade-tolerant species if managed with traditional uneven-aged systems. Two-aged systems or systems where retention blocks are aggregated to allow sufficient sunlight will probably work best in mixed-species forests such as the forests of the western hemlock zone, the mixed-conifer forests of the Sierra Nevada, and the mixed forests at higher elevations in the eastside forests including north Idaho.

29.3.3 Exotic pests

Introduced insects, pathogens, and plants alter forest ecosystems in dramatic and frequently negative ways. Even those pests that do not cause widespread tree mortality create changes in such ecosystem attributes as stand structure, regeneration, and light environment. Restoration of these ecosystems becomes especially problematic where little is known about the ecosystem prior to introduction or about the basic ecology of the pest. The Pacific Northwest and California face new introductions as global trade expands, while continuing to confront hundreds of past introductions.

29.3.3.1 Exotic insects and pathogens

The primary effect of an exotic insect or pathogen is the loss or decline of one or more tree species. A common initial approach to restoration is the search for genetic resistance to

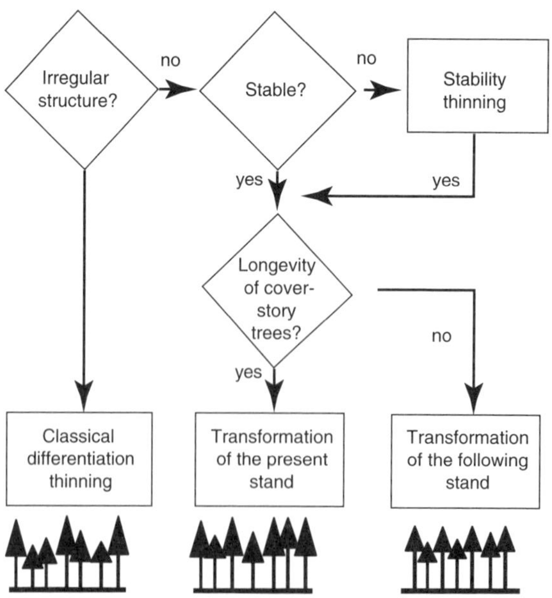

Figure 29.6 Diagram of decision key for transforming even-aged to uneven-aged stands (modified from Schütz, J.-P., *For. Ecol. Manage.* 151, 87, 2001. Reprinted with permission).

exotic pathogens within the host species. One such example involves five-needled white pines and white pine blister rust (caused by *Cronartium ribicola*). Genetic resistance within the white pines, such as sugar pine (*Pinus lambertiana*) and western white pine (*Pinus monticola*), has been coupled with increased knowledge of the ecology of *Cronartium ribicola* and techniques that minimize disease impacts. Specific techniques include regulation of stocking levels, pruning, and careful site and species composition selection. Thus, restoration efforts integrate on-the-ground management tools to provide habitat that is both favorable for the target tree species and unfavorable for the pathogen (Hunt 1998; O'Hara et al. 1995).

Natural Monterey pine (*Pinus radiata*) forests in California provide an example of challenges facing restoration following exotics. Widespread decline of Monterey pine has followed introduction of *Fusarium circinatum*, the causal agent of pine pitch canker disease. The ecology of native Monterey pine stands is not well understood; hence, providing a quantitative basis for restoration may be difficult. Insects transport *F. circinatum*; thus these insects must be incorporated into management and restoration programs. A recent restoration plan for Monterey pine forests included three primary goals: to speed genetic adaptation of the ecosystem to the pine pitch canker disease; continue exotic weed suppression; and allow for natural and surrogate disturbance processes (Storer et al. 2001). Critical components of any restoration system integrate local issues (such as land ownership) with the ecology of both the host system and the exotic. Without this knowledge base, restoration strategies are likely to take longer to complete and may be subject to failure.

Although introduced pests may pose the largest threat, native pests cannot be ignored. Restoration programs that target one pest must incorporate the risk of loss or damage due to native pests following restoration treatments, particularly if the target species for restoration is the primary host for these pests. For example, residual trees may be susceptible to secondary insects such as the red turpentine beetle (*Dendroctonus valens*) or pine engraver (*Ips pini*) if left in poor vigor or near areas, such as downed logs or slash piles, where insect populations may increase dramatically. Disturbance events such as fire may create perfect conditions for the establishment of exotic plants. Land managers must incorporate these risks into the overall restoration treatment plan, while also establishing contingency plans in case of unanticipated effects.

29.3.3.2 Exotic plants

Exotic plants, hundreds of which have established in the Pacific Northwest, create dramatic changes in ecological community composition, structure, and function. Restoration efforts aimed at remediation of other forest issues may actually increase exotic plant problems. For example, thinning and burning to alleviate problems with large-scale fire or insect and disease events may create a suitable seedbed for germination and establishment of weeds. However, prescribed fire may also improve the diversity and production of native plant communities, thus making the forest more resistant to exotic plant invasion or establishment (Harrod 2001). Due to the potentially negative aspects of forest restoration strategies, it is critical to include invasive weed management in a comprehensive forest restoration plan.

Prevention of weed introduction, establishment, and expansion may be a more important aspect than restoration to weed management programs. However, control can be successful and may involve prescribed fire, biological control, herbicides, and manual or mechanical removal. Restoration of invaded sites may involve replanting or seeding native plants, and is often entered into simultaneously with control (Harrod 2001). Additionally, if restoration is aimed at reducing invasive weeds, natural disturbance processes such as flooding or fire can be incorporated into the plan. Desired future conditions may be met through directing plant communities through various development pathways (Sheley et al. 1996). Linking more than one control method may provide expanded possibilities in some cases. For example, spotted knapweed (*Centaurea maculosa*) may best be controlled by first reducing the overall population with herbicides or fire, and then maintaining the low population with biological control insects. Another option is to use herbicides first, and then reintroduce fire when the threat of weed invasion is lower (Rice 2000).

29.3.4 Reintroduction of fire

Reintroducing fire into forests after decades of suppression has gained popularity as a way to restore many fire-dependent ecosystems. This has resulted in new fire policy among public agencies, such as allowing some wildfires to burn and actively using prescribed burning.

Decades of fire suppression have left millions of ha of overstocked stands in the Pacific Northwest and California, mostly in inland areas where unmanaged forests are common or precommercial thinning funds are scarce (Mutch et al. 1993). Past management of many of these forests included partial harvests that removed only commercially valuable trees, leaving smaller trees and high-graded stands. Decades of fire suppression formed stands with high stocking throughout multiple size classes with fuel ladders leading to risk of crown fires. Restoration treatments provide opportunities to produce fire-resistant stands and organize them across landscapes.

29.3.4.1 Prescribed burning

Historical fire intensities in the Pacific Northwest and California ranged from low-severity nonlethal understory burns to high-severity stand replacement fires, depending upon forest ecosystem. This variability allows for the creative use of prescribed burning in forest restoration, and the integration of fire with other silvicultural tools. Examples include widespread use of fire in dry ponderosa pine ecosystems by simulating the historical fire regime of low-intensity, high-frequency fires, sometimes preceded by thinning (Mutch 1994; Smith and Arno 1999). In subalpine zones, current policy allows for lightning-ignited, high-intensity fires that begin within wilderness area boundaries to burn under prescribed conditions, again simulating pre-suppression era regimes of stand-replacement fire every 35 to 300 years. Human-ignited, high-intensity, stand-replacement prescribed burns have become important components within some forest restoration treatments, an event that would have been unlikely just 30 years ago (Hardy et al. 2000; Keane and Arno 1996).

Prescribed burning becomes hazardous when long periods of fire suppression have built up large amounts of fuel (Figure 29.7). In these cases, even an understory burn can lead to stand replacement fires. Lethal understory burns have increased in frequency, while nonlethal understory burns have decreased (as a proportion of total fires), as compared to presuppression fire history (Brown et al. 1994). Simulating some fire effects using surrogate treatments such as thinning are potential alternatives to fire in these situations.

29.3.4.2 Stand-level fuel treatments

Stands resistant and resilient to fire generally have low stocking levels, large trees, high crown base height, absence of vertical fuel ladders, thin litter and duff layers, and low levels of living and dead surface fuels. Variations in species composition can also be important because of differences between species in crown bulk density, which affects the behavior of crown fires. Stands in many western forests commonly develop with shade-intolerant conifers in the canopy above more shade-tolerant species in the understory. These understory species form fuel ladders that facilitate the spread of surface fires into the main canopy. They often have higher crown bulk densities for foliage than less shade-tolerant species (Brown 1978). Low thinning removes many of these understory trees, thereby reducing density, raising live crown heights, and removing more flammable tree species (Graham et al. 1999). Surface fuel treatments are also essential, particularly after thinning. Surface fuel treatments include piling, burning, lopping, crushing, and chipping, and are often done in combination with other fuel treatments. Following timber harvest, these operations are usually part of site preparation treatments. Machine or hand piling treatments often accompany intermediate operations such as thinnings. Chipping for biomass utilization is also an option (Keyes and O'Hara 2002; Witherspoon and Skinner 1995; Graham et al. 1999).

29.3.4.3 Landscape-level organization of stand structures

Restoration of fire-resistant forests also involves landscape-level organization of stands with different structures. Fire regime largely determines landscape vegetation patterns. Low-severity fires generally result in small patches with low amounts of edge (Table 29.1). High-severity fires are typically large and result in large patch size. Moderate-severity fires kill some trees and leave others uninjured, resulting in high spatial variability and

(A)

(B)

Figure 29.7 Accumulated litter under ponderosa pine after decades of fire suppression (left), and after burning charred stem shows depth of litter before burning (right).

Table 29.1 Landscape Characteristics of Western Forest Fire Regimes (Agee 1998)

	Fire Regime		
Landscape Character	Low Severity	Moderate Severity	High Severity
Patch size	Small (<1 ha)	Medium (1–300 + ha)	Large (1–10,000 + ha)
Edge	Low amount	High amount	Moderate amount
Pre–post fire similarity	High	Moderate	Low

Patch size denotes the average area in which tree regeneration will be open-grown. Similarity is the proportion of the total area burned that resembles the prefire forest structure.

high amounts of edge (Agee 1998). Given variations in terrain, forest type, time since last fire, and levels of accumulated fuels, many patterns are possible on any given landscape.

Landscape-level fuelbreaks have gained interest in recent years as a means to defend areas against large wildfires by breaking the continuity of fuels. Current interest is in "shaded fuelbreaks" where stand structures are modified, but not removed, over large areas. Shaded fuelbreaks can be located strategically to modify fire behavior such that the size and severity are reduced (Agee et al. 2000). The Quincy Library Group in the northern Sierra Nevada of California designed stand-level treatments to reduce density and fuel ladders that would result eventually in an uneven-aged forest maintained with single-tree and group-selection treatments. Fuelbreaks would be applied in strips approximately 400 m wide and placed along roads or other defensible locations. These shaded fuelbreaks were not intended to substitute for prescribed burning; instead, they were intended as a precondition to expanded use of prescribed burning and part of a comprehensive strategy to reduce fuels (Weatherspoon and Skinner 1996). The desired future condition of the stands in these landscapes is a structure that is fire-resistant and, coincidentally, corresponds to a presettlement condition.

Restoration of fire-resistant stand structures or presettlement fire regimes will require more than single treatments. Fire-prone forest types that require mechanical treatment or prescribed burning will need periodic treatments on a schedule that resembles natural fire frequency. In dry ponderosa pine forests, for example, initial treatments will be followed by redevelopment of dense stands and new fire ladders. Merely treating them once will only provide a temporary solution. In this light, forest restoration has to be considered a long-term, recurring process.

Forest restoration requires that processes and functions be restored. A knowledge of stand structure development and the function of particular structures in the broader context of ecosystems are critical for designing restoration treatments that consider the dynamic nature of forest ecosystems. Restoration to a particular target condition will result in a structure that will soon develop into another. Thus, restoration programs should target a range of conditions centered on predictable changes or processes over time. Some objectives of forest restoration will conflict with other objectives. For example, the diversity of stand structures that provide habitat for some wildlife and improve aesthetics may also promote high flammability with large amounts of fuel and vertical fuel ladders. Likewise, thinning to reduce fuel hazards will reduce structural diversity. Restoration of fire as an ecological process may be a goal of restoration. Prescribed burning can also meet other goals, such as control of noxious weeds and increasing biodiversity (Walstad et al. 1990).

References

Acker, S.A., Sabin, T.E., Ganio, L.M., and McKee, W.A., Development of old-growth structure and timber volume growth trends in maturing Douglas-fir stands, *For. Ecol. Manage.*, 104, 265, 1998.

Agee, J.K., The landscape ecology of western forest fire regimes, *Northwest Sci.*, 72 (special issue), 24, 1998.

Agee, J.K., Behro, B., Finney, M.A., Omi, P.N., Sapsis, D.B., Skinner, C.N., van Wagtendonk, J.W., and Weatherspoon, C.P., The use of shade fuelbreaks in landscape fire management, *For. Ecol. Manage.*, 127, 55, 2000.

Arno, S.F., Reinhardt, E.D., and Scott, J.H., *Forest Structure and Landscape Patterns in the Subalpine Lodgepole Pine Type: A Procedure for Quantifying Past and Present Conditions*, U.S. Department of Agriculture, Forest Service, General Techincal Report INT-294, 1993.

Aubry, K.B., Amaranthus, M.P., Halpern, C.B., White, J.D., Woodward, B.L., Peterson, C.E., Lagoudakis, C.A., and Horton, A.J., Evaluating the effects of varying levels and patterns of green-tree retention: experimental design of the DEMO study, *Northwest Sci.*, 73 (special issue), 12, 1999.

Bailey, J.D. and Tappeiner, J.C., Effects of thinning on structural development in 40- to 100-year-old Douglas-fir stands in western Oregon, *For. Ecol. Manage.*, 108, 99, 1998.

Barrett, J.W., *Twenty-Year Growth of Ponderosa Pine Saplings Thinned to Five Spacings in Central Oregon*, U.S. Department of Agriculture, Forest Service, Research Paper PNW-301, 1982.

Barrett, J.W., Ed., *Regional Silviculture of the United States*, 3rd ed., John Wiley & Sons, New York, 1995.

Brown, J.K., *Weight and Density of Crowns of Rocky Mountain Conifers*, U.S. Department of Agriculture, Forest Service, Research Paper INT-197, 1978.

Brown, J.K., Arno, S.F., Barrett, S.W., and Menakis, J.P., Comparing the prescribed natural fire program with pre-settlement fires in the Selway-Bitterroot Wilderness, *Int. J. Wildland Fire*, 4, 157, 1994.

Carey, A.B., Experimental manipulation of spatial heterogeneity in Douglas-fir forests: effects on squirrels, *For. Ecol. Manage.*, 152, 13, 2001.

Carey, A.B., Thysell, D.R., and Brodie, A.W., *The Forest Ecosystem Study: Background, Rationale, Implementation, Baseline Conditions and Silvicultural Assessment*, U.S. Department of Agriculture, Forest Service, General Technical Report PNW-GTR-457, 1999.

Chew, J.D., *Development of a System for Simulating Vegetative Patterns and Processes at Landscape Scales*, Ph.D. thesis, University of Montana, Missoula, 1995.

Cissel, J.H., Swanson, F.J., McKee, W.A., and Burditt, A.L., Using the past to plan for the future in the Pacific Northwest, *J. For.*, 92, 30, 1994.

Cissel, J.H., Swanson, F.J., Grant, G.E., Olson, D.H., Gregory, S.V., Garman, S.L., Ashkenas, L.R., Hunter, M.G., Kertis, J.A., Mayo, J.H., McSwain, M.D., Swetland, S.G., Swindle, K.A., and Wallin, D.O., *A Landscape Plan Based on Historical Fire Regimes for a Management Forest Ecosystem: The Augusta Creek study*, U.S. Department of Agriculture, Forest Service, General Technical Report PNW-GTR-422, 1998.

Cochran, P.H. and Barrett, J.W., *Growth of Ponderosa pine Thinned to Different Stocking Levels In Central Oregon: 30-Year Results*, U.S. Department of Agriculture, Forest Service, Research Paper PNW-RP-508, 1999

Curtis, R.O., The role of extended rotations, in *Creating a Forestry for the 21st Century*, Kohm, K.A. and Franklin, J.F., Eds., Island Press, Washington, DC, 1997, chap. 10.

Curtis, R.O. and Carey, A.B., Timber supply in the Pacific Northwest: managing for economic and ecological values in Douglas-fir forests, *J. For.*, 94, 4. 1996.

Curtis, R.O. and Marshall, D.D., Douglas-fir rotations: time for reexamination, *West. J. Appl. For.*, 8, 81, 1993.

Eyre, F.H., Ed., *Forest Cover Types of the United States and Canada*, Society of American Foresters, Washington, DC, 1980.

Fites, J.A., Chapel, M., and Corbin, B., *Preliminary Ecological Old-Growth Definitions for Mixed Conifer (SAF type 243) in California*, Unpublished document, U.S. Department of Agriculture, Forest Service, Pacific Southwest Regional Office, San Francisco, CA, 1992.

Franklin, J.F., Berg, D.R., Thornburg, D.A., and Tappeiner, J.C., Alternative silvicultural approaches to timber harvesting: Variable retention harvest systems, in *Creating a Forestry for the 21st Century*, Kohm, K.A. and Franklin, J.F., Eds., Island Press, Washington, DC, 1997, chap. 7.

Graham, R.T., Harvey, A.E., Jain, T.B., and Tonn, J.R., *The Effects of Thinning and Similar Stand Treatments on Fire Behavior in Western Forests*, U.S. Depfartment of Agriculture, Forest Service, General Technical Report PNW-GTR-463, 1999.

Gray, A.N. and Spies, T.A., Gap size, within-gap position and canopy structure effects on conifer seedling establishment, *J. Ecol.*, 84, 635, 1996.

Hardy, C.C., Keane, R.E., and Stewart, C.A., Ecosystem-based management in the lodgepole pine zone, in *The Bitterroot Ecosystem Management Research Project: What we have Learned*, Smith, H.Y., Ed., U.S. Department of Agriculture, Forest Service, RMRS-P-17, 31, 2000.

Harris, L.D., *The Fragmented Forest: Island Biogeography Theory and the Preservation of Biotic Diversity*, University of Chicago Press 1984.

Harrod, R.J., The effect of invasive and noxious plants on land management in Eastern Oregon and Washington, *Northwest Sci.*, 75, 85, 2001.

Helms, J.A., Ed., *The Dictionary of Forestry*, Society of American Foresters, Bethesda, MD, 1998.

Hobbs, S.D., Tesch, S.D., Owsten, P.W., Stewart, R.E., Tappeiner II, J.C., and Wells, G.E., Eds., *Reforestation Practices in Southwestern Oregon and Northern California*, Forest Research Laboratory, Oregon State University, Corvallis, OR, 1992.

Hunt, R.S., Pruning western white pine in British Columbia to reduce white pine blister rust losses: 10-year results, *West. J. Appl. For.*, 13, 60, 1998.

Keane, R.E. and Arno, S.F., Whitebark pine ecosystem restoration in western Montana, in *The Use of Fire in Forest Restoration*, Hardy, C.C. and Arno, S.F., Eds., U.S. Department of Agriculture, Forest Service, General Technical Report INT-GTR-341, 51, 1996.

Keane, R.E., Long, D.G., Menakis, J.P., Hann, W.J., and Bevins, C.D., *Simulating Coarse-Scale Vegetation Dynamics using the Columbia River Basin Succession Model — CRBSUM*, U.S. Department of Agriculture, Forest Service, General Technical Report INT-GTR-340, 1996.

Keyes, C.R. and O'Hara, K.L., Quantifying stand targets for silvicultural prevention of crown fires, *West. J. Appl. For.*, 17, 101, 2002.

McCarter, J.B., Wilson, J.S., Baker, P.J., Moffett, J.L., and Oliver, C.D., Landscape management through integration of existing tools and emerging technologies, *J. For.*, 96, 17, 1998.

Meyer, W.H., *Growth in Selectively Cut Ponderosa Pine Forests of the Pacific Northwest*, U.S. Department of Agriculture, Technical Bulletin No. 407, 1934.

Morgan, P., Aplet, G.H., Haufler, J.B., Humphries, H.C., Moore, M.M., and Wilson, W.D., Historical range of variability: a useful tool for evaluating ecosystem change, *J. Sustain. For.*, 2, 87, 1994.

Morrison, P.H. and Swanson, F.J., *Fire History and Pattern in a Cascade Range Landscape*, U.S. Department of Agriculture, Forest Service, General Technical Report PNW-GTR-254, 1990.

Mutch, R.W., Fighting fire with prescribed fire: a return to ecosystem health, *J. For.*, 92, 31, 1994.

Mutch, R.W., Arno, S.F., Brown, J.K., Carlson, C.E., Ottmar, R.D., and Peterson, J.L., *Forest Health in the Blue Mountains: A Management Strategy for Fire-Adapted Ecosystems*, U.S. Department of Agriculture, Forest Service, General Technical Report PNW-GTR-310, 1993.

National Research Council, *Environmental Issues in Pacific Northwest Forest Management*, National Academy Press, Washington, DC, 2000.

Newton, M. and Cole, E.C., A sustained yield scheme for old-growth Douglas-fir, *West. J. Appl. For.*, 2, 22, 1987.

O'Hara, K.L., Silviculture for structural diversity: a new look at multi-aged systems, *J. For.*, 96, 4, 1998.

O'Hara, K.L. and Oliver, C.D., A decision system for assessing stand differentiation potential and prioritizing precommercial thinning treatments, *West. J. Appl. For.*, 14, 7, 1999.

O'Hara, K.L., Parent, D.R., and Hagle, S.K., Pruning eastern Cascade and northern Rocky Mountain species: biological opportunities, in *Forest Pruning and Wood Quality of Western North American Conifers*, Institute of Forest Resources Contribution No. 77, 1995.

O'Hara, K.L., Latham, P.A., Hessburg, P., and Smith, B.G., A structural classification for inland northwest forest vegetation, *West. J. Appl. For.*, 11, 97, 1996.

O'Hara, K.L., Seymour, R.S., Tesch, S.D., and Guldin, J.M., Silviculture and our changing profession: leadership for shifting paradigms, *J. For.*, 92, 8, 1994.

Oliver, C.D., Forest development in North America following major disturbance, *For. Ecol. Manage.*, 3, 153, 1980.

Oliver, C.D., Harrington, C., Bickford, M., Gara, R., Knapp, W., Lightner, G., and Hicks, L., Maintaining and creating old-growth structural features in previously disturbed stands typical of the eastern Washington Cascades, *J. Sustain. For.*, 2, 353, 1994.

Oliver, C.D. and Larson, B.C., *Forest Stand Dynamics*, John Wiley & Sons, New York, 1996.

Rice, P., Restoration of native plant communities infested by invasive weeds — Sawmill Creek Research Natural Area, in *The Bitterroot Ecosystem Management Research Project: What we have learned*, Smith, H.Y., Ed., U.S. Department of Agriculture, Forest Service, RMRS-P-17, 29, 2000.

Ross, D.W. and Niwa, C.G., Using aggregation and anti-aggregation pheromones of the Douglas-fir beetle to produce snags for wildlife habitat, *West. J. Appl. For.*, 12, 52, 1997.

Schütz, J.-P., Opportunities and strategies of transforming regular forests to irregular forests, *For. Ecol. Manage.*, 151, 87, 2001.

Sheley, R.L, Svejcar, T.J., and Maxwell, B.D., A theoretical framework for developing successional weed management strategies on rangeland, *Weed Technol.*, 10, 766, 1996.

Smith, H.Y. and Arno, S.F., Eds., *Eighty-Eight Years of Change in a Managed Ponderosa Pine Forest*, U.S. Department of Agriculture, Forest Service, General Technical Report RMRS-GTR-23, 1999.

Spies, T.A. and Franklin, J.F., *The Structure of Natural Young, Mature, and Old-Growth Douglas-Fir Forests in Oregon and Washington, in Wildlife and Vegetation of Unmanaged Douglas-fir Forests*, U.S. Department of Agriculture, Forest Service, General Technical Report PNW-GTR-285, 91, 1991.

Storer, A.J., Wood, D.L., Gordon, T.R., and Libby, W.J., Restoring native Monterey pine forests in the presence of an exotic pathogen, *J. For.*, 14, 2001.

Tappeiner, J.C., Huffman, D., Marshall, D., Spies, T.A., and Bailey, J.D., Density, ages, and growth rates in old-growth and young-growth forests in coastal Oregon, *Can. J. For. Res.*, 27, 638, 1997.

Thornburgh, D.A., Noss, R.F., Angelides, D.P., Olson, C.M., Euphrat, F., and Welsh, Jr., H.H., Managing redwoods, in *The Redwood Forest: History, Ecology and Conservation of the Coast Redwoods*, Noss, R.F., Ed., Island Press, Washington, DC., 229, 2000.

Thysell, D.R. and Carey, A.B., Manipulation of density of *Pseudotsuga menziesii* canopies: preliminary effects on understory vegetation, *Can. J. For. Res.*, 31, 1513, 2001.

Wallin, D.O., Swanson, F.J., Marks, B., Cissel, J.H., and Kertis, J., Comparison of managed and pre-settlement landscape dynamics in forests of the Pacific Northwest, USA, *For. Ecol. Manage.*, 85, 291, 1996.

Walstad, J.D., Radosevich, S.R., and Sandberg, D.V., Eds., *Natural and Prescribed Fires in Pacific Northwest Forests*, Oregon State University, Corvallis, OR, 1990.

Weatherspoon, C.P. and Skinner, C.N., An assessment of factors associated with damage to tree crowns from the 1987 wildfires in northern California, *Forest Sci.*, 41, 430. 1995.

Weatherspoon, C.P. and Skinner, C.N., *Landscape-Level Strategies for Forest Fuel Management*, University of California-Davis, Centers for Water and Wildland Resources, 1471, 1996.

Wilson, J.S. and Oliver, C.D., Stability and density management in Douglas-fir plantations, Can. *J. For. Res.*, 30, 910, 2000.

Wonn, H.T. and O'Hara, K.L., Height:diameter ratios and tree stability relationships for four northern Rocky Mountain tree species, *West. J. Appl. For.*, 16, 87, 2001.

York, R.A., Battles, J.J., and Heald, R.C., Edge effects in mixed conifer group selection openings: tree height response to resource gradients, *For. Ecol. Manage.*, 179, 107, 2003.

chapter thirty

Restoring dry and moist forests of the inland northwestern U.S.

Theresa B. Jain and Russell T. Graham
US Department of Agriculture Forest Service

Contents

30.1 Introduction

The complex topography of the inland northwestern U.S. (58.4 million ha) interacts with continental and maritime air masses to create a highly variable climate, which results in a variety of forest settings. Historically (1850 to 1900), approximately 20% of the area was covered by dry forests (*Pinus ponderosa, Pseudotsuga menziesii*), and an estimated 18% was covered by moist forests (*Pinus monticola, Tsuga heterophylla*). Frequent surface fires burned over 75% of the area of dry forests; today, these fires burn approximately 45% of the area. In the dry forests, successful fire exclusion and harvesting allowed dense stands of *Abies grandis, Pseudotsuga menziesii*, and small *Pinus ponderosa* to develop. Historically, forest canopies (*Pinus ponderosa, Larix occidentalis*) and their nutrients were located well above the soil surface; fine roots and microbial activity were located deep in mineral soils, thus protecting them from wildfire. In contrast, the *Abies*- and *Pseudotsuga*-dominated forests of today contain nutrient-rich crowns that extend to the forest floor. Nutrients and microbial activity are located near the soil surface, increasing their susceptibility to loss from wildfire.

In the moist forests, fire exclusion, harvesting, and the introduction of *Cronartium ribicola* (a stem disease) from Europe are the primary change agents. In the northern Rocky Mountain moist forests, early-seral *Pinus monticola* has nearly been extirpated and mid- to late-seral conifers now dominate. In the moist forest of the eastern Cascades Mountains in Washington and Oregon, an increase in homogeneity of mid-seral forests containing *Abies grandis*, *Tsuga heterophylla*, and *Pseudotsuga menziesii* has occurred, encouraged by the harvesting of *L. occidentalis* and *Pinus ponderosa*. These changes have elevated the risk to large-scale insect and disease epidemics and uncharacteristic wildfires. Successful restoration strategies in both the dry and moist forests should be cognizant of the changes that have occurred not only in the tree component but also those occurring in the soil and across landscapes. The reintroduction of fire alone is not the answer to restoring these forests. Fire should only be used when the trees and soil are in harmony with its reintroduction. Given ever-changing social desires, changes in soil microbial and chemical properties, potential changes in long-term climate, and both native and exotic diseases and insects, a multiscale approach applied over short- and long-term temporal (decades to centuries) and spatial (site to landscape) scales may provide a template for restoring the moist and dry forests of the Inland Northwest.

30.2 Forests of the inland northwest

The inland northwestern U.S. (58.4 million ha) is defined by the Bitterroot, Selkirk, Cabinet, Salmon River, Lemhi, Steens, Purcell, Cascade, and Blue Mountain ranges with elevations over 1,500 m (Figure 30.1). Within these ranges, the valley bottoms can be low (225 m) and the topography steep. This rough and complex topography results in a variety of forest settings ranging from steep slopes, in narrow V-cut canyons, to gentle rolling slopes, in wide river valleys. During the Pleistocene, alpine glaciers shaped the canyons and valleys; today, a mantle of glacial till covers these glaciated landscapes. Much of the

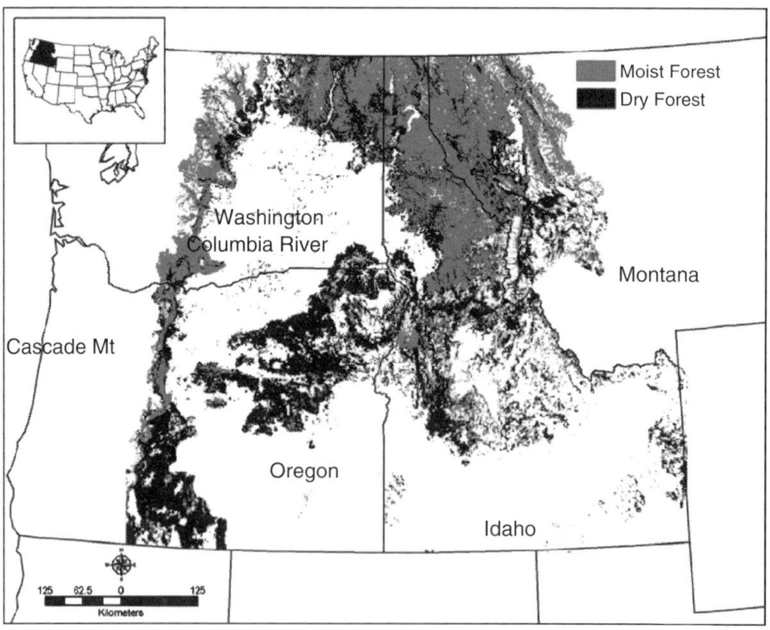

Figure 30.1 There are 58,361,400 ha in the inland northwestern U.S. framed by the Columbia River Basin. The topography is rugged ranging from the Cascade Mountains in the West to the Bitterroot and Salmon River Mountains in Idaho. Elevations in the region range from 225 m to over 3000 m. Dry and moist forests make up 90% of the forests occurring in the inland northwestern U.S. The moist forests occur primarily in northern Idaho, northwestern Montana, and northeastern Washington and along the eastern slopes of the Cascade Mountains in Washington and Oregon. The dry forests are dispersed throughout the region (Hann et al. 1997).

fine silt washed out by the glaciers was redeposited by winds, leaving deep layers of loess over many landscapes. Some 12,000 to 15,000 years ago, Glacial Lake Missoula repeatedly filled and emptied, flooding most of northern Idaho and eastern Washington. The eruption of prehistoric Mt. Mazama 7,500 years ago formed Crater Lake in Oregon and deposited a layer of ash up to 62 cm thick across the area. Disturbance events continually modify the granitic and metasedimentary rocks, ash, and loess deposits, giving rise to diverse soils (Quigley et al. 1996).

Moist marine air originating from the Pacific Ocean moderates temperatures within the Inland Northwest, while continental dry and cold air from the north and east brings cold weather in winter and hot weather in summer. During the summer, these air masses interact and bring convective precipitation, lightning, and cool periods. Dry Arctic air in the winter brings damaging frosts and cold temperatures ($<-5°C$) that alternate with wet warm periods. This highly variable climate interacts with the heterogeneous and rugged topography to create mosaics of compositionally and structurally diverse forests (Franklin and Dyrness 1973; Foiles et al. 1990; Graham 1990; Hann et al. 1997).

Until 1900, forests covered over 47% of the Inland Northwest (Figure 30.1). Dry forests occupied an estimated 11 million ha, dominated by *Pinus ponderosa*, and moist forests covered an estimated 10.5 million ha (18%). The United States Forest Service and the Bureau of Land Management administer more than 50% of both the dry and moist forests (Quigley et al. 1996). Other federal and state agencies administer approximately 5% of these forests, and several industrial and nonindustrial owners manage smaller tracts. Both the moist and dry forests have lost many native structures (large early-seral tree component) and processes (native fire regimes) that were integral in maintaining these systems and the myriad plants, animals, and uses they supported (Quigley et al. 1996).

30.3 *Dry forests*

Dry forests occurred across a wide range of elevations in northeastern Washington, northeastern Oregon, central and southern Idaho, and south-central Oregon (Figure 30.1) (Hann et al. 1997). Soil parent materials include granites, metasedimentaries, glacial tills, and basalts. Vegetation in these forests is usually limited by water availability and is often subject to drought. Nutrient deficiencies develop in eroded areas that can limit forest development. *Pseudotsuga menziesii*, *Pinus ponderosa*, and dry *Abies grandis/Abies concolor* potential vegetation types (PVTs) dominate these settings (Hann et al. 1997). Potential vegetation type is a classification system based on the physical and biological environment characterized by the abundance and presence of vegetation in the absence of disturbance. They are defined by and named for indicator species that grow in similar environmental conditions (Hann et al. 1997). When *L. occidentalis* is present in dry forests, it is always an early-successional species (dominant after disturbance). *Abies grandis/Abies concolor* or *Pseudotsuga menziesii* are late-successional species that are usually more shade-tolerant than the early-seral species they succeed, while *Pinus ponderosa* can play both roles, depending on the PVT (Daubenmire and Daubenmire 1968; Hann et al. 1997). Surface vegetation in the dry forests includes shrubs (*Arctostaphylos uva-ursi*, *Ceanothus* spp., *Purshia tridentata*, *Symphoricarpos albus*, *Physocarpus malvaceus*), grasses (*Calamagrostis rubescens*, *Bromus vulgaris*), and sedges (*Carex* spp.) (Foiles et al. 1990; Hermann and Lavender 1990; Oliver and Ryker 1990).

Fire, insects, diseases, snow, ice, and competition thinned these forests, and surface fires provided opportunities for regeneration (Foiles et al. 1990; Hermann and Lavender 1990; Oliver and Ryker 1990). In concert, these disturbances historically maintained a variety of structural and successional stages (Table 30.1). Approximately 18% of the area was in a grass, forb, and shrub stage for long (hundreds of years) periods and 15% contained

Table 30.1 Historical (1850 to 1900) and Current (1991) Distribution of Forest Structures Within the Dry and Moist Forests of the Inland Northwest (Hann et al. 1997)

Forest Structure	Historical (%)	Current (%)	Change (%)
	Dry Forests		
Grass/forb/shrub	18	1	−17
Early seral intolerant	15	14	−1
Early seral tolerant	3	3	0
Mid-seral intolerant	21	35	+14
Mid-seral tolerant	8	22	+14
Late seral – intolerant single story	21	5	−16
Late seral – tolerant single story	2	3	+1
Late seral – intolerant multistory	9	8	−1
Late seral – tolerant multistory	3	9	+6
	Moist Forests		
Northern Rocky Mountain Region (NRM)			
Early seral – single story	29	20	−9
Mid-seral	41	69	+28
Late seral – single story	11	3	−8
Late seral – multistory	19	8	−11
Eastern Cascade Region (Northern Cascade/Southern Cascade)			
Early seral – single story	23/25	32/15	+9/−10
Mid-seral	37/34	48/37	+11/+3
Late seral – single story	9/9	4/29	−5/+20
Late seral – multistory	31/32	16/19	−15/−13

early-seral *Pinus ponderosa* with diameters ranging from 5 to 80 cm (Meyer 1938). As these forests aged, mid-seral multistoried forest structures developed. Three percent of the area contained stands dominated by late-seral *Pseudotsuga menziesii* and *Abies grandis/Abies concolor* with multiple canopies. Large, widely spaced (~250 trees per ha) *Pinus ponderosa* often dominated 21% of the dry forests, with the plurality of diameters ranging from 30 to 60 cm (Figure 30.2A) (Daubenmire and Daubenmire 1968; Hann et al. 1997). Late-seral single-storied forests containing *Pseudotsuga menziesii* and *Abies grandis/Abies concolor* complexes dominated some settings.

30.3.1 *Dry forest change*

The dry forests were adapted to a wide range of site conditions and short-term climate variation. These characteristics created an ecosystem that appeared to be long-lived and relatively resilient to disturbances (fire, insect, and disease) (Harvey et al. 1994). Since 1900, approximately 8% (600,000 ha) of the dry forests have been converted to agriculture, urbanization, and industry (Hann et al. 1997). Fire exclusion, harvesting, and changes in fire regime altered the composition and structure of the remaining dry forests (Hann et al. 1997). The area burned by surface fires has decreased from an estimated 80% to less than 50% of the area. The mean fire return interval has also increased from less than 20 years to 40 to 80 years. Mixed fires (combination of surface and crown fires) have increased from 5% to an estimated 35% of burned area and the mean fire return interval has increased from 45 to 60 years. A similar increase in crown fires has also occurred (Hann et al. 1997). Mid-seral structures have increased (from an estimated 29 to 57% of the area), often containing dense stands of small *Pinus ponderosa*, *Pseudotsuga menziesii*, or *Abies grandis/Abies concolor* (Table 30.1; Figure 30.2B). The proportion of the dry forests occupied by late-seral single-storied *Pinus ponderosa* has declined from 21 to 5% (Figure 30.2A). In addition,

(A)

(B)

Figure 30.2 A historical (1850 to 1900) (*Pinus ponderosa*) stand exhibiting a lush understory layer of forbs and grasses (A). These conditions were maintained by frequent (<20 year) nonlethal surface fires. Within the dry forests, successful fire exclusion and harvesting have allowed dense stands of vegetation to develop (B) (U.S. Forest Service photos).

small-diameter trees have encroached and now occupy all but 1% of the dry forests that were formerly covered by grasses, forbs, and shrubs (Figure 30.2B). The dominant tree species has changed from *Pinus ponderosa* to *Pseudotsuga menziesii* or *Abies grandis/Abies concolor*, changing the character and canopy architecture of the forest.

The shift in species composition from *Pinus ponderosa*- to *Abies*- and *Pseudotsuga*-dominated forests changed litter type and quantity, which changed soil chemistry, microbial processes, and ectomycorrhizal relationships (Rose et al. 1983). For example, decomposed true firs create white rotten wood, which rapidly disperses into the soil and is quickly consumed by decomposers. In contrast, decomposed *Pinus ponderosa* and *L. occidentalis* create brown rotten wood, which can persist in soil for centuries and has been shown to retain nutrients and hold water (Larsen et al. 1980; Harvey et al. 1987). *L. occidentalis* and *Pinus ponderosa* tend to be deep-rooted, in contrast to the relatively shallow-rooted *Pseudotsuga* and *Abies*, which have abundant feeder roots and ectomycorrhizae in the shallow soil organic layers (Minore 1979; Harvey et al. 1987). *Pinus ponderosa* and *L. occidentalis* forests are generally tall and self-pruning, even in moderately dense stands. They have large branches high in the crowns and the base of the crowns is well above surface fuels. In general, this crown architecture protects the nutrients stored in the canopy from surface fires. In contrast, young- to mid-aged (<150 years) *Pseudotsuga menziesii* and *Abies grandis/Abies concolor* generally do not self-prune. This canopy architecture favors lower crown base heights, higher crown densities, and canopies with higher nutrient (especially potassium) content than occur in *L. occidentalis*- and *Pinus ponderosa*-dominated forests (Figure 30.2B) (Harvey et al. 1999; Minore 1979).

In the dry forests, biological decomposition is more limited than biological production. When fire return intervals reflected historical fire frequencies, the accumulation of thick organic layers was minimized and nutrient storage and nutrient turnover were dispersed in the mineral soils (Marschner and Marschner 1996; Harvey et al. 1999). In the absence of fire, bark slough, needles, twigs, and small branches accumulated on the forest floor allowing ectomycorrhizae and fine roots of all species to concentrate in the surface mineral soil and thick organic layers (Harvey et al. 1994).

Harvesting the *L. occidentalis* and *Pinus ponderosa* and the ingrowth of *Abies grandis/Abies concolor* and *Pseudotsuga menziesii* in the dry forests together facilitated the accumulation of both above- and below-ground biomass and their nutrient content close to the soil surface (Harvey et al. 1986). Even low-intensity surface fires now consume the surface organic layers, killing fine roots, volatilizing nutrients, killing trees, and increasing soil erosion potential (Debano 1991; Hungerford et al. 1991; Ryan and Amman 1996; Robichaud et al. 2000). In addition, fir ingrowth creates nutrient-rich ladder fuels that facilitate crown-fire initiation, increasing the likelihood of nutrient loss (Van Wagner 1977; Minore 1979; Harvey et al. 1999). The risk of nutrient loss is great on infertile sites, because dense stands of late-seral species are more demanding of nutrients and water than the historical stands dominated by widely- spaced early-seral species (Minore 1979; Harvey et al. 1999).

30.4 Moist forests

Moist forests of the inland northwestern U.S. occur in two locations: the eastern Cascade Mountains (east of the Cascade Crest in Washington and Oregon) and the northern Rocky Mountains (northeastern Washington and Oregon, northern Idaho, and western Montana) (Figure 30.1). They grow at elevations ranging from 460 to 1,600 m and occasionally occur at elevations up to 1,800 m (Foiles et al. 1990; Graham 1990; Packee 1990; Schmidt and Shearer 1990; Hann et al. 1997) (Figure 30.1). These forests are influenced by a maritime climate with wet winters and dry summers. Most precipitation occurs during November through May, with amounts ranging from 500 to 2,300 mm (Foiles et al. 1990; Graham

1990; Packee 1990; Schmidt and Shearer 1990). Precipitation comes as snow and prolonged gentle rains, accompanied by cloudiness, fog, and high humidity. Rain-on-snow events are common January through March. A distinct warm and sunny drought period occurs in July and August, with rainfall in some places averaging less than 25 mm per month.

Soils that maintain these forests include, but are not limited to, Spodosols, Inceptisols, and Alfisols. A defining characteristic of the northern Rocky Mountains is the layer of fine-textured ash (up to 62 cm thick) that caps the residual soils. The ash soils and loess deposits throughout the moist forests are continually being modified by disturbance events giving rise to soils with differing levels of productivity (Foiles et al. 1990; Graham 1990; Packee 1990; Schmidt and Shearer 1990). The combination of climate, topography, parent material, soils, weathering, and ash depth (unique to the northern Rocky Mountains) creates the most productive of all forests occurring within the Inland Northwest.

For both locations, the historical vegetation complexes ranged from early- to late-seral, and occurred within a landscape mosaic possessing all possible combinations of species and seral stages. The PVTs in the northern Rocky Mountains include *Thuja plicata, Tsuga heterophylla,* and *Abies grandis* with *Pinus monticola, L. occidentalis, Pinus contorta, Pseudotsuga menziesii,* and *Pinus ponderosa* as the early- and mid-seral species (Daubenmire and Daubenmire 1968; Hann et al. 1997). The eastern Cascades PVTs include *Thuja plicata, Tsuga heterophylla, Abies grandis, Abies amabilis,* and *Abies procera*. The early- and mid-seral species include *Pinus contorta, Pseudotsuga menziesii,* and *Pinus ponderosa* while *Pinus monticola* and *L. occidentalis* are less abundant when compared to the northern Rocky Mountains (Franklin and Dyrness 1973; Lillybridge et al. 1995).

Lush ground-level vegetation is the norm in the moist forests. The vegetation complexes are similar to those occurring on the west side of the Cascade Mountains and in some Pacific coastal areas. Tall shrubs include *Acer circinatum, Achylys triphylla, Acer glabrum, Alnus sinuata, Oplopanax horridum, Rosa* spp., *Ribes* spp., *Vaccinium* spp., and *Salix* spp. Forbs include *Actaea rubra, Adenocaulon bicolor, Asarum caudatum, Clintonia uniflora, Cornus canadensis,* and *Coptis occidentalis*. Phytogeographic evidence indicates that some plant populations on the west side of the Cascade Mountains also occur as disjunct populations in the moist forests. For example, low-elevation riparian areas in northern Idaho contain disjunct populations of *Alnus rubra, Cornus nuttallii, Symphoricarpos mollis, Selaginella douglasii,* and *Physocarpus capitatus* (Foiles et al. 1990; Graham 1990; Packee 1990; Schmidt and Shearer 1990).

Native disturbances (snow, ice, insects, disease, and fire), when combined, created heterogeneity in patch sizes, forest structures, and compositions. Ice and snow created small gaps and openings, thinning forest densities and altering species composition (Figure 30.3A). Native insects (*Dendroctonus* spp.) and diseases (*Armillaria* spp., *Arceuthobium* spp.) infected and killed the very old or stressed individuals, which tended to diversify vegetation communities (Figure 30.3B) (Hessburg et al. 1994). A mixed-fire regime best defines the role fire played in creating a mosaic of forest compositions and structures. Nonlethal surface fires occurred at relatively frequent intervals (15 to 25 years) in a quarter of the area (Figure 30.4A). Lethal crown fires burned about a quarter of the area at intervals of 20 to 150 years but occasionally extended to 300 years (Figure 30.4B). The mixed-fire regime occurred across the rest of the moist forests at 20- to 150-year intervals. Fires typically started burning in July and were usually out by early September (Hann et al. 1997).

30.4.1 Moist forest change

The current distribution of successional-stage forest structure, species composition, and disturbance regimes differs from the historic (1850 to 1900) patterns of the moist forest (Hann et al. 1997). In some settings, the mixed-fire regime maintained closed canopy conditions,

(A)

(B)

Figure 30.3 Ice and snow damage create small gaps and openings, decrease forest densities, and alter species composition (A). Historically (1850 to 1900), in the moist forests, diseases (e.g., *Armillaria* spp., *Arceuthobium* spp.) attacked the very old, unthrifty, or stressed individuals (B). These disturbances stabilized and diversified vegetation communities. Currently, in the changed systems, epidemics of these disturbances often occur (U.S. Forest Service photos).

which allowed for the mid-seral stage to develop into late-seral multistory stages (Hann et al. 1997). The late-seral multistory structure, which typically developed in cool, moist bottoms and basins, has decreased by about half in the last century (Table 30.1). The early-seral single-story stands that once occupied an estimated 25 to 30% of the area now occupy only 9 to 10% of the area, except in the northern Cascades (Washington) where they increased in abundance. The mid-seral stages have generally increased in abundance in the northern Rocky Mountains and to a lesser degree in the eastern Cascades.

Species composition has shifted in the northern Rocky Mountains (Hann et al. 1997; Nuenschwander et al. 1999; Fins et al. 2001); before 1900, *Pinus monticola* (early- to

(A) (B) (C)

Figure 30.4 Historically (1850 to 1900), in the moist forests, nonlethal surface fires (A) occurred at relatively frequent intervals (15 to 25 years) in 25% of the moist forests while lethal crown fires (B and C) occurred over 25% of the forests at 300-year intervals (U.S. Forest Service photos).

mid-seral species) dominated, often representing 15 to 80% of the trees within stands (Figure 30.5) (Fins et al. 2001). This species is resistant to many endemic insects and diseases; it is long-lived (300 years) and can grow across 90% of the moist forest environments. It is a prolific seed producer after age 70 and is the only moist-forest conifer with seed remaining viable for up to 3 years, which allows it to regenerate abundantly after disturbance (Haig et al. 1941; Graham 1990). It is broadly adapted genetically (generalists) to the environment (Rehfeldt et al. 1984) and it is moderately shade-tolerant, allowing it to establish and develop within a wide range of canopy openings (Haig et al. 1941; Graham 1990; Jain 2001). *Pinus monticola* often reaches 30 m in height or greater within 50 years of establishment (Graham 1990). *L. occidentalis* and *Pinus ponderosa* also occurred in the early- and mid-seral structures, but declined along with *Pinus monticola* and were succeeded by *Abies grandis*, *Pseudotsuga menziesii*, and *Tsuga heterophylla* (Hann et al. 1997; Atkins et al. 1999). The eastern Cascades had limited amounts of *Pinus monticola* and *L. occidentalis*; therefore, *Pinus ponderosa*, *Pinus contorta*, and *Pseudotsuga menziesii* played more of a role in occupying the early- to mid-seral successional stages.

Native insects and pathogens occurred in these forests, but recent (1991) activity levels far exceed those of the past (Hessburg et al. 1994). Within the *Abies grandis* PVT, fire maintained landscapes that contained a plurality of early-seral *Pinus ponderosa* and *L. occidentalis*; the insects *Dendroctonus pseudotsugae*, *Choristoneura occidentalis*, and *Orgyia pseudotsugata* were generally endemic. But they are often epidemic in the current forests dominated by *Abies grandis* and *Pseudotsuga menziesii* (Hessburg et al. 1994). Similarly, the diseases *Armillaria* spp. and *Phellinus weirii* were historically endemic, but the current fir-dominated forests make epidemics of these diseases more common (Hessburg et al. 1994; Hann et al. 1997). As in many forest ecosystems in the western U.S., effective fire exclusion contributed to these changes. Historically, 25% of the area had surface fires, 50% mixed fires, and 25% stand-replacing crown fires. Today, crown fires burn approximately 60% of the areas in these forests and only 15% are burned by surface fires and 20% are burned by mixed fires (Hann et al. 1997).

Although fire exclusion played a role in altering forests in the northern Rocky Mountains, introduction of a European stem rust, *Cronartium ribicola*, caused the greatest change (Figure 30.6). The rust infects all five-needle pines, and subsequently decimated the abundant *Pinus monticola*. Because the rust killed so many trees, the majority of surviving pines were harvested under the assumption that they too would succumb to the rust (Ketcham et al. 1968). *Abies grandis* and *Pseudotsuga menziesii* readily filled the niche that *Pinus monticola* once held. In the eastern Cascades, blister rust was less severe since

Figure 30.5 Historically (1850 to 1900), 25 to 50% of the moist forests were dominated by *Pinus monticola* and 15 to 80% of the trees within stands were *Pinus monticola*. This photo shows 150- to 180-year-old *Pinus monticola* (c. 1935) growing in northern Idaho (U.S. Forest Service photo).

(A) (B)

Figure 30.6 Photograph A shows a blister rust canker occurring on a young *Pinus monticola*. Photograph B is a mid-aged (70 to 80 years) stand of *Pinus monticola* experiencing extreme mortality from blister rust (U.S. Forest Service photos).

Pinus monticola was not the dominant species; thus, fire exclusion and harvesting were more important agents in altering these forests (Figure 30.7). Harvesting removed the early-seral, shade-intolerant species (*Pinus ponderosa, L. occidentalis*) that were resistant to fire and other disturbances. Partial canopy removal and minimal soil surface disturbance in these harvests

Figure 30.7 Harvesting of *Pinus monticola* (c. 1935) with an Idaho jammer. In the moist forests, these ground-lead cable-harvesting systems required closely spaced roads (150 m), which led to the creation of a dense road network within many drainages (U.S. Forest Service photo).

were ideal for *Pseudotsuga menziesii* and *Abies grandis* which regenerated aggressively, rather than the shade-intolerant *Pinus* and *Larix* species. Fire exclusion also prevented the creation of canopy openings and receptive seedbeds for the regeneration of *Pinus* and *Larix*. Similar to the dry forests, high canopies (>50 m) of *Pinus monticola, L. occidentalis, Pinus ponderosa,* and other early- and mid-seral species is currently absent. In their place, the present forest structure and composition (*Abies grandis* and *Pseudotsuga menziesii*) favor the compression of nutrients, microbial processes, and root activity toward the soil surface (Harvey et al. 1999). When wildfires occur, surface organic layers can be consumed, decreasing the nutrition and microbial processes important for sustaining these forests.

30.5 Restoration strategies

Changing species composition from late-seral to early- and mid-seral species has been suggested as the key to restoring both moist and dry forests (Everett et al. 1994; Hann et al. 1997; Harvey et al. 1999; Neuenschwander et al. 1999; Finns et al. 2001); by doing this, resilience to epidemics from insects and diseases will increase. Moreover, current altered nutrient and microbial processes would be redirected toward historical conditions when early- to mid-seral species dominated these forests. Returning these forests to a more natural fire regime has also been advocated. Some see forests as they exist today, and because they contain live and green trees, assume that the present forests are typical, resilient, and functioning and should be left alone. In contrast, others conclude that because of the changes that have occurred, the forests of today are less healthy and no longer sustainable. Some see forests as valuable sources of commodities and recreational opportunities and want these benefits maintained. Opinions differ, however, whether these are problems with the Inland Northwest forests and what should be done to restore them. These conflicts have led to widely divergent opinions on the effects that management, fire, or no management have on these forests, and whether intervention can restore them to something resembling past conditions. Most of the general solutions that have been suggested apply a simple fix to a variety of complex problems, often causing mixed beneficial and/or detrimental results that can help or damage a forest, depending on details of history, current conditions, and future trends.

For example, reintroducing ground fire is not a simple matter. Current fuel (live and dead) accumulations and distributions place human, vegetative, and soil resources at high risk (Figure 30.2B). Harvesting to reduce fuels to normal levels has been advocated, but results in more roads with potential soil impacts. Even thinning dense stands, with or without prescribed burning, is problematic, as thinning species susceptible to root disease is likely to set off an epidemic, upset natural selection processes, or worsen fire effects if slash is not appropriately treated (Harvey et al. 1999).

Because these forests are complex and profound changes have occurred, restoration strategies should be applied at the appropriate spatial and temporal scale, both broad (landscape, watershed, centuries) and fine (stand, gap, decades). To date, restored forest conditions (sometimes referred to as reference conditions) are defined by, but not limited to, using estimates of historical structure and composition (Covington and Moore 1994; Kauffman et al. 1994) or characteristics required by wildlife (Reynolds et al. 1992). Large landscape units (2,000 to 4,000 ha) are most applicable when developing restoration strategies. Areas this large probably contain a variety of structures (vegetation mosaics, vegetation patches) and processes (fire, decomposition, management opportunities) not observed at fine spatial scales. Landscape units will likely contain the habitats and food webs of one or more wildlife species (*Accipiter* spp., hawks) that depend on the forests (Reynolds et al. 1992). Multiple land ownerships occur throughout these units, probably with different objectives. Wildfires and the environments they create will still occur because humans are present, but not to the same extent, location, or frequency.

Landscape attributes provide a biophysical template for beginning to set restoration priorities. A step-down process from broad to fine scales for planning restoration activities has been suggested for areas as large as the Interior Columbia River Basin (USDA 2000). Factors such as wide-ranging predators (grizzly bears), anadromous fish (salmon), and social and economic infrastructure (population trends, distribution of cities, towns, and communities) are addressed at the broadest scale. At the mid-scale or river basins, forest composition and distributions can be addressed. For example, in the Coeur d'Alene River Basin (>350,000 ha) of the northern Rocky Mountains, historical patterns of *Pinus monticola* abundance were used to define possible restoration priorities (Jain 2001). At the fine scale, Camp et al. (1997) identified fire refugia as a function of physical landscape attributes in the eastern Cascades. Managers could use these techniques in both dry and moist forests to identify key wildlife habitat, or unique vegetative communities.

Within the priorities established using this step-down process, managers can set restoration targets by defining reference conditions using wildlife indicator species or historical forest structure and composition. Particularly in the dry forests, native surface fires, insects, diseases, and other local disturbances kept vegetation patches relatively small (<1.0 ha). These patches and their characteristic structural stages include live trees, snags, decadent large trees (primarily *Pinus ponderosa* and *L. occidentalis*), coarse woody debris, and organic rich soil. Moreover, these small patches are the preferred habitat for many birds and mammals (Reynolds et al. 1992). An uneven-aged silvicultural system can create and maintain this landscape-scale mosaic in the approximate composition, structure, and landscape patterns existing in dry forests before fundamental changes in natural disturbances regime and forest structure occurred (Reynolds et al. 1992; Long and Smith 2000).

Restoring native fire regimes of frequent surface fires requires approaches that conserve nutrients, microbial activity, and fine roots that developed in the uncharacteristically deep surface organic layers. Decreasing the depth and volume of these organic layers by repeating treatments over a series of years will force the fine roots to migrate down into the mineral soil. Depending on the dry forest setting, it may take one to multiple combinations of mechanical and carefully executed prescribed burns before fuel loads, species composition, and forest structures allow prescribed fires to burn freely in the dry forests.

Burning when moisture is high (>100%) in the lower layers will preserve them but allow the drier upper layers to be burned (Brown et al. 1985). In addition, burning at low temperatures (less than −2°C) when fine root growth is minimal results in little root damage (Kramer and Kozlowski 1979). These conditions occur in early spring in dry forests when snow is present but the forest floor around the base of trees is clear (Figure 30.8). Under

(A)

(B)

Figure 30.8 Because of fire exclusion, deep organic layers surround many old *Pinus ponderosa*. By applying prescribed fire (A) when the lower organic layers are high in moisture (>100%), and soil temperatures are low (−2°C), the depth of these layers can be decreased with minimal damage to the fine roots they contain. In addition, mixing (B) the organic layers (increasing decomposition) also decreases organic layer depth (U.S. Forest Service photos).

high moisture conditions (>100% moisture by weight), nutrient volatilization is minimized (Brown et al. 1985). Allowing fine fuels created by mechanical treatments to overwinter before they are burned provides time for nutrients to leach and minimizes volatilization (Harvey et al. 1999).

Silvicultural systems can be developed that maintain high forest cover in moist forests, yet provide sufficient growing space to establish and develop *Larix* or *Pinus* (Jain 2001) and avoid using large openings (clearcuts >16 ha) and their associated roads. Restoration treatments using this fine-scale approach mimic many native disturbances (windstorms, root disease, ice breakage) (Figure 30.3) and their spatial patterns rather than only mimicking large crown fires (Figure 30.4). However, even-aged silvicultural systems (clearcut, shelterwood, seed-tree) are still appropriate for increasing the abundance of resilient species complexes on many lands (Haig et al. 1941; Graham et al. 1983).

The major issue facing restoring the moist forests of the northern Rocky Mountains is the susceptibility of *Pinus monticola* to *Cronartium ribicola* (Figure 30.7). Fortunately, native populations do contain some natural resistance to the rust, thus providing management opportunities. A breeding program currently produces rust-resistant (estimated at 68%) *Pinus monticola* seedlings for reforestation (Fins et al. 2001) and continued breeding will ensure that rust mutations will not compromise the resistant material. In addition to breeding programs, mass selection presents an opportunity to utilize the rust resistance occurring in natural stands (Hoff and McDonald 1980; Graham et al. 1994). In stands where blister rust has killed the majority (over 70%) of the *Pinus monticola*, approximately 7 to 10% of the progeny, often thousands per ha, of the survivors exhibit rust resistance. Regenerating these wild populations can supplement the genetic diversity contained in breeding programs.

30.6 Economic and social aspects

Ecological information is available to restore the moist and dry forests (Reynolds et al. 1992; Covington and Moore 1994; Hann et al. 1997; Harvey et al. 1999; Nuenschwander et al. 1999; Long and Smith 2000; Finns et al. 2001). Interweaving the social, economic, and political needs of the society presents the greater challenge in restoring these forests. Costs can be specified for restoration activities such as harvesting, thinning, planting, weeding/cleaning, prescribed burning, exotic plant control, riparian area treatments, and planning and analysis. Specific costs will vary depending on site characteristics and stand structure. For example, the cost per unit area ranges from $75 (U.S. dollars) per ha for vegetation management to $750 per ha for riparian treatments, with cleaning and weeding costing $200 per ha and prescribed burning costing $80 per ha (USDA 2000).

These cost estimates do not reflect some of the benefits of restoring moist and dry forests. For example, an estimated 83% of the recreational benefits within the Interior Columbia Basin come from federally administered Forest Service and Bureau of Land Management lands (Phillips and Williams 1998; Reyna 1998). These recreation benefits include trail use, hunting, fishing, camping, boating, wildlife viewing, winter sports, day use, and motor viewing. When restoration positively influences these activities, the benefit may exceed the costs. Converting forests dominated by late-seral structures to forests dominated by early- and mid-seral structures most likely will benefit these recreational activities. In addition, restored forest conditions may improve the habitat of legally threatened or endangered wildlife species, but all restoration treatments have to be implemented to avoid negative impacts on species such as the bull trout, grizzly bear, and Canadian lynx. Nevertheless, a vocal segment of society prefers the status quo and resists actively managing forests.

Restoration activities will probably affect timber-dependent and isolated communities the most. Reyna (1998) and Phillips and Williams (1998) reported that 137 communities specialized in logging and wood products manufacturing in the Interior Columbia River Basin with 64 being isolated. Timber harvesting and wood products manufacturing generally has been important in these communities since the 1800s when many towns were established. Active management and restoring forests near these communities most likely would be a positive benefit. However, timber harvesting is usually a controversial issue and likely will continue to be so into the future. In addition, the infrastructure for manufacturing wood products has decreased in recent years, increasing transportation costs for the harvested material (Haynes 2002). Perhaps the greatest economic challenge is that the small trees (especially *Pinus ponderosa*) available for harvesting in most restoration activities have low value (Lippke 2002; McKetta 2002). However, small-diameter (15 cm to 25 cm in diameter) *Abies grandis* and *Pseudotsuga menziesii* seem to be an exception, and are of value in producing construction materials (McKetta 2002).

In general, private landowners have more flexibility for conducting restoration activities where timber production is often the primary objective (Blatner et al.1994). Unlike their federal counterparts, managers of private lands have fewer requirements for analysis and planning prior to conducting activities, which allows them to respond quickly to insects, diseases, wildfires, and storms, as well as changing markets. Moreover, forests containing *Pinus ponderosa*, *Pseudotsuga menziesii*, *Pinus monticola*, and *L. occidentalis* tend to have high commercial value and are resilient to native disturbances (Nuenschwander et al. 1999; McKetta 2002). However, just species presence would not necessarily indicate that a forest is restored; the entire suite of forest structures (biological and physical properties of vegetation, soils, microbes, and water) and compositions distributed in a mosaic over the landscape would most likely resemble a restored forest (Nuenschwander et al. 1999).

30.7 Conclusion

A combination of harvesting and the introduction of *Cronartium ribicola* greatly impacted the moist forests. Fire exclusion played a role in changing these forests, but not to the same extent that it did in the dry forests. Because of current (recreation, scenic, wildlife habitat) and past (harvesting, road construction) human uses and values, restoring both the dry and moist forests will be challenging, but not impossible. By viewing and developing management strategies using a multiscale approach, areas and treatments can be designed that move these forests on a trajectory toward their historical compositions and structures. Majestic stands of *Pinus monticola*, *Pinus ponderosa*, *L. occidentalis*, *Pseudotsuga menziesii*, and all possible combinations of these species and their associates once populated the forests of the Inland Northwest. Once these systems reflect the historical composition and structure, endemic levels of other disturbances can aid in sustaining these forests into the future. However, because of human presence, the extent and intensity of endemic disturbances plus exotic introductions into these forests will make restoration activities challenging. If society determines that the dry and moist forests should be restored, it will take time, patience, perseverance, and commitment by both public and private individuals and organizations to accomplish the task.

References

Atkins, D., Byler, J., Livingston, L., Rogers, P., and Bennett, D., *Health of Idaho's Forests*, U.S. Department of Agriculture, Forest Service, Northern Region, Missoula, MT, Report No. 99-4, 1999.

Blatner, K.A., Keegan, C.E. III, O'Laughlin, J., and Adams, D.L., Forest health management policy, *J. Sustain. For.*, 2, 317, 1994.

Brown, J.K., Marsden, M.A., Ryan, K.C., and Reinhardt, E.D., *Predicting Duff and Woody Fuel Consumed by Prescribed Fire in the Northern Rocky Mountains*, U.S. Department of Agriculture, Forest Service, Intermountain Research Station, Ogden, UT, Research Paper INT-337, 1985.

Camp, A., Oliver, C., Hessburg, P., and Everett, R., Predicting late-successional fire refugia pre-dating European settlement in the Wenatchee Mountains, *For. Ecol. Manage.*, 95, 63, 1997.

Covington, W.W. and Moore, M.M., Post settlement changes in natural fire regimes and forest structure, in *Assessing Forest Ecosystem Health in the Inland West*, Sampson, R.N. and Adams D.L., Eds., The Haworth Press Inc., New York, 153, 1994.

Daubenmire, R. and Daubenmire, J., *Forest Vegetation of Eastern Washington and Northern Idaho*, Washington State Agricultural Experiment Station, Pullman, WA, Bulletin 60, 1968.

Debano, L.F., Effects of fire on soil properties, in *Proceedings: Management and Productivity of Western-montane Forest Soils*, Harvey, A.E. and Neuenschwander, F.L., Comps., U.S. Department of Agriculture, Forest Service, Intermountain Research Station, Ogden, UT, 65, 1991.

Everett, R.L., Hessburg, P., Jensen, M., and Bormann, B., *Eastside Forest Ecosystem Health Assessment: Volume I, Executive Summary*, U.S. Department of Agriculture, Forest Service, Pacific Northwest Research Station, Portland, OR, General Technical Report PNW-GTR-317, 1994.

Fins, L., Byler, J., Ferguson, D., Harvey, A., Mahalovich, M.F., McDonald, G., Miller, D., Schwandt, J., and Zack, A., *Return of the Giants: Restoring White Pine Ecosystems by Breeding and Aggressive Planting of Blister Rust-resistant White Pines*, University of Idaho, Moscow, ID, 2001.

Foiles, M.W., Graham, R.T., and Olson, D. Jr., *Abies grandis* (Dougl. ex D. Don.) Lindl., in *Silvics of North America, Volume 1, Conifers*, Burns, R.M. and Honkala, B.H., Tech. Coords., U.S. Department of Agriculture, Forest Service, Washington, D.C., Agriculture Handbook 654, 52, 1990.

Franklin, J.F. and Dyrness, C.T., *Natural Vegetation of Oregon and Washington*, Oregon State University Press, Corvallis, OR, 1973.

Graham, R.T., Wellner, C.A., and Ward R., Mixed conifers, western white pine, and western redcedar, in *Silvicultural Systems for the Major Forest Types of the United States*, Burns, R.M., Ed., U.S. Department of Agriculture, Forest Service, Washington, DC., Agriculture Handbook 445, 67, 1983.

Graham, R.T., *Pinus monticola* (Dougl. Ex. D. Don.), in *Silvics of North America, Volume 1, Conifers*, Burns, R.M. and Honkala, B.H., Tech. Coords., U.S. Department of Agriculture, Forest Service, Washington, DC., Agriculture Handbook 654, 385, 1990.

Graham, R.T., Tonn, J.R., and Jain, T.B., Managing western white pine plantations for multiple resource objectives, in *Proceedings — Interior Cedar-hemlock-White Pine Forests: Ecology and Management*, Baumgartner, D.M., Lotan, J.E. and Tonn, J.R., Eds., Washington State University, 357, 1994.

Haig, I.T., Davis, K.P., and Weidman, R.H., *Natural Regeneration in the Western White Pine Type*, U.S. Department of Agriculture, Forest Service, Washington, DC., Technical Bulletin No. 767, 1941.

Hann, W.J., Jones, J.L., Karl, M.G., Hessburg, P.F., Keane, R.E., Long, D., Menakis, J.P., McNicoll, C.H., Leonard, S.G., Gravenmier, R.A., and Smith, B.G., Chapter 3: Landscape dynamics of the Basin, in *An Assessment of Ecosystem Components in the Interior Columbia Basin and Portions of the Klamath and Great Basins: Volume II*, Quigley, T.M. and Arbelbide S. J., Tech. Eds., U.S. Department of Agriculture, Forest Service, Pacific Northwest Research Station, Portland OR, General Technical Report PNW-GTR-405, 338, 1997.

Harvey, A.E., Jurgensen, M.F., Larsen, M.J., and Schlieter, J.A., *Distribution of Active Ectomycorrhizal Short Roots in Forest Soils of the Inland Northwest: Effects of Site and Disturbance*, U.S. Department of Agriculture, Forest Service, Intermountain Research Station, Ogden, UT, Research Paper INT-374, 1986.

Harvey, A.E., Jurgensen, M.F., Larsen, M.J., and Graham, R.T., *Decaying Organic Materials and Soil Quality in the Inland Northwest: A Management Opportunity*, U.S. Department of Agriculture, Forest Service, Intermountain Research Station, Ogden, UT, General Technical Report INT-GTR-225, 1987.

Harvey, A.E., Hessburg, P.F., and Byler, J.W., Health declines in interior forests: Symptoms and solutions, in *Ecosystem Management in Western Interior Forests*, Washington State University, Pullman, WA, 163, 1994.

Harvey, A.E., Graham, R.T., and McDonald, G.I., Tree species composition change: forest soil organism interaction potential effects on nutrient cycling and conservation processes in interior forests, in *Pacific Northwest Forest and Rangeland Soil Organism Symposium*, Meurisse, R., Ypsllantis, W.G., and Seybold, C., Tech. Eds., U.S. Department of Agriculture, Forest Service, Pacific Northwest Research Station, Portland, OR, General Technical Report PNW-GTR-461, 137, 1999.

Haynes, R.W., U.S. timber supply and demand in the United States, 1996 to 2050, in *Small Timber: Resource Management, Manufacturing, and Markets, Symposium Proceedings*, Baumgartner, D.M., Johnson, L.R., and Depuit, E.J., Comps., Washington State University, Pullman, WA, 33, 2002.

Hermann, R.K. and Lavender D.P., *Pseudotsuga menziesii* (Mirb.) Franco, in *Silvics of North America, Volume 1, Conifers*, Burns, R.M. and Honkala, B.H., Tech. Coords., U.S. Department of Agriculture, Forest Service, Washington, DC., Agriculture Handbook 654, 527, 1990.

Hessburg, P.F., Mitchell, R.G., and Filip, G.M., *Historical and Current Roles of Insects and Pathogens in Eastern Oregon and Washington Forested Landscapes*, U.S. Department of Agriculture, Forest Service, Pacific Northwest Research Station, Portland, OR, General Technical Report PNW-GTR-327, 1994.

Hoff, R.J. and McDonald, G.I., *Improving Rust-Resistant Strains of Inland Western White Pine*, U.S. Department of Agriculture, Forest Service, Intermountain Forest and Range Experiment Station, Ogden, UT, Research Paper INT-245, 1980.

Hungerford, R.D., Harrington, M.G., Frandsen, W.H., Ryan, K.C., and Niehoff, G.J., Influence of fire on factors that affect site productivity, in *Proceedings — Management and Productivity of Western-montane Forest Soils*, Harvey, A.E. and Neuenschwander, F.L., Comps., U.S. Department of Agriculture, Forest Service, Intermountain Research Station, Ogden, UT, 32, 1991.

Jain, T.B., *Biophysical Characteristics Influencing Growth and Abundance of Western White Pine (Pinus monticola) across Spatial Scales in the Coeur D'Alene River Basin, Idaho*, Ph.D. dissertation, University of Idaho, Moscow, ID, 2001.

Kaufman, M.R., Graham, R.T., Boyce, D.S., Moir, W.H., Perry, L., Reynolds, R.T., Bassett, R.L., Melhop, P., Edminster, C.B., Block, W.M., and Corn, P.S., *An Ecological Basis for Ecosystem Management*, U.S. Department of Agriculture, Forest Service, Rocky Mountain Forest and Range Experiment Station, Fort Collins, CO, General Technical Report RM-246, 1994.

Ketcham, D.A., Wellner, C.A., and Evans, S. Jr., Western white pine management programs realigned on Northern Rocky Mountain National Forests, *J. For.*, 1968, 66, 329.

Kramer, P.J. and Kozlowski, T.T., *Physiology of Woody Plants*, Academic Press, New York, 1979.

Larsen, M.J., Harvey, A.E., and Jurgensen, M.F., Residue decay processes and associated environmental functions in northern Rocky Mountain forests, in *Environmental Consequences of Timber Harvesting in Rocky Mountain Coniferous Forests: Symposium Proceedings*, U.S. Department of Agriculture, Forest Service, Ogden, UT, General Technical Report INT-90, 157, 1980.

Lillybridge, T.R., Kovalchik, B.L., Williams, C.K., and Smith, B.G., *Field Guide for Forested Plant Associations of the Wenatchee National Forest*, U.S. Department of Agriculture, Forest Service, Pacific Northwest Research Station, Portland, OR, General Technical Report PNW-GTR-359, 1995.

Lippke, B., Technology transfer opportunities for small diameter timber problems, in *Small Timber: Resource Management, Manufacturing, and Markets, Symposium Proceedings*, Baumgartner, D.M., Johnson, L.R., and Depuit, E.J., Comps., Washington State University, Pullman, WA, 53, 2002.

Long, J.N. and Smith, F.W., Restructuring the forest: Goshawks and restoration of southwestern ponderosa pine, *J. For.*, August, 25, 2000.

Marschner, G.E. and Marschner, H., Nutrient and water uptake by roots of forest trees, *Pflanzenernahr Bodenk*, 159, 11, 1996.

McKetta, C., Why grow large trees anyway? A timber grower's perspective, in *Small Timber: Resource Management, Manufacturing, and Markets, Symposium Proceedings*, Baumgartner, D.M., Johnson, L.R., and Depuit, E.J., Comps., Washington State University, Pullman, WA, 133, 2002.

Meyer, W.H., *Yield of Even-aged Stands of Ponderosa Pine*, U.S. Department of Agriculture, Washington, DC, Technical Bulletin 630, 1938.

Minore, D., *Comparative Autecological Characteristics of Northwestern Tree species — A Literature Review*, U.S. Department of Agriculture, Forest Service, Pacific Northwest Forest and Range Experiment Station, Portland, OR, General Technical Report PNW-GTR-87, 1979.

Neuenschwander, L.F., Byler, J.W., Harvey, A.E., McDonald, G.I., Ortiz, D.S., Osborne, H.L., Snyder, G.C., and Zach, A., *White Pine in the American West: A Vanishing Species — Can We Save it?*, U.S. Department of Agriculture, Forest Service, Rocky Mountain Research Station, Ogden, UT, General Technical Report RMRS-GTR-35, 1990.

Oliver, W.W. and Ryker, R.A., *Pinus ponderosa* (Dougl. ex Laws), in *Silvics of North America, Volume 1, Conifers*, Burns, R.M. and Honkala, B.H., Tech. Coords., U.S. Department of Agriculture, Forest Service, Washington, DC., Agriculture Handbook 654, 413, 1990.

Packee, E.C., *Tsuga heterophylla* (Raf.) Sarg., in *Silvics of North America, Volume 1, Conifers*, Burns, R.M. and Honkala, B.H., Tech. Coords., U.S. Department of Agriculture, Forest Service, Washington, DC., Agriculture Handbook 654, 613, 1990.

Phillips, R.H. and Williams, G.W., An estimation of effects of the draft EIS alternatives on communities, in *Economic and Social Conditions of Communities: Economic and Social Characteristics of Interior Columbia Basin Communities and an Estimation of Effects on Communities from the Alternatives of the Eastside and Upper Columbia River Basin Draft Environmental Impact Statements*, U.S. Department of Agriculture, Forest Service, U.S. Department of Interior, Bureau of Land Management, Portland, OR, 83, 1998.

Quigley, T., Haynes, R.W., and Graham, R.T., Tech. Eds., *Integrated Scientific Assessment for Ecosystem Management in the Interior Columbia Basin*, U.S. Department of Agriculture, Forest Service, Pacific Northwest Research Station, Portland, OR, General Technical Report PNW-GTR-382, 1996.

Rehfeldt, G.E., Hoff, R.J., and Steinhoff, R.J., Geographic patterns of genetic variation in *Pinus monticola*, *Bot. Gaz.*, 145, 229, 1984.

Reyna, N.E., Economic and social characteristics of communities in the Interior Columbia Basin, in *Economic and Social Conditions of Communities: Economic and Social Characteristics of Interior Columbia Basin Communities and an Estimation of Effects on Communities from the Alternatives of the Eastside and Upper Columbia River Basin Draft Environmental Impact Statements*, U.S. Department of Agriculture, Forest Service, U.S. Department of Interior, Bureau of Land Management, Portland, OR, 3, 1998.

Reynolds, R.T., Graham, R.T., Reiser, M.H., Bassett, R.L., Kennedy, P.L., Boyce, D.A., Goodwin, G., Smith, R., and Fisher, E.L., *Management Recommendations for the Northern Goshawk in the Southwestern United States*, U.S. Department of Agriculture, Forest Service, Rocky Mountain Forest and Range Experiment Station and Southwestern Region, Fort Collins, CO, General Technical Report RM-217, 1992.

Robichaud, P.R., Byers, J.L., and Neary, D.G., *Evaluating the Effectiveness of Postfire Rehabilitation Treatments*, U.S. Department of Agriculture, Forest Service, Rocky Mountain Research Station, Fort Collins, CO, General Technical Report RMRS-GTR-63, 2000.

Rose, S.L., Perry, D.A., Pilz, D., and Schoeneberger, M.M., Allelopathic effects of litter on the growth and colonization of mycorrhizal fungi, *J. Chem. Ecol.*, 9, 1153, 1983.

Ryan, K.C. and Amman, G.D., Bark beetle activity and delayed tree mortality in the Greater Yellowstone area following the 1988 fires, in *Ecological Implications of Fire in Greater Yellowstone USA, IAWF*, 151, 1996.

Schmidt, W.C. and Shearer, R.C., *Larix occidentalis* (Nutt.), in *Silvics of North America, Volume 1, Conifers*, Burns, R.M. and Honkala, B.H., Tech. Coords., U.S. Department of Agriculture, Forest Service, Washington, DC., Agriculture Handbook 654, 160, 1990.

U.S. Department of Agriculture (USDA), *Interior Columbia Basin Supplemental Draft Environmental Impact Statement*, Vol. 1, U.S. Department of Agriculture, Forest Service and U.S. Department of Interior, Bureau of Land Management, Portland, OR, 2000, chap. 4.

Van Wagner, C.E., Conditions for the start and spread of crown fire, *Can. J. For. Res.*, 7, 23, 1977.

chapter thirty-one

Restoration of ponderosa pine forests in the interior western U.S. after logging, grazing, and fire suppression

Merrill R. Kaufmann and Kevin C. Ryan
US Department of Agriculture Forest Service
Peter Z. Fulé
Northern Arizona University
William H. Romme
Colorado State University

Contents

31.1 Introduction

Ponderosa pine forests are a major forest type found in every state in the western U.S. They are readily accessible and aesthetically appealing, and they have provided valuable economic benefits from logging and grazing. Human activities have contributed to extensive alteration of these forests. While early Native American effects may have been locally significant (Allen 2002; Allen et al. 2002; Baker and Kipfmueller 2001; Barrett and Arno 1982; Kaye and Swetnam 1999), most impacts in much of the western U.S. have probably occurred during the last 130–150 years, since settlement by Euro-Americans.

In many western U.S. regions, early photographs, journals written by explorers and surveyors, and scientific reports supplement physical evidence of historical forest structure. Physical evidence includes stand remnants of historical landscapes, fire scars providing insight into a major stand regulating influence, and residual stumps from early logging. Together, these sources of information provide considerable insight into the structural characteristics of natural ponderosa pine forests and their ecology prior to significant human influence (Moore et al. 1999; Swetnam et al. 1999).

Three varieties of ponderosa pine occur. These are Pacific ponderosa pine (*Pinus ponderosa* var. *ponderosa*) along the West Coast (particularly Oregon and Washington), Arizona ponderosa pine (*Pinus ponderosa* var. *arizonica*) found in southern Arizona and New Mexico, and the Rocky Mountain variety (*Pinus ponderosa* var. *scopulorum*) in the interior western states (Oliver and Ryker 1990). Jeffrey pine (*Pinus jeffreyi*), found in California and Mexico, closely resembles ponderosa pine but is a distinct species. Studies of packrat middens indicate that Rocky Mountain ponderosa pine first occurred in northern New Mexico and Arizona about 10,000 to 12,000 years ago, central Colorado only about 5,500 to 7,000 years ago, and northern Wyoming and western Montana only about 2,000 years ago, suggesting a postglacial northward migration (J. Betancourt, U.S. Geological Survey, pers. comm.).

Ponderosa pine generally occurs at the lower elevational life zone of western forests, above drier nonforest vegetation types such as grasslands, shrublands, and pinyon–juniper woodlands, but below subalpine forests. Because the geographic area naturally occupied by ponderosa pine is very large, significant regional differences exist in climate, soils, and topography where ponderosa pine occurs (Oliver and Ryker 1990). In many areas of Arizona, Washington, Oregon, and California, and in other areas with sufficient precipitation, site productivity is higher than in most of the interior Rocky Mountains where precipitation is lower.

Historically, wildfire was a keystone regulating process in the ponderosa pine zone (Swetnam et al. 1999). Ponderosa pine often dominated this zone because it is considerably better adapted to wildfire than other tree species climatically suited to the same zone. Traits associated with fire resistance include open crowns, self-pruning branches, and thick relatively inflammable bark (Keeley and Zedler 1998; Graham 2003). Thick scales protect buds, and tight needle bunches enclose and protect meristems but open into a loose arrangement that does not favor combustion or propagation of flames. The foliage generally has high moisture content, and the roots are deep. These fire adaptations favored ponderosa pine over fire-sensitive species lacking these features, and wildfire often allowed pure or nearly pure ponderosa pine stands to exist. Fires sometimes damaged but did not kill trees, and the study of fire scars provides valuable insight into the role of past fires in the ecology of these forests (Swetnam and Baisan 1996; Agee 1998).

Human effects on ponderosa pine ecosystems include loss of old-growth forests through logging, introduction of exotic and often noxious weed species, damaging effects of overgrazing, and alteration of natural disturbance patterns such as fire and insect activity. Many forests that had been pure or nearly pure ponderosa pine now include competing tree species such as Douglas-fir (*Pseudotsuga menziesii*) or piñon pine (*Pinus edulis*) and Rocky Mountain juniper (*Juniperus scopulorum*) (Biondi 1996). Nearly all ponderosa pine forests are far denser and more vulnerable to crown fire than they were historically. Thus, there is a clear need for forest restoration, both on publicly and privately owned lands.

In this chapter, we begin by summarizing the natural range of variability in stand structure, landscape characteristics, and disturbance regimes of ponderosa pine forests prior to Euro-American settlement. Next, we examine the changes in these forests that have occurred in the last 150 years. Finally, we explore restoration of sustainable ponderosa pine ecosystems, and cite four specific case studies from across the Rocky Mountain region. We focus on Rocky Mountain ponderosa pine forests, found from western Montana to Arizona and east through the Rocky Mountains and into the Great Plains. The interior variety occurs in the elevational zone between grasslands, desert scrub, or piñon–juniper forests at lower elevations and mixed conifer or subalpine forests at higher elevations. However, much of the content of this chapter is applicable for the Pacific and Arizona varieties and to Jeffrey pine.

31.2 Historical fire regimes, stand structures, and landscape characteristics

31.2.1 Variation in fire regimes and stand structures

While fire historically maintained ponderosa pine forests throughout its range, regional differences in fire regimes had significant effects on the structure of historical forest stands and landscapes. Three categories of factors affect how a fire behaves: fuels or vegetative structure, topography and other physical site characteristics, and weather. Relatively subtle differences in ponderosa pine stand development exist across the geographic range of the species, but in combination with physical site characteristics and weather, significant contrasts exist in fire behavior patterns and assemblages of stands into a landscape structure. In this chapter, we recognize two primary types of historical fire regimes in ponderosa pine forests: surface-fire regimes, in which fires burned at generally low to moderate intensity throughout a stand and caused little canopy mortality; and mixed-severity fire regimes, in which fires burned at variable intensity, producing patches where canopy mortality was nearly complete as well as patches of low canopy mortality.

Regional differences in historical fire behavior probably stemmed in large part from differences in understory productivity and fuel accumulation, which are influenced by precipitation and its seasonality, and crown base height, which may have been influenced by surface fire frequency and flame heights and by site productivity. In the southwestern U.S., frequent low-intensity surface fires were common in historical ponderosa pine forests, with mean fire intervals often less than 10 years (Swetnam and Baisan 1996; Brown et al. 2001; Allen et al. 2002). Stand-replacing fires were rare (Woolsey 1911; Cooper 1960), and tree mortality was more common among seedlings and small saplings than in larger trees. In the southwestern U.S., summer monsoon moisture flow provides enough rainfall for high understory productivity. When dry, the grass and forb biomass provided a continuous fuelbed for the spread of surface fire, particularly following wet El Niño events (Swetnam and Betancourt 1998). Trees grew taller in these sites, and self-pruning of lower branches (perhaps enhanced by surface fires) resulted in high crown base heights that limited the likelihood of crown fires (Figure 31.1). Furthermore, the frequent surface fires minimized the establishment of seedlings, which would have provided subcanopy fuel ladders to move fire from the surface into the canopy.

Fires were less frequent farther north in the southern and central Rocky Mountains, where intervals between landscape-scale fires often exceeded 40 or even 50 years (Goldblum and Veblen 1992; Brown and Seig 1996; Brown et al. 1999, 2000; Brown and Shepperd 2001; Kaufmann et al. 2001; Donnegan et al. 2001). Frequently, fires in these more inland areas were mixed in severity, having a nonlethal surface fire component intermixed with a lethal component that killed the overstory, often in a patchy pattern. Surface fuels may have been inadequate to spread surface fires during most years. In the absence of frequent surface fires, shrub communities often developed in the open forests, and fuels

Figure 31.1 Historical low-density ponderosa pine forest in the southwestern U.S. (U.S. Forest Service file photo).

accumulated beneath the crowns of trees. Furthermore, the crown base height was lower, and in the absence of surface fires small trees probably were more common and created fuel ladders. This resulted in portions of forests becoming vulnerable to crown fires, and in occasional periods of higher precipitation, grasses and forbs may have provided enough surface fuels for fire spread. The result was mixed-severity fires, with portions spreading only on the surface, and in other places killing some trees either by torching or scorching of individual trees or by localized crown fires (Figure 31.2). But in some places, fire completely killed all trees and created new openings having no trees. The type of fire behavior at any site undoubtedly depended on local understory and overstory fuel conditions, topography, and weather. Evidence suggests that the proportions of each fire effect in mixed-severity fires varied widely (Brown et al 1999; Kaufmann et al. 2000a; Veblen et al. 2000; Huckaby et al. 2001). Furthermore, these fires historically burned through much of the year, which also would contribute to variable fire effects.

The surface-fire regime was more common in the southwest, and the mixed-severity fire regime occurred more commonly in the eastern and northern portions of the Rocky Mountain ponderosa pine range. However, local variation in fire regimes undoubtedly occurred, and the transition geographically from surface to mixed-severity fire regimes was probably gradual across latitude and elevation. There is an analytical difficulty in sorting out fire regimes. Required data include not only evidence of fire frequency and mean fire return interval under historical conditions but also evidence of tree ages and spatial patterns of fires and tree survival. The ages of trees help determine specific locations within fire perimeters where fire was stand-replacing and where it was not. Without this critical information, it is often difficult to determine the actual historical fire regime governing landscapes over time. No evidence of a true crown-fire system (Romme and Despain 1989; Turner and Romme 1994) has been found in ponderosa pine-dominated landscapes, except for one study that suggested extensive 19th-century crown fires in part of the Black Hills, SD (Shinneman and Baker 1997).

31.2.2 Historical landscape structure

The surface and mixed-severity fire behavior patterns, in combination with regional differences in site productivity, resulted in very different landscape structures. More uniform forest landscapes developed where surface fires occurred frequently, particularly in areas

Figure 31.2 Historical forest structure in December 1896 after a mixed severity fire in 1880 (from Denver Water photo archives).

with a relatively flat topography (Figure 31.1). These forests were generally low in density and dominated by large overstory trees. Forest stands were locally patchy with small clumps of trees and many small openings (generally considerably less than one ha). At a larger scale, landscapes may have included larger openings that were relatively stable over long periods, perhaps influenced by local variations in soil characteristics (Weaver 1951; Cooper 1960; White 1985; Covington and Moore 1994; Fulé et al. 1997; Fulé et al. 2002a).

In contrast, more complex and dynamic forest landscapes developed in regions where the mixed-severity fire behavior pattern predominated (Figure 31.2). Denser patches of trees may have existed but were not extensive, and open woodlands and openings with no trees were common. Landscapes were patchy and heterogeneous, and periodic fires and sporadic tree recruitment into openings or open woodlands contributed to constant changes in landscape structure over time.

31.3 Current condition of ponderosa pine forests

Ponderosa pine forests occur in relatively accessible terrain, and winter snow packs usually do not limit year-round access. Accordingly, pine forests throughout the west were widely exploited soon after Euro-American settlement, and more than a century of logging and grazing, coupled with nearly a century of fire suppression, resulted in a very different forest structure than occurred historically in most areas (Kaufmann et al. 1992; Covington and Moore 1994). Fire exclusion began in most areas with the onset of extensive, unregulated livestock grazing in the late 1800s, apparently because the animals consumed the fine herbaceous fuels that formerly carried surface fires (Allen et al. 2002). Aggressive fire suppression policies instituted in the early 1900s resulted in continued exclusion of surface fires (Pyne 1982). Logging removed many of the larger and older trees and created an environment for germination and establishment of many new trees (Veblen and Lorenz 1991; Covington and Moore 1994; Kaufmann et al. 2000a; Friederici 2003). Livestock grazing, particularly cattle, reduced understory competition for seedlings and reduced fuels for the spread of surface fire, which further favored germination and survival of new seedlings (Belsky and Blumenthal 1997; Baisan and Swetnam 1997). Examples of forest condition prior to logging and change over time caused by logging and fire suppression are shown in Figure 31.3 and Figure 31.4. Fire suppression removed the major natural disturbance that historically kept tree density low, and large tree regeneration events occurred during climatically favorable periods (Savage et al. 1996). Together, these practices resulted in dense, relatively uniform stands and landscapes dominated by young, small trees.

During the 20th century, millions of ha of ponderosa pine forest developed much higher stand densities and canopy cover, and they became vulnerable to catastrophically large crown fires and damage to watersheds, and massive bark beetle epidemics. For

Figure 31.3 Example of urban encroachment into ponderosa pine forest, creating a wildland/urban interface zone in which property and lives are at risk of fire (photograph by David Steinke, USDA Forest Service).

Figure 31.4 Current condition of ponderosa pine forest on a north aspect near Cheesman Lake in the Colorado Front Range (photo by M.R. Kaufmann).

example, in the Colorado Front Range on the east face of the southern Rocky Mountains, a series of large wildfires occurred since 1996, most exceeding 4,000 ha, each having a large crown-fire component, and each well outside the range of fire severity that occurred historically (Brown et al. 1999; Graham 2003; M. Kaufmann, unpublished data). The 55,000-ha Hayman fire in June 2002 burned the Cheesman Lake landscape, a major research site for studies on historical ponderosa pine landscape characteristics (Brown et al. 1999; Kaufmann et al. 2000b, 2001, 2003; Huckaby et al. 2001; Fornwalt et al. 2002, 2003). Many of these forests also had large amounts of fuel created by a mountain pine bark beetle epidemic in the 1970s, exacerbated by high stand densities and tree crowding.

Where surface-fire regimes occurred historically, the recent crown fires are highly atypical (Woolsey 1911; Weaver 1951). Where mixed-severity fires occurred historically, the current crown-fire patch size appears to be considerably larger than occurred historically because of a greater accumulation of fuels in the past century or more (Kaufmann et al. 2000b, 2001). In both the surface and mixed-severity fire systems, large crown fires are not consistent with ecological sustainability. They result in large areas of charred forest that are vulnerable to excessive erosion and invasion by exotic plant species (Griffis et al. 2001), and they are unsuited for native species that favor historical forest structures. In addition, human population growth and urban encroachment into forests have placed people's lives and property at risk (Figure 31.3). More than half of the forest in many western counties is classified as a wildland/urban interface.

31.4 Restoration goals for ecologically sustainable landscapes

Large crown fires in ponderosa pine forests reset the forest landscape to an unforested condition. Reestablishment of older forests takes at least several centuries, as the old-growth condition does not occur until ponderosa pine trees are about 200 years old (Kaufmann 1996), and the oldest trees in unlogged landscapes are often more than 400 or even 600 years of age (Huckaby et al. 2001). Re-formation of soils and recovery of genetic diversity takes even longer; thus, the negative effects of large crown fires persist for a long time. In contrast, carefully designed forest treatments offer opportunities to improve forest structure, composition, and processes while reducing the likelihood of an ecological reset by large crown fire.

Differences in fire effects of the two natural fire behavior systems described above (surface and mixed-severity fire regimes) provide a basis for different approaches in forest restoration across the range of interior ponderosa pine forests. Where surface fire predominated historically, restored forests would be relatively uniform at a large scale, with fine-scale patchiness (on the order of tenths of a ha) provided by irregular tree distribution and small openings. Where mixed-severity fires occurred, however, the scale of patchiness would be larger (one to tens or hundreds of ha), reflecting the thinning or stand-replacing components of fire and irregular spatial and temporal patterns of tree establishment (Huckaby et al. 2001; Kaufmann et al. 2000a; Kaufmann et al. 2003). Because of differences in historical fire effects on landscape structure, no single set of management recommendations for restoration is appropriate across the entire range of ponderosa pine. Other differences, such as soil characteristics, patterns of nonfire disturbance, climate, or topography, also vary, underscoring the importance of site-specific restoration approaches. Finally, management goals and social issues commonly preclude "strict restoration" practices that are based exclusively on ecological factors (Higgs 1997).

It is noteworthy that restoring forests to regionally specific conditions similar to those occurring historically would address both the sustainability and wildfire hazard issues. Where properties or other human values are at risk, thinning forests and creating fuel breaks having a very low overstory density could reduce wildfire hazards. Such thinning would also lead to improved ecological sustainability where surface fire predominated historically. The restoration of a more complex landscape structure where mixed-severity fires occurred would also require thinning, but restoring dispersed patches of open woodlands and openings would be necessary to provide an ecologically sustainable landscape. A restored complex landscape would also provide considerable reduction in wildfire hazard. Therefore, forest restoration can mitigate wildfire hazards while also achieving ecological goals appropriate for each region.

Spatial scale is important. Forest restoration can be accomplished at nearly any scale, from stand to landscape. Even small-scaled restoration activities near a house may provide ecological benefits. However, landscape restoration requires that forest conditions be improved at a spatial scale of landscape processes such as fire or maintenance of viable species populations. Landscape restoration necessarily is a huge undertaking with many potential social, economic, and ecological issues and impediments.

Ideally, ponderosa pine restoration would yield forests having appropriate tree age and size distributions, properly configured spatially across the landscape to support a trajectory of future self-regulation of the landscape. Historically, ponderosa pine forests had a high percentage of old-growth trees, even in complex landscapes where mixed-severity fires often killed trees and created openings (Huckaby et al. 2001). Many current forests still have at least some old-growth trees, easily recognized by characteristic features of their tree crowns and branches (Kaufmann 1996; Huckaby et al. 2003). This aids in restoring the appropriate tree age distribution in forest landscapes (Mast et al. 1999). Large, young trees approach the size of old trees found in historical forests (Kaufmann et al 2000b; Allen et al. 2002). Where old trees are rare, these trees are candidates for retention. However, retention of all large trees ultimately leads to overstocked stands and increased wildfire hazard in some stands, and where openings should be re-created, such trees clearly must be removed. The number of smaller trees usually far exceeds historical densities, however, and small trees contribute significantly to current wildfire hazards.

31.5 Restoration treatment options

Forest restoration at any scale requires a careful consideration of the result desired and of the social acceptability of restoration alternatives. Land managers have three basic options: continue current fire suppression policies with no other treatment to forests; attempt to restore more sustainable forest structure and dynamics by means of natural or prescribed fire only; and actively restore more natural stand and landscape structures by means of mechanical thinning, prescribed fire, and other techniques. The first two options probably are not feasible. While leaving forests alone or using fires without mechanical thinning is appealing for certain ecological and social reasons (Cole 2001), fire suppression is simply not reliable enough to protect forests from fire. Thus, relying on fire suppression to hold ponderosa pine landscapes in their current condition without attempting restoration is difficult and hazardous.

In certain remote wildlands, a burn-only approach to restoration may be useful (Miller and Urban 2000). In a vast majority of ponderosa pine landscapes, however, high tree densities and biomass have already resulted in catastrophic crown fires well outside the historical patterns. Thus, the best alternative both ecologically and socially is the third option: manipulate ponderosa pine forest structure to align it with conditions that occurred naturally, thereby placing these forests on a trajectory of future self-regulation. Many agencies are carrying out fuel hazard and wildland fire use projects that involve the combination of thinning and prescribed fire, although the scale of the work is inadequate.

Two major types of treatment, mechanical thinning accompanied by removal of biomass, and the use of fire to reduce fuels, have proven useful for reducing forest density and biomass. In many ponderosa pine forests, there are 10 to 20 or more times as many trees as there were historically (Kaufmann et al. 2001). Compare a historical forest in the Colorado Front Range (Figure 31.2) with a current forest only 2 km away (Figure 31.4), for example. Felling trees but leaving slash in the forest does not appear to be a viable option. Subsequent fires are too intense and generally kill the retained trees. Converting trees into mulch with grinding machines is another option considered below.

Thinning from below in forests that historically had a surface fire regime can reduce the density of smaller trees while retaining both old trees and an appropriate diameter class distribution. Where mixed-severity fires occurred, some forest patches require thinning, but in other areas treatment should lead to either openings (complete clearcut) or open woodlands, both of which remove significant numbers of overstory trees regardless of size or even age. In a Colorado Front Range ponderosa pine landscape, 42% of the land area was estimated to have a historical canopy cover of 10% or less, 40% had between 11 and 30%, and only 7% had 35% canopy cover or higher (Kaufmann et al. 2001; Fornwalt et al. 2002; M. Kaufmann, unpublished data). In contrast, only 21% of current forests have a canopy cover of 10% or less, and 45% have a canopy cover of 35% or higher. At canopy covers of 35% or higher, crown fire occurs readily when other conditions are favorable for fire.

Prescribed fire is used to reduce wildfire hazard by reducing forest biomass, especially fine fuels and small trees (Figure 31.5). By choosing conditions under which to use prescribed fire, forest managers can effectively reduce wildfire hazards in many cases, and in rare instances may be able to recreate small openings or low-density patches by mortality of the overstory. However, prescribed burning by itself rarely can accomplish the changes in forest structure needed to make forests ecologically sustainable and safe from catastrophic fire. In many forests, a dense understory of smaller trees makes prescribed burning too likely to progress into a catastrophic crown fire. Furthermore, prescribed fire consumes deep litter accumulated beneath older trees, often producing enough heat to damage roots and kill trees that should be retained (Swezy and Agee 1991; Sackett et al. 1996). In general, it is difficult to remove enough biomass with prescribed fire, or to achieve the appropriate patch and forest structure needed to restore the landscape.

The use of both mechanical thinning and prescribed fire is very promising, however. Thinning can be used to reduce overstory density and canopy closure to levels found

Figure 31.5 Use of prescribed fire in Arizona to reduce understory fuels.

historically, thereby reducing the likelihood of active crown fire (fire spread from crown to crown). One additional option is the use of machines that chop or grind entire trees into chunks and scatter them in the forest. This process converts living biomass to material scattered on the ground, leaving it in a form that can be consumed with prescribed fire but not resulting in flame heights that lead to crown fires. However, such tree mulching is costly and has no potential for economic return, and at least temporary shifts in site nitrogen balance are likely.

Prescribed fire is useful for consuming slash and small trees left after mechanical treatment operations. Furthermore, prescribed burning is a primary tool for treatment in subsequent years, and in inaccessible areas. Frequently, soil disturbance during mechanical treatments and more open stand conditions create favorable conditions for reproduction. In a few decades, treatment benefits are quickly lost. Prescribed fire may be an effective tool for limiting ingrowth after mechanical treatment. In areas having a historical surface-fire regime, relatively frequent prescribed fire could be used to mimic the historical fire effects. In areas having a mixed-severity fire regime, prescribed fire might be timed to kill new seedlings a few years after mechanical treatment, but subsequent prescribed fire may not be required for several decades.

31.6 The concept of a landscape strategy

Millions of ha of ponderosa pine forests are in poor ecological condition and pose high risk to human lives and property; yet on an annual basis, restoration treatments involving mechanical thinning are applied to only thousands of ha. Given the staggering economic and human resources needed for large-scale restoration, a strategy for implementing restoration in stages is needed to optimize the use of limited resources. This strategy should provide the largest ecological and wildfire mitigation benefits over the greatest amount of land in a reasonable time and at a low cost per ha, while minimizing social disruption. Considerable information is available to develop at least parts of such a strategy. Like adaptive management, the development of a restoration strategy should be considered an iterative process that begins with locally relevant scientific information for developing the desired ecological outcome.

The effect of fire behavior in the landscape is one strategic factor. For wildfire management, there is growing scientific evidence that under some circumstances, treating only a portion of a landscape in an appropriate treatment pattern may significantly alter crown fire spread for areas much larger than those treated (Finney 2001). Thus, carefully positioning fuel break treatments in areas having highly erosive soils or where critical habitats need protection may leverage treatment benefits to larger areas. More information of the ecological consequences of alternative treatment patterns across the landscape is needed.

Similarly, mechanical thinning treatments may be made less costly by positioning them in areas already lower in tree density or that are more accessible, and optimizing the use of less costly prescribed fire in concert with mechanical treatments may expand treatment benefits to larger areas at reduced cost. Furthermore, some trees may have economic value as forest products.

31.7 Case examples

Several ongoing projects in Arizona, Colorado, and Montana illustrate that forest restoration of ponderosa pine forests at a landscape scale can be accomplished while accommodating differences in regional fire behavior patterns and in the importance of issues at a local scale. Several features are common among all the case studies and are summarized here. First, all sites have suffered ecologically from the effects of logging, grazing, and fire suppression that led to dense and fairly uniform forests dominated by young, small-diameter trees (Figure 31.3 and Figure 31.4). Second, all are highly vulnerable to large crown fires and are ecologically unsustainable in their present condition. Third, each site had previously provided adequate material to support forest products industries, but industry presence is now limited or nonexistent. Fourth, all sites provide a range of other amenities, including hunting, firewood gathering, and general recreation.

While similar in many ways, these sites vary in other features, such as the degree of urban encroachment, role as municipal watersheds, concern about endangered and invasive species, and especially the approaches taken by partnerships to achieve improvement in ecological condition and reduction in wildfire hazard. They also differ in the potential to reestablish a viable industry that can assist in restoration activities through forest product markets. The case studies below are presented in order of increasing latitude from Arizona to Montana, and generally reflect a transition from surface fire regimes to mixed-severity fire regimes.

31.7.1 *Northern Arizona, U.S.A. — including the Greater Flagstaff Forests Partnership*

The high country of Arizona and New Mexico supports nearly 3 million ha of ponderosa pine and lower mixed conifer forests (Fiedler et al. 2002; O'Brien 2002), distributed among several national forests, national parks, and Native American lands. In New Mexico, over 90% of these forests are considered at moderate or high risk of crown fire due to dense stand structure and accumulated fuels (Fiedler et al. 2002). A long legacy of research has shown that the exclusion of frequent surface fires changed southwestern pine forests from open to dense conditions, dominated by fuel ladders of small, crowded trees (Weaver 1951; Cooper 1960; Covington and Moore 1994; Swetnam et al. 1999). The consequences of increased fuels may have been masked by unusually wet conditions from the mid-1970s to mid-1990s (Salzer in press), but a drought period beginning in 1996 has led to large and severe wildfires such as the Rodeo-Chediski fire of 2002, which covered approximately 190,000 ha. Recent landscape-scale bark beetle outbreaks, previously not seen in the southwest (Sánchez-Martínez and Wagner 2002), may further exacerbate fire behavior.

A pioneering attempt at restoring forest structure and function was initiated in Arizona in 1992 by Covington et al. (1997). Treated areas showed significant improvements in old-growth tree photosynthesis, reduced moisture stress, increased insect resistance (Feeney et al. 1998; Stone et al. 1999), higher understory plant productivity (Covington et al. 1997), and altered nutrient transformations and hydrological responses (Kaye and Hart 1998; Kaye et al. 1999). This set of findings provides a strong scientific basis for viewing treatments as restoring ecosystem-level patterns of structure and function, not simply as ameliorating fuel hazards.

The largest restoration project to date, on Bureau of Land Management lands in northwestern Arizona, covers 1,200 ha (Moore et al. 1999; Waltz et al. 2003). This scale is large enough to address invertebrate (Meyer and Sisk 2001) and vertebrate responses (Germaine and Germaine 2002). Different silvicultural approaches also are being evaluated, including tests of multiple levels of thinning and burn-only treatments (Fulé et al. 2001a, 2001b, 2002b). The degree of thinning was directly related to reduction in potential crown fire behavior, as measured by fire behavior modeling (Scott 1998). Empirical evidence of the effectiveness of thinning in reducing wildfire intensity in Arizona was also reported by Pollet and Omi (2002).

As in other areas of the U.S., the social context of forest restoration has been clouded with disagreement and distrust between some environmental activist groups and governmental agencies such as the U.S. Forest Service. A novel collaboration has attempted to create consensus. The Greater Flagstaff Forests Partnership (2004) was formed after severe wildfires in 1996 by nongovernmental organizations, Northern Arizona University, municipal and county governments, and other stakeholders, to work with the Coconino National Forest to restore forests around the city of Flagstaff, Arizona.

The Partnership and its restoration efforts have not been free of controversy. However, by 2004 the first project was nearly complete, and a series of subsequent projects were also moving forward. The Partnership's work is becoming a useful example of the role of collaboration and its limitations in southwestern forest restoration (Moseley and KenCairn 2001).

31.7.2 Southwestern Colorado — the Ponderosa Pine Partnership

The Ponderosa Pine Partnership in southwestern Colorado began in 1992 with concerns about a dwindling supply of large trees for the local timber industry and overly dense forests of smaller trees vulnerable to destructive wildfires, insect outbreaks, and loss of biodiversity. A public partnership of local national forest administrators, government officials, and representatives of the timber industry sought constructive and sustainable solutions to these problems. A scientific assessment of the ecological history and current state of the landscape (Lynch et al. 2000; Romme et al. 2003) demonstrated natural and human disturbance histories much like those documented in northern Arizona and many other places throughout western North America. Reconstructions of past canopy density in three representative stands ranged from 37 to 59 trees/ha, whereas current densities range from 178 to 338 trees/ha in the same stands (Romme et al. 2003).

The Ponderosa Pine Partnership designed a restoration treatment experiment to test the effectiveness of reaching four goals: reduce canopy density and basal area, leaving clumps of trees interspersed with small openings similar to those of pre-1880 stands; reintroduce low-intensity fire; stimulate productivity and diversity of suppressed herbaceous plants on the forest floor; and provide timber for local sawmills, mostly from small-diameter trees, and to find markets that would allow the companies to harvest small-diameter timber without losing money.

The prescription for achieving these objectives included three stages: retention of the largest and oldest trees and a cluster of trees of all sizes surrounding each large or old tree; fuel reduction with prescribed fire; and ecological and economic monitoring, including a complete accounting of loggers' costs and revenues associated with the project (Romme et al. 2003).

Treatments were conducted from 1995 to 1998, and monitoring was done from 1995 to 2001. The ecological objectives were largely achieved, but monitoring results provided guidelines for improving the results as the project moved forward (Lynch et al. 2000; Romme et al. 2003). Treatments did not significantly reduce the quantity of downed woody fuels on the forest floor, and additional prescribed burning was recommended. The combination of canopy thinning and prescribed burning led to dramatic increases in herbaceous cover and diversity. The economic objective was only partially achieved, the largest obstacle being the lack of suitable markets for small-diameter material (Romme et al. 2003). Numerous creative alternatives for adding value to forest products were explored with only limited success. Overall, the loggers' profit in the entire operation was less than 1% of gross revenues.

In 2001, a scientific review team generally endorsed the approach being taken, and encouraged the partnership to expand restoration treatments from the 20 to 40 ha experimental stands to much larger units. The team recommended creating larger tree clusters and larger open areas than had been accomplished to date, to better emulate the pre-1880 spatial pattern and to provide more appropriate wildlife habitat. The partnership is using these recommendations to develop restoration timber sales for much larger units.

Overall, an ecologically sound restoration prescription has been developed, tested, and refined, and can now be applied more extensively throughout the 75,000-ha ponderosa pine landscape in southwestern Colorado. The local community has supported restoration treatments without legal challenges by environmental or industry groups that often hold up projects of this kind, in large part because of early involvement of researchers and all stakeholder groups in formulating restoration treatments, plus the transparent development and evaluation of the experimental prescription. The largest challenge is financial: the restoration prescription has been only marginally profitable at best. Ultimately, the single greatest economic need for forest restoration projects of this kind is the development of alternative, profitable markets for small-diameter material (Lynch and Mackes 2002). This challenge is being addressed by several institutions, such as the Interior West Center for the Innovative Use of Small Diameter Trees, located at Colorado State University.

31.7.3 East central Colorado — the Upper South Platte Watershed Protection and Restoration Project

The Upper South Platte Project was initiated in the South Platte watershed southwest of Denver, CO, after the 5,000-ha Buffalo Creek fire in 1996. A series of subsequent fires in 2000 and 2002, most spectacularly the 55,000-ha Hayman fire (Graham 2003), contributed to both local and national action on forest restoration. In nearly every case, fire severity was far worse than experienced historically, illustrating a major problem with respect to forest condition in the Colorado Front Range (Kaufmann et al. 2001; Graham 2003). Many structures were burned, and erosion introduced massive amounts of sediment into the metropolitan water supply for Denver, CO. Five firefighters were killed en route to the Hayman fire. Three pilots were killed during aerial suppression activities for a smaller fire near an urban area, and two residents died during postfire flooding.

The continued occurrence of large fires in the Colorado Front Range ponderosa pine zone leads to several conclusions. First, a threshold of forest condition has been reached in which current forests have become particularly vulnerable to large crown fires. Second, there is an urgency to implement restoration and wildland fire hazard mitigation to protect both ecosystems and human values at risk. And third, the scale of treatments must be massive, given the spatial extent of the ponderosa pine zone in the Colorado Front Range and the huge amount of urban encroachment into this zone.

The 1996 Buffalo Creek fire led to the establishment of a partnership to establish goals focused on protecting watersheds and improving the ecological condition of forests. After the 2000 and 2002 fires, a Front Range Fuel Treatment Partnership was formed to address the condition of ponderosa pine forests in the entire Front Range area (several million ha), with an increased focus on the wildland/urban interface.

Research on the historical ecology of Front Range ponderosa pine forests has been a compelling factor guiding restoration efforts. A major portion of this research was conducted on the Cheesman Lake ponderosa pine landscape in the South Platte watershed. While this unlogged 3,000-ha landscape was completely destroyed by the 2002 Hayman fire, it provided exceptional insight into the effects of natural fire behavior and tree recruitment patterns on historical ponderosa pine landscape structure and change over time (Brown et al 1999; Kaufmann et al. 2000b, 2001, 2003; Huckaby et al. 2001, 2003; Fornwalt et al. 2002, 2003). This research demonstrated not only that the ecological condition of current forests was poor, but that characteristics of historical forests could be used to guide treatment selection and implementation.

Initial partners included the U.S. Forest Service (both national forest management and research), Colorado State Forest Service, Denver Water, U.S. Fish and Wildlife Service,

U.S. Geological Survey, and U.S. Natural Resource Conservation Service. More recently, environmental organizations have become constructively engaged in moving restoration activities forward. Efforts have been aided by large-scale landscape assessments to identify high-priority watershed and wildland/urban interface areas for initial restoration treatments on both publicly and privately owned land. Public meetings presenting the scientific information available and demonstration sites aided early acceptance of restoration concepts.

From the inception of the restoration project, activities focused on practicing good ecology by altering forest structure toward conditions found before Euro-American settlement. These activities are reducing the probability of crown fires around structures and in watershed areas most critically in need of protection from erosion. Activities also focused on the protection of two endangered species: the Pawnee montane skipper (a butterfly) and the Prebles meadow jumping mouse. Both mechanical thinning and prescribed fire are being used to improve ecological sustainability and mitigate wildfire hazards. Thinning includes both decreasing tree density and canopy cover and creating openings similar to those found historically. This creates fuel breaks that protect property in the wildland/urban interface and limit the size of new crown fires. It also improves butterfly habitat by creating grassy openings. Prescribed fire is focused on reducing fuels where it is safe to do so and burning slash from mechanical treatments; fire will also be used to limit regeneration. Currently, the combined areas of mechanical thinning and prescribed burning treatments are approaching tens of thousands of ha annually. As with other projects, the low economic value of small-diameter trees is a major limitation for large-scale treatment.

31.7.4 Western Montana — the Lick Creek Demonstration Study

In 1982, George Gruell, a pioneer in the use of repeat photography, and his associates published a series of photographs that documented 70 years of management-era change on the structure and composition of ponderosa pine forests (Gruell et al. 1982). They proposed the use of partial cutting and prescribed burning to restore ecological conditions. This led in 1991 to the initiation of the Lick Creek Demonstration Study on the Bitterroot National Forest in west central Montana, documented in a compendium by Smith and Arno (1999). Prior to 1900, light surface fires, believed to have been caused by lightning, burned the area on average every 7 years (Arno 1976; Gruell et al. 1982). The forest consisted primarily of large mature and old-growth pines with grass and forb understory. Based on reconstruction of stand structure, the pines appear to have been predominantly in the 200- to 400-year range (Arno et al. 1995; Menakis 1994).

Management of the Lick Creek site began in 1906. Gifford Pinchot, the first Chief of the U.S. Forest Service (1905 to 1910) and classically trained at the French National Forestry School in Nancy, France, instructed the marking of the trees for cutting (Smith and Arno 1999). The original stand averaged 124 trees per ha. About half the trees were cut favoring larger-diameter trees (trees over 48 cm), and about two thirds of the volume was removed. In the 1950s and 1960s, multiple harvests that favored cutting of the larger pines, combined with fire suppression, led to substantial loss of ecological integrity and a conversion to a stand replacement fire regime (Figure 31.6) (Gruell et al. 1982; Smith and Arno 1999).

In 1991, a series of replicated treatments was located on 215 ha. Three silvicultural systems were tested and combined with varying fuel treatments and controls, and commercial harvesting was conducted in 1992. Treatment effects on trees, understory vegetation, soil nutrients and microbial populations, wildlife forage, and avian use were monitored and evaluated (Smith and Arno 1999). Silvicultural treatments included retention or

Figure 31.6 Lick Creek restoration site in 1979.

Figure 31.7 Lick Creek restoration site after thinning treatment (U.S. Forest Service file photo).

irregular shelterwood cutting, prescribed for areas that consisted predominantly of second-growth pine 80 to 85 years old, followed by two levels of prescribed burning to treat fuels and rejuvenate understory vegetation; modified individual tree selection cut (Fiedler et al. 1988) with and without prescribed fire on an area of primarily uneven-aged ponderosa pine ranging from seedlings to old-growth trees up to 400 years in age; and commercial thinning in an area where precommercial thinning had taken place in the 1960s, with and without prescribed fire. The cutting and burning prescriptions were designed, in general, to greatly reduce all Douglas-fir and to reduce ponderosa pine primarily in the middle- and smaller-size classes. Additional mortality occurred predominantly in the smaller-size classes 2 to 5 years after treatments, due to fire injury and bark beetle attack. The early effects of the treatments on the various response variables are numerous and too complex to describe in this chapter, but are documented by Smith and Arno (1999). However, the treatments were generally successful at achieving the desired stand structure and species composition and at reducing fuels and fire potential in all of the areas (Figure 31.7). Total cover of understory species surpassed pretreatment levels by the second year, particularly in burned treatments. An increase in the presence of invasive weeds and heavy wildlife foraging, particularly in burned areas, compromised some of the restoration goals. Soil studies indicate few significant effects beyond the second year.

The restoration of Lick Creek was not without public controversy. Despite many public hearings and field trips, the treatments were appealed. The long history of documentation of management activities, the availability of repeat photography, and the efforts of a large number of scientists and educators were fundamental to getting this restoration project off the ground. Sound baseline data combined with active monitoring provide a basis for adaptive management as we move forward with restoration projects throughout the region, projects that enjoy increasing public acceptance because of the Lick Creek Demonstration Area.

31.8 Summary

The condition of almost all ponderosa pine forests in the western U.S. is poor, but not irreversibly so, except where recently burned at high severity. While forest densities are high and in many cases there has been an invasion of more fire-sensitive species, there is ample opportunity for ecological restoration and return of ponderosa pine landscapes to an ecologically sustainable condition. The costs of restoration are enormous, however, because fuel treatments to improve ecological conditions and reduce wildfire hazards generally require subsidies in today's economic environment, even when biomass being removed may have some commercial value. Yet, the ecological costs of not treating forests are equally great and often long-lasting because of the nature of current fires. Large crown fires preclude the restoration of sustainable old-growth ponderosa pine forest for centuries, during which many ecological services such as provision for biodiversity are absent or adversely affected.

Evidence is accumulating that restoration activities can return ponderosa pine systems to a condition in which they can once again self-regulate. Restoring the structure of forests and landscapes can set the stage for protection and recovery of sustainable populations of plants and animals. Restoration activities that lead to more open forest structures similar to those found historically result in protection from severe damage during fire because of reduced fire severity, including avoidance of soils destabilized by severe fire. Protection of human lives and property is dramatically easier where crown fires are reduced by an open forest structure. And finally, experience is teaching that humans can adapt more readily than expected to a more open forest landscape structure that had once been considered undesirable, particularly when scientific evidence shows that the open landscape is more natural and ecologically sustainable, and safer for humans.

References

Agee, J.K., Fire and pine ecosystems, in *Ecology and Biogeography of Pinus*, Richardson, D.M., Ed., Cambridge University Press, Cambridge, U.K., 193, 1998.

Allen, C.D., Lots of lightning and plenty of people: an ecological history of fire in the upland Southwest, in *Fire, Native Peoples, and the Natural Landscape*, Vale, T.R., Ed., Island Press, Washington, D.C., 143, 2002.

Allen, C.D., Savage, M., Falk, D.A., Suckling, K.F., Swetnam, T.W., Schulke, T., Stacey, P.B., Morgan, P., Hoffman, M., and Klingel, J.T., Ecological restoration of Southwestern Ponderosa pine ecosystems: a broad perspective, *Ecol. Appl.*, 12, 1418, 2002.

Arno, S.F., *The Historical Role of Fire on the Bitterroot National Forest*, U.S. Department of Agriculture, Forest Service, Intermountain Forest and Range Experiment Station, Research Paper INT-187, Ogden, UT, 29 pp., 1976.

Arno, S.F., Scott, J.H., and Hartwell, M.G., *Age-class Structure of Old Growth Ponderosa Pine/Douglas-fir Stands and its Relationship to Fire History*, U.S. Department of Agriculture, Forest Service, Intermountain Research Station, Research Paper INT-RP-481, Ogden, UT, 25 pp., 1995.

Baisan, C.H. and Swetnam, T.W., *Interactions of Fire Regimes and Land Use in the Central Rio Grande Valley*, U.S. Department of Agriculture, Forest Service, Rocky Mountain Forest and Range Experiment Station, Research Paper RM-RP-330, Fort Collins, CO, 20 pp., 1997.

Baker, W.L. and Kipfmueller, K.F., Spatial ecology of pre-Euro-American fires in a Southern Rocky Mountain subalpine forest landscape, *Prof. Geogr.*, 53, 248, 2001.

Barrett, S.W. and Arno, S.F., Indian fires as an ecological influence in the Northern Rockies, *J. For.*, 80, 647, 1982.

Belsky, A.J. and Blumenthal, D.M., Effects of livestock grazing on stand dynamics and soils in upland forests of the Interior West, *Conserv. Biol.*, 11, 315, 1997.

Betancourt, J., U.S. Geological Survey, personal communication.

Biondi, F., Decadal-scale dynamics at the Gus Pearson Natural Area: evidence for inverse (a)symmetric competition?, *Can. J. For. Res.*, 26, 1397, 1996.

Brown, P.M. and Seig, C.H., Fire history in interior ponderosa pine communities of the Black Hill, South Dakota, USA, *Int. J. Wildland Fire*, 6, 97, 1996.

Brown, P.M., Kaufmann, M.R., and Shepperd, W.D., Long-term landscape patterns of past fire events in a montane ponderosa pine forest of central Colorado, *Landscape Ecol.*, 14, 513, 1999.

Brown, P.M., Ryan, M.G., and Andrews, T.G., Historical surface fire frequency in ponderosa pine stands in Research Natural Areas, central Rocky Mountains and Black Hills, USA, *Nat. Areas J.*, 20, 133, 2000.

Brown, P.M. and Shepperd, W.D., Fire history and fire climatology along a 5° gradient in latitude in Colorado and Wyoming, USA, *Paleobotanist*, 50, 133, 2001.

Brown, P.M., Kaye, M.W., Huckaby, L.S., and Baisan, C.H., Fire history along environmental gradients in the Sacramento Mountains, New Mexico: influences of local patterns and regional processes, *Ecoscience*, 8, 115, 2001.

Cole, D.N., Management dilemmas that will shape wilderness in the 21st century, *J. For.*, 99, 4, 2001.

Cooper, C.F., Changes in vegetation, structure, and growth of southwestern pine forests since white settlement, *Ecology*, 42, 493, 1960.

Covington, W.W. and Moore, M.M., Southwestern ponderosa forest structure and resource conditions: changes since Euro-American settlement, *J. For.*, 92, 39, 1994.

Covington, W.W., Fulé, P.Z., Moore, M.M., Hart, S.C., Kolb, T.E., Mast, J.N., Sackett, S.S., and Wagner, M.R., Restoration of ecosystem health in southwestern ponderosa pine forests. *J. For.*, 95, 23, 1997.

Donnegan, J.A., Veblen, T.T., and Sibold, J.S., Climatic and human influences on fire history in the Pike National Forest, Central Colorado, *Can. J. For. Res.*, 31, 1526, 2001.

Feeney, S.R., Kolb, T.E., Covington, W.W., and Wagner, M.R., Influence of thinning and burning restoration treatments on presettlement ponderosa pines at the Gus Pearson Natural Area, *Can. J. For. Res.*, 28, 1295, 1998.

Fiedler, C.E., Becker, R.R., and Haglund, S.A., Preliminary guidelines for uneven-aged silvicultural prescriptions in ponderosa pine, in *Ponderosa Pine — the Species and its Management*, Baumgartner, D.M. and J.E., Loten, Comps., Washington State University Cooperative Extension, Pullman, WA, 235, 1988.

Fiedler, C.E., Keegan III, C.E., Robertson, S.H., Morgan, T.A., Woodall, C.W., and Chmelik, J.T., *A Strategic Assessment of Fire Hazard in New Mexico*, Final report to the Joint Fire Science Program, Boise, ID, 2002.

Finney, M.A., Design of regular landscape fuel treatment patterns to modify fire growth and behavior, *For. Sci.*, 47, 219, 2001.

Fornwalt, P.J., Kaufmann, M.R., Huckaby, L.S., and Stoker, J.M., Using the forest vegetation simulator to reconstruct historical stand conditions in the Colorado Front Range, in *Second Forest Vegetation Simulator Conference*, Crookston, N. L. and R. N. Havis, Comps., RMRS-P-25, U.S. Department of Agriculture, Forest Service, Rocky Mountain Research Station, Ogden, UT, 108, 2002.

Fornwalt, P.J., Kaufmann, M.R., Huckaby, L.S., Stoker, J.M., and Stohlgren, T.J., Non-native plant invasions in managed and protected ponderosa pine/Douglas-fir forests of the Colorado Front Range, *For. Ecol. Manage.*, 177, 515, 2003.

Friederici, P., Ed., *Ecological Restoration of Southwestern Ponderosa Pine Forests*, Island Press, Washington, DC, 2003.

Fulé, P.Z., Moore, M.M., and Covington, W.W., Determining reference conditions for ecosystem management in southwestern ponderosa pine forests, *Ecol. Appl.*, 7, 895, 1997.

Fulé, P.Z., Waltz, A.E.M., Covington, W.W., and Heinlein, T.A., Measuring forest restoration effectiveness in hazardous fuels reduction, *J. For.*, 99, 24, 2001a.

Fulé, P.Z., McHugh, C., Heinlein, T.A., and Covington, W.W., Potential fire behavior is reduced following forest restoration treatments, in *Ponderosa Pine Ecosystems Restoration and Conservation: Steps Toward Stewardship*, Vance, G.K., C.B. Edminster, W.W. Covington, and J.A. Blake, Comps., RMRS-P-22, U.S. Department of Agriculture, Forest Service, Rocky Mountain Research Station, Ogden, UT, 28, 2001b.

Fulé, P.Z., Covington, W.W., Moore, M.M., Heinlein, T.A., and Waltz, A.E.M., Natural variability in forests of Grand Canyon, USA, *J. Biogeogr.*, 29, 31, 2002a.

Fulé, P.Z., Covington, W.W., Smith, H.B., Springer, J.D., Heinlein, T.A., Huisinga, K.D., and Moore, M.M., Testing ecological restoration alternatives: Grand Canyon, Arizona, *For. Ecol. Manage.*, 170, 19, 2002b.

Germaine, H.L. and Germaine, S.S., Forest restoration treatment effects on the nesting success of Western Bluebirds (*Sialia mexicana*), *Restor. Ecol.*, 10, 362, 2002.

Goldblum, D. and Veblen, T.T., Fire history of a ponderosa pine/Douglas-fir forest in the Colorado Front Range, *Phys. Geogr.*, 13, 133, 1992.

Graham, R.T., Tech. Ed., *Hayman Fire Case Study*, U.S. Department of Agriculture, Forest Service, Rocky Mountain Research Station, Ogden, UT, General Technical Report RMRS-GTR-114, 396 pp., 2003.

Greater Flagstaff Forest Partnership, Website [Online]. Available at http://www.gffp.org, 2004.

Griffis, K.L., Crawford, J.A., Wagner, M.R., and Moir, W.H., Understory response to management treatments in northern Arizona ponderosa pine forests, *For. Ecol. Manage.*, 146, 239, 2001.

Gruell, G.E., Schmidt, W.C., Arno, S.F., Reich, W.J., *Seventy Years of Vegetative Change in a Managed Ponderosa Pine Forest in Western Montana — Implications for Resource Management*, U.S. Department of Agriculture, Forest Service, Intermountain Forest and Range Experiment Station, Ogden, UT, General Technical Report INT-130, 42 pp., 1982.

Higgs, E.S., What is good ecological restoration? *Conserv. Biol.*, 11, 338, 1997.

Huckaby, L.S., Kaufmann, M.R., Fornwalt, P.J., and Stoker, J.M., Landscape patterns of montane forest age structure relative to fire history at Cheesman Lake in the Colorado Front Range, in *Ponderosa Pine Ecosystems Restoration and Conservation: Steps toward Stewardship*, Vance, R.K., C.B. Edminster, W.W. Covington, and J.A. Blake, Comps. U.S. Department of Agriculture, Forest Service, Rocky Mountain Research Station, Ogden, UT, RMRS-P-22, 19, 2001.

Huckaby, L.S., Kaufmann, M.R., Fornwalt, P.J., Stoker, J.M., and Dennis, C., *Identification and Ecology of Old Ponderosa Pine Trees in the Colorado Front Range*, U.S. Department of Agriculture, Forest Service, Rocky Mountain Research Station, RMRS-GTR-110, 47 pp., 2003.

Kaufmann, M.R., To live fast or not: growth, vigor and longevity of old-growth ponderosa pine and lodgepole pine trees, *Tree Physiol.*, 16, 139, 1996.

Kaufmann, M.R., Moir, W.H., and Covington, W.W., Old-growth forests: what do we know about their ecology and management in the Southwest and Rocky Mountain Regions? in *Old-Growth Forest in the Southwest and Rocky Mountain Regions, Proceedings of a Workshop*, M.R. Kaufmann, W.H. Moir, and R.L. Basset, Tech. Coords., U.S. Department of Agriculture, Forest Service, Rocky Mountain Research Station, General Technical Report RM-213, 1, 1992.

Kaufmann, M.R., Huckaby, L.S., and Gleason, P., Ponderosa pine in the Colorado Front Range: long historical fire and tree recruitment intervals and a case for landscape heterogeneity, in *Proceedings from the Joint Fire Science Conference and Workshop, "Crossing the Millennium: Integrating Spatial Technologies and Ecological Principles for a New Age in Fire Management."* Vol. 1, L.F. Neuenschwander, K.C. Ryan, G.E. Goldberg, and J.D. Greer, Eds., University of Idaho and the International Association of Wildland Fire, Moscow, ID, 153, 2000a.

Kaufmann, M.R., Regan, C.M., and Brown, P.M., Heterogeneity in ponderosa pine/Douglas-fir forests: age and size structure in unlogged and logged landscapes of central Colorado, *Can. J. For. Res.*, 30, 1, 2000b.

Kaufmann, M.R., Fornwalt, P.J., Huckaby, L.S., and Stoker, J., Cheesman Lake — a historical ponderosa pine landscape guiding restoration in the South Platte watershed of the Colorado Front Range, in *Ponderosa Pine Ecosystems Restoration and Conservation: Steps toward Stewardship*, Vance, R.K., C.B. Edminster, W.W. Covington, and J.A. Blake, Comps., U.S. Department of Agriculture, Forest Service, Rocky Mountain Research Station, Ogden, UT, RMRS-P-22, 9, 2001.

Kaufmann, M.R., Huckaby, L.S., Fornwalt, P.J., Stoker, J.M., and Romme, W.H., Using tree recruitment patterns and fire history to guide restoration of an unlogged ponderosa pine/Douglas-fir landscape in the southern Rocky Mountains after a century of fire suppression, *Forestry*, 76, 231, 2003.

Kaye, J.P. and Hart, S.C., Ecological restoration alters nitrogen transformations in a ponderosa pine — bunchgrass ecosystem, *Ecol. Appl.*, 8, 1052, 1998.

Kaye, J.C., Hart, S.C., Cobb, R.C., and Stone, J.E., Water and nutrient outflow following the ecological restoration of a ponderosa pine — bunchgrass ecosystem, *Restor. Ecol.*, 7, 252, 1999.

Kaye, M.W. and Swetnam, T.W., An assessment of fire, climate, and Apache history in the Sacramento Mountains, New Mexico, *Phys. Geogr.*, 20, 305, 1999.

Keeley, J.E. and Zedler, P.H., Evolution of life histories in *Pinus*, in *Ecology and Biogeography of Pinus*, Richardson, D.M., Ed., Cambridge University Press, Cambridge, U.K., 219, 1998.

Lynch, D.L. and Mackes, K.H., *Opportunities for Making Wood Products from Small Diameter Trees in Colorado*, U.S. Department of Agriculture, Forest Service, Research Paper RMRS-RP-37, 2002.

Lynch, D.L., Romme, W.H., and Floyd, M.L., Forest restoration in southwestern ponderosa pine, *J. For.*, 98, 17, 2000.

Mast, J.N., Fulé, P.Z., Moore, M.M., Covington, W.W., and Waltz, A., Restoration of pre-settlement age structure of an Arizona ponderosa pine forest, *Ecol. Appl.*, 9, 228, 1999.

Menakis, J.P., *A Summary of Ponderosa Pine (Pinus ponderosa) Management Activities in the Lick Creek Drainage of the Bitterroot National Forest*, University of Montana, Missoula, MT, thesis, 79 pp., 1994.

Meyer, C.L. and Sisk, T.D., Butterfly response to microclimatic conditions following ponderosa pine regeneration, *Restor. Ecol.*, 9, 453, 2001.

Miller, C. and Urban, D.L., Modeling the effects of fire management alternatives on Sierra Nevada mixed conifer forests, *Ecol. Appl.*, 10, 85, 2000.

Moore, M.M., Covington, W.W., and Fulé, P.Z., Evolutionary environment, reference conditions, and ecological restoration: a southwestern ponderosa pine perspective, *Ecol. Appl.*, 9, 1266, 1999.

Moseley, C. and KenCairn, B., Problem solving or social change? The Applegate and Grand Canyon forest partnerships, in *Ponderosa Pine Ecosystems Restoration and Conservation: Steps Toward Stewardship*, Vance, G.K., C.B. Edminster, W.W. Covington, and J.A. Blake, Comps., U.S. Department of Agriculture, Forest Service, Rocky Mountain Research Station, Ogden, UT, RMRS-P-22, 121, 2001.

O'Brien, R.A., *Arizona's Forest Resources, 1999*, U.S. Department of Agriculture, Forest Service, Rocky Mountain Research Station, Fort Collins, CO, RMRS-RB-2, 2002.

Oliver, W.W. and Ryker, R.A., *Pinus ponderosa Dougl. ex. Laws. Ponderosa Pine. Silvics of North America. Volume 1. Conifers*, U.S. Department of Agriculture, Forest Service, Washington, D.C., Agriculture Handbook 654, 413, 1990.

Pollet, J. and Omi, P.N., Effect of thinning and prescribed burning on crown fire severity in ponderosa pine forests, *Int. J. Wildland Fire*, 11, 1, 2002.

Pyne, S.J., *Fire in America: A Cultural History Of Wildland And Rural Fire*, University of Washington Press, Seattle, 1982.

Romme W.H. and Despain, D.G., Historical perspectives on the Yellowstone fires, 1988, *BioScience*, 39, 695, 1989.

Romme, W.H., Preston, M., Lynch, D.L., Kemp, P., Floyd-Hanna, L., Hanna, D.D., and Burns, S., The ponderosa pine forest partnership: ecology, economics, and community involvement in forest restoration, in *Ecological Restoration of Southwestern Ponderosa Pine Forests*, Friederichi, P., Ed., Island Press, Washington, D.C., 99, 2003.

Sackett, S.S., Haase, S.M., and Harrington, M.G., *Lessons Learned from Fire Use for Restoring Southwestern Ponderosa Pine Ecosystems*, U.S. Department of Agriculture, Forest Service, Rocky Mountain Forest and Range Experiment Station, Fort Collins, CO, General Technical Report RM-GTR-278, 53, 1996.

Salzer, M.W., Dendroclimatology in the San Francisco Peaks region of northern Arizona, USA, Ph.D. dissertation, University of Arizona, Tucson, 211, 2000.

Sánchez-Martínez, G. and Wagner, M.R., Bark beetle community structure under four ponderosa pine forest stand conditions in northern Arizona, *For. Ecol. Manage.*, 170, 145, 2002.

Savage, M, Brown, P.M. and Feddema, J., The role of climate in a pine forest regeneration pulse in the southwestern United States, *EcoScience*, 3, 310, 1996.

Scott, J.H., *Fuel Reduction in Residential and Scenic Forests: A Comparison of Three Treatments in a Western Montana Ponderosa Pine Stand*, U.S. Department of Agriculture, Forest Service, Rocky Mountain Research Station, Ogden, UT, Research Paper RMRS-RP-5, 1998.

Shinneman, D.J. and Baker, W.L., Nonequilibrium dynamics between catastrophic disturbances and old-growth forests in ponderosa pine landscapes of the Black Hills, *Conserv. Biol.*, 11, 1276, 1997.

Smith, H.Y. and Arno, S.F., Eds., *Eighty-eight Years of Change in a Managed Ponderosa Pine Forest*, U.S. Department of Agriculture, Forest Service, Rocky Mountain Research Station, Ogden UT, General Technical Report RMRS-GTR-23, 55 p., 1999.

Stone, J.E., Kolb, T.E., and Covington, W.W., Effects of restoration thinning on presettlement Pinus ponderosa in northern Arizona, *Restor. Ecol.*, 7, 172, 1999.

Swetnam, T.W. and Baisan, C.H., Historical fire regime patterns in the southwestern United States since AD 1700, in *Proceedings of the 2nd La Mesa Fire Symposium*, Allen, C.D., Ed., U.S. Department of Agriculture, Forest Service, Rocky Mountain Forest and Range Experiment Station, Fort Collins, CO, General Technical Report RM-GTR-286, 11, 1996.

Swetnam, T.W. and Betancourt, J.L., Mesoscale disturbance and ecological response to decadal climatic variability in the American Southwest, *J. Climate*, 11, 3128, 1998.

Swetnam, T.W., Allen C.D., and Betancourt, J.L., Applied historical ecology: using the past to manage for the future, *Ecol. Appl.*, 9, 1189, 1999.

Swezy, D.M. and Agee, J.K., Prescribed fire effects on fine-root and tree mortality in old-growth ponderosa pine, *Can. J. For. Res.*, 21, 626, 1991.

Turner, M.G. and Romme, W.H., Landscape dynamics in crown fire ecosystems, *Landscape Ecol.*, 9, 59, 1994.

Veblen, T.T. and Lorenz, D.C., *The Colorado Front Range: A Century of Ecological Change*, University of Utah Press, Salt Lake City 1991.

Veblen, T.T., Kitzberger T., and Donnegan, J., Climatic and human influences on fire regimes in ponderosa pine forests in the Colorado Front Range, *Ecol. Appl.*, 10, 1178, 2000.

Waltz, A.E.M., Fulé, P.Z., Covington W.W., and Moore, M.M., Diversity in ponderosa pine forest structure following ecological restoration treatments, *For. Sci.*, 49, 885, 2003.

Weaver, H., Fire as an ecological factor in the southwestern ponderosa pine forests, *J. For.*, 49, 93, 1951.

White, A.S., Presettlement regeneration patterns in a southwestern ponderosa pine stand, *Ecology*, 66, 589, 1985.

Woolsey, T.S., *Western Yellow Pine in Arizona and New Mexico*, U.S. Department of Agriculture, Forest Service, Washington, D.C., Bulletin 101, 64 pp., 1911.

chapter thirty-two

Restoring longleaf pine forest ecosystems in the southern U.S.

Dale G. Brockway, Kenneth W. Outcalt, and Donald J. Tomczak
US Department of Agriculture Forest Service
Everett E. Johnson
Auburn University

Contents

32.1 Southern forest environment

Longleaf pine (*Pinus palustris*) ecosystems are native to nine states of the southern region of the U.S. Longleaf pine can grow on a variety of site types including wet flatwoods and savannas along the Atlantic and Gulf coastal plain, higher droughty sand deposits from the fall line sandhills to the central ridge of Florida (Stout and Marion 1993), and the montane slopes and ridges of Alabama and northwest Georgia up to 600 m elevation (Boyer 1990b). This region has a humid subtropical climate (Bailey 1995). Maximum July temperatures average 29°C to > 35°C while minima during January range from 0 to 13°C. The mean annual precipitation is 1,040 to 1,750 mm and is well distributed through the year. The growing season is comparatively long, ranging from 300+ days in Florida to 220 days along the northern limit of longleaf. During the late summer and fall, hurricanes can develop over the Atlantic Ocean, move westward, and impact coastal plain forests. Such tropical storms are one of the principal large-scale disturbance agents for longleaf pine forests growing near the seacoast.

Longleaf pine grows on soils derived from marine sediments ranging from deep, coarse, excessively drained sands to finer textured clays (Boyer 1990b). Entisols and Spodosols, two of the major orders occupied by longleaf, are generally sandy, acidic, low in organic matter, and relatively infertile. Quartzipsamments, the most prevalent Entisol on xeric sandhills, are deep sands with weak horizon development. Spodosols, principally Aquods, are found on lower coastal plain flatwoods. These are wet sandy soils with a shallow water table that is at or near the ground surface during the rainy season. Longleaf pine is also found on more fertile clay soils (Ultisols) such as the red hills region of southern Georgia. Typic Paleudults and Plinthic Paleudults are the Ultisols most frequently supporting longleaf pine.

32.2 Longleaf pine ecology

32.2.1 Longleaf pine ecosystems

Longleaf pine forests were once among the most extensive ecosystems in North America (Landers et al. 1995). Prior to European settlement, these forests occupied ~38 million ha in the southeastern U.S. (Frost 1993). Travelers in this region during the late 18th and early 19th centuries reported vast areas of longleaf pine that sometimes covered >90% of the landscape (Bartram 1791; Williams 1837). The native range of longleaf pine (Figure 32.1) encompasses an area along the Gulf and Atlantic Coastal Plains from Texas to Virginia, extending well into central Florida and the Piedmont and mountains of northern Alabama and northwest Georgia (Boyer 1990b; Stout and Marion 1993).

An open, park-like stand structure (Figure 32.2) is a distinguishing characteristic of longleaf pine ecosystems (Schwarz 1907; Wahlenberg 1946). Naturally occurring longleaf pine forests contain numerous embedded special habitats such as stream bottoms, wetlands, and seeps (Brockway and Outcalt 1998; Hilton 1999; Platt and Rathbun 1993). In the western Gulf Coastal Plain, bluestem grasses (*Schizachyrium scoparium* and *Andropogon* spp.) dominate longleaf pine understories. From Florida north and eastward, longleaf pine typically is associated with wiregrass (*Aristida stricta* and *Aristida beyrichiana*), also known as pineland threeawn. Fallen pine needles and understory grasses facilitate the ignition and spread of fire, which limits woody shrubs and hardwood trees (Landers 1991). While such woody plants may be more numerous on mesic sites, their stature is typically limited by frequent burning.

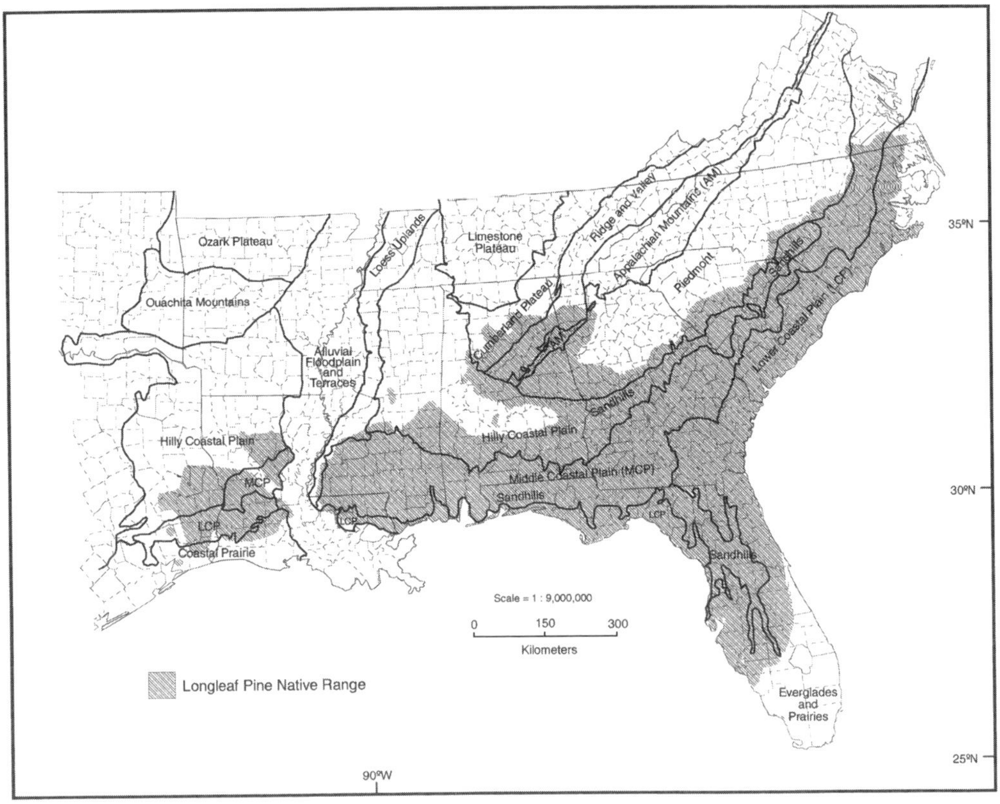

Figure 32.1 Native range of longleaf pine and physiographic provinces of the southeastern U.S. (Little 1971; Miller and Robinson 1995).

Figure 32.2 Longleaf pine bunchgrass ecosystem on xeric sandhills.

At various locations within the native range, *Quercus*, *Ilex*, and *Serenoa* may be common tree and shrub associates. Longleaf pine ecosystems support a great variety of herbaceous plant species. The high diversity of understory plants per unit area makes these ecosystems among the most species-rich plant communities outside the tropics (Peet and Allard 1993).

Longleaf pine is closely associated with frequent surface fires (Brockway and Lewis 1997; Garren 1943; Outcalt 2000). Longleaf pine and bunchgrasses function together as keystone species that facilitate but are resistant to fire (Noss 1989; Platt et al. 1988b). Their longevity and nutrient and water retention ability reinforce their site dominance and minimize change in the plant community following disturbance (Landers et al. 1995). The long and highly flammable needles of longleaf pine together with the living and dead leaves of bunchgrasses constitute a fine-fuel matrix that facilitates the rapid spread of fire (Abrahamson and Hartnett 1990; Landers 1991).

Prior to landscape fragmentation, natural fires occurred every 2 to 8 years throughout much of the region (Abrahamson and Hartnett 1990; Christensen 1981). Longleaf pine dominated this large expanse primarily because it tolerates frequent fire better than seedlings of thinner-barked competitors. Longleaf pine seedlings are susceptible to fire-caused mortality during the first year following germination, but they become increasingly resistant to fire in subsequent years. A unique adaptation of longleaf pine to a fire-prone environment is a seedling "grass stage," during which root growth is favored and the seedling top remains a tuft of needles surrounding and protecting a large terminal bud. The lack of a stem limits exposure to damage from surface fires. When sufficient root reserves have accumulated, grass stage longleaf pine seedlings "bolt" by rapidly growing 1 to 2 m in a short time period, putting their terminal bud beyond the lethal reach of most surface fires. Larger longleaf pine trees have a thick bark that protects cambial tissue from the lethal heating of surface fires (Wahlenberg 1946). Fires assist in the natural pruning of longleaf pine, creating a clear bole between the crown and any accumulated surface fuels. Surface fires are thereby prevented from easily moving into the canopy. Longleaf pine also tends to regenerate more successfully in forest openings than directly beneath mature trees (Brockway and Outcalt 1998), thus keeping ladder fuels away from the crowns of adult trees.

Longleaf pine evolved in an environment influenced by frequent disturbance, principally fire (Engstrom et al. 2001; Palik and Pederson 1996). Damaging tropical storms, such as hurricanes and associated tornadoes, may fell trees over an extensive area and open gaps in the canopy of longleaf pine forests (Croker 1987). Lightning is another important disturbance agent, typically killing individual trees but occasionally striking small groups of trees (Komarek 1968; Palik and Pederson 1996; Taylor 1974). Insect infestations are uncommon; however, annosus root rot (*Heterobasidion annosum*), pitch canker (*Fusarium moniliforme* var. *subglutinans*), and cone rust (*Cronartium strobilinum*) are among the pathogens that may infect longleaf pine (Boyer 1990b). Epidemics of brown-spot disease (*Mycosphaerella dearnessii*) occasionally occur in young longleaf pines; this pathogen is usually fatal unless a surface fire consumes infected needles and cleanses the stand of inoculum (Boyer 1990b).

Longleaf pine is a shade-intolerant tree species and regenerates naturally only in canopy gaps (Wahlenberg 1946). Seedlings developing in gaps at different times result in a network of forest patches at various stages of development dispersed across the landscape (Pickett and White 1985). Such gap-phase regeneration dynamics produce a forest structure commonly observed in natural longleaf pine ecosystems of even-aged patches distributed within an uneven-aged mosaic (Palik et al. 1997).

32.2.2 Ecological significance

The complex natural pattern and disturbance-mediated processes of longleaf pine forests cause extraordinarily high levels of biological diversity in these ecosystems, with as many as 140 species of vascular plants in a 1,000 m^2 area. Counts of more than 40 species per m^2 have

Figure 32.3 Former longleaf pine site invaded and occupied by oak.

been recorded in many longleaf pine communities (Peet and Allard 1993). A large number of these plant species are restricted to, or found principally in, longleaf pine habitats. Not surprisingly, many animal species also depend on longleaf pine ecosystems for much of their habitat, including two increasingly rare animals that are important primary excavators. Tree cavities created by red-cockaded woodpeckers (*Picoides borealis*) and ground burrows dug by gopher tortoises (*Gopherus polyphemus*) provide homes for a variety of secondary users such as insects, snakes, birds, and mammals (Engstrom 2001; Jackson and Milstrey 1989).

The longleaf pine forests and savannas of the southeastern coastal plain are among the most critically endangered natural ecosystems in the U.S., now occupying less than 3% of their original extent (Noss et al. 1995; Ware et al. 1993). Extreme habitat reduction is the primary cause for increasing rarity of 191 taxa of vascular plants and several terrestrial vertebrate species that are endemic to or exist largely in longleaf pine communities (Hardin and White 1989; Walker 1993). Habitat loss principally has resulted from conversion of longleaf pine forests to other land uses (i.e., agriculture, industrial pine plantations, and urban development), landscape fragmentation, and interruption of natural fire regimes (Landers et al. 1995; Wear and Greis 2002). Long-term suppression of fire typically depresses species diversity, and a substantial hardwood understory and midstory develops a thick layer of forest litter (Brockway and Lewis 1997; Kush and Meldahl 2000) (Figure 32.3). Such extraordinary buildup of forest fuel poses a serious wildfire hazard and, rather than naturally occurring surface fires, crown fires with potentially catastrophic effects on rare plants and animals are likely. Safe and effective reintroduction of fire into long-unburned forests remains the critical conservation challenge (Wear and Greis 2002).

Longleaf pine bunchgrass ecosystems are also vital to the maintenance of numerous biotic communities embedded within the southern forest landscape matrix (Landers et al. 1990). Many of these adjacent communities require periodic fire to maintain their ecological structure and health (Kirkman et al. 1998). Wildfires typically begin in longleaf pine forests and spread into adjoining habitats such as seepage slopes, canebrakes, treeless savannas, and sand pine scrub. Without periodic fire, these communities also change in ways that make them less suitable habitats for other fire-adapted plants and animals.

32.3 History of longleaf pine ecosystems

Longleaf pine, moving northward and eastward from its ice age refugia in southern Texas or northern Mexico (Schmidtling and Hipkins 1998), established in the lower coastal plain ~8,000 years ago (Watts et al. 1992) and during the ensuing 4,000 years spread throughout the southeast (Delcourt and Delcourt 1987). Interestingly, this time period coincided with increased population levels of Native Americans throughout the region; their use of fire is thought to be related to the spread of longleaf pine forests (Landers and Boyer 1999; Pyne 1997; Schwartz 1994). Native Americans frequently used fire to manipulate their environment (Anderson 1996; Carroll et al. 2002; Robbins and Myers 1992; Stanturf et al. 2002). Recognizing the benefits of fire on the landscape, early European settlers adopted the practice of periodically burning nearby forests and woodlands to improve forage quality for cattle grazing and discouraged the encroachment of shrubby undergrowth.

European settlement had little impact on longleaf pine forests initially, with harvesting limited to areas near towns and villages for building log structures (Croker 1987). By the 1700s, water-powered sawmills became common, but log transportation was inefficient and remained confined to rivers (Frost 1993). After 1830, removal of the longleaf pine resource accelerated with the arrival of steam railroads, and was quickly followed by steam skidders. By 1880, most of the longleaf pine forests along streams and railroads had been harvested (Frost 1993). During the next 40 years, the great forests of longleaf and other southern pines were harvested, with temporary railroad spur lines laid down every quarter mile (Croker 1987). Timber extraction peaked in 1907, when 39 million m^3 were removed (Wahlenberg 1946). By 1930, nearly all old-growth longleaf pine was harvested and lumber companies migrated west.

Although well adapted to frequent disturbance from surface fires, longleaf pine was not well suited to disturbances brought by European settlement. As a result of cumulative impacts over three centuries of changing landuse, longleaf pine forests declined dramatically. By 1900, logging, extraction for naval stores, and agriculture had reduced the area dominated by longleaf pine by more than half (Frost 1993). Second-growth longleaf pine stands became established on only one third of the sites previously occupied (Wahlenberg 1946). Harvest of these second-growth stands, often followed by conversion to other southern pines or urban development, continued through 1985 (Kelly and Bechtold 1990) until longleaf pine was reduced to less than 5% of its original area (Outcalt and Sheffield 1996).

32.4 Social and political context

Longleaf pine ecosystems have provided raw materials for economic development in the southern U.S. Wild game, forage grasses, wood, and naval stores (chemicals derived from pine resin) were the principal products of these forests (Franklin 1997). During the early 20th century, affluent landowners recognized the value of longleaf pine forests as habitat for bobwhite quail (*Colinus virginianus*) and white-tailed deer (*Odocoileus virginianus*) and acquired large tracts for private hunting reserves. Many large areas of longleaf pine exist today only because of the opportunities they provided for hunting and timber harvest. Nevertheless, economic exploitation has played a major role in the decline of these forests. Recent developments provide hope that these negative trends may be reversed. Conversion of longleaf pine to other tree species has slowed, as numerous federal and state agencies have begun regenerating longleaf pine on their lands following harvest, and they rehabilitate degraded longleaf pine forests with fire and other appropriate techniques (Hilliard 1998; McMahon et al. 1998). Interest in longleaf pine reforestation and afforestation has increased on private lands because of incentives provided by the federal government; from 1998 to 2000, longleaf pine was planted on 68,240 ha across the region.

The southern forestry community has also gained an improved understanding of longleaf pine ecosystems and has come to appreciate the natural heritage that could be lost. No single entity dominates landownership in longleaf pine ecosystems but numerous groups share a sense of urgency, and partnerships have developed. The Nature

Conservancy, Tall Timbers Research Station, Joseph W. Jones Ecological Research Center, USDA Forest Service, USDI Fish and Wildlife Service, U.S. Department of Defense, Cooperative Extension Service, state agencies, private landowners, universities, and forest industry now work together to promote longleaf pine ecosystem restoration. In 1995, the Longleaf Alliance was formed to serve as a regional clearinghouse for a broad range of information on the regeneration, restoration, and management of longleaf pine ecosystems. The Alliance is housed at Auburn University in Alabama and facilitates communication among these groups and provides training for private landowners concerning successful longleaf pine regeneration.

32.5 Restoration perspectives

Longleaf pine still occurs over most of its natural range, albeit in isolated fragments; thus, restoration is feasible (Landers et al. 1995). Restoration to historical authenticity may not be desirable or even possible, but natural authenticity is a reasonable goal, meaning that compositional, structural, and functional components are present within an appropriate physical environment. Thus, ecological processes can be sustained in restored longleaf pine ecosystems, providing for native species perpetuation and evolution, ecosystem resiliency to disturbance and adaptation to long-term environmental change, goods and services for human societies, and safe harbors for rare and endangered species (Clewell 2000).

Because natural longleaf pine forest ecosystems are so variable, the range of conditions that fall within natural variability are correspondingly broad. Overall, full restoration would mean an overstory dominated by longleaf pine, occurring as uneven-aged stands or even-aged patches across an uneven-aged landscape mosaic. Depending on site type and location within the native range, a lesser component of other tree species may be present, such as slash pine (*Pinus elliottii*) or oaks, which may occur singly or in clusters. The midstory should generally be absent or mostly composed of ascending longleaf pines. Native grasses and forbs should dominate the understory, with lesser cover of shrubs and vines. Long-term ecosystem recovery and sustainability will be fostered by properly functioning ecological processes such as periodic surface fires, natural regeneration that leads to normal stand replacement dynamics, nutrient cycling that maintains primary productivity, and suitable habitat that facilitates life cycle completion by numerous native organisms. Augmenting existing longleaf pine fragments and creating new connecting habitat patches will achieve reductions in habitat fragmentation, population isolation, and species rarity.

32.6 Restoration methods

32.6.1 Restoration framework

Despite a wealth of knowledge and experience concerning longleaf pine restoration, much uncertainty still exists, fostering a healthy debate about the best approaches. Desirable changes in longleaf pine communities can be achieved by using a variety of methods, machines, and products, either singly or in combination. Prescribed fire may be used to reduce midstory, understory, and occasionally overstory layers and encourage fire-tolerant plants. Because frequent fire is crucial for ecosystem restoration, other treatments should be planned to facilitate the eventual application of prescribed fire. Physical or mechanical treatments include complete overstory harvest, selective thinning of overstory and midstory trees, and shredding or mowing midstory and understory plant layers. Chemical treatments, principally herbicide application, can be used to selectively induce mortality of undesirable plant groups. In highly degraded ecosystems, biological approaches such as reintroducing extirpated species will likely be required for full restoration.

32.6.2 Selecting techniques

Historical events and changing landuse provide an array of candidate sites in various conditions for restoration of longleaf pine ecosystems. While about 1.2 million ha currently have an overstory of longleaf pine (Outcalt and Sheffield 1996), only 0.5 to 0.8 million ha of these have intact native understories (Noss 1989). Other candidate areas with little overstory longleaf pine have understories that range from having most of the native species to highly altered understories with no native species (Outcalt 2000). This variety of existing vegetation exists across the range of sites that longleaf pine can occupy, from dry sandhills to wet savannas. Suitable restoration techniques depend on the site type and degree of ecosystem degradation (Table 32.1). The types of longleaf pine ecosystems discussed are based on the classification of Peet and Allard (1993), with sandhills corresponding to their xeric and subxeric series, flatwoods and wet lowlands to their seasonally wet series, and uplands to their mesic series. However, we include their Piedmont/upland subxeric woodland community in the uplands rather than sandhills.

32.6.3 Restoration prescriptions

32.6.3.1 Xeric and subxeric sandhills dominated by longleaf pine with native understory

In many existing xeric and subxeric sandhills longleaf pine forests, fire suppression allowed turkey oak (*Q. laevis*), bluejack oak (*Q. incana*), sand live oak (*Q. geminata*), and sand post oak (*Q. stellata* var. *margaretta*) to develop into a scrub oak midstory. Repeated applications of fire during the growing season are effective at restoring these sites, by gradually reducing the density of the midstory oaks (Glitzenstein et al. 1995). Fires stimulate grasses and forbs to produce flowers and seeds (Christensen 1981; Clewell 1989; Outcalt 1994; Platt et al. 1988a), which aid in colonization of newly exposed microsites.

Reintroducing growing-season fires into xeric longleaf pine forests that have not burned for a prolonged period may kill older trees over the 1 to 3-year interval following

Table 32.1 Prescriptions for Restoring Longleaf Pine (LLP) Ecosystems in Varying Stages of Degradation

	Moderately Degraded	Very Degraded	Highly Degraded
Overstory: Understory:	Longleaf pine Native Plants	Other trees Native Plants	Other trees Nonnative Plants
Xeric and subxeric sandhills:	Growing-season fire Dormant-season fire Mechanical removal and herbicide hardwoods	Mechanical harvest Growing-season fire Herbicide sprouts Plant LLP seedlings	Roller-chop twice and burn Herbicide if needed Plant LLP seedlings Sow native understory seed
Montane and mesic uplands:	Growing-season fire Dormant-season fire Mechanical removal and herbicide hardwoods	Growing-season fire Mechanical harvest to create canopy gaps Plant LLP seedlings	Growing-season fire Harvest, chop, harrow Herbicide if needed Plant LLP seedlings Sow native understory seed
Flatwoods and wet lowlands:	Growing-season fire Dormant-season fire at 2-year intervals	Growing-season fire Mechanical harvest Roller-chop once and burn Plant LLP seedlings	Roller-chop twice and burn Herbicide if needed Plant LLP seedlings Sow native understory seed

the initial burn. The precise cause for this is unknown, but seems to be related to excessive forest litter accumulation around the base of larger longleaf pines and damage to roots, cambium, or both from smoldering combustion of this litter. To decrease this mortality, several dormant-season fires should be applied to gradually reduce the accumulated litter before switching to growing-season burning. Duff moisture levels during burns must be high enough to prevent ignition of the lower litter layer at the base of larger longleaf pines.

Usually three or four growing-season fires are sufficient to control scrub oak on these sites, but supplemental treatments can accelerate the restoration process. Mechanical methods such as chainsaw felling, girdling, or chipping on-site can reduce midstory hardwoods (Provencher et al. 2001); following these treatments with prescribed burning will stimulate grasses and forbs and reduce the growth of hardwood sprouts. If woody material from the midstory is not chipped or removed from the site, it should be allowed to decay before introducing the first prescribed burn. Mechanical methods are expensive and are most appropriate for critical areas in need of rapid restoration, such as red-cockaded woodpecker colony sites or along the urban–wildland interface where it is difficult to schedule the series of prescribed fires required for restoration.

Hexazinone herbicide can be useful in accelerating the restoration process compared with burning alone (Brockway et al. 1998). Application rates of 1 to 2 kg active ingredient ha^{-1} liquid formulation in a 2×2 m grid pattern will produce 80 to 90% oak mortality without long-term damage to herbaceous understory species (Brockway and Outcalt 2000). Because hexazinone does impact woody species, desirable nontarget species, such as gopher apple (*Licania michauxii*), may be reduced for a time. During dry periods, liquid hexazinone may photodegrade before sufficient rainfall transports it into the soil for oak roots to absorb it (Berish 1996); hence, application should be timed for periods of periodic rainfall. Granular hexazinone is less subject to this problem, but it potentially causes a greater reduction in the cover of grasses and forbs when it is uniformly applied across the entire site (Brockway et al. 1998).

32.6.3.2 *Xeric and subxeric sandhills dominated by other trees with native understory*

Scrub oaks have captured substantial areas and have become dominant following the harvest of longleaf pine. Although somewhat suppressed in the absence of frequent fire, the understory plant community still contains many native species. Other areas were converted to slash pine plantations following the removal of longleaf pine. Although understory species, especially the important grasses, are susceptible to severe mortality from soil disturbance on dry sandhills sites (Grelen 1962; Outcalt and Lewis 1990), some slash pine plantations have intact understory communities due to less intense site preparation or high soil moisture levels. A third condition is found on extensive areas in western Florida where Choctawhatchee sand pine (*Pinus clausa* var. *immuginata*) invaded former longleaf pine sites following harvest. Unlike slash pine, sand pine is more adapted to dry sites, forming a nearly continuous canopy that severely reduces understory density. However, plant diversity in these stands is generally unaffected, with native species surviving but much reduced in number (Provencher et al. 2001). Restoration under these conditions requires invigorating the herbaceous understory, if present, removing off-site slash pine or sand pine, reducing the scrub oak tree layer, and establishing longleaf pine seedlings.

Areas dominated by scrub oak can be treated with a small (3 to 5 t) single-drum roller-chopper with no offset. Heavier choppers with offset rollers should be avoided, because they can cause excessive soil disturbance that will harm understory plants. The objective of this treatment is to knock down the oaks and compress them into a layer that will carry a prescribed burn after drying. By contrast, slash pine plantations often have enough needle litter to support a prescribed burn. Burning these plantations will invigorate the grasses, allowing them to accumulate root reserves, and thereby increase their ability to recover from disturbance associated with removing the slash pine and establishing longleaf pine seedlings. A second fire following harvest will remove logging slash, help control oak sprouts, and increase the cover of herbaceous species. If slash pine plantations have several scrub oaks, hexazinone can be applied as outlined earlier. Application can be made prior to harvest, in which case the logging activity will knock down many of the standing dead stems, which then serve as additional fuel for prescribed burning. If herbicide is applied after logging, the dead oak stems should be allowed to fall before burning, as this will remove debris and facilitate planting. Sand pine often grows so densely that it must be removed to release surviving understory species. Sites can then be burned to remove logging slash, reduce abundant sand pine seedlings, and consume sand pine seed.

Options for establishing longleaf pine seedlings include manual or machine planting of either bareroot or container seedlings (Barnett et al. 1990; Barnett and McGilvray 1997). Site preparation, other than that discussed above, should be avoided to protect the understory plant community. It is much less expensive to plant additional longleaf pine seedlings to compensate for lower survival than it is to reestablish key understory species lost to excessive soil disturbance. If grass competition is vigorous ($\geq$60% cover) and bareroot seedlings are to be used, a planting machine with a small scalper blade can increase seedling survival (Outcalt 1995). Although this removes a strip of vegetation ~1 m wide, native grasses and forbs will recolonize these strips within 3 to 5 years, as long as invasive woody plants are discouraged by periodic growing-season fire. Planting container, longleaf pine seedlings results in acceptable survival rates without site preparation other than burning, although hexazinone application may increase survival on areas with vigorous scrub oak competition.

32.6.3.3 *Xeric and subxeric sandhills without a native understory*

Highly altered sites that once supported native longleaf pine ecosystems may have no longleaf pine trees and a much altered understory, or longleaf pines may be present but the native understory is not. Most of these sites were once used for agriculture or intensively managed plantations of other pines. Restoration of the understory is a formidable and therefore expensive task. Restoring understory plant communities is also the area where knowledge is most lacking and experience is limited to a few operational-scale restoration projects. In most cases, the first step is removal of trees other than longleaf pine from the overstory. Since there are few understory plants to protect, many options are available for site preparation. Chopping with a double-drum offset roller-chopper effectively controls all competition and produces a clean site for planting (Burns and Hebb 1972). This treatment can be combined with burning if there is sufficient woody residue. Much of the nutrient capital on these sites is in the litter layer and upper soil horizon; therefore, soil and litter movement should be minimized and root raking and shearing, if used, must be carefully applied. Longleaf pine bareroot or container seedlings can be planted after the soil has settled.

Restoration of understory plants is best done simultaneously with replanting longleaf pine seedlings to take advantage of the reduced competition and ease of onsite operability. The most critical part in this process is reestablishing grasses, because of their important role as fuel to support recurrent fire. To date, most work on reestablishment of wiregrass has focused on the eastern portion of the range (Means 1997; Seamon 1998; Mulligan et al. 2002). A planting density of 0.5 to 1 seedlings m^{-2} is recommended for restoration of wiregrass with plugs (Outcalt et al. 1999). To successfully establish wiregrass under existing plantations of longleaf pine, repeated burning, mechanical felling, herbicide application, or some combination must remove any hardwood midstory. A heavy-duty woods-harrow is then used to disk strips between trees. In the spring, wiregrass plugs can be planted in these strips using 1 × 1 m spacing. Applying fertilizer during the second or third growing season will stimulate wiregrass growth (Outcalt et al. 1999), but should be applied only around wiregrass plants to avoid stimulating the growth of competing vegetation. In pastures occupied by bahia grass (*Paspalum notatum*), cultivation to break up the old sod and herbicide to control the bahia grass will improve both the survival and growth of wiregrass (Uridel 1994).

Direct seeding to reestablish wiregrass between rows of trees in newly planted and existing plantations is less expensive than planting seedlings or plugs (Hattenbach et al. 1998). Small quantities of seed can be collected by hand or with a hand-held seed stripper. For larger quantities, a tractor-mounted flail-vac is effective. Seeds can be stored in woven bags or sown immediately, by hand or with a small bale chopper. Rolling seed into the soil can improve wiregrass establishment and survival (Hattenbach et al. 1998). Other grass species are part of the native understory in sandhills longleaf pine forests and should be included in seed mixes. Pineywoods dropseed (*Sporobolus junceus*), for example, is common on many sites and, like wiregrass, will produce seed following fire. Its seed can be collected by hand and mixed with wiregrass seed when sowing restoration sites.

The most extensive direct seeding program of understory species is at Fort Stewart, GA (Table 32.2), where resource managers have collected and sown seed on site-prepared areas since 1997. Seed is collected using a tractor-mounted flail-vac from areas burned during the growing season yielding from 750 to 1,100 kg year^{-1}. At a mean sowing rate of 13.2 kg ha^{-1}, enough seed is collected to sow 57 to 83 ha year^{-1}. Seed is spread using a platform-mounted bale chopper on the back of a farm tractor. Their goal is to restore 8,100 ha of former agricultural fields to functioning longleaf pine ecosystems (Hilliard 1998).

Many understory species on sandhills sites survive extreme disturbance as propagules in the soil, or reinvade sites after the disturbance ends (Hattenbach et al. 1998). In one comparison, understories of remnant xeric longleaf pine stands and 30- to 40-year-old plantations on old-field sites were similar (Smith et al. 2002). Although the remnant stands had higher species diversity, nearly 90% of the understory species in the plantations were native to natural longleaf pine communities. Similar comparisons for the sandhills of South Carolina showed that species abundance was the same in plantations and reference stands, except for wiregrass and dwarf huckleberry (*Gaylussacia dumosa*), which were significantly reduced in plantations (Walker 1998). Thus, restoration does not require that

Table 32.2 Production Rates, Equipment and Costs[a] for Understory Plant Restoration at Fort Stewart, Georgia (Seed Collection Season Varies from Late October to Late December)

	Seed Collection					Seed Sowing			
Year	days	ha	kg	kg day^{-1}	kg ha^{-1}	days	ha	ha day^{-1}	kg ha^{-1}
1998	17	61.9	821	48.3	13.3	12	62.3	5.2	13.2
1999	24	79.5	938	39.1	11.8	14	74.1	5.3	12.7
2000	20	20.2	539	27.0	26.7	9	40.5	4.5	13.3
2001	30	95.1	1096	36.5	11.5	16	97.2	6.1	11.3
2002	17	57.9	746	43.9	12.9	6	40.5	6.8	18.4
Mean	22	62.9	828	37.6	13.2	11	62.9	5.7	13.2

Equipment	Cost	Item	Cost
4-wheel-drive vehicle	$47,594.00	Seed collection labor	$10.74 kg ha^{-1}
Flail-vac seed collector	$11,950.00	Sowing labor	$58.05 ha^{-1}
Small tractor	$35,000.00	Seed cost[a]	$141.77 ha^{-1}
Bale chopper	$3,800.00	Understory restoration cost[a]	$199.82 ha^{-1}

[a]Does not include equipment purchase or operation and maintenance costs.

every plant species be reintroduced. In addition to certain common species that do not easily reinvade or survive, reintroduction of some rare species will likely be required (Glitzenstein et al. 1998, 2001; Walker 1998).

32.6.3.4 *Flatwoods and wet lowlands dominated by longleaf pine with native understory*

Some stands have been degraded by years of fire suppression. Rehabilitation using prescribed burning to reduce woody understory and midstory species and allow grasses and forbs to increase can be effective. Growing-season fires are as useful as dormant-season burns and may be more effective. One or two dormant-season fires will gradually reduce litter buildup and is advisable before the first growing-season burn. Initial burns should be conducted when the Keetch–Byram Drought Index (KBDI) (Keetch and Byram 1968) is less than 250 (Miller and Bossuot 2000). Flatwoods understories dominated by saw palmetto, gallberry (*I. glabra*), waxmyrtle (*Myrica cerifera*), and sweetgum (*Liquidambar styraciflua*) are quite resistant to fire. Only repeated fires at short return intervals over a long period significantly reduce these woody species (Waldrop et al. 1987). Thus, burning every 2 years over a period of 10 to 20 years may be required to readjust the understory composition on wet sites.

Lightweight choppers or heavy-duty mowers may be used to reduce saw–palmetto coverage and dominance (Huffman and Dye 1994). Both methods cause limited soil disturbance and thus do not reduce native grass species. Preliminary findings from research under way at Myakka River State Park in Florida indicate that the chopping treatment is more effective for reducing saw–palmetto cover. Prescribed burning 3 to 6 months before or after these mechanical treatments seems to increase their effectiveness.

32.6.3.5 *Flatwoods and wet lowlands dominated by other trees with native understory*

Longleaf pine overstory on wet sites may have been replaced by other pines, leaving a native understory. Such sites include naturally regenerated stands that were invaded by slash pine and loblolly pine after the removal of longleaf pine, and site-prepared plantations that were planted with other southern pines. Rehabilitation (conversion) requires removal of the loblolly or slash pine overstory and reestablishment of longleaf pine. Prescribed burning 2 years prior to the harvest will reduce woody competition and stimulate the growth of herbaceous understory species. A site-preparation fire following logging may be needed to remove debris and discourage hardwood trees and shrubs. Between the harvest and site preparation burn, chopping may be used to control woody competitors. A single-drum chopper should be used to avoid excessive soil disturbance. Some managers prescribe bedding on these wet sites before planting to increase survival rates of bareroot or container longleaf pine seedlings. Bedding will improve seedling survival during wetter years by about 15%. However, this survival gain comes at a cost, not only of the operation but also from damage to the native groundcover. Bedding may also alter site moisture relations and nutrient distribution for more than 30 years (Schultz 1976). Alternatively, planting additional longleaf pine seedlings during drier seasons can offset lower survival.

32.6.3.6 *Upland and montane sites dominated by longleaf pine with native understory*

Few upland and mountain sites remain in longleaf pine because these were preferred areas for agricultural, urban, and residential development. However, there are upland areas mostly in Alabama, Mississippi, Louisiana, and Texas and montane sites in Alabama and Georgia that have developed unnaturally dense hardwood midstories. Because these are among the most biologically productive longleaf pine sites, they change the most rapidly, quickly developing midstory layers in the absence of frequent fire. In addition to a very dense midstory and a shrub-dominated understory, these sites also accumulate significant quantities of potentially hazardous fuel. Frequent growing-season fires are needed on upland sites with better soils to adequately control competition from woody plants. Like flatwoods sites, frequent growing-season fires over many years are required to reduce the hardwood rootstocks (Boyer 1990a). As noted for other longleaf pine ecosystem types, a series of dormant-season fires may be necessary to gradually reduce fuel levels before growing-season burning begins.

A variant of this ecosystem type where longleaf pine is present but other southern pines are dominants or codominants is common. In addition to prescribed burning as outlined above, these stands may need selective harvesting to reduce the presence of other southern pines and hardwoods in the overstory. The objective is not total elimination of other tree species, but rather a proportional adjustment of overstory composition, recognizing that these other species are part of the natural longleaf pine community. Understory burning should begin prior to selective harvest to control competition from woody plants or they will proliferate and form a shrub thicket in openings created by harvesting. Herbicide application and mechanical reduction of nonmerchantable woody species may accelerate the process of adjusting species composition and dominance (Boyer 1991).

32.6.3.7 *Upland and montane sites dominated by other species*

Only limited research and experience are available to guide restoration on upland sites dominated by other overstory species. The few sites that show no evidence of severe soil disturbance contain scattered natural longleaf pine trees in a mixture dominated by loblolly pine, shortleaf pine (*Pinus echinata*), and hardwoods. Some native understory likely still exists in the soil seed bank or as suppressed individuals (Varner et al. 2000). Therefore, restoration would consist of prescribed burning to reduce fuel and control woody shrubs and hardwood trees. Repeated and prolonged treatment with prescribed burning should eventually reduce the abundance and cover of woody plants in the understory. If timber markets allow, selective harvest can be used to release any native longleaf pine and reduce the hardwood component. Otherwise, thinning would be performed at a financial cost. Other pines may need to be retained onsite to furnish sufficient needlefall for prescribed burning and to avoid release of woody competition. Once prescribed burning and other mechanical or chemical methods have reduced the woody midstory and understory layers, some of these other pines could be removed and replaced with longleaf pine seedlings. This is probably best done by creating canopy gaps in areas where the understory has become dominated by desirable grasses and forbs.

Restoring upland sites with a history of severe soil disturbance from agriculture or intensive forestry will be more challenging. It is unlikely that many native understory grasses and forbs survived intensive soil disturbance; however, there is a large soil seedbank of herbaceous weeds that must be controlled. Restoration techniques being tested include multiple-pass harrowing to reduce weeds followed by planting wiregrass plugs. High survival rates have been obtained with this method, but long-term growth rates are still uncertain (Mulligan and Kirkman 2002).

32.7 *Costs and benefits associated with restoration*

32.7.1 *Estimating restoration costs*

Reestablishing longleaf pine as the dominant tree species on a site is often the first and, in many ways, easiest step in the restoration process. Establishment costs vary according to ambient conditions and the type and amount of site preparation needed. On previously

Table 32.3 Comparative Value of Wood Products among Major Southern Pine Species, Assuming a Mean Stand Volume of ~29 m^3 ha^{-1} at Age 55 (Holliday 2001)

	Sawtimber Price m^{-3}	High-Quality Poles	Value ha^{-1}
Longleaf pine	$264	66%	$8,492
Slash pine	$257	12%	$7,640
Loblolly pine	$254	5%	$7,454

harvested or old-field sites, costs typically range from $370 to 740 ha^{-1}, depending on site conditions and whether bareroot seedlings or container seedlings are used. This range reflects the current costs for site preparation, seedlings, and planting. To control competing vegetation, increase survival, and stimulate early growth, an additional $85 to 100 ha^{-1} might be expended for herbicide application. Despite these expenditures occurring early in a timber rotation, the average internal rate of return for such an investment ranges from 8 to 12% (Busby et al. 1996).

Restoring groundcover plants can be very expensive, with costs sharply rising when quick success is desired. In relatively undisturbed forests, many native plants will respond to reintroduction of fire, particularly growing-season burning, through stimulating residual seed banks and inducing flowering and seed production in existing plants. The cost of fire reintroduction varies with existing site conditions, especially the number of fuel reduction burns needed. Where seed banks are depleted from severe soil disturbance, restoring the plant community is more difficult. Reseeding or replanting selected understory plant species has been accomplished successfully, but at a considerable cost, from several hundred to several thousand dollars per ha. Seed collection, cultivation, distribution, planting techniques, and other steps in the process are being developed and generally focus on pyrophytic graminoids (e.g., wiregrass), species consumed by wildlife (e.g., legumes), and species of special concern due to rarity or endangerment, such as American chaffseed (*Schwalbea americana*).

32.7.2 *Benefits of restored longleaf pine ecosystems*

The material and intangible benefits of restoring longleaf pine ecosystems are substantial. The economic value of longleaf pine forests is considerable and sustainable forest management for commercial products is achievable. Longleaf pine is the most versatile of all the southern pines and provides a variety of highly valued products (Table 32.3). Longleaf pine forests typically produce up to five times more tree stems of sufficient quality to be used as utility poles than stands of slash pine or loblolly pine (Boyer and White 1990). Stumpage values for such poles exceed prices for sawtimber by about 40% in local wood markets. When the high value of pine straw (i.e., fallen needles used as landscaping material which may be harvested from stands as early as age 10) is added, the economic value of longleaf pine forests becomes increasingly obvious. Surveys consistently indicate the value of hunter access to private lands as a tradable commodity throughout the natural range of longleaf pine, and private pine forests are leased for hunting rights for more than the timber value. Where longleaf pine forests are maintained in open park-like condition, the higher quality of this habitat for quail, turkey, and deer brings a premium in hunting leases and related services to private landowners.

Acknowledgments

The authors express their appreciation to Becky Estes for searching the literature to identify numerous relevant publications. We are also grateful to John Stanturf, Palle Madsen,

Dave Haywood, Steve Jack, John Kush, Dave Borland, and Ric Jeffers for helpful comments that improved this chapter.

References

Abrahamson, W.G. and Hartnett, D.C., Pine flatwoods and dry prairies, in *Ecosystems of Florida*, Myers, R.L. and Ewel, J.J., Eds., University of Central Florida Press, Orlando, FL, 103, 1990.

Anderson, K.M., Tending the wilderness, *Restor. Manage. Notes*, 14, 154, 1996.

Bailey, R.G., *Description of the Ecoregions of the United States*, U.S. Department of Agriculture, Forest Service, Miscellaneous Publication 1391, Washington, DC, 108 pp., 1995.

Barnett, J.P., Lauer, D.K., and Brissette J.C., Regenerating longleaf pine with artificial methods, in *Management of Longleaf Pine*, Farrar R.M., Ed., U.S. Department of Agriculture, Forest Service, Southern Forest Experiment Station, New Orleans, LA, General Technical Report SO-75, 72, 1990.

Barnett, J.P. and McGilvray J.M., *Practical Guidelines for Producing Longleaf Pine Seedlings in Containers*, U.S. Department of Agriculture, Forest Service, Southern Research Station, Asheville, NC, General Technical Report SRS-14, 28 pp., 1997.

Bartram, W., *Travel Through North and South Carolina, Georgia and East and West Florida*, Dover Publishers, New York, NY, 414 pp., 1791.

Berish, S.J., *Efficacy of Three Formulations of the Forest Herbicide Hexazinone as an Aid to Reforestation of Longleaf Pine (Pinus palustris) Sandhills at Eglin Air Force Base, Florida*, M.S. thesis, University of Florida, Gainesville, FL, 1996.

Boyer, W.D., *Growing-Season Burns for Control of Hardwoods in Longleaf Pine Stands*, U.S. Department of Agriculture, Forest Service, Southern Forest Experiment Station, New Orleans, LA, Research Paper SO-256, 7 pp., 1990a.

Boyer, W.D., *Pinus palustris*, Mill. longleaf pine, in *Silvics of North America*, Vol. 1, *Conifers*, Burns, R.M. and Honkala, B.H., Tech. Coords., U.S. Department of Agriculture, Forest Service, Washington, D.C., 405, 1990b.

Boyer, W.D., *Effects of a Single Chemical Treatment on Long-term Hardwood Development in a Young Pine Stand*, U.S. Department of Agriculture, Forest Service, Southeastern Forest Experiment Station, Asheville, NC, General Technical Report SE-70, 599, 1991.

Boyer, W.D. and White, J.B., Natural regeneration of longleaf pine, in *Management of Longleaf Pine*, Farrar, R.M., Ed., U.S. Department of Agriculture, Forest Service, Southern Forest Experiment Station, New Orleans, LA, General Technical Report SO-75, 94, 1990.

Brockway, D.G. and Lewis, C.E., Long-term effects of dormant-season prescribed fire on plant community diversity, structure and productivity in a longleaf pine wiregrass ecosystem, *For. Ecol. Manage.*, 96, 167, 1997.

Brockway, D.G. and Outcalt, K.W., Gap-phase regeneration in longleaf pine wiregrass ecosystems, *For. Ecol. Manage.*, 106, 125, 1998.

Brockway, D.G. and Outcalt, K.W., Restoring longleaf pine wiregrass ecosystems: hexazinone application enhances effects of prescribed fire, *For. Ecol. Manage.*, 137, 121, 2000.

Brockway, D.G., Outcalt, K.W., and Wilkins, R.N., Restoring longleaf pine wiregrass ecosystems: plant cover, diversity and biomass following low-rate hexazinone application on Florida sandhills, *For. Ecol. Manage.*, 103, 159, 1998.

Burns, R.M. and Hebb, E.A., *Site Preparation and Reforestation of Droughty, Acid Sands*, U.S. Department of Agriculture, Forest Service, Washington, DC, Agriculture Handbook No. 426, 61 pp., 1972.

Busby, R.L., Thomas, C.E. and Lohrey, R.E., The best kept secret in southern forestry: longleaf pine plantation investments, in *Longleaf Pine: A Regional Perspective of Challenges and Opportunities*, Kush, J.S., Comp., Longleaf Alliance Report No. 1, Solon Dixon Forestry Education Center, Andalusia, AL, 26, 1996.

Carroll, W.D., Kapeluck, P.R., Harper, R.A., and Van Lear, D.H., Historical overview of the southern forest landscape and associated resources, in *The Southern Forest Resource Assessment*, Wear, D.N. and Greis, J.G., Eds., U.S. Department of Agriculture, Forest Service, Southern Research Station, Asheville, NC, General Technical Report SRS-53, 583, 2002.

Christensen, N.L., Fire regimes in southeastern ecosystems, in *Fire Regimes and Ecosystem Properties*, Mooney, H.A., Bonnicksen, T.M., Christensen, N.L., Lotan, J.E., and Reiners, W.A., Eds., U.S. Department of Agriculture, Forest Service, Washington, DC, General Technical Report WO-26, 112, 1981.

Clewell, A.F., Natural history of wiregrass (*Aristida stricta* Michx., Gramineae), *Nat. Areas J.*, 9, 223, 1989.

Clewell, A.F., Restoring for natural authenticity, *Ecol. Rest.*, 18, 216, 2000.

Croker, T.C., *Longleaf Pine: A History of Man and a Forest*, U.S. Department of Agriculture, Forest Service, Southern Region, Atlanta, GA, Forestry Report R8-FR7, 37 pp., 1987.

Delcourt, P.A. and Delcourt, H.R., *Long-term forest dynamics of the temperate zone*, Springer-Verlag Publishers, New York, NY, 439 pp., 1987.

Engstrom, R.T., Red-cockaded woodpeckers: prospects for recovery, in *The Fire Forest: Longleaf Pine-wiregrass Ecosystems*, Wilson, J.R., Ed., Georgia Wildlife, 8, 12, 2001.

Engstrom, R.T., Kirkman, L.K., and Mitchell, R.J., The natural history of the fire forest, in *The Fire Forest: Longleaf Pine-wiregrass Ecosystems*, Wilson, J.R., Ed., Georgia Wildlife, 8, 5, 14, 2001.

Franklin, R.M., *Stewardship of Longleaf Pine Forests: A Guide for Landowners*, Longleaf Alliance Report No. 2, Longleaf Alliance, Solon Dixon Forestry Education Center, Andalusia, AL, 44 pp., 1997.

Frost, C.C., Four centuries of changing landscape patterns in the longleaf pine ecosystem, in *Proceedings of the 18th Tall Timbers Fire Ecology Conference*, Herman, S.M., Ed., Tall Timbers Research Station, Tallahassee, FL, 17, 1993.

Garren, K.H., Effects of fire on vegetation of the southeastern United States, *Bot. Rev.*, 9, 617, 1943.

Glitzenstein, J.S., Platt, W.J., and Streng, D.R., Effects of fire regime and habitat on tree dynamics in north Florida longleaf pine savannas, *Ecol. Monogr.*, 65, 441, 1995.

Glitzenstein, J.S., Streng, D.R., and Wade, D.D., A promising start for a new population of *Parnassia caroliniana* Michx., in *Ecological Restoration and Regional Conservation Strategies*, Kush, J.S., Comp., Longleaf Alliance Report No. 3, Solon Dixon Forestry Education Center, Andalusia, AL, 44, 1998.

Glitzenstein, J.S., Streng, D.R., Wade, D.D., and Brubaker, J., Starting new populations of longleaf pine ground-layer plants in the Outer Coastal Plain of South Carolina, USA, *Nat. Areas J.*, 21, 89, 2001.

Grelen, H.E., Plant succession on cleared sandhills in northern Florida, *Am. Midl. Nat.*, 67, 36, 1962.

Hardin, E.D. and White, D.L., Rare vascular plant taxa associated with wiregrass (*Aristida stricta*) in the southeastern United States, *Nat. Areas J.*, 9, 234, 1989.

Hattenbach, M.J., Gordon, D.R., Seamon, G.S., and Studenmund, R.G., Development of direct-seeding techniques to restore native groundcover in a sandhill ecosystem, in *Ecological Restoration and Regional Conservation Strategies*, Kush, J.S., Comp., Longleaf Alliance Report No. 3, Solon Dixon Forestry Education Center, Andalusia, AL, 64, 1998.

Hilliard, T., Longleaf-wiregrass restoration at Fort Stewart, Georgia: the military's role in restoration, in *Ecological Restoration and Regional Conservation Strategies*, Kush, J.S., Comp., Longleaf Alliance Report No. 3, Solon Dixon Forestry Education Center, Andalusia, AL, 32, 1998.

Hilton, J., Biological diversity in the longleaf pine ecosystem, *Alabama's Treasured For.*, 18, 28, 1999.

Holliday, P.P., *Going, going…saving the longleaf pine ecosystem before it's gone*, Sherpa Guides, ⟨www.sherpaguides.com/georgia/fire_forest/saving_ecosystem/index.html⟩, 9 pp., 2001.

Huffman, J.M. and Dye, R., Summary of wiregrass ecosystem restoration projects at Myakka River State Park, in *Proceedings of the Wiregrass Ecosystem Restoration Workshop*, Clewell, A.F. and Cleckley, W., Eds., State of Florida, Northwest Florida Water Management District, Tallahassee, FL, 21, 1994.

Jackson, D.R. and Milstrey, E.R., The fauna of gopher tortoise burrows, in *Proceedings of the Gopher Tortoise Relocation Symposium*, Diemer, J.E., Ed., State of Florida, Game and Freshwater Fish Commission, Tallahassee, FL, 86, 1989.

Keetch, J.J. and Byram, G.M., *A Drought Index for Forest Fire Control*, U.S. Department of Agriculture, Forest Service, Southeastern Forest Experiment Station, Asheville, NC, Research Paper SE-38, 1968.

Kelly, J.F. and Bechtold, W.A., The longleaf pine resource, in *Management of Longleaf Pine*, Farrar, R.M., Ed., U.S. Department of Agriculture, Forest Service, Southern Forest Experiment Station, New Orleans, LA, General Technical Report SO-75, 11, 1990.

Kirkman, L.K., Drew, M.B., West, L.T. and Blood, E.R., Ecotone characterization between upland longleaf pine/wiregrass stands and seasonally-ponded isolated wetlands, *Wetlands*, 18, 346, 1998.

Komarek, E.V., Lightning and lightning fires as ecological forces, in *Proceedings of the 9th Tall Timbers Fire Ecology Conference*, Tall Timbers Research Station, Tallahassee, FL, 169, 1968.

Kush, J.S. and Meldahl, R.S., Composition of a virgin stand of longleaf pine in south Alabama, *Castanea*, 65, 56, 2000.

Landers, J.L., Disturbance influences on pine traits in the southeastern United States, in *Proceedings of the 17th Tall Timbers Fire Ecology Conference*, Tall Timbers Research Station, Tallahassee, FL, 61, 1991.

Landers, J.L. and Boyer, W.D., *An Old-growth Definition for Upland Longleaf and South Florida Slash Pine Forests, Woodlands and Savannas*, U.S. Department of Agriculture, Forest Service, Southern Research Station, Asheville, NC, General Technical Report SRS-29, 15 pp., 1999.

Landers, J.L., Byrd, N.A. and Komarek, R., A holistic approach to managing longleaf pine communities, in *Management of Longleaf Pine*, Farrar, R.M., Ed., U.S. Department of Agriculture, Forest Service, Southern Forest Experiment Station, New Orleans, LA, General Technical Report SO-75, 135, 1990.

Landers, J.L., Van Lear, D.H. and Boyer, W.D., The longleaf pine forests of the Southeast: requiem or renaissance?, *J. For.*, 93, 39, 1995.

Little, E.L., *Atlas of United States Trees. Volume 1. Conifers and Important Hardwoods*, U.S. Department of Agriculture, Forest Service, Washington, DC, Miscellaneous Publication 1146, 320 pp., 1971.

McMahon, C.K., Tomczak, D.J., and Jeffers, R.M., Longleaf pine ecosystem restoration: the role of the USDA Forest Service, in *Ecological Restoration and Regional Conservation Strategies*, Kush, J.S., Comp., Longleaf Alliance Report No. 3, Solon Dixon Forestry Education Center, Andalusia, AL, 20, 1998.

Means, D.B., Wiregrass restoration: probable shading effects in a slash pine plantation, *Restoration and Manage. Notes*, 15, 52, 1997.

Miller, J.H. and Robinson, K.S., A regional perspective of the physiographic provinces of the Southeastern United States, in *Proceedings of the 8th Biennial Southern Silvicultural Research Conference*, Edwards, M.B., Ed., U.S. Department of Agriculture, Forest Service, Southern Research Station, Asheville, NC, General Technical Report SRS-1, 581, 1995.

Miller, S.R. and Bossuot, W.R., Flatwoods restoration on the St. Johns River Water Management District, Florida: a prescription to cut and burn, in *Proceedings of the 21st Tall Timbers Fire Ecology Conference*, Tall Timbers Research Station, Tallahassee, FL, 212, 2000.

Mulligan, M.K. and Kirkman, L.K., Competition effects on wiregrass (*Aristida beyrichiana*) growth and survival, *Plant Ecol.*, 167, 39, 2002.

Mulligan, M.K., Kirkman, L.K., and Mitchell, R.J., *Aristida beyrichiana* (wiregrass) establishment and recruitment: implications for restoration, *Restor. Ecol.*, 10, 68, 2002.

Noss, R.F., Longleaf pine and wiregrass: keystone components of an endangered ecosystem, *Nat. Areas J.*, 9, 211, 1989.

Noss, R.F., LaRoe, E.T., and Scott, J.M., *Endangered Ecosystems of the United States: A Preliminary Assessment of Loss and Degradation*, U.S. Department of Interior, National Biological Service, Washington, DC, Biological Report 28, 59 pp., 1995.

Outcalt, K.W., Seed production of wiregrass in central Florida following growing season prescribed burns, *Int. J. Wildland Fire*, 4, 123, 1994.

Outcalt, K.W., *Maintaining the Native Plant Community during Longleaf Pine Establishment*, Forestry Research Institute, Rotorua, New Zealand, Bulletin No. 192, 283, 1995.

Outcalt, K.W., Occurrence of fire in longleaf pine stands in the southeastern United States, in *Proceedings of the 21st Tall Timbers Fire Ecology Conference*, Moser W.K. and Moser, C.F., Eds., Tall Timbers Research Station, Tallahassee, FL, 178, 2000.

Outcalt, K.W. and Lewis, C.E., Response of wiregrass (*Aristida stricta*) to mechanical site preparation, in *Wiregrass Biology and Management*, Duever, L.C. and Noss, R.F., Eds., KBN Engineering and Applied Sciences, Gainesville, FL, 60, 1990.

Outcalt, K.W. and Sheffield, R.M., *The Longleaf Pine Forest: Trends And Current Conditions*, U.S. Department of Agriculture, Forest Service, Southern Research Station, Asheville, NC, Resource Bulletin SRS-9, 23 pp., 1996.

Outcalt, K.W., Williams, M.E., and Onokpise, O., Restoring *Aristida stricta* to *Pinus palustris* ecosystems on the Atlantic Coastal Plain, USA, *Restor. Ecol.*, 7, 262, 1999.

Palik, B.J., Mitchell, R.J., Houseal, G., and Pederson, N., Effects of canopy structure on resource availability and seedling responses in a longleaf pine ecosystem, *Can. J. For. Res.*, 27, 1458, 1997.

Palik, B.J. and Pederson, N., Overstory mortality and canopy disturbances in longleaf pine ecosystems, *Can. J. For. Res.*, 26, 2035, 1996.

Peet, R.K. and Allard, D.J., Longleaf pine-dominated vegetation of the southern Atlantic and eastern Gulf Coast region, USA, in *Proceedings of the 18th Tall Timbers Fire Ecology Conference*, Herman, S.M., Ed., Tall Timbers Research Station, Tallahassee, FL, 45, 1993.

Pickett, S.T.A. and White, P.S., *The Ecology of Natural Disturbance and Patch Dynamics*, Academic Press, Orlando, FL, 1985.

Platt, W.J., Evans, G.W., and Davis, M.M., Effects of fire season on flowering of forbs and shrubs in longleaf pine forests, *Oecologia*, 76, 353, 1988a.

Platt, W.J., Evans, G.W., and Rathbun, S.L., The population dynamics of a long-lived conifer (*Pinus palustris*), *Am. Nat.*, 131, 491, 1988b.

Platt, W.J. and Rathbun, S.L., Dynamics of an old-growth longleaf pine population, in *Proceedings of the 18th Tall Timbers Fire Ecology Conference*, Herman, S.M., Ed., Tall Timbers Research Station, Tallahassee, FL, 275, 1993.

Provencher, L., Litt, A.R., Galley, K.E.M., Gordon, D.R., Tanner, G.W., Brennan, L.A., Gobris, N.M., McAdoo, J.P., and Herring, B.J., *Restoration of fire suppressed longleaf pine sandhills at Eglin Air Force Base, Niceville, Florida*, The Nature Conservancy, Science Division, Gainesville, FL, 294 pp., 2001.

Pyne, S.J., *Fire in America: A Cultural History of Wildland and Rural Fire*, Princeton University Press, Princeton, NJ, 654 pp., 1997.

Robbins, L.E. and Myers, R.L., *Seasonal Effects of Prescribed Burning in Florida: A Review*, Tall Timbers Research Station, Tallahassee, FL, Miscellaneous Publication No. 8, 96 pp., 1992.

Schmidtling, R.C. and Hipkins, V., Genetic diversity in longleaf pine (*Pinus palustris*): influence of historical and prehistorical events, *Can. J. For. Res.*, 28, 1135, 1998.

Schultz, R.P., *Environmental Change After Site Preparation and Slash Pine Planting on a Flatwoods Site*, U.S. Department of Agriculture, Forest Service, Southeastern Forest Experiment Station, Asheville, NC, Research Paper SE-156, 20 pp., 1976.

Schwartz, M.W., Natural distribution and abundance of forest species and communities in northern Florida, *Ecology*, 75, 687, 1994.

Schwarz, G.F., *The Longleaf Pine Virgin Forest: A Silvical Study*, John Wiley and Sons, New York, NY, 135 pp., 1907.

Seamon, G., A longleaf pine sandhill restoration in northwest Florida, *Restor. Manage. Notes*, 16, 46, 1998.

Smith, G.P., Shelburne, V.B., and Walker, J.L., Structure and composition of vegetation of longleaf pine plantations compared to natural stands occurring along an environmental gradient at the Savannah River Site, in *Proceedings of the 11th Biennial Southern Silvicultural Research Conference*, Outcalt, K.W., Ed., U.S. Department of Agriculture, Forest Service, Southern Research Station, Asheville, NC, General Technical Report SRS-48, 481, 2002.

Stanturf, J.A., Wade, D.D., Waldrop, T.A., Kennard, D.K., and Achtemeier, G.L., Fire in southern landscapes, in *The Southern Forest Resource Assessment*, Wear, D.N. and Greis, J.G., Eds., U.S. Department of Agriculture, Forest Service, Southern Research Station, Asheville, NC, General Technical Report SRS-53, 607, 2002.

Stout, I.J. and Marion, W.R., Pine flatwoods and xeric pine forests of the southern lower coastal plain, in *Biodiversity of the Southeastern United States: Lowland Terrestrial Communities*, Martin, W.H., Boyce, S.G. and Echternacht, A.C., Eds., Wiley, New York, NY, 373, 1993.

Taylor, A.R., Ecological aspects of lightning in forests, in *Proceedings of the 13th Tall Timbers Fire Ecology Conference*, Tall Timbers Research Station, Tallahassee, FL, 455, 1974.

Uridel, K.W., *Restoration of Native Herbs in Abandoned Paspalum notatum (Bahia grass) Pastures*. M.S. thesis, University of Florida, Gainesville, FL, 1994.

Varner, J.M., Kush, J.S., and Meldahl, R.S., Ecological restoration of an old-growth longleaf pine stand utilizing prescribed fire, in *Proceedings of the 21st Tall Timbers Fire Ecology Conference*, Moser, W.K. and Moser, C.F., Eds., Tall Timbers Research Station, Tallahassee, FL, 216, 2000.

Wahlenberg, W.G., *Longleaf Pine: Its Use, Ecology, Regeneration, Protection, Growth and Management*, C.L. Pack Forestry Foundation and U.S. Department of Agriculture, Forest Service, Washington, DC, 429 pp., 1946.

Waldrop, T.A., Van Lear, D.H., Lloyd, F.T. and Harms, W.R., *Long-Term Studies Of Prescribed Burning In Loblolly Pine Forests Of The Southeastern Coastal Plain*, U.S. Department of Agriculture, Forest Service, Southeastern Forest Experiment Station, Asheville, NC, General Technical Report SE-45, 23 pp., 1987.

Walker, J.L., Rare vascular plant taxa associated with the longleaf pine ecosystem, in *Proceedings of the 18th Tall Timbers Fire Ecology Conference*, Hermann, S.M., Ed., Tall Timbers Research Station, Tallahassee, FL, 105, 1993.

Walker, J.L., Ground layer vegetation in longleaf pine landscapes: an overview for restoration and management, in *Ecological Restoration and Regional Conservation Strategies*, Kush, J.S., Comp., Longleaf Alliance Report No. 3, Solon Dixon Forestry Education Center, Andalusia, AL, 2, 1998.

Ware, S., Frost, C.C. and Doerr, P.D., Southern mixed hardwood forest: the former longleaf pine forest, in *Biodiversity of the Southeastern United States: Lowland Terrestrial Communities*, Martin, W.H., Boyce, S.G. and Echternacht, A.C., Eds., John Wiley and Sons, New York, NY, 447, 1993.

Watts, W.A., Hansen, B.C.S. and Grimm, E.C., Camel Lake: a 40,000-yr record of vegetational and forest history from north Florida, *Ecology*, 73, 1056, 1992.

Wear, D.N. and Greis, J.G., Summary report, in *The Southern Forest Resources Assessment*, Wear, D.N. and Greis, J.G., Eds., U.S. Department of Agriculture, Forest Service, Southern Research Station, Asheville, NC, General Technical Report SRS-53, 114 pp., 2002.

Williams, J.L., *The Territory of Florida*, 1962 Edition, University of Florida Press, Gainesville, FL, 1837.

Part 7

Case Studies: Reclamation

chapter thirty-three

Reclamation of forests degraded by air pollution in central and eastern Europe

Mikołaj Mikułowski and Tadeusz Zachara
Forest Research Institute
Inge Stupak Møller
Danish Centre for Forest, Landscape and Planning, KVL

Contents

33.1 Introduction

Over the past 50 years, unprecedented environmental deterioration in central and eastern Europe (CEE) has induced substantial problems for forest health. Air pollutants from fossil fuel combustion and heavy industry threaten forest productivity; in some places, the actual survival of the forest is at risk. The heaviest damage to forests occurs in the mountains and industrial regions. Emissions of sulfur dioxide (SO_2) and nitrogen oxides (NO_X), and deposition of heavy metals are the main problems (Luttermann and Freedman 2000). The political changes in CEE at the end of the 1980s were followed by significant declines in pollution, especially SO_2 emissions. However, the reductions in the atmospheric concentrations of toxic gases, including SO_2, have not eliminated local and regional forest damage (Vancura et al. 2000). Forest health problems can be observed as a direct damage to vegetation near large point sources, while lower-level, regional effects may be more widespread, but also more difficult to document (Lutterman and Freedman 2000).

1-56670-635-1/05/$0.00+$1.50

Observed symptoms of low-level pollution are nonspecific: leaf discoloration, necrosis, or crown dieback (Vancura et al. 2000).

The CEE countries presently represent an intermediate stage between the immense pollution problems of the Russian Federation and the relatively minor problems in western Europe (Oleksyn and Innes 2000). This chapter focuses especially on the ecological, economic, and socioeconomic aspects of forest reclamation after the occurrence of air pollution damage in the CEE, and it also gives an introduction to the relation between forest health and air pollution (Innes and Oleksyn 2000; Lutterman and Freedman 2000; Vancura et al. 2000; Kozlov et al. 2000). Geographically, the chapter encompasses the following postcommunist CEE countries in transition toward market economies: Belarus, Czech Republic, Estonia, Hungary, Latvia, Lithuania, Poland, Slovakia, Slovenia, and Ukraine. Austria is used as a reference country for comparison; although it has a stable economy, it is close to the CEE countries in terms of forest management, geography, and history.

The area of the Ore and Sudeten Mountains in southern Poland, northern Czechoslovakia, and southeastern Germany provides a regional case study of the difficulty of restoring forests damaged by air pollution. The area was one of the most polluted regions in Europe until the mid-1990s, earning the name the "Black Triangle" (Vancura et al. 2000). The results have been serious: a decrease in the growing stock and permanent deforestation (Paschalis and Zajączkowski 1997).

33.2 Physiography

The CEE countries included in this chapter range from 44 to 60°N and 12 to 40°E, with elevations from –2 to 2655 m asl. Soil, climate, and vegetation cover vary. Quaternary glaciation affected the northern part of CEE (Poland, the Baltic countries, Belarus, and the northwestern Ukraine) and the soils are of two types: more fertile moraine tills and loamy sands (mostly utilized for agricultural purposes), and poor sandy soils where the proportion of forest cover is higher (Wallen 1977). The soils vary more in the central and southern parts of the CEE that were not reached by the glacier. They are rich chernozems (Czech Republic, Ukraine, and Hungary), good-quality brown soils in the drier and lower areas, and stony mountain soils at high elevations, especially in the Sudeten and Carpathian Mountains (Slovakia, Czech Republic, Ukraine, and Poland). Vertical soil zonality prevails in the mountainous areas and some southern parts of the area are characterized by poor soil conditions (glacial outwash sandy soils).

The forests of CEE are situated in a temperate climate that is generally moist and cool. Influenced by the Atlantic Ocean to the west, the region becomes increasingly continental with harsher winters toward the east. This gives rise to greater temperature amplitudes and longer winters, as well as less frequent winds and spring frosts that cause damage to forests. A Mediterranean climate occurs in the south with hot dry summers in Slovenia. As such, four basic climatic zones are distinguished: temperate oceanic, temperate continental, temperate mountain, and temperate steppe, the latter also characterized by dry summers (only part of Ukraine). The Sudeten and Carpathian Mountain ranges physically separate the Czech Republic, Slovakia, and Hungary from the European lowlands to the east. The mountains block air masses from penetrating from the western Atlantic, thus reducing the frequency of strong westerly winds and the amount of precipitation (Wallen 1977). The precipitation level ranges from less than 300 mm year^{-1} in southeastern Ukraine, where forest cover disappears naturally, and below 500 mm year^{-1} in some parts of the Great Hungarian Lowland (Puszta steppe) and western Poland (Wielkopolska Lowland, farmlands, and pine forests), to 1,000 to 1,500 mm at timberline in the Carpathian and Sudeten Mountains. The January mean temperature ranges from -1°C in western Poland to -8°C in the Czech Republic, eastern Belarus, and Ukraine. In July, the mean temperature ranges from 16°C in the Baltic countries to 24 to 25°C in the Hungarian lowlands and southern Ukraine.

In general, CEE lies in the mixed forest zone. Most of the lowland is situated in the temperate continental forest zone, with forest types distributed along local and regional gradients of climate and nutrient availability. Only western Poland belongs to the temperate oceanic forest zone. The mountainous areas belong to the temperate mountain system (Anon. 2001a). Soils and climate determine the occurrence of main forest trees species in CEE. Scots pine (*Pinus sylvestris* L.) is extensive in the whole region, except for Hungary and most of Ukraine. Norway spruce (*Picea abies* [L.] Karst.) occupies the Sudeten and Carpathian Mountains, northern Poland, Belarus, and the Baltic countries. Silver fir (*Abies alba* Mill.) occurs mainly in the Carpathians. Common oak (*Quercus robur* L.) occurs across the entire region except southern Ukraine. The natural range of sessile oak (*Quercus petrea* [Mattus.] Liebl.) extends from eastern Poland to western Ukraine. Beech (*Fagus sylvatica* L.) is an Atlantic-mountain species that occurs within Poland (Dengler 1990). Silver fir and beech are extremely sensitive to SO_2 pollution; pine, spruce, and the oaks are less sensitive species (Akkermann 1987).

33.3 Economical, social, and political context

All CEE countries are postcommunist countries, but they vary considerably in terms of per capita income levels (Table 33.1). Economic transformations that followed political change, such as restructuring of heavy industry, production decline, and technological modernization, were accompanied by declines in pollutant emissions. The CEE countries face daunting tasks of modernizing industry and institutions, as well as land privatization and economic restitution. These processes also concern forests and forest management practices. Following the fall of communism in the 1980s, the CEE countries inherited generally well-managed forests and well-educated staff from the previous regime, but often dilapidated forest industries and infrastructure (Anon. 2001a).

There are three main types of forest owners: the state, other public bodies such as communes and municipalities, and private individuals. The ownership structure of forests varies from country to country, with a temporal trend toward an increasing share of private forest because of the privatization and restitution processes. However, the percentage of public (state-owned) forests in almost all CEE countries is very high (Table 33.1), and the average area of private forests is small. Therefore, state forest holdings, relatively rich in financial and human resources, play a dominant role in forest restoration efforts (Henžlik 1997).

33.4 Air pollution impacts on forests

Air pollution resulting from human activity can be defined as the contamination of the atmosphere by toxic or radioactive gases and particulate matter to a degree where they cause measurable and undesirable effects on organisms and ecosystems (Lutterman and Freedman 2000). Two common air pollutants are sulfur dioxide (SO_2) and nitrogen oxides (NO_x); they are the major cause of so-called acid rain. Metals (lead, zinc, copper, nickel, mercury, aluminum, and others) can impact forests locally, especially around metal smelters, and long-range transport in the atmosphere can occur. In the Czech Republic, for example, an increasing role is played by photochemical pollutants such as hydrogen fluorides and ozone, which can also be transported over long distances from urban sources to forest areas (Müller-Edzards et al. 1997; Vancura et al. 2000).

In the period from 1975 to 1989, Poland, Czech Republic, Slovakia, Ukraine, and Belarus were among those CEE countries whose forests suffered the greatest health impairment induced by air pollution (Gutkowski and Winnicki 1997). In Poland, Czech Republic, and

Table 33.1 Statistical Data from the CEE Countries and Austria

Country	Area	GDP per Capita 1999	Forest Cover (Forest and Woodlands)		Public Forests Proportion of Total Forest Area	Average Size of a Private Forest	Species Distribution (Public and Private Forests)		
							Conifers	Broadleaves	Mixed
	10^3 km^2	US$ year $^{-1}$	10^3 km^2	%	%	ha	%	%	%
Austria	84	24,668[a]	40	48	18	8	68	12	20
Belarus	208	6,880[b]	92	44	100	—	39	22	39
Czech Republic	79	13,027[a]	27	34	84	3	31	13	56
Estonia	45	3,689[c]	23	52	92	11	39	21	40
Hungary	93	11,232[a]	19	20	65	9	11	78	10
Latvia	65	2,815[c]	31	48	56	11	39	19	42
Lithuania	65	6,490[a]	21	33	82	3	46	34	20
Poland	313	8,793[b]	91	29	83	2	67	15	18
Slovakia	49	10,215[a]	21	42	56	31	31	48	22
Slovenia	20	15,900[a]	12	58	30	3	30	38	33
Ukraine	604	33,60[b]	97	16	100	—	39	50	11

[a] Gross domestic product.

[b] Gross national income.

[c] Data for 1997 (Anon. 2001b).

Gross domestic product (GDP) per capita was estimated from various sources as purchasing power parity, current prices (Data from Anon. 2000, 2001b).

Table 33.2 Air Pollutant Emissions (t km^{-2} $year^{-1}$) in Selected CEE Countries and Austria, 1990 and 1997 (Data from Anon. 2001b)

Countries	Year	Sulfur Oxides	Nitrogen Oxides	Carbon Oxide	Carbon Dioxide	Particulates
Austria	1990	1.1[a]	2.3	15.3	703	0.47
	1997	0.7[a]	2.1	11.8[c]	763	0.45
Czech Republic	1990	23.8[a]	9.4	13.4	1,800	8.0
	1997	8.9[a]	5.4	11.1	1,534	1.5
Hungary	1990	10.9[a]	2.6	8.2	731	2.20
	1997	7.1[a]	2.1	7.8	624	1.46
Poland	1990	10.3[a]	4.1[b]		1,116	6.24[d]
	1997	7.0[a]	3.6[b]	15.0	1,119	3.61[d]
Slovakia	1990	11.1	4.6	17.3	1,102	6.20
	1997	4.1	2.5	6.9	776	1.29

[a] SO_2 only.
[b] NO_2 only.
[c] 1996.
[d] Data are only emissions from stationary sources.

Slovakia, a major part of the anthropogenic air pollution originated from obsolete, poorly maintained industrial plants relying upon rich deposits of soft brown coal from the region where the borders of southwestern Poland, Germany, and the Czech Republic meet. This brown coal has a very high sulfur content (5 to 10% by weight), resulting in emissions of harmful acidic gases, particles, and toxic metals. The annual rate of industrial SO_2 emissions from the three greatest power plants (Hirschfelde, Hagenwerder, and Turoszów) increased from 45,000 t in 1957 to 500,000 t in 1980 and further to 900,000 t in the last years of the decade (Balcar et al. 1997). In Slovakia and Slovenia, acidifying and toxic air pollution was emitted from metallurgical or chemical plants. In Estonia, major emissions of SO_2 originated from oil-shale-burning power plants in the northeast part of the country (Anon. 2003b), and immense alkaline emissions from the Kunda cement plant have had significant effects on site characteristics (e.g, bog ecosystems) and forest health, reducing both the production and vitality of conifers (Mandre et al. 1998; Ploompuu 1997). In Belarus and Ukraine, 35,000 km^2 of forests were contaminated with radioactive fallout after the nuclear reactor accident at Chernobyl in northern Ukraine in 1986, causing major forest health problems. All forest activity is forbidden in the 30-km zone around Chernobyl, effectively removing 15 million m^3 of timber from economic use. In the rest of the contaminated area, wood production is possible, but the radioactivity level of nonwood forest products such as berries, mushrooms, medical plants, and wild food exceed permissible threshold values, thus inflicting local communities with considerable economic and health problems (Vancura et al. 2000).

Current levels of acidifying emissions in CEE countries are significantly lower than under communism as a result of the Convention on Long-Range Transboundary Air Pollution and macroeconomic and social factors. The reduction in SO_2 emissions was particularly noticeable, more than 50% in some countries (Table 33.2) (Vancura et al. 2000). Nevertheless, SO_2 remains a significant pollutant in the central CEE countries (Anon. 1999). Emissions of NO_X from stationary sources fell due to an economic recession, but this was partly offset by the sharp growth in the use of private cars, especially in the large cities.

33.5 Forest decline

Attempts to define forest decline and to link it to causal agents have sparked much controversy. Manion (1991) defines forest decline as "an interaction of interchangeable, specifically ordered abiotic and biotic factors, that leads to produce a gradual general deterioration, often ending in death of trees." He divides the possible causing factors into three groups: predisposing, inciting, and contributing factors. Predisposing factors work, more or less, over the long term, for example, climate, site, age, and genetic makeup. These may not lead to obvious problems, but predispose trees to inciting factors that work in the short term, for example, defoliation and frost damage. If not for the predisposing factors, trees would recover quickly, but predisposed trees go into decline and are then vulnerable to the contributing factors: opportunistic fungi and insects like bark beetles, leaf-feeding insects, and Armillaria root rot. They finally kill the trees, but normally would not do so unless the tree was declining. Each decline situation involves one factor from each group (Worrel 2004). Hypotheses that refer to air pollution as causing forest decline are still controversial (Müller-Edzards et al. 1997).

Large-scale effects of pollution in Europe were first detected at the end of the 1960s (Capecki 1969). Systematic monitoring of forest conditions was initiated in 1985 following signing of the Convention on Long-Range Transboundary Air Pollution by all CEE countries. Monitoring was mainly comprised of assessments of crown condition (Anon. 2003a) and large-scale forest health problems were diagnosed by nonspecific symptoms such as defoliation and leaf discoloration. In the 1980s, severe forest decline due to air pollution was estimated at less than 0.5% (8,000 km^2) of the European forest area (Kauppi et al. 1992). This general pattern also applies for the CEE subregion. Air pollution as the triggering cause of forest dieback is usually restricted to different point-source polluters, urban and industrial agglomerations, or both; one of these was the Black Triangle (Vancura et al. 2000).

In spite of the general reduction in atmospheric emissions, since the 1990s, average daily concentrations of SO_2 in "hot spots" can be very high and cause heavy damage to forests, especially when accompanied by unfavorable weather conditions. During the winter of 1995 to 1996, for example, acute injury occurred in large areas of forests in the northeastern parts of the Ore Mountain ridge because of high SO_2 concentrations and unfavorable weather. The damage mechanism involved frost (temperatures as low as −17.4°C), extreme water stress because of foliar transpiration, and high volumes of ice in the crowns (72 days with rime, from November 1995 through January 1996). In Medenec, Czech Republic, peaks in daily levels of SO_2 concentrations between November 1995 and February 1996 ranged from 187 to 2,542 $\mu g\ m^{-3}$ (Vancura et al. 2000); in Saxony, throughfall pH was below 3.0 with a minimum of 2.7 (Raben et al. 1996). Cold and wet weather in the spring and summer of 1995 and 1996 adversely influenced the reproduction of several insect pests, but favored the spread of pathogenic fungi (Zahradnik 1997).

Physiological symptoms, foliar injury, decreases of productivity, and tree mortality are common phenomena in the CEE countries. Damage to forest resources in CEE as measured by defoliation percentages (Figure 33.1) has long been one of the highest on the continent, especially in Poland, the Czech Republic, and Slovakia, despite the small area directly affected by air pollution. By one estimate, 7% or 28,500 km^2 of the forested area of the entire CEE are damaged by air pollution (Anon. 1999). In the Carpathian Mountains, the area in decline is about 13 times higher than the area severely affected by air pollution. Although air pollution is suspected, it cannot be identified as the cause of decline (Niemtur et al.1999). In spite of a clear coincidence between high crown defoliation/discoloration and excessive concentrations and deposition levels for sulfur and nitrogen compounds in central Europe (Müller-Edzards et al. 1997), existing data are insufficient to establish clear relations between large-scale forest decline and air pollution. However, there might be a causal relationship between air pollution and more widespread forest health problems. Apart from damaging trees directly by foliar exposure to elevated concentrations of SO_2, NO_2, NH_3, or some combination, deposition on soil may indirectly

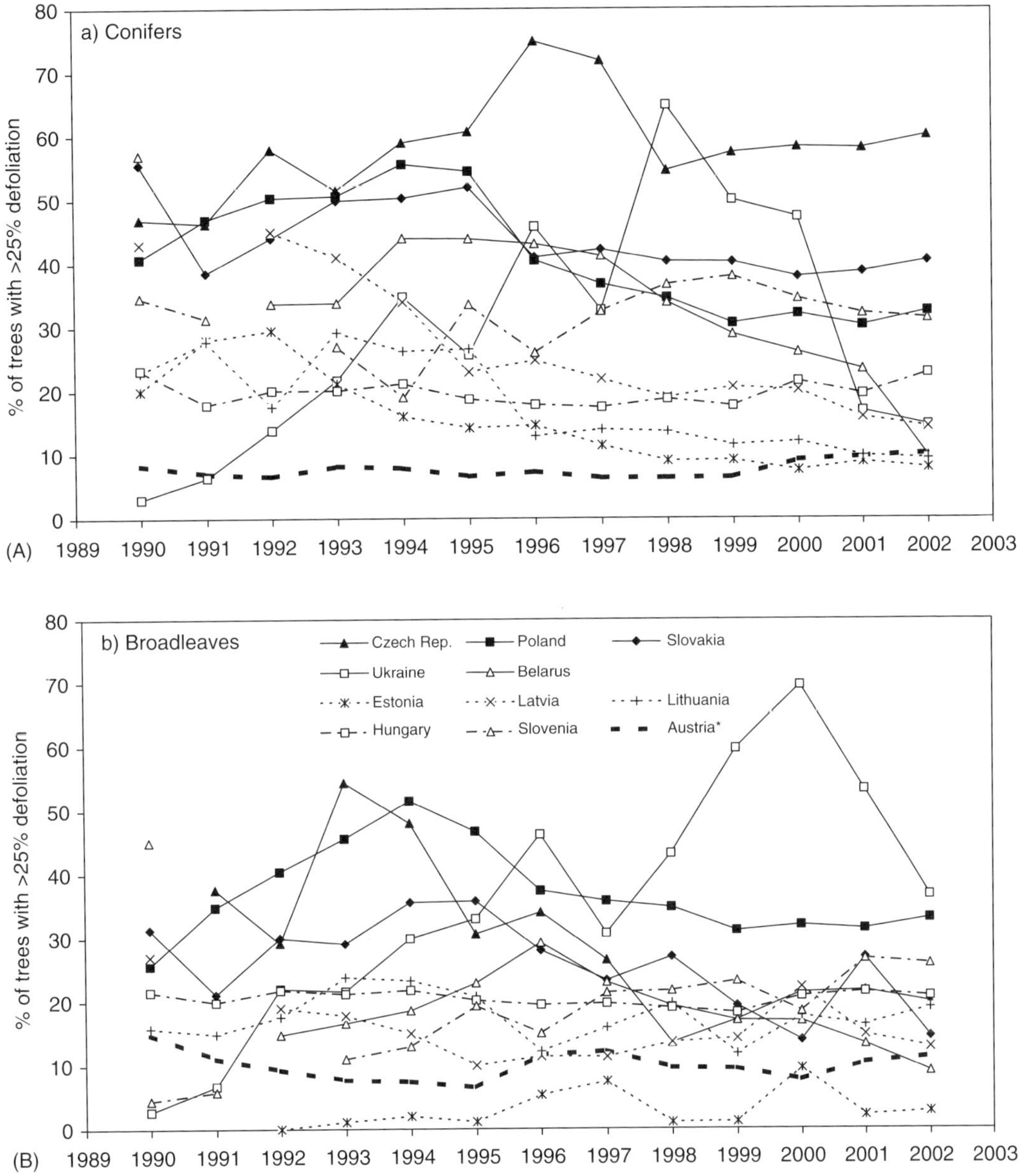

Figure 33.1 Tree defoliation in CEE countries compared to Austria for the years 1990 to 2002 (data from Anon. 2003). Comparisons between countries should be made cautiously because damage surveys were not intercalibrated until 1995. There may still be bias in the data, especially as older trees show more defoliation than younger trees (Müller-Edzards et al. 1997); therefore, age composition of the forests influences the results.

influence vegetation by initiating complex and long-term changes in soil chemistry that negatively influence forest vitality (Müller-Edzards et al. 1997; Vancura et al. 2000). In this scenario, the environment remains contaminated even after the pollution level has been reduced (Ploompuu 1997; Stachurski et al. 1995; Zimka and Stachurski 1996). Imbalanced nutrition of macronutrients (NPK), together with toxic levels of aluminum, lead, and sulfur, reduces foliar chlorophyll concentrations and disturbs the physiological processes of trees (Oleksyn et al. 1996; Stachurski et al. 1995; Ulrich et al. 1980). An indirect effect on soils could account for tree defoliation remaining at high levels in many CEE countries, despite reduced pollutant emissions (Tesař 1998).

33.6 Forest restoration

Forests are managed according to general principles of utilization and restocking over space and time, accounting for differences in site conditions, as well as basic principles of forest protection. Forest decline over extensive areas disrupts the planned spatial and temporal order in a managed forest, and causes serious disturbance in forest management at a regional scale. The overall restoration aim for forest degraded by industrial air pollution is to reconstruct a forest that is maximally resistant to disturbances and at the same time serves the productive, ecological, and social functions normally required by forests (Rozwałka 2003).

Clearcutting systems with artificial regeneration have been widely used in the CEE. Natural regeneration has traditionally been very limited, especially in the Czech Republic (5%), Poland (7%), Hungary (8%), and Slovakia (10%) as compared to European Union (EU) and other western European countries with similar traditions and natural conditions. For example, throughout the 1990s, natural regeneration accounted for approximately 90% of the regeneration in Switzerland, 84% in Austria, and 40% in Germany (Anon. 1999). Not surprisingly, the first comprehensive strategy for forest management in affected areas (developed in 1957) was based on traditional spruce monocultures, with the substitution of Norway spruce by introduced species considered to be more tolerant to air pollution (Kozlov et al. 2000). Efforts to prevent losses of commercial timber in stands damaged by air pollution in the Bohemian part of the Ore Mountains led to clearcuts over large areas; Kozlov et al. (2000) called this "technical restoration." However, the appropriateness of clearcutting as a sustainable management system must be assessed in the specific context of individual forest zones, including natural processes and local site conditions. In the mountain forest, where the most severe damage occurred, the clearcutting system is inappropriate due to the risk of erosion (Anon. 1999). Concerns for erosion, together with the failure of introduced species such as Colorado blue spruce (*Picea pungens* Engelm.) (Tesař and Tichy 1990) and greater appreciation for the ability of autochthonous species to regenerate naturally, led to more ecological restoration strategies (Kozlov et al. 2000). This strategy is based on restoration according to a hierarchy of needs in a deliberately chosen sequence (Lipka-Chudzik and Stępień 1996; Niemtur 1997).

Significant factors for establishing a hierarchy of needs include site factors (altitude, topography, regional and local climate, soil, soil contamination level, pollution loads) and stand characteristics (degree of damage, presence of natural regeneration and weeds, stand age). For forests damaged by air pollution, local climate as affected by altitude and topography is typically the most important site criteria. For example, the greatest temperature amplitudes are noticed in the bottom of mountain valleys and in the upper parts of slopes, whereas the highest snow cover appears in the part of slope just below the peak (Mikułowski 2001). But soil type (moisture, shallowness, fertility) can also affect the choice of restoration method. The most important stand characteristics are the degree of damage and the presence of natural regeneration. In the higher elevations of the Carpathian Mountains (>1000 m), three categories of damage were distinguished and, correspondingly, three different restoration strategies were recommended (Niemtur et al. 1999). These strategies included "forced" restoration, where the stand was completely destroyed by air pollution and any remaining tress were incapable of forming a stand that could recover the original structure and species composition. "Obligatory" restoration was called for where stands exhibited distinct signs of air pollution damage and suffered from fungal diseases and insect attacks. A third strategy was "optional," where stands were under persistent impact of low concentrations of industrial emissions and species composition was not adjusted to habitat conditions, but there were no noticeable signs of decline. The areas requiring obligatory or optional restoration could only be roughly delineated. Stand age (Lipka-Chudzik and Stępień 1996) and the presence of competing vegetation were also important for establishing the hierarchy of needs.

In the case of forced restoration, there is no time for long-term local studies of optimal restoration methods. Thus, the actions undertaken to regenerate the forest ecosystem are usually based on general silvicultural knowledge or on analyses of similar cases. Most research is focused on case studies (Latocha 1990; Niemtur et al. 1999; Pysek 1992; Tesař and Tichy 1990). Restoration of stands damaged by air pollution consists of a combination of several components securing successful stand establishment or conversion, the combination and time sequence differing according to the above-mentioned hierarchy of needs. The components are regeneration method (natural regeneration, planting, supplementary planting), removal of dead or damaged trees (e.g., clearcut, cutting in strips, sheltering), mechanical soil preparation, chemical soil preparation (liming, fertilization), weed control, species choice, genetic regeneration material, plant quality, and insect pest and game control. After successful stand establishment or conversion, appropriate tending methods should be applied.

Natural regeneration for obligatory and optional restoration requires favorable conditions. It enables a gradual restoration with a continuous forest cover (Latocha 1990), and it is particularly desirable on slopes because of decreased erosion risk. Nevertheless, natural regeneration may be inappropriate if the species that regenerates is off-site or of a foreign provenance. For severely damaged mountain forests lacking seed sources or facing severe competition, planting seedlings in clusters has proved successful in restoring disturbed forests at high elevations (Schönenberger et al. 1990).

Selecting the correct species choice is critical. Special attention must be given to ecological site conditions as compared to species requirements, the degree of soil and stand degradation, sensitivity of the species and provenance to air pollution, and severity of competing vegetation (mainly grasses). At high elevation, choices are limited: mainly broadleaved species such as beech, birch, sycamore, and mountain ash, and at the highest altitudes, conifers such as Norway spruce, European larch (*Larix decidua* Mill.), and mountain pine (*Pinus mugo* Turra) (Szymański et al. 1997; Vacek et al. 1997). In areas of heavy pollution damage, the only choice is between birch (*Betula* sp.) and mountain ash (*Sorbus aucuparia* L.), which are resistant to harsh climate conditions (Tesař and Tichy 1990). At lower elevations and in less polluted areas, the possibilities are greater. Species composition can be reconstructed by sequentially introducing native broadleaved species (oaks, lime [*Tilia cordata* Mill., *T. platyphyllos* Scop.], sycamore maple [*Acer pseudoplatanus* L.], or beech) under the canopy of pine or spruce where possible (Latocha 1990). In the most polluted lowlands, nonnative species such as black pine (*Pinus nigra* Arnold), blue spruce, black locust (*Robinia pseudoacacia* L.), and red oak (*Q. rubra* L.) are advisable (Anon. 1988; Rozwałka 2003).

The choice of regeneration material should be optimized and, generally, locally adapted provenances are preferred. However, industrial pollution may reduce the genetic diversity of the populations of the main forest tree species by reducing the ability of populations to adapt to future changes in environmental conditions. At the level of moderate contamination of the environment, the tolerant individuals exhibit increased heterozygosity and genotypic polymorphism, and a higher mean number of alleles per locus. This suggests that genotypes with high heterozygosity and genetic diversity have higher adaptive value than genotypes lacking these attributes (Prus-Głowacki and Wojnicka 1996). Another adaptation strategy under high concentrations of pollutants consists of the selection of genotypes that are specifically adapted to the changed environment, leading to increased homozygosity and a loss of genotypic polymorphism (Pitelka 1988). Thus, *ex situ* conservation of genotypes is important in the attempt to secure the best regeneration material for restoring forests degraded by air pollution. In Poland, the Forest Gene Bank in Kostrzyca, located in the Sudeten Mountain region, has been established to carry out genetic conservation for international needs in general, but particularly for the Black Triangle region.

33.7 Costs

Forest restoration is a long-term process ensuring sustainability of forest production, but in the short term it increases the costs of silviculture. If the natural regeneration is effective, and restoration costs are lowered but without intensive silvicultural measures, the effectiveness of forest restoration is unlikely (Küssner 1997). Recent experiences in the Czech Republic, Slovakia, and Poland indicate that afforestation is expensive; total expenditures range between 1000 and 1300€ per ha for conifers or broadleaves (Anon. 1999). Costs for establishing mixed-species plantations, including protection and cultivation, are greater by nearly 850€ per ha than for monocultures (Bludovsky 1995). This is a serious expenditure, even for large-scale, state-owned forests. For small-scale private-forest holdings, it is impossible to bear. Furthermore, costs are also increased due to complex administrative procedures and legal regulations in the agricultural and environmental sectors (Anon. 1999). Currently, small owners prefer a nonintensive model of forest management, especially because the forest is never their main income source.

Forest owners are beginning to organize into associations to represent their interests to government and other institutions in order to obtain subsidies (Kaliszewski 2004). In some CEE countries, private forest owners can apply for financial support from state or local authorities for afforestation or restoration of degraded forests. For example, Polish forestry obtained about 120 M€ in subsidies in the years 1992 to 1999, and in the Czech Republic total subsidies exceeded 300 M€ for the same period (Kaliszewski 2004). Approximately half of the total sum in both countries was associated with forest management in air-polluted areas. But the system of subsidies for afforestation and reclamation is not stable, and the amount of funding is often insufficient. Private owners can also gain through rural development funds within the EU framework and from World Bank resources (Kaliszewski 2004).

33.8 Regional example — the Black Triangle

The Black Triangle is comprised of northern Bohemia in the Czech Republic, parts of Lower Silesia in Poland, and Saxony in Germany, an area of about 34,500 km^2 (Abraham et al. 2001). The most severe forest health problems occur at higher elevations due to transboundary air pollution. A trilateral approach to solving these problems was begun with a 1991 joint declaration on cooperation, signed by the ministers of environment of (the former) Czechoslovakia, Poland, and Germany. The European Commission was a fourth partner in this initiative and provided financial support that resulted in a Joint Air Monitoring System (JAMS) with 40 measuring stations (Abraham et al. 2001). The involved countries provided the bulk of the funding.

33.8.1 Physiography

The Ore Mountains and Sudeten massifs are of Paleozoic origin, similar in age, lithology, relief, climate, flora, and fauna. They consist mostly of gneiss, crystalline schist, and granite; base-rich carbonate and volcanic rocks are scarce. Such a geologic structure results in low natural soil buffering capacity and susceptibility to acidification, which is one of the major reasons for forest site degradation even at low levels of SO_2 deposition.

A complex orography causes considerable variation in climate with typical annual mean temperatures of about 8 and 3°C in lowland and mountainous parts, respectively. The annual sum of precipitation ranges from 450 to 1400 mm, being unevenly distributed with sharp gradients caused also by the strong influence of the complex orography (Abraham et al. 2001). Westerly winds prevail, transporting air pollutants from the old-technology brown coal power plants to the area. Otherwise, the climate is characterized by a thick and long-lasting snow cover, a high frequency of mists with elevated water content from liquid deposition and hoarfrost, night frosts in terrain depressions throughout the year with draughts, and considerable drops in temperatures (Dore et al. 1999; Migała et al. 2002; Sobik 1991; Sobik and Migała 1993).

Unsuitable silvicultural management practices in the past have produced a disturbed species composition, as compared to the potential plant communities. Forest area in the Sudetes declined by half between the 12th and 18th centuries, with broadleaves and silver fir among the most heavily harvested species. Attempts at forest management emerged in the early 1800s when cuttings were limited in size and seed trees had to be left. A high demand for spruce wood, however, led to systematic harvesting of all hardwood stands and replacement with Norway spruce. The proportion of Norway spruce increased, from 30 to 95%, between 1834 and 1925 (Mazurski 1986; Wilczkiewicz 1982). Natural regeneration of Norway spruce was largely unsuccessful and local seed sources were insufficient. This necessitated artificial regeneration based on seeds imported from Austria. At that time, the significance of the seed source for productivity and resistance to biotic and abiotic factors was not recognized (Vancura et al. 2000). The reliance on artificial regeneration led to the formation of even-aged monocultures of Norway spruce that later proved particularly susceptible to disturbance.

33.8.2 *Primary and secondary decline*

Pollution impacts on the Norway spruce forests of the Sudeten Mountains were noted early in the 20th century, around coal mines and other small industrial plants in the mountain valleys (Capecki 1989). By the end of the 1960s, observers noted large-scale effects of air pollution in forests above 900 m elevation (Capecki 1969). In time, a decline was observed at lower elevations, down to 700 m. The decline was likely a result of high concentrations of several pollutants (mostly SO_2 and NO_x) and their combination with other stresses related to climate (Fisak et al. 2002; Hartig et al. 1998; Kmiec et al. 1995; Mazurski 1986; Pacyna and Tørseth 1997; Paschalis and Zajączkowski 1997). Primary forest decline (Table 33.3) resulted from the weakening of mostly coniferous species that promoted insect outbreaks of larch bud moth (*Zeiraphera griseana* Hb.) and spruce bark beetle (*Ips typographus* L.), which were the direct causes of mortality (Capecski et al. 1996; Grodzki 1995).

From the end of the 1980s until the mid-1990s, forest decline and mortality in the Sudeten Mountain range continued, however at a small scale. As the older, most sensitive stands were replaced, age distribution was skewed toward a greater proportion of younger stands. For example, the average age of stands in Świeradów Forest District was 58 years in 1974 to 1976 and declined to 41 years by 1988 to 1991. In the Szklarska Poręba Forest District, the average stand age over the same period declined from 58 years to 39 years. Further decreases in average stand age were due to restocking the extensive deforested areas. The decline of large areas of older forests has exposed other previously undamaged stands to air pollutants.

Health problems increasingly appear in younger stands. Newly established stands (currently 1 to 20 years) occur at higher elevations where forest decline was the greatest and growth conditions are the most difficult. On windward, west-facing slopes, there is increased damage to the assimilatory apparatus, with the number of needle-year classes in Norway spruce stands growing on exposed slopes greater than in stands not directly

Table 33.3 Groups of Stressors Affecting Forest Decline in the Area of 'Black Triangle' (Manion 1991; Mikulowski 2001)

Stressors	Primary Decline Stressors	Secondary Decline Stressors
Predisposing Historic	• Allocation of more fertile land to agriculture • Improper forest management • Change in species composition • Introduction of foreign provenances	
Predisposing Current	• Wind, frost, snow • Increase in pollution level in the 1970s • Interaction between pollution and climate (heavy precipitation, acid rains, mists) resulting in high deposition levels	• Exposure of slopes due to primary decline
Inciting Current	• Strong wind, droughts, frost • Acute peak levels of pollutants concentrations	• Acute peak levels of air pollution concentrations and abiotic stressors (wind, snow, and frost)
Contributing Current	• Larch bud moth (*Z. glisseana*) and Spruce bark beetle (*I. typographus*)	

exposed to pollutants (Mikułowski and Gil 2003). The rate and extent of the process have been slower for secondary than for the primary decline, and it has not been accompanied by insect invasion. Although not conclusive, this points to air pollution combined with abiotic factors, mainly wind and snow, as the causes of decline in the stands (Mikułowski 2001) (Table 33.3).

33.8.3 Change of ecological preconditions by deforestation

Harsh natural conditions in this region pose significant problems for successfully regenerating plantations, and deforestation caused by air pollution further exacerbates the difficulty. In the degraded areas, soil acidification, lack of ectomycorrhizae, and fine root dieback have been observed (Zimka and Stachurski 1996). Soil pH is considerably lower in the surface and decomposition of organic matter after overstory removal is slower (C/N ratio is 25 to 35 in the deforested area and 20 to 25 in the less-damaged stand). Nitrogen content is lower (total N in the Ofh horizon is 0.8 to 1.2% in the deforested area and 1.0 to 1.5% in the less-damaged stand) as are concentrations of K, Ca, and Mg (Walendzik et al. 1995a, 1995b). Acidophilus grass communities have invaded, particularly *Dechampsia flexuosa* (L.) Trin., *Calamagrostis villosa* (Chaix) J. Gmelin, and *Carex* species, which make the replanting of forest trees extremely difficult or even impossible (Kozlov et al. 2000). Plants characteristic of natural mountain coniferous forests such as *Vaccinium vitis-idaea* (L.) are absent, which may be an effect of site acidification (Fabiszewski and Brej 2000; Fabiszewski and Wojtuń 1994).

33.8.4 Forest restoration

The first step in the restoration process is determining the degree of damage. In the Black Triangle, spruce stands were grouped into three categories: stands consisting of dead trees only, generally at elevations above 900 m; partially damaged stands at various elevations; and weakened stands with clearly reduced vitality caused by air pollution and insect attacks at lower elevations (<700 m), or in some cases at higher elevations on mainly north- and northeast-facing slopes, which were not exposed to air pollutants (Mazurski 1986; Mikułowski 2001).

In the large areas deforested during the period of greatest air pollution impact (1978 to 1989), the primary goal was to secure plant survival during the regeneration phase. Subsequent goals have been protection against insects and wildlife damage and silvicultural options for stabilizing the young stands against wind and snow damage. From 1978 to 1989, most completely damaged spruce stands were clearcut. Standing dead trees were left if sites were inaccessible or harvesting cost more than the value of the salvaged timber. In partially damaged stands, both clearcuts and partial cuts to salvage standing dead trees were utilized. Only dead trees were removed in weakened stands, which is also a protective measure against spruce bark beetles (Mikułowski 2001; Szymański et al. 1997).

Because of the difficult terrain, manual soil preparation was used to scarify patches of about 0.6×0.6 m. This ensured adequate exposure of mineral soil at low costs. Depending on the topography and general accessibility of sites, occasionally equipment such as ploughs and planting borers mounted on tractors were used. Planting without soil preparation was infrequent, but proved to be satisfactory locally (Gorzelak 1995; Szymański et al. 1997; Vacek et al. 1997).

In the Czech Republic, only container seedlings were used for reforestation of degraded sites. In Poland, forest nurseries could not supply sufficient quantities of container seedlings, but bareroot seedlings generally had poor survival and growth. Since the 1990s, container seedlings have been obligatory for replanting deforested sites in the mountains (Szymański et al. 1997). At higher elevations, mainly conifers were planted, including Norway spruce, European larch, and mountain pine (Mikułowski 2001; Szymański et al. 1997; Vacek et al. 1997; Vancura et al. 2000). At lower elevations, the broadleaved species were preferred, including European beech, sycamore maple, and oaks.

Fertilizers were applied prior to or after planting; top dressing applied to individual seedlings was most effective (Walendzik et al. 1995a, 1995b). Chemical control of weeds was limited due to concerns for water protection. Manual weed control consisted of mulching or grasscutting. The mulches used were usually black foil or cardboard sheets spread around seedlings during the growing season, which markedly improved the survival and growth of seedlings in young plantations (Gorzelak 1995).

Natural regeneration of birch occurred on clearcuts on north-facing slopes at elevations below 900 m. The birch was treated as a shelter crop for regeneration of the main tree species of beech, sycamore, and fir. In a number of sites situated above 900 m, self-seeded mountain ash provided shelter to spruce seedlings. Regeneration of beech by self-seeding in partially damaged and weakened stands was noted at elevations up to 700 m. Some limited artificial regeneration under a canopy was considered a supplementary method for forest renewal if natural regeneration was inadequate. In such cases, dead or weakened trees and strips of healthy trees were left as shelter for beech, silver fir, sycamore maple, and Norway spruce seedlings (Szymański et al. 1997; Vacek et al. 1997).

Cervid populations expanded after large areas were opened by canopy mortality due to air pollution. High game pressure in young stands (1 to 40 years) includes browsing during the snowless period and debarking when the snow cover lingers. Commonly used protective measures included repellents, tree shelters, and fencing. Individual shelters to protect broadleaves were successful but expensive. The most successful method for limiting damage by deer was to reduce population size and reestablish natural sex and age ratios (Konca et al. 1997; Szukiel and Nasiadka 1997; Zahradnik 1997).

In even-aged young stands (1 to 20 years) at high elevations, controlling density by tending cuts can increase the resistance of stands and minimize the negative impacts of weather. Two preferable approaches in spruce stands are heavy thinning (Slodičak 1995), which decreases the H/dbh ratio and thereby stabilizes the stand against strong wind; or group thinning (Zajączkowski 1994). Group thinning is an important element in seminatural silvicultural management. It is commonly used in forestry management but is especially recommended for mountainous conditions (Rozwałka 2003), where it ensures a relatively stable stand structure and increases long-term resistance against abiotic factors, particularly wind and snow (Otto 1994a, 1994b; Zajączkowski 1994).

Although there are still problems, forest restoration in the Black Triangle can be viewed as successful. Recent evaluations of the restored areas in the western Sudeten Mountains show that restoration has had an effect. After a period of stagnation, spruce and larch plantations have experienced rapid growth (Mikułowski, unpublished data, 2003). Natural regeneration of spruce can be found under the plantation canopies even at high elevations (>900 m), and invasive grasses have been suppressed by the shade of tree crowns. Restoration was only possible because of the dominant role of state forest holdings, international cooperation, and economic support of international agencies.

33.9 Conclusions

The level of air pollution in the CEE countries has significantly decreased in recent years, but degradation of forests by air pollutants and indirect effects of soil acidification are still serious problems in parts of the CEE. The heaviest damage to forests has been in the mountain regions and industrial centers, especially in the vicinity of coal mines. Special actions have been undertaken, and are still required, to enhance the resistance of forests against complex disturbance factors. The public forestry sector is expected to fulfill an important mission in providing an organizational framework for forest restoration. Integration of small-scale private forest owners into the process is needed to accelerate forest restoration through improved forest management. However, relatively low gross national products in the CEE countries make it difficult to sufficiently fund the needed restoration work. The regions with the greatest ecological collapse need international cooperation and funding in the process of forest reclamation.

References

Abraham, J., Ciechanowicz-Kusztal, R., Drücke, M., Jodłowska-Opyd, G., Kallweit, D., Keder, J., Kulaszka, W., and Novák, J., *Common Report on Air Quality in the Black Triangle Region 2000*, ČHMÚ, WIOŚ, LfUG, UBA, Prague, 113 pp., 2001.

Akkermann, R., *Zur Situation der Baum- und Waldschäden in Nordwestdeutschland — Teil 1*, BSH Verlag, Wardenburg, 293 pp., 1987.

Anon., *Zasady Hodowli Lasu (Principles of Silviculture)*, revised Edition V, Państwowe Wydawnictwo Rolnicze i Leśne, Warszawa, 172 pp., 1988 (in Polish).

Anon., *Conservation and Sustainable Management of Forests in Central and Eastern European Countries*, PHARE Environment Consortium, European Commission PHARE Programme, Brussels, 1999.

Anon., *Forest Resources of Europe, CIS, North America, Australia, Japan and New Zealand (TBFRA-2000)*, Geneva Timber and Forest Study Papers, 17, UN/ECE FAO, New York and Geneva, 445 pp., 2000.

Anon., *Global Forest Resources Assessment 2000*, Main Report, FAO Forestry Paper, 140, FAO, Rome, 481 pp., 2001a.

Anon., *Statistical Yearbook of the Republic of Poland Year LXI*, Central Statistical Office, Warsaw, 2001b.

Anon., *Forest Condition in Europe. Results of the 2002 Large-scale Survey*, Technical Report, UN/ECE and EC, Geneva and Brussels, 114 pp., 2003a.

Anon., The World Fact Book, http://www.cia.gov/cia/publications/factbook/, 2003b.

Balcar, V., Vacek, S., and Henžlik, V., Poškozeni a uhyn lesnich porostou v Sudetskych horach (Injury and decline of forest stands in the Sudeten Mts.), in *Protection of Forest Ecosystems. Selected Problems in Sudeten Mountains*, Paschalis, P. and Zajączkowski, S., Eds., Grant GEF 05/21685 POL Forest Biodiversity Protection Project, Warszawa, 29, 1997 (in Czech; English summary).

Bludovský, Z., *Vliv změny druhové skladby na zvýšeni nákladů (Effect of changes in species composition on expenditure level)*, Les. Prace, 6, 4, 1995 (in Czech).

Capecki, Z., Zagrożenie lasów sudeckich przez szkodniki na tle szkód powodowanych przez huragany i okiść (*Danger to the Sudeten forests by pests on the background of forest decline caused by hurricanes and frosts)*, Sylwan, 3, 57, 1969 (in Polish; English summary).

Capecki, Z., Rejony zdrowotności lasów sudeckich *(Forest condition of different parts of the Sudeten forests)*, Prace Instytutu Badawczego Lesnictwa, Ser. A, 688, 5, 1989 (in Polish; German summary).

Capecki, Z., Grodzki, W., Kosibowicz, M., and Kozioł, M., Valuation of relationship between major pests and industrial pollution of mountain forests, in *Reakcje biologiczne drzew na zanieczyszczenia przemysłowe (Biological Reactions of Trees to Industrial Pollution)*, Siwecki R., Ed., III National Symposium Kórnik 24–26 May 1994, Wyd. Sorus, Poznań, 543, 1996.

Dengler, A., *Waldbau auf ökologischer Grundlage, Vol. 2, Baumartenwahl, Bestandesbegründung und Bestandespflege*, revised by Röhrig, E. and Gussone, H., Parey, Hamburg, 314 pp., 1990.

Dore, A.J., Sobik, M., and Migala, K., Patterns of precipitation and pollutant deposition in the western Sudeten mountains, Poland, *Atmos. Environ.*, 33, 20, 3301, 1999.

Fabiszewski, J. and Brej, T., Contemporary habitat and floristic changes in the Sudeten Mts., *Acta Soc. Bot. Pol.*, 69, 3, 215, 2000.

Fabiszewski, J. and Wojtuń, B., Zjawiska ekologiczne towarzysząńce wymieraniu lasów w Sudetach *(Ecological phenomena accompanying forest decline in Sudeten)*, Prace Instytutu Badawczego Lesnictwa, Ser. B, 21/2, 195, 1994 (in Polish).

Fisak, J., Tesar, M , Rezacova, D., Elias, V., Weignerova, V., and Fottova, D., Pollutant concentrations in fog and low cloud water at selected sites of the Czech Republic, *Atmos. Res.*, 64, 1, 75, 2002.

Gorzelak, A., *Wpływ zabiegów pielęgnacyjnych na wzrost I stabilność upraw w Sudetach Zachodnich, (Effect of tending treatments on growth and stability of plantations in West Sudeten Mountains)*, Prace Instytutu Badawczego Lesnictwa, Ser. B, 25, 277, 1995 (in Polish).

Grodzki, W., *Zanieczyszczenia przemysłowe a gradacje szkodników owadzich w lasach górskich* (*Air pollution and insect outbreaks in mountain forests*), Sylwan, 89, 5, 13, 1995 (in Polish; English summary).

Gutkowski, R.M. and Winnicki, T., Eds., *Restoration of Forests: Environmental Challenges in Central and Eastern Europe*, Proceedings of the NATO Advanced Research Workshop on Science and Technology to Save and Better Utilize Central and Eastern Europe's Forests, Saulberg, Lower Bavaria, Germany, July 15–19, 1996. NATO Science Partnership, Subseries 2, 472 pp., 1997.

Hartig, M., Pfalz, W., and Wollmerstädt, J., *Waldbauliche Behandlung und Umbau immissionsgeschädigter Fichtenwälder im Sächsischen Erzgebirge*, Forst und Holz, 53, 20, 621, 1998.

Henžlik, V., Forests and air pollution in the Czech Republic, in *Restoration of Forests: Environmental Challenges in Central and Eastern Europe*, Gutkowski, R.M., and Winnicki, T., Eds., Proceedings of the NATO Advanced Research Workshop on Science and Technology to Save and Better Utilize Central and Eastern Europe's Forests, Saulberg, Lower Bavaria, Germany, July 15–19, 1996, NATO Science Partnership, Subseries 2, 133, 1997.

Innes, J.L. and Oleksyn, J., Eds., *Forest Dynamics in Heavily Polluted Areas*, Report No. 1 of the IUFRO Task Force on Environmental Change, CABI Publishing, Wallingford, 248 pp., 2000.

Kaliszewski, A., Financing of forestry from public means in the Czech Republic, Estonia, Poland and Slovenia — policy context, organization, and supported activities, *J. For. Sci.*, 50, 181, 2004.

Kauppi, P.E., Melikäinen, K., and Kuusela, K., Biomass and carbon budget in European forests, 1971 to 1990, *Science*, 256, 70, 1992.

Kmiec, G., Kacperczyk, K., Zwozdziak, A., and Zwozdziak, J., Acid pollutants in air and precipitation/deposition at the Sudeten mountains, Poland, *Water Air Soil Pollut.*, 85, 4, 2131, 1995.

Konca, B., Zimny, J. and Michalski, J., Ochrona lasu w Sudetach — czynniki abiotyczne I biotyczne oraz stan obecny i prognozy (Protection of the forest in the Sudety Mts., abiotic and biotic factors nowadays and forecast for protective future), in *Protection of Forest Ecosystems. Selected*

Problems in Sudeten Mountains, Paschalis, P. and Zajączkowski, S., Eds., Grant GEF 05/21685 POL Forest Biodiversity Protection Project, Warszawa, 216, 1997 (in Polish; English summary).

Kozlov, M.V., Haukioja, E., Niemelä, P., Zvereva, E. and Kytö, M., Revitalization and restoration of boreal and temperate forests damaged by air pollution, in *Forest Dynamics in the Heavily Polluted Regions*, Innes, J.L. and Oleksyn, J., Eds., Report No. 1 of the IUFRO Task Force on Environmental Change, CABI Publishing, Wallingford, 2000, chap. 11.

Küssner, R., *Sukzessionale Prozesse in Fichtenbeständen (Picea abies) des Osterzgebirges — Möglichkeiten ihrer waldbaulichen Beeinflussung und ihre Bedeutung für einen ökologisch begründeten Waldumbau*, Forstwissenshcaftliches Centralblatt, 116, 359, 1997.

Latocha, E., Scots pine stand reconstruction possibilities in the Upper Silesia industrial region, in *XIX World Congress of IUFRO, Montreal, Canada, 5–11 August 1990. Proceedings, Division 2*, Oxford Forestry Institute, Oxford, 447, 1990.

Lipka-Chudzik, E. and Stępień, E., *Niektóre problemy organizacji ładu czasowego w przebudowie drzewostanów sosnowych (Some problems with arranging temporal order in restructured pine stands)*, Sylwan, 12, 57, 1996 (in Polish; English summary).

Lutterman, A. and Freedman, B., Risks in heavily polluted regions, in *Forest Dynamics in the Heavily Polluted Regions*, Innes, J.L. and Oleksyn, J., Eds., Report No. 1 of the IUFRO Task Force on Environmental Change, CABI Publishing, Wallingford, 2000, chap. 2.

Mandre, M., Ots, K., Rauk, J., and Tuulmets, L., Impact of air pollution emitted from the cement industry on forest bioproduction, *Oil Shale*, 15, 4, 353, 1998.

Manion, P.D., *Tree Disease Concepts,* Prentice-Hall, Englewood Cliffs, NJ, 409 pp., 1991.

Mazurski, K.R., The destruction of forests in the Polish Sudetes Mountains by industrial emissions, *For. Ecol. Manage.*, 17, 4, 303, 1986.

Migala, K., Liebersbach, J., and Sobik, M., Rime in the Giant Mts. (The Sudetes, Poland), *Atmos. Res.*, 64, 63, 2002.

Mikułowski, M., *Zmiany środowiska leśnego i gospodarki leśnej w polskich Sudetach Zachodnich (Changes in forest ecosystems and forest management in the Sudety Mountains, Poland)*, Prace Instytutu Badawczego Lesnictwa, Ser. A, 3, 103, 2001 (in Polish).

Mikułowski, M., unpublished data, 2003.

Mikułowski, M. and Gil, W., Wind induced damage to Polish forests and the methods of mitigating its effect, in *International Conference: Wind Effects on Trees*, Ruck, B., Kottmeier, C., Mattheck, C., Quine, C. and Wilhelm, G., Eds., 16–18 September 2003, University of Karlsruhe, Germany, 349, 2003.

Müller-Edzards, C., De Vries, W., and Erisman, J.W., Eds., *Ten Years Of Monitoring Forest Condition in Europe, Studies on Temporal Development, Spatial Distribution, Impacts of Natural and Anthropogenic Stress Factors*, Technical Background Report, UN/ECE and EC, Geneva and Brussels, 386 pp., 1997.

Niemtur, S., *Przebudowa drzewostanów uszkodzonych przez imisje przemysłowe w Sudetach Zachodnich i w Górnośląskim Okręgu Przemysłowym (Transformation of forest stands damaged by industrial emissions in Western Sudetes and Upper Silesian Industrial Region)*, Postępy Techniki w Leśnictwie, 61, 51, 1997 (in Polish).

Niemtur, S., Ambroży, S., Gazda, M., and Grodzki, W., *Przebudowa karpackich drzewostanów regla górnego uszkodzonych przez imisje przemysłowe (Transformation of Carpathian stands in the upper mountain zone damaged by industrial emissions)*, Prace Instytutu Badawczego Leśnictwa, Ser. B, 35, 1, 1999 (in Polish).

Oleksyn, J. and Innes, J.L., Air pollution and forests in heavily industrialized regions: an introduction, in *Forest Dynamics in the Heavily Polluted Regions*, Innes, J.L. and Oleksyn, J., Eds., Report No. 1 of the IUFRO Task Force on Environmental Change, CABI Publishing, Wallingford, 2000, chap. 1.

Oleksyn, J., Karolewski, P., Giertych, M.J., Werner, A., Tjoelker, M.G., and Reich, P.B., Altered root growth and plant chemistry of *Pinus sylvestris* seedlings subjected to aluminium in nutrient solution, *Trees*, 10, 135, 1996.

Otto, H.J., Ökologischer Waldbau. Walddynamische Processe bei Fichte und Kiefer in Rahmen waldbaulicher Rationalisierung, *Der Wald*, I, 44, 11, 364, 1994a.

Otto, H.J., Ökologischer Waldbau. Walddynamische Processe bei Fichte und Kiefer in Rahmen waldbaulicher Rationalisierung, *Der Wald*, II, 44, 12, 408, 1994b.

Pacyna, J. and Tørseth, K., Central European hot spots of air pollution, in *Restoration of Forests. Environmental Challenges in Central and Eastern Europe*, Gutkowski, R.M. and Winnicki, T., Eds., Kluwer Academic Publishers, Dordrecht, 15, 1997.

Paschalis, P. and Zajączkowski, S., Preface, in *Protection of Forest Ecosystems. Selected Problems in Sudeten Mountains*, Paschalis, P. and Zajączkowski, S., Eds., Grant GEF 05/21685 POL Forest Biodiversity Protection Project, Warszawa, 7, 1997.

Pitelka, L.F., Evolutionary responses of plants to anthropogenic pollutants, *Trends Ecol. Evol.*, 3, 233, 1988.

Ploompuu, T., *Rabade taimekoosluste teisenemine leeliselise tolmusaaste toimel Kunda ümbruses. (Changes in plant communities of bogs upon the influence of alkaline dust pollution in Kunda region (NE Estonia))*, Kaasaegse ökoloogia probleemid, Ajalised muutused Eesti elusloođuses ja keskkonnas, Tartu, 177, 1997 (in Estonian; English summary).

Prus-Głowacki, W. and Wojnicka, A., Wpływ skażenia przemysłowego na pule genowe populacji drzew leśnych (The influence of industrial pollution on gene pools of forest tree populations), in *Reakcje biologiczne drzew na zanieczyszczenia przemysłowe* (Biological reactions of trees to industrial pollution), Siwecki R., Ed., *III National Symposium Kórnik* 24-26 May 1994 (Wyd. Sorus), Poznań, 391, 1996 (in Polish; English summary).

Pysek, P., Dominant species exchange during succession in reclaimed habitats: a case study from areas deforested by air pollution, *For. Ecol. Manage.*, 54, 27, 1992.

Raben, G., Andrae, H. and Leube, F., Schadstoffbelastungen in sächsischen Waldökosystemen, *Allgemeine Forszeischrift/Der Wald*, 51, 1244, 1996.

Rozwałka, Z., *Zasady Hodowli Lasu (Silvicultural Principles)*, Dyrektor Generalny Lasów Państwowych, 159 pp., 2003 (in Polish).

Schönenberger, W., Frey, W., and Leuenberger, F., Ökologie und Technik der Aufforstung im Gebirge, Anregungen fur die Praxis, *Berichte*, 325, 58 pp., 1990.

Slodičak, M., Thinning regime in stands of Norway spruce subjected to snow and wind damage, in, *Wind and trees*, Coutts, M.P. and Grace, J., Eds., Cambridge University Press, Cambridge, 436, 1995.

Sobik, M., *Precipitation and Cloudwater Acidity in the Karkonosze Mountains*, Alpex Regional Bulletin, Swiss Meteorological Institute, 17, 26, 1991.

Sobik, M. and Migała, K., *The Role of Cloud Water and Fog Deposits on the Water Budget in the Karkonosze Mountains*, Alpex Regional Bulletin, Swiss Meteorological Institute, 21, 13, 1993.

Stachurski, A., Zimka, J.R., and Kwiecień, M., Forest decline in Karkonosze (Poland). I. Chlorophyll, phenols, defoliation index and nutrient status of the Norway spruce (*Picea abies* L.), *Ekologia Polska*, 42, 289, 1995.

Szukiel, E. and Nasiadka, P., Gospodarka łowiecka — stan i dotychczasowe działania oraz prognozy i zalecenia (Game management — the present state and prognoses and recomendations), in *Protection of Forest Ecosystems. Selected Problems in Sudeten Mountains*, Paschalis, P. and Zajączkowski, S., Eds., Grant GEF 05/21685 POL Forest Biodiversity Protection Project, Warszawa, 159, 1997.

Szymański, S., Ceitel, J., and Zientarski, J., Odnowienie lasu (zagospodarowanie terenów poklęskowych) — wyniki dotychczasowych działań I propozycje na przyszłość (Forest regeneration management of disaster areas — results of hitherto activities and proposals for the future), in *Protection of Forest Ecosystems. Selected Problems in Sudeten Mountains*, Paschalis, P. and Zajączkowski, S., Eds., Grant GEF 05/21685 POL Forest Biodiversity Protection Project, Warszawa, 59, 1997 (in Polish; English summary).

Tesař, V., The Impact of Air Pollution and Strategies for Spruce Monoculture Conversion in Central Europe, in *Spruce Monocultures in Central Europe — Problems and Prospects*, Klimo, E., Hager, H., and Kulhavý, J., Eds., EFI Proceedings, 33, 208 pp., 1998.

Tesař, V. and Tichy, J., Results of new objectives in restoring the forest damaged by air pollution in Bohemian mountains, in *XIX World Congress of IUFRO*, Montreal, Canada, 5-11 August 1990. Proceedings, Division 2, Oxford Forestry Institute, Oxford, 455, 1990.

Ulrich, B., Mayer, R., and Khanna, P.K., Chemical changes due to acid precipitation in a loss-derived soil in central Europe, *Soil*, 130, 193, 1980.

Vacek, S., Lokvenc, T., Balcar, V., and Henžlik, V., Obnova a stabilizace lesa v horskych oblastach Sudet (Forest regeneration and stabilization in the Sudeten mountain region), in *Protection of Forest Ecosystems. Selected Problems in Sudeten Mountains*, Paschalis, P. and Zajączkowski, S., Eds., Grant GEF 05/21685 POL Forest Biodiversity Protection Project, Warszawa, 93, 1997 (in Czech; English summary).

Vancura, K., Raben, G., Gorzelak, A., Mikulowski, M., Caboun, V., and Oleksyn, J., Impact of air pollution on the forests of Central and Eastern Europe, in *Forest Dynamics in the Heavily Polluted regions*, Innes, J.L. and Oleksyn, J., Eds., Report No. 1 of the IUFRO Task Force on Environmental Change, CABI Publishing, Wallingford, 2000, chap. 8.

Walendzik, R.J., Szołtyk, G., Olejarski, I., and Goć, K., *Wdrożenie nawożenia mineralnego w Sudetach Zachodnich, (The use of mineral fertilizers in West Sudetes)*, Prace Instytutu Badawczego Lesnictwa, Ser. B, 25, 385, 1995a (in Polish).

Walendzik, R.J., Szołtyk, G., Olejarski, I. and Przyborowska, I., *Rewitalizacja gleb leśnych i drzewostanów na terenach zagrożenia ekologicznego w Sudetach Zachodnich (Revitalization of forest soils and stands in the ecological risk areas in West Sudetes)*, Prace Instytutu Badawczego Lesnictwa, Ser. B, 25, 315, 1995b (in Polish).

Wallen, C.C., Ed., *Climates of Central and Southern Europe*, World Survey of Climatology, 6, Elsevier, Amsterdam, 248 pp., 1977.

Wilczkiewicz, M., *Rys historyczny gospodarki w lasach sudeckich (Historical outline of the management in Sudetic forests)*, Sylwan, CXXVII, 49, 1982 (in Polish).

Worrel, J., Decline diseases, at http://www.forestpathology.org/decline.html, 2004.

Zahradnik, P., Ochrana lesa — abioticke a bioticke factory současny stav a prognoza v oblasti Sudet (Forest protection — abiotic and biotic factors — the present state and prognosis in the area of the Sudeten), in *Protection of Forest Ecosystems. Selected Problems in Sudeten Mountains*, Paschalis, P. and Zajączkowski, S., Eds., Grant GEF 05/21685 POL Forest Biodiversity Protection Project, Warszawa, 274, 1997.

Zajączkowski, J., *Biogrupy drzew w drzewostanach — możliwości i celowość ich wykorzystania przy prowadzeniu trzebieży (Tree biogroups in forest stands — possibilities and rationale for their use at thinnings)*, Prace Instytutu Badawczego Leśnictwa, Ser. A, 778/782, 5, 1994 (in Polish; English summary).

Zimka, J. R. and Stachurski, A., Forest decline in Karkonosze Mts. (Poland). Part II. An analysis of acidity and chemistry of atmospheric precipitation, throughfall and forest streamwaters, *Ekologia Polska*, 44, 153, 1996.

chapter thirty-four

Reclaiming peatlands for forestry: the Irish experience

Florence Renou and E. P. Farrell
University College Dublin

Contents

34.1 Introduction

Lying between 51° and 55° north latitude, the Republic of Ireland covers an area of 70,280 km^2. The country is made up of a broad north central plain enclosed by coastal highlands. Its 3.9 million inhabitants enjoy a mild temperate climate due to the influence of the Gulf Stream, characterized by cool summers, mild winters, high amounts of cloud, high atmospheric humidity, and windiness. Ireland belongs to the temperate deciduous

1-56670-635-1/05/$0.00+$1.50

forest biome. While the country's flora includes a small number of native species, mostly broadleaves, the climate is conducive to the growth of a wide range of exotic species.

34.2 Overview of Irish forestry

34.2.1 Current state

Following thousands of years of exploitation, by the end of the 19th century the forests of Ireland covered only 1% of the land area. The 20th century witnessed the gradual restoration of this lost resource. Despite a very high level of planting in recent years, Ireland has the lowest proportion of land under forest in the European Union (EU), at 9.4% or 610,000 ha (Forest Service 2000). The rate of afforestation peaked at 20,000 ha in 1996 and the annual rate has since ranged from 12,600 to 15,700 ha (ITGA 2002). The government ambitiously plans to double the forest area to 1.2 million ha (i.e., 17% of the land area) by 2030 (Department of Agriculture Food and Forestry 1996). The main impetus for afforestation is the financial incentives provided by the European Union (EU) and the Irish Government. These grants and premiums are inflating the price of land for forestry.

Ireland has a young, intensively managed forest estate, of which 70% is in public ownership, reflecting the dominant role of the state in forestry development. The balance of ownership is changing however, as the private sector (farmers in particular) becomes increasingly involved (Gillmor 1998). The sector is growing at some 10% per annum and employs about 16,000 people. Ireland has moved from a position of timber deficit to one of surplus in less than 100 years.

Ireland has only four native tree species of commercial value. Eighty percent of the forests established to date are exotic conifers, planted in monocultures or mixtures. Half of these plantations are Sitka spruce (*Picea sitchensis* [Bong] *Carr.*), a commercially attractive species that is capable of higher yields over a wider range of site types than other species. The emphasis on quick-growing softwoods reflects the favorable conditions of the Irish climate for these species; growth rates in Ireland are among the highest in Europe. Timber production is still the prime business of Irish forestry. Plantations are managed through planting and thinning; stands are dense and rotations are short (the crop is clearfelled from 35 years of age, depending on species and growth rate). As a result, Irish coniferous forests rarely develop to the open character of mature forests.

The level of broadleaf afforestation was very low up to the early 1990s, averaging less than 500 ha (3 to 4%) of total annual afforestation. Financial incentives brought that level to 20% in 1995, but it has steadily decreased to 13% in 2000 (ITGA 2002). Pockets of seminatural woodland survive around the country and these receive some protection under different schemes. State forests planted on peatlands account for 44%, but the proportion on wet mineral soil has steadily increased.

34.2.2 Historical framework

Temperate deciduous forests once covered a large part of the Irish landscape and from earliest times the forests were exploited. Early migrants came by sea and were both boat builders and craftsmen of some skill. A Viking ship discovered at Roskilde, Denmark, has been shown by tree-ring analysis to be constructed of oak from Ireland (probably Dublin) in 1059 (Olsen and Crumlin-Pedersen 1978). In the 17th century, Sir Arthur Chichester wrote that "the Irish build very good ships and many English merchants choose to build there. Their oak is very good and they have a very good store of it." Another author of the same era claimed that "the forests of Cork and Kerry in the southwest of Ireland were used to cask the wine that France and Spain produced" (McCracken 1971). Exploitation became destruction under the Tudors as the resource was liquidated to build ships for the English Navy. Ireland's remarkable

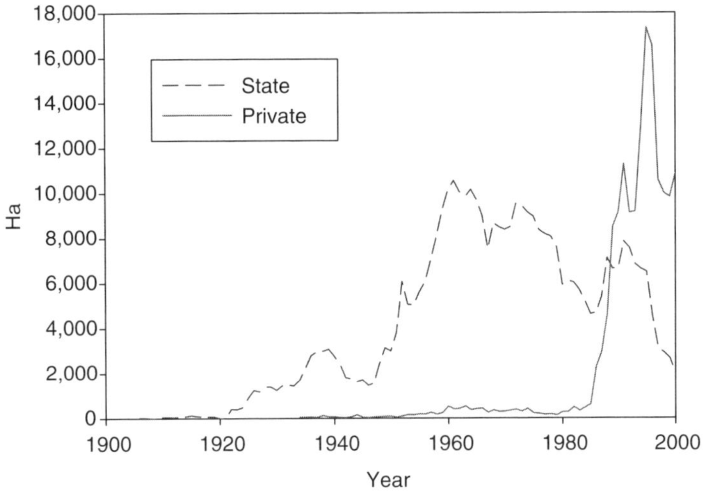

Figure 34.1 Annual planting rate on public and private land in the Republic of Ireland.

demographic and settlement history from the 17th century onward ensured that by the mid-19th century it was among the countries most denuded of timber. According to one 18th-century French observer (de la Latocnaye 1985), "they had not left wood enough to make a toothpick." At the beginning of the 20th century, the total forest area of Ireland was under 50,000 ha or less than 1% of the land area (Neeson 1991).

Historians may one day view the 20th century as the restoration phase of forests in Ireland. Paradoxically, because of the earlier devastation, experimentation was feasible and new techniques could be used without disrupting established traditions. Hence the early interest in conifers. Afforestation began with the establishment of the Forest Service in 1903 and planting increased with passage of the Forestry Act of 1919 (Figure 34.1). Afforestation policy had only one objective: to produce timber. Afforestation relied almost totally on conifers, on the grounds that short-rotation softwood forests would have a greater commercial appeal than the slow-growing, native hardwood, for which demand was weak at that time.

Expansion of forestry was limited by the scarcity of suitable and affordable land. With the Great Famine of the 1840s still haunting memories, priority was given to food production and agricultural crops. Forestry was not perceived to have a role in this scheme and it rapidly became accepted policy that no land fit for agriculture should be afforested. Thus, in the absence of incentives, private landowners refrained from planting forests. The restoration of Irish forests became synonymous with state forestry, which from the early 1950s extended onto very poor soils described as agriculturally "marginal." Foremost within this category were the peatlands.

34.3 Role of peatlands in the afforestation of Ireland

34.3.1 Irish peatlands

After Finland, Ireland has the highest proportion of peat cover of any EU country, with 17% or 1.3 million ha (Hammond 1981). There are two main peatland types in Ireland: fens and bogs. The area of undisturbed fens is small: most were drained and reclaimed for agricultural use long ago. Bogs, or ombrotrophic peatlands, can be subdivided into *raised* and *blanket* bogs.

34.3.1.1 Raised bogs

At the end of the last glacial period 10,000 years ago, shallow lakes left by the retreating ice covered much of central Ireland. Nutrient-rich groundwater derived from calcareous glacial drift fed these lakes (Mitchell and Ryan 1997). Reeds and sedges encroached around the lake edges; their remains fell into the water where they were only partly decomposed, in time forming a thick layer of reed peat. The upward growth of sedges and other plants (bog mosses) and the accumulation of their undecayed remains elevated the bog surface above the level of the surrounding countryside. The result was the deepest raised bogs in Ireland, averaging 6 to 7 m, which were particularly suitable for peat exploitation. Raised bogs are found mainly in the midlands under moderate rainfall, between 750 and 1,000 mm per year. The surface is strongly acid, but beneath the peat are alkaline moraine soils of limestone origin. Raised bogs are *Sphagnum* dominated, with heather (*Calluna vulgaris* [L.] Hull), bog cotton (*Eriophorum angustifolium* Honckeny), and several species of sundew and orchids also present.

34.3.1.2 Blanket bogs

Blanket bogs developed about 4,000 years ago; they are found chiefly in the west of Ireland and on mountaintops throughout the rest of the country. They are most widespread in areas where annual rainfall levels are greater than 1,200 mm and the number of rain days exceeds 225. These bogs are shallow and form a blanket-like layer averaging 2.5 m in depth over an underlying acidic mineral soil. In their natural condition, these areas are dominated by *Eriophorum* species, black bog rush (*Schoenus nigricans* L.), and purple moor-grass (*Molinia caerulea* [L.] Moench).

34.3.1.3 Peatland utilization

Irish bogs are mainly open, windswept, wilderness areas virtually devoid of tree cover due to the wet weather conditions that have prevailed in recent centuries. It is to the past, to periods when bog growth was slower and the surface drier, that one must look to find Irish peatlands covered with pine. Dendrochronology and radiocarbon dating have shown that this forest resource existed for a period of at least 500 years sometime between 2,500 and 1,800 BC. In time, bogs grew and engulfed these ancient forests under several meters of peat. These buried trees were a valuable resource in the 17th century, when standing timber became scarce. The remnants of pine forests can still be seen on harvested peatlands and are sometimes used for firewood, construction, and wood sculpture.

Peat has been a source of fuel since prehistoric times. In the 19th century, peat was the only fuel available to a population that had expanded to more than 8.2 million. The introduction of large-scale mechanized peat extraction in the 1940s and the establishment of the Irish Peat Board (Bord na Móna) as a quasi-state body in 1946 were direct results of government policies introduced to deal with the increasing difficulties associated with importing fuel supplies. By developing Irish peatlands, the government not only redressed the trade imbalance but also stimulated rural economies where the bogs were located.

34.3.2 Birth of peatland forestry and afforestation of blanket bogs

Unlike in Fenno-Scandinavian countries, wetland forestry was not popular in Ireland and little had been done to afforest peatlands prior to the 1940s. Fuel production was still seen as the priority for the raised bogs of the midlands. However, in the early 1950s, the government decided to increase the rate of afforestation significantly. Blanket peatland was seen to offer the possibility of large-scale plantations on treeless land that was virtually unused, apart from some extensive grazing. The increased availability of inexpensive phosphatic fertilizers and the introduction of tracked vehicles that could pull large, heavy-duty ploughs made blanket peatland forestry possible. Peatland afforestation became significant: the annual area planted grew from 225 ha in 1950 to 3,750 ha in 1960, or from 10 to 32% of the annual total. Although probably inaccurate, 200,000 ha of forestry was estimated on peatland in Ireland in 1990, the greater part being on blanket peat in the west of the country (Farrell 1990b). Forestry moved onto blanket peatlands with some difficulty; blanket peat is extremely acid and has very low fertility. Phosphorus is in particularly short supply and nitrogen reserves are often much smaller than in mineral soils. Blanket peat has a high resistance to water movement (saturated hydraulic conductivity of less than 1 cm per day), an obstacle to drainage (Galvin 1976; Gleeson 1985). Blanket bogs can be subdivided by altitude into low-level and high-level (montane) bog (Hammond 1981), which is also relevant to silviculture. Plantations on higher grounds are more prone than low-level plantations to acute nutritional problems, exposure, and higher risk of windthrow (Dickson and Savill 1974; Farrell and McAleese 1972; Farrell and Mullen 1979; Gallagher 1974; Schaible 1992). Nonetheless, the productivity of some high-level bogs and most low-level blanket bogs has greatly exceeded expectations, with yield class estimates for Sitka spruce at 13 (yield class is an index of potential maximum mean annual volume growth increment expressed in m^3 ha^{-1}), and for lodgepole pine (*Pinus contorta* Dougl.) at 10 (Farrell and Boyle 1990). These figures represent a remarkable success for Irish forestry as they indicate a much higher level of productivity than was considered possible at the start of the blanket peatland afforestation program (Anon. 1951). The economic viability of the enterprise, however, has always been marginal despite the relatively high productivity achieved and the availability of EU and state grants.

Management techniques have been developed to overcome establishment problems; yet, government and public support for afforestation of blanket peatland has declined, not only on economic grounds but, more importantly perhaps, because of its potential environmental and aesthetic impacts. Drainage, aerial fertilization, and clearfelling have direct impacts on surface water quality (Cummins and Farrell 2000, 2003), which is a serious issue as blanket peatlands are located where most streams are of particularly high quality for salmonid populations. The oxidation and subsidence of drained peat cast doubt over the long-term sustainability of afforested ecosystems (Byrne and Farrell 1997). Attention has also focused on the aesthetics of planting trees in large open areas such as peatlands and on the wildlife value of peatland habitats. Blanket peatlands are increasingly appreciated as a natural resource and a unique landscape (Foss et al. 2001). In certain cases, rehabilitating the peatlands may be more appropriate than reforesting after clearfelling. Although the Forestry Act of 1946 currently stipulates that these sites should be replanted, the Forest Service realizes that replanting many sites with the same species may not be sustainable. Research into peatland restoration and natural regeneration of lodgepole pine in western blanket peatlands has been initiated and will be integrated in further discussions on the future of blanket peatland forestry.

34.4 Prospects for cutaway peatland forestry

Prospects are limited for further forestry development on both blanket and intact raised peatlands. Smaller areas of raised peatlands were planted at the same time as blanket peatlands in the 1950s, but current government policy does not grant-aid the afforestation of intact raised bogs. National and international efforts are under way to conserve representative samples of these ecosystems. Greater potential exists for afforestation of industrial cutaway peatlands (Table 34.1). Since the late 1950s, afforestation has been considered a feasible option for cutaway areas in several countries (Ferda 1972, 1975, 1986; Kaunisto 1986; Mikola and Mikola 1958; Nilsson 1992; Nilsson et al. 1987). In Ireland, commercial afforestation of cutaway

Table 34.1 Irish Peatlands and Their Forestry Potential

	Raised Bogs	Blanket Bogs		Cutaway Peatlands	
		High Level	Low Level	Sod Peat	Milled Peat
Approximate area (ha)	311,400	565,000	339,000	10,000	90,000
Formation	Deep bog dominated by heather species	Shallow bog dominated by *Eriophorum* species		Peat has been removed from a vertical face, leaving behind sphagnum strippings (up to 0.5 m) over 1 m of minerotrophic peat	Peat has been extracted on the horizontal plane, leaving a varying depth of reed peat over an undulating alkaline mineral soil
Afforestation potential	None	200,000 ha planted; no more State afforestation		1,000 ha planted; most sod peat bogs subsequently have been milled	3,200 ha planted and potential for another 50,000 ha
Yield class (m^3 ha^{-1} $year^{-1}$)		Sitka spruce 13 Lodgepole pine 10		Sitka spruce 18–24 Lodgepole pine 18–20	Potential species: Norway spruce, hybrid larch, Scots pine, alder, birch, oak, Sitka spruce under nurse species
Forestry issues		Windthrow Nutrition Access Landscape aesthetics		Nutrition Waterlogging	Late spring frost Nutrition Competing vegetation Waterlogging

peatlands commenced in the 1980s. In theory, prospects for this type of peatland forestry were promising. Some foresters described these cutaways as homogeneous in botanical origin and in chemical and physical characteristics: drainable and available in large accessible units (up to a thousand ha). These peats were perceived as very productive without any requirements for cultivation. Twenty years' experience shows that this was overly simplistic, and afforestation of cutaway peatlands is a delicate exercise that has so far yielded only mediocre results.

34.4.1 *Reclamation of industrial cutaway peatlands*

The term "industrial cutaway" refers to large areas of former peatlands where peat harvesting for fuel has ceased. Peat harvesting is well developed in Ireland and the largest production company, Bord na Móna, currently manages some 88,000 ha. The harvested peat is used for horticultural products and for energy. Currently, milled peat produces approximately 10% of the Republic of Ireland's electricity requirements, supplying six peat-fired power stations.

Two types of cutaway peatlands can be found in Ireland. The first type is the result of an old method of harvesting peat using the sod-peat extraction process by which peat was removed from a vertical trench (Figure 34.2). Described as the Clonsast Complex (named after the area where it is most commonly found in Ireland) (Hammond 1981), it consists of a disturbed layer of sphagnum strippings overlying a shallow layer of minerotrophic peat. Another method removes deeper peat layers using the milling process (Figure 34.3), where all the peat is extracted except for a residual layer varying in thickness from 0 to over 1 m. Most milled cutaway peatlands are classified as Boora Complex and have a varying depth of peat. This residual peat is generally minerotrophic, nonuniform, classified on the basis of included plant fossils. Boora Complex peats can be a woody fen peat, frequently corresponding with the occurrence of well-developed sub-peat mineral soils. However, the bulk of cutaway peat available for future use is a lake- or reed-peat dominated by *Phragmites australis* (Cav.) Trin., which is found on the deep contours of the bog floor. The peat remaining after the milling process is quite dense, having been compacted for thousands of years under several meters of peat and then for many seasons under heavy machinery. Peat depth can vary greatly over short distances; in cutaway peatland greater than 60 cm in depth, hydraulic conductivity and gas diffusion are low. The pH of the peat varies from 4.8 to 5.8 (Renou 2000; Walsh and Barry 1958). The bottom layer of a cutaway peatland is generally infertile, containing small amounts of mineral nutrients and

Figure 34.2 Sod peat extraction system, also called the bagger method.

Figure 34.3 Milled peat extraction system, also called the Peco milling system. Bord na Móna developed this system.

usually extremely low levels of phosphorus and potassium. Unlike bogs in Finland (Aro and Kaunisto 1998) or Germany that overlie acidic soils, the substrata of raised bogs in the Irish midlands are generally alkaline, varying greatly from pH 7 (silty clay) to pH 9 (marl) (Barry et al. 1973). The challenge to restoration of these cutaway ecosystems is illustrated by their slow recolonization after abandonment.

Industrial harvesting generally ceases for three main reasons:

1. Woody remains in the lower strata impede the removal of the peat by the milling process.
2. The subsoil becomes exposed; fuel production requires no contamination by any mineral subsoil.
3. The site becomes too wet; in some areas, water is pumped away to maintain the water table below the peat surface.

Within the next three to four decades, large areas of industrial cutaway peatland will become available for other land uses. Several options are open, depending on a variety of factors, including the properties of the remaining peat type, its thickness, the geomorphology of sub-peat mineral soils, and hydrological conditions. These options include naturally regenerated ecosystems (including wetlands), grassland, or forestry. Declining demand for grassland as a result of Common Agriculture Policy reform, which favors voluntary set-aside and a decrease in land under production, shifted focus toward forestry as the major after-use for the 88,000 ha of industrial cutaways. Approximately 50,000 ha are likely to become available for afforestation, with the potential to account for 10% of the national planting program over the next 30 years. The phased replacement of peat exploitation by forestry and downstream forest industries is likely to be of crucial importance for economic and social reasons within the Irish midlands.

34.4.2 *Initial research into forestry on cutaway peatlands*

Industrial cutaway peatlands are significantly different from all other types of peatlands (Figure 34.4). Their afforestation thus gives rise to problems different from those of blanket peatland forestry. Research into the afforestation of cutaway peatlands began in the 1950s and focused on sod-cutaways. Results from one experiment (Trench 14) showed that

Figure 34.4 Cutaway peatland in Ireland newly afforested with Norway spruce (*Picea abies* [L.] Karst.).

several coniferous species performed satisfactorily on mixed-peat types varying between 0.5 and 1.5 m in depth. Leading species in terms of yield class were grand fir (*Abies grandis* Lindl.) at 24, Sitka spruce at 18 to 24, and Lodgepole pine at 18 to 20. Good yields and low establishment and management costs (no cultivation was performed on this free-draining site) produced a rate of return of almost 11% (Carey et al. 1985). Unfortunately, results from sod-cutaway cannot be applied to industrial cutaway peatlands that were milled. Based largely on these optimistic projections, however, state forests leased 4,000 ha of cutaway peatlands from Bord na Móna in the 1980s. These vegetation-free sites were planted mainly with Sitka spruce, but also Lodgepole pine without any soil preparation. Bare-root seedlings were slit-planted by hand and fertilized immediately with unground mineral phosphate (350 kg ha^{-1}) and muriate of potash (250 kg ha^{-1}). Results from three surveys in the early 1990s found that a third of the trees were of acceptable size and form, but the rest were small and stunted and unlikely to form a commercial crop (Jones and Farrell 1997a, 1997b; Lynch and McGuire 1993).

34.5 BOGFOR: finding ways to afforest industrial cutaway peatlands

Following the widespread mortality and poor growth of trees planted on milled cutaway peatlands in the 1980s, the BOGFOR Research Programme was initiated in 1998 to focus on the specific problems of milled-cutaway peatlands. Research objectives are to find ways to establish and develop a successful forest resource on the cutaway peatlands of the Irish midlands, providing appropriate silvicultural options to ensure optimum forest productivity compatible with environmental imperatives. The program is multidisciplinary, investigating a range of issues such as cultivation methods, species selection, drainage, vegetation control as well as edaphic and climatic limitations. The results of BOGFOR will assist the Forest Service in drafting a code of best forest practice for the successful establishment of plantations on cutaway peatlands.

34.5.1 Industrial cutaway peatlands: a challenging ecosystem

Poor growth of trees on cutaway peatlands is due to a combination of several factors stemming from the complexity of these degraded ecosystems, but falling under two headings: climatic limitations and imperfections of peat as a substrate for tree growth. Combinations of these factors develop further difficulties with vegetation competition and species suitability.

34.5.1.1 Climatic conditions

Temperatures recorded in the Irish Midlands over the period 1969 to 1990 show that January is the coldest month with an average of 4.1°C. July and August are the warmest months, being on average 14.5°C. Because of their low-lying position in the landscape, cutaway peatlands are susceptible to late spring frost. Temperatures as low as −3°C occur regularly and frosts of −5°C are recorded as late as the end of June (Green et al. 2002). Freezing temperatures at the time of budding and early shoot growth are a major barrier to the successful establishment of Sitka spruce; 50% of trees planted in the 1980s suffered severe frost damage in the form of lateral or leading shoot dieback (Figure 34.5).

Preliminary findings show that vegetation cover and localized frost risk are positively correlated, suggesting a number of establishment and management techniques such as vegetation control or cultivation prior to planting that may reduce the risk of frost damage to trees (Green et al. 2002; Smith et al. 2000). Use of nurse species such as birch (*Betula* spp.) has been suggested (Paavilainen and Päiviänen 1995; Smith et al. 1998) and is being investigated at demonstration scale, in addition to planting Sitka spruce under an existing naturally regenerated birch shelterwood system. Norway spruce (*Picea abies* [L.] Karst) and other species may have less sensitive provenances that can be matched with climatic

Figure 34.5 Columnar shape of a Sitka spruce (*Picea sitchensis* [Bong.] Carr.) is due to late spring frosts that have repeatedly killed lateral shoots.

conditions offering earlier or later dates for bud-burst, thus minimizing the risk of frost damage (Aiken and Hannerz 2001).

34.5.1.2 Soil moisture regime and water table levels

Soil moisture status is critical to the successful establishment of young trees. Both too little and too much water can limit the physiological processes taking place during establishment. Despite drainage for peat production, waterlogging difficulties occur regularly in cutaway peatlands. The water table is noticeably high and fluctuates rapidly (Figure 34.6). A high water table may not kill standing trees but it can stress them and make them more vulnerable to disease. A fluctuating water table can also retard tree growth dramatically, favoring competitors such as *Juncus* that thrive under these conditions. High soil moisture content also impairs mineralization and the release of nutrients. Water management in a typical milled cutaway bog involves dividing it into a series of peat fields that are 15 m wide, separated by a ditch of 1 m width and depth. This drainage network usually deteriorates rapidly after peat harvesting stopped.

Cleaning and deepening the existing ditches are necessary before site preparation for afforestation. Additional ditches may be required to remove the excess water from hollows. Because local conditions are so variable, no precise drainage method can be applied.

In deep cutaway bogs, wet conditions can result in oxygen deficiency in the rooting zone. Site preparation techniques that disturb peat can induce better soil water drainage (Figure 34.6). The natural density of the peat is often higher than is optimal for seedling establishment. Cultivation can improve both the hydraulic conductivity and bulk density of the planting medium. Removing excess water from the rooting zone can also be achieved by creating an elevated planting spot or mound. The Phragmites peat type is unsuitable for mounding, as it is susceptible to cracking and irreversible drying, thus promoting the rapid disintegration of the mound. Ripping and disking techniques are preferred as they help to aerate the medium and also create extra water channels.

An impermeable layer between the peat and the sub-peat mineral soil can exacerbate drainage problems and create long-term surface ponding. Cultivating the peat may not improve these wet conditions but the water table will recede when regenerating vegetation transpires enough to draw the water table down. In addition to *Juncus* spp., other species such as birch, willow (*Salix* spp.), and alder (*Alnus* spp.) can establish relatively quickly on some of these wet sites. Typically seen as competitors, these species are adapted

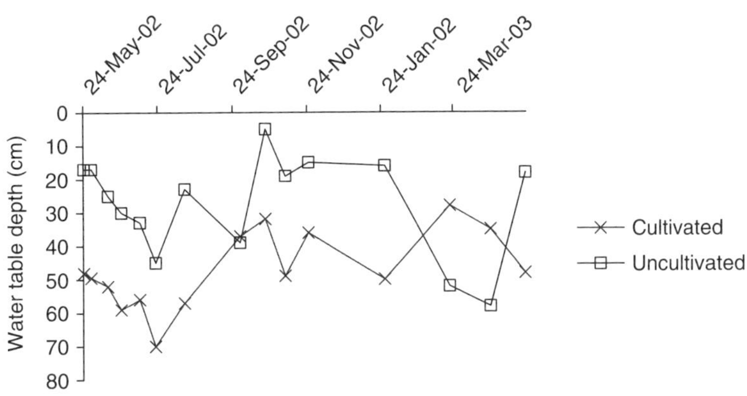

Figure 34.6 Effect of cultivation on water table levels in a newly afforested cutaway peatland, Republic of Ireland.

to growing on wet sites and may benefit underplanted conifer seedlings, lowering the water table through transpiration.

34.5.1.3 Soil nutrient regime and tree nutrition

Most of the remaining peat present on cutaway bogs is infertile and acidic (pH 3 to 5). Fertilization is vital to tree survival at the establishment phase. Within the BOGFOR Program, soil and foliar analysis methods have been used in the ongoing nutritional diagnosis of 1980s Sitka spruce peatland forests, which so far have been below critical thresholds for most macro- and micronutrients (Jones and Farrell 1997a, 1997b, 2000). Deficiencies in nitrogen, phosphorus, potassium, and copper (N, P, K, Cu) were common across different species.

The current Code of Best Forest Practice (Collins et al. 2000) recommends applying 42 kg-P ha^{-1} on afforested cutaway peatland, in the form of granulated rock phosphate. As P is mobile in peat soils, foresters have to consider the potential impact on local aquatic systems. Phosphorus movement from a recently planted and manually fertilized plantation showed an immediate loss of P in surface runoff water (Renou et al. 2000). Attempts to reduce the impact by splitting the application of phosphatic fertilizer into half of the recommended amount at planting and the other half in year 3 significantly decreased P concentration in runoff. This difference became less significant over time, however, because the loss of P depends more on retention capacity of the substrate than the level applied (Renou and Farrell 2001). Splitting the application did not have an impact on the crop trees but reduced the level of vegetation competition, especially on deep peat. Although P concentrations in runoff water clearly increase following fertilization, the complex hydrology of downstream water bodies makes it difficult to assess the real impact of this change. Cultivating the surface of the peat by ripping and disking has shown decreased P concentrations in runoff water and mixing sub-peat mineral soil with the peat will reduce the amount of P leaving the root zone while allowing satisfactory tree growth (Renou and Cummins 2002). New methods of applying and placing phosphatic fertilizers are currently being tested.

Sitka spruce cutaway peatland forests showed K and N deficiencies five years after planting that continue to deteriorate (Figure 34.7). Whether to apply K at planting is still uncertain. Evidence from field observations and foliar analyses justifies application of K before it limits tree growth, that is, before year 4 or 5 (Carey 1977; Jones and Farrell 1997a).

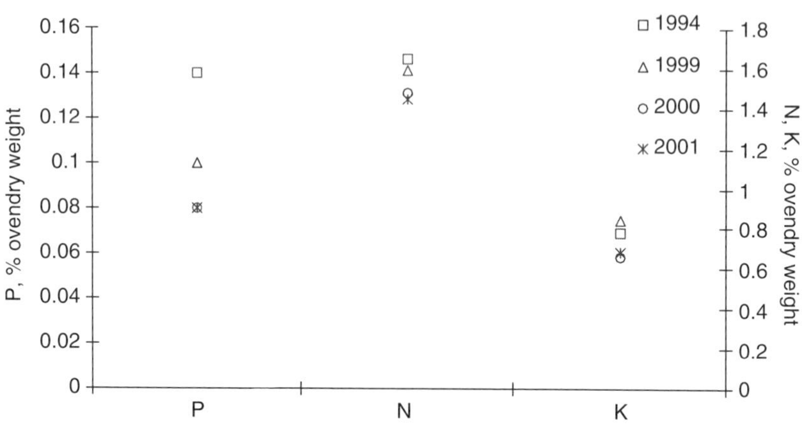

Figure 34.7 Declining nutrition over time of Sitka spruce (*Picea sitchensis* [Bong.] Carr.) planted in 1989 on a cutaway peatland. Foliar concentrations of phosphorus, potassium, and nitrogen show declines over time, which is most pronounced for phosphorus.

Nitrogen can be deficient on these sites (O'Carroll 1962), but fertilization with urea is carried out only if N alone is deficient. If both N and P are deficient, application of P will not only overcome that deficiency but will increase the N mineralization in acid peat and uptake of N by trees (Farrell 1990a; Kaila 1958; Kaunisto 1997). Added N is not essential for establishment on cutaway peatlands except in very high water table environments where anaerobic conditions impede the release of organically bound N in the peat (Adams 1974). Drainage and adequate site preparation may be sufficient to meet N requirements by promoting mineralization of organically bound N through increased microbial activity (Päiviänen and Paavilainen 1996).

A less costly alternative to fertilizers is the use of native pioneer species sown or planted on bare cutaway bogs prior to the commercial crop or in mixture. Alder can contribute to site N capital through symbiotic N fixation (Simard 1996; Wheeler et al. 1986). Birch in mixture with Norway spruce has a positive effect on the tree nutrient status, promoting total stand productivity in the long term (Thelin et al. 2002). These hardwood species supply leaf litter that may improve the nutrient cycle by increasing the rate of litter decomposition (Fyles and Fyles 1993).

34.5.1.4 Vegetation competition

Industrial cutaway peatlands are usually bare sites and competing vegetation is not a problem initially. Depending on environmental conditions, natural invasion will occur more or less rapidly after P fertilizer is applied, and competition from ground vegetation can contribute to the poor performance of some species, particularly Norway spruce (Parr 2000). Acid peat together with high water table levels favor the invasion of *Juncus* species (Pöllänen and Renou 2002). Chemical weed control is less effective on peat than on mineral soil because of high organic matter content with long-term absorbent capacity. Spraying methods are also limited by windy conditions. The use of specially designed equipment such as flail mowers and roto-wipers is successful if carried out at the right time. Nonetheless, invasion by *Juncus* species in very wet areas is particularly difficult to overcome and can only be dealt with manually (McCorry and Renou 2003).

34.5.1.5 Species suitability

Stand establishment by sowing or natural seeding is not possible on cutaway peatland due to wind exposure and the remoteness of tree seed sources. Afforestation is based on 2 to 4 year-old bare-root nursery stock, manually slit-planted. Planting density varies depending on species; the normal density for conifers is approximately 2,500 ha^{-1} to 6,000 ha^{-1} for broadleaves. Single-species plantations have been favored. Fencing against hares and deer is necessary when planting broadleaves and may be considered for conifers if hare populations are high. The Irish hare (*Lepus timidus* L.) is a protected species in Ireland and is considered to be the only mammal that truly belongs to the bog (Feehan and O'Donovan 1996). Cutaway peatland is its preferred habitat.

Over 150 ha of species trials covering 13 coniferous species and 10 broadleaf species have been established within the last six years. It is still too early to draw up a list of the most suitable species to plant on industrial cutaway peatlands. Sitka spruce has been favored, but it is highly susceptible to late spring frost, and a return to lower-yielding Norway spruce may be advisable. Norway spruce is nutritionally demanding, but produces better quality timber that may yield a premium over Sitka spruce. The resilience and tolerance of Norway spruce, as well as its potential to produce reasonable yields, mark it out as a promising candidate. Survival rates for Norway spruce can be reduced in very exposed sites but a shelterbelt can greatly enhance its growth during the establishment phase. Until recently, Lodgepole pine was the second favorite species. Although considered a "low-risk" choice, the incidence of pine shoot moth in the Irish Midlands has moved pine down the list of suitable species. Its susceptibility to basal sweep (curving at the base of the tree) also suggests that its use be reevaluated. In free-draining, frost-free cutaway sites, hybrid larch (*Larix × eurolepis* A. Henry) has been another successful exotic conifer. Native broadleaves, such as pedunculate oak (*Quercus robur* L.) and common alder (*Alnus glutinosa* [L.] Gaertn.), have shown unexpected potential. Although oak seems to suffer from leader dieback on very exposed sites, height and leader growth were greatly enhanced where trees were underplanted in an old recolonized birch cutaway bog. Birch grows extremely well in some cutaway sites, but it is a species of minimal economic value in Ireland. Not all forests established on cutaway bogs will be commercial, however, and two native birch species (*Betula pendula* Roth and *B. pubescens* Ehrh.) are of great interest, firstly as a timber crop, secondly as a nurse species for frost protection, and thirdly in mixture with conifers for biodiversity value.

34.5.1.6 Environmental issues

The afforestation program of the cutaway peatlands is one of the largest reclamation projects in Ireland and as such will be critically evaluated in a wider environmental context. The capacity of forests to influence the greenhouse gas balance has come into sharp focus, especially in the context of Ireland's significant current economic growth and accompanying increase in emissions. The role of peatland forests, and especially cutaway peatland forests, in sequestering atmospheric CO_2 remains uncertain. Following drainage for afforestation, the depth of the aerobic layer in peat increases and methane emissions cease, but decomposition rates and soil CO_2 emissions increase (Byrne and Farrell 1997, 2000; Waddington et al. 2002). Nevertheless, these losses may be offset by CO_2 sequestration in the forest crop (Byrne and Farrell 2000). Afforested peatlands in Scotland have been shown to accumulate more carbon in litter, soil, and forest products than is lost from the peat for 90 to 190 years (Hargreaves et al. 2003). However, future management of cutaway peatland forests, such as rotation lengths, clearfelling, reforestation, and site preparation, will impact upon carbon storage and the greenhouse gas balance.

The first impression of a cutaway peatland is often a bleak one: a flat, bare, windswept dark desolate area lacking microtopography. The replacement of these areas with forests will affect the aesthetics and biodiversity of cutaway landscape of the future. The promotion of biodiversity has become central to the implementation of sustainable forest management, which includes the cutaway peatland afforestation program. There is a strong case to be made for a broadening of the range of tree species planted in Ireland; if plantations on cutaway peatlands were to be considered for objectives other than commercial timber production, then the range of tree species and their associated biodiversity would be considerably wider.

34.6 Conclusions

Conventional establishment techniques have limited applicability and resulted in variable success on cutaway peatlands. No single reclamation measure will be appropriate to all cutaway areas, and successful afforestation requires an understanding of site variation. Soil nutrient and moisture regime, as well as aeration, must be carefully considered in afforesting a cutaway peatland. An extensive ground survey to establish peat type and depth, as well as the nature of the underlying mineral soil, is arguably necessary before planting decisions are made. The need for site preparation including drainage, mechanical cultivation, and fertilization is evident, although the necessary intensity of such operations is not yet clearly established. Species and provenance must be carefully matched with the climatic and soil characteristics of the particular cutaway peatland site. Well-planned forestry, with best management practices as an integral part of management, allows for good commercial production without compromising environmental quality. At the end of the day, the success of these new forests will be measured not only by how well they meet the objective of financial profit but also by how well they meet social, aesthetic, and environmental needs.

Acknowledgments

The BOGFOR Research Programme has received funding from the Department of Communication, Marine and Natural Resources (COFORD, National Development Plan), Bord na Móna and Coillte Teoranta.

References

Adams, S.N., Some relations between forest litter and growth of Sitka spruce on poorly drained soils, *J. Appl. Ecol.*, 11, 761, 1974.

Aiken, S.N. and Hannerz, M., Genecology and gene resource management strategies for conifer cold hardiness, in *Conifer Cold Hardiness*, Brigras, F.J. and Colombo, S.J., Eds., Kluwer Academic Publishers, Dordrecht, 23, 2001.

Anon., *Report on Forestry Mission to Ireland*, Government Publications, Dublin, Ireland, 1951.

Aro, L. and Kaunisto, S., Nutrition and development of 7–17 year old Scots pine and silver birch plantations in cutaway peatlands, in *Peatland Restoration and Reclamation: Techniques and Regulatory Considerations; Proceedings of the 1998 International Peat Symposium*, Malterer, T., Johnson, K., and Stewart, J., Eds., Duluth, Minnesota, 109, 1998.

Barry, T.A., Carey, M.L., and Hammond, R.F., *A Survey of Cutover Peats and Underlying Mineral Soils, Bord na Móna Cnoc Cioluin Group*, Bord na Móna and An Foras Taluntais, Dublin, Ireland, 1973.

Byrne, K.A. and Farrell, E.P., The influence of forestry on blanket peatland, in *Humic Substances, Peat and Sludges*, Hayes, M.H.B. and Wilson, W.S., Eds., The Royal Society of Chemistry, Cambridge, 262, 1997.

Byrne, K.A. and Farrell, E.P., Greenhouse gas balances in peatland forests, in *Carbon Sequestration, Policy, Science and Economics, Proceedings of a COFORD seminar on Carbon Sequestration and Irish Forests*, Hendrick, E. and Ryan, M., Eds., Dublin, Ireland, 15 June 2000, 7, 2000.

Carey, M.L., Nutritional disorders in Sitka spruce in the Republic of Ireland, *Ir. For.*, 34, 40, 1977.

Carey, M.L., Hammond, R.F., and McCarthy, R., Plantation forestry on cutaway raised bogs and fen peats in the Republic of Ireland, *Ir. For.*, 42, 106, 1985.

Collins, K.D., Gallagher, G., Gardiner, J.J., Hendrick, E., and McAree, A., *Code of Best Forest Practice — Ireland*, Forest Service, Department of Marine and Natural Resources, Dublin, 2000.

Cummins, T. and Farrell, E.P., Effects of forest operations on acid peatland streamwater composition, *Verh. Internat. Verein. Limnol*, 27, 1556, 2000.

Cummins, T. and Farrell, E.P., Biogeochemical impacts of clearfelling and reforestation on blanket peatland streams. I. Phosphorus, *For. Ecol. Manage.*, 180, 545, 2003.

Department of Agriculture, Food and Forestry, *Growing for the Future — A Strategic Plan for the Development of the Forestry Sector in Ireland*, Department of Agriculture, Food and Forestry, The Stationery Office, Dublin, Ireland, 1996.

Dickson, D.A. and Savill, P.S., Early growth of *Picea sitchensis* on deep oligotrophic peat in Northern Ireland, *Forestry*, 47, 57, 1974.

Farrell, E.P., Aspects of the nitrogen cycle in peatland and plantation forest ecosystems in western Ireland, *Plant Soil*, 128, 13, 1990a.

Farrell, E.P., Peatland forestry in the Republic of Ireland, in *Biomass Production and Element Fluxes in Forested Peatland Ecosystems*, Hånell, B., Ed., Umea, Sweden, 13, 1990b.

Farrell, E.P. and Boyle, G., Peatland forestry in the 1990s. 1. Low-level blanket bog, *Ir. For.*, 47, 69, 1990.

Farrell, E.P. and McAleese, D.M., The response of Sitka spruce to sulphate of ammonia and ground rock phosphate on peat, *Ir. For.*, 29, 14, 1972.

Farrell, E.P. and Mullen, G.J., Rooting characteristics of *Picea sitchensis* and *Pinus contorta* on blanket peat in Ireland, *J. Life Sci., Royal Dublin Soc.*, 1, 1, 1979.

Feehan, J. and O'Donovan, G., *The Bogs of Ireland*, The Environmental Institute, Dublin, Ireland, 1996.

Ferda, J., Fertilizing and nutrition of forest woody plants in exploited fen deposits, in *4th International Peat Congress*, Dublin, Ireland, 509, 1972.

Ferda, J., Forestry reclamation of extracted peat deposits, *Res. Inst. Land Reclam. Improv. Sci. Monogr.*, 6, 1, 1975.

Ferda, J., Reclamation and utilization of extracted peat deposits for forest planting, in *Socio-Economic Impacts of the Utilization of Peatlands in Industry and Forestry*, International Peat Society, Finn. Nat. Comm., Helsinki, Finland, 227, 1986.

Forest Service, *Irish National Forest Standard*, Stationary Office, Dublin, Ireland, 2000.

Foss, P.J., O'Connell, C.A., and Crushell, P.H., *Bogs and Fens of Ireland Conservation Plan 2005*, Irish Peatland Conservation Council, Dublin, Ireland, 2001.

Fyles, I.H. and Fyles, J.W., Interaction of Douglas-fir with red alder and salal foliage litter during decomposition, *Can. J. For. Res.*, 23, 358, 1993.

Gallagher, G.J., Windthrow in State forests in the Republic of Ireland, *Ir. For.*, 31, 154, 1974.

Galvin, L.F., Physical properties of Irish peats, *Ir. J. Agric. Res.*, 15, 207, 1976.

Gillmor, D.A., Trends and spatial patterns in private afforestation in the Republic of Ireland, *Ir. For.*, 55, 10, 1998.

Gleeson, T.N., Drainage of lowland peats and wet mineral soils, in *Peatland Production Seminar*, Lullymore, Co. Kildare, Ireland, 10, 1985.

Green, C., Renou, F., and Farrell, E.P., Late spring frost damage to Sitka spruce on industrial cutaway peatlands, in *Faculty of Agriculture Research Report 2000–2001*, University College Dublin, Dublin, Ireland, 2002.

Hammond, R.F., *The Peatlands of Ireland*, An Foras Taluntais, Dublin, Ireland, 1981.

Hargreaves, K.J., Milne, R., and Cannell, M.G.R., Carbon balance of afforested peatland in Scotland, *Forestry*, 76, 299, 2003.

ITGA, *Forestry Yearbook 2002*, Irish Timber Grower Association, Dublin, Ireland, 2002.

Jones, S.M. and Farrell, E.P., *Re-Survey of Plantation Forests on Bord na Móna Cutaway Bogs*, Report No. 23, Forest Ecosystem Research Group, Department of Environmental Resource Management, University College Dublin, 1997a.

Jones, S.M. and Farrell, E. P., *Survey of Plantation Forests on Bord na Móna Cutaway Bogs*, Report No. 20, Forest Ecosystem Research Group, Department of Environmental Resource Management, University College Dublin, 1997b.

Jones, S.M. and Farrell, E.P., *Research Programme to Develop a Forest Resource on Industrial Cutaway Peatland in the Midlands*. BOGFOR 3 Project — Phase 1. Final Report, Report No. 52, Forest Ecosystem Research Group, Department of Environmental Resource Management, University College Dublin, 2000.

Kaila, A., Effect of superphosphate on the mobilisation of nitrogen in a peat soil, *J. Sci. Agric. Soc. Finland*, 30, 114, 1958.

Kaunisto, S., Peatlands before and after peat harvesting, in *Socio-economic Impacts of the Utilisation of Peatlands in Industry and Forestry*, International Peat Society, Finn. Nat. Comm., Helsinki, 241, 1986.

Kaunisto, S., Peatland forestry in Finland: problems and possibilities from the nutritional point of view, in *Northern Forested Wetlands: Ecology and Management*, Trettin, C., Jurgensen, M.F., Grigal, D.F., and Jeglum, J.K., Eds., Lewis Publishers, Boca Raton, 387, 1997.

Latocnaye, Chevalier de la [1798], *A Frenchman's Walk through Ireland*, Blackstaff Press, Belfast, 1985.

Lynch, B. and McGuire, A., *Survey Report on Lands Leased From Bord na Móna*, Coillte Teoranta, Dublin, Ireland, 1993.

McCorry, M.J. and Renou, F., *Ecology and Management of* Juncus effusus *on Cutaway Peatlands in the Irish Midlands*, Report No. Forest Ecosystem Research Group Report Number 69, Department of Environmental Resource Management, University College Dublin, 2003.

McCracken, E., *The Irish Woods Since Tudor times*, David and Charles, Newton Abbot, 1971.

Mikola, P. and Mikola, I., Suon metsittäminen pollttoturpeen noston jälkeen (Reforestation of bogs after peat harvesting), *Suo*, 9, 44, 1958.

Mitchell, F. and Ryan, M., *Reading the Irish Landscape*, Town House and Country House, Dublin, Ireland, 1997.

Neeson, Eoin, *A History of Irish Forestry*, Lilliput Press and the Department of Energy, Dublin, Ireland, 1991.

Nilsson, T., Effects of drainage intensity and PK fertilisation on the growth and needle nutrient contents of Scots pine on a cutover peat area in west central Sweden, in *Proceedings of the 9th International Peat Congress*, International Peat Society, Uppsala, Sweden, 315, 1992.

Nilsson, T., Olsson, M., and Lundin, L., Markförbättring och intensive skogsproducktion på en torvmark (Soil improvements and intense forest production at a peatland), *Swed. Univ. Agric. Sci, Rep. For. Ecol. For. Soils*, 58, 1, 1987.

O'Carroll, R.N., The progress of peatland afforestation in the Republic of Ireland, *Ir. For.*, 19, 93, 1962.

Olsen, Olaf and Crumlin-Pedersen, Ole, *Five Viking ships from Roskilde Fjord*, The National Museum, Copenhagen, 1978.

Paavilainen, E. and Päiviänen, J., *Peatland Forestry: Ecology and Principles*, Springer-Verlag, Berlin, 1995.

Päiviänen, J. and Paavilainen, E., Forestry on peatlands, in *Peatlands in Finland*, Vasander, H., Ed., Finnish Peatland Society, Helsinki, Finland, 72, 1996.

Parr, S.L., *Baseline Studies on Ground Vegetation in Plantation Forests on Cutaway Peat*, Department of Environmental Resource Management, University College Dublin, Dublin, Ireland, 2000.

Pöllänen, M. and Renou, F., *Development of Plantation Forest on Industrial Cutaway Peatlands in Relation to Ground Vegetation*, Forest Ecosystem Research Group Report Number 66, Department of Environmental Resource Management, University College Dublin, 2002.

Renou, F., Environmental monitoring, in *Research Programme to Develop a Forest Resource on Industrial Cutaway Peatland in the Midlands. BOGFOR — Phase 1: Final Report*, Forest Ecosystem Research Group Report 52., Jones, S.M. and Farrell, E.P., Eds., Department of Environmental Resource Management, University College Dublin, Dublin, Ireland, 134, 2000.

Renou, F. and Cummins, T., Soil as a key to sustainable forest management, in *Achievement and Challenge Rio + 10 and Ireland*, Feehan, J., and Convery, F., Eds., The Environmental Institute, University College Dublin, Dublin, Ireland, 25, 2002.

Renou, F. and Farrell, E.P., Investigation into the fate of phosphatic fertiliser applied to cutaway peatland forests, in *Environ 2001: 11th Irish Environmental Researchers' Colloquium*, 26–27 January 2001, Dublin City University, 16, 2001.

Renou, F., Jones, S.M., and Farrell, E.P., Leaching of phosphorus fertiliser applied on cutaway peatland forests recently established in central Ireland, in *Sustaining our Peatlands: Proceedings of the 11th International Peat Congress*, Rochefort, L. and Daigle, J.-Y., Eds., Québec, 984, 2000.

Schaible, R., Sitka spruce in the 21st century, establishment and nutrition, *Ir. For.*, 49, 10, 1992.

Simard, S., Mixtures of paper birch and conifers: an ecological balancing act, in *Silviculture of Temperate and Boreal Broadleaf-conifer Mixtures*, Comeau, P.G. and Thomas, K.D., Eds., B.C. Ministry of Forests, British Columbia, 15, 1996.

Smith, E.M., Jones, S.M., and Farrell, E. P., *The Use of Nursing Mixtures with Special Reference to Frost Protection*, Report No. 38, Forest Ecosystem Research Group, Department of Environmental Resource Management, University College Dublin, 1998.

Smith, E.M., Jones, S.M., and Farrell, E.P., *Late Spring Frosts*, Report No. 50, Forest Ecosystem Research Group, Department of Environmental Resource Management, University College Dublin, 2000.

Thelin, G., U., Rosengren, Callesen, I., and Ingerslev, M., The nutrient status of Norway spruce in pure and in mixed-species stands, *For. Ecol. Manage.*, 160, 115, 2002.

Waddington, J.M., Warner, K.D., and Kennedy, G.W., Cutover peatlands: a persistent source of atmospheric CO_2, *Global Biogeochem. C.*, 16, 21, 2002.

Walsh, T. and Barry, T.A., The chemical composition of some Irish peats, *Proc. R. Ir. Acad.*, 59 (Sec. B), 305, 1958.

Wheeler, I.C.T., Hooker, J.E., Crowe, A., and Berrie, A.M., The improvement and utilization in forestry of nitrogen fixation by actinorhizal plants with special reference to *Alnus* in Scotland, *Plant Soil*, 90, 393, 1986.

Index

D

T

U